D1799872

Infectious disease in aquaculture

Related titles:

New technologies in aquaculture: Improving production efficiency, quality and environmental management
(ISBN 978-1-84569-384-8)
With wild stocks declining due to over-fishing, aquaculture will have a more significant role to play in meeting future demand for fresh fish. Developments in research continue to lead to improvements in aquaculture production systems, resulting in increased production efficiency, higher product quality for consumers and a more sustainable industry. *New technologies in aquaculture* reviews essential advances in these areas. Chapters focus on key aspects of genetic improvement, reproduction, diet and husbandry, health and aquaculture systems design. Contributions on environmental issues and farming new species complete the volume.

Shellfish safety and quality
(ISBN 978-1-84569-152-3)
Shellfish are a very popular and nutritious food source worldwide and their consumption has risen dramatically. Because of their unique nature as compared to beef and poultry, shellfish have their own distinct aspects of harvest, processing and handling. Edited by leading authorities in the field, this collection reviews issues of current interest and outlines steps that can be taken by the shellfish industry to improve shellfish safety and eating quality. Opening chapters consider microbial, biotoxin, metal and organic contaminants of shellfish. Techniques to reduce contamination are then discussed, such as mitigation of the effects of harmful algal blooms. Chapters also address approaches to managing disease and other methods to improve quality, such as improved packaging methods and reduction of biofouling.

Improving farmed fish quality and safety
(ISBN 978-1-84569-299-3)
Fish farming enables greater control of product quality, but there have been concerns about the levels of contaminants found in farmed products. Their sensory and nutritional quality can also not equal that of wild-caught fish. This important collection reviews potential negative safety and quality issues in farmed fish and presents methods to improve product characteristics. The first part of the book discusses contaminants, such as persistent organic pollutants and veterinary drug residues and methods for their reduction and control. The second part addresses important quality issues, such as genetic control of flesh characteristics and the effects of feed on product nutritional and sensory quality.

Details of these and other Woodhead Publishing materials books can be obtained by:

- visiting our web site at www.woodheadpublishing.com
- contacting Customer Services (e-mail: sales@woodheadpublishing.com; fax: +44 (0) 1223 832819; tel.: +44 (0) 1223 499140 ext. 130; address: Woodhead Publishing Limited, 80 High Street, Sawston, Cambridge CB22 3HJ, UK)
- in North America, contacting our US office (e-mail: usmarketing@woodheadpublishing.com; tel.: (215) 928 9112; address: Woodhead Publishing, 1518 Walnut Street, Suite 1100, Philadelphia, PA 19102-3406, USA)

If you would like e-versions of our content, please visit our online platform www.woodheadpublishingonline.com. Please recommend it to your librarian so that everyone can benefit from the wealth of content on the site.

Woodhead Publishing Series in Food Science, Technology and Nutrition:
Number 231

Infectious disease in aquaculture

Prevention and control

Edited by
Brian Austin

WP

WOODHEAD
PUBLISHING

Oxford   Cambridge   Philadelphia   New Delhi

Published by Woodhead Publishing Limited,
80 High Street, Sawston, Cambridge CB22 3HJ, UK
www.woodheadpublishing.com
www.woodheadpublishingonline.com

Woodhead Publishing, 1518 Walnut Street, Suite 1100, Philadelphia,
PA 19102-3406, USA

Woodhead Publishing India Private Limited, G-2, Vardaan House, 7/28 Ansari Road,
Daryaganj, New Delhi – 110002, India
www.woodheadpublishingindia.com

First published 2012, Woodhead Publishing Limited
© Woodhead Publishing Limited, 2012
The authors have asserted their moral rights.

British Library Cataloguing in Publication Data
A catalogue record for this book is available from the British Library.

Library of Congress Control Number: 2012932537

ISBN 978-0-08-101633-6 (print)
ISBN 978-0-85709-573-2 (online)
ISSN 2042-8049 Woodhead Publishing Series in Food Science, Technology and Nutrition (print)
ISSN 2042-8057 Woodhead Publishing Series in Food Science, Technology and Nutrition (online)

The publisher's policy is to use permanent paper from mills that operate a sustainable forestry policy, and which has been manufactured from pulp which is processed using acid-free and elemental chlorine-free practices. Furthermore, the publisher ensures that the text paper and cover board used have met acceptable environmental accreditation standards.

Typeset by Toppan Best-set Premedia Limited, Hong Kong

Transferred to Digital Printing in 2016

Contents

Contributor contact details

(* = main contact)

Editor and Chapter 9

Brian Austin
Institute of Aquaculture
University of Stirling
Stirling
FK9 4LA
UK

E-mail: brian.austin@stir.ac.uk;
  baustin5851@gmail.com

Chapter 1

C. J. Secombes* and T. Wang
Scottish Fish Immunology
  Research Centre
University of Aberdeen
Aberdeen
AB24 2TZ
UK

E-mail: c.secombes@abdn.ac.uk;
  t.h.wang@abdn.ac.uk

Chapter 2

Lage Cerenius and Kenneth
  Söderhäll*
Department of Comparative
  Physiology
Uppsala University
Norbyvägen 18A
SE-752 36 Uppsala
Sweden

E-mail: Kenneth.Soderhall@ebc.
  uu.se

Chapter 3

B. Novoa and A. Figueras*
Instituto de Investigaciones
  Marinas (IIM)
CSIC
Eduardo Cabello 6
36208 Vigo
Spain

E-mail: Antoniofigueras@iim.csic.es

Chapter 4

Prof. James F. Turnbull
Institute of Aquaculture
University of Stirling
Stirling
FK9 4LA
UK

E-mail: jft1@stir.ac.uk; j.f.turnbull@
    stir.ac.uk

Chapter 5

A. Adams* and K. D. Thompson
Institute of Aquaculture
University of Stirling
Stirling
FK9 4LA
UK

E-mail: alexandra.adams@stir.ac.uk

Chapter 6

Dr Margaret Crumlish
Institute of Aquaculture
University of Stirling
Stirling
FK9 4LA
UK

E-mail: margaret.crumlish@stir.
    ac.uk; mc3@stir.ac.uk

Chapter 7

Peter Smith
Department of Microbiology
National University of Ireland,
    Galway
University Road
Galway
Ireland

E-mail: peter.smith@nuigalway.ie

Chapter 8

Andrew P. Shinn* and James E.
    Bron
Institute of Aquaculture
University of Stirling
Stirling
FK9 4LA
UK

E-mail: aps1@stir.ac.uk; jeb1@stir.
    ac.uk

Chapter 10

Jarl Bøgwald and Roy A. Dalmo*
Norwegian College of Fishery
    Science
University of Tromsø
N-9037 Tromsø
Norway

E-mail: roy.dalmo@uit.no

Chapter 11

Donald V. Lightner* and Rita M.
  Redman
Department of Veterinary and
  Microbiology
University of Arizona
Tucson
AZ 85721
USA

E-mail: dvl@u.arizona.edu

Chapter 12

C. J. Rodgers*
IRTA-Sant Carles de la Ràpita
Crta. Poble Nou s/n
Apartat de Correus 200
43540 Tarragona
Spain

E-mail: chris.rodgers@irta.cat

E. J. Peeler
Centre for Environment, Fisheries
  and Aquaculture Science
  (CEFAS)
Barrack Rd
Weymouth
DT4 8UB
UK

Chapter 13

S. MacKenzie* and S. Boltaña
Institut de Biotecnologia i de
  Biomedicina
Universitat Autonoma de
  Barcelona
08193 Barcelona
Spain

E-mail: Simon.Mackenzie@uab.cat

B. Novoa and A. Figueras
Instituto de Investigaciones
  Marinas (IIM)
CSIC
Eduardo Cabello 6
36208 Vigo
Spain

F. W. Goetz
School of Freshwater Sciences
University of Wisconsin-Milwaukee
600 E. Greenfield Ave.
Milwaukee
WI 53204
USA

Chapter 14

Chao Ran and Mark R. Liles*
Department of Biological Sciences
Room 316, Rouse Life Sciences
  Building
120 West Samford Avenue
Auburn University
Auburn
AL 36849
USA

E-mail: lilesma@auburn.edu

Abel Carrias and Jeffery S.
  Terhune
Department of Fisheries and Allied
  Aquacultures
203 Swingle Hall
Auburn University
Auburn
AL 36849
USA

Chapter 15

Peter De Schryver, Tom Defoirdt
  and Peter Bossier*
Laboratory for Aquaculture &
  Artemia Reference Center
Faculty of Bio-science Engineering
Ghent University
Rozier 44
9000 Ghent
Belgium

E-mail: Peter.Bossier@UGent.be

Nico Boon and Willy Verstraete
Laboratory for Microbial Ecology
  and Technology
Faculty of Bio-science Engineering
Ghent University
Coupure Links 653
9000 Ghent
Belgium

Chapter 16

Thavasimuthu Citarasu
Centre for Marine Science and
  Technology
Manonmaniam Sundaranar
  University
Rajakkamangalam
Kanyakumari District
Tamilnadu
629502
India

E-mail: citarasu@gmail.com;
  citarasu@msuniv.ac.in

Chapter 17

A. Falco, A. Martinez-Lopez and A.
  Estepa*
IBMC
Miguel Hernández University
03202 Elche
Spain

E-mail: aestepa@umh.es

J. M. Coll
INIA-SIGT–Biotechnology
28040 Madrid
Spain

Chapter 18

Christina Sommerville
Institute of Aquaculture
University of Stirling
Stirling
FK9 4LA
UK

E-mail: cs3@stir.ac.uk

Woodhead Publishing Series in Food Science, Technology and Nutrition

Preface

B. Austin, University of Stirling, UK

Interest in diseases of aquatic animals centres on aquaculture, which in recent years has been seen as a method of replacing the stocks no longer provided by capture fisheries. The reasons for aquaculture include:

- the production of high-quality protein for mass human populations, so allowing the dwindling capture fisheries to regenerate
- the provision of high-value edible species for the middle/upper classes and for export
- providing specimens for restocking to augment dwindling natural stocks and/or to provide stock for sports fisheries
- the highly lucrative ornamental (= pet) fish trade
- the provision of specialist products, such as pearls and compounds for biotechnology
- the establishment of lagoonaria, for example in Tahiti, whereby lagoons are closed off and the organisms within allowed to flourish. This combines ecotourism with habitat protection and the preservation of biodiversity.

Aquaculture production is certainly increasing in most areas of the world with yearly increase of approximately 9% in most countries except sub-Saharan Africa. In 2008, the total worldwide production of animals and plants was >52 million tonnes and >15 million tonnes, respectively, of which China remains the biggest single producer by contributing >32 million tonnes and >9 million tonnes of animals and plants, respectively (FAO, 2008). Thus aquaculture production is clearly dominated by Asia. The number one product was silver carp (*Hypophthalmichthys molitrix*) of which >3 million tonnes were produced in 2008. This contributed to >20 million tonnes of cyprinids that were produced in 2008 alone (FAO, 2008).

Disease remains a significant constraint of aquaculture. The range of pathogens is ever increasing, and industry is crying out for better diagnostic and disease control methods. *Francisella* has emerged as a significant pathogen, initially in Norway, and already two new species of bacterial fish pathogens have been recognized and described, namely *Francisella asiatica* and *noatunensis*. Organisms have been seen in pathological material, e.g. summer enteritic syndrome of rainbow trout, but for which culturing has not been achieved. In this case, the causal agent was coined *Candidatus* Arthromitus. Researchers have been quick to embrace new molecular technologies, which have been applied to both diagnostic procedures and vaccine development. In contrast, the use of antibiotics is losing favour as concerns increase about antibiotic resistance and tissue residues. Alternative disease-control strategies including the use of natural plant products are being embraced widely. There is a clear need for continued research in fish diseases. Against this background, this book was developed to provide the reader with up-to-date information about approaches to mitigate the effects of infectious disease in aquaculture. I am grateful to all the authors who responded to the challenge of producing manuscripts often within a tight timescale.

Reference

FAO (2008), *FAO Yearbook; Fishery and Aquaculture Statistics*. Rome, Italy. Food and Agricultural Organization of the United Nations.

Part I

Immune responses in fish and shellfish and their implications for disease control

1

The innate and adaptive immune system of fish

C. J. Secombes and T. Wang, University of Aberdeen, UK

Abstract: This chapter describes what is known about the main components and responses of the innate and adaptive immune system of fish. The chapter first reviews the organs, cells and molecules of the immune system known in a few economically important or model fish species. Molecular evidence suggests a similar immune system exists throughout the jawed vertebrates yet marked differences are also apparent. The innate parameters are at the forefront of fish immune defence and are a crucial factor in disease resistance. The adaptive response of fish is commonly delayed but is essential for long lasting immunity and a key factor in successful vaccination.

Key words: fish immune organs, pattern recognition receptors (PRR), innate immune responses, adaptive immune responses, immune regulation.

1.1 Introduction

Fish possess innate and adaptive immune defence systems. The innate parameters are at the forefront of immune defence and are a crucial factor in disease resistance. The adaptive response of fish is commonly delayed but is essential for long-lasting immunity and is a key factor in successful vaccination. The massive increase in aquaculture in recent decades has put greater emphasis on studies of the fish immune system and defence against diseases commonly associated with intensive rearing of a few economically important species. Such research has helped define the optimum conditions for maintaining immunocompetent fish in culture, for selection of fish stock (breeding), as well as developing and improving prophylactic measures such as vaccination, and use of probiotics and immunostimulation in the aqua-cultured species.

However, there is great variation in disease susceptibility and immune defence between different fish species, a reflection of the extended time the present day teleosts have been separated during the evolution of this fish group. Thus the immune response described in one species may not be the same in other species. Indeed, the immune system is largely unknown in most fish species, especially in newly aquacultured species, limiting the development of immune control strategies against infectious disease. This chapter will describe what is known about the main components of the innate and adaptive immune system of fish.

1.2   Overview of immune cells and organs in fish

Vertebrates live in an environment containing a great variety of infectious agents – viruses, bacteria, fungi, protozoa and multicellular parasites – that can cause disease, and if they multiply unchecked they will eventually kill their host. Thus vertebrates have evolved effective immune responses that initially recognize the pathogens or other foreign molecules (antigens), triggering pathways that subsequently elicit effector mechanisms to attempt to eliminate them. The immune responses elicited fall into two main categories: innate (or non-specific) immune responses and adaptive (or specific) immune responses.

Immune responses are mediated by a variety of cells and secreted soluble mediators. Leucocytes are central to all immune responses, and include lymphocytes (T cells, B cells, large granular lymphocytes), phagocytes (mononuclear phagocytes, neutrophils and eosinophils) and auxiliary cells (basophils, mast cells, platelets). Other cells in tissues also participate in the immune responses by signalling to the leucocytes and responding to the soluble mediators (cytokines) released by leucocytes such as T cells and macrophages.

The cells involved in the immune responses are organized into tissues and organs in order to perform their functions most effectively. These structures are collectively referred to as the lymphoid system, and are arranged into either discretely encapsulated organs or accumulations of diffuse lymphoid tissue. The major lymphoid organs and tissues are classified as either primary (central) or secondary (peripheral). Lymphocytes are produced in the primary lymphoid organs and function within the secondary lymphoid organs and tissues.

In mammals, the thymus, foetal liver and bone marrow are the primary lymphoid organs, where lymphocytes differentiate from lymphoid stem cells, proliferate and mature into functional cells. T cells mature in the thymus whereas B cells mature in the foetal liver and bone marrow. In the primary lymphoid organs, lymphocytes acquire their repertoire of specific antigen receptors, i.e. T cell receptor (TCR) and B cell receptor (BCR), in order to cope with antigenic challenges that individuals receive during their lifespan, with cells having receptors for autoantigens mostly eliminated early in development. For example, in the thymus, T cells learn to recognize self-MHC (major histocompatibility complex) molecules but if they react to self-antigens presented by these molecules they are eliminated. It is worth noting that some lymphocytes develop outside the primary lymphoid organs (Alitheen *et al.*, 2010; Peaudecerf and Rocha, 2011).

The generation of lymphocytes in primary lymphoid organs is followed by their migration into peripheral secondary lymphoid tissues. In mammals, the secondary lymphoid tissues comprise well-organised, encapsulated organs, such as the spleen and lymph nodes (systemic organs) and non-encapsulated accumulations of lymphoid tissues. The spleen is responsive to blood-borne antigens and lymph nodes protect the body from antigens from skin or from internal surfaces. The lymphoid tissue found in association with mucosal surfaces is called the mucosal associated lymphoid tissue (MALT), and includes GALT (gut-associated lymphoid tissue) in the intestinal tract, BALT (bronchus-associated lymphoid tissue) in the respiratory tract, and lymphoid tissue in the genitourinary tract (Randall, 2010; Suzuki *et al.*, 2010).

In the secondary lymphoid organs, germinal centres (GC) are unique structures in birds and mammals where the collaboration between proliferating antigen-specific B cells, T follicular helper cells (Tfh), and the specialized follicular dendritic cells (FDC) produces high-affinity antibody-secreting plasma cells and memory B cells that ensure sustained immune protection and rapid recall responses against previously encountered foreign antigens (Gatto and Brink, 2010). GCs develop in the B cell follicles of secondary lymphoid tissues during T cell-dependent (TD) antibody responses. The mature GC is divided into the dark and light zones on the basis of their histological appearance. The antigen-specific B cells proliferate in these locations (Hauser *et al.*, 2007) and undergo somatic hypermutation, antibody class-switch recombination, and are then selected by the FDC and Tfh. Therefore, the GC response endows a population of antigen-activated B cells that secrete antibodies (or immunoglobulins, Ig) with a high affinity for the antigen and with a relevant Ig isotype, resulting in a more efficient clearance of the antigen (Good-Jacobson and Shlomchik, 2010).

1.2.1   The thymus

The term 'fish' refers to a heterogeneous group of organisms that include the Agnathans (lampreys and hagfish – jawless vertebrates), Chondrichthyes (sharks and rays) and Osteichthyes (bony fish, that include the largest group of fish the teleosts) (Nelson, 1994). In this chapter the term fish will be used to refer to bony fish unless otherwise specified. As in birds and mammals, fish have cellular and humoral immune responses, and central organs (Fig. 1.1) whose main function is involved in immune defence.

The thymus is considered a key organ of the immune system in jawed vertebrates. It is thought to have evolved in early fish species as a thickening in the epithelium of the pharyngeal area of the gastro-intestinal tract (Bowden *et al.*, 2005), and is identifiable in the Chondrichthyes and the

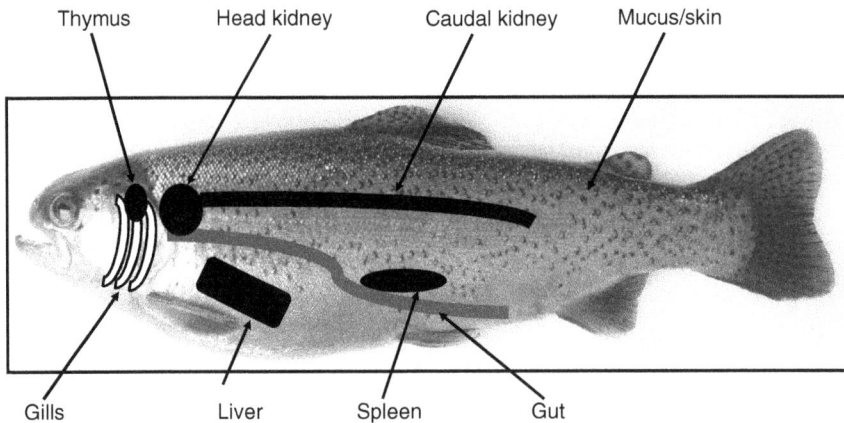

Fig. 1.1   Immune tissues in teleost fish. The approximate sites of immune tissues
are superimposed onto a rainbow trout (*Oncorhynchus mykiss*).

Osteichthyes. It generally develops in the lamina propria of the gastrointestinal tract in pouches located at the base of the gill arches, and subsequently migrates to the underlying mesenchyme during ontogeny. In most
teleosts the thymus is located near the gill cavity and is closely associated
with the pharyngeal epithelium (Zapata *et al.*, 1996). Although usually
found as a paired organ in most vertebrates, the thymus can appear as more
than one pair of organs in teleosts. For example, each gill chamber of cling-
fish, *Sicyases sanguineus*, has a pair of thymus glands with one taking up a
superficial position and the second located close to the gill epithelium
(Gorgollon, 1983).

    The cells in the mammalian thymus can be divided into hematopoietic
cells (CD45$^+$ cells, which include thymocytes, dendritic cells (DC), macrophages and B cells) that are transient passengers and resident stromal cells
(CD45$^-$ cells). CD45$^-$ cells include two lineages: thymus epithelial cells
(Keratin$^+$) that originate from the pharyngeal pouch endoderm (third pouch
in the mouse) and mesenchymal cells (Keratin$^-$), which are a mixture of cell
types that contribute to various structures of the thymus such as the capsule
or vasculature (Rodewald, 2008). The thymus is organized into the inner,
morphologically lighter zone, the medulla, and the outer, morphologically
darker zone, the cortex. In mammals T-cell progenitors enter through the
cortico-medullary blood vessels and can differentiate into NK cells, DC and
T cell lineages (De and Pal, 1998). Within the cortex are lymphocytes in a
stroma of cells of epithelial morphology, and macrophages. The structure of
the fish thymus is highly variable between species and within a species in
an age-dependent manner. In many fish species there is no clear cortico-
medullary differentiation as would normally be seen in mammals. Zonation

of the thymus has been observed in turbot (*Scophthalmus maximus* L.) and halibut (*Hippoglossus hippoglossus* L.) but not in salmonids (Tatner and Manning, 1982; Fournier-Betz *et al.*, 2000).

Although zonation is absent in the young carp thymus, later on a complex intermingling of cortex into the medulla occurs in developing carp, with zonation becoming visible during the fourth week post-fertilisation (Romano *et al.*, 1999). The size of lymphocytes also can vary with species. A comparative study of three fish species revealed that lymphocytes are typically basophilic and 3–5 μm in diameter, whilst populations of darker staining small lymphocytes (2–2.5 μm) are observed in later development (Chantanachookin *et al.*, 1991).

The thymus, kidney and spleen are the major (non-mucosal) lymphoid organs of fish. In freshwater fish, the thymus is the first organ to become lymphoid, although prior to this the kidney can contain hematopoietic precursors but not lymphocytes. However, in marine fish the order in which the major lymphoid organs develop is kidney, spleen and then the thymus (Zapata *et al.*, 2006). Early development of the thymus in fish has been studied in many diverse teleost species and has shown that the development timeframe can differ from species to species even when accounting for temperature effects on growth (Bowden *et al.*, 2005). The relationship between growth and development can be dynamic and physiological age expressed as degree-days does not factor out all differences. Thus at 5 days pre-hatching at 14°C, the rainbow trout embryo already possesses the rudiments of a thymus (Grace and Manning, 1980), whereas a thymus is only seen at 28 days post-hatching in Atlantic cod (Schrøder *et al.*, 1998).

1.2.2   The bone marrow equivalent: the head kidney

Hematopoietic stem cells (HSCs) found in the mammalian bone marrow (BM) are crucial throughout life for their ability to differentiate and generate all hematopoietic lineages while maintaining the capacity for self-renewal. This remarkable ability can be demonstrated in mice where a single HSC can reconstitute all the immune cells of a lethally irradiated animal, thereby maintaining a functional immune system throughout life. The bone marrow in mammals is the site where B-cells originate and develop from HSCs via progression through downstream multipotent progenitors, lymphoid primed multipotent progenitors, common lymphoid progenitors, B-cell progenitor intermediates and finally naive B cells expressing rearranged surface bound Ig (see below) (Santos *et al.*, 2011). The bone marrow is absent in fish but the cephalic portion of the kidney (head kidney or pronephros) is considered analogous to mammalian bone marrow, at least in terms of hematopoiesis (Zapata, 1979). The trunk kidney (mesonephros) is also hematopoietic, although it also contains renal tissue.

The kidney in fish is often a Y-shaped organ that is placed along the body axis above the swim bladder (Fig. 1.1). The lower part is a long structure situated parallel to the vertebral column, most of which works as a renal system. The active immune part, the head kidney, is formed by the two arms, which penetrate under the gills. The head kidney has a reticulo-endothelial stroma consisting of sinusoidal cells (endothelial and adventitial cells) and reticular cells (macrophage-type reticulum and fibroblast-like reticular cells) similar to those of the mammalian bone marrow (Meseguer et al., 1995). The macrophage-type reticulum cells are characterized by their cytoplasmic processes and acid phosphatase positive lysosomes. The fibroblast-like reticular cells are peroxidase negative and acid and alkaline phosphatase, glucose-6-phosphatase, beta-glucuronidase and ATPase positive, and are joined by desmosomes and form an extensive network between the hematopoietic parenchyma. The types of hematopoiesis described within the fish head kidney include erythropoiesis, granulopoiesis, thrombopoiesis, monopoiesis and lymphoplasmopoiesis (Abdel-Aziz et al., 2010). Erythropoiesis includes a number of developmental stages, including proerythroblasts, basophilic erythroblasts, polychromatic erythroblasts, acidophilic erythroblasts and young and mature erythrocytes. The granulopoietic series consists of cells with variable shape and size depending on the stage of maturity, from myeloblasts to mature granulocytes. The lymphopoietic cells include lymphoblasts, large lymphocytes, small lymphocytes and active and inactive plasma cells, whilst the thrombopoietic series consists of thromboblasts, prothromboblasts and thrombocytes. Melano-macrophage centres (MMC) are also present in the head kidney, and are thought to function as primordial GCs (Agius and Roberts, 2003; Saunders et al., 2010).

The fish head kidney is also an important endocrine organ, homologous to mammalian adrenal glands, and contains aminergic chromaffin and interrenal steroidogenic cells. The adrenal homologue and hematopoietic tissues can be mixed, adjacent or completely separated, with the former lining the endothelium of the venous vessels or being located in close proximity to them (Gallo and Civinini, 2003). The fish adrenal homologue is under hormonal and neuronal control. Interrenal cells secrete corticosteroids and other hormones that may play an important role in modulating stress responses, osmoregulation and the immune response. Thus, the head kidney is an important organ with key regulatory functions and the central organ for immune–endocrine interactions and even neuroimmunoendocrine cross-talk.

The fish head kidney appears to be the primary organ for antibody production (Tian et al., 2009). The first appearance of antibody secreting lymphocytes varies considerably among fish species. The first appearance of B cells, as defined by the expression of Ig, is later in marine species compared to freshwater species, with larvae being 20–30 mm in length when Ig is first expressed, about a week after hatching in the case of rainbow trout and channel catfish (Magnadottir et al., 2005).

1.2.3 The spleen

The spleen in mammals is the largest secondary immune organ in the body and is responsible for initiating immune reactions to blood-borne antigens and for filtering the blood of foreign material and old or damaged red blood cells. These functions are carried out by the two main compartments of the spleen, the white pulp (including the marginal zone) and the red pulp, which are vastly different in their architecture, vascular organization and cellular composition (Cesta, 2006). The spleen is also a major secondary lymphoid organ in fish, although absent in Agnathans where spleen-like lymphohematopoietic tissues occur in the intestine (Fänge and Nilsson, 1985; Press and Evensen, 1999). It contains the same elements as the other vertebrates: blood vessels, ellipsoids, red pulp and white pulp. However, the red and white pulp in fish is less clearly defined than in homeothermic vertebrates. The pulp occupies the majority of the organ, and consists of a reticular cell network supporting blood-filled sinusoids that hold diverse cell populations, including macrophages and lymphocytes. The white pulp is often poorly developed and typically has two main components: the melano-macrophage accumulations and the ellipsoids. The spleen can also be a major reservoir of disease and there is much interest in trying to understand its role in protection against bacterial infection (Hadidi *et al.*, 2008) as well as in red blood cell regulation. The populations of lymphocytes and macrophages capable of mounting an immune response are situated close to sites of antigen trapping and often associated with accumulations of melano-macrophages.

The melano-macrophages may form MMCs, bound by a thin argyrophilic capsule and surrounded by white pulp, often in association with thin-walled, narrow blood vessels (Agius, 1980; Press and Evensen, 1999). Fish MMCs are typically located in the stroma of the haematopoietic tissue of the spleen and kidney, and in the liver in some species, and may be primitive analogues of the GCs of mammals and birds. GCs contain specialized FDC that interact with antigen-specific B cells to produce high-affinity antibody-secreting plasma cells and memory B cells (Gatto and Brink, 2010). An antibody, CNA-42, usually employed for labelling FDC of higher vertebrates, can label free melano-macrophages and splenic MMCs, and the key initiator of antibody affinity maturation (activation-induced cytidine deaminase, AID) is expressed in cells that co-locate with melano-macrophages, suggesting an evolutionary relationship between fish MMCs and mammalian GCs (Vigliano *et al.*, 2006; Saunders *et al.*, 2010). The MMCs can retain antigens for long periods, possibly in the form of immune-complexes (Agius, 1980; Press and Evensen, 1999), and increase in size or frequency in conditions of environmental stress and during infection (De Vico *et al.*, 2008; Suresh, 2009).

The ellipsoids terminate in arterioles with a narrow lumen that runs through a sheath of reticular fibres, reticular cells and macrophages, the ellipsoid. Ellipsoids appear to have a specialized function for plasma filtration and the trapping of blood-borne substances, particularly immune complexes (Secombes *et al.*, 1982; Press and Evensen, 1999). Blood-borne substances are retained in the ellipsoidal wall and taken up by the rich population of macrophages surrounding these vessels. The subsequent migration of antigen-laden macrophages to MMCs has been described (Press and Evensen, 1999). Ellipsoids occur in most fish but may be indistinct or lacking in certain species (Fänge and Nilsson, 1985). Similarly, splenic lymphoid tissue is poorly developed in some fish where diffuse layers of lymphocytes surround arteries and MMCs, with scattered lymphocytes within the whole parenchyma. In contrast, in the icefish (*Chaenocephalus aceratus*), a teleost which possesses practically no erythrocytes, the dominant cells of the spleen parenchyma are lymphocytes and macrophages (Walvig, 1958). Plasma cells secreting Ig are scattered throughout the white pulp and isolated spleen lymphocytes stimulated *in vitro* produce plasma cells (Bromage *et al.*, 2004).

1.2.4   The gills

The fish gill is a multifunctional organ involved in gas exchange, ionoregulation, osmoregulation, acid–base balance, ammonia excretion, hormone production, modification of circulating metabolites and immune defence (Rombough, 2007). In filter-feeding species, such as the sardine (*Sardina pilchardus*), the gills may also perform a feeding function. Agnathan hagfishes have primitive gill pouches, while lampreys have arch-like gills similar to the higher fishes. In lampreys and elasmobranchs, the gill filaments are supported by a complete interbranchial septum and water exits via external branchial slits or pores. In contrast, the teleost interbranchial septum is much reduced, leaving the ends of the filaments unattached, and the multiple gill openings are replaced by the single caudal opening of the operculum (Wilson and Laurent, 2002). The basic functional unit of the gill is the filament, which supports rows of plate-like lamellae. The lamellae are designed for gas exchange with a large surface area and a thin epithelium surrounding a well-vascularized core of pillar cell capillaries. The lamellae are positioned for the blood flow to be counter-current to the water flow over the gills. The lamellar gas-exchange surface is covered by squamous pavement cells, while large, mitochondria-rich, ionocytes and mucocytes are found in greatest frequency in the filament epithelium.

Fish pathogens readily spread in the water, and the thin respiratory epithelium of the gills represents an obvious entry for pathogens. For example, infectious salmon anaemia (ISA) virus infection is believed to be established first in the gills before spreading to other organs (Rimstad and Mjaaland, 2002). The physical barrier of the fish gills consists of the gill epithelium, a glycocalyx layer and a mucus layer. The gill is a major organ for antibody secreting cell production following direct immersion immunization (Dos Santos *et al.*, 2001). Lymphocyte accumulations have been identified recently on the caudal edge of the interbranchial septum, at the base of the gill filaments in salmonid fish (Haugarvoll *et al.*, 2008). Flow cytometry analysis reveals high numbers of T cells (CD3ε$^+$) in the gills, as well as in thymus and intestine (Koppang *et al.*, 2010). The interbranchial T cells are embedded in a meshwork of epithelial cells.

1.2.5   The gut

The gastrointestinal tract (the gut) plays dual roles in mammals: digestion/ uptake of nutrients and immune homeostasis, the latter to protect the body from potentially harmful microbes but also to induce a tolerogenic response to innocuous food, commensals and self-antigens. The mammalian GALT is composed of aggregations of lymphoid follicles called Peyer's patches and diffusely dispersed effector cells at sites such as the lamina propria and the intraepithelial lymphocyte (IEL) compartment. The separation of these sites serves to limit and control immune responses (Mason *et al.*, 2008). In addition to its distinct architecture, the gut has specialized immune cells i.e. dendritic cells and M cells that transport antigens and pathogens into the lymphoid tissues. M cells have numerous microfolds on their luminal surface and contain deep invaginations of the baso-lateral plasma membrane which form pockets containing B cell, T cells, DCs and macrophages. M cells are positioned above the Peyer's patches and transcytose antigens and micro-organisms into the pocket and to the underlying lymphoid tissue, resulting in IgA class switching and the secretion of high amounts of dimeric IgA at the effector sites (Cerutti, 2008). M cells are not exclusive to Peyer's patches and are also found in epithelia associated with lymphoid cell accumulations at antigen sampling areas in other mucosal sites. The secreted IgA is sub-sequently bound by the polymeric Ig receptor (pIgR) and transcytosed to the intestinal lumen or to the bile in the liver. The extracellular part of the receptor is then cleaved off and secreted as the secretory component together with the IgA at the mucus site (Rombout *et al.*, 2010).

Although Peyer's patches and IgA are not reported in teleost fish and lymph nodes are absent, fish do have a local mucosal defence in the gut to sample antigens (Fuglem *et al.*, 2010) and produce local Ig responses (Hamuro *et al.*, 2007; Zhang *et al.*, 2010). In stomachless species the bile and pancreatic ducts immediately enter the gut just posterior of the oesophagus and the so-called intestinal bulb. In nearly all species investigated the intestine can be subdivided into three segments based on the microscopical anatomy of their mucosa, especially their enterocytes. The enterocytes of the first segment (60–75% of the total gut length, dependent on the species) can be considered as absorptive. The enterocytes of the second segment (15–30% of the gut length), are characterized by the presence of large supranuclear vacuoles, irregular microvilli zone and high pinocytotic activity at the apical part and strongly take up macromolecules. The enterocytes of the third segment (5–15% of the gut length) have been less studied but are thought to have an osmoregulatory function (Stroband *et al.*, 1979; Rombout *et al.*, 2010).

Leucocytes are abundantly present in the lamina propria and intestinal epithelium of the fish gut. However, the lack of suitable antibodies in fish has hampered the distinction of subpopulations within the GALT. However, it is likely that the level of GALT organization in teleosts is lower than in mammals, with fish having a more diffusely organized immune system in their gut, containing lymphocytes, macrophages, eosinophilic and neutrophilic granulocytes. M cells and DCs are important for sampling antigens in the GALT of mammals and in fish a small number of DC-like cells and epithelial cells are located in the mucosal folds in the second segment that can take up gold bovine serum albumin (BSA). These gold-positive epithelial cells display diverging and electron-dense microvilli with channels intruding into their cytoplasm and have a characteristic lectin binding property typical of mammalian M cells (Fuglem *et al.*, 2010).

The first indications that fish could have local and/or mucosal responses came from the detection of specific antibodies in mucosal secretions after intestinal or immersion immunization of a variety of fish species, which were rarely detectable after systemic immunization (Rombout *et al.*, 2010). Intravenous administration of radiolabelled Ig never reached the mucosal secretions and therefore the Ig in mucosal secretions was suggested to be the result of local synthesis (Lobb and Clem, 1981; Rombout *et al.*, 1993b). Biochemical analysis of cutaneous mucus of a variety of teleost species revealed only IgM-like molecules (see below), in tetrameric, dimeric and monomeric forms (Rombout *et al.*, 1993b). Whilst fish lack IgA, it appears that a second Ig isotype (IgT) is specialized for mucosal immunity, and in trout the IgT response to a gut parasite is restricted to the intestine (Zhang *et al.*, 2010). The pIgR is also an essential component for mucosal immunity in mammals. It is expressed by mucosal epithelia and hepatocytes, binds IgA and IgM, and can transcytose these Igs to the luminal surface. A pIgR-like molecule has been described in a few fish species (Hamuro *et al.*, 2007; Rombout *et al.*, 2008; Feng *et al.*, 2009), although the teleost pIgR only consists of two instead of five Ig-like domains relative to the mammalian molecule.

Ig$^+$ B cells and Ig$^-$ T cells are abundantly present in gut of fish (Rombout *et al.*, 1993a; Abelli *et al.*, 1997) but only limited data are available on their functional relevance. T cell receptor (TCRβ chain – see below) transcripts of rainbow trout IEL are highly diverse and polyclonal in adult naive individuals. The trout TCRβ repertoire is significantly modified upon a systemic rhabdovirus infection, but no specific differences between the trout IEL TCRβ repertoire can be detected compared with the spleen and head kidney repertoires, questioning whether a distinct IEL compartment really exists in teleosts (Bernard *et al.*, 2006). IELs show non-specific, MHC-independent cytotoxic activity, and IEL isolated from trout intestine are spontaneously cytotoxic against a mouse tumour cell line (McMillan and

Secombes, 1997). Lymphocytes purified from the sea bass intestinal mucosa also exhibit significant cytotoxic activity against xenogeneic (different species) or allogeneic (genetically different individual of same species) cell targets (Picchietti et al., 2010). Mucosal T cells in mammals are notoriously heterogeneous with regard to their phenotypes and functions, and two major subsets can be distinguished based on the type of TCR and MHC co-receptor expressed. The first group, or 'type a' cells, consists of TCR$\alpha\beta^+$ MHC class II-restricted CD4$^+$ and TCR$\alpha\beta^+$ MHC class I-restricted CD8$\alpha\beta^+$ lymphocytes, that resemble conventional thymus-selected antigen-experienced T cells also found in the blood, spleen and other secondary lymphoid organs. The second group, or 'type b' cells, express either TCR$\alpha\beta$ or TCR$\gamma\delta$ and they frequently express CD8$\alpha\alpha$ molecules but lack expression of the typical TCR co-receptors CD4 or CD8$\alpha\beta$ (van Wijk and Cheroutre, 2009). High ratios of CD8α^+ cells are found in trout intestine as well as in thymus and gill, but have a relatively low abundance in the head kidney, spleen and blood (Takizawa et al., 2011). Whether the CD8α^+ cells are CD8$\alpha\beta^+$ or CD8$\alpha\alpha^+$ is unknown.

1.2.6 Other immune relevant organs: the liver and the integumentary surface

As an aquatic organism, the surface of fish is subjected to continuous contact with many different types of microorganisms. The first barrier against pathogens, the integumentary surface, is equipped with mechanisms to protect against pathogen entry. One of the most important is the secretion of mucus containing a diverse group of antibacterial molecules. The latter are peptide-based molecules that act both directly and indirectly on components of the bacterial cell wall, resulting in lysis. Mucus is an important barrier in fish, and covers most of the external surfaces, especially the skin. It provides a substrate in which antimicrobial mechanisms may act, and contains many immune molecules including antibacterial agents (Kumari et al., 2011), anti-viral components (Raj et al., 2011) and interlectins (Rajan et al., 2011; Tsutsui et al., 2011). The production of mucus is significantly increased when a fish is subjected to a stressful situation, and it is apparent that most freshwater species have a higher production of mucus compared with marine species.

The liver is responsible for protein, carbohydrate and lipid metabolism, bile secretion and detoxification. With such critical metabolic functions it is often forgotten that it is also an important immune organ. In mammals the liver is a mediator of systemic and local innate immunity and is an important site of immune regulation. Hepatic cells of the myeloid lineage are present, including Kupffer cells and DCs. Intrahepatic lymphocytes are also present but distinct in both phenotype and function from their counterparts in other organs and include both conventional T cells, B cells, NK cells and nonconventional lymphoid cells (natural killer T (NKT) cells, $\gamma\delta$TCR$^+$ T cells, CD4⁻CD8⁻ T cells) (Nemeth *et al.*, 2009). It appears that the liver is the only non-lymphoid organ able to retain and activate naive CD8⁺ T cells in an antigen-specific manner and this is associated with the induction of T cell tolerance (Holz *et al.*, 2010). The immune relevance of fish liver is understudied but the large impact on immune gene expression after bacterial infection (Martin *et al.*, 2010; Millán *et al.*, 2011) suggests that the fish liver is actively involved in immune defence.

1.3   Fish innate immune response

1.3.1   Pattern recognition receptors (PRRs)

The innate immune response is initiated by detection of infectious agents by pathogen recognition receptors (PRRs) (Palm and Medzhitov, 2009). Vertebrates have evolved a vast array of PRRs, in both the extracellular and intracellular compartments, for detecting and responding to pathogen-associated molecular patterns (PAMPs), or to danger-associated molecular patterns (DAMPs) which are endogenous molecules released by damaged or stressed host cells (Hansen *et al.*, 2011). Bacterial PAMPs include components of the bacterial cell wall such as lipopolysaccharide (LPS) and peptidoglycan, flagellin, and DNA or RNA structures that are unique to bacteria (Boltaña *et al.*, 2011). Viral PAMPs survey the cytosolic compartment for viral genome amplification and/or mRNA metabolism and viral protein expression, including both double-stranded and single-stranded non-capped RNA (Wilkins and Gale, 2010). Fungal PAMPs can also be detected and are associated with early germ tube formation and hyphal forms that express high levels of zymosan.

PRRs are germline-encoded and can be classified into at least five major groups: the C-type lectins (CLRs), the Toll-like receptors (TLRs); retinoic acid inducible gene I (RIG-I)-like receptors (RLRs); the nucleotide-binding domain, leucine-rich repeat containing proteins (NLRs, also known as nucleotide binding oligomerization domain (NOD)-like receptors); and the newly classified HIN200/PYHIN family members that have recently been designated absent in melanoma (AIM)-like receptors (ALRs) (Hansen *et al.*, 2011).

The basic characteristics of PRRs are the same; they all possess a protein domain for recognizing PAMPs (or DAMPs) coupled (sometimes with an intervening domain) to a protein domain that interacts with downstream signalling molecules. A limited collection of functionally analogous protein domains for both PAMP recognition and signal propagation is shared between PRRs. Leucine-rich repeat (LRR) domains are essential for the specificity of PAMP recognition by both TLRs and NLRs. LRR are also present in the variable lymphocyte receptors of Agnathans that are analogous to antigen receptors in fish, indicating the specificity and usage of LRR in recognizing non-self is not limited to PRR (Pancer *et al.*, 2004). ALRs use the unique hematopoietic interferon-inducible nuclear proteins that contain a 200-amino acid repeat (HIN-200) domain(s) for PAMP recognition (Hornung *et al.*, 2009), whereas RLRs bind PAMPs with both helicase and regulatory domains (RD) (Yoneyama and Fujita, 2008). After sensing PAMPs, PRRs deliver a signal via their effector or signalling domain. In TLRs, the effector domain is a Toll/interleukin-1 receptor (TIR) domain. However, NLRs, RLRs and ALRs use either a caspase recruitment domain (CARD) or the related PYRIN domain (PYD) to interact with other proteins.

For the most part, CLRs, TLRs, RLRs, NLRs and ALRs are expressed by first responder cells of the immune system, including monocytes, macrophages, DCs and neutrophils. In addition, some non-immune cells express PRRs, including tissue-specific epithelial and endothelial cells as well as cells of the nervous system. The cellular location of PRRs varies; they are expressed on the cell surface (TLRs), in endosomes (TLRs) or within the cytosol (NLRs, RLRs and ALRs) of host cells.

C-type lectin receptors (CLRs)
CLRs contain a highly conserved carbohydrate recognition domain named a C-type lectin-like domain (CTLD) that binds carbohydrates in a Ca^{2+}-dependent manner (den Dunnen *et al.*, 2010). They interact with pathogens primarily through recognition of distinct carbohydrates, such as mannose, fucose or glucan structures. Most pathogens express carbohydrate structures on their surface that function as a so-called sugar fingerprint and are recognized by specific CLRs. Carbohydrate recognition by CLRs enables antigen presenting cells to recognize the major pathogen classes, i.e. mannose specificity allows recognition of viruses, fungi and mycobacteria, whereas fucose structures are more specifically expressed by particular bacteria and helminths, and glucan structures are present on mycobacteria and fungi. It is becoming clear that C-type lectins are important PRRs that recognize carbohydrate structures. Following pathogen binding, CLRs trigger distinct signalling pathways that induce the expression of specific cytokines which determine T cell polarization fates. Some CLRs induce signalling pathways that directly activate the transcription factor NF-κB, whereas other CLRs affect signalling by Toll-like receptors (Geijtenbeek and Gringhuis, 2009).

Many CLRs have been described in various fish, e.g. mannose-binding lectin (MBL), galectins and DC-SIGN. Mammalian MBL is a member of the collectin family, whose members possess a collagen-like domain in the N-terminus and C-type lectin-like domain in the C-terminus. MBLs form an oligomer in the collagen-like domain, and then form complexes with MBL-associated serine proteases (MASPs). By forming a complex with MASPs, MBLs activate the lectin pathway of the complement system (Kondo *et al.*, 2007). Two carp MBLs and three trout MBLs have been identified. The trout MBLs are synthesized in spleen, HK and are co-expressed with MASP in the liver (Kania *et al.*, 2010). The carp MBLs are associated with MASP2 and cleave human C4. Interestingly, one carp MBL has galactose specificity, whilst the other has mannose specificity (Nakao *et al.*, 2006). This suggests that bony fish have developed a diverged set of MBL homologues that recognize carbohydrates from different pathogens in the lectin complement activation pathways.

CLRs are either secreted proteins or transmembrane proteins. Galectins are synthesized and stored in the cytoplasm, but upon infection-initiated tissue damage and/or following prolonged infection, cytosolic galectins are either passively released by dying cells or actively secreted by inflammatory activated cells through a non-classical pathway, the 'leaderless' secretory pathway. Once exported, galectins act as PRRs, as well as immunomodulators in the innate response to some infectious diseases (Sato *et al.*, 2009). As galectins are predominantly found in lesions where pathogen-initiated tissue damage signals appear, this lectin family is also considered as a potential DAMP candidate that may orchestrate innate immune responses alongside the PAMP system. A sea bass (*Dicentrarchus labrax*) galectin is upregulated by a nodavirus infection and decreases respiratory burst activity (oxygen free radical production) in head kidney leucocytes (Poisa-Beiro *et al.*, 2009). The galectin protein is expressed in the brain in virus-infected fish, and decreases the expression of IL-1β, TNF-α and Mx in the brain when co-injected with nodavirus, suggesting a potential anti-inflammatory, protective role of galectin during viral infection.

Mammalian DC-SIGN contains a carbohydrate recognition domain (CRD), a repeat neck region, and a transmembrane region followed by a cytoplasmic tail containing recycling and internalization motifs. DC-SIGN ligation can result in transmission of intracellular signalling and this has been associated with the presence of a di-leucine motif and a tyrosine residue in the cytoplasmic tail (Svajger *et al.*, 2010). DC-SIGN is preferentially expressed on myeloid DCs and is considered a DC-specific phenotypic marker. DC-SIGN recognizes several pathogens, contributing to generation of pathogen-tailored immune responses. A zebrafish DC-SIGN is associated with various antigen presenting cells, including macrophages, B cells and DC-like cells. Its expression in immune-related tissues is up-regulated by exogenous antigens and IL-4, and it is involved in T cell activation, antibody (IgM) production, and bacterial vaccine-elicited immunoprotection (Lin *et al.*, 2009).

Toll-like receptors (TLRs)
TLRs are type I transmembrane proteins that consist of three major domains: (1) a leucine-rich repeat (LRR) extracellular domain, (2) a transmembrane domain, and (3) a cytoplasmic TIR domain. There are 10 human TLRs (TLR1–10) and 13 murine TLRs (TLR1–13, although TLR10 is not functional in mice because of a retroviral insertion) that each have a different PAMP specificity. TLR1–10 are conserved between humans and mice, but TLR11–13 are not present in humans. Thus, despite some

species-specific receptors, many TLR members are conserved in mammals (Yamamoto and Takeda, 2010). TLRs can be divided into extracellular TLRs (e.g. human TLR1, 2, 4, 5, 6 and 10) that recognize their ligands on the cell surface; and intracellular TLRs (e.g. human TLR3, 7, 8, and 9) that are expressed in endosomes. In each case, the TIR moiety residues are in the cell cytoplasm while the LRR domain is positioned to detect either extracellular PAMPs (outside the cell) or during sampling within endosomes (Hansen *et al.*, 2011).

TLRs form hetero- or homodimers to provide different PAMP specificity and this probably facilitates dimerization of the cytoplasmic TIR domain to activate intracellular signalling. The extracellular LRR domain is composed of 19–25 tandem copies of the 'xLxxLxLxx' motif that confer TLR ligand specificity (Yamamoto and Takeda, 2010). A number of PAMPs from various microbes are detected by TLRs. For example, lipoprotein is detected by TLR1, 2 and 6, LPS by TLR4, flagellin by TLR5, double-stranded (ds) RNA by TLR3, single-stranded (ss)RNA by TLR7–8, and CpG DNA by TLR9. Upon engagement of their LRRs with ligands, TLR engage specific adaptor molecules using their cytoplasmic TIR domains and trigger inflammatory cascades that lead to the production of cytokines, which are critical for innate responses such as phagocytosis and the respiratory burst, and for activation of the adaptive immune system.

TLRs are an ancient family of PRRs and even non-vertebrate genomes encode multiple TLRs. Thus the amphioxus genome has 28 predicted TLR genes composed of combinations of LRR, TM and TIR domains, with over 200 TLRs in the sea urchin and at least 8 unique TLRs in *Drosophila* (Huang *et al.*, 2008). There are about 10 distinct TLR genes in birds (Temperley *et al.*, 2008) and approximately 20 TLRs in amphibians (Ishii *et al.*, 2007). A total of 17 distinct TLRs have been identified in more than a dozen different fish species (Rebl *et al.*, 2010). Of these, several are direct structural orthologues of mammalian TLRs (e.g. TLRs 1–5, 7–9, 12–13) while others have no predicted mammalian orthologues (e.g. TLRs 14 and 18–23). It is worth noting that mammalian TLR6 and TLR10 are absent in teleosts, and some piscine TLRs are encoded by duplicated genes, for example salmonid TLR22. Functional studies reveal that stimulation with viruses, or Gram-positive and Gram-negative bacteria may regulate the expression of certain TLRs in fish (Stafford *et al.*, 2003; Hirono *et al.*, 2004; Rebl *et al.*, 2007; Meijer *et al.*, 2004; Matsuo *et al.*, 2008).

Divergent TLRs require divergent downstream binding partners. All mammalian TLRs, except TLR3, interact with myeloid differentiation primary response gene-88 (MyD88) to activate NF-κB, MAP kinase, or IRF signalling pathways; many also interact with a second adaptor, while TLR4 interacts with several adaptors. Many TLR signalling proteins, including MyD88, are structurally and functionally conserved in teleost fish (Sullivan *et al.*, 2007). Numerous studies have revealed that specific piscine TLRs share functional properties with their mammalian counterparts (Boltaña *et al.*, 2011). Nevertheless, distinct features of teleost TLR cascades have been discovered.

Mammalian TLR4 is mainly responsible for recognition of LPS, a characteristic cell wall component of Gram-negative bacteria that provokes strong immune reactions in mammals, which in extreme cases culminate in septic shock (Gutsmann *et al.*, 2007). In contrast, fish are often resistant to the toxic effects of LPS (Swain *et al.*, 2008). A key step in LPS detection in mammals is the transportation of LPS aggregates to the cell surface, mediated by LPS-binding protein (LBP), in order to form a ternary complex with CD14. This facilitates the transfer of monomeric LPS to TLR4 and another co-stimulatory molecule known as myeloid differentiation protein 2 (MD2) (Dauphinee and Karsan, 2006). Whilst LBP-like molecules are present in a number of different piscine species (Solstad *et al.*, 2007), so far neither CD14 nor MD2 have been isolated from fish. Indeed, the TLR4 gene has only been identified in cyprinids to date, including Chinese rare minnow (*Gobiocypris rarus*) and zebrafish and has not been identifiable in other genome sequenced fish, such as fugu (*Takifugu rubripes*), tetraodon (*Tetraodon nigroviridis*) and stickleback (*Gasterosteus aculeatus*) (Boltaña *et al.*, 2011). The expression of rare minnow TLR4 can be induced by bacterial and viral infection (Su *et al.*, 2009). Unexpectedly, the extracellular region (LRR) of zebrafish TLR4 cannot sense LPS (Sullivan *et al.*, 2009) and zebrafish TLR4 negatively regulates the MyD88-dependent TLR pathway in embryos (Sepulcre *et al.*, 2009).

Although the fish TLRs of mammalian orthologues may have different roles from those seen in mammals, the roles of the non-mammalian TLRs will be particularly important to determine. Genomic analysis has revealed that TLR22 is present in fish and amphibians but not in birds and mammals, and thus is an aquatic animal-specific gene presumably lost during vertebrate evolution (Matsuo *et al.*, 2008). Fish TLR22 expression is induced by Poly I:C and induces interferon (IFN) production as seen with mammalian TLR3, but unlike TLR3, TLR22 localizes to the cell surface (Hirono *et al.*, 2004). When cells expressing TLR22 are exposed to dsRNA or dsRNA viruses, the cells induce IFN responses and acquire resistance to virus infection. In addition, TLR3 and TLR22 can discriminate the size of dsRNA. TLR3 preferentially recognizes short dsRNA (<1 kbp), whereas TLR22 prefers long dsRNA (>1 kbp) (Matsuo *et al.*, 2008). Thus fish have two non-redundant RNA sensors, TLR22 and TLR3, on the cell surface and intracellularly, respectively, for effective protection of fish from RNA virus infection in the water (Oshiumi *et al.*, 2008).

RIG-I-like receptors (RLRs)
The RLRs are crucial to the RNA virus triggered interferon response. They consist of three members, retinoic acid-inducible gene I (RIG-I), melanoma differentiation-associated gene-5 (MDA5) and laboratory of genetics and physiology-2 (LGP2) (Zou *et al.*, 2009; Matsumiya *et al.*, 2011). RLRs are

structurally related and contain a DExD/H or closely related type III restriction enzyme (Res III) domain, a helicase domain, and a C-terminal regulatory domain (RD). All the molecules have a common functional RNA helicase domain near the C terminus specifically binding to RNA molecules of viral origin. The ATP-dependent DExD/H domain contains a conserved motif Asp-Glu-X-Asp/His (DExD/H) which is involved in ATP-dependent RNA or DNA unwinding. The C-terminal RD domain binds viral RNA in a 5′-triphosphate-dependent manner and may confer ligand specificity (Cui et al., 2008). For example, whilst RIG-I can recognize viral ssRNA as well as short dsRNA, MDA5 recognizes long dsRNA (Takahasi et al., 2009). In addition both RIG-I and MDA-5 have two caspase recruitment domains (CARD), which are essential for the interaction with an adaptor molecule and the ensuing antiviral responses.

Under resting conditions, the CARDs are enveloped in other domains of RIG-I and MDA5. Once the ligand is bound by the RD, a conformational shift is thought to orientate the CARD domains of the RLR to interact with their specific adaptor molecules, leading to the activation of antiviral signalling molecules including IRF-3 and NF-κB (Cui et al., 2008). Mitochondrial antiviral signalling (MAVS) protein is one of the adaptor molecules responsible for antiviral signalling triggered by RLRs. LGP2, which lacks a CARD domain, has been shown to negatively regulate RLR signalling via sequestration of dsRNA, complexing with MAVS, or by directly binding RIG-I (Komuro et al., 2008). It can also potentiate IFN production in response to viral infection (Satoh et al., 2010).

The origins of the RLR system of viral detection predate the emergence of the Gnathostomes and RIG-1, MDA5, LGP2 and MAVS have been identified in teleost fish (Zou et al., 2010; Chang et al., 2011a; Hansen et al., 2011). The salmonid orthologues have the same domain structures as seen in mammals. Teleost MDA5 and LGP2 are both up-regulated in a rapid but transient manner in lymphoid tissues during virus infection in grass carp, indicating a potential involvement in the early anti-viral response (Huang et al., 2010; Su et al., 2010). Over-expression of either full-length teleost RIG-I or the RIG-I CARD domain alone leads to significant induction of an antiviral state, as measured by plaque reduction assays and the induction of IFN stimulated genes including the RLRs and IFN itself (Biacchesi et al., 2009). Over-expression of teleost MAVS protects cells from infection by both DNA and RNA viruses by inducing IFN stimulated genes, such as IRF-3, Mx and Vig-1, as well as type I IFN (Biacchesi et al., 2009; Lauksund et al., 2009; Simora et al., 2010). These findings suggest that teleost fish possess a functional RLR anti-viral pathway.

NOD-like receptors (NLRs)
NLRs are a family of molecules that sense a wide range of ligands within the cytoplasm of cells. This family comprises 23 members in humans and approximately 34 in mice. These sensors comprise three domains: the

C-terminal LRR domain which is thought to be involved in the recognition of microbial PAMPs; an N-terminal effector domain and an intermediate domain consisting of nucleotide-binding and oligomerization (NACHT) domains, which are required for ligand-induced, ATP-dependent oligomerization of the sensors and formation of active receptor complexes for activation of downstream signalling (Kumar et al., 2011). Ligand sensing by the LRR results in NACHT oligomerization, which shifts the receptor conformation such that the N-terminal effector domains are accessible for interacting with signalling molecules or molecular adaptors to induce the production of inflammatory cytokines, or activate a multiprotein-complex, the 'inflammasome', which either initiates the proteolytic cleavage (or maturation) of various caspases resulting in the maturation and production of inflammatory cytokines, such as IL-1β and IL-18, or initiates cell death (Kumar et al., 2011).

The function of several NLR members, e.g. the NLRC family members NOD1–2 and NLRP family members, are well characterized in mammals and have also been discovered in fish (Stein et al., 2007; Laing et al., 2008; Chang et al., 2011b). In addition, teleost fish possess a unique group of NLRs of several hundred genes, many of which are predicted to encode a C-terminal B30.2 domain (Laing et al., 2008). Mammalian NOD1 and NOD2 (also known as CARD4 and CARD15, respectively) are mainly expressed in the cytosol of various cells and comprise C-terminal LRRs, a central NACHT domain and an N-terminal domain containing either one (NOD1) or two (NOD2) CARDs. They discriminate bacterial pathogens via detection of bacterial cell wall components; they recognize diaminopimelic acid (DAP, restricted to Gram-negative bacteria) and muramyl dipeptide (MDP, found in Gram-negative and positive bacteria) respectively using their distinct LRRs (Hansen et al., 2011). PAMP recognition initiates oligomerization of these sensors, which subsequently recruit a CARD-containing adaptor protein known as RIP2 (RICK) via CARD–CARD interactions, and activate NF-κB and MAP kinases to induce the transcription of inflammatory cytokines. Teleost NOD1 and NOD2 have the same structure as their mammalian orthologues and their transcripts are up-regulated during bacterial and viral infection, or stimulation with Poly I:C (Sha et al., 2009; Chen et al., 2010a; Chang et al., 2011b). Overexpression of the CARD domain of trout NOD2 significantly induces pro-inflammatory genes, including IL-1β, presumably through interaction with RIP2 kinase (Chang et al., 2011b).

AIM2-like receptors (ALRs)
ALRs are a small gene family that includes four members in humans and six in mice that share a characteristic hallmark IFI200 domain (also known as HIN200, Pfam domain 02760) and is currently characterized only in mammals. They were named after the founding member of this gene family, absent in melanoma 2 (AIM2) (Unterholzner et al., 2010). AIM2 possesses

an N-terminal PYD coupled to the IFI200 domain whereas the γ-IFN-inducible protein 16 (IFI16) encodes PYD coupled to two tandem IFI200 domains. AIM2 and IFI16 have recently been directly implicated as PRRs to detect intracellular microbial DNA leading to IFN production (Rathinam et al., 2010; Unterholzner et al., 2010). The presence of this family of PRRs in fish is waiting to be defined (Hansen et al., 2011).

The existence of multiple types of PRRs in different cellular compartments, some of which sense similar ligands, highlights the synergistic complexities possible in innate immune responses in a single organism. As well as being the first line of defence, the innate immune system plays an essential role in inducing and manipulating adaptive immunity. By working together or antagonistically, the different outcomes induced by TLRs, RLRs and NLRs/ALRs in the context of infection can allow for a high degree of specificity in the overall response to particular microbes. The intersection of these PRR pathways and downstream effects is an exciting area of future research and may reveal how different animals evolve unique solutions to the pressures posed by pathogenic microbes.

1.3.2   Antimicrobial peptides (AMPs)

Fish continually fight against pathogens by secreting a wide range of antimicrobial peptides (AMPs) as an innate defence mechanism. AMPs, also known as host defence peptides, play major roles in the innate immune system, and protect against a wide variety of bacterial, viral, fungal and other pathogenic infections by disruptive 'lytic' or pore-forming 'ionophoric' actions (Smith et al., 2010). In general, AMPs are secreted in the saliva, mucus, circulatory system and other areas which are high-risk pathogen targets (Noga et al., 2011). In addition to their direct microbicidal effects, AMPs have other roles in inflammatory responses, including recruitment of neutrophils and fibroblasts, promotion of mast cell degranulation, enhancement of phagocytosis and decreasing fibrinolysis (Plouffe et al., 2005).

Many of the fish AMPs isolated from the epidermal cells or secretions of the skin, gills and intestine in early studies have high sequence homology to segments of other proteins (particularly histone or histone-like molecules) indicating that they may in fact be cleavage products of larger molecules (e.g. Park et al., 1998; Fernandes et al., 2004). For example, a number of specific antimicrobial molecules characterized in rainbow trout, called oncorhyncins, have been found to be very similar to chromosomal proteins (Fernandes et al., 2004). A number of AMP genes, including genes for liver expressed AMPs (LEAPs), β-defensins, cathelicidins and piscidins have also been cloned recently that will be discussed in detail below.

LEAPs

LEAP-1 or hepcidin, is a highly disulphide bonded (rich in Cys) β-sheet AMP. Since the first identification of hepcidin from human liver, hepcidin genes have been identified from various vertebrates and fish species, including Japanese flounder, tilapia, zebrafish and salmonids (Douglas *et al.*, 2003; Chen *et al.*, 2005; Hirono *et al.*, 2005; Huang *et al.*, 2007). Some fish (e.g. Japanese flounder and tilapia) possess multiple paralogues of hepcidin genes.

The Japanese flounder hepcidin JF2 peptide (26-aa) has antimicrobial activities against various Gram-negative (*Escherichia coli*) and Gram-positive (*Staphylococcus aureus* and *Lactococcus garvieae*) bacterial species but does not show antimicrobial function against the Gram-negative *Edwardsiella tarda*. Transgenic zebrafish expressing tilapia hepcidin show resistance to infection by the Gram-negative bacterium *Vibrio vulnificus* (Hsieh *et al.*, 2010). In mammalian systems, hepcidin is involved in both iron-homeostasis and host-defence functions and its expression is induced by IL-6 only in hepatocytes but not macrophages. However, rainbow trout hepcidin can be induced by IL-6 in macrophages and IL-6 is also induced in macrophages following infection (Costa *et al.*, 2011). Thus, in fish during sepsis patrolling macrophages express increased IL-6 that induces hepcidin expression, and in this way may act to reduce iron availability and lead to iron deficiency, as a means to limit the spread of infection. It has been hypothesized that the Japanese flounder JF1 peptide may function in iron homeostasis, while JF2 is involved in antimicrobial functions (Hirono *et al.*, 2005) although additional work is needed to confirm this. The tilapia hepcidin TH1–5 molecule has antiviral activity against fish nervous necrosis virus (NNV) infection confirming that AMPs are not only antibacterial (Chia *et al.*, 2010). A second LEAP, LEAP-2, has also been discovered in fish (Zhang *et al.*, 2004; Bao *et al.*, 2006; Liu *et al.*, 2010). It is generally most highly expressed in the liver, but can be induced elsewhere upon infection. Thus, grass carp infected by *Aeromonas hydrophila* have significant up-regulation of LEAP-2 in gill, skin, muscle, spleen, blood, head kidney, heart and intestine, whilst in channel catfish infected with *Edwardsiella ictaluri*, the causative agent of enteric septicemia of catfish, up-regulation is seen in the spleen.

β-Defensins (BD)

Defensins are small (3.5–4.5 kDa), cationic and amphipathic peptides with six conserved cysteine residues. They can be classified as α-, β- or θ-defensins depending on the position of the cysteines and topology of the disulphide bonds. Only β-defensin (BD) genes have been found so far in fish (Cuesta *et al.*, 2011), first identified in zebrafish, tetraodon and fugu aided by their sequenced genomes (Zou *et al.*, 2007). BD were later cloned in rainbow trout and other fish species (Casadei *et al.*, 2009; Falco *et al.*, 2008; Zhao *et al.*, 2009; Jin *et al.*, 2010; Nam *et al.*, 2010; Cuesta *et al.*, 2011).

Four BD (BD-1–4) have been cloned in rainbow trout to date. They are expressed in most tissues and are up-regulated in head kidney leucocytes by Poly I:C stimulation. After bacterial infection *in vivo*, only BD-2 (in gut) and BD-3 (in gills) genes were significantly up-regulated (Casadei *et al.*, 2009). The recombinant BD-1 peptide increases Mx1 gene expression and has antiviral activity against viral hemorrhagic septicaemia virus (VHSV) (Falco *et al.*, 2008). Five Japanese flounder (*Paralichthys olivaceus*) BD genes have been identified by expressed sequence tag (EST) analysis and are found clustered at the same locus with a conserved gene organization (Nam *et al.*, 2010). The flounder BD-1 mRNA is expressed constitutively in early developmental stages after hatching, and is induced in the head kidney of juvenile fish by pathogen challenge. The recombinant flounder BD-1 protein has antimicrobial activity against *E. coli* (Nam *et al.*, 2010). Interestingly, in grouper (*Epinephelus cocoides*) a β-defensin has been identified recently that is exclusively expressed in the pituitary gland and testis with antimicrobial and antiviral functions, suggesting a role in the reproductive and endocrine systems (Jin *et al.*, 2010).

Cathelicidins
Cathelicidins comprise a large number of precursors of AMPs in vertebrate species, which typically contain a conserved N-terminal sequence ('cathelin' domain) and a C-terminal antimicrobial domain of varied sequence and length. The C-terminal peptide expresses antimicrobial activity after cleavage from the prepropeptide by neutrophil elastase or other proteases (Bals and Wilson, 2003). They exhibit broad-spectrum antimicrobial activity against a wide range of microorganisms, and possess the ability to neutralize endotoxin. Cathelicidin genes have been identified in mammals and birds. Fish cathelicidins were first described in Atlantic hagfish (*Myxine glutinosa*) where three potent broad-spectrum antimicrobial peptides isolated from intestinal tissues are encoded by three cathelicidin genes (Uzzell *et al.*, 2003). Teleost fish cathelicidin genes were subsequently cloned in salmonids, Atlantic cod (*G. morhua*) and ayu (*P. altivelis*) (Chang *et al.*, 2005, 2006; Maier *et al.*, 2008; Scocchi *et al.*, 2009; Broekman *et al.*, 2011; Lu *et al.*, 2011; Shewring *et al.*, 2011).

The Atlantic cod cathelicidin is expressed early in development and is modulated by different feeding regimes (Broekman *et al.*, 2011). The expression of cathelicidins can be induced by bacterial infection, as seen in salmonids, Atlantic cod and ayu, indicating a role of these proteins in fish innate immunity. Indeed, synthesized peptides from the mature peptide N-terminal region from different species have been shown to have antimicrobial activity. A 36-residue peptide corresponding to the core part of the fish cathelicidin exhibits various antibacterial activities against all 10 different microorganisms studied, including Gram-negative and Gram-positive bacteria, and has a low hemolytic effect (Chang *et al.*, 2005). Whilst the full-length ayu cathelicidin has no antimicrobial activity detected by inhibition zone assay, the mature peptide exhibits an antimicrobial capability against all tested bacteria and has the strongest activity against *A. hydrophila* (Lu *et al.*, 2011).

Piscidins

Piscidins are a family of AMPs found in a fish species and show broad-spectrum activity against bacteria, fungi and viruses (Cole *et al.*, 1997; Yin *et al.*, 2006; Fernandes *et al.*, 2010). These molecules have a marked amphipathic character, due to well-defined hydrophobic and hydrophilic regions, and most are linear peptides with less than 26 residues and a high proportion of basic amino acids (phenylalanine and isoleucine). They are unstructured in water but have a high α-helix content in dodecylphosphocholine (DPC) micelles, a structure similar to those determined for other cationic peptides involved in permeabilization of bacterial membranes (Campagna *et al.*, 2007).

Pleurocidin, a 25-residue linear antimicrobial peptide, was the first piscidin isolated, from the skin secretions of winter flounder (*Pleuronectes americanus*) (Cole *et al.*, 1997). The pleurocidin gene comprises four exons encoding for a 68-residue prepropeptide that undergoes proteolytic cleavage of its amino-terminal signal and carboxy-terminal anionic propiece to form the active, mature peptide. Pleurocidin is localized in mucin granules of skin and intestinal goblet cells (Cole *et al.*, 2000).

The piscidins of the hybrid striped bass (*Morone chrysops* female × *Morone saxatilis* male) were the first to be shown to reside within mast cells (Silphaduang and Noga, 2001). This phenomenon is now known to be widespread in a number of fish species (Fernandes *et al.*, 2010). Four piscidins (1–4) have been isolated from hybrid striped to date, with piscidin-4 being unusual in being twice as long (44 amino acids) as typical members of the piscidin family (Park *et al.*, 2011).

Epinecidin-1 is another member of the fish piscidin family of antimicrobial peptides, isolated from the grouper (*E. coioides*). The epenecidin-1 transcript is highly expressed in the head kidney, intestine and skin, and is up-regulated by stimulation with LPS and Poly I:C (Yin *et al.*, 2006). Three genes, with either a short or a long 5′-untranslated region (UTR), have been isolated that potentially encode for three epinecidin-1 peptides (Pan *et al.*, 2008). Most recently piscidins (Gaduscidin-1 and 2) have been isolated from Atlantic cod, where they are highly expressed constitutively in immune tissues and are induced in the spleen following injection with killed bacteria (Browne *et al.*, 2011).

1.3.3 Fish complement system

Components of the fish complement system

The complement system is a major component of the innate defences. In mammals it is composed of about 30 distinct plasma proteins and membrane-associated proteins (Table 1.1), responsible for various immune effecter functions. These include elimination of invading pathogens, promotion of inflammatory responses, clearance of apoptotic cells and necrotic cell debris, and modulation of the adaptive immune responses (Nakao

Table 1.1 Complement components identified in teleost fish, with rainbow trout and carp/zebrafish examples

Human	Trout	Carp or zebrafish
C1q		Zebrafish C1qA, C1qB, C1qC (Hu *et al.*, 2010)
C1r and C1s	C1r/C1s (Wang and Secombes, 2003)	C1r/C1s/MASP2-like (Nakao *et al.*, 2001)
MBL	MBL-H1,H2,H3 (Kania *et al.*, 2010)	MBL, GalBL (Vitved *et al.*, 2010; Nakao *et al.*, 2006)
MASP1		
MASP2		MASP2 (Nakao *et al.*, 2006)
MASP3		MASP3 (Endo *et al.*, 1998)
sMAP-like		MRP (Nagai *et al.*, 2000)
C3	C3-1, C3-2, C3-3, C3-4 (Nonaka *et al.*,1984, 1985; Sunyer *et al.*, 1996; Zarkadis *et al.*, 2001)	C3-H1, C3-H2, C3-S, C3-Q1 and C3-Q2 (Nakao *et al.*, 2000)
C4	C4-1, C4-2 (Wang and Secombes, 2003; Boshra *et al.*, 2004a)	C4-1, C4-2 (Mutsuro *et al.*, 2005)
C5	C5 (Franchini *et al.*, 2001)	C5-1, C5-2 (Kato *et al.*, 2003)
B and C2	Bf1, Bf2 (Sunyer *et al.*, 1998)	B/C2-A1,2,3, B/C2-B (Nakao *et al.*, 1998, 2002)
C6	C6 (Chondrou *et al.* (2006b)	
C7	C7-1, C7-2 (Zarkadis *et al.*, 2005; Papanastasiou and Zarkadis, 2005)	
C8α	C8α (Papanastasiou and Zarkadis, 2006a)	C8α (Uemura *et al.*, 1996)
C8β	C8β (Kazantzi *et al.*, 2003)	C8β (Uemura *et al.*, 1996)
C8γ	C8γ (Papanastasiou and Zarkadis, 2006b)	C8γ (Uemura *et al.*, 1996)
C9	C9 (Chondrou *et al.*, 2006a)	C9 (Uemura *et al.*, 1996)
D	D (Boshra *et al.*, 2004a)	D (Yano and Nakao, 1994)
I	I (Anastasiou *et al.*, 2011)	FI-A, FI-B (Nakao *et al.*, 2003)
H or C4bp	H (Anastasiou *et al.*, 2011)	Zebrafish CFH and CFHL1-4 (Sun *et al.*, 2010)
MCP or DAF		
CR1 or CR2		
CR3	CR3 (Mikrou *et al.*, 2009)	
CR4		
P	Pfc1, Pfc2, Pfc3 (Chondrou *et al.*, 2008)	
C3aR	C3aR (Boshra *et al.*, 2005)	
C5aR	C5aR (Fujiki *et al.*, 2003; Boshra *et al.*, 2004b)	
C1 inhibitor	C1 inhibitor (Wang and Secombes, 2003)	
CD59	CD59-1, CD59-2 (Papanastasiou *et al.*, 2007)	
Clusterin	Clusterin-1, Clusterin-2 (Londou *et al.*, 2008)	
Vitronectin	Vitronectin (Marioli and Zarkadis, 2008)	

et al., 2011). The identification of fish complement components was aided by application of homology cloning techniques, and then much accelerated by utilization of genomic and transcriptomic data as it became available for several teleost species. It is now evident that almost all of the mammalian complement components have homologues in teleost fish, as shown in Table 1.1. Thus, the teleost complement system is equivalent or comparable to the mammalian system from both a structural and functional viewpoint (Boshra *et al.*, 2006).

As seen in Table 1.1, several components, such as C3, C4, C5, C7, MBL, factor B/C2 and factor I, have multiple paralogues, as a result of genome or gene duplications. Some duplicates are only seen in pseudotetraploid species such as carp and salmonids, indicating that they most likely arose by tetraploidization. However, some occur in diploid species such as zebrafish, medaka and fugu (Boshra *et al.*, 2006), and may be attributable to a genome duplication event that is believed to have occurred in the common ancestor of the teleost lineage, although other tandem gene duplication events may also contribute.

The expression of fish complement components

In mammals, the liver is recognized as the primary site for production of the majority of the complement components, with exceptions being C1q and properdin which are expressed mainly in macrophages. However, local extrahepatic expression of complement components can play a significant role in regulation of local immune responses including crosstalk with adaptive immunity as well as damage and repair of tissues (K. Li *et al.*, 2007). In teleost fish, mRNAs encoding complement components show a substantially wider tissue distribution, and are found in the head kidney, renal (body) kidney, intestine, gill, skin, brain and gonads (Løvoll *et al.*, 2007). These results suggest that the teleost complement system is not only operating in blood, lymph and body fluids but is also present at the local interface with the environment, potentially ready for invading pathogens. The extrahepatic production of complement components may also play a significant role in the clearance of damaged host cells, organ morphogenesis and tissue regeneration in fish as reported for mammals (Ricklin *et al.*, 2010).

Many complement components have been shown to be present in eggs and to be of maternal origin. Thus these proteins are present in the embryo before they can start to be synthesized during ontogeny, as reported for C3, C4, C5, Bf and C7 in rainbow trout. In general, complement transcript levels increase steadily from day 28 post-fertilization to hatching, followed by a decrease during yolk-sac resorption (Løvoll *et al.*, 2006). In zebrafish, immunization with *A. hydrophila* increases the levels of C3 and B in female fish and the early developmental stages of their offspring. It is particularly noteworthy that the maternal transfer of these up-regulated complement components can also transfer protection against the same bacterial species

(Wang *et al.*, 2009). If this maternally derived protection is attributable to enhancement of innate immunity including the complement system, then this maternal transfer may have value practically, to help improve the survival of fry against a wide range of pathogens in aquaculture.

Some teleost complement components have been recognized as acute-phase proteins as seen in mammals, suggesting they have an important role in immediate (first-line) defences against microbial infection. Thus, C3 behaves as a major acute-phase protein upon bacterial, viral and parasite infection (Nakao *et al.*, 2011). After bath exposure of zebrafish to viral hemorrhagic septicemia virus (VHSV) transcripts of the complement components (C3b, C8 and C9) are more highly increased in fins than in internal organs (spleen, head kidney and liver), suggesting local synthesis of the complement components at the infection site may play a key role in natural defence against VHSV infection (Encinas *et al.*, 2010).

The activation pathways of the fish complement system
The mammalian complement system has three activation pathways, classical, alternative and lectin, which merge at the proteolytic activation step of C3, the central component of the complement system. C3 is equipped with a unique intra-molecular thioester bond which is exposed to the molecular surface upon activation and forms a covalent bond with invading microorganisms (Endo *et al.*, 2006). The tagging of invading microorganisms by C3 activation products enhances phagocytosis of pathogens via C3 receptors on phagocytes, and contributes to the activation of the late complement components, C5–C9, which form a cytolytic complex (the so-called membrane attack complex). The classical pathway is activated by antibody–antigen complexes and is a major effector of antibody-mediated immunity. The lectin pathway activates complement following the recognition of microbial carbohydrate patterns by either mannose-binding lectin (MBL) or ficolins, and subsequently activates the associated unique enzymes MASPs. The alternative pathway is initiated by the covalent binding of a small amount of C3 molecules to hydroxyl or amine groups in cell surface molecules of microorganisms and does not involve specific recognition. This pathway also functions to amplify C3 activation (amplification loop) (Nakao *et al.*, 2011).

Although the above effector functions play crucial roles for successful host defence, excessive complement activation and activation on host cells can potentially cause serious damage to host tissue. Therefore, complement activation is controlled at multiple points in the pathway by several kinds of regulatory factors, such as C1-inhibitor, carboxypeptidase N, C4-binding protein (C4bp), factor H, decay-accelerating factor (DAF), membrane-cofactor protein (MCP), cluterin, vitronectin and CD59 (Nakao *et al.*, 2011).

The complement system of bony fish and cartilaginous fish appears to be fully equipped with the three C3-activation pathways and the cytolytic pathway, and shows many of the effector activities known from the mammalian complement system, such as target cell killing, opsonization and anaphylatoxic leucocyte stimulation (Boshra *et al.*, 2006).

The anaphylatoxins produced after fish complement activation
After activation, the target-bound C3b and its cleavage fragment iC3b, are recognized by various leucocytes, including neutrophils and macrophages and function as opsonins. On the other hand, C3a and C5a peptides released from C3 and C5, respectively, show potent physiological activities that induce leucocyte chemotaxis and degranulation, and promote inflammatory and allergic reactions (Walport, 2001).

Teleost fish, unlike any other known vertebrate group, contain multiple forms of the C3a anaphylatoxin as a result of the multiple isoforms of C3. All of these molecules are functionally active and play a prominent role in inducing oxygen free radical (superoxide) production from fish leucocytes. The C5a anaphylatoxin has also been characterized in fish, and as in mammals plays an important role in leucocyte chemotaxis and in triggering the respiratory burst of leucocytes (Sunyer *et al.*, 2005). Interestingly, it has been shown that rainbow trout anaphylatoxins play an unexpected role in enhancing phagocytosis of particles (Li *et al.*, 2004).

1.3.4   Cellular components of fish innate immunity
Innate immune responses lack 'immunological memory' and rapid 'secondary' responses following repeated exposure to the same pathogen that characterize adaptive immune responses, but can be stimulated by interaction with cells of the adaptive immune response and their products (Secombes, 1996). A broad range of key cell types are involved in the innate defences of teleost fish, including monocytes/macrophages, non-specific cytotoxic cells (NCC), NK-like cells and granulocytes (e.g. neutrophils). Some teleosts have been reported to have both acidophilic and basophilic granulocytes in peripheral blood in addition to neutrophils, but in other species only the latter cell type has been found (Ainsworth, 1992).

Monocytes/macrophages
The mononuclear phagocytic system consists of endothelial cells and macrophages which line small blood vessels and eliminate an array of soluble macromolecular physiologic and foreign waste products from the circulation by receptor-mediated endocytosis and phagocytosis (Whyte, 2007). Whilst the scavenger endothelial cells in all vertebrate classes are geared to express scavenger receptors for endocytosis of major physiologic waste products and are able to clear all major categories of biological macromolecules, macrophages can phagocytose particulate material (Seternes *et al.*, 2002). These scavenger endothelial cells are differentially positioned in vertebrate classes, i.e. gills in Agnatha and Chondrichthyes;

heart or kidney in Osteichthyes; and liver in Amphibia, Reptilia, Aves and Mammalia.

Macrophages arise from hematopoietic progenitors which differentiate directly, or via circulating monocytes, into subpopulations of tissue macrophage that, in mammals, include Kupffer cells in the liver, alveolar macrophages in the lung, microglia cells in the central nervous system, osteoclasts in bone tissue and specialized macrophages in the spleen (Forlenza *et al.*, 2011). All these types of macrophage are important for the maintenance of homeostasis, including the immune response to pathogens. Macrophages express various receptors on their cell surface, including TLRs, scavenger receptors, CLRs and complement receptors, and are able to phagocytose pathogens and exert major effects through the production of cytokines and growth factors. Stimulation of fish macrophages *in vitro* with microbial stimuli, such as LPS, lipoteichoic acid (LTA) or peptidoglycan (PGN), flagellin and Poly I:C, leads to increased respiratory burst activity and associated production of oxygen radicals, increased phagocytosis and production of pro-inflammatory cytokines.

Macrophages are prodigious phagocytic cells and display remarkable plasticity in their physiology in response to environmental cues. These changes can give rise to different populations of cells with distinct functions. Based on three different homeostatic activities – host defence, wound healing and immune regulation – macrophages are classified as classically activated, wound healing and regulatory macrophages (Mosser and Edwards, 2008). IFN-γ and TNF can give rise to classically activated macrophages that produce high levels of IL-12 and modest levels of IL-10 and promote the differentiation of Th1 cells (see below). By contrast, regulatory macrophages produce high levels of IL-10 and low levels of IL-12 and are therefore involved in the regulation of pro-inflammatory responses and in the dampening of inflammatory reactions. Regulatory macrophages can be generated in the presence of TLR ligands in combination with a second signal such as immune complexes, prostaglandins, apoptotic cells, glucocorticoids, etc., or IL-10 alone. Wound-healing (alternatively activated) macrophages arise in response to IL-4, which can be produced during an adaptive immune response by Th2 cells or during an innate immune response by granulocytes. Wound-healing macrophages produce low levels of these cytokines but express resistin-like molecule-α intracellularly, a marker that is not expressed by the other macrophage populations, and play an important role in the protection of the host by decreasing inflammation and promoting tissue repair (Varin and Gordon, 2009).

Most of the signals for macrophage activation, including IFN-γ, TNF, an IL-4 like molecule, IL-10 and TLRs, are present in teleost fish. However, the pathways leading to the activation of fish macrophages is largely unknown and a definite involvement of these signals in fish macrophage activation is still to be proven. Nevertheless, *in vivo* studies on immune responses of carp to the parasites *Trypanoplasma borreli* and *Trypanosoma carassii*, that induce fundamentally different immune responses in their host, suggest the presence of classically activated, alternatively activated and regulatory macrophages (Forlenza *et al.*, 2011).

Cells involved in non-specific cell-mediated cytotoxicity (CMC)
NCC and NK-like cells have been reported to be involved in non-specific cell-mediated cytotoxicity (CMC) in fish. NCCs were originally described in channel catfish and are the most extensively studied killer cell population in teleosts. NCCs spontaneously kill a wide variety of target cells including tumour cells, virally transformed cells and protozoan parasites, and express components of the granule exocytosis pathway of CMC similar to mammalian cytotoxic lymphocytes (Froystad *et al.*, 1998; Praveen *et al.*, 2004). Catfish NCCs are small agranular lymphocytes that express a novel type III membrane protein termed the NCC receptor protein 1 (NCCRP-1) (Evans *et al.*, 1984). NCC activity has been described in other fish species including rainbow trout, carp, damselfish and tilapia (Shen *et al.*, 2002).

NK-like cells that lack the markers that define T cells (e.g. TCR), neutrophils, monocytes/macrophages and NCCs, have also been first described in catfish (Shen *et al.*, 2004). They were cloned from alloantigen-stimulated blood leucocytes and kill not only the stimulating allogeneic cells but also unrelated allogeneic targets by a perforin/granzyme-mediated apoptosis pathway. Catfish NK-like cells are heterogeneous in terms of target specificities and cell surface phenotype. The relationship between NCCs and NK-like cells in channel catfish is currently unknown except that the source of cells differs, i.e. NCCs are organ-derived cells while NK-like cells are isolated from blood leucocytes.

Mammalian NK cells are lymphocytes that are distinct from the T and B lineages and mediate an alternative form of innate immunity, which is triggered through receptors that recognize anomalies on the surfaces, e.g. changes in cell surface glycoproteins or presence of bacterial and viral molecules (Yoder and Litman, 2011). Mammalian NK receptors (NKRs) are Ig and/or lectin-type receptors that can be classified into two groups, the inhibitory and activating NKRs, based on the physiological function of signalling motifs. Inhibitory NKRs typically encode a cytoplasmic immunoreceptor tyrosine-based inhibition motif (ITIM) whilst activating NKRs typically possess a positively charged residue within the transmembrane domain that associates physically with an adaptor protein, e.g. DAP12, which encodes a negatively charged residue within the transmembrane domain. The adaptor protein has a cytoplasmic immunoreceptor tyrosine based activation motif (ITAM). Upon ligand recognition, activating NKRs initiate a signalling cascade that leads to the production of cytokines and chemokines and the release of cytolytic granules. In contrast, the signalling initiated by inhibitory NKRs counters that of the activating NKRs (Lanier, 2005). Numerous receptors have been identified in fish that share structural relationships with mammalian NKRs. The potential fish NKR candidates include the novel immune-type receptors (NITR), the novel Ig-like transcripts (NILTs) and the leucocyte immune-type receptors (LITRs) (Yoder and Litman, 2011; Cortes *et al.*, 2012).

Other innate immune cells: neutrophils, mast cells and rodlet cells
Neutrophils are key components of the inflammatory immune response against a variety of bacterial, viral, protozoan and fungal pathogens. As one of the first cells recruited to an inflammatory site, neutrophils possess a formidable armoury of responses that in most cases efficiently remove the invading pathogens. Neutrophils can phagocytose, produce toxic reactive oxygen and nitrogen intermediates, degranulate and release neutrophil extracellular traps (NETs) in response to invading pathogens (Lamas and Ellis, 1994; Kemenade *et al.*, 1996; Katzenback and Belosevic, 2009).

Mammalian mast cells are known to play pivotal roles in maintaining a healthy physiology, in wound healing and angiogenesis, and defence against a whole host of pathogens, participating in both innate and adaptive immunity. They are also involved in the inflammatory process, attracting different leucocyte subsets to the site of injury, and in allergy and allergic diseases (Weller *et al.*, 2011). Fish mast cells show marked diversity in their staining properties, with both basophilic and acidophilic components in their granules. In some fish families the eosinophilic component dominates, and they are termed eosinophilic granular cells (EGCs), whereas in others such as the pike the granules are strongly basophilic and show the metachromatic staining more typical of mast cell granules (Reite and Evensen, 2006). Teleost mast cells are localized in the vicinity of blood vessels in the intestine, gills and skin, and may play an important role in the inflammatory response because they express a number of functional proteins, including alkaline and acid phosphatases, arylsulphatase, 5′-nucleotidase, lysozyme and AMPs, which act against a broad-spectrum of pathogens (Silphaduang and Noga 2001; Dezfuli *et al.*, 2010). An increase in the number of mast cells in various tissues and organs of teleosts seems to be linked to a wide range of stressful conditions, such as exposure to heavy metals (cadmium, copper, lead and mercury), exposure to herbicides, parasitic infections and chronic inflammation (Lauriano *et al.*, 2011). Intraperitoneal injection of inactivated *A. salmonicida* in salmonids causes mast cell degranulation and release of mediators of inflammation, which is followed by an inflammatory reaction and vasodilation (Reite, 1997).

Rodlet cells are unique in fish and are characterized morphologically by their typical cytoplasmic inclusions, the so-called rodlets, and a thickened capsule-like cell border. They show a clear association with epithelial organized tissues or organs (Siderits and Bielek, 2009). It is suggested that the rodlet cell may represent a type of eosinophilic granulocyte that populates the tissues when immature and that mature in response to appropriate stimuli, in a way similar to that of mast cell precursors (Reite and Evensen, 2006). There is a close relationship between the presence of helminths or other noxious agents and the presence of rodlet cells. Massive aggregations of such cells can be seen in affected epithelia (e.g. gills or intestinal tract) and points to a functional role for the rodlet cells in teleost host defence against parasites.

1.4   An overview of the adaptive immune response in fish

The mammalian type of adaptive immune response is a relatively recent development in evolutionary time, assumed to have appeared in early jawed vertebrates (i.e. Gnathostomata) about 400–500 million years ago. The key elements in the evolution of the adaptive immune system are the appearance of the thymus, B and T lymphocytes and the RAG (recombination activation gene) enzymes, which allow the gene rearrangements necessary to generate great diversity of immunoglobulin superfamily (IgSF) antigen receptors on B and T cells (Hirano et al., 2011). Thus, the adaptive immune response is mediated by the actions of these two major groups of lymphocytes, B cells that mediate antibody (humoral) responses and T cells that mediate cell-mediated immune responses. Jawless vertebrates also have an adaptive immune system consisting of B- and T-like cells. However, the antigen receptors (also called variable lymphocyte receptors, VLRs) are generated through recombinatorial usage of a large panel of highly diverse leucine-rich-repeat (LRR) sequences (Mariuzza et al., 2010).

In antibody responses, B cells are activated to secrete antibodies, which are soluble forms of their surface Ig antigen receptor. The antibodies circulate in the bloodstream and permeate the other body fluids, where they bind specifically to the foreign antigen that stimulated their production. Binding of antibody inactivates viruses and microbial toxins (such as tetanus toxin or diphtheria toxin) by blocking their ability to bind to receptors on host cells. Antibody binding also marks invading pathogens for destruction, mainly by cells with cell surface receptors for the Ig molecules, such as phagocytic cells of the innate immune system (Schroeder and Cavacini, 2010).

In cell-mediated immune responses, antigen-specific T cells are activated that react directly against a foreign antigen that is presented to them on the surface of a host cell. Unlike B cells, T cells can only recognize antigen that has been processed and presented by antigen-presenting cells via their MHC proteins. Thus T cells can kill, for example, virus-infected host cells that have viral antigens on their surface, thereby eliminating the infected cells before the virus has had a chance to replicate. In other cases, the T cell produces cytokines that activate the innate defences to destroy the invading microbes (Laing and Hansen, 2011).

Both types of lymphocytes respond in a highly specific manner, and have diverse rearranging and adapting antigen receptors (Litman et al., 2010). A key feature of the adaptive immune system is immunological memory.

Repeat infections by the same virus or bacteria are met immediately with a strong and specific response that usually stops the infection and has less reliance on the innate system. Vaccination against infection is possible due to this immune memory. The first adaptive response against an infection, called the primary response, often takes days to mature. In contrast, a memory response develops within hours of infection. Memory is maintained by a subset of B and T lymphocytes called memory cells, which can potentially survive for years in the body. Memory cells remain ready to respond rapidly and efficiently to a subsequent encounter with a pathogen, giving rise to stronger and faster so-called secondary responses (Litman *et al.*, 2010).

1.4.1   The humoral adaptive immune response in fish

Immunoglobulins

A typical Ig molecule consists of two heavy (H) and two light (L) chains, each of which contain one amino-terminal variable (V) IgSF domain and one (in the L chain) or more (in the H chain) carboxyl-terminal constant (C) IgSF domains. The V domains are created by means of a complex series of gene rearrangement events and can then be subjected to somatic hypermutation after exposure to antigen to allow affinity maturation. Each V domain can be split into three regions of sequence variability termed the complementarity-determining regions (CDRs) and four regions of relatively constant sequence termed the framework regions. The three CDRs of the H chain are paired with the three CDRs of the L chain to specify the antigen-binding site, whilst the H chain C domains define effector functions. The C domains of the heavy chain can be switched to allow altered effector function while maintaining antigen specificity. Five classes of H chain C domains are known in mammals, that define the IgM, IgG, IgA, IgD and IgE isotypes (Schroeder and Cavacini, 2010; Zhang *et al.*, 2011).

Three isotypes, IgM, IgD and IgT (also called IgZ in zebrafish) have been identified in almost all studied species belonging to the main orders of teleost fish. The main exception is that IgT has not been found thus far in the channel catfish (*I. punctatus*) (Edholm *et al.*, 2010, 2011). IgM is the most ancient and the only isotype functionally conserved in all jawed vertebrates. IgD has been found in all jawed vertebrate groups except birds, indicating that it is also a primordial antibody class despite its highly plastic structure and unclear function in evolution (Edholm *et al.*, 2010, 2011). Other Ig isotypes have evolved to play specialized roles either within mucosal or systemic compartments. In mammals and birds, the IgM, IgG and IgY isotypes have major roles in systemic responses, while IgA is the main player in mucosal areas. In amphibians, IgM and IgY play a prevalent role in systemic immunity whereas IgX is an isotype chiefly expressed in the gut (Flajnik, 2010). The fish IgT is a mucosal-epithelial Ig preferentially expressed in the gut, bound to resident bacteria and induced specifically by a mucosal pathogen (Zhang et al., 2010).

The fish IgH loci in zebrafish and rainbow trout are organized as VH-DJCτ (elements for τ IgH chain) and VH-DJCμCδ (elements for μ and δ IgH chains) (Danilova et al., 2005; Hansen et al., 2005). This organization is strikingly similar to that of the mouse TRd-TRa locus encoding the TCRδ and TCRα chains. In both loci, upstream V segments rearrange either to DJCτ (or DJCδ) to encode τ (IgT) (or TCRδ) or to DJCμ (IgM) (or DJCα) to encode μ (or TCRα). As the messenger RNAs for the H chains of IgM (μ) and IgD (δ) are generated by alternative splicing the recombined VDJ to either Cμ or Cδ in the IgH loci thus allows IgM and IgD to be expressed in the same cells. The cells expressing IgM delete the IgT-encoding locus, thus fish B cells appear to express either IgT or IgM (Zhang et al., 2011). It is worth noting that fish may have multiple IgH loci for both IgT and IgM. For example, two IgH loci were discovered in Atlantic salmon, in each of which three or five VHDτJτCτ clusters are upstream of one copy of VHDμJμCμCδ, encoding for three putative functional IgT subclasses (Yasuike et al., 2010; Tadiso et al., 2011).

It is established that IgM is the predominant Ig isotype found in teleost blood/serum, with IgD and IgT in lesser amounts. In most teleosts, serum IgM is expressed as a tetramer, although IgM monomers have been described in some fish species. In contrast, serum IgT is expressed as a monomer in rainbow trout serum, and a tetramer in gut mucus (Zhang et al., 2010). To date serum IgD has only been described in catfish, by western blot analysis, and whether it exists as a monomer has yet to be determined (Edholm et al., 2011). Both IgT and IgM can be expressed from very early developmental stages (as early as 4 days post-fertilization) in teleost fish, with the expression of IgT increasing more rapidly when compared with that of IgM, perhaps suggesting that IgT plays a significant role in protecting fish larvae. In most adult fish, IgT and IgM can be expressed constitutively in both primary and secondary lymphoid organs, with IgM always being the dominant isotype but in some situations (i.e. after parasite infection) IgT is more highly expressed at mucosal sites (Zhang et al., 2011).

B cells

In humans (and mice) IgM and IgD are expressed on the surface of naive mature B cells by alternative splicing of a long primary RNA transcript. These double positive B cells (IgM$^+$/IgD$^+$) make up the majority of the peripheral B cells and upon antigen binding IgD expression is down-regulated. In contrast, the majority of long-lived memory B cells have class switched, undergone somatic hypermutation and express high affinity IgG, IgE or IgA. IgM$^-$/IgD$^+$ (IgD-only) cells can also be produced in humans by class switching (Chen et al., 2009). As fish lack IgG, IgE and IgA, and as IgT and IgM are not co-expressed in the same B cell, fish only have the potential to make IgM$^+$IgD$^+$, IgM$^+$IgD$^-$, single IgD$^+$ and IgT$^+$ B cells.

It appears that rainbow trout have IgM$^+$IgD$^+$ B cells (IgM$^+$/IgD$^+$/IgT$^-$) and IgT$^+$ B cells (IgM$^-$/IgD$^-$/IgT$^+$) (Zhang *et al.*, 2010), whilst catfish have IgM$^+$/IgD$^+$, IgM$^+$/IgD$^-$ and IgM$^-$/IgD$^+$ subsets in the absence of IgT (Edholm *et al.*, 2010). Thus, each teleost fish species appears to contain two or more major subsets of B cells although the presence of the IgT$^+$ and IgD$^+$ B cell subsets needs to be confirmed in other fish species. The rainbow trout IgT$^+$ B cell subset represents 16–27% of all B cells in the main systemic lymphoid organs, whereas IgM$^+$ B cells comprise about 72–83% of all B cells. In contrast, in gut the percentage of IgT$^+$ and IgM$^+$ lymphocytes is around 54% and 46% of all B cells, respectively (Zhang *et al.*, 2010). Catfish IgD-only B cells can represent as much as 60–80% of total peripheral blood B cells, depending on the individual fish (Edholm *et al.*, 2011). This highlights the complexity and fish species-specific differences in B cell composition.

B cell immune response
The specific immune response to infection and vaccination has been documented in fish by measuring IgM responses, and more recently IgT responses. Agglutinating and/or neutralizing antibodies can be induced against viral, bacterial and parasitic diseases and vaccines (Sanmartín *et al.*, 2008; Peñaranda *et al.*, 2011; Raida *et al.*, 2011). However, it appears that fish Ig isotypes are compartmentalized into mucosal IgT and systemic IgM in response to pathogenic challenge. IgT$^+$ B cells accumulate in the gut of rainbow trout that survive an infection of *Ceratomyxa shasta*, an intestinal parasite, whereas the number of IgM$^+$ B cells does not change relative to control fish (Zhang *et al.*, 2010). More critically, gut mucus from fish surviving intestinal *C. shasta* infection contains parasite-specific IgT titres but no specific IgM titre. Conversely, the serum of surviving fish contains parasite-specific IgM but not IgT titres (Zhang *et al.*, 2010). Whether IgT/IgM compartmentalization is present in all fish species and whether IgT in the gut can confer protection against other parasites will require further work.

It is generally believed that the antibody repertoire of fish is more restricted compared with mammalian species, due to limitations in the genetic machinery, the slower metabolic rate of poikilothermic animals, the lack of lymph nodes and GCs, and absence of Ig class switching (Magnadottir, 2010). However, the adaptive system of fish deviates from that in mammals in some interesting ways, which may to some extent compensate for these limitations. For example, fish IgM is not a static molecule but plastic in both structure and affinity. In rainbow trout, structural diversity of the 800 kDa IgM is generated by non-uniform disulphide cross-linking of the native tetramer. Denaturing, non-reducing electrophoretic analysis of the salmonid serum Ig produces four dominant species on gels: monomers (200 kDa), dimers (400 kDa), trimers (600 kDa), and fully cross-linked tetramers (800 kDa) (Costa *et al.*, 2012). The structural heterogeneity of IgM could also diversify the antibody's capabilities by producing molecules with specialized effector functions (Kaattari *et al.*, 1998). Fish IgM can also be present in relatively high concentrations in fish serum as natural antibodies, which are believed to have PRR properties (Magnadottir, 2010). Another interesting deviation from the mammalian adaptive system is the apparent phagocytic activity of B cells described in some fish, which may indicate that fish lymphocytes have multiple roles (Li *et al.*, 2006).

1.4.2   The cellular adaptive immune response in fish
T cell receptors (TCR) and co-receptors
Conventional mammalian T cells share some basic characteristics once they reach maturity. They all possess a TCR by which they recognize linear antigens presented by MHC molecules; they all possess the CD3 complex of signalling molecules through which the antigen – TCR interaction propagates cellular activation; they all possess the same basic machinery of co-stimulatory (e.g. CD28) and co-inhibitory (e.g. CTLA-4) surface molecules, as well as transmembrane (e.g. CD45) and intracellular enzymes (e.g. Lck and ZAP70) that ensure the correct balance of activation or inhibition; they express the TCR co-receptor CD8 or CD4 which drives their specificity for MHC class I or MHC class II presented antigens, respectively; and they all have the potential to form immunological memory in preparation for a future pathogenic insult (Laing and Hansen, 2011).

The TCR in mammals comes in two forms: a heterodimer composed of the TCR-α and TCR-β chains is found on the surface of conventional circulating αβ-T cells, and a heterodimer composed of the TCR-γ and TCR-δ chains is found on the 'more primitive' mucosa-associated γδ-T cells. In mammals the TCR genes, TRA, TRB, TRG and TRD, encoding for the TCR- α, β, γ and δ chains, respectively, have the same system as Ig genes for receptor diversification, using V-(D)-J genomic rearrangement. All the TCR genes have been identified in multiple fish species and are well conserved. The structural constraints of the TCR/peptide/MHC complex seem to constitute an evolutionary lock that efficiently prevents further isotypic diversification (Castro *et al.*, 2011).

The TRA and TRD genes are encoded in the same locus in tetraodon, Japanese flounder, Atlantic salmon and probably all teleosts, as seen in other vertebrates. In the Atlantic salmon, the complete TRA/TRD locus covers 900 kb and contains 292 TRAV/TRDV genes and 123 TRAJ/TRDJ genes, of which 1 TRDJ, 113 TRAJ and 128 TRAV/TRDV genes appear to be functional (Yazawa *et al.*, 2008a). The fish TRB locus seems to be organized as in humans and mice. For example, in rainbow trout, multiple Vβ are followed by a TRBD gene, 11 TRBJ genes and a TRBC (De Guerra and Charlemagne, 1997). The zebrafish has the same general TRB locus organization with 51 TRBV genes, a single TRBD gene, 27 TRBJ1 genes, a single TRBJ2 gene and two TRBC genes (Meeker *et al.*, 2010). Two different TRG loci have been discovered in Atlantic salmon, with the first, TRG1, spanning 260 kbp and containing four tandemly repeated clusters each of which consists of 1–4 TRGV segments, 1–2 sets of a TRGJ and a TRGC (Yazawa *et al.*, 2008b).

The CD8 molecule has two chains, CD8α and CD8β, which form either heterodimers (CD8αβ) or homodimers (CD8αα) (Cole and Gao, 2004). Mature mammalian TCR-αβ-expressing cytotoxic T cells (CTL) generally possess CD8αβ while most γδ-T cells lack CD8 (Kabelitz *et al.*, 2000). However, a small population of intestinal epithelial αβ- and γδ-T cells express CD8αα (Jarry *et al.*, 1990). Orthologues of both CD8α and CD8β have been described in multiple species of teleost fish (Hansen and Strassburger, 2000; Moore *et al.*, 2005; Suetake *et al.*, 2007). Both CD8 chains are expressed by thymocytes and lymphocytes, supporting their role as co-receptors on teleost T cells.

The CD4 molecule in mammals is a single protein with four extracellular Ig-like domains and an intracellular tail. Teleost fish possess a structurally conserved orthologue of mammalian CD4 (Laing *et al.*, 2006; Edholm *et al.*, 2007; Sun *et al.*, 2007; Buonocore *et al.*, 2008; Moore *et al.*, 2009). In addition, a second CD4-like (CD4L or CD4REL) gene exists in teleost fish that encodes a protein with either two (Dijkstra *et al.*, 2006; Laing *et al.*, 2006) or three (Edholm *et al.*, 2007) Ig-like domains. This latter protein fits the predicted structure of a primordial CD4 molecule from which the four Ig-domain CD4 molecule may have arisen following duplication of the two Ig-domain form (Laing *et al.*, 2006). Both forms of CD4 are expressed in lymphoid tissues, IgM⁻ lymphocytes and clonal T cell lines, akin with their function on T cells (Laing *et al.*, 2006; Edholm *et al.*, 2007). CD4 transcripts are found in TCR⁺ cultured lymphocytes that lack CD8α, CD8β and CD4L (Yamaguchi *et al.*, 2011), and CD4L (but not CD4) expression is associated with teleost Treg-like cells (Wen *et al.*, 2011), supporting a distinction between CD8 and CD4 expressing T cells in fish, and suggesting CD4L and CD4 expressing cells may differ functionally. Which version of CD4, if any, will act as a true CD4 orthologue has still to be verified. Interestingly, a study on catfish CD4 showed that the two CD4 molecules are also expressed on the cytotoxic TS32.17 cell line, suggesting that fish may possess CD4⁺ CTLs in addition to conventional CD8⁺ CTL and CD4⁺ helper cells.

T cells
Different populations of T cells are deemed either αβ- or γδ-T cells depending on which TCR they express. αβ-T cells are the more abundant T cell type found in lymphoid organs and blood in mammals. γδ-T cells differ from αβ-T cells in their ability to recognize antigen and their main site of residence. These cells appear to recognize unprocessed antigen in a manner similar to that of pattern recognition receptors; thus γδ-T cells are more like innate immune cells with less dependence on MHC presentation and are most abundant in epithelial and mucosal tissues (Bonneville *et al.*, 2010). Based on the expression of the co-receptors, CD8 or CD4, that stabilize the

interaction of the TCR with MHC molecules thus enhancing TCR activation through the CD3 tyrosine phosphorylation pathway, two main sets of T cells are distinguished – CD4$^+$ and CD8$^+$ T cells. CD8 marks CTL that recognize antigenic peptides associated with MHC class I molecules on the surface of antigen presenting cells, and whose main function is the direct killing of target cells. In contrast, CD4 marks T helper cells (Th cells) that recognize peptides associated with MHC class II and orchestrate many aspects of the immune response by release of cytokines.

The isolation of key T cell markers found in mammals suggests that fish possess a T cell paradigm in common with its mammalian counterparts. It is believed that the fish T cell primary differentiation takes place in the thymus, from where mature T cells migrate to secondary lymphoid organs such as the head kidney and spleen. T cells can also be found in the gills, liver, olfactory pit and gut (Hansen and Zapata, 1998). In zebrafish, the initiation of rearrangements of the TR genes coincides with the expression of RAG 1 and 2, and terminal deoxynucleotidyl transferase. However, little is known about the details of the T cell differentiation processes in fish. CD8$^+$ and CD4$^+$ mRNAs are highly expressed in the thymus of naive fish, as well as in kidney and spleen tissues, with a lower degree of expression in gill and gut. The expression pattern of these molecules is correlated with the expression of other T cell markers such as LAG3 or the TCR. The expression of these markers can be modulated after infection, or after *in vitro* exposure to T cell mitogens (Somamoto *et al.*, 2006; Araki *et al.*, 2008; Buonocore, *et al.*, 2008; Hetland *et al.*, 2010).

Ginbuna crucian carp CD8α$^+$ T cells mediate alloantigen-specific killing in a similar manner to CTL in mammals (Toda *et al.*, 2009). Using monoclonal antibodies (mAbs) against rainbow trout CD8α, high ratios of CD8α$^+$ cells are found in trout thymus, gill and intestine, but a relatively low abundance is seen in head kidney, spleen and blood. In secondary lymphoid tissues, CD8α$^+$ lymphocytes (which do not react with anti-thrombocyte or anti-IgM mAbs) express CD8α, CD8β and TCRα, while Ig and CD4 transcripts were found in CD8α$^-$ lymphocytes (Takizawa *et al.*, 2011). In contrast, considerable CD4 expression is seen in CD8α$^+$ thymocytes and suggests the presence of double-positive early T cells in this site, that likely represent an early developmental stage in T cell development, as in other vertebrates. CD4$^+$ T cells proliferate *in vitro* in mixed leucocyte cultures (the so-called mixed leucocyte reaction – MLR) and antigen-specific proliferation of CD4$^+$ T cells occurs after *in vitro* sensitization with ovalbumin (Toda *et al.*, 2011). Collectively, these preliminary findings using mAbs against fish CD4 and CD8α suggest the basic characteristics of CD4$^+$ and CD8α$^+$ lymphocytes seem similar in teleost fish and mammals. The final definition of T cell subsets in fish will require the use of multiple mAbs specific to other markers (e.g. CD3) in addition to CD4 and CD8, which are lacking at the moment but should be available in the coming years.

T cell immune response

It has long been suggested that antibody responses are less diverse in fish than in mammals, although they can efficiently protect the animals against pathogens (Du Pasquier, 1982). The expressed Ig diversity in fish may be restricted compared with the repertoire available to a given antigenic determinant. This restriction is probably more pronounced in species of small size, such as zebrafish, where a limited number of lymphocytes are present in the organism at a given time. However, the T cell available repertoire seems large enough to support the selection of diverse responses to pathogens, as shown by immunoscope analysis of the T cell response against VHSV infection in rainbow trout (Boudinot *et al.*, 2001).

In channel catfish, three functionally distinct leucocyte subpopulations can be isolated from peripheral blood, surface (s) Ig⁺ and sIg⁻ lymphocytes, and macrophages. Antibody responses to T-dependent antigens in catfish requires the co-culture of the three cell types and antibody production by fish B cells is dependent upon the presentation of the antigen and on T cell help, as seen in mammals (Miller *et al.*, 1985). Interestingly, the magnitude of the primary response to T-independent antigens and the secondary response to T-dependent antigens is relatively independent of *in vitro* culture temperature. In contrast, the magnitude of primary responses to T-dependent antigens is suppressed at lower *in vitro* temperatures (Miller and Clem, 1984).

Specific cell-mediated cytotoxicity was first described in catfish, and later in several other fish species (Somamoto *et al.*, 2000; Stuge *et al.*, 2000; Utke *et al.*, 2007). Such assays require the use of fish and/or cell lines that are genetically identical (syngeneic) to the responder cells, in order to ensure the MHC molecules present appropriate peptides to the TCR, since MHC molecules are highly polymorphic within a population. An autologous cell-mediated response against a syngenic cell line infected with infectious pancreatic necrosis virus was described in the ginbuna crucian carp (Somamoto *et al.*, 2000). In the same species, virus-specific responsive cells from carp haematopoietic necrosis virus (CHNV)-infected fish proliferated after stimulation *in vitro* with CHNV-infected syngeneic stimulator cells. After *in vitro* stimulation, TCR αβ+ effector cells lysed CHNV-infected syngeneic cells, but not CHNV-infected allogeneic cells or cells infected by other different viruses (Somamoto *et al.*, 2009). In rainbow trout, specific cell-mediated cytotoxicity occurs when leucocytes from VHSV-infected clonal fish are mixed with infected MHC compatible RTG2 fibroblasts (Utke *et al.*, 2007). Interestingly, peripheral blood leucocytes from fish immunized with a DNA vaccine encoding the VHSV G protein efficiently lysed VHSV-infected but not infectious haematopoietic necrosis virus (IHNV)-infected histocompatible targets (Utke *et al.*, 2008). Taken together, these observations strongly suggest that teleost fish possess specific T cell-mediated cytotoxicity as reported in humans and mice.

1.5   Immune regulation: the cytokine network in fish

1.5.1   The cytokines and transcription factors involved in mammalian Th cell development

Mammalian CD4[+] T cells, also known as Th cells, play an important role in orchestrating adaptive immune responses to various infectious agents. Upon activation by TCR- and cytokine-mediated signalling, naive CD4[+] T cells may differentiate into at least four major types of Th cells, Th1, Th2, Th17 and inducible T-regulatory (iTreg) cells, that play a critical role in orchestrating adaptive immune responses to various microorganisms. They can be distinguished by their unique cytokine production profiles, the expression of unique transcription factors and their functions (Fig. 1.2). Th1 cells predominantly produce IFN-γ, and are important for protective immune responses to intracellular viral and bacterial infections. Th2 cells, by producing IL-4, IL-5, IL-9, IL-13 and IL-25, are critical for expelling extracellular parasites such as helminths. Th17 cells are responsible for controlling extracellular bacteria and fungi through their production of IL-17A, IL-17F, IL-21 and IL-22, whilst inducible T-regulatory (iTreg) cells,

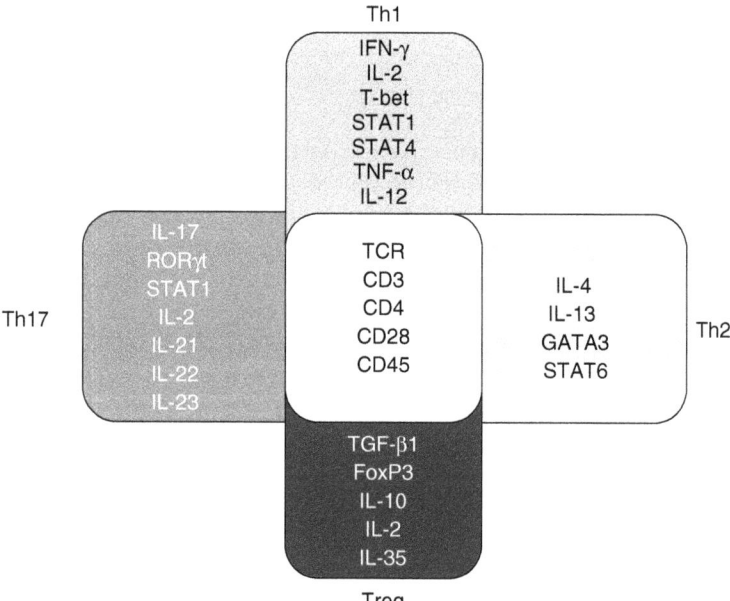

Fig. 1.2 Shared and private components of Th1, Th2, Th17 and regulatory T cells (Treg). Mammalian Th cells share surface molecules such as the TCR, CD3, CD28, and CD45 but each of the Th lineages can be distinguished by their unique cytokine production profiles, the expression of unique transcription factors and cytokines inducing them.

together with naturally occurring T-regulatory (nTreg) cells, are important in maintaining immune tolerance, as well as in regulating lymphocyte homeostasis, activation and function (Zhu and Paul, 2010a). An important principle for Th cell differentiation is that one of the signature cytokines produced by each differentiated cell type also plays a critical role in the induction of such cells, potentially providing a powerful positive feedback loop. These 'feedback' cytokines are IFN-γ for Th1, IL-4 for Th2, IL-21 for Th17, and TGF-β for iTregs.

The major determinant for Th cell differentiation is the cytokine milieu at the time of antigen encounter, although the nature of cognate antigen and its affinity to the TCR as well as the available co-stimulants, many of which regulate initial cytokine production, can influence Th cell fate (Zhu and Paul, 2010a,b). IL-12 and IFN-γ are two important cytokines for Th1 differentiation. For Th2 differentiation, cytokines including IL-2, IL-4, IL-7 and thymic stromal lymphopoietin (TSLP) may be involved. TGF-β induces Th17 differentiation in the presence of IL-6, but also promotes iTreg cell differentiation when IL-2 rather than IL-6 is available. In general, more than one cytokine is required for differentiation to any particular phenotype and cytokines that promote differentiation to one lineage may suppress adoption of other lineage fates. Thus, Th cell differentiation involves a complex cytokine network.

Transcription factors are critical for Th cell differentiation and cytokine production. Cell fate determination in each lineage requires at least two types of transcription factors, the master regulators and the STAT proteins. Some STAT proteins are responsible for inducing the expression of master regulators. In addition, the same STATs and the master regulators often collaborate in regulating cytokine production by directly acting on cytokine genes. The essential transcription factors of Th lineages are T-bet / STAT4 (Th1), GATA3 / STAT5 (Th2), RORγt / STAT3 (Th17), and Foxp3 / STAT5 (iTreg), respectively (Zhu and Paul, 2010a,b).

1.5.2   The Th cell cytokines in fish

The critical cytokines involved in mammalian $CD4^+$ T cell differentiation are IFN-γ, TGF-β, the γ-chain cytokines (IL-2, 4, 7, 9 and 21), IL-10 family cytokines (IL-10 and IL-22), IL-12 family cytokines (IL-12, 23, 27 and 35), IL-17 family cytokines (IL-17A, E and F) and IL-6.

IFN-γ is the effector and feedback cytokine of Th1 cells, and is structurally and functionally conserved in teleost fish. Unlike higher vertebrates teleost fish possess two forms of IFN-γ (IFN-γ and IFN-γrel) that likely arose by tandem gene duplication during teleost evolution (Igawa et al., 2006; Milev-Milovanovic et al., 2006; Stolte et al., 2008; Purcell et al., 2009; Savan et al., 2009; Chen et al., 2010a). Whilst teleost IFN-γrel lacks a nuclear localization signal critical to the function of mammalian IFN-γ, its tertiary structure is conserved and there appears to be overlapping function with

IFN-γ (Zou *et al.*, 2005; Chen *et al.*, 2010a; Lopez-Munoz *et al.*, 2011). Recombinant (r)IFN-γ from zebrafish, trout and goldfish can increase the expression of anti-viral and inflammatory genes, the production of reactive oxygen and nitrogen species by macrophages, and the phagocytic activity of macrophages (Zou *et al.*, 2005; Martin *et al.*, 2007; Lopez-Munoz *et al.*, 2009; Grayfer *et al.*, 2010). T cells and NK-like cells are the predominant secretors of teleost IFN-γ, similar to mammalian IFN-γ, although fish macrophages and B-cells also produce IFN-γ (Milev-Milovanovic *et al.*, 2006; Stolte *et al.*, 2008) in response to viral or bacterial infection and exposure to related TLR agonists (Igawa *et al.*, 2006; McBeath *et al.*, 2007; Stolte *et al.*, 2008; Purcell *et al.*, 2009; Chen *et al.*, 2010b). Two IFN-γ genes have been found in some fish species, and the IFN-γ receptor has begun to be elucidated (Gao *et al.*, 2009; Grayfer and Belosevic, 2009; Aggad *et al.*, 2010).

TGF-β is a pleiotropic cytokine with potent regulatory and inflammatory activity, and is best known for its induction of peripheral tolerance. TGF-β orchestrates the differentiation of both Treg and Th17 cells in a concentration-dependent manner, with the latter requiring the presence of IL-6. In addition, TGF-β in combination with IL-4 promotes the differentiation of IL-9- and IL-10-producing T cells, which lack suppressive function and also promote tissue inflammation (Sanjabi *et al.*, 2009). The trout TGF-β1 sequence was probably the first cytokine gene described in fish (Hardie *et al.*, 1998), and to date at least three TGF-βs have been found in fish. TGF-β can significantly induce leucocyte migration (Cai *et al.*, 2010) and proliferation of fibroblasts (Haddad *et al.*, 2008), and can modulate gene expression in leucocytes (un-published). An additional TGF-β1 gene has been cloned recently in rainbow trout (acc. no. FN822750), and unlike the first gene where expression is stable in most situations, the second trout TGF-β1 is modulated by inflammatory stimulants.

The cytokines that signal through the common γ-chain receptor (IL-2, IL-4, IL-7, IL-9, IL-15 and IL-21) have a critical role in the survival and homeostatic proliferation of lymphocyte populations (Wang *et al.*, 2011c). IL-2 is a key T cell growth factor responsible for the clonal expansion of T cells following activation by specific antigen, and for the differentiation and survival of effector T cells and NK cells. The fish IL-2 gene has been identified recently in fugu by genome analysis and the bioactivity of a trout rIL-2 has been reported. Rainbow trout IL-2 expression is induced by the T cell mitogens PHA and PMA, by the mixed leucocyte reaction, and by bacterial infection. The trout rIL-2 increases the expression of two transcription factors, STAT5 and Blimp-1, as well as IFN-γ, γIP and IL-2 itself (Diaz-Rosales *et al.*, 2009). IL-15 together with IL-7 plays a key role in the generation and maintenance of memory CD8[+] T cells. Thus, IL-7 is required for the survival of both naive and memory CD8[+] T cells, while IL-15 is essential for their proliferation (Tanel *et al.*, 2009). A fish IL-7 gene that shares limited homology to mammalian IL-7 has been described in fugu, with assignment of this molecule to IL-7 supported by synteny analysis (Kono *et al.*, 2008). A trout IL-7 gene has also been cloned recently that shares high identity to fugu IL-7. Following intraperitoneal injection of the trout rIL-7, the expression of TNF, IL-1β and IL-11 are increased in head kidney and spleen (unpublished). Fish have two paralogues of mammalian IL-15, IL-15 and IL-15-like (Bei *et al.*, 2006; Fang *et al.*, 2006; Gunimaladevi *et al.*, 2007). The fish IL-15 bioactivity has been described only in rainbow trout to date. Trout rIL-15 increases the expression of IFN-γ in splenic leucocytes from healthy fish, but not in head kidney leucocytes (Wang *et al.*, 2007). Curiously, the expression of trout IL-15 is itself up-regulated by trout rIFN-γ, suggesting a positive feedback loop. IL-4 is the main effector cytokine of Th2 cells and inhibits the Th1 response. Mammalian genes encoding IL-4 and IL-13 are located in tandem within a cluster of cytokine genes. Although no clear authentic homologues of IL-4 or IL-13 have been identified in fish, two IL-4/13 like molecules have been discovered (Bird and Secombes, 2006; J. H. Li *et al.*, 2007), termed IL-4/13A and IL-4/13B by Ohtani *et al.* (2008). The trout IL-4/13A molecule has been cloned recently (Takizawa *et al.*, 2011), where it is expressed highly in thymus, skin and Ig⁻ gill cells. The receptor subunits for mammalian IL-4 and IL-13 signalling have also been identified in fish (Wang *et al.*, 2011a). IL-21 is a key regulator of the cellular immune response and affects the growth, survival and activation of B cells, T cells and NK cells. In particular IL-21 has been shown to promote differentiation of naive CD8⁺ T cells into cells with a highly effective cytolytic activity in response to TCR and co-stimulatory signals. IL-21 is the signature cytokine of Tfh (Sallusto and Lanzavecchia, 2009) and a feedback cytokine for Th17 cells. Fish IL-21 has a six exon/five intron structure in contrast to the five exon/four intron structure in mammals. Trout IL-21 is expressed in immune tissues and is induced by bacterial and viral infection and the T cell stimulant PHA. Trout rIL-21 induces a rapid and long-lasting (4–72 h) induction of expression of IFN-γ, IL-10 and IL-22 *in vitro*, and maintains the expression of CD8α, CD8β and IgM at a late stage of stimulation when their expression is significantly decreased in controls *in vitro*. Trout rIL-21 also increases the expression of CD4, T-bet and GATA3, Th cell markers (see below), suggesting a regulatory role in fish T cell development. JAKs/STAT3, Akt1/2 and PI3K pathways are at least partially responsible for rIL-21 action (Wang *et al.*, 2011b).

The IL-12 family cytokines (IL-12, IL-23, IL-27 and IL-35) are heterodimers, with IL-12 and IL-23 important growth factors for Th1 and Th17 cells respectively, whilst IL-27 has broad inhibitory effects on all effector Th subsets whilst expanding inducible Tregs (Yoshida *et al.*, 2009; Jankowski *et al.*, 2010). In contrast IL-35 is produced by Tregs and serves to suppress inflammatory responses but induces Treg proliferation (Castellani *et al.*, 2010). The peptide chains that form these interleukins include p19, p28, p35, p40 and Epstein–Barr virus induced gene 3 (EBI3). Thus, IL-12 is composed of p35 and p40, IL-23 of p19 and p40, IL-27 of p28 and EBI3 and IL-35 of p35 and EBI3. In addition it is known that the p40 homodimer is biologically active and acts as an IL-12 antagonist (Jana *et al.*, 2009). All the peptide chains except p28 have been discovered in fish but none of these interleukins have been produced as biologically active proteins in fish to date (Secombes *et al.*, 2011).

The IL-17 cytokine subfamily in mammals is composed of six members, IL-17A to IL-17F, with IL-17E also known as IL-25. IL-17A and IL-17F are gene neighbours and are highly homologous, suggesting they have arisen from a recent gene duplication event. In teleost fish several IL-17 family members have been discovered, but homology to mammalian genes has not always been easy to assign. Thus, several IL-17A/F-like genes are known, and teleost IL-17C in fact looks to be related to both IL-17C and IL-17E (Wang *et al.*, 2010c; Secombes *et al.*, 2011), suggesting an ancestral IL-17C/E gene may have duplicated and diverged later in vertebrate evolution (Secombes *et al.*, 2011). To date no IL-17B gene has been found outwith the tetrapods.

IL-6 is a pleiotropic cytokine involved in the physiology of virtually every organ system. IL-6 induces the development of Th17 cells from naive T cells together with TGF-β but in contrast, IL-6 inhibits TGF-β-induced Treg differentiation (Kimura and Kishimoto, 2010). The IL-6 gene has been cloned from a few fish species, but the bioactivity has only been described for rainbow trout. Trout IL-6 expression in macrophages is induced by pro-inflammatory agents (LPS, Poly I:C and IL-1β), and trout rIL-6 rapidly induces STAT3 phosphorylation and expression of suppressors of cytokine signalling (SOCS1–3, CISH) and a transcription factor involved in interferon signalling (IRF-1), as seen in mammals. The trout rIL-6 also promotes the growth of macrophages *in vitro* and induces the expression of AMP genes but down-regulates the expression of the pro-inflammatory cytokines IL-1β and TNF, suggesting an important immune regulatory role of fish IL-6 (Costa *et al.*, 2011).

IL-10 is an anti-inflammatory cytokine that was initially shown to be produced by Th2 cell clones in the mouse, but it is now known to be synthesized by almost all types of leucocytes, including Th1, Th2, Th17 and regulatory T cells (Sabat *et al.*, 2010). IL-10 has been identified in several species of teleost fish (Lutfalla *et al.*, 2003; Savan *et al.*, 2003; Zou *et al.*, 2003; Inoue *et al.*, 2005; Zhang *et al.*, 2005; Pinto *et al.*, 2007; Seppola *et al.*, 2008), where only a single IL-10 gene has been found in each species except rainbow trout which has two (Harun *et al.*, 2011). The bioactivity of fish IL-10 has been reported recently, where goldfish rIL-10 reduced the expression of the pro-inflammatory cytokines IL-1β, TNF, IL-8 and IL-10 itself but increased SOCS3 expression (Grayfer *et al.*, 2011).

Lastly, IL-22 is a cytokine released by Th-17 cells in mammals and crucial for protection against extracellular microbes and at mucosal sites (Kolls, 2010). IL-22 has been cloned in several fish species but only trout IL-22 has been produced as a bioactive recombinant protein. Trout rIL-22 expression is induced in the spleen upon infection of fish with the Gram-negative bacterium *Y. ruckeri* (Monte *et al.*, 2011) and in haddock IL-22 expression is correlated with disease resistance in vaccinated fish (Corripio-Miyar *et al.*, 2009). *E. coli* produced trout rIL-22 enhances the expression of a number of AMPs in spleen cells, and suggests that fish IL-22 promotes host innate immunity against microbes and has a biological similarity with its mammalian counterpart (Monte *et al.*, 2011).

1.5.3   The critical Th cell transcription factors in fish

Th1 cells express three lineage-specific transcription factors: STAT1, STAT4 and T-box expressed in T cells (T-bet), with T-bet representing the hallmark transcription factor for this subset (Szabo *et al.*, 2003). T-bet has been identi-fied by searching teleost genomes and ESTs and its presence confirmed in [Ig] lymphocytes (Takizawa *et al.*, 2008a; Mitra *et al.*, 2010; Wang *et al.*, 2010a). STAT1, which is also associated with Th17 cells, has also been described in teleost and Agnathan fish (Oates *et al.*, 1999; Suzuki *et al.*, 2004; Zhang and Gui, 2004; Collet *et al.*, 2008; Jia and Zhou, 2010), while STAT4 has been reported only in teleost fish at this time (Yap *et al.*, 2005; Guo *et al.*, 2009). Th2 cells express STAT6 and GATA3 (Nurieva *et al.*, 2008). GATA3 is conserved in teleost fish (Kumari *et al.*, 2009; Wang *et al.*, 2010a), where it is expressed and regulated in lymphoid tissues and [Ig] lymphocytes (Neave *et al.*, 1995; Heicklen-Klein *et al.*, 2005; Takizawa *et al.*, 2008b; Kumari *et al.*, 2009; Wang *et al.*, 2010a). STAT6 structure, expression and tyrosine-based phosphorylation is also conserved between fish and mammals (Sung *et al.*, 2010).

The transcription factor profile of Th17 cells includes a splice variant of nuclear RAR-related orphan receptor gamma (RORγt) and STAT1 (Wei *et al.*, 2007). Whilst RORγ is also conserved in fish, to date there is no evidence that a splice variant exists (Flores *et al.*, 2007). Lastly, Tregs are recognized by the cell surface expression of CD25 and the intracellular presence of the FoxP3 transcription factor. Recently, Helios (typically asso-ciated with T cell development) was identified as a further transcription factor associated with a subset of Tregs (Getnet *et al.*, 2010; Thornton *et al.*, 2010), that regulate Treg activity by directly binding FoxP3. A CD25-like gene has been identified in teleost fish that appears to be ancestral to both CD25 (also known as IL-2Rα) and IL-15Rα. The teleost CD25-like protein is detectable on lymphocytes that express CD4L (Wen *et al.*, 2011). FoxP3 also exists in teleost fish, and as with the mammalian equivalent, can inter-act with the transcription factors NF-κB and NFAT and is most highly expressed in tissues associated with T cells, such as thymus and spleen (Mitra *et al.*, 2010; Quintana *et al.*, 2010; Wang *et al.*, 2010b). The FoxP3 gene shows reciprocal expression to IL-17 expression in zebrafish embryos, con-sistent with a presumed Th17 inhibition by Tregs (Quintana *et al.*, 2010), and has been found in immune suppressing lymphocytes that react with polyclonal antibodies to CD4L and a CD25-like molecule (Wen *et al.*, 2011).

Helios is also conserved in fish (John *et al.*, 2009), and thus conservation of Tregs and their functions in early vertebrates seems likely.

1.5.4   Immune regulation

In humans and mice, CD4$^+$ T cells play a central role in initiating and maintaining diverse immune responses by recruitment and activation of other immune cells, including CD8$^+$ T cells, B cells, macrophages and granulocytes. Initial studies by Coffman and Mosmann (Coffman and Carty, 1986; Mosmann *et al.*, 1986) provided the first clues as to how Th cells orchestrate appropriate innate and adaptive responses to microbial pathogens based on the Th1/Th2 paradigm of T cell activation. It is now known that CD4$^+$ T cells can be classified into four major lineages, Th1, Th2, Th17 and Treg cells, based on their functions, pattern of cytokine secretion and their expression of specific transcription factors, although other Th lineages may exist (Zhu and Paul, 2010b). However, each CD4$^+$ lineage shows heterogeneity and plasticity, and subsets of the same lineage may express different effector cytokines, reside at different locations or give rise to cells with different fates, and cells from different lineages may secrete common cytokines, such as IL-2 and IL-10. In addition, the pattern of cytokine secretion may switch from that of one lineage towards another under certain circumstances.

Typically, Th1 cells are thought to regulate cellular immunity via production of IL-2 and IFN-γ, and their response is elicited by intracellular bacteria and viral infections. The differentiation process is promoted by IL-12, IL-18, IFN-α and IFN-γ, and controlled by the transcription factors T-bet, STAT1 and STAT4. All these molecules exist in fish. Some of the recombinant proteins have also been produced recently, including salmonid IFN-γ and IL-2, and show functional activities closely related with the mammalian homologues (Zou *et al.*, 2005; Díaz-Rosales *et al.*, 2009). In addition, T-bet, STAT1 and STAT4 transcription factors are present in fish genomes (see above), supporting the concept that a complete Th1 pathway may exist in fish. This notion has also been supported in a parasitic infection model, where *T. bryosalmonae* infection induced a biased gene expression profile with the expression of T-bet, IFN-γ and IL-2 markedly increased (Wang *et al.*, 2010a).

In mammals, Th2 cells regulate humoral immunity via production of IL-4, IL-5, IL-9 and IL-13, and they are associated with the response to parasitic helminths and other extracellular pathogens. IL-4 directs Th2 differentiation, governed by GATA3 through STAT6 up-regulation. In contrast to the Th1 pathway, no clear counterparts of the Th2 key molecules have been found in fish but two Th2 related cytokine genes are described and encoded on two paralogous loci (IL-4/13A and B). Although both transcription factors (GATA3 and STAT6) are found in several fish genomes, additional data is needed before it is possible to speculate whether true Th2-like responses exist in fish.

Th17 cells have been identified in humans and mice, characterized by the production of the cytokines IL-17A, IL-17F, IL-21 and IL-22. Th17 cells represent an important component of the adaptive immune response against certain extracellular bacterial and fungal infections. Different combinations of cytokines (TGF-β, IL-6, IL-1, IL-21, IL-23) have been shown to activate STAT3 phosphorylation, leading to induction of the transcription factor RORγt, the master regulator of Th17 differentiation. All cytokines primarily related with differentiation of the Th17 pathway and the cytokines produced by Th17 cells exist in teleosts, as well as the ROR family of transcription factors. However, there is still no direct evidence for the presence of Th17 responses in fish, although some recent expression data is interesting in this regard (Ribeiro *et al.*, 2010).

Human and mouse CD4[+] T cells can also develop into a variety of Treg cell populations defined by expression of the transcription factor FoxP3 and by their capacity to produce cytokines such as TGF-β and IL-10. FoxP3, TGF-β and IL-10 are present in the genomes of several fish species, although none of them have yet been associated with a defined Treg cell subset.

Thus a plethora of classical T cell markers, T cell specific cytokines and transcription factors have been reported to date in several species, suggesting that the fish T cell system has many characteristics in common with its mammalian counterparts. However, the presence of additional molecules (e.g. CD4L, IFN-γrel) and putative common ancestors (e.g. IL-17C/E, CD25 like) clearly show that fish T cell markers and T cell functions are likely to be different relative to mammals, and even the number and types of T cell subsets will potentially vary.

1.6   Conclusions

As we learn more about the immune system of fish, the more we realize just how complex it is! Molecular evidence suggests a similar immune system exists throughout the jawed vertebrates yet marked differences are also apparent. The molecular tools are now well established to measure gene expression changes in fish but analysis of proteins and cell types is still in its infancy. Thus there is still a long way to go to really understand how fish defences are regulated and the different mechanisms that can contribute to protection against disease. Nevertheless, whilst future studies still have to overcome many significant challenges, the knowledge generated will help inform the development of novel disease control measures for fish, including vaccinaiton. Although many efficacious vaccines exist for fish, there are still a lot of diseases for which no successful treatment is available. The reasons why some vaccines are effective and others fail are many-fold, and include factors such as the immunogenicity of the antigens and the 'type' of immune response elicited. The immunogenicity of antigens can potentially be enhanced by adjuvants, and the type of immune response can be manipulated by immune regulatory factors (such as cytokines or PAMPs) given at the time of immunization. Thus, a greater understanding of fish immune responses could lead to new or improved disease control strategies in this manner. Another issue in vaccine developmement is the need for *in vitro* assays to determine the effectiveness of pilot vaccines. Currently there is an almost complete lack of established correlates of disease resistance, with disease challenge trials increasingly expensive and giving limited information. It seems likely that as we learn more about factors important for fish disease resistance such assays will naturally emerge and help future vaccine trials.

1.7   References

ABDEL-AZIZ EL-SH, ABDU SB, ALI TEL-S, FOUAD HF (2010), 'Haemopoiesis in the head kidney of tilapia, *Oreochromis niloticus* (Teleostei: Cichlidae): a morphological (optical and ultrastructural) study', *Fish Physiol Biochem*, **36**, 323–36.

ABELLI L, PICCHIETTI S, ROMANO N, MASTROLIA L, SCAPIGLIATI G (1997), 'Immunohistochemistry of gut-associated lymphoid tissue of the sea bass *Dicentrarchus labrax* (L.)', *Fish Shellfish Immunol*, **7**, 235–45.

AGGAD D, STEIN C, SIEGER D, MAZEL M, BOUDINOT P, HERBOMEL P, LEVRAUD JP, LUTFALLA G, LEPTIN M (2010), '*In vivo* analysis of IFN-gamma 1 and IFN-gamma 2 signaling in zebrafish', *J Immunol*, **185**, 6774–82.

AGIUS C (1980), 'Phylogenetic development of melano-macrophage centres in fish', *J Zool London*, **191**, 11–31.

AGIUS C, ROBERTS RJ (2003) 'Melano-macrophage centres and their role in fish pathology', *J Fish Dis*, **26**, 499–509.

AINSWORTH AJ (1992), 'Fish granulocytes: morphology, distribution, and function', *Ann Rev Fish Diseases*, **2**, 123–48.

ALITHEEN NB, MCCLURE S, MCCULLAGH P (2010), 'B-cell development: one problem, multiple solutions', *Immunol Cell Biol*, **88**, 445–50.

ANASTASIOU V, MIKROU A, PAPANASTASIOU AD, ZARKADIS IK (2011), 'The molecular identification of factor H and factor I molecules in rainbow trout provides insights into complement C3 regulation', *Fish Shellfish Immunol*, **31**, 491–9.

ARAKI K, AKATSU K, SUETAKE H, KIKUCHI K, SUZUKI Y (2008), 'Characterization of CD8+ leukocytes in fugu (*Takifugu rubripes*) with antiserum against fugu CD8alpha', *Dev Comp Immunol*, **32**, 850–8.

BALS R, WILSON JM (2003), 'Cathelicidins – a family of multifunctional antimicrobial peptides', *Cell Mol Life Sci*, **60**, 711–20.

BAO BL, PEATMAN E, XU P, LI P, ZENG H, HE CB, LIU ZJ (2006), 'The catfish liver-expressed antimicrobial peptide 2 (LEAP-2) gene is expressed in a wide range of tissues and developmentally regulated', *Mol Immunol*, **43**, 367–77.

BEI JX, SUETAKE H, ARAKI K, KIKUCHI K, YOSHIURA Y, LIN HR *et al.* (2006), 'Two interleukin (IL)-15 homologues in fish from two distinct origins', *Mol Immunol*, **43**, 860–9.

BERNARD D, SIX A, RIGOTTIER-GOIS L, MESSIAEN S, CHILMONCZYK S, QUILLET E, BOUDINOT P, BENMANSOUR A (2006), 'Phenotypic and functional similarity of gut intraepithelial and systemic T cells in a teleost fish', *J Immunol*, **176**, 3942–9.

BIACCHESI S, LEBERRE M, LAMOUREUX A, LOUISE Y, LAURET E, BOUDINOT P, BREMONT M (2009), 'Mitochondrial antiviral signaling protein plays a major role in induction of the fish innate immune response against RNA and DNA viruses', *J Virol*, **83**, 7815–27.

BIRD S, SECOMBES CJ (2006), '*Danio rerio* partial mRNA for interleukin-4', GenBank Accession No. AM403245.

BOLTAÑA S, ROHER N, GOETZ FW, MACKENZIE SA (2011), 'PAMPs, PRRs and the genomics of gram negative bacterial recognition in fish', *Dev Comp Immunol*, DOI: 10.1016/j.dci.2011.02.010

BONNEVILLE M, O'BRIEN RL, BORN WK (2010), 'Gammadelta T cell effector functions: a blend of innate programming and acquired plasticity', *Nat Rev Immunol*, **10**, 467–78.

BOSHRA H, GELMAN AE, SUNYER JO (2004a), 'Structural and functional characterization of complement C4 and C1s-like molecules in teleost fish: insights into the evolution of classical and alternative pathways', *J Immunol*, **173**, 349–59.

BOSHRA H, LI J, PETERS R, HANSEN J, MATLAPUDI A, SUNYER JO (2004b), 'Cloning, expression, cellular distribution, and role in chemotaxis of a C5a receptor in rainbow trout: the first identification of a C5a receptor in a nonmammalian species', *J Immunol*, **172**, 4381–90.

BOSHRA H, WANG T, HOVE-MADSEN L, HANSEN J, LI J, MATLAPUDI A, SECOMBES CJ, TORT L, SUNYER JO (2005), 'Characterization of a C3a receptor in rainbow trout and *Xenopus*: the first identification of C3a receptors in nonmammalian species', *J Immunol*, **175**, 2427–37.

BOSHRA H, LI J, SUNYER JO (2006), 'Recent advances on the complement system of teleost fish', *Fish Shellfish Immunol*, **20**, 239–62.

BOUDINOT P, BOUBEKEUR S, BENMANSOUR A (2001), 'Rhabdovirus infection induces public and private T cell responses in teleost fish', *J Immunol*, **167**, 6202–9.

BOWDEN TJ, COOK P, ROMBOUT JH (2005), 'Development and function of the thymus in teleosts', *Fish Shellfish Immunol*, **19**, 413–27.

BROEKMAN DC, FREI DM, GYLFASON GA, STEINARSSON A, JÖRNVALL H, AGERBERTH B, GUDMUNDSSON GH, MAIER VH (2011), 'Cod cathelicidin: isolation of the mature peptide, cleavage site characterisation and developmental expression', *Dev Comp Immunol*, **35**, 296–303.

BROMAGE ES, KAATTARI IM, ZWOLLO P, KAATTARI SL (2004), 'Plasmablast and plasma cell production and distribution in trout immune tissues', *J Immunol*, **173**, 7317–23.

BROWNE MJ, FENG CY, BOOTH V, RISE ML (2011), 'Characterization and expression studies of Gaduscidin-1 and Gaduscidin-2; paralogous antimicrobial peptide-like transcripts from Atlantic cod (*Gadus morhua*)', *Dev Comp Immunol*, **35**, 399–408.

BUONOCORE F, RANDELLI E, CASANI D, GUERRA L, PICCHIETTI S, COSTANTINI S, FACCHI-ANO AM, ZOU J, SECOMBES CJ, SCAPIGLIATI G (2008), 'A CD4 homologue in sea bass (*Dicentrarchus labrax*): molecular characterization and structural analysis', *Mol Immunol*, **45**, 3168–77.

CAI Z, GAO C, LI L, XING K (2010), 'Bipolar properties of red seabream (*Pagrus major*) transforming growth factor-beta in induction of the leucocytes migration', *Fish Shellfish Immunol*, **28**, 695–700.

CAMPAGNA S, SAINT N, MOLLE G, AUMELAS A (2007), 'Structure and mechanism of action of the antimicrobial peptide piscidin', *Biochemistry*, **46**, 1771–8.

CASADEI E, WANG T, ZOU J, GONZÁLEZ VECINO JL, WADSWORTH S, SECOMBES CJ (2009), 'Characterization of three novel beta-defensin antimicrobial peptides in rainbow trout (*Oncorhynchus mykiss*). *Mol Immunol*, **46**, 3358–66.

CASTELLANI ML, ANOGEIANAKI A, FELACO P, TONIATO E, DE LUTIIS MA, SHAIK B, FULCHERI M, VECCHIET J, TETÈ S, SALINI V, THEOHARIDES TC, CARAFFA A, ANTINOLFI P, FRYDAS I, CONTI P, CUCCURULLO C, CIAMPOLI C, CERULLI G, KEMPURAJ D (2010), 'IL-35, an anti-inflammatory cytokine which expands CD4+CD25+ Treg Cells', *J Biol Regul Homeost Agents*, **24**, 131–5.

CASTRO R, BERNARD D, LEFRANC MP, SIX A, BENMANSOUR A, BOUDINOT P (2011), 'T cell diversity and TcR repertoires in teleost fish', *Fish Shellfish Immunol*, **31**, 644–54.

CERUTTI A (2008), 'The regulation of IgA class switching', *Nat Rev Immunol*, **8**, 421–34.

CESTA MF (2006), 'Normal structure, function, and histology of the spleen', *Toxicol Pathol*, **34**(5), 455–65.

CHANG CI, PLEGUEZUELOS O, ZHANG YA, ZOU J, SECOMBES CJ (2005), 'Identification of a novel cathelicidin gene in the rainbow trout, *Oncorhynchus mykiss*', *Infect Immun*, **73**, 5053–64.

CHANG CI, ZHANG YA, ZOU J, NIE P, SECOMBES CJ (2006), 'Two cathelicidin genes are present in both rainbow trout (*Oncorhynchus mykiss*) and Atlantic salmon (*Salmo salar*)', *Antimicrob Agents Chemother*, **50**, 185–95.

CHANG M, COLLET B, NIE P, LESTER K, CAMPBELL S, SECOMBES CJ, ZOU J (2011a), 'Expression and functional characterization of the RIG-1 like receptors MDA5 and LGP2 in rainbow trout *Oncorhynchus mykiss*', *J Virol*, **85**, 8403–12.

CHANG M, WANG T, NIE P, ZOU J, SECOMBES CJ (2011b) 'Cloning of two rainbow trout nucleotide-binding oligomerization domain containing 2 (NOD2) splice variants and functional characterization of the NOD2 effector domains', *Fish Shellfish Immunol*, **30**, 118–27.

CHANTANACHOOKIN C, SEIKAI T, MASARU T (1991), 'Comparative study of the lymphoid organs in three species of marine fish', *Aquaculture*, **99**, 143–55.

CHEN K, XU W, WILSON M, HE B, MILLER NW, BENGTÉN E, EDHOLM ES, SANTINI PA, RATH P, CHIU A, CATTALINI M, LITZMAN J, B BUSSEL J, HUANG B, MEINI A, RIESBECK K, CUNNINGHAM-RUNDLES C, PLEBANI A, CERUTTI A (2009), 'Immunoglobulin D enhances immune surveillance by activating antimicrobial, proinflammatory and B cell-stimulating programs in basophils', *Nat Immunol*, **10**, 889–98.

CHEN SL, XU MY, JI XS, YU GC, LIU Y (2005), 'Cloning, characterization, and expression analysis of hepcidin gene from red sea bream (*Chrysophrys major*), *Antimicrob Agents Chemother*, **49**, 1608–12.

CHEN WQ, XU QQ, CHANG MX, NIE P, PENG KM (2010a), 'Molecular characterization and expression analysis of nuclear oligomerization domain proteins NOD1 and NOD2 in grass carp *Ctenopharyngodon idella*', *Fish Shellfish Immunol*, **28**, 18–29.

CHEN WQ, XU QQ, CHANG MX, ZOU J, SECOMBES CJ, PENG KM, NIE P (2010b), 'Molecular characterization and expression analysis of the IFN-gamma related gene (IFN-gammarel) in grass carp *Ctenopharyngodon idella*', *Vet Immunol Immunopathol*, **134**, 199–207.

CHIA TJ, WU YC, CHEN JY, CHI SC (2010), 'Antimicrobial peptides (AMP) with antiviral activity against fish nodavirus', *Fish Shellfish Immunol*, **28**, 434–9.

CHONDROU M, PAPANASTASIOU AD, SPYROULIAS GA, ZARKADIS IK (2008), 'Three isoforms of complement properdin factor P in trout: cloning, expression, gene organization and constrained modeling', *Dev Comp Immunol*, **32**, 1454–66.

CHONDROU MP, LONDOU AV, ZARKADIS IK (2006a), 'Expression and phylogenetic analysis of the ninth complement component (C9) in rainbow trout', *Fish Shellfish Immunol*, **21**, 572–6.

CHONDROU MP, MASTELLOS D, ZARKADIS IK (2006b), 'cDNA cloning and phylogenetic analysis of the sixth complement component in rainbow trout', *Mol Immunol*, **43**, 1080–7.

COFFMAN RL, CARTY JA (1986) 'T cell activity that enhances polyclonal IgE production and its inhibition by interferon-gamma', *J Immunol*, **136**, 949–54.

COLE AM, WEIS P, DIAMOND G (1997), 'Isolation and characterization of pleurocidin, an antimicrobial peptide in the skin secretions of winter flounder', *J Biol Chem*, **272**, 12008–13.

COLE AM, DAROUICHE RO, LEGARDA D, CONNELL N, DIAMOND G (2000), 'Characterization of a fish antimicrobial peptide: gene expression, subcellular localization, and spectrum of activity', *Antimicrob Agents Chemother*, **44**, 2039–45.

COLE DK, GAO GF (2004), 'CD8: adhesion molecule, co-receptor and immuno-modulator', *Cell Mol Immunol*, **1**, 81–8.

COLLET B, BAIN N, PREVOST S, BESINQUE G, MCBEATH A, SNOW M, COLLINS C (2008), 'Isola-tion of an Atlantic salmon (*Salmo salar*) signal transducer and activator of tran-scription STAT1 gene: kinetics of expression upon ISAV or IPNV infection', *Fish Shellfish Immunol*, **25**, 861–7.

CORRIPIO-MIYAR Y, ZOU J, RICHMOND H, SECOMBES CJ (2009), 'Identification of interleu-kin-22 in gadoids and examination of its expression level in vaccinated fish', *Mol Immunol*, **46**, 2098–106.

CORTES HD, MONTGOMERY BC, VERHEIJEN K, GARCÍA-GARCÍA E, STAFFORD JL (2012), 'Examination of the stimulatory signaling potential of a channel catfish leukocyte immune-type receptor and associated adaptor', *Dev Comp Immunol*, **36**, 62–73.

COSTA MM, MAEHR T, DIAZ-ROSALES P, SECOMBES CJ, WANG T (2011), 'Bioactivity studies of rainbow trout (*Oncorhynchus mykiss*) interleukin-6: effects on macrophage growth and antimicrobial peptide gene expression', *Mol Immunol*, **48**, 1903–16.

COSTA G, DANZ H, KATARIA P, BROMAGE E (2012), 'A holistic view of the dynamisms of teleost IgM: A case study of *Streptococcus iniae* vaccinated rainbow trout (*Oncorhynchus mykiss*)', *Dev Comp Immunol*, **36**, 298–305.

CUESTA A, MESEGUER J, ESTEBAN MA (2011), 'Molecular and functional characterization of the gilthead seabream β-defensin demonstrate its chemotactic and antimicro-bial activity', *Mol Immunol*, **48**, 1432–8.

CUI S, EISENACHER K, KIRCHHOFER A *et al.* (2008), 'The C-terminal regulatory domain is the RNA 5′-triphosphate sensor of RIG-I', *Mol Cell*, **29**, 169–79.

DANILOVA N, BUSSMANN J, JEKOSCH K, STEINER LA (2005), 'The immunoglobulin heavychain locus in zebrafish: identification and expression of a previously unknown isotype, immunoglobulin Z', *Nat Immunol*, **6**, 295–302.

DAUPHINEE SM, KARSAN A (2006), 'Lipopolysaccharide signaling in endothelial cells', *Lab Invest*, **86**, 9–22.

DE GUERRA A, CHARLEMAGNE J (1997), 'Genomic organization of the TcR beta-chain diversity (Dbeta) and joining (Jbeta) segments in the rainbow trout: presence of many repeated sequences', *Mol Immunol*, **34**, 653–62.

DE SK, PAL SG (1998), 'Studies on a Gobiid fish thymus', *J Freshwater Biol*, **10**, 63–7.

DE VICO G, CATALDI M, CARELLA F, MARINO F, PASSANTINO A (2008), 'Histological, histo-chemical and morphometric changes of splenic melanomacrophage centers (SMMCs) in Sparicotyle-infected cultured sea breams (*Sparus aurata*)', *Immuno-pharmacol Immunotoxicol*, **30**, 27–35.

DEN DUNNEN J, GRINGHUIS SI, GEIJTENBEEK TB (2010), 'Dusting the sugar fingerprint: C-type lectin signaling in adaptive immunity', *Immunol Lett*, **128**, 12–16.

DEZFULI BS, PIRONI F, GIARI L, NOGA EJ (2010), 'Immunocytochemical localization of piscidin in mast cells of infected seabass gill', *Fish Shellfish Immunol*, **28**, 476–82.

DÍAZ-ROSALES P, BIRD S, WANG TH, FUJIKI K, DAVIDSON WS, ZOU J, SECOMBES CJ (2009), 'Rainbow trout interleukin-2: cloning, expression and bioactivity analysis', *Fish Shellfish Immunol*, **27**, 414–22.

DIJKSTRA JM, SOMAMOTO T, MOORE L, HORDVIK I, OTOTAKE M, FISCHER U (2006), 'Identi-fication and characterization of a second CD4-like gene in teleost fish', *Mol Immunol*, **43**, 410–19.

DOS SANTOS NMS, TAVERNE-THIELE JJ, BARNES AC, VAN MUISWINKEL WB, ELLIS AE, ROMBOUT JHWM (2001), 'The gill is a major organ for antibody secreting cell pro-duction following direct immersion of sea bass (*Dicentrarchus labrax*, L.) in a *Photobacterium damselae* ssp. *piscicida* bacterin: an ontogenetic study', *Fish Shell-fish Immunol*, **11**, 65–74.

DOUGLAS SE, GALLANT JW, LIEBSCHER RS, DACANAY A, TSOI SC (2003), 'Identification and expression analysis of hepcidin-like antimicrobial peptides in bony fish', *Dev Comp Immunol*, **27**, 589–601.

DU PASQUIER L (1982), 'Antibody diversity in lower vertebrates – why is it so restricted?', *Nature*, **296**, 311–13.

EDHOLM ES, STAFFORD JL, QUINIOU SM, WALDBIESER G, MILLER NW, BENGTEN E, WILSON M (2007), 'Channel catfish, *Ictalurus punctatus*, CD4-like molecules', *Dev Comp Immunol*, **31**, 172–87.

EDHOLM ES, BENGTÉN E, STAFFORD JL, SAHOO M, TAYLOR EB, MILLER NW, WILSON M (2010), 'Identification of two IgD+ B cell populations in channel catfish, *Ictalurus punctatus*', *J Immunol*, **185**, 4082–94.

EDHOLM ES, BENGTEN E, WILSON M (2011), 'Insights into the function of IgD', *Dev Comp Immunol*, DOI: 10.1016/j.dci.2011.03.002

ENCINAS P, RODRIGUEZ-MILLA MA, NOVOA B, ESTEPA A, FIGUERAS A, COLL J (2010), 'Zebrafish fin immune responses during high mortality infections with viral haemorrhagic septicemia rhabdovirus. A proteomic and transcriptomic approach', *BMC Genomics*, **11**, 518.

ENDO Y, TAKAHASHI M, NAKAO M, SAIGA H, SEKINE H, MATSUSHITA M, NONAKA M, FUJITA T (1998), 'Two lineages of mannose-binding lectin-associated serine protease (MASP) in vertebrates', *J Immunol*, **161**, 4924–30.

ENDO Y, TAKAHASHI M, FUJITA T (2006), 'Lectin complement system and pattern recognition', *Immunobiology*, **211**, 283–93.

EVANS DL, HOGAN KT, GRAVES SS (1984), 'Nonspecific cytotoxic cells in fish (*Ictalurus punctatus*). III. Biophysical and biochemical properties affecting cytolysis', *Dev Comp Immunol*, **8**, 599–610.

FALCO A, CHICO V, MARROQUÍ L, PEREZ L, COLL JM, ESTEPA A (2008), 'Expression and antiviral activity of a beta-defensin-like peptide identified in the rainbow trout (*Oncorhynchus mykiss*) EST sequences', *Mol Immunol*, **45**, 757–65.

FANG W, XIANG LX, SHAO JZ, WEN Y, CHEN SY (2006), 'Identification and characterization of an interleukin-15 homologue from *Tetraodon nigroviridis*', *Comp Biochem Physiol B*, **143**, 335–43.

FÄNGE R, NILSSON S (1985), 'The fish spleen: structure and function', *Experientia*, **41**, 152–8.

FENG LN, LU DQ, BEI JX, CHEN JL, LIU Y, ZHANG Y et al. (2009), 'Molecular cloning and functional analysis of polymeric immunoglobulin receptor gene in orange-spotted grouper (*Epinephelus coioides*)', *Comp Biochem Physiol B*, **154**, 282–9.

FERNANDES JMO, MOLLE G, KEMP GD, SMITH VJ (2004), 'Isolation and characterization of oncorhyncin II, a histone H1-derived antimicrobial peptide from skin secretions of rainbow trout, *Oncorhynchus mykiss*', *Dev Comp Immunol*, **28**, 127.

FERNANDES JMO, RUANGSRI J, KIRON V (2010), 'Atlantic cod piscidin and its diversification through positive selection' *PLoS One*, **5**, e9501.

FLAJNIK MF (2010), 'All GOD's creatures got dedicated mucosal immunity', *Nat Immunol*, **11**, 777–9.

FLORES MV, HALL C, JURY A, CROSIER K, CROSIER P (2007), 'The zebrafish retinoid-related orphan receptor (ror) gene family', *Gene Expression Patterns*, **7**, 535–43.

FORLENZA M, FINK IR, RAES G, WIEGERTJES GF (2011), 'Heterogeneity of macrophage activation in fish', *Dev Comp Immunol*, in press.

FOURNIER-BETZ V, QUENTEL C, LAMOUR F, LEVEN A (2000), 'Immunocytochemical detection of Ig-positive cells in blood, lymphoid organs and the gut associated lymphoid tissue of the turbot (*Scophthalmus maximus*)', *Fish Shellfish Immunol*, **10**, 187–202.

FRANCHINI S, ZARKADIS IK, SFYROERA G, SAHU A, MOORE WT, MASTELLOS D, LAPATRA SE, LAMBRIS JD (2001), 'Cloning and purification of the rainbow trout fifth component of complement (C5)', *Dev. Comp. Immunol.*, **25**, 419–30.

FRØYSTAD MK, RODE M, BERG G, GJØEN T (1998), 'A role for scavenger receptors in phagocytosis of protein-coated particles in rainbow trout head kidney macrophages', *Dev Comp Immunol*, **22**, 533–49.

FUGLEM B, JIRILLO E, BJERKÅS I, KIYONO H, NOCHI T, YUKI Y, RAIDA M, FISCHER U, KOPPANG EO (2010), 'Antigen-sampling cells in the salmonid intestinal epithelium', *Dev Comp Immunol*, **34**, 768–74.

FUJIKI K, LIU L, SUNDICK RS, DIXON B (2003) 'Molecular cloning and characterization of rainbow trout (*Oncorhynchus mykiss*) C5a anaphylatoxin receptor', *Immunogenetics*, **55**, 640–6.

GALLO VP, CIVININI A (2003), 'Survey of the adrenal homolog in teleosts', *Int Rev Cytol*, **230**, 89–187.

GAO Q, NIE P, THOMPSON KD, ADAMS A, WANG T, SECOMBES CJ, ZOU J (2009), 'The search for the IFN-gamma receptor in fish: functional and expression analysis of putative binding and signalling chains in rainbow trout *Oncorhynchus mykiss*', Dev Comp Immunol, **33**, 920–31.

GATTO D, BRINK R (2010), 'The germinal center reaction', *J Allergy Clin Immunol*, **126**, 898–907.

GEIJTENBEEK TB, GRINGHUIS SI (2009), 'Signalling through C-type lectin receptors: shaping immune responses', *Nat Rev Immunol*, **9**, 465–79.

GETNET D, GROSSO JF, GOLDBERG MV, HARRIS TJ, YEN HR, BRUNO TC, DURHAM NM, HIPKISS EL, PYLE KJ, WADA S, PAN F, PARDOLL DM, DRAKE CG (2010), 'A role for the transcription factor Helios in human CD4(+)CD25(+) regulatory T cells', *Mol Immunol*, **47**, 1595–600.

GOOD-JACOBSON KL, SHLOMCHIK MJ (2010), 'Plasticity and heterogeneity in the generation of memory B cells and long-lived plasma cells: the influence of germinal center interactions and dynamics', *J Immunol*, **185**, 3117–25.

GORGOLLON P (1983), 'Fine structure of the thymus in the adult cling fish *Sicyases sanguineus* (Pisces, Gobiesocidae)', *J Morphol*, **177**, 25–40.

GRACE MF, MANNING MJ (1980), 'Histogenesis of the lymphoid organs in rainbow trout, *Salmo gairdneri* Richardson. 1836', *Dev Comp Immunol*, **4**, 255–64.

GRAYFER L, BELOSEVIC M (2009), 'Molecular characterization of novel interferon gamma receptor 1 isoforms in zebrafish (*Danio rerio*) and goldfish (*Carassius auratus* L.)', *Mol Immunol*, **46**, 3050–9.

GRAYFER L, GARCIA EG, BELOSEVIC M (2010), 'Comparison of macrophage antimicrobial responses induced by type II interferons of the goldfish (*Carassius auratus* L.)', *J Biol Chem*, **285**, 23537–47.

GRAYFER L, HODGKINSON JW, HITCHEN SJ, BELOSEVIC M (2011), 'Characterization and functional analysis of goldfish (*Carassius auratus* L.) interleukin-10', *Mol Immunol*, **48**, 563–71.

GUNIMALADEVI I, SAVAN R, SATO K, YAMAGUCHI R, SAKAI M (2007), 'Characterization of an interleukin-15 like (IL-15L) gene from zebrafish (*Danio rerio*)', *Fish Shellfish Immunol*, **22**, 351–62.

GUO CJ, ZHANG YF, YANG LS, YANG XB, WU YY, LIU D, CHEN WJ, WENG SP, YU XQ, HE JG (2009), 'The JAK and STAT family members of the mandarin fish *Siniperca chuatsi*: molecular cloning, tissues distribution and immunobiological activity', *Fish Shellfish Immunol*, **27**, 349–59.

GUTSMANN T, SCHROMM AB, BRANDENBURG K (2007), 'The physicochemistry of endotoxins in relation to bioactivity', *Int J Med Microbiol*, **297**, 341–52.

HADDAD G, HANINGTON PC, WILSON EC, GRAYFER L, BELOSEVIC M (2008), 'Molecular and functional characterization of goldfish (*Carassius auratus* L.) transforming growth factor beta', *Dev Comp Immunol*, **32**, 654–63.

HADIDI S, GLENNEY GW, WELCH TJ, SILVERSTEIN JT, WIENS GD (2008), 'Spleen size predicts resistance of rainbow trout to *Flavobacterium psychrophilum* challenge', *J Immunol*, **180**, 4156–65.

HAMURO K, SUETAKE H, SAHA NR, KIKUCHI K, SUZUKI Y (2007), 'A teleost polymeric Ig receptor exhibiting two Ig-like domains transports tetrameric IgM into the skin', *J Immunol*, **178**, 5682–9.

HANSEN JD, ZAPATA AG (1998), 'Lymphocyte development in fish and amphibians', *Immunol Rev*, **166**, 199–220.

HANSEN JD, STRASSBURGER P (2000), 'Description of an ectothermic TCR coreceptor, CD8 alpha, in rainbow trout', *J Immunol*, **164**, 3132–9.

HANSEN JD, LANDIS ED, PHILLIPS RB (2005), 'Discovery of a unique Ig heavy-chain isotype (IgT) in rainbow trout: implications for a distinctive B cell developmental pathway in teleost fish', *Proc Natl Acad Sci USA*, **102**, 6919–24.

HANSEN JD, VOJTECH LN, LAING KJ (2011), 'Sensing disease and danger: a survey of vertebrate PRRs and their origins', *Dev Comp Immunol*, **35**, 886–97.

HARDIE LJ, LAING KJ, DANIELS GD, GRABOWSKI PS, CUNNINGHAM C, SECOMBES CJ (1998), 'Isolation of the first piscine transforming growth factor beta gene: analysis reveals tissue specific expression and a potential regulatory sequence in rainbow trout (*Oncorhynchus mykiss*)', *Cytokine*, **10**, 555–63.

HARUN NO, COSTA MM, SECOMBES CJ, WANG T (2011), 'Sequencing of a second interleukin-10 gene in rainbow trout *Oncorhynchus mykiss* and comparative investigation of the expression and modulation of the paralogues *in vitro* and *in vivo*', *Fish Shellfish Immunol*, **31**, 107–17.

HAUGARVOLL E, BJERKÅS I, NOWAK BF, HORDVIK I, KOPPANG EO (2008), 'Identification and characterization of a novel intraepithelial lymphoid tissue in the gills of Atlantic salmon', *J Anat*, **213**, 202–9.

HAUSER AE, JUNT T, MEMPEL TR, SNEDDON MW, KLEINSTEIN SH, HENRICKSON SE, VON ANDRIAN UH, SHLOMCHIK MJ, HABERMAN AM (2007), 'Definition of germinal-center B cell migration *in vivo* reveals predominant intrazonal circulation patterns', *Immunity*, **26**, 655–67.

HEICKLEN-KLEIN A, MCREYNOLDS LJ, EVANS T (2005), 'Using the zebrafish model to study GATA transcription factors', *Semin. Cell Dev Biol*, **16**, 95–106.

HETLAND DL, JØRGENSEN SM, SKJØDT K, DALE OB, FALK K, XU C, MIKALSEN AB, GRIMHOLT U, GJØEN T, PRESS CM (2010), '*In situ* localisation of major histocompatibility complex class I and class II and CD8 positive cells in infectious salmon anaemia virus (ISAV)-infected Atlantic salmon', *Fish Shellfish Immunol*, **28**, 30–9.

HIRANO M, DAS S, GUO P, COOPER MD (2011), 'The evolution of adaptive immunity in vertebrates', *Adv Immunol*, **109**, 125–57.

HIRONO I, TAKAMI M, MIYATA M, MIYAZAKI T, HAN HJ, TAKANO T, ENDO M, AOKI T (2004), 'Characterization of gene structure and expression of two toll-like receptors from Japanese flounder, *Paralichthys olivaceus*', *Immunogenetics*, **56**, 38–46.

HIRONO I, HWANG JY, ONO Y, KUROBE T, OHIRA T, NOZAKI R et al. (2005), 'Two different types of hepcidins from the Japanese flounder *Paralichthys olivaceus*', *FEBS J*, **272**, 5257–64.

HOLZ LE, BOWEN DG, BERTOLINO P (2010), 'Mechanisms of T cell death in the liver: to Bim or not to Bim?', *Dig Dis*, **28**, 14–24.

HORNUNG V, ABLASSER A, CHARREL-DENNIS M, BAUERNFEIND F, HORVATH G, CAFFREY DR, LATZ E, FITZGERALD KA (2009), 'AIM2 recognizes cytosolic dsDNA and forms a caspase-1-activating inflammasome with ASC', *Nature*, **458**, 514–18.

HSIEH JC, PAN CY, CHEN JY (2010), 'Tilapia hepcidin (TH)2–3 as a transgene in transgenic fish enhances resistance to *Vibrio vulnificus* infection and causes variations in immune-related genes after infection by different bacterial species', *Fish Shellfish Immunol*, **29**, 430–9.

HU YL, PAN XM, XIANG LX, SHAO JZ (2010), 'Characterization of C1q in teleosts: insight into the molecular and functional evolution of C1q family and classical pathway', *J Biol Chem*, **285**, 28777–86.

HUANG PH, CHEN JY, KUO CM (2007), 'Three different hepcidins from tilapia, *Oreochromis mossambicus*: analysis of their expressions and biological functions', *Mol Immunol*, **44**, 1922–34.

HUANG S, YUAN S, GUO L, YU Y, LI J, WU T, LIU T, YANG M, WU K, LIU H, GE J, HUANG H, DONG M, YU C, CHEN S, XU A (2008), 'Genomic analysis of the immune gene repertoire of amphioxus reveals extraordinary innate complexity and diversity', *Genome Res*, **18**, 1112–26.

HUANG T, SU J, HENG J, DONG J, ZHANG R, ZHU H (2010) 'Identification and expression profiling analysis of grass carp *Ctenopharyngodon idella* LGP2 cDNA', *Fish Shellfish Immunol*, **29**, 349–435.

IGAWA D, SAKAI M, SAVAN R (2006), 'An unexpected discovery of two interferon gamma-like genes along with interleukin (IL)-22 and -26 from teleost: IL-22 and -26 genes have been described for the first time outside mammals', *Mol Immunol*, **43**, 999–1009.

INOUE Y, KAMOTA S, ITO K, YOSHIURA Y, OTOTAKE M, MORITOMO T, NAKANISHI T (2005), 'Molecular cloning and expression analysis of rainbow trout (*Oncorhynchus mykiss*) interleukin-10 cDNAs', *Fish Shellfish Immunol*, **18**, 335–44.

ISHII A, KAWASAKI M, MATSUMOTO M, TOCHINAI S, SEYA T (2007), 'Phylogenetic and expression analysis of amphibian *Xenopus* Toll-like receptors', *Immunogenetics*, **59**, 281–93.

JANA M, DASGUPTA S, PAL U, PAHAN K (2009), 'IL-12 p40 homodimer, the so-called biologically inactive molecule, induces nitric oxide synthase in microglia via IL-12R beta1', *Glia*, **57**, 1553–65.

JANKOWSKI M, KOPIŃSKI P, GOC A (2010), 'Interleukin-27: biological properties and clinical application', *Arch Immunol Ther Exp (Warsz)*, **58**, 417–25.

JARRY A, CERF-BENSUSSAN N, BROUSSE N, SELZ F, GUY-GRAND D (1990), 'Subsets of CD3+ (T cell receptor alpha/beta or gamma/delta) and CD3-lymphocytes isolated from normal human gut epithelium display phenotypical features different from their counterparts in peripheral blood', *Eur J Immunol*, **20**, 1097–103.

JIA W, ZHOU X (2010), 'Molecular structural and functional characterization of STAT1 gene regulatory region in teleost *Channa argus*', *Vet Immunol Immunopathol*, **135**, 146–51.

JIN JY, ZHOU L, WANG Y, LI Z, ZHAO JG, ZHANG QY, GUI JF (2010), 'Antibacterial and antiviral roles of a fish β-defensin expressed both in pituitary and testis', *PLoS One*, **5**, e12883.

JOHN LB, YOONG S, WARD AC (2009), 'Evolution of the Ikaros gene family: implications for the origins of adaptive immunity', *J Immunol*, **182**, 4792–9.

KAATTARI S, EVANS D, KLEMER J (1998), 'Varied redox forms of teleost IgM: an alternative to isotypic diversity?', *Immunol Rev*, **166**, 133–42.

KABELITZ D, GLATZEL A, WESCH D (2000), 'Antigen recognition by human gammadelta T lymphocytes', *Int Arch Allergy Immunol*, **122**, 1–7.

KANIA PW, SORENSEN RR, KOCH C, BRANDT J, KLIEM A, VITVED L, HANSEN S, SKJODT K (2010), 'Evolutionary conservation of mannan-binding lectin (MBL) in bony fish: identification, characterization and expression analysis of three bona fide collectin homologues of MBL in the rainbow trout (*Onchorhynchus mykiss*)', *Fish Shellfish Immunol*, **29**, 910–20.

KATO Y, NAKAO M, MUTSURO J, ZARKADIS IK, YANO T (2003), The complement component C5 of the common carp (*Cyprinus carpio*): cDNA cloning of two distinct isotypes that differ in a functional site', *Immunogenetics*, **54**, 807–15.

KATZENBACK BA, BELOSEVIC M (2009), 'Isolation and functional characterization of neutrophil-like cells, from goldfish (*Carassius auratus* L.) kidney', *Dev Comp Immunol*, **33**, 601–11.

KAZANTZI A, SFYROERA G, HOLLAND MC, LAMBRIS JD, ZARKADIS IK (2003), 'Molecular cloning of the beta subunit of complement component eight of rainbow trout', *Dev Comp Immunol*, **27**, 167–74.

KEMENADE BML, DALY JG, GROENEVELD A, WIEGERTJES GF (1996), 'Multiple regulation of carp (*Cyprinus carpio* L.) macrophages and neutrophilic granulocytes by serum factors: influence of infection with atypical *Aeromonas salmonicida*', *Vet Immunol and Immunopath*, **51**, 189–200.

KIMURA A, KISHIMOTO T (2010), 'IL-6: regulator of Treg/Th17 balance', *Eur J Immunol*, **40**, 1830–5.

KOLLS JK (2010), 'Th17 cells in mucosal immunity and tissue inflammation', *Semin Immunopathol*, **32**, 1–2.

KOMURO A, BAMMING D, HORVATH CM (2008), 'Negative regulation of cytoplasmic RNA-mediated antiviral signaling', *Cytokine*, **43**, 350–8.

KONDO H, YEU TZEH AG, HIRONO I, AOKI T (2007), 'Identification of a novel C-type lectin gene in Japanese flounder, *Paralichthys olivaceus*', *Fish Shellfish Immunol*, **23**, 1089–94.

KONO T, BIRD S, SONODA K, SAVAN R, SECOMBES CJ, SAKAI M (2008), 'Characterization and expression analysis of an interleukin-7 homologue in the Japanese pufferfish, *Takifugu rubripes*', *FEBS J*, **275**, 1213–26.

KOPPANG EO, FISCHER U, MOORE L, TRANULIS MA, DIJKSTRA JM, KÖLLNER B, AUNE L, JIRILLO E, HORDVIK I (2010), 'Salmonid T cells assemble in the thymus, spleen and in novel interbranchial lymphoid tissue', *J Anat*, **217**, 728–39.

KUMAR H, KAWAI T, AKIRA S (2011), 'Pathogen recognition by the innate immune system', *Int Rev Immunol*, **30**, 16–34.

KUMARI J, BOGWALD J, DALMO RA (2009), 'Transcription factor GATA-3 in Atlantic salmon (*Salmo salar*): molecular characterization, promoter activity and expression analysis', *Mol Immunol*, **46**, 3099–107.

KUMARI U, NIGAM AK, MITIAL S, MITIAL AK (2011), 'Antibacterial properties of the skin mucus of the freshwater fishes, *Rita rita* and *Channa punctatus*', *Eur Rev Med Pharmacol Sci*, **15**, 781–6.

LAING KJ, HANSEN JD (2011), 'Fish T cells: recent advances through genomics', *Dev Comp Immunol*, DOI: 10.1016/j.dci.2011.03.004

LAING KJ, ZOU JJ, PURCELL MK, PHILLIPS R, SECOMBES CJ, HANSEN JD (2006), 'Evolution of the CD4 family: teleost fish possess two divergent forms of CD4 in addition to lymphocyte activation gene-3', *J Immunol*, **177**, 3939–51.

LAING KJ, PURCELL MK, WINTON JR, HANSEN JD (2008), 'A genomic view of the NOD-like receptor family in teleost fish: identification of a novel NLR subfamily in zebrafish', *BMC Evol Biol*, **8**, 42.

LAMAS J, ELLIS AE (1994), 'Electron microscopic observations of the phagocytosis and subsequent fate of *Aeromonas salmonicida* by Atlantic salmon neutrophils *in vitro*', *Fish Shellfish Immunol*, **4**, 539–46.

LANIER LL (2005), 'NK cell recognition', *Annu Rev Immunol*, **23**, 225–74.

LAUKSUND S, SVINGERUD T, BERGAN V, ROBERTSEN B (2009), 'Atlantic salmon IPS-1 mediates induction of IFNa1 and activation of NF-kappaB and localizes to mitochondria', *Dev Comp Immunol*, **33**, 1196–204.

LAURIANO ER, CALÒ M, SILVESTRI G, ZACCONE D, PERGOLIZZI S, LO CASCIO P (2011), 'Mast cells in the intestine and gills of the sea bream, *Sparus aurata*, exposed to a polychlorinated biphenyl, PCB 126', *Acta Histochem*, DOI: 10.1016/j.acthis.2011.04.004

LI J, PETERS R, LAPATRA SE, VAZZANA M, SUNYER JO (2004), 'Anaphylatoxin-like molecules generated during complement activation induce a dramatic enhancement of particle uptake in rainbow trout phagocytes', *Dev Comp Immunol*, **28**, 1005–21.

LI J, BARREDA DR, ZHANG YA, BOSHRA H, GELMAN AE, LAPATRA S, TORT L, SUNYER JO (2006), 'B lymphocytes from early vertebrates have potent phagocytic and microbicidal abilities', *Nat Immunol*, **7**, 1116–24.

LI JH, SHAO JZ, XIANG LX, WEN Y (2007), 'Cloning, characterization and expression analysis of pufferfish IL-4 cDNA: the first evidence of Th2-type cytokine in fish', *Mol Immunol*, **44**, 2078–86.

LI K, SACKS SH, ZHOU W (2007), 'The relative importance of local and systemic complement production in ischaemia, transplantation and other pathologies', *Mol Immunol*, **44**, 3866–74.

LIN AF, XIANG LX, WANG QL, DONG WR, GONG YF, SHAO JZ (2009), 'The DC-SIGN of zebrafish: insights into the existence of a CD209 homologue in a lower vertebrate and its involvement in adaptive immunity', *J Immunol*, **183**, 7398–410.

LITMAN GW, RAST JP, FUGMANN SD (2010), 'The origins of vertebrate adaptive immunity', *Nat Rev Immunol*, **10**, 543–53.

LIU F, LI JL, YUE GH, FU JJ, ZHOU JF (2010), 'Molecular cloning and expression analysis of the liver-expressed antimicrobial peptide 2 (LEAP-2) gene in grass carp', *Vet Immunol Immunopathol*, **133**, 133–43.

LOBB CJ, CLEM LW (1981), 'The metabolic relationships of the immunoglobulins in fish serum, cutaneous mucus and bile', *J Immunol*, **127**, 1525–9.

LONDOU A, MIKROU A, ZAKADIS IK (2008), 'Cloning and characterization of two clusterin isoforms in rainbow trout', *Mol Immunol*, **45**, 470–8.

LOPEZ-MUNOZ A, ROCA FJ, MESEGUER J, MULERO V (2009), 'New insights into the evolution of IFNs: zebrafish group II IFNs induce a rapid and transient expression of IFN-dependent genes and display powerful antiviral activities', *J Immunol*, **182**, 3440–9.

LOPEZ-MUNOZ A, SEPULCRE MP, ROCA FJ, FIGUERAS A, MESEGUER J, MULERO V (2011), 'Evolutionary conserved pro-inflammatory and antigen presentation functions of zebrafish IFN gamma revealed by transcriptomic and functional analysis', *Mol Immunol*, **48**, 1073–83.

LØVOLL M, KILVIK T, BOSHRA H, BØGWALD J, SUNYER JO, DALMO RA (2006), Maternal transfer of complement components C3-1, C3-3, C3-4, C4, C5, C7, Bf, and Df to offspring in rainbow trout (*Oncorhynchus mykiss*)', *Immunogenetics*, **58**, 168–79.

LØVOLL M, DALMO RA, BØGWALD J (2007), 'Extrahepatic synthesis of complement components in the rainbow trout (*Oncorhynchus mykiss*)', *Fish Shellfish Immunol*, **23**, 721–31.

LU XJ, CHEN J, HUANG ZA, SHI YH, LV JN (2011), 'Identification and characterization of a novel cathelicidin from ayu, *Plecoglossus altivelis*', *Fish Shellfish Immunol*, **31**, 52–7.

LUTFALLA G, ROEST CROLLIUS H, STANGE-THOMANN N, JAILLON O, MOGENSEN K, MONNERON D (2003), 'Comparative genomic analysis reveals independent expansion of a lineage-specific gene family in vertebrates: the class II cytokine receptors and their ligands in mammals and fish', *BMC Genomics*, **4**, 29.

MAGNADOTTIR B (2010), 'Immunological control of fish diseases', *Mar Biotechnol (NY)*, **12**, 361–79.

MAGNADOTTIR B, LANGE S, GUDMUNDSDOTTIR S, BØGWALD J, DALMO RA (2005), 'Ontogeny of humoral immune parameters in fish', *Fish Shellfish Immunol*, **19**, 429–39.

MAIER VH, DORN KV, GUDMUNDSDOTTIR BK, GUDMUNDSSON GH (2008), 'Characterisation of cathelicidin gene family members in divergent fish species', *Mol Immunol*, **45**, 3723–30.

MARIOLI DJ, ZARKADIS IK (2008), 'The vitronectin gene in rainbow trout: cloning, expression and phylogenetic analysis', *Fish Shellfish Immunol*, **24**, 18–25.

MARIUZZA RA, VELIKOVSKY CA, DENG L, XU G, PANCER Z (2010), 'Structural insights into the evolution of the adaptive immune system: the variable lymphocyte receptors of jawless vertebrates', *Biol Chem*, **391**, 753–60.

MARTIN SA, MOHANTY BP, CASH P, HOULIHAN DF, SECOMBES CJ (2007), 'Proteome analysis of the Atlantic salmon (*Salmo salar*) cell line SHK-1 following recombinant IFN-gamma stimulation', *Proteomics*, **7**, 2275–86.

MARTIN SAM, DOUGLAS A, HOULIHAN DF, SECOMBES CJ (2010), 'Starvation alters the liver transcriptome of the innate immune response in Atlantic salmon (*Salmo salar*)', *BMC Genomics*, **11**, 418.

MASON KL, HUFFNAGLE GB, NOVERR MC, KAO JY (2008), 'Overview of gut immunology', *Adv Exp Med Biol*, **635**, 1–14.

MATSUMIYA T, IMAIZUMI T, YOSHIDA H, SATOH K (2011), 'Antiviral signaling through retinoic acid-inducible gene-I-like receptors', *Arch Immunol Ther Exp (Warsz)*, **59**, 41–8.

MATSUO M, OSHIUMI H, TSUJITA T, MITANI H, KASAI H, YOSHIMIZU M, MATSUMOTO M, SEYA T (2008), 'Teleost Toll-like receptor 22 recognizes RNA duplex to induce IFN and protect cells from Birnaviruses', *J Immunol*, **181**, 3474–85.

MCBEATH AJ, SNOW M, SECOMBES CJ, ELLIS AE, COLLET B (2007), 'Expression kinetics of interferon and interferon-induced genes in Atlantic salmon (*Salmo salar*) following infection with infectious pancreatic necrosis virus and infectious salmon anaemia virus', *Fish Shellfish Immunol*, **22**, 230–41.

MCMILLAN DN, SECOMBES CJ (1997), 'Isolation of rainbow trout (*Oncorhynchus mykiss*) intestinal intraepithelial lymphocytes (IEL) and measurement of their cytotoxic activity', *Fish Shellfish Immunol*, **7**, 527–41.

MEEKER ND, SMITH AC, FRAZER JK, BRADLEY DF, RUDNER LA, LOVE C, TREDE NS (2010), 'Characterization of the zebrafish T cell receptor beta locus', *Immunogenetics*, **62**, 23–9.

MEIJER AH, GABBY KRENS SF, MEDINA RI, HE S, BITTER W, EWA SNAAR-JAGALSKA B, SPAINK HP (2004), 'Expression analysis of the Toll-like receptor, TIR domain adaptor families of zebrafish', *Mol Immunol*, **40**, 773–83.

MESEGUER J, LÓPEZ-RUIZ A, GARCÍA-AYALA A (1995), 'Reticulo-endothelial stroma of the head-kidney from the seawater teleost gilthead seabream (*Sparus aurata* L.): an ultrastructural and cytochemical study', *Anat Rec*, **241**, 303–9.

MIKROU A, MARIOLI D, PAPANASTASIOU AD, ZARKADIS IK (2009), 'CR3 complement receptor: cloning and characterization in rainbow trout', *Fish Shellfish Immunol*, **26**, 19–28.

MILEV-MILOVANOVIC I, LONG S, WILSON M, BENGTEN E, MILLER NW, CHINCHAR VG (2006), 'Identification and expression analysis of interferon gamma genes in channel catfish', *Immunogenetics*, **58**, 70–80.

MILLÁN A, GÓMEZ-TATO A, PARDO BG, FERNÁNDEZ C, BOUZA C, VERA M, ALVAREZ-DIOS JA, CABALEIRO S, LAMAS J, LEMOS ML, MARTÍNEZ P (2011), 'Gene expression profiles of the spleen, liver, and head kidney in turbot (*Scophthalmus maximus*) along the infection process with *Aeromonas salmonicida* using an immune-enriched oligo-microarray', *Mar Biotechnol (NY)*, **13**, 1099–114.

MILLER NW, CLEM LW (1984), 'Temperature-mediated processes in teleost immunity: differential effects of temperature on catfish *in vitro* antibody responses to thymus-dependent and thymus-independent antigens', *J Immunol*, **133**, 2356–9.

MILLER NW, SIZEMORE RC, CLEM LW (1985), 'Phylogeny of lymphocyte heterogeneity: the cellular requirements for in vitro antibody responses of channel catfish leukocytes', *J Immunol*, **134**, 2884–8.

MITRA S, ALNABULSI A, SECOMBES CJ, BIRD S (2010), 'Identification and characterization of the transcription factors involved in T-cell development, t-bet, stat6 and foxp3, within the zebrafish, *Danio rerio*', *FEBS J*, **277**, 128–47.

MONTE MM, ZOU J, WANG T, CARRINGTON A, SECOMBES C (2011), 'Cloning, expression analysis and bioactivity studies of rainbow trout (*Oncorhynchus mykiss*) interleukin-22', *Cytokine*, **55**, 62–73.

MOORE LJ, SOMAMOTO T, LIE KK, DIJKSTRA JM, HORDVIK I (2005), 'Characterisation of salmon and trout CD8alpha and CD8beta', *Mol Immunol*, **42**, 1225–34.

MOORE LJ, DIJKSTRA JM, KOPPANG EO, HORDVIK I (2009), 'CD4 homologues in Atlantic salmon', *Fish Shellfish Immunol*, **26**, 10–18.

MOSMANN TR, CHERWINSKI H, BOND MW, GIEDLIN MA, COFFMAN RL (1986), 'Two types of murine helper T cell clone. I. Definition according to profiles of lymphokine activities and secreted proteins', *J Immunol*, **136**, 2348–57.

MOSSER DM, EDWARDS JP (2008), 'Exploring the full spectrum of macrophage activation', *Nat Rev Immunol*, **8**, 958–69.

MUTSURO J, TANAKA N, KATO Y, DODDS AW, YANO T, NAKAO M (2005), 'Two divergent isotypes of the fourth complement component from a bony fish, the common carp (*Cyprinus carpio*)', *J Immunol*, **175**, 4508–17.

NAGAI T, MUTSURO J, KIMURA M, KATO Y, FUJIKI K, YANO T, NAKAO M (2000), 'A novel truncated isoform of the mannose-binding lectin-associated serine protease (MASP) from the common carp (*Cyprinus carpio*)', *Immunogenetics*, **51**, 193–200.

NAKAO M, FUSHITANI Y, FUJIKI K, NONAKA M, YANO T (1998), 'Two diverged complement factor B/C2-like cDNA sequences from a teleost, the common carp (*Cyprinus carpio*)', *J Immunol*, **161**, 4811–18.

NAKAO M, MUTSURO J, OBO R, FUJIKI K, NONAKA M, YANO T (2000), 'Molecular cloning and protein analysis of divergent forms of the complement component C3 from a bony fish, the common carp (*Cyprinus carpio*): presence of variants lacking the catalytic histidine', *Eur J Immunol*, **30**, 858–66.

NAKAO M, OSAKA K, KATO Y, FUJIKI K, YANO T (2001), 'Molecular cloning of the complement (C1r/C1s/MASP2-like serine proteases from the common carp (*Cyprinus carpio*)', *Immunogenetics*, **52**, 255–63.

NAKAO M, MATSUMOTO M, NAKAZAWA M, FUJIKI K, YANO T (2002), 'Diversity of complement factor B/C2 in the common carp (*Cyprinus carpio*): three isotypes of B/C2-A expressed in different tissues', *Dev Comp Immunol*, **26**, 533–41.

NAKAO M, HISAMATSU S, NAKAHARA M, KATO Y, SMITH SL, YANO T (2003), 'Molecular cloning of the complement regulatory factor I isotypes from the common carp (*Cyprinus carpio*)', *Immunogenetics*, **54**, 801–6.

NAKAO M, KAJIYA T, SATO Y, SOMAMOTO T, KATO-UNOKI Y, MATSUSHITA M, NAKATA M, FUJITA T, YANO T (2006), 'Lectin pathway of bony fish complement: identification of two homologs of the mannose-binding lectin associated with MASP2 in the common carp (*Cyprinus carpio*)', *J Immunol*, **177**, 5471–9.

NAKAO M, TSUJIKURA M, ICHIKI S, VO TK, SOMAMOTO T (2011), 'The complement system in teleost fish: progress of post-homolog-hunting researches', *Dev Comp Immunol*, in press.

NAM BH, MOON JY, KIM YO, KONG HJ, KIM WJ, LEE SJ, KIM KK (2010), 'Multiple beta-defensin isoforms identified in early developmental stages of the teleost *Paralichthys olivaceus*', *Fish Shellfish Immunol*, **28**, 267–74.

NEAVE B, RODAWAY A, WILSON SW, PATIENT R, HOLDER N (1995), 'Expression of zebrafish GATA 3 (gta3) during gastrulation and neurulation suggests a role in the specification of cell fate', *Mech Dev*, **51**, 169–82.

NELSON JS (1994), *Fishes of the world*, 3rd edition. New York, Wiley & Sons.

NEMETH E, BAIRD AW, O'FARRELLY C (2009), 'Microanatomy of the liver immune system', *Semin Immunopathol*, **31**, 333–43.

NOGA EJ, ULLAL AJ, CORRALES J, FERNANDES JM (2011), 'Application of antimicrobial polypeptide host defenses to aquaculture: exploitation of downregulation and upregulation responses', *Comp Biochem Physiol D Genom Proteom*, **6**, 44–54.

NONAKA M, IWAKI M, NAKAI C, NOZAKI M, KAIDOH T, NONAKA M, NATSUUME-SAKAI S, TAKAHASHI M (1984), 'Purification of a major serum protein of rainbow trout (*Salmo gairdneri*) homologous to the third component of mammalian complement', *J Biol Chem*, **259**, 6327–33.

NONAKA M, NONAKA M, IRIE M, TANABE K, KAIDOH T, NATSUUME-SAKAI S, TAKAHASHI M (1985), 'Identification and characterization of a variant of the third component of complement (C3) in rainbow trout (*Salmo gairdneri*) serum', *J Biol Chem*, **260**, 809–15.

NURIEVA RI, CHUNG Y, HWANG D, YANG XO, KANG HS, MA L, WANG YH, WATOWICH SS, JETTEN AM, TIAN Q, DONG C (2008), 'Generation of T follicular helper cells is mediated by interleukin-21 but independent of T helper 1, 2, or 17 cell lineages', *Immunity*, **29**, 138–49.

OATES AC, WOLLBERG P, PRATT SJ, PAW BH, JOHNSON SL, HO RK, POSTLETHWAIT JH, ZON LI, WILKS AF (1999), 'Zebrafish stat3 is expressed in restricted tissues during embryogenesis and stat1 rescues cytokine signaling in a STAT1-deficient human cell line', *Dev Dyn*, **215**, 352–70.

OHTANI M, HAYASHI N, HASHIMOTO K, NAKANISHI T, DIJKSTRA JM (2008), 'Comprehensive clarification of two paralogous interleukin 4/13 loci in teleost fish', *Immunogenetics*, **60**, 383–97.

OSHIUMI H, MATSUO A, MATSUMOTO M, SEYA T (2008), 'Pan-vertebrate Toll-Like receptors during evolution', *Current Genomics*, **9**, 488–93.

PALM NW, MEDZHITOV R (2009), 'Pattern recognition receptors and control of adaptive immunity', *Immunol Rev*, **227**, 221–33.

PAN CY, CHEN JY, NI IH, WU JL, KUO CM (2008), 'Organization and promoter analysis of the grouper (*Epinephelus coioides*) epinecidin-1 gene', *Comp Biochem Physiol B Biochem Mol Biol*, **150**, 358–67.

PANCER Z, AMEMIYA CT, EHRHARDT GR, CEITLIN J, GARTLAND GL, COOPER MD (2004), 'Somatic diversification of variable lymphocyte receptors in the agnathan sea lamprey', *Nature*, **430**, 174–80.

PAPANASTASIOU AD, ZARKADIS IK (2005), 'Gene duplication of the seventh component of complement in rainbow trout', *Immunogenetics*, **57**, 703–8.

PAPANASTASIOU AD, ZARKADIS IK (2006a), 'Cloning and phylogenetic analysis of the alpha subunit of the eighth complement component (C8) in rainbow trout', *Mol Immunol*, **43**, 2188–94.

PAPANASTASIOU AD, ZARKADIS IK (2006b), 'The gamma subunit of the eighth complement component (C8) in rainbow trout', *Dev Comp Immunol*, **30**, 485–91.

PAPANASTASIOU AD, GEORGAKA E, ZARKADIS IK (2007), 'Cloning of a CD59-like gene in rainbow trout. Expression and phylogenetic analysis of two isoforms', *Mol Immunol*, **44**, 1300–6.

PARK IY, PARK CB, KIM MS, KIM SC, PARASIN I (1998), 'An antimicrobial peptide derived from histone H2A in the catfish, *Parasilurus asotus*', *FEBS Lett*, **437**, 258–62.

PARK NG, SILPHADUANG U, MOON HS, SEO JK, CORRALES J, NOGA EJ (2011), 'Structure–activity relationships of piscidin 4, a piscine antimicrobial peptide', *Biochemistry*, **50**, 3288–99.

PEAUDECERF L, ROCHA B (2011), 'Role of the gut as a primary lymphoid organ', *Immunol Lett*, **140**, 1–6.

PEÑARANDA MM, LAPATRA SE, KURATH G (2011), 'Specificity of DNA vaccines against the U and M genogroups of infectious hematopoietic necrosis virus (IHNV) in rainbow trout (*Oncorhynchus mykiss*)', *Fish Shellfish Immunol*, **31**, 43–51.

PICCHIETTI S, GUERRA L, BERTONI F, RANDELLI E, BELARDINELLI MC, BUONOCORE F *et al.* (2010), 'Cell-mediated intestinal immunity in *Dicentrarchus labrax* (L.): gene expression and functional studies', *First EOFFI Symposium, Viterbo (Italy)*, p. 59.

PINTO RD, NASCIMENTO DS, REIS MI, DO VALE A, DOS SANTOS NM (2007), 'Molecular characterization, 3D modelling and expression analysis of sea bass (*Dicentrarchus labrax* L.) interleukin-10', *Mol Immunol*, **44**, 2056–65.

PLOUFFE DA, HANINGTON PC, WALSH JG, WILSON EC, BELOSEVIC M (2005), 'Comparison of select innate immune mechanisms of fish and mammals', *Xenotransplantation*, **12**, 266–77.

POISA-BEIRO L, DIOS S, AHMED H, VASTA GR, MARTÍNEZ-LÓPEZ A, ESTEPA A, ALONSO-GUTIÉRREZ J, FIGUERAS A, NOVOA B (2009), 'Nodavirus infection of sea bass (*Dicentrarchus labrax*) induces up-regulation of galectin-1 expression with potential anti-inflammatory activity', *J Immunol*, **183**, 6600–11.

PRAVEEN K, EVANS DL, JASO-FRIEDMANN L (2004), 'Evidence for the existence of granzyme-like serine proteases in teleost cytotoxic cells', *J Mol Evol*, **58**, 449e59.

PRESS CMCL, EVENSEN Ø (1999), 'The morphology of the immune system in teleost fish', *Fish Shellfish Immunol*, **9**, 309–18.

PURCELL MK, LAING KJ, WOODSON JC, THORGAARD GH, HANSEN JD (2009), 'Characterization of the interferon genes in homozygous rainbow trout reveals two novel genes, alternate splicing and differential regulation of duplicated genes', *Fish Shellfish Immunol*, **26**, 293–304.

QUINTANA FJ, IGLESIAS AH, FAREZ MF, CACCAMO M, BURNS EJ, KASSAM N, OUKKA M, WEINER HL (2010), 'Adaptive autoimmunity and Foxp3-based immunoregulation in zebrafish', *PLoS One*, **5**, e9478.

RAIDA MK, NYLÉN J, HOLTEN-ANDERSEN L, BUCHMANN K (2011), 'Association between plasma antibody response and protection in rainbow trout *Oncorhynchus mykiss* immersion vaccinated against *Yersinia ruckeri*', *PLoS One*, **6**, e18832.

RAJ VS, FOURNIER G, RAKUS K, RONSMANS M, OUYANG P, MICHEL B, DELFORGES C, COSTES B, FARNIR F, LEROY B, WATTIEZ R, MELARD C, MAST J, LIEFFRIG F, VANDERPLASSCHEN A (2011), 'Skin mucus of *Cyprinus carpio* inhibits cyprinid herpesvirus 3 binding to epidermal cells', *Vet Res*, **42**, 92.

RAJAN B, FERNANDES JM, CAIPANG CM, KIRON V, ROMBOUT JH, BRINCHMANN MF (2011), 'Proteome reference map of the skin mucus of Atlantic cod (*Gadus morhua*) revealing immune competent molecules', *Fish Shellfish Immunol*, **31**, 224–31.

RANDALL TD (2010), 'Bronchus-associated lymphoid tissue (BALT) structure and function', *Adv Immunol*, **107**, 187–241.

RATHINAM VA, JIANG Z, WAGGONER SN, SHARMA S, COLE LE, WAGGONER L, VANAJA SK, MONKS BG, GANESAN S, LATZ E, HORNUNG V, VOGEL SN, SZOMOLANYI-TSUDA E, FITZGERALD KA (2010), 'The AIM2 inflammasome is essential for host defense against cytosolic bacteria and DNA viruses', *Nat Immunol*, **11**, 395–402.

REBL A, SIEGL E, KOLLNER B, FISCHER U, SEYFERT HM (2007), 'Characterization of twin toll-like receptors from rainbow trout (*Oncorhynchus mykiss*): evolutionary relationship, induced expression by *Aeromonas salmonicida salmonicida*', *Dev Comp Immunol*, **31**, 499–510.

REBL A, GOLDAMMER T, SEYFERT HM (2010), 'Toll-like receptor signaling in bony fish', *Vet Immunol Immunopathol*, **134**, 139–50.

REITE OB (1997), 'Mast cells/eosinophilic granule cells of salmonids: staining properties and responses to noxious agents', *Fish Shellfish Immunol*, **7**, 567–84.

REITE OB, EVENSEN O (2006), 'Inflammatory cells of teleostean fish: a review focusing on mast cells/eosinophilic granule cells and rodlet cells', *Fish Shellfish Immunol*, **20**, 192–208.

RIBEIRO CMS, PONTES MJSL, BIRD S, CHADZINSKA M, SCHEER M, VERBURG VAN KEMENADE L, SAVELKOUL HFJ, WIEGERTJES GF (2010), 'Trypanosomiasis-induced Th17-like immune responses in carp', *Plos ONE*, **5**, e13012.

RICKLIN D, HAJISHENGALLIS G, YANG K, LAMBRIS JD (2010), 'Complement: a key system for immune surveillance and homeostasis', *Nat Immunol*, **11**, 785–97.

RIMSTAD E, MJAALAND S (2002), 'Infectious salmon anaemia virus. An orthomyxovirus causing an emerging infection in Atlantic salmon', *APMIS*, **110**, 273–82.

RODEWALD HR (2008), 'Thymus organogenesis', *Annu Rev Immunol*, **26**, 355–88.

ROMANO N, TAVERNE-THIELE AJ, FANELLI M, BALDASSINI MR, ABELLI L, MASTROLIA L *et al.* (1999), 'Ontogeny of the thymus in a teleost fish, *Cyprinus carpio* L.: developing thymocytes in the epithelial microenvironment', *Dev Comp Immunol*, **23**, 123–37.

ROMBOUGH P (2007), 'The functional ontogeny of the teleost gill: which comes first, gas or ion exchange?', *Comp Biochem Physiol A Mol Integr Physiol*, **148**, 732–42.

ROMBOUT JH, TAVERNE-THIELE AJ, VILLENA MI (1993a), 'The gut associated lymphoid tissue (GALT) of carp (*Cyprinus carpio* L.): an immunocytochemical analysis', *Dev Comp Immunol*, **17**, 55–66.

ROMBOUT JHWM, TAVERNE N, VAN DE KAMP M, TAVERNE-THIELE AJ (1993b), 'Differences in mucus and serum immunoglobulin of carp (*Cyprinus carpio* L.)', *Dev Comp Immunol*, **17**, 309–17.

ROMBOUT JH, VAN DER TUIN SJ, YANG G, SCHOPMAN N, MROCZEK A, HERMSEN T, TAVERNE-THIELE JJ (2008), 'Expression of the polymeric Immunoglobulin Receptor (pIgR) in mucosal tissues of common carp (*Cyprinus carpio* L.)', *Fish Shellfish Immunol*, **24**, 620–8.

ROMBOUT JH, ABELLI L, PICCHIETTI S, SCAPIGLIATI G, KIRON V (2010), 'Teleost intestinal immunology', *Fish Shellfish Immunol*, **31**, 616–26.

SABAT R, GRÜTZ G, WARSZAWSKA K, KIRSCH S, WITTE E, WOLK K *et al.* (2010), 'Biology of interleukin-10', *Cytokine Growth Factor Rev*, **21**, 331–44.

SALLUSTO F, LANZAVECCHIA A (2009), 'Heterogeneity of CD4+ memory T cells: functional modules for tailored immunity', *Eur J Immunol*, **39**, 2076–82.

SANJABI S, ZENEWICZ LA, KAMANAKA M, FLAVELL RA (2009), 'Anti-inflammatory and pro-inflammatory roles of TGF-beta, IL-10, and IL-22 in immunity and autoimmunity', *Curr Opin Pharmacol*, **9**, 447–53.

SANMARTÍN ML, PARAMÁ A, CASTRO R, CABALEIRO S, LEIRO J, LAMAS J, BARJA JL (2008), 'Vaccination of turbot, *Psetta maxima* (L.), against the protozoan parasite *Philasterides dicentrarchi*: effects on antibody production and protection', *J Fish Dis*, **31**, 135–40.

SANTOS P, ARUMEMI F, PARK KS, BORGHESI L, MILCAREK C (2011), 'Transcriptional and epigenetic regulation of B cell development', *Immunol Res*, **50**, 105–12.

SATO S, ST-PIERRE C, BHAUMIK P, NIEMINEN J (2009), 'Galectins in innate immunity: dual functions of host soluble beta-galactoside-binding lectins as damage-associated molecular patterns (DAMPs) and as receptors for pathogen-associated molecular patterns (PAMPs)', *Immunol Rev*, **230**, 172–87.

SATOH T, KATO H, KUMAGAI Y, YONEYAMA M, SATO S, MATSUSHITA K, TSUJIMURA T, FUJITA T, AKIRA S, TAKEUCHI O (2010), 'LGP2 is a positive regulator of RIG-I- and MDA5-mediated antiviral responses', *Proc Natl Acad Sci USA*, **107**, 1512–17.

SAUNDERS HL, OKO AL, SCOTT AN, FAN CW, MAGOR BG (2010), 'The cellular context of AID expressing cells in fish lymphoid tissues,' *Dev Comp Immunol*, **34**, 669–76.

SAVAN R, IGAWA D, SAKAI M (2003), 'Cloning, characterization and expression analysis of interleukin-10 from the common carp, *Cyprinus carpio* L.', *Eur J Biochem*, **270**, 4647–54.

SAVAN R, RAVICHANDRAN S, COLLINS JR, SAKAI M, YOUNG HA (2009), 'Structural conservation of interferon gamma among vertebrates', *Cytokine Growth Factor Rev*, **20**, 115–24.

SCHRØDER MB, VILLENA AJ, JØRGENSEN TO (1998), 'Ontogeny of lymphoid organs and immunoglobulin producing cells in Atlantic cod (*Gadus morhua* L.)', *Dev Comp Immunol*, **22**, 507–17.

SCHROEDER HW JR, CAVACINI L (2010), 'Structure and function of immunoglobulins', *J Allergy Clin Immunol*, **125**, S41–52.

SCOCCHI M, PALLAVICINI A, SALGARO R, BOCIEK K, GENNARO R (2009), 'The salmonid cathelicidins: a gene family with highly varied C-terminal antimicrobial domains', *Comp Biochem Physiol B Biochem Mol Biol*, **152**, 376–81.

SECOMBES CJ (1996), 'The nonspecific immune system: cellular defenses', In: Iwama G, Nakanishi T, editors. *The fish immune system: organism, pathogen and environment*. Academic Press, UK.

SECOMBES CJ, MANNING MJ, ELLIS AE (1982), 'Localization of immune complexes and heat aggregated immunoglobulin in the carp *Cyprinus carpio* L.', *Immunology*, **47**, 101–5.

SECOMBES CJ, WANG T, BIRD S (2011), 'The interleukins of fish', *Dev Comp Immunol*, DOI: 10.1016/j.dci.2011.05.001

SEPPOLA M, LARSEN AN, STEIRO K, ROBERTSEN B, JENSEN I (2008), 'Characterisation and expression analysis of the interleukin genes, IL-1beta, IL-8 and IL-10, in Atlantic cod (*Gadus morhua* L.)', *Mol Immunol*, **45**, 887–97.

SEPULCRE MP, ALCARAZ-PEREZ F, LOPEZ-MUNOZ A, ROCA FJ, MESEGUER J, CAYUELA ML, MULERO V (2009), 'Evolution of lipopolysaccharide (LPS) recognition, signaling: fish TLR4 does not recognize LPS, negatively regulates NF-kappaB activation', *J Immunol*, **182**, 1836–45.

SETERNES T, SØRENSEN K, SMEDSRØD B (2002), 'Scavenger endothelial cells of vertebrates: a nonperipheral leukocyte system for high-capacity elimination of waste macromolecules', *Proc Natl Acad Sci USA*, **99**, 7594–7.

SHA Z, ABERNATHY JW, WANG S, LI P, KUCUKTAS H, LIU H, PEATMAN E, LIU Z (2009), 'NOD-like subfamily of the nucleotide-binding domain and leucine-rich repeat containing family receptors and their expression in channel catfish', *Dev Comp Immunol*, **33**, 991–9.

SHEN L, STUGE TB, ZHOU H, KHAYAT M, BARKER KS, QUINIOU SM, WILSON M, BENGTÉN E, CHINCHAR VG, CLEM LW, MILLER NW (2002), 'Channel catfish cytotoxic cells: a mini-review', *Dev Comp Immunol*, **26**, 141–9.

SHEN L, STUGE TB, BENGTÉN E, WILSON M, CHINCHAR VG, NAFTEL JP, BERNANKE JM, CLEM LW, MILLER NW (2004), 'Identification and characterization of clonal NK-like cells from channel catfish (*Ictalurus punctatus*)', *Dev Comp Immunol*, **28**, 139–52.

SHEWRING DM, ZOU J, CORRIPIO-MIYAR Y, SECOMBES CJ (2011), 'Analysis of the cathelicidin 1 gene locus in Atlantic cod (*Gadus morhua*)', *Mol Immunol*, **48**, 782–7.

SIDERITS D, BIELEK E (2009), 'Rodlet cells in the thymus of the zebrafish *Danio rerio* (Hamilton, 1822)', *Fish Shellfish Immunol*, **27**, 539–48.

SILPHADUANG U, NOGA E (2001), 'Peptide antibiotics in mast cells of fish', *Nature*, **414**, 268–9.

SIMORA RM, OHTANI M, HIKIMA J, KONDO H, HIRONO I, JUNG TS, AOKI T (2010), 'Molecular cloning and antiviral activity of IFN-β promoter stimulator-1 (IPS-1) gene in Japanese flounder, *Paralichthys olivaceus*', *Fish Shellfish Immunol*, **29**, 979–86.

SMITH VJ, DESBOIS AP, DYRYNDA EA (2010), 'Conventional and unconventional antimicrobials from fish, marine invertebrates and micro-algae', *Mar Drugs*, **8**, 1213–1262.

SOLSTAD T, STENVIK J, JORGENSEN TO (2007), 'mRNA expression patterns of the BPI/LBP molecule in the Atlantic cod (*Gadus morhua* L.)', *Fish Shellfish Immunol*, **23**, 260–271.

SOMAMOTO T, NAKANISHI T, OKAMOTO N (2000), 'Specific cell-mediated cytotoxicity against a virus-infected syngeneic cell line in isogeneic ginbuna crucian carp', *Dev Comp Immunol*, **24**, 633–40.

SOMAMOTO T, YOSHIURA Y, SATO A, NAKAO M, NAKANISHI T, OKAMOTO N et al. (2006), 'Expression profiles of TCRbeta and CD8alpha mRNA correlate with virus-specific cell-mediated cytotoxic activity in ginbuna crucian carp', *Virology*, **348**, 370–7.

SOMAMOTO T, OKAMOTO N, NAKANISHI T, OTOTAKE M, NAKAO M (2009), '*In vitro* generation of viral-antigen dependent cytotoxic T-cells from ginbuna crucian carp, *Carassius auratus langsdorfii*', *Virology*, **389**, 26–33.

STAFFORD JL, ELLESTAD KK, MAGOR KE, BELOSEVIC M, MAGOR BG (2003), 'A toll-like receptor (TLR) gene that is up-regulated in activated goldfish macrophages', *Dev Comp Immunol*, **27**, 685–98.

STEIN C, CACCAMO M, LAIRD G, LEPTIN M (2007), 'Conservation and divergence of gene families encoding components of innate immune response systems in zebrafish', *Genome Biol*, **8**, R251.

STOLTE EH, SAVELKOUL HF, WIEGERTJES G, FLIK G, LIDY VERBURG-VAN KEMENADE BM (2008), 'Differential expression of two interferon-gamma genes in common carp (*Cyprinus carpio* L.)', *Dev Comp Immunol*, **32**, 1467–81.

STROBAND HWJ, VAN DER MEER H, TIMMERMANS LPM (1979), 'Regional functional differentiation in the gut of the grass carp, *Ctenopharyngodon idella* (Val)', *Histochemistry*, **64**, 235–49.

STUGE TB, WILSON MR, ZHOU H, BARKER KS, BENGTÉN E, CHINCHAR G, MILLER NW, CLEM LW (2000), 'Development and analysis of various clonal alloantigen-dependent cytotoxic cell lines from channel catfish', *J Immunol*, **164**, 2971–7.

SU J, YANG C, XIONG F, WANG Y, ZHU Z (2009), 'Toll-like receptor 4 signaling pathway can be triggered by grass carp reovirus, *Aeromonas hydrophila* infection in rare minnow *Gobiocypris rarus*', *Fish Shellfish Immunol*, **27**, 33–9.

SU J, HUANG T, DONG J, HENG J, ZHANG R, PENG L (2010), 'Molecular cloning and immune responsive expression of MDA5 gene, a pivotal member of the RLR gene family from grass carp *Ctenopharyngodon idella*', *Fish Shellfish Immunol*, **28**, 712–18.

SUETAKE H, ARAKI K, AKATSU K, SOMAMOTO T, DIJKSTRA JM, YOSHIURA Y, KIKUCHI K, SUZUKI Y (2007), 'Genomic organization and expression of CD8alpha and CD8beta genes in fugu *Takifugu rubripes*', *Fish Shellfish Immunol*, **23**, 1107–18.

SULLIVAN C, POSTLETHWAIT JH, LAGE CR, MILLARD PJ, KIM CH (2007), 'Evidence for evolving Toll-IL-1 receptor-containing adaptor molecule function in vertebrates', *J Immunol*, **178**, 4517–27.

SULLIVAN C, CHARETTE J, CATCHEN J, LAGE CR, GIASSON G, POSTLETHWAIT JH, MILLARD PJ, KIM CH (2009), 'The gene history of zebrafish tlr4a and tlr4b is predictive of their divergent functions', *J Immunol*, **183**, 5896–908.

SUN G, LI H, WANG Y, ZHANG B, ZHANG S (2010), 'Zebrafish complement factor H and its related genes: identification, evolution, and expression', *Funct Integr Genomics*, **10**, 755–87.

SUN XF, SHANG N, HU W, WANG YP, GUO QL (2007), 'Molecular cloning and characterization of carp (*Cyprinus carpio* L.) CD8beta and CD4-like genes', *Fish Shellfish Immunol*, **23**, 1242–55.

SUNG SC, CHENG CH, CHOU CM, CHU CY, CHEN GD, HWANG PP, HUANG FL, HUANG CJ (2010), 'Expression and characterization of a constitutively active STAT6 from *Tetraodon*', *Fish Shellfish Immunol*, **28**, 819–28.

SUNYER JO, ZARKADIS IK, SAHU A, LAMBRIS JD (1996), 'Multiple forms of complement C3 in trout that differ in binding to complement activators', *Proc Natl Acad Sci USA*, **93**, 8546–51.

SUNYER JO, ZARKADIS I, SARRIAS MR, HANSEN JD, LAMBRIS JD (1998), 'Cloning, structure, and function of two rainbow trout Bf molecules', *J Immunol*, **161**, 4106–14.

SUNYER JO, BOSHRA H, LI J (2005), 'Evolution of anaphylatoxins, their diversity and novel roles in innate immunity: insights from the study of fish complement', *Vet Immunol Immunopathol*, **108**, 77–89.

SURESH N (2009), 'Effect of cadmium chloride on liver, spleen and kidney melanomacrophage centres in *Tilapia mossambica*', *J Environ Biol*, **30**, 505–8.

SUZUKI K, KAWAMOTO S, MARUYA M, FAGARASAN S (2010), 'GALT: organization and dynamics leading to IgA synthesis', *Adv Immunol*, **107**, 153–85.

SUZUKI T, SHIN IT, KOHARA Y, KASAHARA M (2004), 'Transcriptome analysis of hag-fish leukocytes: a framework for understanding the immune system of jawless fishes', *Dev Comp Immunol*, **28**, 993–1003.

SVAJGER U, ANDERLUH M, JERAS M, OBERMAJER N (2010), 'C-type lectin DC-SIGN: an adhesion, signalling and antigen-uptake molecule that guides dendritic cells in immunity', *Cell Signal*, **22**, 1397–405.

SWAIN P, NAYAK SK, NANDA PK, DASH S (2008), 'Biological effects of bacterial lipopolysaccharide (endotoxin) in fish: a review', *Fish Shellfish Immunol*, **25**, 191–201.

SZABO SJ, SULLIVAN BM, PENG SL, GLIMCHER LH (2003), 'Molecular mechanisms regulating Th1 immune responses', *Annu Rev Immunol*, **21**, 713–58.

TADISO TM, LIE KK, HORDVIK I (2011), 'Molecular cloning of IgT from Atlantic salmon, and analysis of the relative expression of tau, mu, and delta in different tissues', *Vet Immunol Immunopathol*, **139**, 17–26.

TAKAHASI K, KUMETA H, TSUDUKI N, NARITA R, SHIGEMOTO T, HIRAI R, YONEYAMA M, HORIUCHI M, OGURA K, FUJITA T, INAGAKI F (2009), 'Solution structures of cytosolic RNA sensor MDA5 and LGP2 C-terminal domains: identification of the RNA recognition loop in RIG-I-like receptors', *J Biol Chem*, **284**, 17465–74.

TAKIZAWA F, ARAKI K, KOBAYASHI I, MORITOMO T, OTOTAKE M, NAKANISHI T (2008a), 'Molecular cloning and expression analysis of T-bet in ginbuna crucian carp (*Carassius auratus langsdorfii*)', *Mol Immunol*, **45**, 127–36.

TAKIZAWA F, MIZUNAGA Y, ARAKI K, MORITOMO T, OTOTAKE M, NAKANISHI T (2008b), 'GATA3 mRNA in ginbuna crucian carp (*Carassius auratus langsdorfii*): cDNA cloning, splice variants and expression analysis', *Dev Comp Immunol*, **32**, 898–907.

TAKIZAWA F, DIJKSTRA JM, KOTTERBA P, KORYTÁŘ T, KOCK H, KÖLLNER B, JAUREGUIBERRY B, NAKANISHI T, FISCHER U (2011), 'The expression of CD8α discriminates distinct T cell subsets in teleost fish', *Dev Comp Immunol*, **35**, 752–63.

TANEL A, FONSECA SG, YASSINE-DIAB B, BORDI R, ZEIDAN J, SHI Y et al. (2009), 'Cellular and molecular mechanisms of memory T-cell survival', *Expert Rev Vaccines*, **8**, 299–312.

TATNER MF, MANNING MJ (1982), 'The morphology of the trout, *Salmo gairdneri* Richardson, thymus: some practical and theoretical considerations', *J Fish Biol*, **21**, 27–32.

TEMPERLEY ND, BERLIN S, PATON IR, GRIFFIN DK, BURT DW (2008), 'Evolution of the chicken Toll-like receptor gene family: a story of gene gain and gene loss', *BMC Genomics*, **9**, 62.

THORNTON AM, KORTY PE, TRAN DQ, WOHLFERT EA, MURRAY PE, BELKAID Y, SHEVACH EM (2010), 'Expression of Helios, an Ikaros transcription factor family member, differentiates thymic-derived from peripherally induced Foxp3+ T regulatory cells', *J Immunol*, **184**, 3433–41.

TIAN JY, XIE HX, ZHANG YA, XU Z, YAO WJ, NIE P (2009), 'Ontogeny of IgM-producing cells in the mandarin fish Siniperca chuatsi identified by *in situ* hybridisation', *Vet Immunol Immunopathol*, **132**, 146–52.

TODA H, SHIBASAKI Y, KOIKE T, OHTANI M, TAKIZAWA F, OTOTAKE M, MORITOMO T, NAKANISHI T (2009), 'Alloantigen-specific killing is mediated by CD8-positive T cells in fish', *Dev Comp Immunol*, **33**(4), 646–52.

TODA H, SAITO Y, KOIKE T, TAKIZAWA F, ARAKI K, YABU T, SOMAMOTO T, SUETAKE H, SUZUKI Y, OTOTAKE M, MORITOMO T, NAKANISHI T (2011), 'Conservation of characteristics and functions of CD4 positive lymphocytes in a teleost fish', *Dev Comp Immunol*, **35**, 650–60.

TSUTSUI S, KOMATSU Y, SUGIURA T, ARAKI K, NAKAMURA O (2011), 'A unique epidermal mucus lectin identified from catfish (*Silurus asotus*): first evidence of intelectin in fish skin slime', *J Biochem*, **150**, 501–14.

UEMURA T, YANO T, SHIRAISHI H, NAKAO M (1996), 'Purification and characterization of the eighth and ninth components of carp complement', *Mol Immunol*, **33**, 925–32.

UNTERHOLZNER L, KEATING SE, BARAN M, HORAN KA, JENSEN SB, SHARMA S, SIROIS CM, JIN T, LATZ E, XIAO TS, FITZGERALD KA, PALUDAN SR, BOWIE AG (2010), 'IFI16 is an innate immune sensor for intracellular DNA', *Nat Immunol*, **11**, 997–1004.

UTKE K, BERGMANN S, LORENZEN N, KOLLNER B, OTOTAKE M, FISCHER U (2007), 'Cell-mediated cytotoxicity in rainbow trout, *Oncorhynchus mykiss*, infected with viral haemorrhagic septicaemia virus', *Fish Shellfish Immunol*, **22**, 182–96.

UTKE K, KOCK H, SCHUETZE H, BERGMANN SM, LORENZEN N, EINER-JENSEN K et al. (2008), 'Cell-mediated immune responses in rainbow trout after DNA immunization against the viral hemorrhagic septicemia virus', *Dev Comp Immunol*, **32**, 239–52.

UZZELL T, STOLZENBERG ED, SHINNAR AE, ZASLOFF M (2003), 'Hagfish intestinal anti-microbial peptides are ancient cathelicidins', *Peptides*, **24**, 1655–67.

VAN WIJK F, CHEROUTRE H (2009), 'Intestinal T cells: facing the mucosal immune dilemma with synergy and diversity', *Semin Immunol*, **21**, 130–8.

VARIN A, GORDON S (2009), 'Alternative activation of macrophages: immune function and cellular biology', *Immunobiology*, **214**, 630–41.

VIGLIANO FA, BERMÚDEZ R, QUIROGA MI, NIETO JM (2006), 'Evidence for melano-macrophage centres of teleost as evolutionary precursors of germinal centres of higher vertebrates: an immunohistochemical study', *Fish Shellfish Immunol*, **21**, 467–71.

VITVED L, HOLMSKOV U, KOCH C, TEISNER B, HANSEN S, SALOMONSEN J, SKJØDT K (2010), 'The homologue of mannose-binding lectin in the carp family cyprinidae is expressed at high level in spleen, and the deduced primary structure predicts affinity for galactose', *Immunogenetics*, **51**, 955–64.

WALPORT MJ (2001), 'Complement. First of two parts', *N Engl J Med*, **344**, 1058–66.

WALVIG F (1958), 'Blood and parenchymal cells in the spleen of the icefish *Chaeno-cephalus aeeratus* (Lönnberg)', *Nytt Magasin Zool* (Norway), **6**, 111–20.

WANG T, SECOMBES CJ (2003), 'Complete sequencing and expression of three comple-ment components, C1r, C4 and C1 inhibitor, of the classical activation pathway of the complement system in rainbow trout *Oncorhynchus mykiss*', *Immunogene-tics*, **55**, 615–28.

WANG T, HOLLAND J, BOLS N, SECOMBES CJ (2007), 'Molecular and functional characteri-sation of interleukin-15 in rainbow trout *Oncorhynchus mykiss*: a potent inducer of interferon-gamma expression in spleen leucocytes', *J Immunol*, **179**, 1475–88.

WANG T, HOLLAND JW, MARTIN SA, SECOMBES CJ (2010a), 'Sequence and expression analysis of two T helper master transcription factors, T-bet and GATA3, in rainbow trout *Oncorhynchus mykiss* and analysis of their expression during bac-terial and parasitic infection', *Fish Shellfish Immunol*, **29**, 705–15.

WANG T, MONTE MM, HUANG W, BOUDINOT P, MARTIN SA, SECOMBES CJ (2010b), 'Identi-fication of two FoxP3 genes in rainbow trout (*Oncorhynchus mykiss*) with dif-ferential induction patterns', *Mol Immunol*, **47**, 2563–74.

WANG T, MARTIN SA, SECOMBES CJ (2010c), 'Two interleukin-17C-like genes exist in rainbow trout *Oncorhynchus mykiss* that are differentially expressed and modu-lated', *Dev Comp Immunol*, **34**, 491–500.

WANG T, HUANG W, COSTA MM, MARTIN SA, SECOMBES CJ (2011a), 'Two copies of the genes encoding the subunits of putative interleukin (IL)-4/IL-13 receptors, IL-4Ralpha, IL-13Ralpha1 and IL-13Ralpha2, have been identified in rainbow trout (*Oncorhynchus mykiss*) and have complex patterns of expression and modula-tion', *Immunogenetics*, **63**, 235–53.

WANG T, DIAZ-ROSALES P, COSTA MM, CAMPBELL S, SNOW M, COLLET B, MARTIN SA, SEC-OMBES CJ (2011b), 'Functional characterization of a nonmammalian IL-21: rainbow trout *Oncorhynchus mykiss* IL-21 upregulates the expression of the Th cell sig-nature cytokines IFN-gamma, IL-10, and IL-22', *J Immunol*, **186**, 708–21.

WANG T, HUANG W, COSTA MM, SECOMBES CJ (2011c), 'The gamma-chain cytokine/receptor system in fish: more ligands and receptors', *Fish Shellfish Immunol*, **31**, 673–87.

WANG Z, ZHANG S, TONG Z, LI L, WANG G (2009), 'Maternal transfer and protective role of the alternative complement components in zebrafish *Danio rerio*', *PloS One*, **4**, e4498.

WEI L, LAURENCE A, ELIAS KM, O'SHEA JJ (2007), 'IL-21 is produced by Th17 cells and drives IL-17 production in a STAT3-dependent manner', *J Biol Chem*, **282**, 34605–10.

WELLER CL, COLLINGTON SJ, WILLIAMS T, LAMB JR (2011), 'Mast cells in health and disease', *Clin Sci (Lond)*, **120**, 473–84.

WEN Y, FANG W, XIANG LX, PAN RL, SHAO JZ (2011), 'Identification of Treg-like cells in Tetraodon: insight into the origin of regulatory T subsets during early vertebrate evolution', *Cell Mol Life Sci*, **68**, 2615–26.

WHYTE SK (2007), 'The innate immune response of finfish – a review of current knowledge', *Fish Shellfish Immunol*, **23**, 1127–51.

WILKINS C, GALE M JR (2010), 'Recognition of viruses by cytoplasmic sensors', *Curr Opin Immunol*, **22**, 41–7.

WILSON JM, LAURENT P (2002), 'Fish gill morphology: inside out', *J Exp Zool*, **293**, 192–213.

YAMAGUCHI T, KATAKURA F, SHITANDA S, NIIDA Y, TODA H, OHTANI M, YABU T, SUETAKE H, MORITOMO T, NAKANISHI T (2011), 'Clonal growth of carp (*Cyprinus carpio*) T cells *in vitro*', *Dev Comp Immunol*, **35**, 193–202.

YAMAMOTO M, TAKEDA K (2010), 'Current views of toll-like receptor signaling pathways', *Gastroenterol Res Pract*, **2010**, 240365.

YANO T, NAKAO M (1994), 'Isolation of a carp complement protein homologous to mammalian factor D', *Mol Immunol*, **31**, 337–42.

YAP WH, YEOH E, TAY A, BRENNER S, VENKATESH B (2005), 'STAT4 is a target of the hematopoietic zinc-finger transcription factor Ikaros in T cells', *FEBS Lett*, **579**, 4470–8.

YASUIKE M, DE BOER J, VON SCHALBURG KR, COOPER GA, MCKINNEL L, MESSMER A *et al.* (2010), 'Evolution of duplicated IgH loci in Atlantic salmon, *Salmo salar*', *BMC Genomics*, **11**, 486.

YAZAWA R, COOPER GA, HUNT P, BEETZ-SARGENT M, ROBB A, CONRAD M *et al.* (2008a), 'Striking antigen recognition diversity in the Atlantic salmon T-cell receptor alpha/delta locus', *Dev Comp Immunol*, **32**, 204–12.

YAZAWA R, COOPER GA, BEETZ-SARGENT M, ROBB A, MCKINNEL L, DAVIDSON WS *et al.* (2008b), 'Functional adaptive diversity of the Atlantic salmon T-cell receptor gamma locus', *Mol Immunol*, **45**, 2150–7.

YIN ZX, WEI H, CHEN WJ, YAN JH, YANG JN, CHAN SM *et al.* (2006), 'Cloning, expression and antimicrobial activity of an antimicrobial peptide, epinecidin-1, from the orange-spotted grouper, *Epinephelus coioides*', *Aquaculture*, **31**, 204–11.

YODER JA, LITMAN GW (2011), 'The phylogenetic origins of natural killer receptors and recognition: relationships, possibilities, and realities', *Immunogenetics*, **63**, 123–41.

YONEYAMA M, FUJITA T (2008), 'Structural mechanism of RNA recognition by the RIG-I-like receptors', *Immunity*, **29**, 178–81.

YOSHIDA H, NAKAYA M, MIYAZAKI Y (2009), 'Interleukin 27: a double-edged sword for offense and defense', *J Leukoc Biol*, **86**, 1295–303.

ZAPATA A (1979), 'Ultrastructural study of the teleost fish kidney', *Dev Comp Immunol*, **3**, 55–65.

ZAPATA A, DIEZ B, CEJALVO T, GUTIÉRREZ-DE FRÍAS C, CORTÉS A (2006), 'Ontogeny of the immune system of fish', *Fish Shellfish Immunol*, **20**, 126–36.

ZAPATA AG, CHIBA A, VARAS A (1996), 'Cells and tissues of the immune system of fish', In: G. Iwama and T. Nakanishi, Editors, *The fish immune system: organism, pathogen, and environment*, San Diego, Academic Press.

ZARKADIS IK, SARRIAS MR, SFYROERA G, SUNYER JO, LAMBRIS JD (2001), 'Cloning and structure of three rainbow trout C3 molecules: a plausible explanation for their functional diversity', *Dev Comp Immunol*, **25**, 11–24.

ZARKADIS IK, DURAJ S, CHONDROU M (2005), 'Molecular cloning of the seventh component of complement in rainbow trout', *Dev Comp Immunol*, **29**, 95–102.

ZHANG DC, SHAO YQ, HUANG YQ, JIANG SG (2005), 'Cloning, characterization and expression analysis of interleukin-10 from the zebrafish (*Danio rerio*)', *J Biochem Mol Biol*, **38**, 571–6.

ZHANG Y, GUI J (2004), 'Molecular characterization and IFN signal pathway analysis of *Carassius auratus* CaSTAT1 identified from the cultured cells in response to virus infection', *Dev Comp Immunol*, **28**, 211–27.

ZHANG YA, ZOU J, CHANG CI, SECOMBES CJ (2004), 'Discovery and characterization of two types of liver-expressed antimicrobial peptides 2 (LEAP-2) genes in rainbow trout', *Vet Immunol Immunopathol*, **101**, 259–69.

ZHANG YA, SALINAS I, LI J, PARRA D, BJORK S, XU Z, LAPATRA SE, BARTHOLOMEW J, SUNYER JO (2010), 'IgT, a primitive immunoglobulin class specialized in mucosal immunity', *Nat Immunol*, **11**, 827–35.

ZHANG YA, SALINAS I, ORIOL SUNYER J (2011), 'Recent findings on the structure and function of teleost IgT', *Fish Shellfish Immunol*, **31**, 627–34.

ZHAO JG, ZHOU L, JIN JY, ZHAO Z, LAN J, ZHANG YB, ZHANG QY, GUI JF (2009), 'Antimicrobial activity-specific to Gram-negative bacteria and immune modulation-mediated NF-kappaB and Sp1 of a medaka beta-defensin', *Dev Comp Immunol*, **33**, 624–37.

ZHU J, PAUL WE (2010a), 'Heterogeneity and plasticity of T helper cells', *Cell Res*, **20**, 4–12.

ZHU J, PAUL WE (2010b), 'Peripheral CD4+ T-cell differentiation regulated by networks of cytokines and transcription factors', *Immunol Rev*, **238**, 247–62.

ZOU J, WANG T, HIRONO I, AOKI T, INAGAWA H, HONDA T et al. (2002), 'Differential expression of two tumor necrosis factor genes in rainbow trout, *Oncorhynchus mykiss*', *Dev Comp Immunol*, **26**, 161–72.

ZOU J, CLARK MS, SECOMBES CJ (2003), 'Characterisation, expression and promoter analysis of an interleukin 10 homologue in the puffer fish, *Fugu rubripes*', *Immunogenetics*, **55**, 325–35.

ZOU J, CARRINGTON A, COLLET B, DIJKSTRA JM, YOSHIURA Y, BOLS N, SECOMBES C (2005), 'Identification and bioactivities of IFN-gamma in rainbow trout *Oncorhynchus mykiss*: the first Th1-type cytokine characterized functionally in fish', *J Immunol*, **175**, 2484–94.

ZOU J, MERCIER C, KOUSSOUNADIS A, SECOMBES C (2007), 'Discovery of multiple beta-defensin like homologues in teleost fish', *Mol Immunol*, **44**, 638–47.

ZOU J, CHANG M, NIE P, SECOMBES CJ (2009), 'Origin and evolution of the RIG-I like RNA helicase gene family', *BMC Evol Biol*, **9**, 85.

ZOU J, BIRD S, SECOMBES CJ (2010), 'Antiviral sensing in teleost fish', *Current Pharmaceutical Design*, **16**, 4185–93.

2

Crustacean immune responses and their implications for disease control

L. Cerenius and K. Söderhäll, Uppsala University, Sweden

Abstract: This chapter reviews recent advances in our knowledge of crustacean immunity. Emphasis is given to shrimp due to their importance in aquaculture and trade and to freshwater crayfish since they serve as model organisms for research in crustacean immunology. Crustaceans lack antibodies, interferon and some other components from the mammalian immune arsenal but can still mount an efficient defence against many potential pathogens. Crustacean innate immunity relies on a combination of efficient hemocyte and humoral reactions carried out by plasma proteins.

Key words: shrimp, crayfish, hematopoiesis, prophenoloxidase activation, clotting protein, transglutaminase, pattern recognition protein, peroxinectin.

2.1 Introduction

Crustaceans, in particular tropical shrimp, are important commodities subjected to intensive farming. In fact, for crustaceans farming is today as important as catch from wild stocks in terms of revenue. Losses due to diseases are enormous but difficult to estimate: old estimates for the 1990s point to an annual loss due to diseases in the shrimp industry of about 1 billion US dollars. Viruses such as white spot syndrome virus (WSSV), yellowhead virus and taura virus are the most serious causes of disease in aquculture. Also bacterial diseases caused by e.g. *Vibrio* species are serious threats to the well-being of cultured crustaceans. Although relatively simple measures such as good hygiene and the use of brood-stock free of certain pathogens can reduce losses substantially it is probably safe to state that a deeper knowledge of the crustacean immune system is a prerequisite for further improvements of the health status in aquaculture. This chapter reviews present knowledge of the immune system of decapod crustaceans. Emphasis will be on tropical shrimps due to their commercial importance and on freshwater crayfish, the latter being the major model organism for unravelling crustacean defence reactions. Recent progress will be given prominence; for earlier work references to some key review papers will instead be provided.

2.2   Cellular defence

Invertebrate immunity is traditionally divided into cellular and humoral (plasma) reactions. For convenience this division will be used here, although it must be stressed that the more we learn about these immune reactions the more evidence for their interdependence comes to light. A particular pathogen may be dealt with in a number of ways, and the relative importance of a specific immune reaction may vary for a number of reasons: pathogen strain or presence of additional pathogens, environment in general, host developmental stage (young or adult), different aspects of host physiology such as circadian rhythms, etc.

2.2.1   Hemocytes

Hemocytes are instrumental in crustacean immunity. Their cell membrane contains receptors with capacity to recognise and bind pathogen-derived molecules. Alternatively plasma proteins recognising, for example, microbial cell-wall components may bind these and the complex formed may engage hemocyte receptors (see Lee and Söderhäll, 2001, for a detailed example of such a reaction). There are three main types of crustacean circulating hemocytes, namely hyaline cells (HC), semigranular cells (SGC) and granular cells (GC). There is a division of labour between these cell types: phagocytosis is performed by HC and SGC, encapsulation is the result of a joint effort by SGC and GC whereas release of components needed for melanisation mainly takes place after degranulation of GC. This release of granular contents can be initiated by, for example, microbial cell wall components. The main granular proteins released by exocytosis from GC have been identified by proteomic methods (Sricharoen et al., 2005). Among the immune components released from GC are several serine proteinase homologues, prophenoloxidase, properoxinectin and several antimicrobial peptides.

Thus, circulating hemocytes serve as transport vehicles for an array of factors with the potential to kill an intruder. When needed, these components are released into plasma to carry out different immunological reactions. Circulating hemocytes may phagocytise pathogens small enough to be placed inside the cell or several hemocytes may cooperate to encapsulate objects too large to be accommodated by a single hemocyte. In similar reaction, aggregates of bacteria may become surrounded by hemocyes in a reaction termed nodulation. In addition, there are fixed phagocytising cells in several tissues such as the arterioles of the hepatopancreas or (in penaeid shrimp) the lymphoid organ (Rusaini and Owens, 2010). Recent data on other invertebrates clearly indicate that immediate reactions, mainly phagocytosis, coagulation and melanisation, are crucial for quickly removing bacteria that gain access to the body, for example through a wound. The great majority of such intruders may be removed within minutes. In most cases very few bacteria survive this initial onslaught and, in most cases, the survivors are taken care of by antimicrobial proteins (AMPs, see below). However, some crustacean pathogens such as virulent *Aeromonas hydrophila* strains can overcome this defence by destroying the hemocytes by the use of secreted toxins (Jiravanichpaisal *et al.*, 2009). Very little is known about phagocytosis in crustaceans, although some opsonins, i.e. proteins enhancing phagocytosis such as lectins and peroxinectin, have been found (see below). These opsonins are relatively unspecific but recently it was shown in two insect models that the Down syndrome cell adhesion molecule (Dscam) can act as an opsonin. Further, there are thousands of splicing variants produced, a fact opening up the possibility that different Dscam variants may be capable of distinguishing between different bacteria. Recently, it was demonstrated in *P. leniusculus* that more than 22,000 unique Dscam isoforms may be produced. Experiments with two bacterial species, *Escherichia coli* and *Staphylococcus aureus*, showed that the bacteria stimulated the synthesis of different Dscam isoforms. Confirmatory tests with different recombinant Dscam demonstrated that the isoforms formed in response to one bacterium were more efficient in promoting uptake of that particular bacterial species (Watthanasururot *et al.*, 2011a). These experiments suggest that crayfish may mount an immune response that is adapted to the particular bacterial species that is present.

2.2.2 Hemocyte adhesion and peroxinectin

Cell adhesion is essential in multicellular animals for many physiological processes including immune reactions. Hemocyte adherence is crucial in many localised responses to a pathogen and a prerequisite for encapsulation and nodulation of pathogens. The first hemocyte adhesion molecule to be purified and characterised in detail was peroxinectin (Johansson and Söderhäll, 1988; Johansson *et al.*, 1995). It is stored in secretory granules in a proform devoid of cell-adhesion activity. Once outside the hemocyte it gains cell adhesion activity by limited proteolysis concomitant with activation of the prophenoloxidase cascade (Johansson and Söderhäll, 1988; Lin *et al.*, 2007), another important innate immune reaction. Active peroxinectin binds to integrin receptors on the cell surface to stimulate further degranulation in a positive feedback loop. Hereby, further storage components from the hemocyte granules are released (Sricharoen *et al.*, 2005). Peroxinectin is a strong cell adhesion ligand and promotor of cell spreading and in this capacity is important for capsule formation. The protein is multifunctional with several additional functions, such acting as an opsonin and possessing peroxidase activity (its name, peroxinectin, refers to this dual capacity of mediator of cell adhesion and peroxidase activity). Peroxinectin interacts with a cell-surface CuZn superoxide dismutase and possibly produces hypohalic acids in the vicinity of the bound microorganism. This would mean that peroxinectin has dual roles in immunity: to promote contact between pathogen and hemocyte and to contribute to the production of antimicrobial products. Although shrimp peroxinectins have not been as thoroughly characterised as their crayfish counterparts there is plenty of indirect evidence for their occurrence (and importance) in shrimps and other invertebrates (C. Liu *et al.*, 2007; Zamocky *et al.*, 2008).

2.3   Hematopoiesis

Circulating hemocytes rarely, if ever, divide and need to be replenished continuously by production of new such cells, i.e. by hematopoiesis. Therefore, this process is crucial in mounting a successful defence against many pathogens. Recently, substantial insights into how invertebrate blood cells are formed have been obtained in the freshwater crayfish. This has been made possible through the successful establishment of primary cultures of crayfish hematopoietic tissue (Hpt) that are capable of undergoing hemocyte differentiation and the isolation of some key regulatory proteins. Two lineages leading to SGC and GC, respectively, have been identified (Fig. 2.1). Less data are available for differentiation of the third crustacean hemocyte cell type, the hyaline cell. Specific marker proteins for Hpt cells (a proliferating cell nuclear antigen, PCNA), SGC (a specific Kazal-type proteinase inhibitor) and GC (a superoxide dismutase) have been discovered and these have made it possible to monitor in detail the development of different precursor cells into mature hemocytes (Wu *et al.*, 2008). A group of cytokines, named astakines, have recently been shown to stimulate and regulate hemocyte proliferation and maturation (Söderhäll *et al.*, 2005). Astakines are prokineticin homologues that were first identified in freshwater crayfish and later shown to be present in other crustaceans as well as in several other invertebrate groups (Lin *et al.*, 2010). Among these crayfish, astakine 1 specifically induces production of SG whereas astakine 2 is involved in formation of GC (Fig. 2.1). Crayfish astakines lack the amino terminal motif ('AVIT') responsible for receptor binding in vertebrate prokineticins. Instead crayfish astakine 1 mediates its effect via binding to the β-subunit of a cell membrane located ATP synthase (Lin *et al.*, 2009). Furthermore, native crayfish astakine 1 is an oligomer whereas astakine 2 is a monomeric protein (Lin *et al.*, 2010). The receptor for astakine 2 is not characterised but it is separate from the astakine 1 receptor. Thus, at least in crayfish different astakines exhibit different primary and quaternary structure, receptor identity and biological activities. Although astakines are

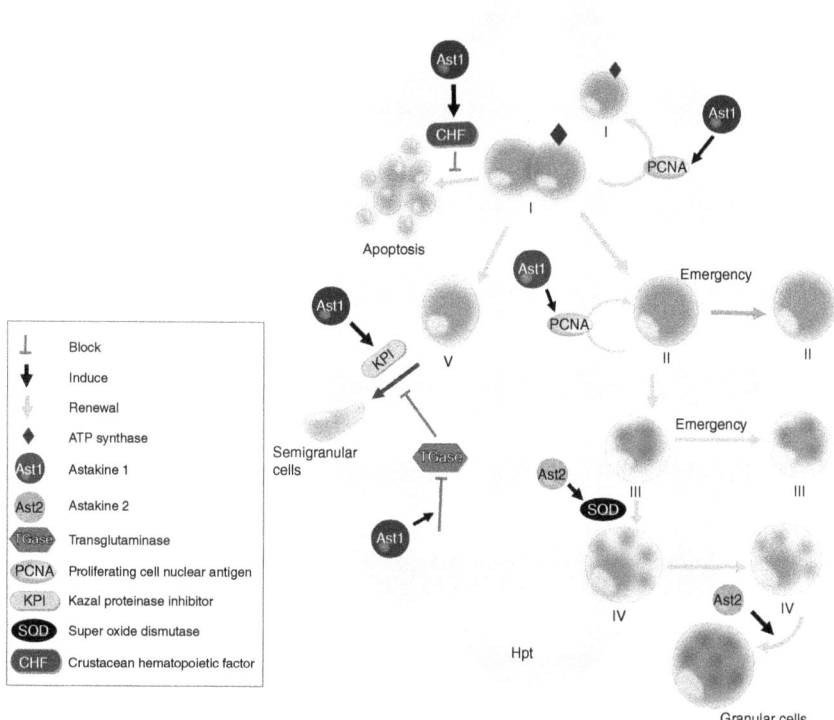

Fig. 2.1 Crustacean hematopoiesis. Shown here is a representation of the hematopoietic tissue from crayfish. There are two main lineages leading to semigranular and granular cells, respectively. Astakine 1 (Ast1) promotes proliferation and differentiation of hematopoietic cells into semigranular cells. Ast1 prevents apoptosis via crustacean hematopoietic factor (CHF) and Ast1 reduces transglutaminase (TGase) and cell surface ATP synthase (black diamond) activities. Astakine 2 (Ast2) stimulates formation of granular cells. Included in the figure are some useful markers for some of the cell types: Proliferating cell nuclear antigen (PCNA) is present in hematopoietic cells but not in mature hemocytes, KPI (a Kazal proteinase inhibitor) is a marker for semigranular cells and SOD (superoxide dismutase) for granular cells.

present in shrimp, chelicerates and several insect orders (Lin *et al.*, 2010), their mode of action has so far only been studied in detail in crayfish. Recently, however, an endogenous astakine 2 was demonstrated to stimulate *in vivo* hemocyte proliferation in *Penaeus monodon* (Hsaio and Song 2010). Thus, further data on astakine activities from especially shrimps are therefore eagerly awaited.

Downstream astakine 1, another protein named crustacean hematopoietic factor (CHF) is critical for the survival of the hemocytes and its removal will cause apoptosis (Lin *et al.*, 2011). CHF is a small cysteine-rich protein with sequence similarities to mammalian morphogenetic proteins. Its transcription in the Hpt is stimulated by astakine 1 but not by astakine 2. It seems likely that CHF is instrumental in allowing proliferation of SGC by decreasing apoptosis in the SGC lineage among the Hpt cells. Astakine stimulation of hemocyte proliferation is accompanied by a decrease in transglutaminase (TGase) activity (Lin *et al.*, 2008). TGase seems to be crucial for keeping the Hpt cells in an undifferentiated state within the tissue. Hemocyte maturation and release from the Hpt follow a decrease in TGase activity and it has been suggested that this enzyme helps to keep the cells within the Hpt tissue to prevent a premature release into the hemolymph. In accordance with this, TGase RNAi treatment of Hpt *in vitro* leads to cell spreading in a way very similar to the effects on these cells by astakine 1 addition. TGase gene silencing in shrimp increases the mortality suggesting that this enzyme is important for survival (Maningas *et al.*, 2008). The direct importance of hemocyte proliferation and astakine synthesis on immune capacity is demonstrated by the strong correlation between astakine expression, hemocyte numbers and survival to an infection by a pathogenic *Pseudomonas* strain (Watthanasurorot *et al.*, 2011b). Interestingly, it was found that astakine expression and thus immune capacity had a circadian rhythm. Potentially this may have large practical implications with respect to, for example, which time of day to choose when handling animals and thus inflicting stress upon them. Another small peptide, hemocyte homeostasis-associated protein (HHAP), has been shown to be important in maintaining the integrity for hemocytes in shrimp and if silenced by RNAi the animals die (Prapavororat *et al.*, 2010). However, silencing of the HHAP gene in crayfish does not have this drastic effect (Prapavororat *et al.*, 2010).

2.4   Defence against viruses

Despite virally caused diseases being of such great concern in the industry, relatively little is known about mechanisms behind viral infections in crustaceans, although many proteins have been implicated (in most cases the mechanism has not been deciphered) to participate in some way in antiviral reactions. For recent reviews on this subject see Liu *et al.* (2009) and Flegel and Sritunyalucksana (2011). A clearer picture on this area is difficult to obtain, since most likely different viruses employ separate pathways to enter the host cell and establish themselves intracellularly, thus complicating efforts to unravel mechanisms. A further complication is that viruses may be present without causing overt disease but infected animals may be capable of passing the virus to more susceptible animals (Flegel, 2007). The white spot syndrome virus (WSSV), can survive in crayfish for several months at low temperature (below 14 °C) but if the temperature goes up the crayfish die of the virus (Jiravanichpaisal *et al.*, 2004). This means that crayfish can function both as a vector and a repository for this virus.

By analogy to better-known viruses infecting mammalian cells one may assume that the ability to interfere with the apoptotic machinery is important for the outcome of a viral infection also in crustacean cells. In contrast to mammals interferon is apparently not present in crustaceans. In the shrimp *Penaeus monodon* for example a caspase has been shown to be a target of a WSSV anti-apoptosis protein (Leu *et al.*, 2008). There are experimental indications of the existence of both a defence based on RNAi against viruses in general and specific defence against different virus types. Several putative components of the RNAi machinery such as an Argonaute homologue have been found in crustaceans (Dechklar *et al.*, 2008). The injection of dsRNA provides a general but incomplete protection against viruses, as first shown by Robalino *et al.* (2004). Any double-stranded RNA (dsRNA) provided it is at least 50 bp long seems to trigger this protection upon its injection into shrimps (Labreuche *et al.*, 2010).

The administration of specific dsRNAs coding for viral genes provides a stronger and specific antiviral response against a particular virus (Robalino *et al.*, 2005). Also the administration of viral envelope proteins or DNA expressing viral proteins have been shown to provide protection presumably by blocking viral uptake into host cells (for an overview see Johnson *et al.*, 2008). For example, oral introduction of WSSV envelope proteins such as VP19, VP28 or VP292 into shrimp (Witteveldt *et al.*, 2004) or crayfish (Xu *et al.*, 2006) will provide a partial but not lasting protection against the virus. Experiments where plasmid DNA encoding some WSSV viral proteins was used (Rout *et al.*, 2007) indicate that protection is lengthened, but more data are certainly needed. The mechanism(s) for the protection obtained in the different treatments mentioned is not known, but it is possible that the virus-derived proteins will compete with the uptake of virus into the host cells. These data suggest that it is possible in principle to interfere with viral propagation by injecting appropriate viral components into the animals. The effect of this treatment may not last long and there are of course practical difficulties with the large-scale administration of any 'vaccine' into shrimp. Perhaps methods to introduce more long-lasting antiviral compounds into eggs need to be developed in the future.

Besides being explored as an antiviral tool, the crustacean RNAi machinery is now widely used for manipulating gene expression in crustaceans (see an overview by Hirono *et al.*, 2011). Gene specific silencing is in most cases readily achieved by injecting dsRNA into to hemolymph and this technique has in a few years become an indispensable tool in crustacean research (Watthanasurorot *et al.* 2010).

2.5   Pattern recognition in crustaceans

Components from (potential) pathogens are known to trigger immune reactions in various organisms from plants to humans. To serve as pattern-recognition molecules they must be present in pathogens but not produced by host tissues. Preferably pattern recognition ought to be present on outer surfaces of the pathogen, secreted by it or in other ways easily becoming accessible for the innate immune system. Among such components are dsRNA, unmethylated CpG DNA, β-1,3-glucans, lipopolysaccharides (LPS) and peptidoglycans (PG). So-called pattern-recognition proteins are, through their capacity for binding these molecules, crucial for initiation of subsequent reactions aimed at the pathogen. Potentially compounds capable of engaging pattern-recognition protein could be used to stimulate crustacean immunity and increase the resistance towards pathogens in these animals. Several crustacean proteins with the capacity to bind microbial cell wall components and subsequently boost immunity have been characterised in detail. Details regarding the detection of dsRNA and non-host forms of DNA in these animals are still awaited. In mammals several receptors for aberrant nucleic acids are present inside immune active cells so it is possible that corresponding crustacean receptors are placed within hemocytes and fixed phagocytic cells in gills, lymphoid organs and elsewhere.

2.5.1   Recognition of β-1,3-glucans

β-1,3-Glucans (BGs) are well-known inducers of immune reactions in many animals. They are present in the cell walls of fungi and oomycetes. In 1977 it was demonstrated using defined oligosaccharides that BGs specifically and in extremely low quantities were capable of activating the proPO-cascade (Unestam and Söderhäll, 1977). The main vehicle for mediating BG-induced immune reactions in crustaceans is the lipopolysaccharide- and glucan-binding protein (LGBP) (Lee et al., 2000). After binding the polysaccharides the complex mediates activation of the proPO cascade. As indicated below LGBP is involved in binding to LPS and peptidoglycans as well and thus this protein is crucial in mediating pattern-recognition and trigger defence reactions against a wide variety of different microorganisms. In insects, LGBP homologues will trigger the Toll pathway for production of antimicrobial peptides. A similar pathway remains to be firmly established in crustaceans, although Toll-like receptors are present in crustaceans (Wang et al., 2010). Another protein, the beta-glucan binding protein (BGBP) is present in high quantities in plasma. It is a major plasma lipid-carrier but will in addition specifically bind BGs, albeit with less affinity than LGBP. The BG–BGBP complex binds specifically to the hemocyte surface to induce degranulation and further defence reactions. It is possible that BGBP is responsible for the removal of excess amounts of BG. This could reduce the effect of using large amounts of BGs in therapies aimed at boosting the general immune capacity of the animals.

2.5.2   Recognition of lipopolysaccharides (LPS)

LPS is present in cell wall of Gram-negative bacteria. Two crustacean PRPs with LPS-binding activities have been studied in detail. These are the LGBP mentioned above and a serine proteinase homologue, the masquerade (mas)-like protein which both upon engagement with LPS mediate defence reactions. The mas-like protein will upon binding the bacteria be proteolytically processed into four subunits, where one of these with a mass of 33 kDa is acting as cell-adhesion factor (Lee and Söderhäll, 2001). The mas-like protein hereby acts as an opsonin and will enhance phagocytosis.

A recent detailed study using *Aeromonas hydrophila* mutants with defects on either different steps in the biosynthesis of LPS or defective with respect to other putative virulence factors flagella, secretion systems, porins, siderophores, etc. has demonstrated the role of LPS in triggering melanisation. In particular, the O:34 antigen and the outer core of the LPS are crucial for both virulence and ability to induce melanisation in crayfish (Noonin *et al.*, 2010).

2.5.3   Recognition of peptidoglycans (PG)

PG are present in most bacterial cell walls. The known ability of PG to specifically trigger defence in crustaceans has been enigmatic; so far no PG recognition proteins (PGRPs) have been detected in any crustacean EST collection or the only complete crustacean genome so far available (i.e. *Dapnia pulex*). PGRP genes are present in many other animals from insects to mammals. A crustacean PG pattern-recognition protein was recently identified in freshwater crayfish by its PGN-binding capability and the recombinant protein produced to study its characteristics (Liu *et al.*, 2011). The protein, a 46 kDa serine proteinase homologue (named Pl-SPH2) is proteolytically processed to a 30 kDa fragment. The 30 kDa fragment binds PG together with another proteinase homologue (Pl-SPH1) and LGBP. If any of these are removed by RNAi, PG-triggered proPO-activation is abolished. This suggests that a complex consisting of SPH1, SPH2 and LGBP bound to the PG is responsible for mediating pattern recognition of this important bacterial cell constituent in crayfish. This means that LGBP has a key role in pattern recognition, being involved in recognition of microbial components present in fungi, Gram-positive as well as Gram-negative bacteria.

2.5.4   Lectins

Lectins are proteins that non-covalently bind specific sugars without catalytically reacting with them. Lectin-binding to carbohydrates on cells, viral particles, etc. may cause their agglutination and therefore aid the immune system. Potentially some lectins could participate in pattern-recognition by virtue of their specific recognition of carbohydrate patterns. A large number of crustacean lectins has been molecularly characterised (for details see Cerenius et al., 2010b). So far, few of them have been assigned a specific role in the immune system. In some cases agglutination of different bacteria (Zhang et al., 2009) or binding to envelope proteins of the WSSV (Zhao et al., 2009) by different lectins have been demonstrated indicative of an immunological function by these proteins. One LPS-binding lectin, PmLec was shown by Luo et al. (2006) to act both as an opsonin to enhance bacterial phagocytosis and to agglutinate bacteria. More functional studies on the numerous crustacean lectins isolated and cloned are needed before their immunological role can be mapped and exploited.

2.6   The prophenoloxidase (proPO) cascade

The prophenoloxidase (proPO) cascade is an important part of the crustacean immune system and has proven to be instrumental in fighting several pathogens including bacteria and parasites (Söderhäll 1982; Söderhäll and Cerenius, 1998; Cerenius and Söderhäll, 2004). For recent updates the reader is referred to Cerenius et al., 2008, 2010c). In short this system is stored in an inactive zymogenic form in the hemocytes (mainly the GCs) and released by exocytosis upon stimulation of the cells. Once released a proteolytic cascade of activation reactions are triggered, resulting finally in the activation of proPO into PO (Fig. 2.2). The latter enzyme carries out the melanisation reaction. All crustaceans investigated so far possess one or two proPO genes in contrast to some dipteran insects that may have a dozen different proPOs. Very little is known about why there are several different proPOs in some species. Cellular defence reactions are in most cases accompanied by melanisation and encapsulated parasites or bacteria-containing nodules are often surrounded by melanin. Melanisation reactions can also occur in the cuticle, presumably by PO being transported there from cells below the cuticle as has been demonstrated in insects. Activation of the cascade is regulated tightly by proteinase inhibitors. In crustaceans, the most effective inhibitor of the proPO-activating proteinase (ppA) is pacifastin, a high molecular weight plasma protein with a light chain with proteinase inhibitor activity and a heavy transferrin chain (Liang et al., 1997). The pacifastin family encompasses a large number of mainly low molecular weight proteinase inhibitors in, for example, insects and some of these inhibitors are involved in regulating the insect melanisation cascade. Since the pacifastin contains a tranferrin subunit capable of binding two iron atoms it is possible that an additional function of pacifastin may be sequestering iron from invading bacteria as suggested by the results from Noonin et al. (2010). Several studies have shown that knocking-down proPO

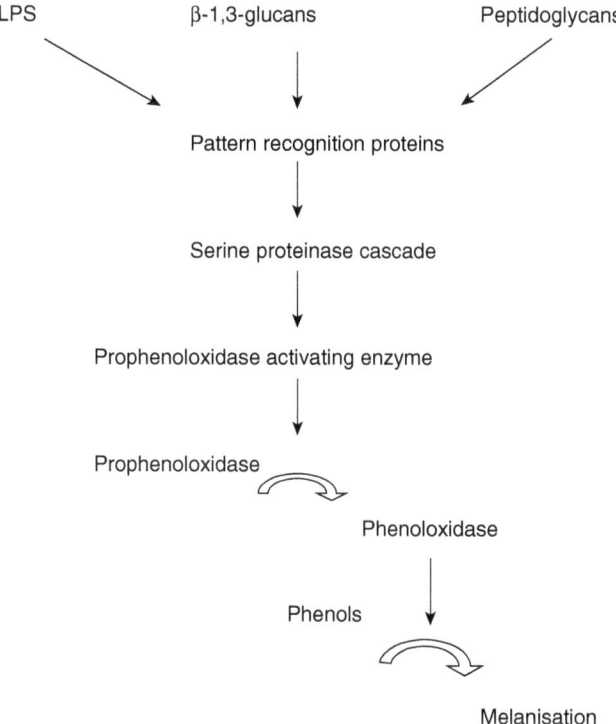

Fig. 2.2 Simplified schema of the prophenoloxidase activating (melanisation) cascade. Pattern recognition protein will, after binding to compounds (e.g. lipopolysaccharide, β-1,3-glucans or peptidoglycans) typical for microorganisms, trigger activation of upstream components of serine proteinase cascade. The terminal proteinase will cleave prophenoloxidase to the catalytically active enzyme phenoloxidase. Different phenolic substances will give rise to melanin as well as toxic short-lived intermediates.

expression will be detrimental for the defence against bacteria (H. Liu *et al.*, 2007; Amparyup *et al.*, 2009; Charoensapsri *et al.*, 2009; Fagutao *et al.*, 2009). There is strong circumstantial evidence as well for the importance of the proPO-system to keep crustacean parasites such as the crayfish plague *Aphanomyces astaci* at bay (Cerenius *et al.*, 2003). Also the antibacterial and antifungal effects of insect and crustacean PO reactions products *in vitro* provide additional support for the immunological role of the proPO-system (Zhao *et al.*, 2007; Cerenius *et al.*, 2010a). There are, however, some pathogenic bacteria that survive with ease the onslaught by the melanisation and other immune reactions aimed at them (Noonin *et al.*, 2010).

The quinones produced by the active PO are potentially very harmful for the host. To regulate undesired production of such compounds, crustaceans (and other invertebrates) employ melanisation inhibitor proteins

(MIPs) that are capable of restricting melanin synthesis by PO. The MIP primary structure differs among taxa but a common essential Asp-rich sequence motif has been identified as crucial for their activities by mutational analyses (Söderhäll *et al.*, 2009). In crayfish and shrimp MIP transcript levels are relatively high in many tissues but the MIP concentration in the plasma is normally very low. When needed, e.g. during a bacterial infection, large amounts of MIP are secreted into the hemolymph, presumably to counteract any dangerous side-effects of the active antibacterial PO (Söderhäll *et al.*, 2009; Angthong *et al.*, 2010). MIP is consumed during the process, and plasma MIP levels will eventually return to a very low level. The ability of different bacteria to trigger melanisation varies and some strains will neither induce melanisation nor cause MIP release into plasma, indicating that ways to block this arm of the innate immune system may develop among some pathogens.

2.7   Coagulation

Coagulation is a crucial event in preventing hemolymph loss upon wounding. Because of its open circulatory system a crustacean risks quickly losing a major part of its hemolymph content unless the wound is quickly sealed by the clotting reaction. The clot of plasma proteins will in addition trap microorganisms that manage to enter through the wound and hinder their further dissemination throughout the body. The crustacean clotting protein (CP) was first cloned from *Pacifastacus leniusculus* in 1999 and the corresponding proteins were subsequently characterised in several shrimp species (Hall *et al.*, 1999; Yeh *et al.*, 2007 and references therein). The crustacean CP circulates in the plasma as homodimer of 210 kDa subunits linked together by disulphide bonds. CP is related to but distinct from vitellogenin, a protein involved in egg formation. Thus, the crustacean clotting system is very different from its counterpart in some other arthropods. In horseshoe crab (a chelicerate) the clotting system consists of a proteolytic cascade released from the hemocytes and TGase is not as in crustaceans necessary for producing a clot (Cerenius *et al.*, 2010c). Crustacean CP polymerises during coagulation and this reaction is catalysed by a TGase released from the hemocytes. The clot consists of long chains of multimeric CP units held together by covalent crosslinks between lysine and glutamine residues on separate CP molecules. In the shrimp species *Marsupenaeus japonicus* experimental RNAi silencing of the CP and TGase genes resulted into defective clotting system but increased susceptibility to bacterial and viral infections as well (Maningas *et al.*, 2008). Whether this increase is due to direct effects (e.g. the clot entrapment of the pathogens is reduced) or indirect effects of the clotting system on other arms of the immune system is at present unclear. So far, little is known about how TGase is released from circulating hemocytes (see, however, above regarding TGase during hematopoiesis) and whether this release is coupled to induction of other immune reactions. CP gene expression seems to be partly regulated by a feed back mechanism since experimental bleedings (and thus blood losses) triggers an increased synthesis of CP in tiger shrimp (Yeh *et al.*, 2007).

2.8 Antimicrobial proteins

In holometabolous insects antimicrobial peptides (AMPs) are very important to take care of bacteria becoming exposed to the body fluids during metamorphosis. In crustaceans a relatively large number of AMPs have been cloned (see Rosa and Barracco, 2010, for a thorough review of crustacean AMPs), although much less is known to what extent they are instrumental in fending off infections. In contrast to insects where the fat body is the major source of hemolymph AMP are crustacean AMPs mainly produced by SGs and GCs. Two major groups of crustacean AMPs that have been discerned are crustins and penaeidins. Crustins have been found to be widely distributed among different crustacean groups since it was first found (and named carcinin) in the shore crab *Carcinus maenas* (see review by Smith *et al.*, 2008) and they are characterised by the presence of a whey acidic protein domain. This domain consists of about 50 amino acids with a so-called four-disulphide core structure containing four disulphide bridges. Crustins are cationic small proteins assumed to be antimicrobial effectors against mainly Gram-positive bacteria. Relatively few crustins, however, have been thoroughly investigated. In *P. leniusculus* recombinant Plcustin1 and 2 both expressed bactericidal activities towards Gram-positive bacteria (Donpudsa *et al.*, 2010). They are mainly, in contrast to most insect AMPs, predominantly constitutively expressed and often present in large amounts although their expression can be further stimulated by the presence of bacteria (Jiravanichpaisal *et al.*, 2007a). Other physiological roles for them besides fighting bacteria and other microorganisms have therefore been proposed (Smith *et al.*, 2008).

Penaeidins are proline/arginine-rich peptides hitherto found exclusively in shrimps (Cuthbertson *et al.*, 2008). Recently, a penaeidin-like protein was, however, characterised from the spider crab *Hyas araneus* and shown to possess antimicrobial activities towards a range of different organisms (Sperstad *et al.*, 2009, 2010). This AMP, named hyastatin, in addition to its antimicrobial activities possessed a strong chitin-binding activity, which perhaps indicates other physiological roles for this protein as well. In the Pacific blue shrimp *Litopenaeus stylirostris* a novel type of AMPs coined stylicins have recently been characterised (Rolland *et al.*, 2010). This protein lacks any apparent sequence similarities with hitherto found AMPs and has a strong antifungal activity against, for example, the shrimp pathogenic species *Fusarium oxysporium*.

Another interesting group of AMPs are the antilipopolysaccharide factors (ALFs). They were first isolated from the large granules of the horseshoe crab (*Limulus*). They are basic proteins around 100 amino acids large. In the horseshoe crab they regulate the coagulation cascade (see Section 2.7) but in crustaceans they have other functions. Crustacean ALFs have been shown by *in vitro* experiments to exhibit bactericidal activity (Somboonwiwat *et al.*, 2005) and in cell cultures and in live animals this protein interferes with WSSV replication (Liu *et al.*, 2006; Tharntada *et al.*, 2009). Knock-down experiments on the *ALF* gene in the shrimp *Litopenaeus vannamei* showed that decreased ALF levels resulted in increased susceptibility to both fungal and bacterial pathogens (de la Vega *et al.*, 2008).

In insects, AMP gene expression is inducible and partly regulated by the Toll and imd pathways. This has led to frequent claims that a similar regulation of AMP expression by especially Toll-like receptors occurs in shrimps. Experimental proof for this is still relatively weak although the genes for Toll-like receptors have been identified in several shrimp species (Mekata *et al.*, 2008; Labreuche *et al.*, 2009; Wang *et al.*, 2010). Evidence at protein level for the involvement of these receptors in AMP induction or for any immunological role in a crustacean is still missing. RNA interference of shrimp Toll homologues indicates though that Toll is involved in crustacean immunity. In the shrimp *L. vannamei* reduced expression of one Toll gene increased the susceptibility of the animal towards the bacterium *Vibrio harveyei* (Wang *et al.*, 2010). It is not known if reduced AMP levels were the cause of the response recorded. In another study using the same shrimp species reduced crustin levels achieved by specific RNAi treatment made the animals more vulnerable to a *V. penaceida* infection (Shockey *et al.*, 2009). This is one of very few *in vivo* studies of AMP importance for the immune capacity of a crustacean. As noted by the authors of the study, it raises new questions since *V. penaceida* is a Gram-negative bacterium and *in vitro* studies of crustins strongly suggest they are effective against Gram-positive bacteria. Certainly, more data are needed in order to assess the role of AMPs in crustacean immunity.

2.9   Future trends and conclusions

Understanding of crustacean immunity is, in spite of the impressive progress in deciphering mechanistic details for some of the immune reactions, still in its infancy. It is likely that very soon the sequencing of the genomes of some commercially important crustaceans and the development of more efficient techniques for making transgenic animals will open new avenues. However, it must be stressed that so far no good candidate genes or proteins to improve disease resistance in shrimps and other crustaceans have been found for these molecular techniques to explore. In addition, there is considerable resistance among consumers regarding the use of transgenic animals as well as widespread scepticism to the likelihood of finding practical and economical techniques for genetic modification that can be put in practice. It is possible that more simple measures involving good husbandry, great attention to immune status of eggs and larva (Jiravanichpaisal *et al.*, 2007b; Zhang *et al.*, 2010) and cost-effective and environmentally friendly use of probiotics (reviewed by Ninawe and Salvin, 2009) will be the ways of the future in crustacean aquaculture. If so, we need to continue to deepen our knowledge of immunity in young animals at different developmental stages (the immune system of young animals differs considerably from that of adults) and to further explore the fascinating interactions between the microflora and the immune system of the stomach and gut (Soonthornchai *et al.*, 2010). Hopefully, research in such directions will be encouraged by governmental agencies and the industry in the future and there will be a pay-off in terms of continuing possibilities for expansion of crustacean aquaculture that is so important in providing revenues for many developing economies.

2.10   References

AMPARYUP P, CHAROENSAPSRI W and TASSANAKAJON A (2009), 'Two prophenoloxidases are important for the survival of *Vibrio harveyi* challenged shrimp *Penaeus monodon*', *Dev Comp Immunol*, **33**, 247–56.

ANGTHONG P, WATTHANASUROROT A, KLINBUNGA S, RUANGDEJ U, SÖDERHÄLL I and JIRAVANICHPAISAL P (2010), 'Cloning and characterization of a melanization inhibition protein (PmMIP) of the black tiger shrimp, *Penaeus monodon*', *Fish Shellfish Immunol*, **29**, 464–8.

CERENIUS L and SÖDERHÄLL K (2004), 'The prophenoloxidase activating system in invertebrates', *Immunol Rev*, **198**, 116–26.

CERENIUS L, BANGYEEKHUN E, KEYSER P, SÖDERHÄLL I and SÖDERHÄLL K (2003), 'Host prophenoloxidase expression in freshwater crayfish is linked to increased resistance to the crayfish plague fungus, *Aphanomyces astaci*', *Cell Microbiol*, **5**, 353–7.

CERENIUS L, LEE B L and SÖDERHÄLL K (2008), 'The proPO-system: pros and cons for its role in invertebrate immunity', *Trends Immunol*, **29**, 263–71.

CERENIUS L, BABU R, SÖDERHÄLL K and JIRAVANICHPAISAL P (2010a), '*In vitro* effects on bacterial growth of phenoloxidase reaction products', *J Invertebr Pathol*, **103**, 21–3.

CERENIUS L, JIRAVANICHPAISAL P, LIU H and SÖDERHÄLL I (2010b), 'Crustacean immunity' in Söderhäll K, *Invertbrate Immunity*, Landes Bioscience and Springer Verlag, 239–69.

CERENIUS L, KAWABATA S, LEE B L, NONAKA M and SÖDERHÄLL K (2010c), 'Proteolytic cascades and their involvement in invertebrate immunity', *Trends Biochem Sci*, **35**, 575–83.

CHAROENSAPSRI W, AMPARYUP P, HIRONO I, AOKI T and TASSANAKAJON A (2009), 'Gene silencing of a prophenoloxidase activating enzyme in the shrimp, *Penaeus monodon*, increases susceptibility to *Vibrio harveyi* infection', *Dev Comp Immunol*, **33**, 811–20.

CUTHBERTSON B J, DETERDING L J, WILLIAMS JG, TOMER K B, ETIENNE K, BLACKSHEAR P J, BULLESBACH E E and GROSS P S (2008), 'Diversity in penaeidin antimicrobial peptide form and function', *Dev Comp Immunol*, **32**, 167–81.

DE LA VEGA E, O'LEARY N A, SHOCKEY J E, ROBALINO J, PAYNE C, BROWDY C L, WARR G W
and GROSS P S (2008), 'Anti-lipopolysaccharide factor in *Litopenaeus vannamei*
(LvALF), a broad spectrum antimicrobial peptide essential for shrimp immunity
against bacterial and fungal infection', *Mol Immunol*, **45**, 1916–25.

DECHKLAR M, UDOMKIT A and PANYIM S (2008), 'Characterization of Argonaute cDNA
from *Penaeus monodon* and implication of its role in RNA interference', *Biochem
Biophys Res Commun*, **367**, 768–74.

DONPUDSA S, RIMPHANITCHAYAKIT V, TASSANAKAJON A, SÖDERHÄLL I and SÖDERHÄLL K
(2010), 'Characterization of two crustin antimicrobial peptides from the freshwater crayfish *Pacifastacus leniusculus*', *J Invertebr Pathol*, **104**, 234–8.

FAGUTAO F F, KOYAMA T, KAIZU A, SAITO-TAKI T, KONDO H, AOKI T and HIRONO I (2009),
'Increased bacterial load in shrimp hemolymph in the absence of prophenoloxidase', *FEBS J*, **276**, 5298–306.

FLEGEL T W (2007), 'Update on viral accommodation, a model for host–viral interaction in shrimp and other arthropods', *Dev Comp Immunol*, **31**, 217–31.

FLEGEL T W and SRITUNYALUCKSANA K (2011), 'Shrimp molecular responses to viral
pathogens', *Mar Biotechnol*, **13**, 587–607.

HALL M, WANG R, ANTWERPEN R, SOTTRUP-JENSEN L and SÖDERHÄLL K (1999) 'The
crayfish plasma clotting protein: a vitellogenin-related protein responsible for clot
formation in crustacean blood', *Proc Natl Acad Sci USA*, **96**, 1965–70.

HIRONO I, FAGUTAO F F, KONDO H and AOKI T (2011), 'Uncovering the mechanisms of
shrimp innate immune response by RNA interference', *Mar Biotechnol*, **13**, 622–8.

HSIAO C Y and SONG Y L (2010), 'A long form of shrimp astakine transcript: molecular
cloning, characterization and functional elucidation in promoting hematopoiesis',
Fish Shellfish Immunol, **28**, 77–86.

JIRAVANICHPAISAL P, SÖDERHÄLL K and SÖDERHÄLL I (2004) 'Effect of water temperature on the immune response and infectivity pattern of white spot syndrome virus
(WSSV) in freshwater crayfish', *Fish Shellfish Immunol*, **17**, 265–75.

JIRAVANICHPAISAL P, LEE S Y, KIM Y A, ANDRÉN T and SÖDERHÄLL I (2007a), 'Antibacterial
peptides in hemocytes and hematopoietic tissue from freshwater crayfish
Pacifastacus leniusculus: characterization and expression pattern', *Dev Comp
Immunol*, **31**, 441–55.

JIRAVANICHPAISAL P, PUANGLARP P, PETKON S, DONNUEA S, SÖDERHÄLL I and SÖDERHÄLL
K (2007b), 'Expression of immune-related genes in larval stages of the giant tiger
shrimp, *Penaeus monodon*', *Fish Shellfish Immunol*, **23**, 815–24.

JIRAVANICHPAISAL P, ROOS S, EDSMAN L, LIU H and SÖDERHÄLL K (2009), 'A highly virulent pathogen, *Aeromonas hydrophila*, from the freshwater crayfish *Pacifastacus
leniusculus*', *J Invertebr Pathol*, **101**, 56–6.

JOHANSSON M W and SÖDERHÄLL K (1988), 'Isolation and purification of a cell adhesion factor from crayfish blood cells', *J Cell Biol*, **106**, 1795–803.

JOHANSSON M W, LIND M I, HOLMBLAD T, THÖRNQVIST P O and SÖDERHÄLL K (1995),
'Peroxinectin, a novel cell adhesion protein from crayfish blood', *Biochem Biophys
Res Commun*, **216**, 1079–87.

JOHNSON K N, VAN HULTEN M C W and BARNES A C (2008), '"Vaccination" of shrimp
against viral pathogens: phenomenology and underlying mechanisms', *Vaccine*,
26, 4885–92.

LABREUCHE Y, O'LEARY N A, DE LA VEGA E, VELOSO A, GROSS P S, CHAPMAN R W, BROWDY
C L and WARR G W (2009), 'Lack of evidence for *Litopenaeus vannamei* Toll receptor (lToll) involvement in activation of sequence-independent antiviral immunity
in shrimp', *Dev Comp Immunol*, **33**, 806–10.

LABREUCHE Y, VELOSO A, DE LA VEGA E, GROSS P S, CHAPMAN R W, BROWDY C L and WARR
G W (2010), 'Non-specific activation of antiviral immunity and induction of RNA
interference may engage the same pathway in the Pacific white leg shrimp *Litopenaeus vannamai*', *Dev Comp Immunol*, **34**, 1209–18.

LEE S Y and SÖDERHÄLL K (2001), 'Characterization of a pattern recognition protein, a masquerade-like protein, in the freshwater crayfish *Pacifastacus leniusculus*', *J Immunol*, **166**, 7319–26.

LEE S Y, WANG R and SÖDERHÄLL K (2000), 'A lipopolysaccharide- and beta-1,3-glucan-binding protein from hemocytes of the freshwater crayfish *Pacifastacus leniusculus* – purification, characterization, and cDNA cloning', *J Biol Chem*, **275**, 1337–43.

LEU J H, KUO Y C, KOU G H and LO C F (2008), 'Molecular cloning and characterization of an inhibitor of apoptosis protein (IAP) from the tiger shrimp, *Penaeus monodon*', *Dev Comp Immunol*, **32**, 121–33.

LIANG Z, SOTTRUP-JENSEN L, ASPÁN A, HALL M and SÖDERHÄLL K (1997), 'Pacifastin, a novel 155-kDa heterodimeric proteinase inhibitor containing a unique transferrin chain', *Proc Natl Acad Sci USA*, **94**, 6682–7.

LIN X, CERENIUS L, LEE B L and SÖDERHÄLL K (2007), 'Purification of properoxinectin, a myeloperoxidase homologue and its activation to a cell adhesion molecule', *Biochim Biophys Acta*, **1770**, 87–93.

LIN X, SÖDERHÄLL K and SÖDERHÄLL I (2008), 'Transglutaminase activity in the hematopoietic tissue of a crustacean, *Pacifastacus leniusculus*, importance in hemocyte hemostasis', *BMC Immunol*, 9e58.

LIN X, KIM Y A, LEE B L, SÖDERHÄLL K and SÖDERHÄLL I (2009), 'Identification and properties of a receptor for the invertebrate cytokine astakine, involved in hematopoiesis', *Exp Cell Res*, **315**, 1171–80.

LIN X, NOVOTNY M, SÖDERHÄLL K and SÖDERHÄLL I (2010), 'Ancient cytokines, the role of astakines as hematopoietic growth factors', *J Biol Chem*, **285**, 28577–86.

LIN X, SÖDERHÄLL K and SÖDERHÄLL I (2011), 'Invertebrate hematopoiesis: an Astakine-dependent novel hematopoietic factor', *J Immunol*, **186**, 2073–9.

LIU C, YEH S, HSU P and CHENG W (2007), 'Peroxinectin gene transcription of the giant freshwater prawn, *Macrobrachium rosenbergii* under intrinsic immunostimulant and chemotherapeutic influences', *Fish Shellfish Immunol*, **22**, 408–17.

LIU H, JIRAVANICHPAISAL P, SÖDERHÄLL I, CERENIUS L and SÖDERHÄLL K (2006), 'Antilipopolysaccharide factor interferes with white spot syndrome virus replication *in vitro* and *in vivo* in the crayfish *Pacifastacus leniusculus*', *J Virol*, **80**, 10365–71.

LIU H, JIRAVANICHPAISAL P, CERENIUS L, LEE B L, SÖDERHÄLL I and SÖDERHÄLL K (2007), 'Phenoloxidase is an important component of the defense against *Aeromonas hydrophila* infection in a crustacean, *Pacifastacus leniusculus*', *J Biol Chem*, **282**, 33593–8.

LIU H, SÖDERHÄLL K and JIRAVANICHPAISAL P (2009), 'Antiviral immunity in crustaceans', *Fish Shellfish Immunol*, **27**, 79–88.

LIU H, WU C, MATSUDA Y, KAWABATA S, LEE B L, SÖDERHÄLL K and SÖDERHÄLL I (2011), 'Peptidoglycan activation of the proPO-system without a peptidoglycan protein (PGRP)?', *Dev Comp Immunol*, **35**, 51–61.

LUO T, YANG H J, LI F, ZHANG X B and XU X (2006), 'Purification, characterization and cDNA cloning of a novel lipopolysaccharide-binding lectin from the shrimp *Penaeus monodon*', *Dev Comp Immunol*, **30**, 607–17.

MANINGAS M B B, KONDO H, HIRONO I, SAITO-TAKI T and AOKI T (2008), 'Essential function of transglutaminase and clotting protein in shrimp immunity', *Mol Immunol*, **45**, 1269–75.

MEKATA T, KONO T, YOSHIDA T, SAKAI M and ITAMI T (2008), 'Identification of a cDNA encoding Toll receptor , MjToll gene from kurama shrimp, *Marsupenaeus japonicus*', *Fish Shellfish Immunnol*, **24**, 122–33.

NINAWE A S and SELVIN J (2009), 'Probiotics in shrimp aquaculture: avenues and challenges', *Crit Rev Microbiol*, **35**, 43–66.

NOONIN C, JIRAVANICHPAISAL P, SÖDERHÄLL I, MERINO S, TOMAS J M AND SÖDERHÄLL K (2010), 'Melanization and pathogenicity in the insect, *Tenebrio molitor*, and the crustacean, *Pacifastacus leniusculus*, by *Aeromonas hydrophila* AH-3', *PLoS ONE*, **5**(12), e15728.

PRAPAVORARAT A, VATANAVICHARN T, SÖDERHÄLL K and TASSANAKAJON A (2010), 'A novel viral responsive protein is involved in hemocyte homeostasis in the black tiger shrimp, *Penaeus monodon*', *J Biol Chem*, **285**, 21467–77.

ROBALINO J, BROWDY C L, PRIOR S, METZ A, PARNELL P, GROSS P and WARR G (2004), 'Induction of antiviral immunity by double-stranded RNA in a marine invertebrate', *J Virol*, **78**, 10442–8.

ROBALINO J, BARTLETT T, SHEPARD E, PRIOR S, JARAMILLO G, SCURA E, CHAPMAN R W, GROSS P S, BROWDY C L and WARR G W (2005), 'Double-stranded RNA induces sequence-specific antiviral silencing in addition to nonspecific immunity in a marine shrimp, convergence of RNA interference and innate immunity in the invertebrate antiviral response?', *J Virol*, **79**, 13561–71.

ROLLAND J L, ABDELOUAHAB M, DUPONT J, LEFEVRE F, BACHÉRE E and ROMESTAD B (2010), 'Stylicins, a new family of antimicrobial peptides from the Pacific blue shrimp *Litpenaeus stylirostris*', *Mol Immunol*, **47**, 1269–77.

ROSA R D and BARRACCO M A (2010), 'Antimicrobial peptides in custaceans', *Invertebr Surviv J*, **7**, 262–84.

ROUT N, KUMAR S, JAGANMOHAN S and MURUGAN V (2007), 'DNA vaccines encoding viral envelope proteins confer protective immunity against WSSV in black tiger shrimp', *Vaccine*, **25**, 2778–86.

RUSAINI and OWENS L (2010), 'Insight into the lymphoid organ of penaeid prawn: a review', *Fish Shellfish Immunol*, **29**, 367–77.

SHOCKEY J E, O'LEARY N A, DE LA VEGA E, BROWDY C L, BAATZ J E and GROSS P S (2009), 'The role of crustins in *Litopenaeus vannamei* in response to infection with shrimp pathogens: an *in vivo* approach', *Dev Comp Immunol*, **33**, 668–73.

SMITH V J, FERNANDES J M O, KEMP G D and HAUTON C (2008), 'Crustins: enigmatic WAP-domain-containing antibacterial proteins from crustaceans', *Dev Comp Immunol*, **32**, 758–72.

SÖDERHÄLL K (1982), 'Prophenoloxidase activating system and melanization – a recognition mechanism of arthropods? A review', *Dev Comp Immunol*, **6**, 601–11.

SÖDERHÄLL K and CERENIUS L (1998), 'Role of the prophenoloxidase activating system in invertebrate immunity', *Curr Opin Immunol*, **10**, 23–38.

SÖDERHÄLL I, KIM Y A, JIRAVANICHPAISAL P, LEE S Y and SÖDERHÄLL K (2005), 'An ancient role for a prokineticin domain in invertebrate hematopoiesis', *J Immunol*, **174**, 6153–60.

SÖDERHÄLL I, WU C, NOVOTNY M, LEE B L and SÖDERHÄLL K (2009), 'A novel protein acts as a negative regulator of prophenoloxidase activation and melanization in the freshwater crayfish *Pacifastacus leniusculus*', *J Biol Chem*, **284**, 6301–10.

SOMBOONWIWAT K, MARCOS M, TASSANAKAJON A, KLINBUNGA S, AUMELAS A, ROMESTAND B, GUEGUEN Y, BOZE H, MOULIN G and BACHERE E (2005), 'Recombinant expression and anti-microbial activity of anti-lipopolysaccharide factor (ALF) from the black tiger shrimp *Penaeus monodon*', *Dev Comp Immunol*, **29**, 841–51.

SOONTHORNCHAI W, RUNGRASSAMEE W, KAROONNUTHAISIRI N, JARAYABHAND P, KLINBUNGA S, SÖDERHÄLL K and JIRAVANICHPAISAL P (2010), 'Expression of immune-related genes in the digestive organ of shrimp, *Penaeus monodon*, after oral infection by *Vibrio harveyi*', *Dev Comp Immunol*, **34**, 19–28.

SPERSTAD S V, HAUG T, VASSKOG T and STENSVÅG K (2009), 'Hyastatin, a glycine-rich multi-domain antimicrobial peptide isolated from the spider crab (*Hyas araneus*) hemocytes', *Mol Immunol*, **46**, 2604–12.

SPERSTAD S V, SMITH V J and STENSVÅG K (2010), 'Expression of antimicrobial peptides from *Hyas araneus* haemocytes following bacterial challenge', *Dev Comp Immunol*, **34**, 618–24.

SRICHAROEN S, KIM J J, TUNKIJJANUKIJ S and SÖDERHÄLL I (2005), 'Exocytosis and proteomic analysis of the vesicle content of granular hemocytes from a crayfish', *Dev Comp Immunol*, **29**, 1017–31.

THARNTADA S, PONPRATEEP S, SOMBOONWIWAT K, LIU H, SÖDERHÄLL I, SÖDERHÄLL K and TASSANAKAJON A (2009), 'Role of anti-lipopolysaccharide factor from the black tiger shrimp, *Penaeus monodon*, in protection from white spot syndrome virus infection', *J Gen Virol*, **90**, 1491–8.

UNESTAM T and SÖDERHÄLL K (1977), 'Soluble fragments from fungal cell walls elicit defence reactions in crayfish', *Nature*, **267**, 45–6.

WANG H, TSENG C, LIN H, CHEN I, CHEN Y, CHEN Y, CHEN T and YANG H (2010), 'RNAi knock-down of the *Litopenaeus vannamei* Toll gene (LvToll) significantly increases mortality and reduces bacterial clearance after challenge with *Vibrio harveyi*', *Dev Comp Immunol*, **34**, 49–58.

WATTHANASUROROT A, JIRAVANICHPAISAL P, SÖDERHÄLL I and SÖDERHÄLL K (2010), 'A gC1qR prevents white spot syndrome virus replication in the freshwater crayfish *Pacifastacus leniusculus*', *J Virol*, **84**, 10844–51.

WATTHANASUROROT A, JIRAVANICHPAISAL P, LIU H, SÖDERHÄLL I and SÖDERHÄLL K (2011a), 'Bacteria-induced Dscam isoforms of the crustacean, *Pacifastacus leniusculus*', *PLoS Pathogens* **7**(6), e1002062.

WATTHANASUROROT A, SÖDERHÄLL K, JIRAVANICHPAISAL P and SÖDERHÄLL I (2011b), 'An ancient cytokine, astakine, mediates circadian regulation of invertebrate hematopoiesis', *Cell Mol Life Sci*, **68**, 315–23.

WITTEVELDT J, CIFUNTES C C, VLAK J M and VAN HULTEN M C (2004), 'Protection of *Penaeus monodon* against white spot syndrome virus by oral vaccination', *J Virol*, **78**, 2057–61.

WU C, SÖDERHÄLL I, KIM YA, LIU H and SÖDERHÄLL K (2008), 'Hemocyte-lineage marker proteins in a crustacean, the freshwater crayfish, *Pacifastacus leniusculus*', *Proteomics*, **8**, 4226–35.

XU Z R, DU H H, XU Y X, SUN J Y and SHEN J (2006), 'Crayfish *Procambarus clarkii* protected against white spot syndrome virus by oral administration of viral proteins expressed in silk worms', *Aquaculture*, **253**, 179–83.

YEH M, HUANG C, CHENG J and TSAI I (2007), 'Tissue-specific expression and regulation of the haemolymph clottable protein of tiger shrimp (*Penaeus monodon*)', *Fish Shellfish Immunol*, **23**, 272–9.

ZAMOCKY M, JAKOPITSCH C, FURTMULLER P G, DUNAND C and OBINGER C (2008), 'The peroxidase-cyclooxygenase superfamily: reconstructed evolution of critical enzymes of the innate immune system', *Proteins Struct Funct Bioinform*, **72**, 589–605.

ZHANG X, XU W, WANG X, MU Y, ZHAO X, YU X and WANG J (2009), 'A novel C-type lectin with two CR domains from Chinese shrimp *Fenneropenaeus chinensis* functions as a pattern recognition protein', *Mol Immunol*, **46**, 1626–37.

ZHANG Y, SÖDERHÄLL I, SÖDERHÄLL K and JIRAVANICHPAISAL P (2010), 'Expression of immune-related genes in one phase of embryonic development of freshwater crayfish, *Pacifastacus leniusculus*', *Fish Shellfish Immunol*, **28**, 649–53.

ZHAO P, LI J, WANG Y and JIANG H (2007), 'Broad-spectrum antimicrobial activity of the reactive compounds generated *in vitro* by *Manduca sexta* phenoloxidase', *Insect Bichem Mol Biol*, **37**, 952–9.

ZHAO Z, YIN Z, XU X, WENG S, RAO X, DAI Z, LUO Y, YANG G, GUAN H, LI Z, CHAN S, YU X and HE J (2009), 'A novel C-type lectin from the shrimp *Litopenaeus vannamei* possesses anti-white spot syndrome virus activity', *J Virol*, **83**, 347–56.

3

Immune responses in molluscs and their implications for disease control

B. Novoa and A. Figueras, Instituto de Investigaciones Marinas (IIM), CSIC, Spain

Abstract: Bivalves are particularly sensitive to their environment since they gather a large number of microorganisms due to their filter-feeding habits. As invertebrates, bivalves only present an innate immune system. Their immune cells, the haemocytes, are able to recognize, phagocytose and destroy non-self particles. Although in the last few years significant efforts have been made to identify protein-encoding genes that are putatively related to the immune system, more studies are needed in this area. Further research is also required to determine the pathogenicity of newly identified organisms and their effects on the bivalve immune system and to develop disease resistance selection programmes.

Key words: shellfish pathology, haemocytes, immune genes, selection programmes, pathogens.

3.1 Introduction

Bivalves are particularly sensitive to their environment and can suffer severe mortalities caused by pathogens (Bower *et al.*, 1994; Figueras and Novoa, 2004). They gather a large number of microorganisms due to their filter-feeding habits. Similar to other invertebrates, bivalves rely on their innate immune system to respond to external aggressions. After an initial contact with a pathogen, they lack the immune memory and antibody response that is seen in vertebrates, which allows for a stronger and quicker response during a second encounter with the same pathogen. Vaccination is therefore not an option in invertebrates. Moreover, using drugs to fight or control these diseases is not feasible because molluscs are usually farmed in the open marine environment.

Although several biologically active molecules have been reported in the haemolymph of bivalve molluscs, a comprehensive view of their immune mechanisms is lacking and few genes involved in immunity have been identified. The shell and mucus are the primary defence barriers. They cover the

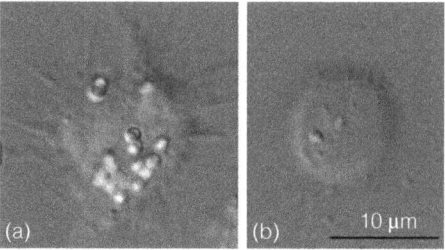

Fig. 3.1   A representation of the clam (*Ruditapes decussatus*) granulocytes (a) and hyalinocytes (b) phagocytosing refringent latex particles.

entire body and contain enzymes with bactericidal activity (Glinski and Jarros, 1999; Ratcliffe *et al.*, 1985). The innate system is composed of humoral and cellular components. Haemocytes, the cellular components, are involved in physiological functions such as digestion and excretion; however, their major role is in the bivalve immune response. Although there are several hypotheses postulating whether different haemocyte types are associated with specific roles, a precise classification of these cells in hyalinocytes, where there is a lack or very few granules in the cytoplasm, and in granulocytes, where there are granules in the cytoplasm, remains vague (Fig. 3.1). Several techniques have been applied to separate different subpopulations of bivalve haemocytes, but there is little agreement on the designation and function of the different bivalve haemocyte types (Cheng *et al.*, 1980; Bachere *et al.*, 1988; Dyrynda *et al.*, 1997; Pipe *et al.*, 1997; Xue *et al.*, 2000, 2001; Allam *et al.*, 2002; Hegaret *et al.*, 2003; Lambert *et al.*, 2003; Goedken and De Guise, 2004; Buggé *et al.*, 2007; García-García *et al.*, 2008).

It is well known that the bivalve immune response is influenced by different factors, such as environmental components, including stress, salinity fluctuations, temperature, food availability, the presence of infectious agents, genetic factors and pollution (Chu *et al.*, 2002). Marine bivalves, such as mussels, are often employed as sentinel organisms of aquatic pollution because, although they are resistant to high doses of contaminants, the induced physiological alterations can be used for pollution monitoring (Cajaraville *et al.*, 2000). Contaminants such as heavy metals or polychlorinated biphenyls (PCBs) can affect the signal transduction pathways (Viarengo, 1989; Canesi *et al.*, 2001, 2003; Marchi *et al.*, 2004) or the production of reactive oxygen species (ROS) (Winston *et al.*, 1996). Recently, genomic approaches have been used to determine the effect of these environmental factors on the gene expression profile of mussels (Dondero *et al.*, 2006; Venier *et al.*, 2006). This chapter provides an overview of the molluscan immune system and major methods of disease control in bivalves.

3.2   The molluscan immune system

Although many of the components and molecules that are involved in the bivalve immune response were identified long ago, the recent use of molecular techniques, such as cDNA libraries and high-throughput sequencing, has allowed researchers to confirm the presence of molecules that were identified indirectly by their biological activity. Here, we describe the main components of the immune system.

3.2.1   The immune cells: haemocytes

Phagocytosis is one of the most important functions of bivalve haemocytes, which identify 'non-self' particles, ingest, destroy and expel them. Bivalve phagocytosis has been described in several species, including scallops (Mortensen and Glette, 1996), clams (Tripp, 1992; López *et al.*, 1997b) and mussels (Carballal *et al.*, 1997). It is likely that haemocytes are attracted to the small peptides released by the potential pathogens or the components of their cellular walls (Howland and Cheng, 1982; Fawcett and Tripp, 1994). However, there is little evidence elucidating which receptors are responsible for sensing pathogens. A strong haemocyte infiltration into tissues is usually observed when pathological processes are present (Fig. 3.2). After recognition and adhesion, the haemocytes internalise foreign particles by extending pseudopods that engulf microorganisms or strange particles (Fig. 3.1) (Ratcliffe *et al.*, 1985). Consequently, phagocytic vacuoles, known as phagosomes, are produced and interact with lysosomes present in haemocytes (Fisher, 1986). The interior of the lysosome is full of hydrolytic enzymes that attack the recently formed vacuoles by releasing their contents while inside, consequently destroying the pathogen. Once the pathogen is destroyed, the remaining materials are released into the environment through a process known as exocytosis (Fisher, 1986).

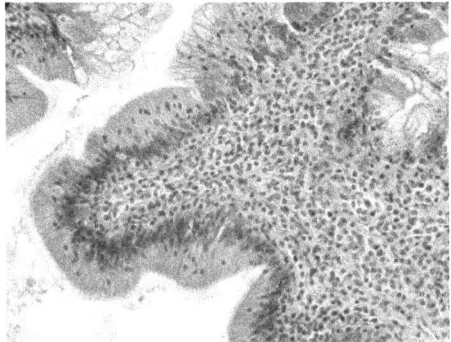

Fig. 3.2   Mussel haemocytes infiltrating the tissue surrounding the digestive gland. Magnification 40×.

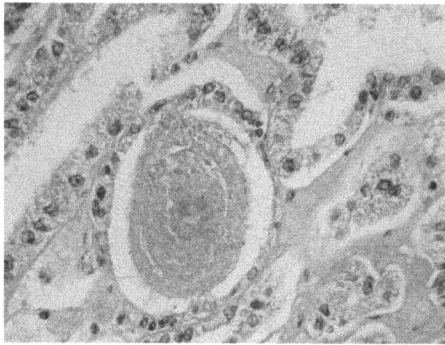

Fig. 3.3   Mussel haemocytes encapsulating bacteria. Magnification 40×.

Aggregation and encapsulation, which involve the formation of cellular aggregates that surround pathogens with a posterior deposit of melanin, also frequently coincide with phagocytosis (Fig. 3.3). When the foreign bodies are bigger than the haemocytes, recruited fibroblasts deposit mucopolysaccaride residues and fibrous material to form a glycoprotein-associated reticulum (Ratcliffe *et al.*, 1985).

During phagocytosis, oxygen radicals are produced, a process known as the respiratory burst, which is caused by the action of the multiprotein complex known as NADPH–oxidase. After a haemocyte encounters a foreign particle, the cell membrane undergoes small changes that activate the NADPH–oxidase system (Torreilles *et al.*, 1996), which reduces the molecular oxygen reactive species and produce reactive oxygen intermediates (ROIs). ROIs are effector molecules that can act on their own or with the hydrolytic enzymes that are released during phagocytosis (Adema *et al.*, 1991). ROI production has been described in several bivalve molluscs, such as the American oyster (*Crassostrea virginica*) (Anderson, 1994; Austin and Paynter, 1995), the Japanese oyster (*Crassostrea gigas*) and the flat oyster (*Ostrea edulis*) (Bachere *et al.*, 1991), the European scallop (*Pecten maximus*) (Le Gall *et al.*, 1991), the hard clam (*Mercenaria mercenaria*) (Buggé *et al.*, 2007), the blue mussel (*Mytilus edulis*) (Pipe, 1992) and the Mediterranean mussel (*M. galloprovincialis*) (Torreilles *et al.*, 1996; Ordás *et al.*, 2000).

Bivalve haemocytes also release nitrogen radicals. Most living beings are able to produce small quantities of nitric oxide (NO) from arginine via nitric oxide synthase (NOS) (Nathan and Xie, 1994). The activity of this enzyme, and the consequent release of nitrogen radicals, has been shown in haemocytes of several species, including *C. gigas* (Nakayama and Maruyama, 1998; Torreilles and Romestand, 2001), *C. virginica* (Villamil *et al.*, 2007), *Ruditapes decussatus* (Tafalla *et al.*, 2003), *M. edulis* (Ottaviani *et al.*, 1993) and *M. galloprovincialis* (Arumugan *et al.*, 2000; Gourdon *et al.*, 2001; Torreilles and Romestand, 2001; Tafalla *et al.*, 2002; Novas *et al.*, 2004).

The release of NO can be modulated by live organisms such as the proto-zoan *Perkinsus marinus*, which is found in the American oyster, *C. virginica* (Villamil *et al.*, 2007).

3.2.2   Complement pathways

Although complement pathways have not been extensively described in bivalves, there are several lines of evidence that support the presence of this defence mechanism in these organisms. The complement system is composed of more than 30 soluble plasma proteins that collaborate to distinguish and eliminate non-self particles. C3 is the central component system. In vertebrates, it is proteolytically activated by a C3 convertase through the classic, lectins and alternative routes (Nonaka and Yoshizaki, 2004). In the classic activation route, immunoglobulins are recognised by the C1 protein, whereas activation is initiated by the recognition of mem-brane carbohydrates by lectins and ficolins in the lectin route (Nonaka and Kimura, 2006).

In the American oyster *C. virginica*, some expressed sequence tags (ESTs) with similarity to proteins that function in the complement pathway, such as the precerebellin-like protein, which possesses a homologous C1q domain, and the mannose-binding lectin (MBL), have been detected (Jenny *et al.*, 2002). The recent analysis of the MBL from the tropical clam *Codakia orbicularis* (Gourdine and Smith-Ravin, 2007) or the C1q domain-containing protein from the Zhikong scallop *Chlamys farreri* and the mussel *M. galloprovincialis* (Zhang *et al.*, 2008; Gestal *et al.*, 2010) illustrate the putative presence of this system in bivalve molluscs. The extensive intra- and inter-individual diversity present in the sequences of mussel MgC1q at both the DNA and cDNA levels indicates the complexity of this system. EST annotation of a suppression subtractive hybridisation (SSH) prepared with bacteria-infected clam (*R. decussatus*) tissues has allowed the identifi-cation of novel C1q adiponectin-like and C3 and factor B-like proteins (Prado-Alvarez *et al.*, 2009a, b).

3.2.3   Pattern recognition receptors (PRRs)

Lectins

Lectins are a family of carbohydrate-recognition proteins that play crucial roles in innate immunity. They have been described as being soluble pro-teins in the haemolymph or associated with cell membranes (Vasta *et al.*, 1982, 1984). C-type lectins exhibit calcium-dependent carbohydrate-binding activity and show characteristic carbohydrate-recognition domains (CRDs). Lectins have been purified from several bivalve species (Tatsumi *et al.*, 1982; Dam *et al.*, 1994; Puanglarp *et al.*, 1995; Tunkijjanukij *et al.*, 1997) and they have also been cloned or purified from clam, *Ruditapes philippinarum*, (Kang *et al.*, 2006; Kim *et al.*, 2008a), oyster *C. gigas* (Yamaura *et al.*, 2008) and scallops *C. farreri* and *A. irradians* (Wang *et al.*, 2007; Zheng *et al.*, 2008; H. Zhang *et al.*, 2009; Zhu *et al.*, 2009). Galectins, which constitute one family of lectins, are characterised by a conserved sequence motif in their carbo-hydrate CRD and a specific affinity for β-galactosides. They have been shown to be involved in cell attachment and inflammatory responses. In bivalves, they have been characterised in several species: in eastern oyster *C. virginica*, a galectin with a unique domain organisation seems to be a receptor for the protistan parasite *P. marinus* (Tasumi and Vasta, 2007) and in Manila clam, a tandem-repeat galectin is induced upon infection with *Perkinsus olseni* (Kim *et al.*, 2008b). A galectin has also been identified in *C. gigas*, although it is not induced after a bacterial infection (Yamaura *et al.*, 2008).

β-Glucan recognition proteins
Although these receptors have been partially sequenced in several bivalves, only one group has published a complete description in scallop (Su *et al.*, 2004). Their results indicated that this lipopolysaccharide and β-1,3-glucan-binding protein (LGBP) was a constitutive and inducible acute-phase protein that may critically participate in the scallop–pathogen interaction.

Peptidoglycan recognition proteins (PGRP)
This family of proteins specifically binds peptidoglycans (PGN), a major component of bacterial cell wall. They were first identified in scallop *C. farreri* and *A. irradians* (Ni *et al.*, 2007; Su *et al.*, 2007) and the Pacific oyster *C. gigas*, from which four different types of PGRPs were identified. CgPGRPs have also been classified as short-type PGRPs. Although phylo-genetic analysis indicated that CgPGRPs are closely related, they exhibit different tissue expression patterns (Itoh and Takahashi, 2008). A PGRP detected in *C. gigas* was classified into the short PGRP group, although its estimated molecular weight of 54 kDa indicates a greater similarity to long PGRP groups (Itoh and Takahashi, 2009).

Toll-like receptors
Toll-like receptors (TLRs) are an ancient family of pattern recognition receptors with homology to the *Drosophila* Toll protein that plays key roles in detecting several non-self substances and initiating and activating the immune system. The first bivalve TLR was identified in the Zhikong scallop, *C. farreri*. It presents an extracellular domain with a potential signal peptide, 19 leucine-rich repeats (LRR), two LRR-C-terminal (LRRCT) motifs, and a LRR-N-terminal (LRRNT) motif that is followed by a transmembrane segment of 20 amino acids and a cytoplasmic region of 138 amino acids containing the Toll/IL-1R domain (TIR) (Qiu *et al.*, 2007). Moreover, several ESTs with similarity to TLRs have been identified in the American oyster *C. gigas* (Tanguy *et al.*, 2004) and in the scallop *A. irradians* (Song *et al.*, 2006).

3.2.4   Protease inhibitors

Pathogens use proteases to fight the host's defence mechanisms (La Peyre and Faisal, 1995; Oliver *et al.*, 1999; Ordás *et al.*, 2001; Brown and Reece, 2003). These proteases facilitate infection, diminish the activity of lysozymes and quench the agglutination capacity of haemocytes (La Peyre *et al.*, 1996). Many organisms have developed enzymes, termed protease inhibitors, that regulate the activities of pathogen proteases as well as endogenous proteases (Laskowski and Kato, 1980; Faisal *et al.*, 1998). Protease inhibitors play important roles in invertebrate immunity. However, information on protease inhibitors in marine molluscs, including in commercially relevant bivalve species, is limited (Zhu *et al.*, 2006; Xue *et al.*, 2006). *C. gigas* has been thought to be significantly more resistant to *P. marinus* than *C. virginica* due to the enhanced protease inhibitory activity against the parasite proteases (Meyers *et al.*, 1991). This inhibitory activity could represent the key event in resistance to parasite infection (Romestand *et al.*, 2002).

Recently, a tissue inhibitor of a metalloproteinase homologue was cloned from the Pacific oyster *C. gigas* (Montagnani *et al.*, 2001), and the genes encoding a Kazal-type serine protease inhibitor were identified in *A. irradians* (Zhu *et al.*, 2006) and *C. farreri* (Wang *et al.*, 2008). In addition, new serine protease inhibitors, cvSI-1 and cvSI-2, were characterised in the eastern oyster *C. virginica*, which likely represent a novel family of serine protease inhibitors in bivalve molluscs (Xue *et al.*, 2009; La Peyre *et al.*, 2010).

3.2.5   Lysozyme

Lysozyme is a widely distributed anti-bacterial molecule present in numerous animals including bivalves. It is a hydrolytic protein that can break the glycosidic union of the peptidoglycans of the bacteria cell wall. It is present in the haemolymph and in cells, and its release is associated with physiological changes, age, stress (López *et al.*, 1997a; Cronin *et al.*, 2001) and bacterial infections (Li *et al.*, 2008). Although lysozyme activity was first reported in molluscs over 30 years ago, complete sequences were published only recently, including those from *M. edulis*, *M. galloprovincialis*, and two different forms in *C. virginica*, *C. gigas*, *O. edulis* and *C. farreri* (Bachali *et al.*, 2002; Itoh and Takahasi, 2007; Li *et al.*, 2008). Bivalve lysozymes belong to the invertebrate type (i-type) lysozyme family, which differs from the insect c-type lysozyme but is similar to the vertebrate c-type (Nilsen and Myrnes, 2001).

3.2.6   Antimicrobial peptides

Antimicrobial peptides (AMPs) are small, gene-encoded cationic peptides that constitute important innate immune effectors from organisms spanning most of the phylogenetic spectrum. Their action mechanism begins with its union to the microorganisms' plasmatic membranes, altering their permeability and causing cellular lysis (Zasloff, 2002; Patrzykat and Douglas, 2005).

In bivalves, they were first purified from mussel haemocyte granules (Charlet *et al.*, 1996; Hubert *et al.*, 1996) and classified into three families: defensins related to arthropod defensins, mytilins and myticins. These three AMPs are present in different quantities. Myticin is expressed 300-fold more than defensin and 30-fold more than mytilin (Gestal *et al.*, 2008). Moreover, the genes are differentially regulated according to the challenging bacteria (Cellura *et al.*, 2007). In *M. galloprovincialis*, all of the peptides possess eight cysteines arranged in specific conserved arrays (Mitta *et al.*, 2000a,b). The typical structure of these peptides is shared by the new myticin class, myticin C, recently described in *M. galloprovincialis* (Pallavicini *et al.*, 2008). The genes encoding these proteins showed a high polymorphic variability (Padhi and Verguese, 2008), which has been suggested to account for the high resistance of the Mediterranean mussel to disease (Costa *et al.*, 2009). This variability was also observed in clam and mussel mytilins (Gestal *et al.*, 2007; Parisi *et al.*, 2009). In the oyster *C. gigas*, a molecular biological approach revealed the presence of two isoforms of a defensin-like protein (Gueguen *et al.*, 2006; Gonzalez *et al.*, 2007). In contrast to the one from the mussel, the oyster defensin was not altered after a bacterial challenge (Gueguen *et al.*, 2006).

3.2.7   Prophenoloxidase (proPO)

The prophenoloxidase (ProPO) system is the origin of melanin production and serves as an innate defence mechanism in several bivalve species, including *C. gigas*, *M. edulis*, *Perna viridis*, *Illex argentinus* and *Perna perna*, *Saccostrea glomerata* and *R. phillipinarum* (Coles and Pipe, 1994; Asokan *et al.*, 1997; Cong *et al.*, 2005; Aladaileh *et al.*, 2007; Hellio *et al.*, 2007). In these species, proPO has been detected in both the haemolymph and in haemocytes (Ashida *et al.*, 1983; Saul *et al.*, 1987; Hernández-Lopez *et al.*, 1996). The proPO, an inactive proenzyme, can be activated by lipopolysaccharide (LPS), β-1,3-glucans and peptidoglycans (Söderhall and Hall, 1984) through a signal cascade mediated by serine proteases. Moreover, proPO expression can be modulated by pathogens such as bacteria and parasites (Hong *et al.*, 2006; Muñoz *et al.*, 2006; Bezemer *et al.*, 2006). Phenoloxidase was found to be involved in the tolerance to *Marteilia sydneyi* of *S. glomerata* strains (Bezemer *et al.*, 2006; Aladaileh *et al.*, 2007; Butt and Raftos, 2008), and this was correlated with the levels of expression of genes coding for antioxidant enzymes in the selected strains for their tolerance to the parasite (Green *et al.*, 2009). Significant differences in phenoloxidase activity have been observed in bivalves exposed to environmental stressors or pollution (Kuchel *et al.*, 2010; Luna-Acosta *et al.*, 2010). Recently, phenoloxidase activity has been detected in different developmental stages of the Pacific oyster *C. gigas*, and it has been suggested that it plays an important function in non-self recognition and host immune reactions in the early life stages of bivalves (Thomas-Guyon *et al.*, 2009).

3.2.8   Heat shock proteins (hsp)

Heat shock proteins (hsp) are critical in mediating protein–protein interactions, including folding, assembly and degradation, by acting as molecular chaperones. Of these proteins, the hsp70s have been studied extensively. They have been linked to developmental changes, such as gametogenesis, embryogenesis and metamorphosis in *C. virginica* (Ueda and Boettcher, 2009). Their overexpression also represents an important mechanism in stress response (Ivanina *et al.*, 2009). Members of the hsp70 group have been described in scallops (Song *et al.*, 2006), *C. gigas* (Gourdon *et al.*, 2000; Boutet *et al.*, 2003b), *C. virginica* (Rathinam *et al.*, 2000), *O. edulis* (Boutet *et al.*, 2003a; Piano *et al.*, 2004, 2005), *M. edulis* (Luedeking and Koehler, 2004) and *M. galloprovincialis* (Cellura *et al.*, 2007). The expression of hsp70 has been reported to increase after a bacterial challenge (Cellura *et al.*, 2006, 2007), and injection of *Vibrio anguillarum*, but not *V. splendidus* or *Micrococcus lysodeikticus*, induced the overexpression of the mussel hsp70 gene (Cellura *et al.*, 2006). However, a recent study did not show a change in hsp70 gene expression following a bacterial challenge, in contrast to previously reported work (Li *et al.*, 2010).

3.2.9   Other immune molecules

Recently, several genes related to the inflammatory response against LPS stimulation have been detected in bivalves. This is the case for the LPS-induced TNF-α factor (LITAF), which is a novel transcription factor that critically regulates the expression of TNF-α and various inflammatory cytokines in response to LPS stimulation. It has been described in pearl oyster *Pinctada fucata* (D. Zhang *et al.*, 2009), *C. gigas* (Park *et al.*, 2008) and scallop *C. farreri*, where it is up-regulated in LPS-challenged, but not peptidoglycan (PGN)-challenged, haemocytes (Yu *et al.*, 2007).

Other TNF-related genes have been identified in the Zhikong scallop, such as a TNFR homologue (Li *et al.*, 2009) and a tumor necrosis factor receptor-associated factor 6 (TRAF6), which is a key signalling adaptor molecule common to the TNFR superfamily and to the IL-1R/TLR family, which is involved in PGN signalling (Qiu *et al.*, 2009).

3.3   Disease control in bivalves

Bivalves are typically cultured on the open sea, which precludes the use of treatments to control disease. Bivalves can be treated only at the larval stage, in the hatcheries. Moreover, as previously discussed, bivalves, similar to all invertebrates, lack an immune response that would allow vaccination against the most virulent pathogens. Here we discuss several methods to fight bivalve diseases.

3.3.1   Sanitary control of trade by efficient and specific pathogen diagnosis

Theoretically, eradication of a pathogen from a specific area is possible but has failed in practice (Van Banning, 1988, 1991). The reservoir or carriers of most bivalve pathogens, usually protozoan parasites, remain unknown, precluding their permanent eradication. Additionally, some of them may have a free living developmental stage in their life cycles. The compartmentalisation of the litoral area into free and infected zones, strictly controlling bivalve movements among areas, has been used. For this reason, disease-causing shellfish pathogens that are included in the lists of the European Union or the World Organisation for Animal Health (OIE) not only impact the populations of susceptible species but also impact their trade and restrict their transfer.

In this context, pathogen definition and identification are of primary importance. Initial descriptions of many pathogens, such as *Marteilia*, *Bonamia* and *Perkinsus*, were based on histological and ultrastructural techniques. However, these methods do not allow for the identification of parasites beyond the genus level. Furthermore, bacterial and viral pathogens cannot be identified with this technique. Other methods, including molecular-based techniques such as PCR, are now widely incorporated in routine diagnosis of mollusc pathogens.

Effective diagnostic techniques and epidemiological data can contribute to the management of the bivalve cultures. In fact, they allow growers to manipulate the disease by harvesting bivalves before they die. This has been done with *C. virginica* infected with *Haplosporidium nelsoni* and *P. marinus* on the east coast of the United States (Andrews and Ray, 1988; Ford and Haskin, 1988). The harvest of flat oysters at a smaller size was proposed to ameliorate the deleterious effects of *Bonamia ostreae* (Figueras, 1991).

3.3.2   Substitution of one species with another

Substituting one susceptible species with a resistant one has been used by the industry to avoid the economic losses caused by massive mortalities. This was the case for introducing *C. gigas* in France. Growers believed that this species was more resistant than *Crassostrea angulata* or *O. edulis* against iridoviruses *Marteilia* and *Bonamia*, respectively (Grizel and Héral, 1991). Recently, due to the *Crassostrea virginica* mortalities caused by *P. marinus* and *H. nelsoni*, the possibility of introducing *C. ariakensis* and *C. gigas* to the US was considered. However, *C. gigas* was shown to be susceptible to both infections, and *C. ariakensis* was shown to be susceptible to *P. marinus* and *Bonamia* (Calvo *et al.*, 1999, 2001; Burreson *et al.*, 2004; Paynter *et al.*, 2008). This industry practice entails high risks that should be taken into account, such as the introduction of new pathogens or competitors that could modify the receptor ecosystem. For this reason, these activities should be implemented following the ICES recommendations on introductions of foreign species. In the last few years, *C. gigas* has been dramatically reduced in France by the infection by a new strain of herpes virus that has caused high economic losses to oyster growers (Segarra *et al.*, 2010).

3.3.3   Identification of genes involved in resistance

Recently, several cDNA or subtractive hybridised libraries have been constructed in stimulated or infected bivalves to identify protein-encoding genes that are putatively related to the immune system (Saavedra and Bachère, 2006; Hedgecock *et al.*, 2007). In general, the percentage of well-characterised immune-related genes is very low in these libraries. For example, only 19 ESTs involved in immunity and cell communication were identified in response to a *P. marinus* challenge in *C. virginica* and *C. gigas* (Tanguy *et al.*, 2004). This low percentage was also observed in *C. gigas* exhibiting summer mortalities. However, this technology led to the identification of antimicrobial peptides in oysters that belonged to the defensin family (Gueguen *et al.*, 2003, 2006; Peatman *et al.*, 2004). In mussels, genes encoding AMP myticins were found to be highly expressed in the cDNA and subtracted libraries (Venier *et al.*, 2003; Pallavicini *et al.*, 2008).

In scallops, 131 host defence-related gene sequences, such as lectins, defensins, proteases, protease inhibitors, heat shock proteins, antioxidants and TLRs, were identified (Song *et al.*, 2006). In clams, a cDNA library constructed in *P. olseni*-infected *R. philippinarum* allowed for the identification of lectins that were the largest group of immune function ESTs, suggestive of a role in manila clam innate immunity for these molecules. Other libraries were also constructed in carpet shell clams (*R. decussatus*) and other bivalves (*C. gigas*, *M. edulis*, *R. decussatus* and *Bathymodiolus azoricus*) after stimulation with dead bacteria (Gestal *et al.*, 2007) or with *P. olseni* (Prado-Alvarez *et al.*, 2009a) and by using numerous tissues and physiological conditions.

The impact of these molecules in disease resistance remains unknown, but a better understanding of the molecular basis of bivalve immune response may allow these genes to be used as markers for disease resistance selection programmes.

3.3.4   Bivalve selection programmes, genetic maps and quantitative trait loci (QTL) selection

Efforts to identify bivalve stocks resistant to a variety of pathogens have been made over the last decades. Selection programmes based on breeding of disease survivors have been used to generate *C. virginica* resistance against the protozoans *H. nelsoni* and *P. marinus* (Ford and Haskin, 1987; Gaffney and Bushek, 1996; Ragone Calvo *et al.*, 2003). The results obtained have been promising, but the progress has been slow because long periods of time were needed to obtain substantial improvements in disease resistance rates.

Recently, the use of advanced molecular biology techniques has helped to improve our knowledge of this subject. Quantitative trait loci (QTLs) are stretches of DNA closely linked to the genes associated with a particular trait of interest. Disease-resistant QTLs were identified by mapping markers that showed significant frequency shifts after disease-inflicted mortalities. In *C. virginica*, Guo *et al.* (2006) reported the development of physical mapping and characterisation of chromosomes. Over 600 amplified fragment length polymorphism (AFLP) markers were developed, and four moderately dense linkage maps were constructed. Mapping analysis in two families identified 12 putative *Perkinsus*/summer mortality resistance QTLs.

Summer mortality suffered by *C. gigas* likely results from a complex interaction between the host, pathogens and environmental factors. Huvet *et al.* (2004) reported that genetic variability in the host is a major determinant in its sensitivity to summer mortality. The Ostreid Herpes virus type 1 (OsHV-1) has been associated with summer mortality. Sauvage *et al.* (2010) conducted a QTL analysis for survival and OsHV-1 load using five F_2 full-sib families from a divergent selection experiment for resistance to summer mortality. A consensus linkage map was built using 29 SNPs and 51 microsatellite markers. Five significant QTLs were identified and assigned to linkage groups. Lallias *et al.* (2009) identified QTLs in the flat oyster *O. edulis* associated with resistance to *Bonamia ostreae*. An F_2 family, from a cross between a wild oyster and an individual from a family selected for resistance to bonamiosis, was cultured with wild oysters injected with the parasite, leading to 20% cumulative mortality. Selective genotyping of the F_2 progeny was performed using 20 microsatellites and 34 AFLP primer pairs, where several QTLs were identified. Sokolova *et al.* (2006) conducted an arbitrary fragment length polymorphism analysis to search for potential genetic markers associated with resistance to infection by *Vibrio vulnificus* and *P. marinus* in oysters. Forty-eight arbitrary fragment length polymorphism markers were used, where two of these markers exhibited significant associations with the incidence of infection of *P. marinus*. Additionally, two separate markers were associated with the magnitude of *V. vulnificus* infection.

3.3.5   The use of triploids

The use of triploids to improve *Perkinsus* resistance has been assessed. In a five-month comparison study, 200 1-year-old market-size triploid *C. ariakensis* were compared with 200 3-year-old diploid *C. virginica*. After five months of exposure, the prevalence of *P. marinus* in *C. virginica* was 65% with a relative intensity of 1.50 on a 0–7 scale, whereas *C. ariakensis* exhibited no signs of infection. Although the non-native *C. ariakensis* may be attractive due to its high meat yield and disease resistance, rapid triploid growth results in very thin shells, which may sensitise them to blue crab predation (McLean and Abbe, 2006).

3.4   Conclusions

Despite recent increases in efforts to investigate bivalve pathology and immunology, a series of issues warrant further research. These include the pathogenicity of newly identified organisms and their effects on the bivalve immune system, disease resistance selection programmes and the molecular basis of bivalve immune defences. Further work also needs to be carried out to validate new diagnostic methods. The formation of transnational multidisciplinary research groups will facilitate research in these areas and others, with the aim of addressing the many unresolved questions in bivalve pathology.

3.5   Acknowledgements

Our work on mollusc immunology is supported by the Spanish Ministry of Science and Innovation (Project AGL2008-05111/ACU) and by the EU Projects REPROSEED (245119) and BIVALIFE (266157).

3.6   References

ADEMA C M, VAN DER KNAAP W P W, SMINIA T (1991), Molluscan hemocyte-mediated cytotoxicity: the role of reactive oxygen intermediates. *Rev Aquat Sci*, **4**, 201–223.

ALADAILEH S, RODNEY P, NAIR S V, RAFTOS D A (2007), Characterization of phenoloxidase activity in Sydney rock oysters (*Saccostrea glomerata*). *Comp Biochem Physiol B Biochem Mol Biol*, **148**, 470–480.

ALLAM B, ASHTON-ALCOX K A, FORD S E (2002), Flow cytometric comparison of haemocytes from three species of bivalve molluscs. *Fish Shellfish Immunol*, **13**, 141–158.

ANDERSON R S (1994), Hemocyte-derived reactive oxygen intermediate production in four bivalve mollusks. *Dev Comp Immunol*, **18**, 89–96.

ANDREWS J D, RAY S M (1988), Management strategies to control the disease caused by *Perkinsus marinus. Am Fish Soc Spec Publ*, **18**, 257–264.

ARUMUGAM M, ROMESTAND B, TORREILLES J, ROCH P (2000), *In vitro* production of superoxide and nitric oxide (as nitrite and nitrate) by *Mytilus galloprovincialis* haemocytes upon incubation with PMA or laminarin or during yeast phagocytosis. *Eur J Cell Biol*, **79**, 513–519.

ASHIDA M, ISHIZAKI Y, IWAHANA H (1983), Activation of pro-phenoloxidase by bacterial cell walls or beta-1,3-glucans in plasma of the silkworm, *Bombyx mori*. *Biochem Biophys Res Commun*, **113**, 562–568.

ASOKAN R, ARUMUGAM M, MULLAINADHAN P (1997), Activation of prophenoloxidase in the plasma and haemocytes of the marine mussel *Perna viridis* Linnaeus. *Dev Comp Immunol*, **21**, 1–12.

AUSTIN K A, PAYNTER K T (1995), Characterization of the chemiluminiscence measured in hemocytes of the eastern oyster, *Crassostrea virginica*. *J Exp Zool*, **273**, 461–471.

BACHALI S, JAGER M, HASSANIN A, SCHOENTGEN F, JOLLÈS P, FIALA-MEDIONI A, DEUTSCH J S (2002), Phylogenetic analysis of invertebrate lysozymes and the evolution of lysozyme function. *J Mol Evol*, **54**, 652–654.

BACHERE E, CHAGOT D, GRIZEL H (1988), Separation of *Crassotrea gigas* hemocytes by density gradient centrifugation and counterflow centrifugal elutriation. *Dev Comp Immunol*, **12**, 549–559.

BACHERE E, HERVIO D, MIALHE E (1991), Luminol-dependent chemiluminiscence by hemocytes of 2 marine bivalves, *Ostrea edulis* and *Crassostrea gigas*. *Dis Aquat Org*, **11**, 173–180.

BEZEMER B, BUTT D, NELL J, ADLARD R, RAFTOS D (2006), Breeding for QX disease resistance negatively selects one form of the defensive enzyme, phenoloxidase, in Sydney rock oysters. *Fish Shellfish Immunol*, **20**, 627–636.

BOUTET I, TANGUY A, MORAGA D (2003a), Organisation and nucleotide sequence of the European flat oysters *Ostrea edulis* heat shock cognate 70 (hsc70) and heat shock protein 70 (hsp) genes. *Aquat Toxicol*, **65**, 221–225.

BOUTET I, TANGUY A, ROUSSEAU S, AUFFRET M, MORAGA D (2003b), Molecular identification and expression of heat shock cognate 70 (hsc70) and heat shock protein 70 (hsp70) genes in the Pacific oyster *Crassostrea gigas*. *Cell Stress Chaperones*, **8**, 76–85.

BOWER S M, MCGLADDERY S E, PRICE I M (1994), Synopsis of infectious diseases and parasites of commercially exploited shellfish. *Ann Rev Fish Dis*, **4**, 1–199.

BROWN G D, REECE K S (2003), Isolation and characterization of serine protease gene(s) from *Perkinsus marinus*. *Dis Aquat Organ*, **57**, 117–126.

BUGGÉ D M, HÉGARET H, WIKFORS G H, ALLAM B (2007), Oxidative burst in hard clam (*Mercenaria mercenaria*) haemocytes. *Fish Shellfish Immunol*, **23**, 188–196.

BURRESON E, STOKES N, CARNEGIE R, BISHOP M (2004), *Bonamia* sp. (Haplosporidia) found in nonnative oysters *Crassostrea ariakensis* in Bogue Sound, North Carolina. *J Aquat Anim Health*, **16**, 1–9.

BUTT D, RAFTOS D (2008), Phenoloxidase-associated cellular defence in the Sydney rock oyster, *Saccostrea glomerata*, provides resistance against QX disease infections. *Dev Comp Immunol*, **32**, 299–306.

CAJARAVILLE M P, BEBIANNO M J, BLASO J, PORTE C, SARASQUETE C, VIARENGO A (2000), The use of biomarkers to assess the impact pollution in coastal environments of the Iberian Peninsula: a practical approach. *Sci Total Environ*, **247**, 295–311.

CALVO G W, LUCKENBACH M W, ALLEN S K JR, BURRESON E M (1999), A comparative field study of *Crassostrea gigas* (Thunberg 1793) and *Crassostrea virginica* (Gmelin 1791) in relation to salinity in Virginia. *J Shellfish Res*, **18**, 465–473.

CALVO G W, LUCKENBACH M, ALLEN S K J R, BURRESON E M (2001), A comparative field study of *Crassostrea ariakensis* (Fujita 1913) and *Crassostrea virginica* (Gmelin 1791) in relation to salinity in Virginia. *J Shellfish Res*, **20**, 221–229.

CANESI L, BETTI M, CIACCI C, GALLO G (2001), Insulin-like effect of zinc in mytilus digestive gland cells: modulation of tyrosine kinase-mediated cell signaling. *Gen Com Endocrinol*, **122**, 60–66.

CANESI L, CIACCI C, BETTI M, SCARPATO A, CITTERIO B, PRUZZO C, GALLO G (2003), Effects of PCB congeners on the immune function of *Mytilus* hemocytes: alterations of tyrosine kinase-mediated cell signaling. *Aquat Toxicol*, **63**, 293–306.

CARBALLAL M J, LÓPEZ C, AZEVEDO C, VILLALBA A (1997), *In vitro* study of phagocytic ability of *Mytilus galloprovincialis* Lmk. hemocytes. *Fish Shellfish Immunol*, **7**, 403–416.

CELLURA C, TOUBIANA M, PARRINELLO N, ROCH P (2006), HSP70 gene expression in *Mytilus galloprovincialis* hemocytes is triggered by moderate heat shock and *Vibrio anguillarum*, but not by *V. splendidus* or *Micrococcus lysodeikticus*. *Dev Comp Immunol*, **30**, 984–997.

CELLURA C, TOUBIANA M, PARRINELLO N, ROCH P (2007), Specific expression of antimicrobial peptide and HSP70 genes in response to heat-shock and several bacterial challenges in mussels. *Fish Shellfish Immunol*, **22**, 340–350.

CHARLET M, CHERNYSH S, PHILIPPE H, HÈTRU C, HOFFMANN D (1996), Innate immunity. Isolation of several cysteine-rich antimicrobial peptides from the blood of a mollusc, *Mytilus edulis*. *J Biol Chem*, **271**, 21808–21813.

CHENG T C, HUANG J V, KARADOGAN H, RENWRANTZ L R, YOSHINO T P (1980), Separation of oyster hemocytes by density gradient centrifugation and identification of their surface receptors. *J Inverteb Pathol*, **36**, 35–40.

CHU F L, VOLETY A K, HALE R C, HUANG Y (2002), Cellular responses and disease expression in oysters (*Crassostrea virginica*) exposed to suspended field contaminated sediments. *Ma Env Res*, **53**, 17–35.

COLES J A, PIPE R K (1994), Phenoloxidase activity in the haemolymph and haemocytes of the marine mussel *Mytilus edulis*. *Fish Shellfish Immunol*, **4**, 337–352.

CONG R, SUN W, LIU G, FAN T, MENG X, YANG L, ZHU L (2005), Purification and characterization of phenoloxidase from clam *Ruditapes philippinarum*. *Fish Shellfish Immunol*, **18**, 61–70.

COSTA MM, DIOS S, ALONSO-GUTIERREZ J, ROMERO A, NOVOA B, FIGUERAS A (2009), Myticin C variability in mussel: ancient defence mechanism or self/non self discrimination? *Dev Comp Immunol*, **33**, 162–170.

CRONIN M A, CULLOTY S C, MULCAHY M F (2001), Lysozyme activity and protein concentration in the haemolymph of the flat oyster *Ostrea edulis* (L.). *Fish Shellfish Immunol*, **11**, 611–622.

DAM T K, BANDYOPADHYAY P, SARKAR M, GHOSAL J, BHATTACHARYA A, CHOUDHURY A (1994), Purification and partial characterization of a heparin-binding lectin from the marine clam *Anadara granosa*. *Biochem Biophys Res Commun*, **203**, 36–45.

DONDERO F, PIACENTINI L, MARSANO F, REBELO M, VERGANI L, VENIER P, VIARENGO A (2006), Gene transcription profiling in pollutant exposed mussels (*Mytilus* spp.) using a new low-density oligonucleotide microarray. *Gene*, **376**, 24–36.

DYRYNDA E A, PIPE R K, RATCLIFFE N A (1997), Sub-populations of haemocytes in the adult and developing marine mussel, *Mytilus edulis*, identified by use of monoclonal antibodies. *Cell Tissue Res*, **289**, 527–536.

FAISAL M, MACINTYRE E A, ADHAM K G, TALL B D, KOTHARY M H, LA PEYRE J F (1998), Evidence for the presence of protease inhibitors in eastern (*Crassostrea virginica*) and Pacific (*Crassostrea gigas*) oysters. *Comp Biochem Physiol B*, **121**, 161–168.

FAWCETT L B, TRIPP M R (1994), Chemotaxis of *Mercenaria mercenaria* hemocytes to bacteria *in vitro*. *J Invertebr Pathol*, **63**, 275–284.

FIGUERAS A (1991), *Bonamia* status and its effects in cultured flat oysters in the ria de Vigo, Galicia (NW Spain). *Aquaculture*, **93**, 225–233.

FIGUERAS A, NOVOA B (2004), What has been going on in Europe in bivalve pathology? *Bull Eur Fish Pathol*, **24**, 16–21.

FISHER W S (1986), Structure and functions of oyster hemocytes. In: Brehelin M, Editor, 1986. *Immunity in Invertebrates*, Springer-Verlag, Berlin/Heidelberg, pp. 25–35.

FORD S E, HASKIN H H (1987), Infection and mortality patterns in strains of oysters *Crassostrea virginica* selected for resistance to the parasite *Haplosporidium nelsoni* MSX. *J Parasitol*, **73**, 368–376.

FORD S E, HASKIN H H (1988), Management strategies for MSX (*Haplosporidium nelsoni*) disease in eastern oysters. *Am Fish Soc Spec Publ*, **18**, 249–256.

GAFFNEY P M, BUSHEK D (1996), Genetic aspects of disease resistance in oysters. *J Shellfish Res*, **15**, 135–140.

GARCÍA-GARCÍA, E, PRADO-ÁLVAREZ M, NOVOA B, FIGUERAS A, ROSALES C (2008), Immune responses of mussel hemocyte subpopulations are differentially regulated by enzymes of the PI 3-K, PKC, and ERK kinase families. *Dev Comp Immunol*, **32**, 637–653.

GESTAL C, COSTA M M, FIGUERAS A, NOVOA B (2007), Analysis of differentially expressed genes in response to bacterial stimulation in hemocytes of the carpet-shell clam *Ruditapes decussatus*: identification of new antimicrobial peptides. *Gene*, **406**, 134–143.

GESTAL C, ROCH P, RENAULT T, PALLAVICINI A, PAILLARD C, NOVOA B, OUBELLA R, VENIER P, FIGUERAS A (2008), Study of diseases and the immune system of bivalves using molecular biology and genomics. *Rev Fish Sci*, **16**, 131–154.

GESTAL C, PALLAVICINI A, VENIER P, NOVOA B, FIGUERAS A (2010), MgC1q, a novel C1q-domain-containing protein involved in the immune response of *Mytilus galloprovincialis*. *Dev Comp Immunol*, **34**, 926–934.

GLINSKI Z, JARROS J (1999), Molluscan immune defenses. *Arch Immunol Ther Exp (Warsz)*, **45**, 149–155.

GOEDKEN M, DE GUISE S (2004), Flow cytometry as a tool to quantify oyster defence mechanisms. *Fish Shellfish Immunol*, **16**, 539–552.

GONZALEZ M, GUEGUEN Y, DESSERRE G, DE LORGERIL J, ROMESTAND B, BACHÈRE E (2007), Molecular characterization of two isoforms of defensin from hemocytes of the oyster *Crassostrea gigas*. *Dev Comp Immunol*, **31**, 332–339.

GOURDINE J P, SMITH-RAVIN E J (2007), Analysis of a cDNA-derived sequence of a novel mannose-binding lectin, codakine, from the tropical clam *Codakia orbicularis*. *Fish Shellfish Immunol*, **22**, 498–509.

GOURDON I, GRICOURT L, KELLNER K, ROCH P, ESCOUBAS J M (2000), Characterization of a cDNA encoding a 72 kDa heat shock cognate protein (Hsc72) from the Pacific oyster, *Crassostrea gigas*. *DNA Seq*, **11**, 265–270.

GOURDON I, GUÉRIN M C, TORREILLES J, ROCH P (2001), Nitric oxide generation by hemocytes of the mussel *Mytilus galloprovincialis*. *Nitric Oxide*, **5**, 1–6.

GREEN T J, DIXON T J, DEVIC E, ADLARD R D, BARNES A C (2009), Diferential expression of genes encoding anti-oxidant enzymes in Sydney rock oysters, *Saccostrea glomerata* (Gould) selected for disease resistance. *Fish Shellfish Immunol*, **26**, 799–810.

GRIZEL H, HÉRAL M (1991), Introduction into France of the Japanese oyster (*Crassostrea gigas*). *Journal du Conseil International pour l'Exploration de la Mer*, **47**, 399–403.

GUEGUEN Y, CADORET J P, FLAMENT D, BARREAU-ROUMIGUIÈRE C, GIRARDOT A L, GARNIER J, HOAREAU A, BACHÈRE E, ESCOUBAS J M (2003), Immune gene discovery by expressed sequence tags generated from hemocytes of the bacteria-challenged oyster, *Crassostrea gigas*. *Gene*, **303**, 139–145.

GUEGUEN Y, HERPIN A, AUMELAS A, GARNIER J, FIEVET J, ESCOUBAS J M, BULET P, GONZALEZ M, LELONG C, FAVREL P, BACHÈRE E (2006), Characterization of a defensin from the oyster *Crassostrea gigas*. Recombinant production, folding, solution structure, antimicrobial activities and gene expression. *J Biol Chem*, **281**, 313–323.

GUO X, WANG Y, YU Z, WANG L, LEE J H (2006), Genome mapping in the eastern Oyster (*Crassostrea virginica* gmelin). *J Shellfish Res*, **25**, 733.

HEDGECOCK D, LIN J Z, DECOLA S, HAUDENSCHILD C D, MEYER E, MANAHAN D T, BOWEN B (2007), Transcriptomic analysis of growth heterosis in larval Pacific oysters (*Crassostrea gigas*). *Proc Nat Acad Sci USA*, **104**, 2313–2318.

HEGARET H, WIKFORS G H, SOUDANT P (2003), Flow cytometric analysis of haemocytes from eastern oysters, *Crassostrea virginica*, subjected to a suden temperature elevation II. Haemocyte functions: aggregation, viability, phagocytosis, and respiratory burst. *J Exp Mar Bio Ecol*, **293**, 249–265.

HELLIO C, BADO-NILLES A, GAGNAIRE B, RENAULT T, THOMAS-GUYON H (2007), Demonstration of a true phenoloxidase activity and activation of a ProPO cascade in Pacific oyster, *Crassostrea gigas* (Thunberg) *in vitro*. *Fish Shellfish Immunol*, **22**, 433–440.

HERNÁNDEZ-LÓPEZ J, GOLLAS-GALVÁN B T, VARGAS-ALBORES F (1996), Activation of the prophenoloxidase system of the brown shrimp *Penaeus californiensis*. *Holmes Comp Biochem Physiol C*, **113**, 61–66.

HONG X T, XIANG L X, SHAO J Z (2006), The immunostimulating effect of bacterial genomic DNA on the innate immune responses of bivalve mussel, *Hyriopsis cumingii* Lea. *Fish Shellfish Immunol*, **21**, 357–364.

HOWLAND K H, CHENG T (1982), Identification of bacterial chemoattractants for oyster (*Crassostrea virginica*) hemocytes. *Invertebr Pathol*, **39**, 123–132.

HUBERT F, NOËL T, ROCH P (1996), A member of the arthropod defensin family from edible Mediterranean mussels (*Mytilus galloprovincialis*). *Eur J Biochem*, **240**, 302–306.

HUVET A, HERPIN A, DÉGREMONT L, LABREUCHE Y, SAMAIN J F, CUNNINGHAM C (2004), The identification of genes from the oyster *Crassostrea gigas* that are differentially expressed in progeny exhibiting opposed susceptibility to summer mortality. *Gene*, **343**, 211–220.

ITOH N, TAKAHASHI K G (2007), cDNA cloning and *in situ* hybridization of a novel lysozyme in the Pacific oyster, *Crassotrea gigas*. *Comp Biochem Physiol B*, **148**, 160–166.

ITOH N, TAKAHASHI K G (2008), Distribution of multiple peptidoglycan recognition proteins in the tissues of Pacific oyster, *Crassostrea gigas*. *Comp Biochem Physiol B Biochem Mol Biol*, **150**, 409–417.

ITOH N, TAKAHASHI K G (2009), A novel peptidoglycan recognition protein containing a goose-type lysozyme domain from the Pacific oyster, *Crassostrea gigas*. *Mol Immunol*, **46**, 1768–1774.

IVANINA A V, TAYLOR C, SOKOLOVA I M (2009), Effects of elevated temperature and cadmium exposure on stress protein response in eastern oysters *Crassostrea virginica* (Gmelin). *Aquat Toxicol*, **91**, 245–254.

JENNY M J, RINGWOOD A H, LACY E R, LEWITUS A J, KEMPTON J W, GROSS P S, WARR G W, CHAPMAN R W (2002), Potential indicators of stress response identified by expressed sequence tag analysis of hemocytes and embryos from the American oyster, *Crassostrea virginica*. *Mar Biotechnol*, **4**, 81–93.

KANG Y S, KIM Y M, PARK K I, CHO S K, CHOI K S, CHO M (2006), Analysis of EST and lectin expressions in hemocytes of Manila clams (*Ruditapes philippinarum*) (Bivalvia: Mollusca) infected with *Perkinsus olseni*. *Dev Comp Immunol*, **30**, 1119–1131.

KIM J Y, ADHYA M, CHO S K, CHOI K S, CHO M (2008a), Characterization, tissue expression, and immunohistochemical localization of MCL3, a C-type lectin produced by *Perkinsus olseni*-infected Manila clams (*Ruditapes philippinarum*). *Fish Shellfish Immunol*, **25**, 598–603.

KIM J Y, KIM Y M, CHO S K, CHOI K S, CHO M (2008b), Noble tandem-repeat galectin of Manila clam *Ruditapes philippinarum* is induced upon infection with the protozoan parasite *Perkinsus olseni*. *Dev Comp Immunol*, **32**, 1131–1141.

KUCHEL R P, RAFTOS D A, NAIR S (2010), Immunosuppressive effects of environmental stressors on immunological function in *Pinctada imbricata*. *Fish Shellfish Immunol*, **29**, 930–936.

LALLIAS D, GOMEZ-RAYA L, HALEY C S, ARZUL I, HEURTEBISE S, BEAUMONT A R, BOUDRY P, LAPÈGUE S (2009), Combining two-stage testing and interval mapping strategies to detect QTL for resistance to bonamiosis in the european flat oyster *Ostrea edulis*. *Mar Biotechnol*, **11**, 570–584.

LAMBERT C, SOUDANT P, CHOQUET G, PAILLARD C (2003), Measurement of *Crassostrea gigas* hemocyte oxidative metabolism by flow cytometry and the inhibiting capacity of pathogenic vibrios. *Fish Shell Immunol*, **15**, 225–240.

LA PEYRE J F, FAISAL M (1995), *Perkinsus marinus* produces extracellular proteolytic factor (s). *Bull Eur Assoc Fish Pathol*, **15**, 1–4.

LA PEYRE J F, YARNALL H A, FAISAL M (1996), Contribution of *Perkinsus marinus* extracellular products in the infection of eastern oysters (*Crassostrea virginica*). *J Invertebr Pathol*, **68**, 312–313.

LA PEYRE J F, XUE Q G, ITOH N, LI Y, COOPER R K (2010), Serine protease inhibitor cvSI-1 potential role in the eastern oyster host defense against the protozoan parasite *Perkinsus marinus*. *Dev Comp Immunol*, **34**, 84–92.

LASKOWSKI M J R, KATO I (1980), Protein inhibitors of proteinases. *Annu Rev Biochem*, **49**, 593–626.

LE GALL G, BACHÈRE E, MIALHE E (1991), Chemiluminiscence analysis of the activity of *Pecten maximus* haemocytes stimulated with zymosan and host-specific Rickettsiales-like organisms. *Dis Aquat Org*, **11**, 181–186.

LI H, PARISI M G, TOUBIANA M, CAMMARATA M, ROCH P (2008), Lysozyme gene expression and haemocyte behaviour in the Mediterranean mussel, *Mytilus galloprovincialis*, after injection of various bacteria or temperature stresses. *Fish Shellfish Immunol*, **25**, 143–152.

LI H, VENIER P, PRADO-ALVÁREZ M, GESTAL C, TOUBIANA M, QUARTESAN R, BORGHESAN F, NOVOA B, FIGUERAS A, ROCH P (2010), Expression of *Mytilus* immune genes in response to experimental challenges varied according to the site of collection. *Fish & Shellfish Immunol*, **28**, 640–648.

LI L, QIU L, SONG L, SONG X, ZHAO J, WANG L, MU C, ZHANG H (2009), First molluscan TNFR homologue in Zhikong scallop: molecular characterization and expression analysis. *Fish Shellfish Immunol*, **27**, 625–632.

LÓPEZ C, CARBALLAL M J, AZEVEDO C, VILLALBA A (1997a), Enzyme characterization of the circulating haemocytes of the carpet shell clam, *Ruditapes decussatus* (Mollusca: bivalvia). *Fish Shellfish Immunol*, **7**, 595–608.

LÓPEZ C, CARBALLAL M J, AZEVEDO C, VILLALBA A (1997b), Morphological characterization of the hemocytes of the clam, *Ruditapes decussatus* (Mollusca: Bivalvia). *J Invertebr Pathol*, **69**, 51–57.

LUEDEKING A, KOEHLER A (2004), Regulation of expression of multixenobiotic resistance (MXR) genes by environmental factors in the blue mussel *Mytilus edulis*. *Aquat Toxicol*, **69**, 1–10.

LUNA-ACOSTA A, BUSTAMANTE P, GODEFROY J, FRUITIER-ARNAUDIN I, THOMAS-GUYON H (2010), Seasonal variation of pollution biomarkers to assess the impact on the health status of juvenile Pacific oysters *Crassostrea gigas* exposed in situ. *Environ Sci Pollut Res Int*, **17**, 999–1008.

MARCHI B, BURLANDO B, MOORE M N, VIARENGO A (2004), Mercury-and copper-induced lysosomal membrane destabilisation depends on $[Ca^{2+}]_i$ dependent phospholipase A2 activation. *Aquat Toxicol*, **66**, 197–204.

MCLEAN R I, ABBE G R (2006), Comparison of *Crassostrea ariakensis* and *C. virginica* in the discharge area of a nuclear power plant in central Chesapeake Bay. *J Shellfish Res*, **25**, 754.

MEYERS J A, BURRESON E M, BARBER B J, MANN R (1991), Susceptibility of diploid and triploid pacific oysters *Crassostrea gigas* (Thunberg, 1793) and eastern oysters, *Crassostrea virginica* (Gmelin, 1791) to *Perkinsus marinus*. *J Shellfish Res*, **10**, 433–437.

MITTA G, VANDENBULCKE F, HUBERT F, SALZET M, ROCH P (2000a), Involvement of mytilins in mussel antimicrobial defence. *J Biol Chem*, **275**, 12954–12962.

MITTA G, HUBERT F, DYRYNDA E A, BOUDRY P, ROCH P (2000b), Mytilin B and MGD2, two antimicrobial peptides of marine mussels: gene structure and expression analysis. *Dev Comp Immunol*, **24**, 381–393.

MONTAGNANI C, LE ROUX F, BERTHE F, ESCOUBAS J M (2001), Cg-TIMP, an inducible tissue inhibitor of metalloproteinase from the Pacific oyster *Crassostrea gigas* with a potential role in wound healing and defense mechanisms. *FEBS Lett*, **500**, 64–70.

MORTENSEN S H, GLETTE J (1996), Phagocytic activity of scallop (*Pecten maximus*) haemocytes maintained *in vitro*. *Fish Shellfish Immnunol*, **6**, 111–121.

MUÑOZ P, MESEGUER J, ESTEBAN M A (2006), Phenoloxidase activity in three commercial bivalve species. Changes due to natural infestation with *Perkinsus atlanticus*. *Fish Shellfish Immunol*, **20**, 12–19.

NAKAYAMA K, MARUYAMA T (1998), Differential production of active oxygen species in photo-symbiotic and non-symbiotic bivalves. *Dev Comp Immunol*, **22**, 151–159.

NATHAN C, XIE Q W (1994), Regulation of biosynthesis of nitric oxide. *J Biol Chem*, **269**, 13725–13728.

NI D, SONG L, WU L, CHANG Y, YU Y, QIU L, WANG L (2007), Molecular cloning and mRNA expression of peptidoglycan recognition protein (PGRP) gene in bay scallop (*Argopecten irradians*, Lamarck 1819). *Dev Comp Immunol*, **31**, 548–558.

NILSEN I W, MYRNES B (2001), The gene of chlamysin, a marine invertebrate-type lysozyme, is organized similar to vertebrate but different from invertebrate chicken-type lysozyme genes. *Gene*, **269**, 27–32.

NONAKA M, KIMURA A (2006), Genomic view of the evolution of the complement system. *Immunogenetics*, **58**, 701–713.

NONAKA M, YOSHIZAKI F (2004), Primitive complement system of invertebrates. *Immunol Rev*, **198**, 203–215.

NOVAS A, CAO A, BARCIA R, RAMOS-MARTÍNEZ J I (2004), Nitric oxide release by hemocytes of the mussel *Mytilus galloprovincialis* Lmk was provoked by interleukin-2 but not by lipopolysaccharide. *Int J Biochem Cell Biol*, **36**, 390–394.

OLIVER J L, LEWIS T D, FAISAL M, KAATTARI S L (1999), Analysis of the effects of *Perkinsus marinus* proteases on plasma proteins of the eastern oyster (*Crassostrea virginica*) and the Pacific oyster (*Crassostrea gigas*). *J Invertebr Pathol*, **74**, 173–183.

ORDÁS M C, NOVOA B, FIGUERAS A (2000), Modulation of the chemiluminescence response of Mediterranean mussel (*Mytilus galloprovincialis*) haemocytes. *Fish Shellfish Immunol*, **10**, 611–622.

ORDÁS M C, NOVOA B, FAISAL M, MCLAUGHLIN S, FIGUERAS A (2001), Proteolytic activity of cultured *Pseudoperkinsus tapetis* extracellular products. *Comp Biochem Physiol B Biochem Mol Biol*, **130**, 199–206.

OTTAVIANI E, PAEMEN L R, CADET P, STEFANO G B (1993), Evidence for nitric oxide production and utilization as a bacteriocidal agent by invertebrate immunocytes. *Eur J Pharmacol Environ Toxicol Pharmacol Sec*, **248**, 319–324.

PADHI A, VERGHESE B (2008), Molecular diversity and evolution of myticin-C antimicrobial peptide variants in the Mediterranean mussel, *Mytilus galloprovincialis*. *Peptides*, **29**, 1094–1101.

PALLAVICINI A, COSTA M M, GESTAL C, DREOS R, FIGUERAS A, VENIER P, NOVOA B (2008), High sequence variability of myticin transcripts in hemocytes of immune-stimulated mussels suggests ancient host–pathogen interactions. *Dev Comp Immunol*, **32**, 213–226.

PARISI, M G, LI H, TOUBIANA M, PARRINELLO N, CAMMARATA M, ROCH P (2009), Polymorphism of mytilin B mRNA is not translated into mature peptide. *Mol Immunol*, **46**, 384–392.

PARK E M, KIM Y O, NAM B H, KONG H J, KIM W J, LEE S J, KONG I S, CHOI T J (2008), Cloning, characterization and expression analysis of the gene for a putative lipopolysaccharide-induced TNF-alpha factor of the Pacific oyster, *Crassostrea gigas*. *Fish Shellfish Immunol*, **24**, 11–17.

PATRZYKAT A, DOUGLAS S E (2005), Antimicrobial peptides: cooperative approaches to protection. *Protein Peptide Lett*, **12**, 19–25.

PAYNTER K T, GOODWIN J D, CHEN M E, WARD N J, SHERMAN M W, MERITT D W, ALLEN S K (2008), *Crassostrea ariakennsis* in Chesapeake Bay: growth, disease and mortality in shallow subtidal environments. *J Shellfish Res*, **27**, 509–515.

PEATMAN E J, WEI X, FENG J, LIU L, KUCUKTAS H, LI P, HE C, ROUSE D, WALLACE R, DUNHAM R, LIU Z (2004), Development of expressed sequence tags from eastern oyster (*Crassostrea virginica*): lessons learned from previous efforts. *Mar Biotechnol*, **6**, S491–496.

PIANO A, VALBONESI P, FABBRI E (2004), Expression of cytoprotective proteins, heat shock protein 70 and metallothioneins, in tissues of *Ostrea edulis* exposed to heat and heavy metals. *Cell Stress Chaperones*, **9**, 134–142.

PIANO A, FRANZELLITTI S, TINTI F, FABBRI E (2005), Sequencing and expression pattern of inducible heat shock gene products in the European flat oyster, *Ostrea edulis*. *Gene*, **361**, 119–126.

PIPE R K (1992), Generation of reactive oxygen metabolites by the haemocytes of the mussel *Mytilus edulis*. *Dev Comp Immunol*, **16**, 111–122.

PIPE R K, FARLEY S R, COLES J A (1997), The separation and characterisation of haemocytes from the mussel *Mytilus edulis*. *Cell Tissue Res*, **289**, 537–545.

PRADO-ALVAREZ M, GESTAL C, NOVOA B, FIGUERAS A (2009a), Differentially expressed genes of the carpet shell clam *Ruditapes decussatus* against *Perkinsus olseni*. *Fish Shellfish Immunol*, **26**, 72–83.

PRADO-ALVAREZ, M, ROTLLANT J, GESTAL C, NOVOA B, FIGUERAS A (2009b), Characterization of a C3 and a factor B-like in the carpet-shell clam, *Ruditapes decussatus*. *Fish Shellfish Immunol*, **26**, 305–315.

PUANGLARP N, OXLEY D, CURRIE G J, BACIC A, CRAIK D J, YELLOWLEES D (1995), Structure of the N-linked oligosaccharides from tridacnin, a lectin found in the haemolymph of the giant clam *Hippopus hippopus*. *Eur J Biochem*, **232**, 873–880.

QIU L, SONG L, XU W, NI D, YU Y (2007), Molecular cloning and expression of a Toll receptor gene homologue from Zhikong scallop, *Chlamys farreri*. *Fish Shellfish Immunol*, **22**, 451–466.

QIU L, SONG L, YU Y, ZHAO J, WANG L, ZHANG Q (2009), Identification and expression of TRAF6 (TNF receptor-associated factor 6) gene in Zhikong scallop *Chlamys farreri*. *Fish Shellfish Immunol*, **26**, 359–367.

RAGONE-CALVO L M, CALVO G W, BURRESON E M (2003), Dual disease resistance in a selectively bred eastern oyster, *Crassostrea virginica*, strain tested in Chesapeake Bay. *Aquaculture*, **220**, 69–87.

RATCLIFFE N A, ROWLEY A F, FITZGERALD S W, RHODES C P (1985), Invertebrate immunity: basic concepts and recent advances. *Int Rev Cytol*, **97**, 183–350.

RATHINAM A V, CHEN T T, GROSSFELD R M (2000), Cloning and sequence analysis of a cDNA for an inducible 70 kDa heat shock protein (Hsp70) of the American oyster (*Crassostrea virginica*). *DNA Seq*, **11**, 261–264.

ROMESTAND B, CORBIER F, ROCH P (2002), Protease inhibitors and haemagglutinins associated with resistance to the protozoan parasite, *Perkinsus marinus*, in the Pacific oyster, *Crassostrea gigas*. *Parasitology*, **125**, 323–329.

SAAVEDRA C, BACHÈRE E (2006), Bivalve genomics. *Aquaculture*, **256**, 1–14.

SAUL S J, BIN L, SUGUMARAN M (1987), The majority of prophenoloxidase in the hemolymph of *Manduca sexta* is present in the plasma and not in the hemocytes. *Dev Comp Immunol*, **11**, 479–485.

SAUVAGE C, BOUDRY P, DE KONING, D J, HALEY CS, HEURTEBISE S, LAPÈGUE S (2010), QTL for resistance to summer mortality and OsHV-1 load in the Pacific oyster (*Crassostrea gigas*). *Anim Genet*, **41**, 390–399.

SEGARRA A, PEPIN J F, ARZUL I, MORGA B, FAURY N, RENAULT T (2010), Detection and description of a particular Ostreid herpesvirus 1 genotype associated with massive mortality outbreaks of Pacific oysters, *Crassostrea gigas*, in France in 2008. *Virus Res*, **153**, 92–100.

SÖDERHALL K, HALL L (1984), Lipopolysaccharide-induced activation of prophenoloxidase activating system in crayfish haemocyte lysate. *Biochem Biophys Acta*, **797**, 99–104.

SOKOLOVA I M, OLIVER J D, LEAMY L J (2006), An AFLP approach to identify genetic markers associated with resistance to *Vibrio vulnificus* and *Perkinsus marinus* in eastern oysters. *J Shellfish Res*, **25**, 95–100.

SONG L, XU W, LI C, LI H, WU L, XIANG J, GUO X (2006), Development of expressed sequence tags from the bay scallop *Argopecten irradians irradians*. *Mar Biotechnol*, **8**, 161–169.

SU J, SONG L, XUW, WU L, LI H, XIANG J (2004), cDNA cloning and mRNA expression of the lipopolysaccharide- and beta-1,3-glucan recognition protein gene from scallop, *Chlamys farreri*. *Aquaculture*, **239**, 69–80.

SU J, NI D, SONG L, ZHAO J, QIU L (2007), Molecular cloning and characterization of a short type peptidoglycan recognition protein (CfPGRP-S1) cDNA from Zhikong scallop *Chlamys farreri*. *Fish Shellfish Immunol*, **23**, 646–656.

TAFALLA C, NOVOA B, FIGUERAS A (2002), Production of nitric oxide by mussel (*Mytilus galloprovincialis*) hemocytes and effect of exogenous nitric oxide on phagocytic functions. *Comp Biochem Physiol B Biochem Mol Biol*, **132**, 423–431.

TAFALLA C, GÓMEZ-LEÓN J, NOVOA B, FIGUERAS A (2003), Nitric oxide production by carpet shell clam (*Ruditapes decussatus*) hemocytes. *Dev Comp Immunol*, **27**, 197–205.

TANGUY A, GUO X, FORD S E (2004), Discovery of genes expressed in response to *Perkinsus marinus* challenge in eastern (*Crassostrea virginica*) and Pacific (*C. gigas*) oysters. *Gene*, **338**, 121–131.

TASUMI S, VASTA G R (2007), A galectin of unique domain organization from hemocytes of the eastern oyster (*Crassostrea virginica*) is a receptor for the protistan parasite *Perkinsus marinus*. *J Immunol*, **179**, 3086–3098.

TATSUMI M, ARAI Y, ITOH T (1982), Purification and characterization of a lectin from the shellfish, *Saxidomus purpuratus*. *J Biochem*, **91**, 1139–1146.

THOMAS-GUYON H, GAGNAIRE B, BADO-NILLES A, BOUILLY K, LAPÈGUE S, RENAULT T (2009), Detection of phenoloxidase activity in early stages of the Pacific oyster *Crassostrea gigas* (Thunberg). *Dev Comp Immunol*, **33**, 653–659.

TORREILLES J, ROMESTAND B (2001), *In vitro* production of peroxynitrite by haemocytes from marine bivalves: C-ELISA determination of 3-nitrotyrosine level in plasma proteins from *Mytilus galloprovincialis* and *Crassostrea gigas*. *BMC Immunol*, **2**, 1.

TORREILLES J, GUERIN M C, ROCH P (1996), Espèces oxygénées réactives et systèmes de défense des bivalves marins. *CR Acad Sci Paris*, **319**, 209–218.

TRIPP M R (1992), Phagocytosis by hemocytes of the hard clam, *Mercenaria mercenaria*. *J Invertebr Pathol*, **59**, 222–227.

TUNKIJJANUKIJ S, MIKKELSEN H V, OLAFSEN J A (1997), A heterogeneous sialic acid-binding lectin with affinity for bacterial LPS from horse mussel (*Modiolus modiolus*) hemolymph. *Comp Biochem Physiol*, **117B**, 273–286.

UEDA N, BOETTCHER A (2009), Differences in heat shock protein 70 expression during larval and early spat development in the eastern oyster, *Crassostrea virginica* (Gmelin, 1791). *Cell Stress Chaperones*, **14**, 439–443.

VAN BANNING P (1988), Management strategies to control diseases in the Dutch culture of edible oysters. In: W S Fisher, Editor, *Disease Processes in Marine Bivalve Molluscs. Am Fish Soc, Spec Publ*, **18**, 243–245.

VAN BANNING P (1991), Observations on bonamiosis in the stock of the European flat oyster, *Ostrea edulis*, in the Netherlands, with special reference to the recent developments in Lake Grevelingen. *Aquaculture*, **93**, 205–211.

VASTA G R, SULLIVAN J T, CHENG T C, MARCHALONIS J J, WARR G W (1982), A cell membrane associated lectin of the oyster hemocyte. *J Invertebr Pathol*, **40**, 367–377.

VASTA G R, CHENG T C, MARCHALONIS J J (1984), A lectin on the hemocyte membrane of the oyster (*Crassotrea virginica*). *Cell Immunol*, **88**, 475–502.

VENIER P, PALLAVICINI A, DE NARDI B, LANFRANCHI G (2003), Towards a catalogue of genes transcribed in multiple tissues of *Mytilus galloprovincialis*. *Gene*, **314**, 29–40.

VENIER P, DE PITTÀ C, PALLAVICINI A, MARSANO F, VAROTTO L, ROMUALDI C, DONDERO F, VIARENGO A, LANFRANCHI G (2006), Development of mussel mRNA profiling: can gene expression trends reveal coastal water pollution? *Mutat Res*, **602**, 121–134.

VIARENGO A (1989), Heavy metals in marine invertebrates: mechanisms of regulation and toxicity at the cellular level. *CRC Crit Rev Aquat Sci*, **1**, 295–317.

VILLAMIL L, GÓMEZ-LEÓN J, GÓMEZ-CHIARRI M (2007), Role of nitric oxide in the defenses of *Crassostrea virginica* to experimental infection with the protozoan parasite *Perkinsus marinus*. *Dev Comp Immunol*, **31**, 968–977.

WANG B, ZHAO J, SONG L, ZHANG H, WANG L, LI C, ZHENG P, ZHU L, QIU L, XING K (2008), Molecular cloning and expression of a novel Kazal-type serine proteinase inhibitor gene from Zhikong scallop *Chlamys farreri*, and the inhibitory activity of its recombinant domain. *Fish Shellfish Immunol*, **24**, 629–637.

WANG H, SONG L, LI C, ZHAO J, ZHANG H, NI D, XU W (2007), Cloning and characterization of a novel C-type lectin from Zhikong scallop *Chlamys farreri*. *Mol Immunol*, **44**, 722–731.

WINSTON G W, MOORE M N, KIRCHIN M A, SOVERCHIA P (1996), Production of reactive oxygen species by hemocytes from the marine mussel, *Mytilus edulis*: lysosomal localization and effect of xenobiotics. *Comp Biochem Physiol*, **113C**, 221–229.

XUE Q, RENAULT T, COCHENNEC N, GERARD A (2000), Separation of European flat oyster, *Ostrea edulis*, haemocytes by density gradient centrifugation and SDS-PAGE characterisation of separated haemocyte sub-populations. *Fish Shellfish Immunol*, **10**, 155–165.

XUE Q, RENAULT T, CHILMONCZYK S (2001), Flow cytometric assessment of haemocyte sub-populations in the European flat oyster, *Ostrea edulis*, haemolymph. *Fish Shellfish Immunol*, **11**, 557–567.

XUE Q G, WALDROP G L, SCHEY K L, ITOH N, OGAWA M, COOPER R K, LOSSO J N, LA PEYRE J F (2006), A novel slow-tight binding serine protease inhibitor from eastern oyster (*Crassostrea virginica*) plasma inhibits perkinsin, the major extracellular protease of the oyster protozoan parasite *Perkinsus marinus*. *Comp Biochem Physiol B Biochem Mol Biol*, **145**, 16–26.

XUE Q, ITOH N, SCHEY K L, COOPER R K, LA PEYRE J F (2009), Evidence indicating the existence of a novel family of serine protease inhibitors that may be involved in marine invertebrate immunity. *Fish Shellfish Immunol*, **27**, 250–259.

YAMAURA K, TAKAHASHI K G, SUZUKI T (2008), Identification and tissue expression analysis of C-type lectin and galectin in the Pacific oyster, *Crassostrea gigas*. *Comp Biochem Physiol Part B: Biochem Mol Biol*, **149**, 168–175.

YU Y, QIU L, SONG L, ZHAO J, NI D, ZHANGM Y, XU W (2007), Molecular cloning and characterization of a putative lipopolysaccharide-induced TNF-alpha factor (LITAF) gene homologue from Zhikong scallop *Chlamys farreri*. *Fish Shellfish Immunol*, **23**, 419–429.

ZASLOFF M (2002), Antimicrobial peptides of multicellular organisms. *Nature*, **415**, 389–395.

ZHANG D, JIANG J, JIANG S, MA J, SU T, QIU L, ZHU C, XU X (2009), Molecular characterization and expression analysis of a putative LPS-induced TNF-alpha factor (LITAF) from pearl oyster *Pinctada fucata*. *Fish Shellfish Immunol*, **27**, 391–396.

ZHANG H, SONG L, LI C, ZHAO J, WANG H, QIU L, NI D, ZHANG Y (2008), A novel C1q-domain-containing protein from Zhikong scallop *Chlamys farreri* with lipopolysacharide binding activity. *Fish Shellfish Immunol*, **25**, 281–289.

ZHANG H, WANG H, WANG L, SONG L, SONG X, ZHAO J, LI L, QIU L (2009), Cflec-4, a multidomain C-type lectin involved in immune defense of Zhikong scallop *Chlamys farreri*. *Dev Comp Immunol*, **33**, 780–788.

ZHENG P, WANG H, ZHAO J, SONG L, QIU L, DONG C, WANG B, GAI Y, MU C, LI C, NI D, XINGM K (2008), A lectin (CfLec-2) aggregating *Staphylococcus haemolyticus* from scallop *Chlamys farreri*. *Fish Shellfish Immunol*, **24**, 286–293.

ZHU L, SONG L, CHANG Y, XU W, WU L (2006), Molecular cloning, characterization and expression of a novel serine proteinase inhibitor gene in bay scallops (*Argopecten irradians*, Lamarck 1819). *Fish Shellfish Immunol*, **20**, 320–331.

ZHU L, SONG L, XU W, QIAN P Y (2009), Identification of a C-type lectin from the bay scallop *Argopecten irradians*. *Mol Biol Rep*, **36**, 1167–1173.

4

Stress and resistance to infectious diseases in fish

J. F. Turnbull, University of Stirling, UK

Abstract: It has been long recognized that stress increases disease susceptibility. Therefore disease is not just the presence of pathogens but results from a combination of factors. As a result we have to be wary of concentrating solely on the role of pathogens in the aetiology of infectious diseases. Since the 1980s a great deal of work has been conducted on the relationship between stress and disease susceptibility, which has resulted in a much improved understanding of stress and disease. This includes the nature and evolutionary context of the immune and the stress responses. Despite a more profound understanding of the internal systems we must still be aware of the direct and indirect effects of stressors in order to understand and control diseases in fish. Although individual variability and population level effects make it difficult to predict the outcome in specific circumstances, general principles have important implications for aquaculture health management.

Key words: fish, stress, disease susceptibility, immunity.

4.1 Introduction: disease is not just the presence of pathogens

The idea that disease is more complex than the simple combination of host and pathogen is not new. More than 50 years ago Dubos (1955) eloquently explained that diseases require more than the presence of a pathogen. Since Sniezko (1973) most fish health textbooks have explained that hosts, environment and pathogens all play a role in fish diseases. However, the implications of this simple concept are not always understood or applied in the control of fish diseases. Although Sniezko's Venn diagram is useful it is very simplistic. Similarly Koch's postulates, while a useful tool, are too focused on the pathogen and far too simplistic for many diseases. They do not take account of more subtle aspects of causality, such as dose, duration, non-agent factors (e.g. age, behaviour), carrier states, multiple agents resulting in a single condition or a single agent resulting in multiple clinical presentations. As a result Koch's postulates have largely been superseded by other criteria for determining causality such as Bradford-Hill (Hill, 1965) but these alternatives are not yet widely applied in the study or control of fish diseases.

Even these brief comments on causality clearly indicate that one agent one disease is far too simplistic and pathogen identification is definitely not the same as diagnosis. However, there are many examples of people still proposing techniques such as polymerase chain reaction (PCR) for initial diagnosis. Although such pathogen identification techniques have a role in research and surveillance (and other areas), they very rarely, if ever, confirm the infectious cause of a disease and never provide any information on the rest of the causal web.

An understanding of the complexity of disease is not only important for identifying causes but also controlling diseases. Disease control is often dependent on reducing or eliminating stressors, which requires an appreciation of the nature of the interaction between stressors and disease susceptibility.

Most experienced fish keepers have encountered the relationship between stress and disease first hand. Figure 4.1 shows mortalities in a population of fish following transportation (unpublished data). The fish, which came from an apparently healthy population, suffered a temperature shock during transportation. There were immediate losses, probably as a result of fish failing to cope with the acute thermal shock; however, on day 7 mortalities started again and followed a typical infectious epidemic curve

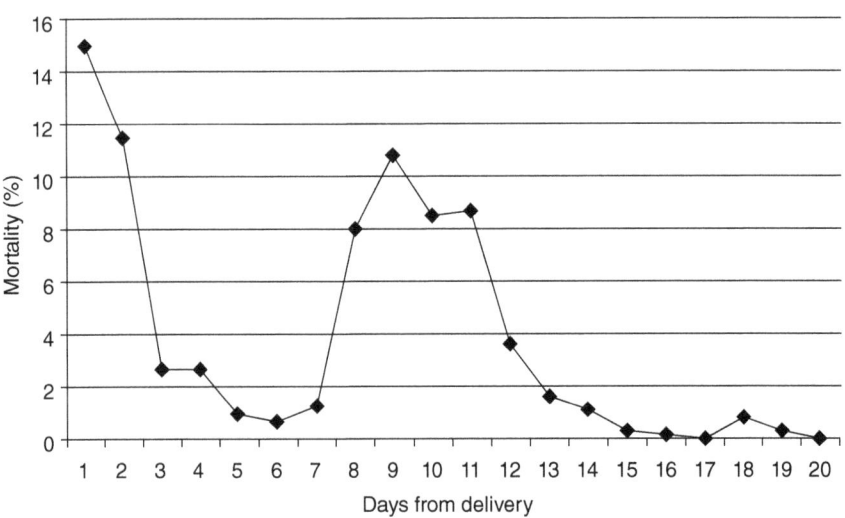

Fig. 4.1 Mortalities expressed as a percentage of the remaining population following a stressful transportation and delivery on day 0.

but no specific pathogens were recovered. All of the clinical evidence supported the hypothesis of an outbreak of an opportunist pathogen in the population as a result of stress-related immuno-suppression. This pattern is predictable and in our institute if stock-fish suffer any significant mortalities following transportation we consider antimicrobial therapy, following standard veterinary criteria (Mateus *et al.*, 2011).

There are many research publications that describe reduced immune response and increased disease susceptibility associated with severe or chronic stress responses and therefore it is clear that in general diseases are exacerbated by stress. Many papers have reported specific effects of stressors on aspects of the immune system or susceptibility to specific diseases (Walters & Plumb, 1980; Pickering & Duston, 1983; Angelidis *et al.*, 1987; Houghton & Matthews, 1990; Johnson & Albright, 1992; Wise *et al.*, 1993; Bilodeau *et al.*, 2003). The general area has also been the subject of several reviews (e.g. Wedemeyer, 1970; Ellis, 1981; Wedemeyer *et al.*, 1981; Wood *et al.*, 1983; Rottmann *et al.*, 1992; Schreck *et al.*, 1993; Iwama *et al.*, 1997; Pickering, 1997).

The increased disease susceptibility associated with the stress response is not solely mediated through the immune system. Cortisol also impairs healing of epithelium (Roubal & Bullock, 1988; Johnson & Albright, 1992) and the stress response can lead to changes in behaviour. The nature of the association between stressors and disease susceptibility is complex even in individual fish and extrapolation from observed changes in immune response to disease susceptibility can be difficult. In the past there has been much unexplained variation among populations and individuals (e.g. Walters & Plumb, 1980) and only recently have we started to understand the mechanisms underlying individual and population variability (Blanford *et al.*, 2003; Lafferty & Holt, 2003; MacKenzie *et al.*, 2009).

4.1.1 Advances in understanding of stress and disease in fish

For a dramatic demonstration of the advances in knowledge over the last 30 years readers should refer to Ellis (1981) and Korte *et al.* (2005). The paper by Ellis dealt specifically with fish but was forced to draw on literature from other vertebrates; the paper by Korte *et al.*, although referring to some examples in fish, is primarily concerned with birds and mammals. The first paper by the late Tony Ellis was a seminal work and much cited in aquatic animal health literature. However, many of the mechanisms involved were poorly understood at that time and Ellis was forced to describe many of the processes as complex but due to lack of information was unable to explain them. Korte *et al.* (2005) dealt with many of the same mechanisms but with a wealth of genetic, biochemical and conceptual information was able to describe and explain many of the complexities.

Although a large body of research has greatly improved our understanding of the interdependent nature of many systems in the body of mammals, similar studies in fish are still in their early stages. It is, however, apparent that in fish, as well as mammals, the neuroendocrine system, including the perception of external events, the hypothalamic-pituitary-adrenal (HPA) stress response and the immune system are all mutually regulatory. Both inflammatory stimuli and challenges to the immune system elicit a HPA stress response (Shanks & Lightman, 2001). In artificial lipopolysaccharide challenges MacKenzie et al. (2006) reported that cortisol did not simply antagonise the response of macrophages but appeared to induce a complex series of up- and down-regulation in a wide range of genes. The response to a stressor is not just a physical phenomenon; the predictability of an event affects the perceived level of threat and as a result the magnitude of the stress response. There are many examples of this, for example sudden changes in light levels can cause a panic and a stress response in fish but if switching the lights occurs at a regular time of day and is followed by feeding it can become a positive stimulus (Stien et al., 2007). In humans and other animals central nervous, immune and stress interactions have been studied for more than three decades. It was recognized in 1975 that an antibody response to antigen and the immunosuppressive effects of a drug can be classically conditioned (Ader & Cohen, 1975).

The lack of information relevant to fish experienced by Ellis in the early 1980s is still a problem today. Many of the more profound studies have been conducted on terrestrial vertebrates. Although we are still forced to extrapolate from birds, mammals and others to fish at least we have a better understanding of the evolutionary and functional context of that extrapolation.

4.2   Fish immune and stress responses

4.2.1   Immune response

We now understand in more detail that fish immunity is at the 'crossroads' between reliance of the innate and adaptive immunity (Tort et al., 2003; Huntingford et al., 2006). Some fish have the capacity to produce specific antibodies but the range of antibodies produced is far less diverse when compared with mammals or birds (Press, 1998). One relatively common misconception among non-specialists is that this reliance on the innate immunity is in some way less effective than a reliance on specific antibodies. The prevalence and diversity of invertebrates (98% of animal species) and fish (more than half of all vertebrates), in addition to their persistence through evolutionary history (Coelacanth relatively unchanged for 400 million years; Johanson et al., 2006), demonstrate beyond any reasonable doubt that the innate immune system is very effective.

Although the neural endocrine response to stress in fish is similar to that in mammals and is mediated through the HPA (or in fish hypothalamic-pituitary-inter-renal) axis, we now know that some aspects of the immune system are not as easily interpreted in fish when compared with mammals. For example, although fish have an antibody response, estimating titres using enzyme-linked immunosorbent assay (ELISA), agglutination or precipitation does not necessarily produce meaningful results. Carp immunized against *Ichthyophthirius multifilis* were more susceptible to the parasite after injection of corticosteroid even though their serum antibody levels remained high (Houghton & Matthews, 1990). If a particular redox state is required for antibody function, absolute titre may be very poorly correlated with activity (Tort *et al.*, 2003), with high titres in the wrong form being relatively ineffective and low titres in the correct form being highly effective.

Lysozyme has been a component of many studies on the immune system of fish. It is present in all fish species studied, and although variable among tissues and species it is influenced by stress (Grinde *et al.*, 1988). It would also appear that acute stress causes lysozyme to be up-regulated and that chronic stress results in down-regulation. Although lysozyme may be correlated with disease susceptibility, it is not a major component of the fish defensive mechanisms and therefore it is difficult to interpret the functional significance of these observations (Möck & Peters, 1990).

Many of the observable changes in circulating leucocytes associated with the stress response may be due to movement or redistribution of cells between tissues rather than proliferation or activation (Felten *et al.*, 1987; Murray *et al.*, 1993). The immune capacity of a fish is not necessarily a simple reflection of circulating leucocyte numbers. Even lymphopenia may be associated with a functional increased capacity to fight off infections as a result of cell redistribution (Dhabhar & McEwen, 1996), whereas if lymphocytes are reduced in number and also do not have the capacity to activate macrophages then lymphopenia may indicate a serious immunosuppression (Ellsaesser & Clem, 1986). This cell trafficking is the result of several hormones. Pickering (1997) reported 14 hormones thought to be affected by stress.

4.2.2 Stress response
Reproductive success drives evolution, not the immediate health or welfare of the individual (Nesse, 1999, 2000, 2001). The stress response has to be viewed not just in relation to the individual but also in the context of benefits to the population and the species. The stress response may invoke physiological and behavioural changes that benefit long-term reproductive success but result in severe adverse consequences for the individual. This is most dramatically illustrated in some Pacific salmon, which individually suffer exhaustion, starvation and eventually death in order to reproduce as a species (McEwen & Lasley, 2002).

The stress response was previously viewed as a means to maintain homeostais; however, the concept of homeostasis has largely been superseded by the concept of allostasis. Allostasis implies actively maintaining stability through change (Sterling & Eyer, 1988). This is a refinement and improvement on the concepts of homeostasis and the general adaptive syndrome. In many circumstances the stress response is adaptive, providing resources to deal with the stressor and its immediate consequence and at the same time the stress response has maladaptive consequences.

It may benefit the animal to have a link between acute stress and immunity providing a brief immuno-stimulation to cope with injury or infectious challenge (Peters, 1991; Demers & Bayne, 1997; Ruis & Bayne, 1997; Weyts *et al.*, 1998). The stress response itself is part of the mechanism to fight infection with a peak of cortisol occurring at time of maximum bacterial clearance following experimental challenge with bacteria (Bilodeau *et al.*, 2003). However, if the stress is repeated, persistent or mismanaged then it will result in an allostatic load or cost leading to immuno-suppression (Peters & Schwarzer, 1985; Maule *et al.*, 1989; Schreck, 1996; McEwen, 2003; McEwen & Wingfield, 2003). The initial immune enhancement may be a noradrenaline response rather than a cortisol response; however, noradrenaline release may be suppressed by cortisol. This has led some authors to describe the seeds of immuno-supression being present in the initial immuno-stimulation associated with acute stress (Narnaware & Baker, 1996).

4.2.3   Direct and indirect effects of stressors

Stressors not only induce a stress response through HPA activation and elevated cortisol but can also cause direct injury or damage, resulting in increased disease susceptibility. Many environmental contaminants even at low levels of exposure have been linked to reduction in many aspects of the defence mechanisms (Khan & Thulin, 1991; Weeks *et al.*, 1992; Arkoosh *et al.*, 1998 cited in MacKinnon, 1998). For example, physical or chemical challenges are stressful but also damage the epithelium of skin or gills reducing their efficacy as a barrier and adversely affecting ionic regulation (e.g. Post, 1987; Klontz, 1993 cited in Conte, 2004).

Fish are commonly challenged by temperature extremes or rapid fluctuations; these not only result in a stress response but have a profound effect on most metabolic processes (Ndong *et al.*, 2007).

Several authors have demonstrated that the relationship between temperature and immunity or disease resistance is complex. It is not simply a case of thermal (high or low) challenge reducing immuno-competence. Carlson *et al.* (1995) demonstrated that temperatures higher or lower than optimal can have a negative effect on aspects of the immune system but the effects are not necessarily the same for temperatures above or below the optimal, suggesting some non-linear or complex process.

Results produced by Aranishi *et al.* (1998), although somewhat contradictory, demonstrated that exposure to high temperature and external bacterial challenge in the Japanese eel (*A. japonica*) led to changes in skin protective bacteriolytic enzymes. However, there was no effect on their response to internal bacterial challenge nor was the effect observed with a low temperature challenge. Even the effect of increased temperature on a single aspect of the immune system is not necessarily straightforward. Nikoskelainen *et al.* (2004 cited in Ndong *et al.*, 2007) reported increased respiratory burst with increasing temperature in rainbow trout (*Oncorhynchus mykiss*), whereas a study of hybrid striped bass demonstrated higher or lower temperature reduced respiratory burst (Carlson *et al.*, 1995).

It would appear that fish have more tolerance to lower than higher temperature; however, below certain temperatures many immune functions will be reduced (Bly & Clem, 1992; Vasta & Lambris, 2002 cited in Tort *et al.*, 2003). In addition thermal shock may often be combined with other challenges such as maturation or migration. Under these circumstances the fish may suffer from the effects of temperature compounded by the action of sex steroids or pre-existing cortisol elevation and a stress response due to the temperature challenge (Magnadóttir *et al.*, 1999).

Interactions with conspecifics is another example of a complex challenge. Aggressive competition for space, food, position in the hierarchy or other resources has profound behavioural and neuro-endocrine consequences for winners and losers. The associated stress may in turn influence behaviour in addition to reducing immuno-competence (Peters & Schwarzer, 1985; Cooper *et al.*, 1989; Peters, 1991). Behavioural responses to stressful circumstances can be very diverse and dependent on the nature of the circumstances and behavioural strategy of the animal (see Section 4.3.1). Such responses include hiding, shoaling, change of colour, changes in posture and swimming patterns, change in feeding patterns (Huntingford *et al.*, 2006). The behavioural response can in turn affect disease susceptibility through physical damage, changes in uptake of chemotherapeutants, routes or rates of pathogen transmission.

4.3   Individual variability and population level effects

4.3.1   Individual variability

Although a chronic stress response or allostatic load will result in immuno-suppression, not all individuals respond in a similar manner. Several studies have indicated genetic variation in disease resistance in Atlantic salmon (*Salmo salar*) (MacKinnon, 1998; Røed *et al.*, 2002). However, differences in susceptibility to disease and cortisol responses between strains of fish are poorly understood (Bilodeau *et al.*, 2003). Rainbow trout (*O. mykiss*) (Fevolden & Røed, 1993) and Atlantic salmon (Fevolden *et al.*, 1993) have been selected for high or low cortisol response. While it is not clear whether the fish have reduced sensitivity, reduced production or increased removal of cortisol, the high-stress response fish had lower immuno-competence.

However, the association between the cortisol response and disease susceptibility is far from clear.

Not all of that variability in response to a stressor is explained by genotype, life stage or circumstances. The magnitude and nature of the response are also partly determined by behavioural strategy. These have also been referred to as behavioural syndromes, personalities or temperaments.

Evolution tends to sustain a range of behavioural strategies rather than drift towards homogeneity (Smith, 1982). The extremes on the spectrum of behavioural strategies have been referred to as pro-active/reactive, bold/shy, hawks/doves and others. These terms reflect some aspects of these varying strategies. For example, in some species hawks would tend to respond with a fight or flight response, take risks, be bold and aggressive, lack behavioural flexibility and explore new environments rapidly and superficially. Doves tend to a freeze and hide response, avoid risks, be cautious and non-aggressive, behaviourally flexible and explore new environments cautiously and thoroughly. Hawks tend to have a lower HPA cortisol response than doves to a given stimulus and tend to suffer less immunosuppression as a result of HPA activation (Korte *et al.*, 2005).

These strategies are not fixed and can be influenced by the experiences of the fish, for example transportation (Ruiz-Gomez *et al.*, 2008). In mammals and birds early life experiences or even the experiences of the parents can also influence the nature of the offspring through epigenetic mechanisms (e.g. Weinstock, 1997; Gluckman *et al.*, 2007; Rutherford *et al.*, 2009). In fish maternal condition does have various effects on offspring (e.g. Schreck *et al.*, 2001; Tierney *et al.*, 2009); however, the few studies that have examined such effects in fish have found no effect on susceptibility to infectious disease (Contreras-Sanchez *et al.*, 1998).

Behavioural strategies are associated with profound differences from the level of gene expression through the metabolome to the neuroendocrine, immune and other systems (MacKenzie et al., 2009). These differences are not only at the level of the neuroedocrine and immune systems but may be mediated through perception since hawks are inherently less responsive to environmental changes. In both hawks and doves an event can be more stressful if it is unpredictable.

4.3.2 Population level effects

Diseases are the expression of host, pathogen and stressor interaction and over the last two decades there has been increasing evidence that diseases play a significant role in population and community dynamics (Norman *et al.*, 1994; Holmes, 1996). The possible interactions are complex and multilayered with effects at the level of genes, cells, individuals, communities, populations and species (Lafferty & Holt, 2003). Changes in stressors in natural or lightly managed populations can range from the disease becoming more prevalent and damaging to the extinction of the pathogen.

Understanding the effects of stress on diseases in populations is also important for disease management in aquaculture.

The impact of pathogens on the host populations is dependent on both the transmission rate, the life history of the pathogen (vectors, secondary hosts, reservoir populations) and the effect on host proliferation or population growth. In general, circumstances that reduce the host population size, or contact rate, or adversely affect pathogens or vectors can reduce the prevalence and impact of diseases. However, the opposing effect of increased host susceptibility makes the outcome difficult to predict (Lafferty & Holt, 2003).

If the impact of the pathogen on the host is close to or just higher than the growth rate of the host population a stable state of chronic infection can occur. Pathogens that are not highly damaging may demonstrate periodic outbreaks with no demonstrable effect on host populations while highly damaging pathogens may drive host population oscillations or in the most extreme case extinctions (Anderson & May, 1981). The disease crayfish plague associated with the Oomycete pseudofungus *Aphanomyces astici* is carried by North American signal crayfish (*Pacifastacus leniusculus*) and was introduced to Europe with them. It is highly damaging to native European crayfish including the white-clawed crayfish (*Austropotamobius pallipes*) and has led to local extinctions and may ultimately lead to extinctions of species (Reynolds, 1988). Stress, depending on its nature and effect, can move pathogen and host populations from one of these states to another.

Although the complexity of the interactions makes outcomes difficult to predict in specific circumstances, Lafferty and Holt (2003) predicted that in general the impact of host-specific diseases would decline with stress while the impact of non-specific infections would increase. In farmed populations non-specific diseases could be defined as diseases where there is an uncontrolled source of infection, either from vectors, wild hosts or other farmed populations.

It is tempting to concentrate on the adverse effects of stressors on hosts; however, adverse conditions, especially the presence of chemicals, can frequently be more damaging for pathogens than hosts. This is the basis of all chemotherapy for the control of infectious diseases. There are also examples outwith chemotherapy. Selenium is more toxic to some cestodes or tapeworms than their fish hosts (Riggs *et al.*, 1987) and there are many other examples of pollutants reducing the survival of free-living stages of pathogens (e.g. Evans, 1982; Siddall & des Clers, 1994). Unpublished data (Shinn, Morris, Turnbull) suggested that pollution from mining affected the local distribution of Gyrodactylids.

Although most of the modelling has been done on wild populations the population level effects discussed above also have implications for aquaculture. In-feed medication and vaccines have a similar effect when controlling infectious diseases in aquaculture, that is, they reduce the number of susceptible individuals in the population. This reduces transmission rate or the capacity for the infection to propagate. Vaccines in aquaculture are not effective just because they protect most individuals. Vaccines work through reducing the transmission rate to the point where an outbreak is unlikely; however, a stressor may reduce the capacity of the fish to respond in the face of an infection. Stress can lead to outbreaks through a relatively small change in the proportion of susceptible individuals. Antibiotics in aquaculture do not treat sick fish. Only actively feeding fish consume medicated, or any other feed and the first sign of ill health in most fish is loss of appetite. Therefore the benefit must be achieved through fish that are not yet sick. Taking a bacterial infection as an example, if the host population is stressed a smaller proportion of the uninfected fish may be feeding and therefore more individuals will remain susceptible, increasing the probability of an outbreak. In addition the stressor may also make the individuals more susceptible to infection.

Farmers and regulators also use population level effects to control disease by separating farm sites physically and other means of reducing transmission. One of these strategies is to apply a very severe form of stressor affecting only the hosts. Slaughtering stock and fallowing allows a farmer, industry or other authority to precipitate the extinction of the pathogen. This is the theory on which eradication schemes are based; remove the potential hosts and drive the pathogen to extinction in a defined geographic area. However, this does not work when the pathogen has access to alternative hosts e.g. wild reservoir populations.

There have been other examples where infected individuals have been removed to reduce the potential for an outbreak. For example in Thailand in the early 1990s tiger shrimp (*P. monodon*) affected with white spot disease were selectively removed from the culture ponds to prevent pond level outbreaks. Unfortunately evidence for the efficacy of this strategy is largely anecdotal.

4.4   Conclusions

The principle that stress makes disease worse has been accepted for a long time and we are beginning to understand the underpinning mechanisms. Although it is still difficult to predict the effect of a stressor on the health of population under specific circumstances we can predict general behaviour. In some cases, despite this weight of evidence, far too much attention is devoted to the pathogen when attempting to control diseases in aquaculture. In theory, the absence of pathogens could reduce the occurrence of disease even if the fish are stressed but in practice there is always something around to take advantage of sick, damaged or stressed fish. Diseases are complex and to effect sustainable control we need to understand that complexity but develop strategies that are sufficiently simple to be implemented and sufficiently robust to deal with real world variability.

4.5   References

ADER, R. & COHEN, N. (1975) Behaviorally conditioned immunosuppression. *Psychosomatic Medicine.* **37**, 333–340.

ANDERSON, R. M. & MAY, R. M. (1981) The population dynamics of microparasites and their invertebrate hosts. *Philosophical Transactions of the Royal Society of London Series B, Biological Sciences.* 451–524.

ANGELIDIS, P., BAUDIN-LAURENCIN, F. & YOUINOU, P. (1987) Stress in rainbow trout, *Salmo gairdneri*: effects upon phagocyte chemiluminescence, circulating leucocytes and susceptibility to *Aeromonas salmonicida*. *Journal of Fish Biology.* **31**, 113–122.

ARANISHI, F., MANO, N., NAKANE, M. & HIROSE, H. (1998) Epidermal response of the Japanese eel to environmental stress. *Fish Physiology and Biochemistry.* **19**, 197–203.

ARKOOSH, M. R., CASILLAS, E., CLEMONS, E., KAGLEY, A. N., OLSON, R., RENO, P. & STEIN, J. E. (1998) Effect of pollution on fish diseases: potential impacts on salmonid populations. *Journal of Aquatic Animal Health.* **10**, 182–190.

BILODEAU, A. L., SMALL, B. C. & WOLTERS, W. R. (2003) Pathogen loads, clearance and plasma cortisol response in channel catfish, *Ictalurus punctatus* (Rafinesque), following challenge with *Edwardsiella ictaluri*. *Journal of Fish Diseases.* **26**, 433–437.

BLANFORD, S., THOMAS, M. B., PUGH, C. & PELL, J. K. (2003) Temperature checks the Red Queen? Resistance and virulence in a fluctuating environment. *Ecology Letters.* **6**, 2–5.

BLY, J. E. & CLEM, L. W. (1992) Temperature and teleost immune functions. *Fish & Shellfish Immunology.* **2**, 159–171.

CARLSON, R. E., BAKER, E. P. & FULLER, R. E. (1995) Immunological assessment of hybrid striped bass at three culture temperatures. *Fish & Shellfish Immunology.* **5**, 359–373.

CONTE, F. S. (2004) Stress and the welfare of cultured fish. *Applied Animal Behaviour Science.* **86**, 205–223.

CONTRERAS-SANCHEZ, W. M., SCHRECK, C. B., FITZPATRICK, M. S. & PEREIRA, C. B. (1998) Effects of stress on the reproductive performance of rainbow trout (*Oncorhynchus mykiss*). *Biology of Reproduction.* **58**, 439–447.

COOPER, E. L., PETERS, G., FAISAL, M., AHMED, I. I. & GHONEUM, M. (1989). Aggression in Tilapia affects immunocompetent leucocytes. *Aggressive Behavior.* **15**, 13–22.

DEMERS, N. E. & BAYNE, C. J. (1997) The immediate effects of stress on hormones and plasma lysozyme in rainbow trout. *Developmental & Comparative Immunology.* **21**, 363–373.

DHABHAR, F. S. & MCEWEN, B. S. (1996) Stress-induced enhancement of antigen-specific cell-mediated immunity. *The Journal of Immunology.* **156**, 2608–2615.

DUBOS, R. J. (1955) Second thoughts on the germ theory. *Scientific American.* **192**, 31–35.

ELLIS, A. E. (1981) Stress and the modulation of defense mechanisms in fish. In: *Stress and Fish.* Pickering, A. D. (Ed). Academic Press. pp147–169.

ELLSAESSER, C. F. & CLEM, L. W. (1986) Haematological and immunological changes in channel catfish stressed by handling and transport. *Journal of Fish Biology.* **28**, 511–521.

EVANS, N. A. (1982) Effects of copper and zinc on the life cycle of *Notocotylus attenuatus* (Digenea: Notocotylidae). *International Journal for Parasitology.* **12**, 363–369.

FELTEN, D. L., FELTEN, S. Y., BELLINGER, D. L., CARLSON, S. L., ACKERMAN, K. D., MADDEN, K. S., OLSCHOWKI, J. A. & LIVNAT, S. (1987) Noradrenergic sympathetic neural interactions with the immune system: structure and function. *Immunological Reviews.* **100**, 225–260.

FEVOLDEN, S. E. & RØED, K. H. (1993) Cortisol and immune characteristics in rainbow trout (*Oncorhynchus mykiss*) selected for high or low tolerance to stress. *Journal of Fish Biology.* **43**, 919–930.

FEVOLDEN, S. E., NORDMO, R. & REFSTIE, T. (1993) Disease resistance in Atlantic salmon (*Salmo salar*) selected for high or low responses to stress. *Aquaculture.* **109**, 215–224.

GLUCKMAN, P. D., HANSON, M. A. & BEEDLE, A. S. (2007) Early life events and their consequences for later disease: a life history and evolutionary perspective. *American Journal of Human Biology.* **19**, 1–19.

GRINDE, B., JOLLÈS, J. & JOLLÈS, P. (1988) Purification and characterization of two lysozymes from rainbow trout (*Salmo gairdneri*). *European Journal of Biochemistry.* **173**, 269–273.

HILL, A. B. (1965) Environment and disease: association or causation? *Proceedings of the Royal Society of Medicine.* **58**, 295–300.

HOLMES, J. C. (1996) Parasites as threats to biodiversity in shrinking ecosystems. *Biodiversity and Conservation.* **5**, 975–983.

HOUGHTON, G. & MATTHEWS, R. A. (1990) Immunosuppression in juvenile carp, *Cyprinus carpio* L.: the effects of the corticosteroids triamcinolone acetonide and hydrocortisone 21-hemisuccinate (cortisol) on acquired immunity and the humoral antibody response to *Ichthyophthirius multifiliis* Fouquet. *Journal of Fish Diseases.* **13**, 269–280.

HUNTINGFORD, F. A., ADAMS, C., BRAITHWAITE, V. A., KADRI, S., POTTINGER, T. G., SANDØE, P. & TURNBULL, J. F. (2006) Current issues in fish welfare. *Journal of Fish Biology.* **68**, 332–372.

IWAMA, G. K., PICKERING, A. D., SUMPTER, J. P. & SCHRECK, C. B. (Eds) (1997) *Fish Stress and Health in Aquaculture.* Cambridge University Press.

JOHANSON, Z., LONG, J. A., TALENT, J. A., JANVIER, P. & WARREN, J. W. (2006) Oldest coelacanth, from the Early Devonian of Australia. *Biology letters.* **2**, 443–446.

JOHNSON, S. C. & ALBRIGHT, L. J. (1992) Comparative susceptibility and histopathology of the response of naive Atlantic, chinook and coho salmon to experimental infection with *Lepeophtheirus salmonis* (Copepoda: Caligidae). *Diseases of Aquatic Organisms.* **14**, 179–193.

KHAN, R. A. & THULIN, J. (1991) Influence of pollution on parasites of aquatic animals. *Advances in Parasitology.* **30**, 201–238.

KLONTZ, G. W. (1993) Environmental requirements and environmental diseases of salmonids. In: *Fish Medicine.* Stoskopf, M. (Ed.). Saunders. pp333–342.

KORTE, S. M., KOOLHAAS, J. M., WINGFIELD, J. C. & MCEWEN, B. S. (2005) The Darwinian concept of stress: benefits of allostasis and costs of allostatic load and the trade-offs in health and disease. *Neuroscience & Biobehavioral Reviews.* **29**, 3–38.

LAFFERTY, K. D. & HOLT, R. D. (2003) How should environmental stress affect the population dynamics of disease? *Ecology Letters.* **6**, 654–664.

MACKENZIE, S., ILIEV, D., LIARTE, C., KOSKINEN, H., PLANAS, J. V., GOETZ, F. W., MÖLSÄ, H., KRASNOV, A. & TORT, L. (2006) Transcriptional analysis of LPS-stimulated activation of trout (*Oncorhynchus mykiss*) monocyte/macrophage cells in primary culture treated with cortisol. *Molecular immunology.* **43**, 1340–1348.

MACKENZIE, S., RIBAS, L., PILARCZYK, M., CAPDEVILA, D. M., KADRI, S. & HUNTINGFORD, F. A. (2009) Screening for coping style increases the power of gene expression studies. *PLoS One.* **4**, e5314.

MACKINNON, B. M. (1998) Host factors important in sea lice infections. *ICES Journal of Marine Science: Journal du Conseil.* **55**, 188–192.

MAGNADÓTTIR, B., JÓNSDÓTTIR, H. & HELGASON, S. (1999) Humoral immune parameters in Atlantic cod (*Gadus morhua* L.): I. The effects of environmental temperature. *Comparative Biochemistry and Physiology Part B: Biochemistry and Molecular Biology.* **122**, 173–180.

MATEUS, A., BRODBELT, D. & STÄRK, K. (2011) Antimicrobial use: evidence-based use of antimicrobials in veterinary practice. *In Practice.* **33**, 194–202.

MAULE, A. G., TRIPP, R. A., KAATTARI, S. L. & SCHRECK, C. B. (1989) Stress alters immune function and disease resistance in chinook salmon (*Oncorhynchus tshawytscha*). *Journal of Endocrinology.* **120**, 135–142.

MCEWEN, B. & LASLEY, E. N. (2002) *The end of stress as we know it.* National Academies Press.

MCEWEN, B. S. (2003) Interacting mediators of allostasis and allostatic load: towards an understanding of resilience in aging. *Metabolism.* **52**, 10–16.

MCEWEN, B. S. & WINGFIELD, J. C. (2003) The concept of allostasis in biology and biomedicine. *Hormones and Behavior.* **43**, 2–15.

MÖCK, A. & PETERS, G. (1990) Lysozyme activity in rainbow trout, *Oncorhynchus mykiss* (Walbaum), stressed by handling, transport and water pollution. *Journal of Fish Biology.* **37**, 873–885.

MURRAY, D. R., POLIZZI, S. M., HARRIS, T., WILSON, N., MICHEL, M. C. & MAISEL, A. S. (1993) Prolonged isoproterenol treatment alters immunoregulatory cell traffic and function in the rat. *Brain, Behavior and Immunity.* **7**, 47–62.

NARNAWARE, Y. K. & BAKER, B. I. (1996) Evidence that cortisol may protect against the immediate effects of stress on circulating leukocytes in the trout. *General and Comparative Endocrinology.* **103**, 359–366.

NDONG, D., CHEN, Y. Y., LIN, Y. H., VASEEHARAN, B. & CHEN, J. C. (2007) The immune response of tilapia *Oreochromis mossambicus* and its susceptibility to *Streptococcus iniae* under stress in low and high temperatures. *Fish & Shellfish Immunology.* **22**, 686–694.

NESSE, R. (1999) Proximate and evolutionary studies of anxiety, stress and depression: synergy at the interface. *Neuroscience & Biobehavioral Reviews.* **23**, 895–903.

NESSE, R. M. (2000) Is depression an adaptation? *Archives of General Psychiatry.* **57**, 14–20.

NESSE, R. M. (2001) On the difficulty of defining disease: a Darwinian perspective. *Medicine, Health Care and Philosophy.* **4**, 37–46.

NIKOSKELAINEN, S., BYLUND, G. & LILIUS, E. M. (2004) Effect of environmental temperature on rainbow trout (*Oncorhynchus mykiss*) innate immunity. *Developmental & Comparative Immunology.* **28**, 581–592.

NORMAN, R., BEGON, M. & BOWERS, R. G. (1994) The population dynamics of microparasites and vertebrate hosts: the importance of immunity and recovery. *Theoretical Population Biology.* **46**, 96–119.

PETERS, G. (1991) Social stress induces structural and functional alterations of phagocytes in rainbow trout (*Oncorhynchus mykiss*). *Fish & Shellfish Immunology.* **1**, 17–31.

PETERS, G. & SCHWARZER, R. (1985) Changes in hemopoietic tissue of rainbow trout under influence of strcss. *Diseases of Aquatic Organisms.* **1**, 1–10.

PICKERING, A. D. (1997) Husbandry and stress. In: *Furunculosis Multidisciplinary Fish Disease Research.* Bernoth, E-M., Ellis, A. E., Midtlyng, P. J. & Olivier, G. (Eds). Academic Press. pp178–202.

PICKERING, A. D. & DUSTON, J. (1983) Administration of cortisol to brown trout, *Salmo trutta* L., and its effects on the susceptibility to Saprolegnia infection and furunculosis. *Journal of Fish Biology.* **23**, 163–175.

POST, G. B. (1987) *Text Book of Fish Diseases.* TFH Publications.

PRESS, C. M. (1998) Immunology of fishes. In: *Handbook of Vertebrate Immunology.* Pastoret, P-P., Griebel, P., Bazin, H. & Govaerts, A. (Eds). Academic Press. pp3–62.

REYNOLDS, J. D. (1988) Crayfish extinctions and crayfish plague in central Ireland. *Biological Conservation.* **45**, 279–285.

RIGGS, M. R., LEMLY, A. D. & ESCH, G. W. (1987) The growth, biomass, and fecundity of *Bothriocephalus acheilognathi* in a North Carolina cooling reservoir. *The Journal of Parasitology.* **73**, 893–900.

RØED, K. H., FEVOLDEN, S. E. & FJALESTAD, K. T. (2002) Disease resistance and immune characteristics in rainbow trout (*Oncorhynchus mykiss*) selected for lysozyme activity. *Aquaculture.* **209**, 91–101.

ROTTMANN, R. W., FRANCIS-FLOYED, R. & DURBOROW, R. (1992) *The Role of Stress in Fish Disease.* Southern Regional Aquaculture Center. Publication number 474.

ROUBAL, F. R. & BULLOCK, A. M. (1988) The mechanism of wound repair in the skin of juvenile Atlantic salmon, *Salmo salar* L., following hydrocortisone implantation. *Journal of Fish Biology.* **32**, 545–555.

RUIS, M. A. W. & BAYNE, C. J. (1997) Effects of acute stress on blood clotting and yeast killing by phagocytes of rainbow trout. *Journal of Aquatic Animal Health.* **9**, 190–195.

RUIZ-GOMEZ, M. D., KITTILSEN, S., HOGLUND, E., HUNTINGFORD, F. A., SORENSEN, C., POT-TINGER, T. G., BAKKEN, M., WINBERG, S., KORZAN, W. J. & OVERLI, O. (2008). Behavioral plasticity in rainbow trout (*Oncorhynchus mykiss*) with divergent coping styles: when doves become hawks. *Hormones and Behaviour.* **54**, 534–538.

RUTHERFORD, K., ROBSON, S. K., DONALD, R. D., JARVIS, S., SANDERCOCK, D. A., SCOTT, E. M., NOLAN, A. M. & LAWRENCE, A. B. (2009) Pre-natal stress amplifies the immediate behavioural responses to acute pain in piglets. *Biology Letters.* **5**, 452–454.

SCHRECK, C. B. (1996) Immunomodulation: endogenous factors. In: *The Fish Immune System: Organism, Pathogen and Environment.* Iwama, G. & Nakanishi T. (Eds). Academic Press. pp311–337.

SCHRECK, C., MAULE, A. G. & KAATTARI, S. L. (1993) Stress and disease resistance. In: *Stress and Fish. Recent Advances in Aquaculture* Vol 4. Muir, J.F. & Roberts, R.J. (Eds). Blackwell Scientific Publications. pp170–175.

SCHRECK, C. B., CONTRERAS-SANCHEZ, W. & FITZPATRICK, M. S. (2001) Effects of stress on fish reproduction, gamete quality, and progeny. *Aquaculture.* **197**, 3–24.

SHANKS, N. & LIGHTMAN, S. L. (2001) The maternal-neonatal neuro-immune interface: are there long-term implications for inflammatory or stress-related disease? *Journal of Clinical Investigation.* **108**, 1567–1574.

SIDDALL, R. & DES CLERS, S. (1994) Effect of sewage sludge on the miracidium and cercaria of *Zoogonoides viviparus* (Trematoda: Digenea). *Helminthologia.* **31**, 143–153.

SMITH, J. M. (1982) *Evolution and the Theory of Games.* Cambridge University Press.

SNIEZKO, S. (1973) Recent advances in scientific knowledge and developments pertaining to diseases of fishes. *Advances in Veterinary Science and Comparative Medicine.* **17**, 291–314.

STERLING, P. & EYER, J. (1988) Allostasis: a new paradigm to explain arousal pathology. In: *Handbook of Life Stress, Cognition and Health.* Fisher, S. & Reason, J. (Eds). John Wiley & Sons. pp629–649

STIEN, L. H., BRATLAND, S., AUSTEVOLL, I., OPPEDAL, F. & KRISTIANSEN, T. S. (2007) A video analysis procedure for assessing vertical fish distribution in aquaculture tanks. *Aquacultural engineering.* **37**, 115–124.

TIERNEY, K. B., PATTERSON, D. A. & KENNEDY, C. J. (2009) The influence of maternal condition on offspring performance in sockeye salmon *Oncorhynchus nerka*. *Journal of Fish Biology.* **75**, 1244–1257.

TORT, L., BALASCH, J. C. & MACKENZIE, S. (2003) Fish immune system. A crossroads between innate and adaptive responses. *Inmunología.* **22**, 277–286.

VASTA, G. R. & LAMBRIS, J. D. (2002) Innate immunity in the Aegean: ancient pathways for today's survival. *Developmental & Comparative Immunology.* **26**, 217–225.

WALTERS, G. R. & PLUMB, J. A. (1980) Environmental stress and bacterial infection in channel catfish, *Ictalurus punctatus* Rafinesque. *Journal of Fish Biology*. **17**, 177–185.

WEDEMEYER, G. (1970) *The Role of Stress in the Disease Resistance of Fishes*. American Fisheries Society Special Publication. **5**, 30–35.

WEDEMEYER, G. A., WOOD, J. W. & FISH, U. S. (1981) *Stress as a Predisposing Factor in Fish Diseases*. US Dept. of the Interior, Fish and Wildlife Service, Division of Fishery Research. FDL-38.

WEEKS, B. A., ANDERSON, D. P., DUFOUR, A. P., FAIRBROTHER, A., GOVEN, A. J., LAHVIS, G. P. & PETERS, G. (1992) Immunological biomarkers to assess environmental stress. In: Biomarkers: *Biochemical, physiological, and histological markers of anthropogenic stress*. Huggett, R. J., Kimerle, R. A., Merle Jr. P. M. & Bergman H. L. (Eds). Lewis Publishers. pp211–234.

WEINSTOCK, M. (1997) Does prenatal stress impair coping and regulation of hypothalamic-pituitary-adrenal axis? *Neuroscience and Biobehavioral Reviews*. **21**, 1–10.

WEYTS, F. A. A., FLIK, G. & KEMENADE, B. M. L. (1998) Cortisol inhibits apoptosis in carp neutrophilic granulocytes. *Developmental & Comparative Immunology*. **22**, 563–572.

WISE, D. J., SCHWEDLER, T. E. & OTIS, D. L. (1993) Effects of stress on susceptibility of naive channel catfish in immersion challenge with *Edwardsiella ictaluri*. *Journal of Aquatic Animal Health*. **5**, 92–97.

WOOD, C. M., TURNER, J. D. & GRAHAM, M. S. (1983) Why do fish die after severe exercise? *Journal of Fish Biology*. **22**, 189–201.

Part II

Advances in disease diagnostics, veterinary drugs and vaccines

5

Advances in diagnostic methods for mollusc, crustacean and finfish diseases

A. Adams and K. D. Thompson, University of Stirling, UK

Abstract: Diagnostic methods for mollusc, crustacean and finfish infectious diseases have evolved differently depending on the availability of reagents, constraints on cell lines for culturing pathogens, the extent of laboratory facilities, and expertise of available staff to perform the tests. Traditional and immunodiagnostic methods are still very important and although widely used, these have very much taken a 'back seat' in many countries where molecular technologies now prevail. Recently focus in clinical medicine has been on the development of multiplex assays so that large numbers of samples for a series of pathogens can be rapidly processed. Some of these methods are currently too expensive for use in aquaculture, but clearly have potential for the future. This chapter provides an insight into the recent advances in rapid diagnostic methods for use in aquaculture, discuss how appropriate these techniques are for this sector and look towards future potential technologies.

Key words: rapid diagnostics, diseases in aquaculture, immunodiagnostics, molecular diagnostics, nanotechnology, multiplex diagnostics.

5.1 Introduction

Diagnostic methods in aquaculture fall into three categories, i.e. field, clinical and agent detection/identification techniques. Field diagnostic methods involve the observation of the animal and its environment (i.e. clinical and behavioural signs), while clinical diagnostic methods focus on pathological effects of the agent on the host (i.e. gross pathology, clinical chemistry, microscopic pathology, wet mounts, smears, fixed sections and transmission electron microscopy (TEM)/cytopathology). Agent detection and identification involves techniques that directly detect/identify the pathogen (i.e. microscopic methods, agent isolation and identification) and also serological methods, which are classed as indirect (to detect pathogen-specific host antibody response). These methods are listed in detail in the OIE Manual of Diagnostic Tests for Aquatic Animals (OIE, 2011) to provide an internationally agreed standardised approach to the diagnosis of OIE-listed diseases for finfish, molluscs and crustacean disease (Aquatic Animal Health Code) (OIE, 2010). The aim is to facilitate safe international trade in aquatic animals and their products by ensuring the quality of diagnostic testing.

Not all diseases of aquatic animals are of course listed in the OIE manual, and rapid advancement in methods has mainly been in the third category of diagnostic methods (i.e. agent detection/identification), very often termed 'rapid' diagnostic methods. Details of diagnostic methods for non-OIE-listed diseases tend to be published in scientific publications with little standardisation or validation of methods between groups. This can lead to a variation in methods for a single test, e.g. different primers for polymerase chain reaction (PCR) to detect a specific pathogen making the literature a minefield for scientists to select which to use, and results dependent on which protocol is used.

Diagnostic methods for mollusc, crustacean and finfish diseases (Table 5.1) have evolved differently depending on the availability of reagents, constraints on cell lines for culturing pathogens, extent of laboratory facilities, and expertise of available staff to perform the tests. In general there has been a rapid expansion of rapid diagnostic methods for use in aquaculture over the last decade, as techniques used in clinical and veterinary medicine have been adapted and commercial reagents and kits have become available (Adams and Thompson, 2011). Traditional and immunodiagnostic methods are still very important and although widely used, these have very much taken a 'back seat' in many countries where molecular technologies, in particular PCR-based methods, now prevail. Modern methods have permitted real progress in the understanding of the taxonomic relationships of many fish pathogens (Austin, 2011), and have led to an understanding of the epidemiology of diseases of aquatic animals (Snow, 2011). Recently focus in clinical medicine has been on the development of multiplex assays so that large numbers of samples for a series of pathogens can be rapidly processed. Some of these methods are currently too expensive for use in aquaculture, but clearly have potential for the future.

This chapter will provide an insight into the recent advances in rapid diagnostic methods for use in aquaculture, how appropriate these techniques are for this sector and a look towards future potential technologies.

5.2   Mollusc disease diagnostic methods

5.2.1   Mollusc diseases

Molluscs represent an important contribution of global aquaculture production (approximately one-quarter) and diseases of molluscs (in both farmed and wild populations) are a primary concern for the development and sustainability of the sector. Paramyxeans, belonging to the genera *Marteilia* and *Marteiliodes*, are considered an important group of parasites for molluscs, with *Marteilia refringens* and *M. sydneyi* highlighted as having a

Table 5.1 Rapid diagnostic methods reported for the detection and identification of pathogens causing disease in finfish, molluscs and crustaceans

Finfish	Molluscs	Crustaceans
Immunodiagnostics	**Immunodiagnostics**	**Immunodiagnostics**
Agglutination[1]	IHC	IHC[1]
Enzyme-linked immuno sorbent assay (ELISA)[1]		FAT/IFAT[1]
Immunohistochemistry (IHC)[1]		Rapid kits[1]
Fluorescence antibody test/ indirect fluorescence antibody test (FAT/IFAT)[1]		Dot blot (WSSV and IHHNV)
Western blot		
Dot blot		
Rapid kits[1]		
Serology	N/A	N/A
ELISA		
Molecular	**Molecular**	**Molecular**
PCR/rtPCR/nested PCR	PCR/rtPCR/nested PCR	PCR/rtPCR/nested PCR
Quantitative PCR (real time PCR)	Restriction fragment length polymorphism (RFLP)	DNA viruses: WSSV, IHHNV, HPV, MBV, BP, SMV
Random amplified polymorphic DNA (RAPD)	Quantitative PCR (real time PCR)	RNA viruses: TSV, YHV, IMNV, GAV, PuNV, MoV, LSNV
Nucleic acid based sequence amplification (NASBA)	In situ hybridisation	Bacteria: NHP-B
Loop-mediated isothermal amplification (LAMP)[1]	Sequencing	Quantitative PCR (real time PCR)
In situ hybridisation		DNA viruses: WSSV, IHHNV, HPV, MBV
Sequencing		RNA viruses: TSV, YHV, IMNV
		RAPD
		NASBA
		LAMP
		Quantitative PCR
		In situ hybridisation
		Sequencing
Multiplex assays	**Multiplex assays**	**Multiplex assays**
Multiplex PCR	None	Multiplex PCR
Bead array (Luminex xMAP™)		DNA array
DNA array		
Others	**Others**	
Microfluid	Flow cytometry	
Matrix-assisted laser desorption ionisation time-of-flight (MALDI-TOF)		

[1] Commercial reagents kits available.

negative impact on mollusc aquaculture globally (Berthe *et al.*, 2004). Marteiliosis is also known as Aber disease (caused by *M. refringens*) and QX disease (caused by *M. sydneyi*) with *M. refringens* currently listed as notifiable by the OIE. Bonamiosis caused by *Bonamia ostreae* and *B. exitiosa* represent important parasitic problems in oysters and these need to be diagnosed and differentiated from Microcytosis (*Microcytos mackani*) and Perkinsosis (*Perkinsus marinus* and *P. olseni*). *Haplosporidium nelsoni*, the agent of MSX, multinucleated sphere unknown (Burreson and Ford, 2004), was previously listed by the OIE, but has very recently been removed based on its restricted geographical distribution of susceptible species (Berthe, 2008). In contrast, the significant abalone diseases often resulting in large scale mortalities are caused by bacterial (*Xenohaliotis californiensis*) and viral pathogens (abalone herpes-like virus, AbHV or abalone viral ganglioneuritis) (Chang *et al.*, 2005; Hooper *et al.*, 2007).

5.2.2   Use of molecular methods in mollusc disease diagnostics

The development of PCR-based molecular methods has, without doubt, led to a significant improvement in mollusc disease diagnosis. Although, histology is still used (Howard *et al.*, 2004) and recommended by the OIE as a standard screening method, for surveillance more specific methods are required to differentiate many of the closely related pathogens causing diseases in molluscs. Molecular methods are therefore increasingly being used as well as electron microscopy (EM) for specific identification. Longshaw *et al.* (2001) reported ultrastructural characterisation of *Marteilia* species (Paramyxea) from *Ostrea edulis*, *Mytilus edulis* and *M. galloprovincialis* in Europe, while Le Roux *et al.* (2001) provided molecular evidence for the existence of the two species. Molecular methods include PCR (e.g. used to detect *Bonamia ostreae* (Carnegie *et al.*, 2000); *Perkinsus marinus* (Audemard *et al.*, 2004), PCR-RFLP (restriction fragment length polymorphism), e.g. for *Perkinsus* species (Abollo *et al.*, 2006), sequencing for example to discriminate between two *Perkinsus* species from soft shell clams (Kotob *et al.*, 1999), and *in situ* hybridisation, for example to detect *Marteilia sydneyi* and *M. refringens* (Kleeman *et al.*, 2002). Both EM (Tan *et al.*, 2008) and a variety of molecular methods have been developed and validated, including conventional PCR (Renault *et al.*, 2000) and qPCR (Corbeil *et al.*, 2010), which have been used successfully to identify herpes-like virus of abalone (*Haliotis* spp.).

5.2.3   Use of antibody-based methods in mollusc disease diagnostics

Antibody-based methods have also been used, e.g. immunohistochemistry and electron immunohistochemical assays on parasitic infections in molluscs (e.g. Anderson *et al.*, 1994; Boulo *et al.*, 1989; Carnegie and Cochennec-Laureau, 2004; Dungan and Roberson, 1993; Mialhe *et al.*, 1988). Clearly serological methods cannot be used because molluscs do not produce antibodies and therefore identification of disease agents is restricted to direct detection of the pathogen. In the case of the soft shell clam, *Mya arenaria*, where diagnosis of haemic neoplaia (HN) is usually performed using hematocytology and histology, flow cytometry cell cycle analysis has also been applied (Delaporte *et al.*, 2008) to determine the percentage of haemocyte tetraploid cells (i.e. neoplastic cells). This provides a high-throughput method for HN status assessment, which can be at either an individual or a population level. Further research is now focusing on specific biomarkers of HN (Delaporte *et al.*, 2008). Flow cytometry has also been used for the analysis of neoplastic cells in cockles (da Silva *et al.*, 2005).

5.3   Crustacean disease diagnostic methods

5.3.1   Crustacean diseases

The global crustacean aquaculture industry is worth more than US$10 billion. A growing number of crustacean species (including crabs, lobsters and prawns) are intensively farmed, and increased production and movement of live products have led to the emergence of several internationally important crustacean diseases. In the past 15 years losses due to disease have been estimated to be in the region of $15 billion, of which 60% of losses were attributed to viruses and 20% to bacteria (Flegel *et al.*, 2008).

In contrast to finfish and shellfish, aquaculture production of crustaceans within Europe accounts for around 2000 Mt/annum, with a total value of $3m (Stentiford *et al.*, 2009). Three of the globally significant crustacean diseases are currently listed in Europe under EC Council Directive 2006/88/EC adopted during 2008; these being white spot disease (WSD), yellowhead disease (YHD) caused by yellowhead virus (YHV), and Taura syndrome (TS) caused by TSV. WSD is currently listed as a 'non-exotic' pathogen to the EU, based on its reported occurrence in penaeid shrimp farms in southern Europe (see Stentiford *et al.*, 2009), while YHD and TS are listed as exotic due to their absence from the EU. The listing of these diseases is in recognition of their global importance in causing significant economic losses and the potential for their international transfer via the transboundary trade in live animals and their products (Stentiford *et al.*, 2009). Other crustacean diseases listed as notifiable by the OIE are crayfish plague (*Aphanomyces astaci*) and infectious hypodermal and hematopoietic necrosis virus (IHHNV).

Diagnostic methods include the traditional methods of gross pathology, histopathology, classical microbiology, animal bioassay, antibody-based methods, and molecular methods using DNA probes and DNA amplification.

5.3.2   Use of molecular methods in crustacean disease diagnostics

The development of diagnostic tests for crustaceans almost immediately followed the molecular route with the use of PCR and *in situ* hybridisation, by-passing many of the traditional methods because cell lines were not available to culture the major pathogens, i.e. viruses. Despite considerable efforts no continuous shrimp cell lines, or lines from other crustaceans, have yet been developed. A DNA hybridisation assay for the parvovirus IHHNV was the first molecular test developed for a shrimp disease (Mari *et al.*, 1993), followed by the first PCR test for monodon baculovirus (MBV) (Chang *et al.*, 1993; Lu *et al.*, 1993), an important baculovirus disease of shrimp, while a PCR able to detect the virus from different geographical regions was developed later in 2005 (Surachetpong *et al.*, 2005). Shrimp disease diagnostic laboratories routinely use molecular tests for diagnostic and surveillance purposes for most of the important penaeid shrimp diseases (Lightner, 1996; Lightner *et al.*, 2006).

WSD is still considered to be the most serious threat to the shrimp industry globally, with economic losses estimated at $10 billion since 1993 (European Community Reference Laboratory, 2008). A PCR test became available to detect white spot syndrome virus-1 (WSSV) in 1996. This method is recommended for surveillance with a variety of methods listed for confirmation (histology, transmission electron microscopy (TEM), *in situ* hybridisation (Nunan and Lightner, 1997), PCR and sequencing (Claydon *et al.* 2004). It is interesting that the PCR is still cited and recommended by OIE using the protocol of Lo *et al.* (1996, 1997), with cloning and sequencing the nested PCR product (Claydon *et al.*, 2004) or direct sequencing of the PCR amplicon, used for confirmation purposes. Crayfish plague (*Aphanomyces astaci*) is confirmed by positive PCR and sequencing of PCR products while IHHNV is confirmed on positive results for any two of the following tests: dot-blot hybridisation, *in situ* hybridisation (ISH) positive histology, PCR and sequencing of PCR specific products if required to determine the genotype of the IHHNV. Other viral diseases including MBV mentioned above, as well as YHV, taura syndrome virus (TSV); infectious myonecrosis virus (IMN) and *Macrobrachium rosenbergii* nodavirus (*Mr*NV) are also mainly diagnosed using PCR or related methods as outlined in the OIE health manual (OIE, 2011). A variety of commercial kits are also available for this testing.

5.3.3   Use of commercial kits in crusctacean disease diagnosis

Molecular methods, in particular PCR and, reverse transcriptase polymerase chain reaction (RT-PCR) have been very important in helping to control the spread of major shrimp disease agents (Flegel *et al.*, 2008) and this accelerated the availability of commercial kits (PCR kits, LAMP kits, ELISA, IFAT/FAT, IHC, rapid kits) for use in aquaculture (Table 5.1). On site, user friendly, relatively inexpensive lateral flow kits have also been developed. It is important that kits are fully validated for use as a variety of genotypes can occur for viruses. Thus, it may be important to detect more than one genotype or some of the genotypes may be more virulent than others. Both could lead to false results being reported. For example, as reviewed by Flegel *et al.* (2008) it is crucial to be able to distinguish between three genotypes of YHV in Thailand. Lateral-flow (Sithigorngul *et al.* 2006; Takahashi *et al.*, 2003; Wang and Zhan, 2006) and flow through immunoassay (Wang *et al.*, 2006) systems have been developed for rapid field-level detection of WSSV in shrimp. A commercial lateral-flow test for WSSV, Shrimple test kit, is available from EnBioTec Laboratories Co., Ltd. Tokyo, Japan. The flow through immunoassay for WSSV is available as a rapid test kit called 'RapiDot' for the early detection of WSSV under field conditions (Patil *et al.*, 2011).

5.4   Finfish disease diagnostic methods

5.4.1   Finfish diseases

The number of diseases reported in finfish and the methods used to detect the causative agents is vast (reviewed by Adams, 2009). The diseases currently listed as notifiable by the OIE include epizootic haematopoietic necrosis (EHN); epizootic ulcerative syndrome (EUS); gyrodactylosis (*Gyrodactylus salaris*); infectious haematopoietic necrosis (IHN); infectious salmon anaemia (ISA); koi herpesvirus disease (KHVD); red sea bream iridoviral disease (RBID); spring viraemia of carp (SVC); viral haemorrhagic septicaemia (VHS) (OIE, 2011).

5.4.2   Use of traditional and antibody-based methods in finfish disease diagnostics

As with molluscs and crustaceans, molecular diagnostics is extensively used and commercial reagents and kits have become more widely available (Adams, 2009; Adams and Thompson, 2006, 2008; Cunningham, 2004; Karunasagar *et al.*, 1997; Wilson and Carson, 2003). Traditional bacteriology, whereby the pathogen is isolated and identified biochemically (e.g. using API strips), and observation of histological sections from disease fish are still used by many laboratories, and immunodiagnostic standardised methods (e.g. immunohistochemistry) and commercial products are available. Such techniques have been combined with cell culture with regard to virus identification, whereby the viral pathogens are cultured in cell lines and then quickly identified using fluorescently labelled antibodies (FAT or IFAT), whereas in the past observation by EM would have been used. Immunodiagnostic methods are used more widely in finfish diagnostics than mollusc or crustacean diagnostics, and the sandwich ELISA in particular has been developed for many finfish pathogens (Austin and Austin, 2007), and commercial kits exist (Table 5.1). Because of the higher sensitivity of PCR and related methods many laboratories have moved to molecular diagnostics.

Serological methods (i.e. analysing serum by ELISA for the presence of host antibodies to specific pathogens) are an indirect method to indicate infection, and have proved very useful in fish diagnostics. The ELISA is flexible and for serology is set up with the pathogen attached to the ELISA plate making the test simple to run. Serological methods are vital in clinical and veterinary medicine (Fournier and Raoult, 2003; Palmer-Densmore et al., 1998; Yuce et al., 2001) and have potential for use in aquaculture. To date few have been validated and it is likely that they will be most useful for viral infections (Adams, 2009). Serology has been used effectively for detecting exposure to KHV (Adams and Thompson, 2008; Adkison et al., 2005) and nervous necrosis virus (NNV) (Breuil and Romstand, 1999).

5.4.3   Use of molecular methods in finfish disease diagnostics

The development and use of molecular methods to detect and identify pathogens in finfish have increased exponentially over the last 5 years with commercial kits and service laboratories making the technology more accessible. These highly sensitive methods are ideal for detecting low levels of pathogens and characterising pathogens such as mycobacteria to species level (Pourahmed et al., 2008; Puttinaowarat et al., 2000), although for some species polygenic sequencing was the only way of fully characterising the isolates (Pourahmed, 2007). In addition, they are extremely useful from an epidemiological point of view as individual strains can be differentiated from closely related ones and traced back to point of origin, e.g. VHSV, ISAV and salmon alphavirus (Snow, 2011). Although a wide range of molecular methods have been described for detecting pathogens in finfish (reviewed by Adams and Thompson, 2011), without a doubt real-time PCR or qPCR is now the most widely used. This method offers high sample throughput and quantification, and gets around much of the previous contamination with PCR that led to false positive results. TaqMan and Cyber green are the two of most common qPCR methods used (Zähringer, 2009). TaqMan probes have a fluorescent reporter dye attached to the 5′ end of the molecule and a quencher moiety attached to the 3′ end. In the unhybridised state, the close proximity of the fluorescent report to the quench molecules prevents a fluorescent signal being produced. However, during the PCR reaction, the 5′-nuclease activity of the Taq polymerase (used in the PCR) cleaves the oligonucleotide hybridised to the complementary DNA. As a result the fluorescent reporter is released and an increase in the fluorescent signal is seen over each cycle, proportional to the amount of probe cleavage. SYBR Green, on the other hand, is the most economical option for real-time PCR. It binds to double-stranded DNA, eliminating the need to design a specific probe, and as the product accumulates, so does the amount of fluorescence. Care needs to be taken to produce well-designed primers because SYBR Green will bind to any double-stranded DNA present in the reaction (e.g. primer-dimers or other non-specific products), leading to overestimation of the PCR product.

Such a powerful test lends itself to very practical applications such as monitoring specific pathogen (*Flavobacterium psychrophilum*) levels on fish farms (Orieux *et al.*, 2011) and demonstrating a positive correlation between a *Rickettsia*-like organism and severity of strawberry disease lesions in rainbow trout (Lloyd *et al.*, 2011). Validation of new methods such as qPCR is clearly important if these are to be used for diagnostics and efforts are being made to address this. For example, a reverse transcription qPCR was recently developed and validated for universal detection of viral hemorrhagic septicaemia virus (Garver *et al.*, 2011). The assay performed equivalent to the traditional detection method of virus isolation via cell culture with the advantage of faster turnaround times and high throughput capacity, indicating suitability for use in a diagnostics.

5.4.4 Recently developed rapid diagnostic methods for use in finfish disease diagnostics

A variety of other novel rapid diagnostic methods are currently being developed that have potential for future application in the diagnosis of aquatic animal health. Some of these were recently reviewed by Adams and Thompson (2008), including LAMP and lateral flow. Both methods have the advantage of being very sensitive, quick to perform and the results can be read by eye. The former has been developed to detect *Edwardsiella tarda*, *E. ictaluri*, *Nocardia seriolae*, *Tetracapsuloides bryosalmonae* and infectious haematopoietic necrosis virus (IHNV) (El-Matbouli and Soliman, 2005; Gunimaladevi *et al.*, 2005; Itano *et al.*, 2006; Savan *et al.*, 2005). Lateral flow tests on the other hand have only so far been developed to detect a few fish pathogens, including ISAV and NNV (Godoy *et al.*, 2008; Rega Tec Inc, 2011). This simple technology is ideal for field tests, while LAMP is more complex to perform. A number of research groups are also developing DNA and oligo microarray technology for diagnostics to enable simultaneous detection of pathogens (González *et al.*, 2004; Nash *et al.*, 2006). Bead arrays also have potential using Luminex technology (Adams and Thompson, 2011) and nanotechnology using magnetic beads in immunomagnetic reduction (IMR) is currently being developed in Taiwan for use in fish diseases with regard to NNV (Adams and Thompson, 2011). The magnetic nano-particles are coated with antibody and a high transition temperature superconductive quantum interference device is used to sense the immunomagnetic reduction of the reagents on reaction with virus. This again follows on from clinical diagnostics work where the method was developed to detect HIV (Yang *et al.*, 2008).

5.5   Future trends

The number of diagnostic methods potentially available appears endless, as technologies in clinical medicine expand, but not all are suited for application in aquaculture, mainly due to cost and complexity. Recently, focus in clinical medicine has been on the development of multiplex assays so that large numbers of samples can be analysed. This is also an active area of research with regard to aquaculture, and although such methods are currently expensive, the cost will come down and these technologies will become more accessible for application in aquatic animal diagnostics. Lievens *et al.* (2011) recently described a method based on multiplex and broad-range PCR amplification combined with DNA array hybridisation for the simultaneous detection and identification of all cyprinid herpesviruses (CyHV-1, CyHV-2 and CyHV-3) targeting DNA polymerase and helicase genes, while for bacterial identification *Flavobacterium* species, including *F. branchiophilum*, *F. columnare* and *F. psychrophilum*, were differentiated using a ribosomal RNA gene. The power of the array for sensitive pathogen detection and identification in complex samples such as infected tissue was effectively demonstrated in this study.

Methods previously developed but not used for diagnostics purposes, are also being revisited. For example, it has recently been suggested (Tae Sung Jung, personal communication) that MALDI-TOF mass spectrometry analysis of pathogens could be a powerful tool for use in aquaculture. Although this technology was proposed more than 30 years ago (Anhalt and Fenselau, 1975; Claydon *et al.*, 1996; Krishnamurthy and Ross, 1996) it has only recently been shown to be a cost-effective, accurate method for routine identification of bacterial isolates in clinical diagnostics (Seng *et al.*, 2009). The technique enabled identification to be completed within one hour and correlation with traditional methods was 95%. One disadvantage, however, is the complexity of the mass spectrophotometer and to establish if the method is applicable for routine diagnostics in aquaculture. Certainly there is potential for application in research in the analysis of closely related isolates or for the identification of antibiotic resistant isolates in aquaculture.

Microfluid technologies also offer much potential for the future and many have already been developed for clinical diagnostics in developing countries. This enables the use of small reaction volumes, and channel architecture can be designed to perform multiple tests or experimental steps on one integrated, automated platform resulting in sensitive, specific, inexpensive, rapid, integrated, and automated tests (McCalla and Tripathi, 2011). Microfluidics technology has recently been developed for use in aquaculture using RT-PCR (Lien *et al.*, 2009) to detect NNV, Iridovirus, and *Vibrio anguillarum* and RT-LAMP to detect NNV in grouper (Wang *et al.*, 2010).

5.6   Sources of further information

5.6.1   Websites for databases, organisations and documents

ASEM Aquaculture Platform Aquatic Animal Health: EU project building partnerships between Asia and Europe to improve cultured aquatic animal health and welfare including diagnosis, preventative measures and treatments. http://www.asemaquaculturehealth.net/

EC FP7 KBBE Projects in Fisheries & Aquaculture: Catalogue of under the EU FP7 COOPERATION – THEME 2 Fisheries, Aquaculture, Food safety & quality, and Marine Biotechnology projects (2007–2009). http://eceuropaeu/research/agriculture/pdf/marine_v6.pdf

European Association of Fish Pathologists: A society to promote the exchange of knowledge, and assistance in the coordination of research related to fish and shellfish pathology. http://eafp.org/

European Medicines Agency: The European Medicines Agency is a decentralised body of the European Union responsible for the scientific evaluation of medicines developed by pharmaceutical companies for use in the European Union. http://www.ema.europa.eu/

European Union Reference Laboratory for Crustacean Diseases: Cefas Weymouth Laboratory is the designated European Union Reference Laboratory (EURL) for Crustacean Diseases. http://www.crustaceancrl.eu/

European Union Reference Laboratory for Fish Diseases: The European Union Reference Laboratory (EURL) is funded by the European Commission and is concerned with harmonising diagnostic procedures for notifiable fish diseases in Europe. http://www.crl-fish.eu/

Fish Health Section of the Asian Fisheries Society: Society aimed at improving knowledge on fish health management among Asian aquaculturists. http://www.fhs-seafdec.org.ph

Fisheries and Aquaculture Department, Food and Agriculture Organization of the United Nations: FAO of the United Nations serves both developed and developing countries and acts as a neutral forum for nations to negotiate agreements and policy. FAO is also a source of knowledge and information. http://www.fao.org/documents/en/detail/69879; http://www.fao.org/fishery/publications/en

International Database on Aquatic Animal Diseases (IDAAD): Database containing OIE listed aquatic animal disease outbreaks (maintained by CEFAS, UK). http://www.cefas.defra.gov.uk/idaad/

Network of Aquaculture Centres in Asia-Pacific: Network of Aquaculture Centre in Asia Pacific News and Resources. http://www.enaca.org/

OIE Aquatic Animal Health Code: Guidelines covering aquatic animal disease management. http://www.oie.int/international-standard-setting/aquatic-code/access-online/, http://www.oie.int/fileadmin/Home/eng/Support_to_OIE_Members/docs/pdf/Good_vet_governance.pdf

OIE Aquatic Animal Health Manual: Manual of diagnostic tests for aquatic animals. http://www.oie.int/en/international-standard-setting/aquatic-manual/access-online

Pescalex Diagnostic Tool: Online information to help communication of disease-related problems between farmers and fish health professionals. http://www.pescalextool.org/pescalex.html

Registry of Aquatic Pathology: Includes hundreds of examples of disease conditions and parasites from aquarium, cultured and wild fish, and bivalve molluscs and crustacea from freshwater and marine environments around the world. http://www.cefas.defra.gov.uk/our-science/animal-health-and-food-safety/aquatic-animal-disease/registry-of-aquatic-pathology.aspx

Synopsis of Infectious Diseases and Parasites of Commercially Exploited Shellfish: Synopsis of infectious diseases and parasites of commercially exploited shellfish. http://www.pa.dfo-mpo.gc.ca/science/species-especes/shellfish-coquillages/diseases-maladies/index-eng.htm

The European Union Reference Laboratory for Mollusc Diseases: One of the main activities of the EURL is to assist in the diagnosis of disease outbreaks for confirmatory diagnosis, characterisation and epizootic studies. http://www.ifremer.fr/crlmollusc

VetBact: Swedish database containing information about bacteria of importance in veterinary bacteriology These species belong to 80 genera with a search facility. http://www.vetbact.org/vetbact/

World Animal Health Information Database (WAHID) Interface: The WAHID Interface providing access to data held within OIE's new World Animal Health Information System (WAHIS) including aquatic animals. http://web.oie.int/wahis/public.php?page=home

World Aquatic Veterinary Medical Association: WAVMA offers support, continuing education and ongoing development to aquatic veterinary medicine, private practitioners and WAVMA members. http://www.wavma.org/contact-WAVMA

5.6.2   Journals

Diseases of Aquatic Organisms, Inter-Research. http://www.int-rescom/journals/dao/

Journal of Aquatic Animal Health, journal from the American Fisheries Society, http://www.afsjournals.org/

Journal of Fish and Shellfish Immunology, Elsevier, http://www.sciencedirect.com/science/journal/10504648

Journal of Fish Diseases, Wiley/Blackwell, http://www.wiley.com/bw/journal.asp?ref=0140-7775&site1

The EAFP Bulletin, European Association of Fish pathologists, http://eafp.org/

5.6.3   Books and manuals

Austin B and Austin D A (2007), *Bacterial Fish Pathogens: Disease of Farmed and Wild Fish*, Springer Praxis Books.

Bruno D W and Woo P T K (2011), *Fish Diseases and Disorders*, Volume 3: *Viral, Bacterial and Fungal Infections*, Cabi Publishing.

Buchmann K and Woo P T K (2011), *Fish Parasites*, Cabi Publishing.

Buchmann K, Bresciani J, Pedersen K and Ariel E and Dalsgaard I (2009), *Fish Diseases: An introduction*, Biofolia.

Buller N B (1996), *Bacteria from Fish and Other Aquatic Animals: A practical identification manual*, Cabi Publishing.

Chanratchakool P, Turnbull J F, Funge-SmithS J, MacRae I H and Limsuan C (1998), *Health Management in Shrimp Ponds*, Third Edition. Aquatic Animal Health Research Institute, Department of Fisheries, Kasetsart University Campus Jatujak, Ladyao, Bangkok, Thailand.

Des Roza Z, Koesharyani I, Johnny F and Yuasa K (1998), *Manual for Fish Diseases Diagnosis: Marine fish and crustacean diseases in Indonesia*. Gondol Research Station for Coastal Fisheries, PO Box 140 Singaraja, Bali, Indonesia.

Eiras J, Segner H, Wahli T and Kapoor B G (2008), *Fish Diseases*, Volume 2, Science Publishers.

Herfort A and Rawlin G (1999), *Australian Aquatic Animal Disease – Identification field guide*. Agriculture, Fisheries and Forestry – Australia, GPO Box 858, Canberra, Australia.

Koesharyani I, Des Roza Z, Mahardika K, Johnny F and Yuasa K (2001) *Manual for Fish Disease Diagnosis – II: Marine fish and crustacean diseases in Indonesia*. Gondol Research Station for Coastal Fisheries, PO Box 140, Singaraja, Bali, Indonesia.

Lavilla-Pitogo C R, Lio-Po G D, Cruz-Lacierda E R, Alapide-Tendencia E V and de la Pena L D (2000), *Diseases in Penaeid Shrimps in the Philippines*, Second Edition, Fish Health Section, SEAFDEC Aquaculture Department Tigbauan, Iloilo 5021, Philippines.

Leatherland J F and Woo P T K (2010), *Fish Diseases and Disorders, Volume 2: Non-infectious Disorders*, Cabi Publishing.

Leung K Y (2004) *Current Trends in the Study of Bacterial and Viral Fish and Shrimp*, World Scientific Publishing Company.

Noga E J (2010) *Fish Disease: Diagnosis and treatment*, John Wiley and Sons.

Plumb J A and Hanson L A (2010), *Health Maintenance and Principal Microbial Diseases of Cultured Fishes*, John Wiley and Sons.

Roberts R J (2001), *Fish Pathology*, Elsevier Health Sciences.

Thorne T (1999), *Fish Health for Fish Farmers*, Fisheries Western Australia, 3rd Floor, SGIO Atrium, 186 St Georges Terrace, Perth WA 6000, Australia.

Wakabayashi H (ed.) (1994) *Fish Health Bibliography III*. Fish Health Special Publication No 3/Japanese Society of Fish Pathology, Fisheries Society, Manila, Philippines.

Woo P T K (2006) *Fish Diseases and Disorders*, Volume 1: *Protozoan and metazoan Infections*, Cabi Publishing.

Woo P T K (2011) *Fish Diseases and Disorders*, 2nd Edition, Cabi Publishing.

5.7   References

ABOLLO E, CASAS S M, CESCHIA G and VILLALBA A (2006), Differential diagnosis of *Perkinsus* species by polymerase chain reaction-restriction fragment length polymorphism assay, *Mol Cell Probes*, **20**, 323–329.

ADAMS A (2009), Advances in disease diagnosis, vaccine development and other emerging methods to control pathogens (Chapter 7), in Burnell G and Allan G (eds), *New Technologies in Aquaculture, Improving Production, Efficiency, Quality and Environmental Management*. Woodhead Publishing Ltd, pp 197–211.

ADAMS A and THOMPSON K D (2006), Biotechnology offers revolution to fish health management, *Trends Biochem Sci*, **24**(5), 201–205.

ADAMS A and THOMPSON K D (2008), Recent applications of biotechnology to novel diagnostics for aquatic animals, *Rev Sci Tech OIE*, **27**(1), 197–209.

ADAMS A and THOMPSON K D (2011), Development of diagnostics for aquaculture: challenges and opportunities, *Aqua Res*, **42**(s1), 93–102.

ADKISON M A, GILAD O and HEDRICK R P (2005), An enzyme linked immunosorbent assay (ELISA) for detection of antibodies to the koi herpesvirus (KHV) in the serum of koi *Cyprinus carpio*, *Fish Pathol*, **40**, 53–62.

ANDERSON T J, MCCAUL T F, BOULO V, ROBLEDO J A F and LESTER R J G (1994), Light and electron immunohistochemical assays on paramyxean parasites, *Aquat Living Resour*, **7**, 47–52.

ANHALT J P and FENSELAU C (1975), Identification of bacteria using mass spectrometry, *Anal Chem*, **47**, 219–225.

AUDEMARD C, REECE K S and BURRESON E M (2004), real-time PCR for the detection and quantification of the protistan parasite *Perkinsus marinus* in environmental waters, *Appl Environ Microbiol*, **70**, 6611–6618.

AUSTIN B (2011), Taxonomy of bacterial fish pathogens, *Vet Res*, **42**, 20.

AUSTIN B and AUSTIN D A (2007), Chapter 6 Diagnostics, in *Bacterial Fish Pathogens: Disease of Farmed and Wild Fish*, Springer Praxis Books, Environmental Sciences, 185.

BERTHE F C J (2008), New approaches to effective mollusc health management, in Bondad-Reantaso M G, Mohan C V, Crumlish M and Subasinghe R P, *Diseases in Asian Aquaculture VI*, Fish Health Section, Asian Fisheries Society, Manila, Philippines.

BERTHE F C J, LE ROUX F, ADLARD R D and FIGUERAS A J (2004), Marteiliosis of molluscs: a review, *Aquat Living Resour*, **17**(4), 433–448.

BOULO V, MIALHE E, ROGIER H, PAOLUCCI F and GRIZEL H (1989), Immunodiagnosis of *Bonamia ostreae* (Ascetospora) infection of *Ostrea edulis* L. and subcellular identification of epitopes by monoclonal antibodies, *J Fish Dis*, **12**, 257–262.

BREUIL G and ROMESTAND B (1999), A rapid ELISA method for detecting specific antibody level against nodavirus in the serum of the sea bass, *Dicentrarchus labrax* (L.): application to the screening of spawners in a sea bass hatchery, *J Fish Dis*, **22**(1), 45–52.

BURRESON E M and FORD S E (2004), A review of recent information on the Haplosporidia, with special reference to *Haplosporidium nelsoni* (MSX disease), *Aquat Living Resour*, **17**(4), 499–517.

CARNEGIE R B and COCHENNEC-LAUREAU N (2004), Microcell parasites of oysters: recent insights and future trends, *Aquat Living Resour*, **17**, 519–528.

CARNEGIE R B, BARBER B J, CULLOTY S C, FIGUERAS A J and DISTEL D L 2000, Development of a PCR assay for detection of the oyster pathogen *Bonamia ostreae* and support for its inclusion in the Haplosporidia, *Dis Aquat Org*, **42**, 199–206.

CHANG P H, KUO S T, LAI S H, YANG H S, TING Y Y, HSU C L and CHEN H C (2005), Herpes-like virus infection causing mortality of cultured abalone *Halitotis diversicolor supertexta* in Taiwan, *Dis Aquat Org*, **65**, 23–27.

CHANG P S, LO C F, KOU G H, LU C C and CHEN S N (1993), Purification and amplification of DNA from *Penaeus monodon*-type baculovirus (MBV), *J Invertebr Pathol*, **62**, 116–120.

CLAYDON K, CULLEN B and OWENS L (2004), OIE white spot syndrome virus PCR gives false-positive results in *Cherax quadricarinatus*, *Dis Aquat Organ*, **62**(3), 265–268.

CLAYDON M A, DAVEY S N, EDWARDS-JONES V and GORDON D B (1996), The rapid identification of intact microorganisms using mass spectrometry, *Nat Biotechnol*, **14**, 1584–1586.

CORBEIL S, COLLING A, WILLIAMS L M, WONG F Y K, SAVIN K, WARNER S, MURDOCH B, COGAN N O I, SAWBRIDGE T I, FEGAN M, *et al.* (2010), Development and validation of a TaqMan PCR assay for the Australian abalone herpes-like virus, *Dis Aquat Org*, **91**, 1–10.

CUNNINGHAM C O (2004), Use of molecular diagnostic tests in disease control: making the leap from laboratory to field application, in Leung K-Y, *Current Trends in the Study of Bacterial and Viral Fish and Shrimp Diseases*, Molecular Aspects of Fish and Marine Biology, World Scientific Publishing Co, vol 3, 292–312.

DA SILVA P M, SOUDANT P, CARBALLA M J, LAMBERT C and VILLALBA A (2005), Flow cytometric DNA content analysis of neoplastic cells in haemolymph of the cockle *Cerastoderma edule*, *Dis Aquat Org*, **67**, 133–139.

DELAPORTE M, SYNARD S, PARISEAU J, MCKENNA P, TREMBLAY R, DAVIDSON J and BERTHE F C J (2008), Assessment of haemic neoplasia in different soft shell clam *Mya arenaria* populations from eastern Canada by flow cytometry, *J Invert Pathol*, **98**(2), 190–197.

DUNGAN C F and ROBERSON B S (1993), Binding specificities of mono- and polyclonal antibodies to the protozoan oyster pathogen *Perkinsus marinus*, *Dis Aquat Org*, **15**, 9–22.

EL-MATBOULI M and SOLIMAN H (2005), Rapid diagnosis of *Tetracapsuloides bryo-salmonae*, the causative agent of proliferative kidney disease (PKD) in salmonid fish by a novel DNA amplification method, loop-mediated isothermal amplification (LAMP), *Parasitol Res*, **96**(5), 277–284.

EUROPEAN COMMUNITY REFERENCE LABORATORY (2008), European Community Reference Laboratory for Crustacean Diseases leaflet: White Spot Disease, available from http://www.crustaceancrl.eu/diseases/WhiteSpot.pdf [accessed 10 October 2011].

FLEGEL T W, LIGHTNER D V, LO C F and OWENS L (2008), Shrimp disease control: past, present and future, in Bondad-Reantaso M G, Mohan C V, Crumlish M and Subasinghe R P, *Diseases in Asian Aquaculture VI*, Fish Health Section, Asian Fisheries Society, Manila, Philippines, 355–378.

FOURNIER P-E and RAOULT D (2003), Comparison of PCR and serology assays for early diagnosis of acute Q fever, *J Clin Micro*, **41**, 5094–5098.

GARVER K A, HAWLEY L M, MCCLURE C A, SCHROEDER T, ALDOUS S, DOIG F, SNOW M, EDES S, BAYNES C and RICHARD J (2011), Development and validation of a reverse transcription quantitative PCR for universal detection of viral hemorrhagic septicemia virus, *Dis Aquat Org*, **95**, 97–112.

GODOY M G, AEDO A, KIBENGE M J T, GROMAN D B, YASON C V, GROTHUSEN H, LISPERGUER A, CALBUCURA M, AVENDAÑO F, IMILÁN M, JARPA M and KIBENGE F S B (2008), First detection, isolation and molecular characterization of infectious salmon anaemia virus associated with clinical disease in farmed Atlantic salmon (*Salmo salar*) in Chile, *BMC Vet Res*, **4**, 28.

GONZÁLEZ S, KRUG M, NIELSON M, SANTOS Y and CALL D R (2004), Simultaneous detection of marine fish pathogens using multiplex PCR and DNA microarrays, *J Clin Microbiol*, **42**, 1414–1419.

GUNIMALADEVI I, KONO T, LAPATRA SE and SAKAI M (2005), A loop mediated isothermal amplification (LAMP) method for detection of infectious hematopoietic necrosis virus (IHNV) in rainbow trout (*Oncorhynchus mykiss*), *Arch Virol*, **150**(5), 899–909.

HOOPER C, HARDY-SMITH P and HANDLINGER J (2007), Ganglioneuritis causing high mortalities in farmed Australian abalone (*Haliotis laevigata* and *Haliotis rubra*), *Aus Vet J*, **85**, 188–193.

HOWARD D W, LEWIS E J, KELLER J and SMITH C S (2004), Histological techniques for marine bivalve molluscs and crustaceans, *NOAA Technical Memorandum NOS NCCOS* 5.

ITANO T, KAWAKAMI H, KONO T and SAKAI M (2006), Detection of fish nocardiosis by loop-mediated isothermal amplification, *J Appl Microbiol*, **100**(6), 1381–1387.

KARUNASAGAR I, NAYAK B B and KARUNASAGAR I (1997), Rapid detection of *Vibrio parahaemolyticus* from fish by polymerase chain reaction, In Flegel T W and MacRae L H, eds. Diseases in Asian Aquaculture III. Asian Fisheries Society, Manila, pp 119–122.

KLEEMAN S N, LE ROUX F, BERTHE F and ADLARD R D (2002), Specificity of PCR and *in situ* hybridization assays designed for detection of *Marteilia sydneyi* and *M. refringens*, *Parasitology*, **125**, 131–141.

KOTOB S, MCLAUGHLIN SM, VAN BERKUM P and FAISAL M (1999), Discrimination between two *Perkinsus* spp isolated from the softshell clam, *Mya arenaria*, by sequence analysis of two internal transcribed spacer regions and the 58S ribosomal RNA gene, *Parasitology*, **119**, 363–368.

KRISHNAMURTHY T and ROSS P L (1996), Rapid identification of bacteria by direct matrix-assisted laser desorption/ionization mass spectrometric analysis of whole cells, *Rapid Commun Mass Sp*, **10**, 1992–1996.

LE ROUX F, LORENZO G, PEYRET P, AUDEMARD C, FIGUERAS A, VIVARES C, GOUY M and BERTHE F (2001), Molecular evidence for the existence of two species of *Marteilia* in Europe, *J Eukaryot Microbiol*, **48**, 449–454.

LIEN K-Y, LEE S-H, TSAI T-J, CHEN T-Y and LEE G-B (2009), A microfluidic-based system using reverse transcription polymerase chain reactions for rapid detection of aquaculture diseases, *Microfluid Nanofluid*, **7**, 795–806.

LIEVENS B, FRANS I, HEUSDENS C, JUSTÉ A and JONSTRUP S P (2011), Rapid detection and identification of viral and bacterial fish pathogens using a DNA array-based multiplex assay, *J Fish Dis*, **34**, 11, 861–875.

LIGHTNER D V (1996), *A Handbook of Shrimp Pathology and Diagnostic Procedures for Diseases of Cultured Penaeid Shrimp*, World Aquaculture Society, Baton Rouge, LA7.

LIGHTNER D V, POULOS B T, TANG-NELSON K F, PANTOJA C R, NUNAN L M, NAVARRO S A, REDMAN R M and MOHNEY L L (2006), Application of molecular diagnostic methods to penaeid shrimp diseases: advances of the past 10 years for control of viral diseases in farmed shrimp, *Dev Biol*, **226**, 117–22.

LLOYD S J, LAPATRA S E, SNEKVIK K R, CAIN K D and CALL D R (2011), Quantitative PCR demonstrates a positive correlation between a *Rickettsia*-like organism and severity of strawberry disease lesions in rainbow trout, *Oncorhynchus mykiss* (Walbaum), *J Fish Dis*, **34**, 701–709.

LO C F, LEU J H, HO C H, CHEN C H, PENG S E, CHEN Y T, CHOU C M, YEH P Y, HUANG C J, CHOU H Y, WANG C H and KOU G H (1996), Detection of baculoviruses associated with white spot syndrome (WSBV) in penaeid shrimps using polymerase chain reaction, *Dis Aquat Organ*, **27**, 215–225.

LO C F, HO C H, CHEN C H, LIU K F, CHIU Y L, YEH P Y, PENG S E, HSU H C, LIU H C, CHANG C F, SU M S, WANG C H and KOU G H (1997), Detection and tissue tropism of white spot syndrome baculovirus (WSBV) in captured brooders of *Penaeus monodon* with a special emphasis on reproductive organs, *Dis Aquat Organ*, **30**, 53–72.

LONGSHAW M, FEIST S W, MATTHEWS R A and FIGUERAS A (2001), Ultrastructural characterization of *Marteilia* species (Paramyxea) from *Ostrea edulis*, *Mytilus edulis* and *Mytilus galloprovincialis* in Europe, *Dis Aquat Org*, **44**, 137–142.

LU C C, TANG K F J, KOU G H and CHEN S N (1993), Development of a *Penaeus monodon*-type baculovirus (MBV) DNA probe by polymerase chain reaction and sequence analysis, *J Fish Dis*, **16**, 551–559.

MARI J, BONAMI J R and LIGHTNER D (1993), Partial cloning of the genome of infectious hypodermal and hematopoietic necrosis virus, an unusual parvovirus pathogenic for penaeid shrimps; diagnosis of the disease using a specific probe, *J Gen Virol*, **74**, 2637–2643.

MCCALLA S E and TRIPATHI A (2011), Microfluidic reactors for diagnostics applications, *Ann Rev Biomed Eng*, **13**, 321–343.

MIALHE E, BOULO V, ELSTON R, HILL B, HINE M, MONTES J, VAN BANNING P and GRIZEL H (1988), Serological analysis *Bonamia* in *Ostrea edulis* and *Tiostrea lutaria* using polyclonal and monoclonal antibodies, *Aquat Living Res*, **1**, 67–69.

NASH J H E, FINDLAY W A, LUEBBERT C C, MYKYTCZUK O L, FOOTES J, TABOADA E N, CARRILLO C D, BOYD J M, COLQUHOUN D J, REITH M E and BROWN L L (2006), Comparative genomics profiling of clinical isolates of *Aeromonas salmonicida* using DNA microarrays, BMC *Genomics*, **7**, 43.

NUNAN L M and LIGHTNER D V (1997), Development of a non-radioactive gene probe by PCR for detection of white spot syndrome virus (WSSV), *J Virol Methods*, **63**, 193–201.

OIE (2010), Aquatic Animal Health Code, available from http://www.oie.int/international-standard-setting/aquatic-code/access-online/ [accessed 30 July 2011].

OIE (2011), Manual of Diagnostic Tests for Aquatic Animals, available from http://www.oie.int/en/international-standard-setting/aquatic-manual/access-online/ [accessed 30 July 2011].

ORIEUX N, BOURDINEAUD J-P, DOUET D G, DANIEL P AND' M LE HÉNAFF (2011), Quantification of *Flavobacterium psychrophilum* in rainbow trout, *Oncorhynchus mykiss* (Walbaum), tissues by qPCR, *J Fish Dis*, **34**, 811–821.

PALMER-DENSMORE M L, JOHNSON A F and SABARA M I (1998), Development and evaluation of an ELISA to measure antibody responses to both the nucleocapsid and spike proteins of canine coronavirus, *J Immunoassay*, **19**, 1–22.

PATIL R, SHANKAR K M, KULKARNI A, PATIL P, KUMAR N and MOGER N (2011), Development of a monoclonal antibody based flow through immunoassay (FTA) for detection of white spot syndrome virus (WSSV) in black tiger shrimp *Penaeus monodon*, *J Fish Dis* (in press).

POURAHMAD F (2007), Molecular detection and identification of aquatic mycobacteria PhD thesis, University of Stirling.

POURAHMAD F, THOMPSON K D, TAGGART J B, ADAMS A and RICHARDS R H (2008), Evaluation of INNO-LiPA mycobacteria v2 assay for identification of aquatic mycobacteria, *J Fish Dis*, **31**(12), 931–940.

PUTTINAOWARAT S, THOMPSON K D and ADAMS A (2000), Mycobacteriosis: detection and identification of aquatic *Mycobacterium* species, *Fish Vet J*, **5**, 6–21.

REGA TEC INC (2011), Necrosis Virus NNV One-Step Rapid Test, available from http://www.regatecinc.com/Aqua%20NNV.pdf [accessed 10 October 2011].

RENAULT T, LE DEUFF R M, LIPART C and DELSERT C (2000), Development of a PCR procedure for the detection of a herpes-like virus infecting oysters in France, *J Virol Method*, **88**, 41–50.

SAVAN R, KONO T, ITAMI T and SAKAI M (2005), Loop-mediated isothermal amplification: an emerging technology for detection of fish and shellfish pathogens, *J Fish Dis*, **28**, 573–581.

SENG P, DRANNCOURT M, GOURIET F, LA SCOLA B, FOURNIER P-E, ROLAIN J M and RAOULT D (2009), Ongoing revolution in bacteriology: routine identification of bacteria by matrix-assisted lazer desorption ionization time of flight mass spectrometry, *Clin Infect Dis*, **49**, 543–551.

SITHIGORNGUL W, RUKPRATANPORN S, PECHARABURANIN N, LONGYANT S, CHAIVISUTHANGKURA P and SITHIGORNGUL P (2006), A simple and rapid immuno-chromatographic test strip for detection of white spot syndrome virus (WSSV) of shrimp, *Dis Aquat Org*, **72**, 101–106.

SNOW M (2011), The contribution of molecular epidemiology to the understanding and control of viral diseases of salmonid aquaculture, *Vet Res*, **42**, 56.

STENTIFORD G D, BONAMI J R and ALDAY-SANZ V (2009), A critical review of susceptibility of crustaceans to Taura syndrome, yellowhead disease and white spot disease and implications of inclusion of these diseases in European legislation, *Aquaculture*, **291**, 1–2, 1–17.

SURACHETPONG W, POULOS B T, TANG K F J and LIGHTNER D V (2005), Improvement of PCR method for the detection of *Monodon baculovirus* (MBV) in penaeid shrimp, *Aquaculture*, **249**, 69–75.

TAKAHASHI Y, FUKUDA K, KONDO M, CHONGTHALEONG A, NISHI K, NISHIMURA M, OGATA K, SHINYA I, TAKISE K, FUJISHIMA Y and MATSUMAURA M (2003), Detection and prevention of WSSV infection in cultured shrimp, *Asian Aquaculture Magazine*, **Nov/Dec**, 25–27.

TAN J, LANCASTER M, HYATT A, VAN DRIEL R, WONG F and WARNER S (2008), Purification of a herpes-like virus from abalone (*Haliotis* spp) with ganglioneuritis and detection by transmission electron microscopy, *J Virol Methods*, **149**, 338–341.

WANG C-H, LIEN K-Y, WANG T-Y, CHEN T-Y and LEE G-B (2010), Integrated microfluidic loop-mediated-isothermal-amplification system for rapid diagnosis of aquaculture virus. 14th International Conference on Miniaturized Systems for Chemistry and Life Sciences, Groningen, The Netherlands, pp. 812–814.

WANG X and ZHAN W (2006), Development of an immunochromatographic test to detect white spot syndrome virus of shrimp, *Aquaculture*, **255**, 196–200.

WANG X, ZHAN W and XING J (2006), Development of dot-immunogold filtration assay to detect white spot syndrome virus of shrimp, *J Virol Methods*, **132**, 212–215.

WILSON T and CARSON J (2003), Development of sensitive, high-throughput one-tube RT-PCR-enzyme hybridisation assay to detect selected bacterial fish pathogens, *Dis Aquatic Org*, **54**, 127–134.

YANG S-Y, CHIEH J J, WANG W C, YU C Y, LAN C B, CHEN J H, HORNG H E, HONG C-Y, YANG H-C and HUANG W (2008), Ultra-highly sensitive and wash-free bio detection of H5N1 virus by immunomagnetic reduction assays, *J Virol Methods*, **153**, 250–252.

YUCE A, YUCESOY M, GENC S, SAYAN M and UCAN E S (2001), Serodiagnosis of tuberculosis by enzyme immunoassay using A60 antigen, *Clin Microbiol Infect*, **7**, 372–376.

ZÄHRINGER H (2009), Product survey: Real-time PCR kits:TaqMan or SYBR Green, *Lab Times*, (1), 47–52, available from http://www.labtimes.org/labtimes/issues/lt2009/lt01/lt_2009_01_47_52.pdf [accessed 10 October 2011].

6

Quality assurance in aquatic disease diagnostics

M. Crumlish, University of Stirling, UK

Abstract: Quality assurance (QA) is becoming more important within the field of aquaculture and aquatic disease diagnosis, but at present a single one-size-fits-all approach does not exist. There is a need to raise awareness for a simple, cost-effective QA approach, which could be implemented by even the most basic aquatic diagnostic facility. This chapter will describe the various aspects required for an effective QA system within an aquatic disease diagnosis facility and discuss the benefits derived when working within a quality management scheme.

Key words: quality assurance, validation, aquatic diagnosis, accreditation.

6.1 Introduction

By definition, quality assurance (QA) is 'the maintenance of a desired level of quality in a service or product' (*Oxford English Dictionary*, www.oxforddictionaries.com). QA is becoming more important within the field of aquaculture and aquatic disease diagnosis generally. Implementation of a QA scheme is not easy, and a single one-size-fits-all approach does not currently exist for any diagnostic facility, especially within aquaculture. Some laboratory facilities do not operate under any QA system, whereas others have opted to follow national or international fully accredited schemes. The uptake of a QA system within any diagnostic facility is complicated and may often be considered prohibitively expensive. However, as technology advances and our understanding of aquatic disease outbreaks, pathogens and epidemics increases, there is a recognised need for improved QA systems to be applied within aquatic diagnostic laboratories. There is a need to raise awareness for a simple, cost-effective QA approach, which could be implemented by even the most basic aquatic diagnostic facility. This chapter will describe QA systems within aquatic disease diagnosis and highlight the benefits for such schemes to be implemented.

6.2   The importance of valid and reliable aquatic disease diagnosis and the role of quality assurance systems

Disease diagnosis, whether medically or veterinary based, is a complicated matter and is generally divided into a consideration of infectious or non-infectious diseases. Ultimately the subject is examined by a medical or veterinary practitioner, who will initially perform a differential diagnosis. This is a systematic process of elimination to try to identify what the disease condition may be. If the condition is thought to be infectious or pathogen-driven then the initial diagnosis will often include the collection of both information and clinical material, which will then be analysed using laboratory-based tests. The practitioner is relying on the test results to help with the initial diagnosis; therefore, these laboratory tests are a fundamental part of the disease diagnostic process. Confidence in the laboratory results is essential for the diagnosis and management of infectious diseases. To ensure the validity and reliability of the laboratory test results, a QA scheme should be implemented.

Within medical diagnostic laboratories there are often legally required QA procedures in place that have been well tested and provide confidence in the results of the tests performed. QA has become an integral part of the global market structure where consumers are much more aware of the importance of maintaining quality, particularly within the manufacturing sector. The implementation of a QA system within any sector provides confidence for the consumer, facilitates global trade and the test results can provide a reliable dataset useful for monitoring health and safety as well as provision of environmental legislation. These are some of the benefits described by the International Organization for Standardization (ISO) (www.iso.org/iso/home.html).

Aquaculture is an animal and plant production sector that is continuing to grow globally (Austin and Austin, 2007). Increased consumer demand for aquatic food products has meant that fish and other aquatic animals play a significant contribution to maintaining food security at an international level. Not only can these products provide low cost and high quality protein sources, they also contribute towards employment and supply numerous additional benefits for local and international trade through both the food and ornamental fish industry. However, infectious disease outbreaks continue to threaten the development of aquaculture (Austin and Austin, 2007). Our understanding of diseases and their impact is increasing daily through scientific research. During infectious disease outbreaks great emphasis is placed on the initial disease diagnosis to provide the cause of the infection and ultimately the treatment to reduce the spread of the disease. There is an assumption that the same level and standards of aquatic diagnosis is being followed in all parts of the world; however, this is not necessarily true. The benefits of applying a QA system with an aquatic disease diagnostic facility are plentiful.

The diagnostic laboratory may play various roles within aquatic health management. Some diagnostic facilities have their own in-house laboratories and others rely on clinical samples being sent to an independent laboratory for histopathology, microbiology or serology testing. Clinical samples are more commonly processed for histopathology and microbiology within traditional disease diagnosis. The histology and microbiology results are the foundation for the initial disease diagnosis to be performed. These are often considered as the *routine diagnostic disciplines* but the laboratory may also be involved in screening clinical samples for health certification purposes or checking that the population is free from specific pathogens. These are more commonly associated with local or international shipments and are a prerequisite for aquatic trade in live fish movements (OIE, 2000a, 2000b, 2000c). To complicate matters more, samples may be sent to the diagnostic laboratory from a wide range of clients including the farmer or the health manager for that production site, hobbyists, international companies/traders, veterinarians and government officials. Whilst the individual tests may differ depending on the client's request the manner with which the tests are performed and reported should follow whatever QA system has been adopted by the test laboratory.

Such systems should include good management practices as well as valid test protocols, calibration methods, quality control and quality assurance. If all of these can be applied this would provide a very robust quality management system. Primarily the aim of any QA system is to provide confidence in the validity of the range of diagnostic tests or procedures conducted and the services provided (Bellamy and Olexson, 2000). Whilst the individual components of the QA system used may vary between laboratories any QA or quality management system for an aquatic disease diagnostic facility should include the quality goals of the individual laboratory. These may be internal and/or external QA testing, provision of standard operating procedures (SOP) for each of the tests or assays performed and full traceability and transparency of all documentation including the test results. Provision of laboratory-based results where the diagnostician has confidence in the range of tests performed will enable a rapid and reliable health management strategy.

6.2.1   QA management systems for disease diagnosis

There are numerous examples of the different types of QA management systems available for disease diagnosis. Within aquatic diagnostic facilities this may include a range of options where the level of QA can differ depending on the purpose of the tests, but it may also be influenced by the type of test being performed. Also the number of samples being submitted can guide which level of QA is required, as it is simpler to implement an internationally accredited QA system if there are large volumes of samples being submitted on a regular basis. In general, the QA system may range from a simple cost-effective internal 'check' often performed in-house, to the uptake of national or international quality control (QC) procedures performed at accredited levels. Although research data is improving our understanding of diseases and their impact in aquatic environments, it may also contribute towards the production of improved QA standards for the detection of various pathogens. The aquatic disease diagnostic facility needs to make some careful considerations prior to implementing any QA system. These are discussed within the scope of aquatic disease diagnosis.

6.3   Sampling submission and handling for aquatic disease diagnosis

6.3.1   Traceability and client confidentiality

Within any diagnostic facility the utmost care must be given to protect client confidentiality. This is the same whether it is a medical or veterinary practice. Provision of a unique identification number for any clinical material received by the diagnostic facility is important to protect client confidentiality and will allow full traceability of any biological samples. The sample submission stage is a very important first step in the QA system and this is often not considered a valuable stage in the diagnostic process. Recording the receipt of the samples, results of each test performed and the storage of clinical samples or the data should be part of the quality management system employed by the diagnostic service. In accredited facilities records are kept from the initial receipt of the material to the final result. This process is often referred to as the *chain of command*.

6.3.2   Sample handling

Most aquatic disease laboratories receive clinical material for analyses that may include samples of organs or whole animals. During a fish disease outbreak tissue samples would normally be required for histopathology, where good quality samples are essential. These should always be taken quickly but in a systematic manner so that a full set of organs is available for the pathologist to read. This supplies the best results enabling a clinical description of the disease and ensures that the appropriate management regime or treatment strategy is applied. If the samples are provided by the client the laboratory has little influence on the sampling process but can install technical rigour with the subsequent processing of the samples, following the laboratory management system.

Not all aspects of the QA system need be costly. A simple necropsy sheet can be provided to staff taking the clinical material. The staff can follow the necropsy sheet and take the samples quickly and with confidence. This reduces post-mortem artefact, which in severe cases can render histology slides unreadable. If infectious disease is suspected, then further samples for microbiological analyses will be taken for pathogen recovery and identification. This is where aseptic technique is crucial. Without proper staff training samples can easily become contaminated or mislabelled or the wrong samples are processed. Investing in staff training and provision of necropsy sheets or pre-labelled containers for tissue samples is a very low cost but effective part of a QA management system. This may be considered basic but it is something that is not only vital to ensure the quality of the samples but it could very easily be adopted by all aquatic diagnostic facilities. Checking aseptic technique is an integral part of any QA system within a diagnostic establishment.

Purity checks are something that may not always be seen of value within a diagnostic facility but are crucial to have confidence in the test outcome. In some aquaculture diagnostic facilities, samples have sometimes been described as 'nearly' pure, which shows a fundamental lack of understanding in aseptic technique and the principles of disease diagnosis. If there is any doubt about the purity of the samples then this should be reported and measures taken to correct it. Purification steps within aquatic bacteriology are relatively common but it is important that the diagnostic staff record the number of subcultures required to purify the sample prior to identification. Working with pure bacterial cultures is essential within aquatic disease diagnosis and a simple method of recording the number of subcultures to purity is valuable for any diagnostic facility. This is important for the identification process as some bacterial characteristics may be lost or impaired during multiple subcultures, making it difficult to provide a bacterial species identification.

Previously, shelf-life and stock control may have been an issue within the production of reagents, particularly in microbiology laboratories. However, the increasing availability of commercially produced agar has reduced both errors and wastage of reagents. Most aquatic bacterial pathogens can be recovered on what is considered *general-purpose* agar with the addition of salt if required with marine organisms. However, the more fastidious aquatic bacterial pathogens require selective agar for growth, and in these cases care must be taken by the laboratory to have a good labelling system which clearly identifies the different reagents required to make the agar, expiry dates and storage of the ingredients within the agar. Appropriate storage of the agar plates once made also helps reduce errors. It is not always essential to purchase the agar commercially and laboratories can reduce the cost by making their own agar plates following published recipes. This may be more cost effective for smaller laboratories or those dealing with limited numbers of samples. It is irrelevant where the agar originates from, as the laboratory should still randomly perform validity testing of the agar for the growth of the bacteria. This is especially important if the agar is to be used for the primary recovery of pathogens during the initial disease diagnosis. The laboratory staff should have a clear understanding of what should and should not grow on the selective agar plate prior to use. These are small but very effective measures, which any diagnostic facility can follow with easy checks contributing towards a simple QA system.

6.4   The importance of assay performance in aquatic disease diagnosis

Ultimately the aim of any aquatic diagnostic laboratory is to provide the client with accurate and reliable data whether this is a clinical outbreak investigation, or health certification of freedom from specific pathogens. All of these outcomes can be investigated through the diagnostic laboratory as long as the assay methods are reliable, reproducible and performed to the highest level. All laboratory tests need dependable protocols for trained staff to follow. Such protocols exist for histopathology, microbiology and serology testing and many of the protocols used within aquatic disease diagnosis originated from those applied in human or terrestrial veterinary practice. Regardless of the level of disease diagnosis performed, the laboratory must establish the diagnostic specificity and sensitivity of each test that they perform. This will allow the diagnostic suitability of the test to be established (Greiner and Gardner, 2000).

Most aquatic histopathology laboratories follow a simple and standard approach to the fixation and staining of the clinical material. These follow established assays from the medical and veterinary disciplines (Drury and Wallington, 1980). In aquatic bacteriology some protocols have not changed since they were first developed, e.g. the Gram stain to differentiate between the different bacterial types has not changed since it was first described by H. C. Gram in 1884 (English translation in Brock, 1999). However, with the advances in molecular biology and serology a wide range of protocols have been established and applied for the detection of aquatic pathogens. This can be quite confusing when trying to determine which protocol or assay is the best for the laboratory. One source of information is the OIE manual for aquatic disease diagnosis, which has a list of standard operating procedures (SOP). These SOP are usually restricted to notifiable diseases or those of significance within global aquaculture. Prior to release of the SOP all have been validated and tested by different laboratories and with free on-line access available for any laboratory working in the field of aquatic disease diagnosis (www.oie.int/international-standard-setting/aquatic-manual/access-online).

Whether the laboratory uses their own in-house protocols or performs their diagnostic assays as described in OIE or accredited schemes, there are key considerations. The main factors to consider within assay development for any disease diagnostic laboratory are provided in Table 6.1 and were developed using the OIE principles and validation methods as a guide. All of these will already have been covered if they are part of an internationally recognised accreditation system and checks will be provided through the

Table 6.1 Checklist for development of standard operating procedures (SOP)

Assay term	Term definition	Laboratory checklist
Correctness	The correlation between the mean from a set of tests and the individual data.	This can be achieved by using a certified reference sample, e.g. a standard control.
Repeatability	Measure of variance between results from testing identical samples with the same methods.	Repeated testing conducted by the same staff with the same sample, usually within a defined time period, e.g. 1 day or 1 week, depending on the assay.
Reproducibility	Comparing test results between different laboratories using the same sample and materials.	The laboratory staff can perform internal reproducibility (similar to repeatability) or can engage in proficiency testing with other laboratories. This can be considered further under *ring testing*.
Range of detection	Are the detection levels the same between laboratories following the same protocol with the same samples?	The laboratory staff must determine this from the internal reproducibility tests or from the results between the laboratories.
Linearity	This determines how linear your test results are.	This is particularly important with serology methods or any methods that involve titration of the samples.
Selectivity	Determining if the correct reagents are being used for the protocol.	This is important for microbiology and serology testing, where selective agars may be used or antisera in viral titration assays.
Sensitivity	Measurement of true positives within the samples tested. Often considered as statistical analyses of the test and can be determined by the number of true positive results divided by the combined number of true positive + false negatives.	This is important for molecular or serology testing where the whole organism is not used within the test. The screening of primer sets and antiserum or antibodies should be carefully evaluated.
Specificity	This is a test to detect true negatives within the samples. Often considered as the number of true negatives divided by the combined number of true negative + false positives.	Laboratory staff should determine the specificity of the test prior to uptake as the higher the specificity of the test then the higher chance that a positive result does indicate the presence of the disease or the pathogen correctly.
Robustness	This is a way to determine the failure of the test.	It is not easy to calculate this but staff can investigate if wrong concentrations are used, incubator temperatures are too high or too low, or if steps are omitted. This is critical to determine why the test has failed.

QA. Every laboratory should aim for the highest standards of rigour when following a standard protocol and much work must be invested prior to calling any protocol a SOP. At the very least, the laboratory staff using the assay should know the specificity and sensitivity of the assay as well as what steps to take if the assay fails.

The lack of appropriate control samples is a significant constraint during assay validity testing or the development/refining of standard protocols. This is particularly relevant within emerging diseases in aquatic systems, where there may not always be an appropriate control available. Nevertheless if the assay is to be performed properly and be part of the quality management system then adequate control samples must be included in the assay. In some accredited schemes controls can be provided but other sources can be used. In histology apparently normal samples can be used for a comparison with the affected fish tissues or reference archived material from a previous confirmed disease outbreak may act as controls. Including a negative control is not a substitute for a positive control but it may be considered better than no control. Consideration should be given to both positive and negative control during the assay development and these can be used for staff training as well as monitoring purposes.

More and more diagnosis is performed using commercially available kits in both human and veterinary medicine and molecular-based kits are gaining popularity within aquatic disease diagnosis. These molecular-based assays or kits are more commonly applied in aquatic disease diagnosis for parasite or viral pathogens, where recovery or growth of the aetiological agent can be difficult. These diagnostic kits are marketed as rapid and cheaper to use than more traditional diagnostic methods. Whilst technological advances should be welcomed within aquatic disease diagnosis, care should be taken when relying on kits for the initial diagnosis. They are not a substitute for a traditional disease diagnosis and because they provide only a 'snap-shot' of the disease situation within the population, care must be given with the interpretation of the results from kits. If disease diagnostic facilities wish to use these then it would be wise to conduct their own validation prior to implementing them within the facility. Appropriate checks and inclusion of positive or negative control samples should be screened with the kits or the molecular assay as part of the laboratory QA system prior to implementation. Otherwise this may cause diagnostic inaccuracy leading to false positive or false negative results (Morris et al., 2002): both can be equally damaging for the final disease diagnosis.

6.5   Validation of quality assurance systems for aquatic diagnostic facilities

Validation is an integral part of any QA system. The importance of the validation step is to determine and prove the quality of the test and so the confidence level must be decided. In accredited tests this is predetermined by those providing the scheme but in a standard aquatic disease diagnosis laboratory those working within the laboratory must establish these levels. Aquatic diagnostic facilities now use molecular and serological assays as part of the routine microbiology protocols. Validation of these assays prior to implementation is essential.

All tests performed within the disease diagnostic facility should be valid, irrespective of their purpose. The validation process determines that the assay is *fit-for-purpose*, it ensures that it has been properly developed, optimised and standardised for the intended purpose. This is described in more detail in chapter 1.1.2 in the OIE manual for diagnostic tests in aquatic animals (www.oie.int). Emphasis is placed on the purpose of the assay, which should be clearly defined by the diagnostic laboratory, irrespective of where this has been developed.

Selection of the various laboratory tests under the quality management system will influence the validity of the results obtained. This process is similar whether the laboratory is part of a fully accredited scheme or not. Importance can be placed on the internationally approved but also the scientific acceptance of the different laboratory tests employed. The methodology does not always have to be the most modern method used and in fact, often traditional, simple tests can be more cost effective within a rural aquatic disease diagnostic laboratory.

The diagnostic sensitivity and specificity of a laboratory test are vital and contribute towards the validation of the test selected. This is sometimes referred to as the *diagnostic accuracy* where emphasis is placed on the quality of the information (Zweig and Campbell, 1993). Although this is still limited within aquatic diagnosis, a good review of the objectives, methods and limitation of different veterinary diagnostic tests can be found (Greiner and Gardner, 2000). This paper explores the variability in diagnostic tests at the population level. With any new tests the laboratory should optimise the assay, which will include establishment of critical specification and inclusion of performance standards. The problems and benefits of these have already been described.

Quality assessment within the microbiology laboratory has also been described as laboratory proficiency (Barrow and Feltham, 1993) which acts as a quality control (QC) for the laboratory procedures. This may be driven internally or externally but the principle is the same. The QA method allows the introduction of known but undisclosed identification to be processed by the laboratory staff for examination and reporting. Both national and international QA systems are available for microbiology laboratories, where the United Kingdom National External Quality Assessment Scheme (UKNEQAS) for microbiology is one of the schemes administered by the Public Health Laboratory Service (Barrow and Feltham, 1993). This particular UK-based scheme is quite comprehensive, covering QA testing of pathogens as well as microscopy, serology and antibiotic sensitivity tests.

Although significant progress has been made in aquatic vaccine development bacterial disease outbreaks continue to cause significant animal losses and there is a continued need to prescribe antibiotics. There are currently no universal anti-biogram standards for aquaculture but performance standards are available for veterinary use (CLSI, 2006). These standards describe the methods, reagents and appropriate positive and negative control to test the antibiotic sensitivity of various pathogens. For any aquatic disease diagnostic facility establishing the antibiotic sensitivity testing method is important. As is the validation and assurance that this method is accurate and reliable. Alderman and Smith (2001) produced draft protocols for reference methods of antibiotic sensitivity testing of bacteria recovered from fish disease outbreaks. This is an excellent start in the production of global standards for aquaculture practices.

6.5.1   Ring testing and reference laboratories

Validation methods can be internal or external depending on the level of QA required by the diagnostic facility. One external method is described as *ring testing*. The purpose of this approach is to verify the laboratory methods using different test laboratories that all receive the same sample and follow a single protocol. It allows common problems such as contamination of samples or reagents, mislabelling, sensitivity and specificity of the tests to be evaluated. If a laboratory passes the ring test then this adds confidence that the tests are being performed appropriately and so the results are correct. This can be used for health certification laboratories as well as provision of data in emerging diseases, and in surveillance and monitoring programmes for spread of diseases.

Reference laboratories are important resources within animal disease diagnosis. These are internationally recognised facilities that have a combination of skilled personnel, validated diagnostic tests, reagents and appropriate archive material. The European Union Reference Laboratory (EURL-FISH) is funded by the EU Commission and is located in the national veterinary institute, Aarhus, Denmark. The main purpose of the EURL is to ensure high quality of all fish diagnostics performed in the member states of the EU. Furthermore this aids in the harmonisation of the protocols and methodologies applied for the various fish diseases, as described in Council Directive 2006/88/EC. The EURL is primarily concerned with notifiable viral diseases of fish (www.crl-fish.eu).

6.6   Interpreting and reporting results under a quality assurance system

Once the laboratory is established and protocols are in place then the laboratory can start to generate results. Staff training in performing the tests and also in the interpretation of the results is vital for an efficient aquatic disease diagnostic facility. Part of the quality management scheme should include not only the training of the staff but also evaluating their progress. This is an essential part of QA within the management system. Human error is part of daily life and in the laboratory it counts for a high percentage of any problems encountered. Mistakes happen and cannot always be avoided. To alleviate or reduce human error then all staff must have a good level of scientific understanding of the test they are performing and the limits of these tests.

In aquaculture, the results from the diagnostic tests will be interpretive and provided to the client. This may be the confirmation of the disease during an infectious outbreak where management strategies or treatments may be required to reduce animal losses. If the test results have not been performed as required this can invalidate the test; however, more worrying is if the personnel performing the test do not understand the intended outcomes of the laboratory test and so cannot interpret the results correctly. This may give false negative or false positive results, both of which can be equally damaging to the client and affect the health management decision to control the infection at the farm level.

Reporting results is a crucial step in aquatic disease diagnosis. Training staff to perform the tests can be done at almost any level but experience and scientific knowledge should never be undervalued or replaced with modern techniques/tests. Having skilled technical staff who can identify problems and solve these in a systematic manner is vital to any QA system. It is a poor management system that does not invest in staff training and assessment of staff competency.

6.7   Accreditation and auditing of disease diagnostic laboratories

Various QA management systems and their benefits have been discussed, where adoption of an accreditation system is a viable option for any disease diagnostic laboratory, particularly if it has a high input of samples. This may be considered as a more formal recognition of QA. There are various types or levels of accreditation but in general a third party is employed to verify and check that the laboratory is conforming to the selected standards which are part of the accreditation scheme chosen by the laboratory. A list of the most common accreditation bodies within aquatic animal health is provided in Table 6.2.

One of the most well-recognised accreditation bodies is the ISO. This is the world's largest developer and publisher of international standards with a current total of 18 500 international standards in the portfolio. This organisation comprises a network with the national standards of 162 countries, and is a non-governmental body that aims to provide a bridge between the

Table 6.2 List of the main international standards or accredited systems within aquatic animal health research and diagnosis

Type of accreditation	Source
Good Laboratory Practice (GLP)	www.legislation.gov.uk
International Organization for Standardization (ISO)	www.iso.org
World Orgnisation for Animal Health (OIE)	www.oie.int
Asia-Pacific Laboratory Accreditation co-operation (APLAC)	www.aplac.org
InterAmerican Accreditation Cooperation (IAAC)	www.iaac.org.mx
European Co-operation for Accrediation (EA)	www.european-accreditation.org

public and private sectors. The standards are always developed by a respected technical group of individuals. Whilst ISO standards are available for different aspects of aquaculture these seem to be more related to water quality testing and enumeration of microbes from fish flesh rather than the application of ISO standards for disease diagnosis *per se* (www.iso.org). Good laboratory practice (GLP) is another accredited quality control system whose principles have been applied within aquatic disease diagnostic facilities. GLP is often considered as a framework within which the laboratory studies can be conducted (OECD, 1998). The set of principles are similar to those described within other accredited bodies but in aquaculture GLP studies are research-based laboratory or aquarium studies investigating chemotherapeutant or vaccine products prior to market application. GLP is not often linked with aquatic disease diagnosis.

International accreditation may be considered as a 'gold standard' but the actual type of accreditation adopted should be linked with the running competence of the laboratory. Uptake of an accredited QA system is not something that should be taken on lightly as it can be expensive to run the diagnostic facility to the level required to meet the accreditation standards. When a disease diagnostic facility becomes accredited it means that not only is it competent to perform the various tests but that there are valid test methods in place, there are qualified and trained staff, and all equipment and facilities are sufficient to perform the tests at that standard. There is also a high level of technical expertise required to acquire and maintain accreditation status. More emphasis is placed on the staff having the technical ability to not only foresee possible problems/constraints but to be able to solve these and document them correctly. Whilst there are obvious benefits for the laboratory to work to international accredited levels, there are also other considerations, which is why such a decision should be carefully considered prior to implementation. The biggest stumbling block for many diagnostic facilities is the financial burden on the facility to maintain this QA level. The working costs for accredited status are much higher than for non-accredited schemes, resulting in greater overheads. Ultimately this often means that the final costs to process clinical samples from an accredited laboratory may be perceived as high for some clients. Whilst such QA schemes have their place, in the aquatic domain not every facility needs to run to such a high level.

Whether the aquatic disease diagnostic laboratory adopts a formal QA scheme or prefers to follow an in-house QA system, audits should be included. Ideally these should include internal but also external audits and in formal accreditation schemes they are usually performed annually. It is the role of the auditor to check the implementation of the QA and QC aspects of the quality management system and verify that all documents are correct. This should be from the initial sample submission to the reporting and archiving of any material/information related to disease diagnosis. External auditors are often part of a professionally recognised body or from an official accreditation agency. Although internal auditors can be members of staff within the diagnostic facility, they should not be primarily involved in the area they are auditing. In either case, a report is produced and action points provided to maintain or improve the level of QA. This is an important and often stressful event for the disease diagnostic staff. Normally the action points are there to assist the facility to improve the QA but in extreme cases if the QA level is found to be low then the formal accreditation can be withdrawn.

6.8 Conclusions

There are many aspects to QA to consider within any facility and more often than not, QA may be viewed negatively or thought to be tiresome and boring or restrictive. However, there are some significant benefits of providing a QA system within any aquatic diagnostic facilities. Good QA systems can indirectly improve the diagnostic capabilities of the facility, thus providing high quality results that the aquaculture sector can have confidence in.

6.9 References

ALDERMAN, D.J. and SMITH, P. 2001. Development of draft protocols of standard reference methods for antimicrobial agent suspectibility testing of bacteria associated with fish diseases. *Aquaculture* **196**, 211–243.
AUSTIN, B. and AUSTIN, D.A. 2007. *Bacterial fish pathogens: disease of farmed and wild fish*. Chichester: Springer, published in association with Praxis Publishing, 4th edition.

BARROW, G.I. and FELTHAM, R.K.A. (eds) 1993. *Cowan and Steel's Manual for the Iden-tifcation of Medical Bacteria*, 3rd edition. Cambridge: Cambridge University Press, 184–187.

BELLAMY, J.E.C. and OLEXSON, D.W. 2000. *Quality Assurance Handbook for Veterinary Laboratories*. Ames, IA: Iowa State University Press.

BROCK, T.D. 1999. *Milestones in Microbiology 1546–1940* (2nd Edition). ASM Press, 215–218.

CLSI 2006. *Performance standards for antimicrobial disk and dilution susceptibility tests for bacterial isolates from animals; Approved Standard*, 3rd edition M31-A3, Vol. 28, No 8.

DRURY, R.A.B. and WALLINGTON, E.A. 1980. General staining procedures. In: *Carelton's Histological Techniques*, ed: Drury, R.A.B. and Wallington, E.A. Oxford Medical Publication, Oxford University Press, 125–150.

GREINER, M. and GARDNER, I.A. 2000. Epidemiological issues in the validation of vet-erinary diagnostic tests. *Preventative Veterinary Medicine*, **45**, 3–22.

MORRIS, D.C., MORRIS, D.J. and ADAMS, A. 2002. Development of improved PCR to prevent false positives and false negatives in the detection of *Tetracapsula bryo-salmonae*, the causative agent of profilerative kidney disease. *Journal of Fish Diseases*, **25**, 483–490.

OECD 1998. *Principles of Good Laboratory Practise*, OECD Environmental Health and Safety Publications (OECD) 1.

OIE 2000a. *International Animal Health Code*, Ninth edition. Office International des Epizooties, Paris.

OIE 2000b. *International Animal Health Code*, Third edition. Office International des Epizooties, Paris.

OIE 2000c. *Diagnostic Manual for Aquatic Animal Diseases*, Third edition. Office International des Epizooties, Paris.

ZWEIG, M.H. and CAMPBELL, G. 1993. Receiver-operating characteristics (ROC) plots: a fundamental evaluation tool in clinical medicine. *Clinical Chemistry*, **39**(4), 561–577.

7

Antibiotics in aquaculture: reducing their use and maintaining their efficacy

P. Smith, National University of Ireland, Galway, Ireland

Abstract: The use of antibiotic therapy in aquaculture inevitably leads to the emergence of resistance in the target bacteria and a reduction of the efficacy of this method of disease control. It is argued that a better understanding of the aetiology of aquatic animal diseases and the potential of antibiotics to control disease losses is essential if we are to achieve a more prudent use of this therapy. The ability to detect resistance in target bacteria will be central if we are to use these agents wisely; the methods available to measure bacterial susceptibility are critically analysed.

Key words: antibiotic therapy, prudent use, disease aetiology, susceptibility testing, interpretive criteria.

7.1   Introduction

Before starting on the body of the chapter it is worthwhile presenting some definition of what, in this chapter, will be meant when the term antibiotic is used. In discussing the issue of definitions, FAO (2005) suggested that an antibiotic is 'a drug of natural or synthetic origin, with the capacity to inhibit the growth of or to kill microorganisms. Antibiotics that are sufficiently non-toxic to the host are used as chemotherapeutic agents in the treatment of infectious diseases of man, animals and plants.' For the purpose of this chapter a slightly tighter definition will be used: the term antibiotic will be confined to drugs of natural or synthetic origin, with the capacity to inhibit the growth of or to kill bacteria and that are sufficiently non-toxic to the host to be used as chemotherapeutic agents in the treatment of those infectious diseases of humans, animals and plants that have a bacterial aetiology. As a consequence, agents for the control of eukaryotic microorganisms and substances classed as disinfectants or antiseptics will not be discussed.

7.1.1   The scope of this chapter

This chapter will focus on the reduction of antibiotic use in aquaculture. The primary reason why we should be interested in reducing this use relates to the remarkable ability of bacteria to adapt to changes in their environment. When the environment contains chemicals, such as antibiotics, that are inhibitory to bacteria, strains of bacteria will emerge that are resistant to those chemicals. This has the consequence that the therapeutic use of antibiotics is always subject to a negative feedback. The more an antibiotic is used the greater will be the selection pressure for the emergence of resistant strains. The higher the frequency of resistance in the target bacteria the less likely it is that the antibiotic will be therapeutically effective. Therefore, the more an antibiotic is used, the less effective it is likely to be. It is important to recognise that this is an absolute rule. The rate at which resistance will emerge may vary depending on the agent and the target bacteria, and the time and extent of the response may not be predictable, but the use of antibiotics will always lead to resistance in the target bacterium. Smith *et al.* (1994) have argued that resistance in these target bacteria is the one negative consequence of aquacultural use of antibiotics for which there is hard, direct and compelling evidence.

The risk that aquacultural antibiotic use might impact on the frequencies of resistance in human pathogens has also resulted in calls for the reduction in this use (FAO, 2005; WHO, 2006; Heuer *et al.*, 2009). This is a complex issue and a systematic and critical review of the available evidence is overdue, but it will not be attempted here. We will confine ourselves to noting our view that the conclusions of some studies in this area are not supported by the research they quote. A critical analysis of one much quoted review (Cabello, 2006) is provided by Smith (2010). The conclusions of other papers (Angulo, 1999, 2000; Angulo and Griffin, 2000) are critically reviewed in Smith (2007) and Smith (2008d).

The risks to humans resulting from the presence of residues of therapeutic antibiotics in aquacultural products presented to consumers have also been cited as a concern (WHO, 2006). From a regulator's perspective residues can be dealt with by setting maximum residue levels (MRL) that are compatible with consumer safety. For producers, problems with unacceptable residues can be dealt with by withholding animals from the market for a sufficient time, the withdrawal time, after a period of therapy. There are problems with setting both MRL and withdrawal times. These are mainly bureaucratic and financial in the case of MRL but in the case of withdrawal times the problems relate to the diversity of aquaculture. Once an MRL has been set, the withdrawal time is largely a function of pharmacokinetics (PK) (EAEM/CVMP/026/95). However, the PK of antibiotics varies with the host (Samuelsen, 2007), the environment and in some cases the laboratory doing the analysis (Smith, 2008a). The issues concerning setting withdrawal times will not be discussed further in this chapter. However, it should be noted that, although there is no evidence that residues have ever resulted in adverse effects in consumers, the imposition of MRL has had an effect on patterns of antibiotic use in aquaculture (Gräslund and Bengtsson, 2001). Anecdotal evidence (Alday-Sanz, personal communication) suggests that fear of the economic consequences of market rejection as a consequence of products containing in excess of the MRL has resulted in large reductions in antibiotic use in some industries.

This chapter will focus on the aspects of the therapeutic use of antibiotics that can lead to a reduction in that use and, therefore, to maintenance of their efficacy as therapeutic agents. Whatever our motives, if we are to reduce antibiotic use in aquaculture we must have a good understanding of the diseases we are trying to control, the ways antibiotics are used in aquaculture and the potential and limitation of those uses. In developing this understanding, it is important that we remember that aquaculture is a commercial activity. In aquaculture, antibiotic therapy is primarily seen by aquaculturalists as a method of protecting profits rather than protecting aquatic animals (Smith, 2010). It will be a source of error if we limit our evaluation of antibiotic therapy in aquaculture purely to its ability to combat bacteria.

7.2 The epidemiology and aetiology of bacterial diseases in aquatic animals

To understand the potential of antibiotic therapy to control disease losses we need to develop an understanding of the aetiology of those diseases. It is clear that antibiotics, as we have defined them, can have no beneficial effect when the infectious agent involved in the disease is a virus or a eukaryotic microorganism. They can have a role in the control of losses consequent on diseases only if the disease is associated with a bacterial infection. The most important breakthrough in our theoretical understanding of these infectious diseases was the development of the germ theory at the end of the nineteenth century. Unfortunately a rather simplistic and erroneous understanding of this theory has gained popularity with the general public and more disturbingly with many scientists. What the theory actually stated was that specific bacteria were associated with the aetiology of specific diseases. What it did not state was the bacteria were *the* cause of these diseases. The difference between these two statements may appear to be minor but it has a major influence on our understanding of how and when to use antibiotics.

As Smith (1997) has observed, many standard textbooks on fish diseases and many scientific papers contain comments of the type '*Aeromonas salmonicida*, the causative agent of furunculosis' or 'furunculosis, the disease caused by *Aeromonas salmonicida*'. The mind-set induced by statements of this kind can, all too easily, lead to an excessively simplistic view of therapy. For example, laboratory examination of fish from a disease outbreak may result in the isolation of a number of bacteria that are related to groups that are known to have been associated with disease. On the basis of these tests the bacteria may be presumed to be *the* cause of the disease outbreak and, therefore, antibiotic therapy would be initiated to reduce the losses associated with the disease outbreak. At first glance there might appear to be nothing wrong with arguments of this type.

Our current understanding of infectious diseases, however, is that they always have a multi-factorial aetiology and that bacteria are never more than one of the factors contributing to the aetiology of a particular disease outbreak (Smith, 1997; Thrusfield, 1986). The bacterial component of the aetiology of a disease may vary from major to minor. It is reasonable to suggest that the ability of an antibiotic therapy to reduce losses in a particular disease outbreak will show some proportionality to the degree that the bacterium is a major causal factor of the losses in that outbreak. There will be disease situations where, although a bacterial involvement is demonstrated, that involvement is so minor or secondary to the major causal factors that the removal of these bacteria, by antibiotic treatment, would have little or no effect on the losses consequent on the disease. In investigating a disease we need to establish not only whether a specific bacterium is involved but also achieve some understanding of the importance of its involvement.

It should be noted that the isolation of a particular bacterium could not, of itself, provide information as to the relative importance of that bacterium in the disease process. Some indirect indication of its possible importance might be obtained from previous studies of the virulence of previously studied members of the group to which the isolate belongs. For example, studies of *Vibrio alginolyticus* have suggested it has low virulence with a median lethal dose (LD_{50}) of 10^7–10^8 cfu for large yellow croakers (Qian *et al.*, 2008). This information might provide some guidance as to the appropriate interpretation of laboratory reports of the isolation of *V. alginolyticus* from moribund fish sampled during a disease outbreak. The low virulence of the bacterium would strongly suggest that additional, non-bacterial, causal factors should be considered as making a major contribution to the aetiology of the outbreak. In such situations, any antibiotic therapy might result in only a small and short-term reduction in losses. However, if the non-bacteriological factors are not addressed, the underlying weakness and susceptibility of the host will continue. In all probability the losses will resume after the period of therapy and possibly another opportunistic bacterium, with an equally low virulence, will be found by laboratory investigation.

The knowledge that infectious diseases of aquatic animals have a multi-factorial aetiology means that caution would always have to be exercised in interpreting the importance of any isolation of low-virulence bacteria. It cannot, however, be assumed that the need for caution can be less when a laboratory reports the isolation of a bacterium with high virulence.

A. salmonicida has been shown to be highly virulent with a LD_{50} of less than 1 cfu following intraperitoneal (i.p.) challenge (Drinan, 1985). Despite this, Smith (1997) has developed a number of disease scenarios where the bacterium would be isolated from moribund fish while at the same time playing only a minor role in the aetiology of the outbreak. In the situations represented by these scenarios, it is probable that the administration of antibiotics would have little or no impact on the losses incurred and would not be prudent or economically justifiable.

The information that can be gained from laboratory investigations regarding the aetiology of a disease outbreak will always be partial. Excessive reliance on such data would automatically result in an overemphasis on the bacterial component of the aetiology. Such overemphasis on the bacterial component of the aetiology will inevitably result, on occasions, in recommendations that antibiotics should be administered in situations where they can have little value. A full picture of the epizootiological factors involved in a disease outbreak can be gained only when laboratory data are used in conjunction with on-site observations by an experienced health-care professional. The development of the skills of these professionals is essential if imprudent use of antibiotics is to be avoided.

This problem has been made more acute by the increasing reliance on high-tech molecular tools such as polymerase chain reaction (PCR), by laboratories involved in diagnosis. Hiney and Smith (1998) have reviewed the significant problems associated with validating these tools in a diagnostic context. However, even if these problems are set aside there are further problems encountered when results of PCR analysis are used to inform decisions concerning the appropriateness of antibiotic therapy. These can be illustrated using, as an example, a recently developed PCR system for the detection of *Vibrio anguillarum* (Hong *et al.*, 2007). The authors claimed that this bacterium was a fish pathogen, and the cause of vibriosis in fish, on the basis of an LD_{50} value of 5.4×10^9 cfu in flounder. The PCR was shown to have the ability to detect $<10^3$ *V. anguillarum* per gram. With this tool a diagnostic laboratory would be capable of producing evidence of the 'causative agent of vibriosis' from the examination of fish that contain *V. anguillarum only* at a concentration several orders of magnitude less than its LD_{50}. If the data were not interpreted in the broader context supplied by on-site epizootiological investigations, the potential for inappropriate recommendations concerning antibiotic therapy is obvious.

7.3   The use of antibiotics in aquaculture

7.3.1   Antimicrobial use in populations; prophylaxis and metaphylaxis

The vast majority of antibiotic treatments in aquaculture are administered to populations. Although there is some use of bath administration in aquaculture, particularly in crustacean and shellfish hatcheries, oral administrations, involving the presentation of antibiotic-containing feed, are by far the most frequent means by which such population treatments are delivered in finfish farms and crustacean grow-out facilities. Antibiotic treatments can be classified as prophylactic if the populations treated do not contain any infected individuals. However, rarely, if ever, are all members of a population to be treated experiencing an infection at the time of treatment, therefore, the term therapeutic, as used in classifications of individual therapies, cannot be correctly used with respect to treatments of populations. When treatments are given to populations that contain infected individuals they should be classified as metaphylactic.

Prophylaxis
Prophylactic treatments of populations are undertaken for a variety of reasons or to address a variety of real or perceived problems. These can be usefully designated as specific risk prophylaxis, general risk prophylaxis and growth promotion.

Specific risk prophylaxis can usefully describe treatments given when it is possible to predict, with some confidence, that a population will face a risk of a specific bacterial infection at a specific time. It should be noted that the protection of the host population will be only for the period of the therapy and, possibly, for a few days after. This form of prophylaxis, therefore, is likely to be effective if and only if the occurrence and particularly the timing of the infection and, therefore, the disease risk, can be accurately predicted. Provided these conditions are met, however, specific risk prophylaxis can be considered as prudent and may be economically justifiable.

General risk prophylaxis describes treatments given when a population is entering a phase of production where they can be predicted to experience non-specific, normally low-level, but economically significant mortalities. The use of antibiotics in this way is rare in finfish culture and in crustacean grow-out facilities. It is however, reported to be more widespread in crustacean and shellfish hatcheries (Gräslund *et al.*, 2002). Despite the fact that there are few studies of the efficacy and long-term consequences of this form of prophylaxis, the prudence of antibiotic use in these situations must be seriously questioned. General principles would suggest that the chronic low-level mortalities it is used to combat would have a very large environmental component in their aetiology. Therefore, it is highly probable that investing in improvement in management and husbandry would represent a better option than continual reliance on antibiotics.

Growth promotion is a form of prophylactic treatment, often involving sub-therapeutic dosages, that is undertaken not to address a specific or general disease risk but in the hope that it will improve the growth rate of the host. Historically, growth promotion has been a phenomenon of land-based agriculture. In these animals increases in growth are thought to result from decreases in animal gut mass, increased intestinal absorption of nutrients and energy sparing (FAO, 2005). There are no data demonstrating that these effects are seen in aquatic animals and use of antibiotics as growth promoters is generally not thought to play a significant role in aquaculture.

Metaphylaxis

In aquaculture antibiotic treatments are initiated most frequently after the isolation of a potential target bacterium or after the disease signs have allowed the assumption of the presence of such a bacterium. They are, therefore, metaphylactic in character. An important aspect of metaphylactic population treatments is heterogeneity of the populations that receive them (Smith, 2008a). Individuals in a treated population range across a wide spectrum. At one end there are those that are fully healthy, feeding at the optimal rate and who never will face an infection challenge. At the other end there are individuals in whom the infection is well established, who are in the terminal stages of the disease and who are, in all probability, not feeding at all. We have little or no experimental data that allows us to decide whether metaphylactic treatments work by preventing *de novo* infections in the uninfected portion of the population or by combating the infections in those already infected. The data produced by Coyne *et al.* (2004a,b, 2006) would suggest that prevention of new infections is probably the more significant. Having examined nearly 200 salmonid fish, sampled at the end of five different therapies in commercial fish farms, they found only one that contained both the infecting bacterium and significant concentrations of the agent. In general they reported that the healthy fish they examined contained antibiotic but no target bacteria and the moribund fish they examined contained bacteria but little or no antibiotic. Studies in Norway (Horseberg, personal communication) and the UK (Alderman, personal communication) have also been interpreted as evidence that the primary function of treatments to populations is to prevent the initiation of infections in the healthy fish in that population. If these conclusions are correct, and they appear to be consistent with the intuitive but experience-based assumptions of many veterinarians, then the treatments should be considered as metaphylactic at a population level but prophylactic at the individual level.

These considerations become particularly important when the issue of placing orally administered antibiotic therapy on a rational foundation is addressed. It is difficult to develop rational, prudent and optimised therapies when we do not know in which sub-population we expect the antibiotic to exert its beneficial effect.

7.3.2 The role of antibiotic therapy in the control of disease losses

There are remarkably few publications that demonstrate the efficacy of antibiotic therapies in commercial aquaculture. However, this is more likely to be an artefact of the conventions of scientific publication than evidence of the fact that such efficacy is rarely observed. Given the long history of the use of these agents, it must be assumed that antibiotic therapy can both combat bacterial infections of aquatic animals and reduce, at least in some cases, the economic losses consequent on diseases. What cannot be assumed, however, is that antibiotic therapy is, either in the short term or the long term, the most efficient and the most cost-effective approach to reducing losses in aquaculture.

There are essentially four ways in which losses consequent on infectious diseases can be reduced. The first is to limit the contact between the infectious agents and the hosts. In aquaculture this is normally achieved by controlling the movement of infected animals into aquaculture facilities. The second is to raise the non-specific resistance of the hosts to infection. In aquaculture, this can be achieved by improving their nutrition, the environment in which they are reared and the husbandry practices in their rearing facilities. The third is to raise the specific resistance of the hosts to infection by particular agents. In aquaculture, vaccines can achieve this when they are available and when the host is capable of mounting a protective immune response. All three of these control measures are prophylactic. They can be implemented before any disease outbreak occurs and their aim is to prevent infection of the hosts. However, should these attempts fail, the only course of action open to farmers or their heath-care advisors is therapy. This fourth way of attempting to control losses, antibiotic therapy, is applicable only in those situations where bacterial infection plays a significant role in the aetiology of the disease process. Although it is the only course of action available once a disease outbreak has started, it must always be remembered that the occurrence of that outbreak represents a failure of one or all of the three prophylactic control measures.

From the perspective of this chapter, the question is to what extent can the various aquaculture industries optimise their application of the prophylactic control measures and, thereby, minimise their reliance on antibiotic therapy. We can predict that, in some situations, failures of these prophylactic measures to control losses are almost unavoidable. Failure is particularly likely when new and emerging diseases are encountered, where vaccines have yet to be developed or where the exploitation of new aquatic animals is being explored (Smith. 2008b). In these situations some reliance on antibiotic therapy is to be expected. However, salmon farming in Norway provides an example of a mature industry that can and does operate without the need for antibacterial therapy. Norwegian scientists (Grave et al., 1996, 1999; Markestad and Grave, 1997; Lillehaug et al., 2003) attribute the reduction in antibiotic use to the availability of effective vaccines and not, as claimed by Cabello (2006), to the implementation of regulation limiting that use. Hiney and Smith (2000) have, however, argued that improvements in husbandry may have also played a significant role. Whatever the reason for the decline, the Norwegian experience demonstrates that major epizootics can be effectively controlled by non-chemotherapeutic prophylaxis. It also demonstrates that attempts to deal with them by the administration of antibiotics were ineffective and economically and environmentally unsustainable.

Some idea of the relative importance of antibiotic therapy as a disease control measure in populations can be gained from an analysis of the patterns of human infectious disease mortality. During the last century deaths to infectious diseases showed a dramatic decline in developed countries. This decline is popularly believed to have been a result of developments in scientific medicine and, in particular, of the application of antibiotics. This belief is not, however, supported by the detailed analysis of mortality data for England and Wales reported by McKeown (1976). He observed that deaths to both viral and bacterial diseases decreased by approximately 90% in the period before any chemotherapeutic agents became available or any vaccine use (with the exception of that designed to control smallpox), was widespread. He concluded, 'immunization and treatment contributed little to the reduction of deaths from infectious diseases before 1935 and over the whole period since the cause of death was first registered (1835) they were much less important than other influences'. These analyses demonstrate that, even in densely populated areas and contrary to popular mythology, a dramatic decline in deaths to infectious diseases can and did occur without reliance on chemotherapy.

7.4   Efficacy of antimicrobial therapy in aquaculture

The efficacy of any antibiotic treatment depends on pharmacokinetics (PK) and pharmacodynamics (PD) properties of the agent administered. These are, in turn, contingent on the environment in which the agent is administered. Thus, the huge diversity of aquaculture presents real problems for attempts to develop rational, efficacious, evidence-based and prudent antibiotic therapies (Smith *et al.*, 2008a). Even within treatments of one species treatments can be administered under widely differing environmental conditions. These diversities also present problems for attempts to regulate antimicrobial use via marketing authorisations. The labelling requirements of these authorisations specify the purposes and the specific conditions for which the products may be sold. Unfortunately the diversity of aquaculture has the consequence that much antibiotic use will always remain 'off-label'.

In commercial aquaculture, the efficacy of any course of antibiotic therapy is dependent on the ability of the dosage regimen employed to deliver a concentration of the agent to the appropriate site in the host that is sufficient to inhibit the infecting bacterium. Thus, the major factors that influence efficacy are the dose regimen used and the susceptibility or resistance of the target bacterium.

7.4.1   Dose regimen

In human medicine very sophisticated models, based on PK and PD data, have been developed to facilitate dose optimisation both with respect to clinical outcomes and the minimising of the selective pressure for resistance (Drusano, 2004; Ambrose *et al.*, 2007; Lees *et al.*, 2006; Schmidt *et al.*, 2008). In aquaculture, largely owing to our lack of appropriate PK data and our failure to understand the true nature of metaphylactic treatments, these modelling approaches are not, as yet, applicable (Smith, 2008a). In aquaculture the dose regimens that have been recommended are empirical. In the main they are not even the product of systematic comparative studies of the efficacy of various possible dose regimens. They are simply recommended because they have been shown, under certain circumstances, to work. A reading of the extensive database of fish pharmacokinetics produced by Reimschuessel *et al.* (2008) suggests that there is considerable variation in dose regimen employed and that very few scientific papers have reported on the efficacy of the regimen they studied. Even fewer papers have compared the relative efficacies of different dose regimens in commercial farms. The study of Darwish *et al.* (2002) is a notable exception.

The data requirements for obtaining a market authorisation include the demonstration of clinical efficacy. Therefore we can be assured that, even if much of it is not in the public domain, data for the efficacy of some agents, under some conditions, do exist. We can be less certain, however, that the data, even in the limited conditions to which they apply, allow the establishment of optimal dose regimen. We cannot be certain that the dose regimens associated with market authorisations provide accurate guides as to the minimum amount of antibiotic that can be used to achieve a given clinical goal. We must always be mindful of the fact that pharmaceutical companies have performed or provided the funds for much of the research that led to the dose regimens that have been recommended. The demonstration that the source of funding significantly influences the conclusions drawn from any research (Kjaergard and Als-Nielsen, 2002) suggests that more, independent, research in this area might be productive.

Of particular interest here is the duration of the therapy period. Lambert (1999) has argued that, even for human treatments, there is little empirical data to justify the duration of recommended therapies. He suggests that, for many human infections, much shorter treatments are just as effective as the longer treatments currently recommended. In aquaculture there appears to have been totally inadequate investigation of the influence of the length of therapies on their clinical efficacy.

7.4.2   Susceptibility and resistance to antibiotics

Mechanisms that reduce the susceptibility of bacteria

There is a variety of ways in which bacteria can reduce their susceptibility to antibiotics *in vitro*. These vary in their molecular mechanisms, the location of the genes encoding the mechanisms, in their specificity and in the degree to which they reduce susceptibility. As a general comment it can be stated that, with respect to the mechanisms by which their susceptibility to antibiotics can be reduced, aquatic bacteria show no differences to terrestrial bacteria.

Possibly the most commonly studied resistance mechanisms are those resulting from the possession of genes encoding specific, positive-function resistance (Sørum, 2006). The phenotype conferred by these genes normally results in a very significant reduction in susceptibility. Although occasionally chromosomally located, these genes cannot be acquired by mutation. They are acquired as a result of the transfer of genetic material from other, frequently unrelated, bacteria. They are often located on genetic elements, such as plasmids, that encode their own transfer. Each gene is normally specific to one antibiotic or class of antibiotics but is frequently located on genetic structures that contain multiple resistance genes with separate specificities. Current evidence suggests that the genes of this type that have been detected in bacteria associated with aquatic animals are identical or closely related to those found in the bacteria associated with human disease (Sørum, 2006).

A second mechanism whereby susceptibility may be reduced is as a result of chromosomal mutations in genes encoding the proteins that are the targets of the antibiotic (Giraud et al., 2004). Such mutations result in changes in susceptibility that are specific to a particular antibiotic or class of antibiotics. They may result in major changes in susceptibility but are frequently associated with only minor changes. Importantly these mutated genes are not transferred horizontally but are inherited only vertically by progeny cells.

There are a number of mechanisms that result in reduced susceptibility that is not specific to a single antibiotic class. Poole (2005) and Li and Nikaido (2009) have provided reviews of the efflux pump systems that mediate multi-drug resistance. These are generally chromosomal and, therefore, are not self-transmissible. They frequently result in moderate or small reductions in susceptibility but are often found in conjunction with other resistance mechanisms (Giraud et al., 2004). Other, less characterised, mechanisms may also confer multiple drug-resistance phenotypes (Nikaido, 1989). Barnes et al. (1990) and Wood et al. (1986) have both reported on the multiple, low-level resistances associated with outer membrane changes. Smith et al. (1994) have argued that some of the reduced susceptibilities to antibiotics that have been detected in clinical isolates may be mediated by non-genetic persistence mechanisms (Bryan, 1989: Balaban et al., 2004) that are inherently unstable. The data of Tsoumas et al. (1989) and Rodgers (2001) also suggests that, at least in some cases, such reduced susceptibility might be unstable on storage.

The important issue for susceptibility testing is to note that resistance can be achieved by a number of mechanisms and that these mechanisms result in changes in susceptibility that vary from major to the barely detectable. It is also important to note that, although the molecular biology of these mechanisms has been studied in detail, there have been few, if any, studies that have reported on the clinical significance of the changes in susceptibility they produce.

The need to determine the susceptibility of target bacteria
The efficacy of any dose regimen will be modified by any changes in the susceptibility of the target bacterium. Information on that susceptibility is essential in making the prediction of the clinical outcome of any proposed therapy that is required by the constraints of prudent use. Thus, as a result of the need for prudence, it is essential that the determination of the susceptibility of the target bacterium be performed for every therapy. Ideally, this should be completed before the initiation of therapy. However, performing these tests takes time and there is evidence that any delay in the initiation of therapy reduces its efficacy. A delay of seven days in the initiation of therapy can result in a ten-fold increase in mortalities. In practice, therefore, clinical and commercial considerations may require the initiation of therapy before the necessary susceptibility data are available. In that situation, the choice of agent has to be based on predictions of the probable susceptibility of the target bacterium. These predictions can be informed by the results of previous testing of that bacterium either at the farm or in the region. It is, however, essential that when therapies are initiated based on predictions of probable susceptibility, these predictions be experimentally confirmed by performance of susceptibility tests at the earliest opportunity.

7.5   Laboratory detection of clinical resistance

The laboratory detection of clinical resistance is composed of two separate but interlinked processes. One, the determining of a measure of *in vitro* susceptibility, is simple and the other, determining what clinical meaning can be given to that measure, is very complex. Discussion of this issue has not been facilitated by confusion in the exact meaning of the terms that have been used.

7.5.1   Susceptibility and susceptibility testing
Various *in vitro* studies of the bacterium can be used to generate a quantitative measure of the concentration of an antimicrobial agent required to kill or, more frequently, to inhibit the growth of a bacterium in a particular laboratory environment. In this chapter the methods used in such studies are termed susceptibility tests and the parameter they measure is bacterial susceptibility.

The most commonly used methods fall into one of two general categories (FAO, 2005) that can be termed either direct or indirect. The direct methods measure the concentration of an agent that needs to be added to laboratory media, broth or agar, to prevent bacterial growth. They generate results, termed the minimum inhibitory concentration (MIC), in unit of mg/l. It is important to realise that the relationship between a bacterium and an antibiotic is context dependent. Thus, an MIC determined in one environment would not be the same as an MIC determined in another (Barker *et al.*, 1995). In particular, a laboratory-determined MIC could not be taken as indicating the concentration needed to inhibit growth *in vivo* (McCabe and Treadwell, 1986). Indirect methods include disc diffusion studies where the results, the diameters of the zones of inhibition, are obtained in terms of mm. The limited studies that have been made suggest that there are no differences in the accuracy of MIC and disc diffusion methods (Jones, 1992). However, probably for logistical reasons, the vast majority of front-line laboratories investigating the susceptibility of clinical isolates employ indirect, disc diffusion methods (Smith, 2006).

7.5.2   Resistance and clinical resistance

There is a long tradition and vast literature which uses the term resistant to refer to bacteria that have been shown, in susceptibility tests, to be less susceptible than other members of the group or species to which they belong. It is important to note that resistance, as defined here, is a measure of an *in vitro* bacterial phenotype. It is a formal error to consider that it has a necessary relationship to the clinical outcome of any proposed therapy. Defined in this way resistance is a very useful term for those whose primary interest is in bacterial phenotypes and the molecular and genetic mechanisms that underlie them. It is, however, less useful and can, in fact, be misleading, for those whose primary interest is in predicting the possible outcomes of a proposed antibiotic therapy.

In the application of antibiotics in aquaculture a central problem faced by veterinarians or health-care professionals is to establish whether or not a proposed course of therapy has a chance of working. They need to make a treat/not treat decision. One major reason why a proposed therapy would not work and the 'not treat' option should be taken, is that the target bacterium is, in a clinical sense, resistant. These considerations have led Smith (2008a) to offer the following, clinically relevant, definition of resistance: 'A bacterium should be considered as (clinically) resistant if, as a result of its reduced susceptibility to an antimicrobial agent, it would be able to continue to contribute to morbidity and mortality in a population during and after a particular administration of that agent to the population'. In this chapter the term clinical resistance will be used in this sense. Defined in this way clinical resistance would have direct relevance to those engaged in therapy. If a target bacterium were to be classified as clinically resistant with respect to a proposed therapy, then that therapy would have no chance of working and the initiation of that therapy would be neither prudent nor defensible.

It should be noted that clinical resistance is a parameter that cannot be established using data gained simply from the study of the *in vitro* susceptibility of a bacterium in the laboratory. Establishing whether a bacterium is clinically resistant requires information about the bacterium to be interpreted in terms of quantitative data relating to the properties of both the therapy itself and the infectious process being treated.

7.6   The Clinical and Laboratory Science Institute (CLSI) approach to susceptibility testing in aquaculture

With respect to susceptibility testing and the formulation of criteria to interpret the data generated by such tests, there are major differences between the situation in human and veterinary medicine on the one hand and the situation in aquaculture on the other.

In human and veterinary medicine there are numerous standardised and validated methods each associated with their own interpretive criteria. The main challenge here is, therefore, to harmonise various well-established but differing methods (Kahlmeter *et al.*, 2003). In aquaculture, however, the situation is different. The guidelines M42-A (CLSI, 2006a) and M49-A (CLSI, 2006b) developed by the Clinical and Laboratory Science Institute (CLSI) from the original proposals by Alderman and Smith (2001), present what are essentially the only standardised and validated protocols available for susceptibility testing of bacteria associated with aquatic animals. In aquaculture the challenge is not to harmonise a number of different protocols but to coordinate the studies needed to further develop the one set that has a real chance of gaining worldwide acceptance.

The underlying aim of the CLSI guidelines is to provide universal, validated, laboratory-independent criteria for the clinical interpretation of susceptibility test data. The central problem with achieving this aim is the relatively low inter-laboratory precision of susceptibility testing methods (NicGabhainn *et al.*, 2003). To minimise this inter-laboratory variation, the CLSI guidelines provide detailed specification of the test protocols to be used. However, even when standard protocols are used, there is still significant variation between the data generated by different laboratories (Smith, 2008c). Therefore, the CLSI guidelines also specify obligatory quality control measures designed to further reduce inter-laboratory variation. For each set of test conditions, appropriate control strains are identified and the acceptable range of results that can be obtained with the control strains are specified. Laboratories can claim to be in compliance with the CLSI guidelines only if they meet these obligatory quality control requirements. Only laboratories that are in compliance can apply the interpretive criteria published in the guidelines.

7.6.1   Current status of the CLSI guidelines

To develop susceptibility test guidelines that cover all the bacterial species that are encountered in aquaculture is a mammoth task. Significant progress has been made but much is still to be done (Table 7.1). Essentially these guidelines can be thought of as comprising three sections covering the test protocols, the quality control requirements and the interpretive criteria.

Test conditions

It would appear that approximately a dozen different test conditions would be necessary to facilitate the testing of all the various bacteria that have been associated with aquatic animal diseases. The CLSI has been able to recommend or, at least, to suggest test conditions for the majority of these bacteria.

Quality control requirements

With respect to developing the requisite quality control procedures, the first step is to establish the appropriate control organisms for each of the test conditions. Suitable control strains (*Escherichia coli* ATCC 25922 and *A. salmonicida* ATCC 33658) have been set for tests performed under the conditions recommended for bacteria that yield results on unmodified Mueller-Hinton (MH) agar in 48 h at either 22 °C or 28 °C (Group 1 organisms, Table 7.1). Data showing that the same control strains can be used with the diluted MH media required for the testing of flavobacteria (Group 3) are currently under consideration. The standard control organisms that will be required for the other test conditions have not yet been set.

The second step is the setting of acceptable ranges for control strains. Acceptable ranges have been set for MIC and disc diffusion tests of control strains tested under conditions recommended for the Group 1 bacteria (Table 7.1). With respect to the conditions proposed for MIC tests of Group 3 bacteria, studies of control strain performance have been completed that should lead to the establishment of acceptable ranges. For the other groups, work on setting the other appropriate acceptable ranges cannot even commence until the appropriate test conditions and control strains are identified. The procedures for setting the acceptable ranges for control strains have been specified in the CLSI guidelines M37-A3 (CLSI, 2007) and M31-A3 (CLSI, 2008) and involve the cooperative work of at least seven, multinational laboratories. It should be noted that, until these acceptable ranges have been set for a particular set of test conditions, laboratories cannot claim that tests carried out under those conditions were performed in compliance with CLSI guidelines. Importantly, laboratories cannot contribute to setting interpretive criteria unless they can demonstrate that they are in full compliance.

Table 7.1 Current progress with susceptibility testing reported in M42-A (CLSI, 2006a), M49-A (CLSI, 2006b) and M42/49-S1 (CLSI, 2010)

Species	Test conditions[a]	Quality control strains and acceptable ranges	Interpretive criteria	
			Epidemiological cut-off values	Clinical breakpoints
Group 1 Non-fastidious species				
Aeromonas hydrophila	Set	Set	5 agents disc	2 agents disc
Aeromonas salmonicida subsp. *salmonicida*	Set	Set	2 agents MIC	2 agents MIC
Aeromonas salmonicida (atypical)	Set[a]	Set		
Aeromonas caviae	Set	Set		
Aeromonas sobria	Set	Set		
Edwardsiella ictaluri	Set	Set		
Edwardsiella tarda	Set	Set		
Pseudomonas anguilliseptica	Set	Set		
Vibrio anguillarum	Set	Set		
Vibrio harveyi	Set	Set		
Vibrio ordalii	Set	Set		
Vibrio parahemolyticus	Set	Set		
Vibrio vulnificus	Set	Set		
Yersinia ruckeri	Set	Set		
Group 2 Obligate halophiles				
Obligate halophilic Vibrios[b]	Suggested			
Photobacterium damselae subsp. *piscicida*	Suggested			
Photobacterium damselae subsp. *damselae*	Suggested			

Group 3 Gliding bacteria[c]

Species		
Flavobacterium branchiophilum	Set/suggested	Set/suggested
Flavobacterium columnare	Set/suggested	Set/suggested
Flavobacterium psychrophilum	Set/suggested	Set/suggested

Group 4 Streptococci

Species		
Lactococcus garvieae	Set/suggested	
Streptococcus agalactiae	Suggested	
Streptococcus iniae	Suggested	
Streptococcus phocae	Suggested	
Vagococcus salmoninarum	Suggested	

Group 5 Miscellaneous and fastidious bacteria

Species		
Francisella asiatica		
Francisella piscicida		
Moritella viscose	Suggested	
Piscirickettsia salmonis		
Renibacterium salmoninarum		
Tenacibaculum maritimum	Suggested	
Vibrio salmonicida	Suggested	

[a] In this column 'suggested' indicates that the probable test condition have been identified but that further evidence is required to confirm their suitability for susceptibility testing.

[b] Current evidence suggests that most Vibrios will generate reliable test results with Muller-Hinton media and can, therefore, be treated as non-fastidious species. There may, however, be exceptions that require an addition of 1% NaCl.

[c] For the Group 3 species the test condition for MIC tests have been set but the test condition for disc diffusion tests have only been suggested.

Interpretive criteria

The informational supplement M42/49-S1 (CLSI, 2010) contains two clinical breakpoints (CB) for oxytetracycline and oxolinic acid and five epidemiological cut-off values (ECO) for the interpretation of *A. salmonicida* disc diffusion susceptibility data. With respect to MIC data for this species the numbers are 2 and 2 respectively. No other interpretive criteria have been set and, therefore, at present, the CLSI guidelines cannot be used to interpret the susceptibility test data relevant to any bacteria other than *A. salmonicida*.

7.6.2   Approaches to setting interpretive criteria

The primary aim of interpretive criteria is to allow the determination of clinical resistance and, therefore, facilitate an interpretation of susceptibility measures in terms that can be validly applied to predicting the outcome of a particular therapy. The most effective way this can be achieved is by setting protocol-specific but laboratory-independent clinical CB.

In the approach of CLSI to setting CB (CLSI, 2007, 2008) two primary data sets are used. The first are clinical outcome data and the second are the distribution of *in vitro* susceptibility data. These can be used to generate cut-off values CO_{cl} and CO_{wt}, respectively. CO_{cl} values are the smallest susceptibility inhibition zone (largest MIC) measures that are determined for bacteria associated with successful therapies. CO_{wt} are the lower limit of the distribution of *in vitro* inhibition zones (upper limit MIC) of fully susceptible strains. If these two cut-off values are in agreement they can be used to set CB. Cut-off values (CO_{pd}) can also be generated from a consideration of pharmacokinetic data and pharmacodynamic data. CLSI (2007) recommend that CO_{pd} be used in the setting of CB only to resolve disagreements, if any, between CO_{cl} and CO_{wt}.

In practice clinical outcome data have rarely been produced from commercial aquaculture and even more rarely have they been associated with data on the susceptibility of the target bacterium using CLSI methods. The authority that can be given to estimates of CO_{cl} is proportional to the number of studies of clinical outcomes that were used to generate them. Thus, the paucity of data of the type needed to set CO_{cl} raises serious questions as to whether precise CB can, in the short or medium term, be set using the procedures outlined in M37-A3 (CLSI, 2007).

Smith (2008a) has analysed the use of PK/PD data in setting CO_{pd} relevant to the use of antibiotics, particularly the quinolones, in aquaculture. There have been major advances in this use of PK/PD data in human medicine in the last decade (Turnidge and Paterson, 2007) and European authorities have given this approach a dominant role in setting CB (Kahlmeter *et al.*, 2003). However, Smith (2008a) concluded that the many unresolved theoretical and practical problems, mainly those associated with metaphylactic population treatments, means that this approach to setting CB relevant to aquaculture is unlikely to be cost-effective.

Considerations of the difficulties in estimating cut-off values from either clinical outcome or PK/PD data suggests that the only relevant data will be available, in any reasonable timeframe, is the distribution of *in vitro* susceptibility measures. The only cut-off values for some time will be CO_{wt}.

Estimates of CO_{wt} for one species or closely related group can be made by visual examination of the distribution of the data obtained from testing a sufficient number of them (Miller and Reimschuessel, 2006). This method is subjective and is liable to be influenced by the existence of strains manifesting a low-level resistance that result in a small reduction in their susceptibility. A more sophisticated approach has been developed by Kronvall (2003) and used in aquaculture by Smith *et al.* (2007, 2009), Ruane *et al.* (2007), Douglas *et al.* (2007) and Avendaño-Herrera *et al.* (2011). This approach, normalised resistance interpretation (NRI), relies on a statistical approach and has the added advantage that it is not influenced by the existence of strains with low-level resistance (Kronvall, 2003). Recently Kronvall (2010) has demonstrated that NRI can also be applied to MIC data.

Whether they are analysed visually or by NRI analysis, an important question is the number of strains that are required to generate a reasonably precise estimate of CO_{wt}. The data of Smith *et al.* (2009) have shown that the precision of any estimate of CO_{wt} is proportional to the log of the number of strains examined. They have suggested that reasonable estimates can be made from strain sets that contain as little as 20–40 members.

Also of importance is the issue of how broad a group of bacteria can be covered by a single CO_{wt} value. Are these values species-specific? If so, the work needed to set all the required values will be huge. On the other hand, if CO_{wt} values are applied to multi-species groups or even multi-generic groups, is there a loss of precision? In setting CO_{wt} values are we going to have to strike a balance between a reasonable workload and a reasonable precision?

7.6.3   The use of epidemiological cut-off values

It is important to note that the meaning that can be attributed to CO_{wt} is limited. Importantly, as the properties of any therapy are not involved in their determination, they cannot be used to set CB or to determine clinical resistance. However, as CO_{wt}, also known as microbiological or epidemiological cut-off values (ECO), are all we have, it is worth considering their value and limitations as interpretive criteria. Within a species or group of related bacteria CO_{wt} allow us to identify two groups. One group includes those that are fully susceptible, also termed the wild type (WT) group (EUCAST, 2000). The susceptibility measures for members of this WT group should show a normal distribution (Kronvall, 2003). All other members of the species, those whose susceptibility measure is clearly, and preferably statistically, outside the normal WT distribution, can be classified as not fully susceptible or non-wild type (NWT). The question is, therefore what meaning, in a clinical context, can be given to the categories WT and NWT. Although it is clearly not legitimate to consider them as synonyms for clinically sensitive and clinically resistant, we can extract some clinical meaning from these categories.

If a clinical isolate is classified as NWT we can state only that susceptibility testing suggests that the initiation of therapy may be imprudent. If a clinical isolate is classified as WT, we can state that susceptibility testing has revealed no reason why therapy should not be initiated. It should always be remembered that there might be other reasons why the particular therapy would be inappropriate. Particular care must be taken in interpreting WT classifications of intracellular bacteria.

The special case of intracellular bacteria
Infections involving intracellular bacteria are an emerging problem in aquaculture. In finfish, salmonid rickettsial septicemia (SRS) caused by the intracellular bacterium *Piscirickettsia salmonis* is considered the main cause of mortality in farmed coho salmon in Chile (Bravo and Midtlyng, 2007). *Francisella piscicida*, has been reported as an emerging problem in cold water cod farming (Ottem *et al.*, 2008) and *Francisella asiatica* has been associated with major epizootics in warm water finfish (Soto *et al.*, 2010). In shrimp both necrotising hepatopancreatitis (NHP) a bacterial disease associated with infection by a rickettsial-like organism (Cuéllar-Anjel *et al.*, 2010) and the emerging disease, streptococcosis (Hasson *et al.*, 2009), both involve bacteria with an intracellular life stage.

Studies in humans have revealed major differences between the extracellular and intracellular PK and PD of antibiotics (Carryn *et al.*, 2003; Barcia-Macay *et al.*, 2006; Sandberg *et al.*, 2009). Thus, it is not valid to assume because a particular therapeutic regimen has been shown to be efficacious against an extracellular bacterium it will also be efficacious against a bacterium, of the same susceptibility, that spends a considerable portion of its infection in an intracellular location. These differences in PK/PD parameters may affect the meaning that can be given to a classification of an intracellular bacterium as WT. It may well be that an intracellular bacterium may, because of the reduced intracellular antibiotic concentration and/or activity, be correctly classified as both WT and clinically resistant with respect to a standard dose regimen.

7.7   Questions concerning the CLSI approach

The goal of the CLSI approach is to generate interpretive criteria that are of universal applicability. The availability of a standard set of interpretive criteria that can be used by all laboratories around the world would represent a major breakthrough and the advantages of this approach are self-evident. However some questions have been raised as to whether this goal is achievable (Smith, 2008c).

7.7.1   The issue of time

To achieve prudent use of antibiotics laboratories have to be able to make reasonable predictions as to the probable success of a proposed therapy. This in turn requires that they have access to experimentally determined CB or ECO values relevant to the species they are dealing with. In the CLSI approach such values cannot be established until the appropriate quality control strains have been identified for a specific test condition and, importantly, until the acceptable range for those control strains have been established by multi-laboratory studies. It is inevitable that such a process will take time. In the meantime, frontline clinical labs have no guidance as to what interpretive criteria to apply to their susceptibility data. Smith (2006) has demonstrated that there are strong grounds for believing that, in the absence of such guidance, major errors are common (Table 7.2). The sensitivity cut-off values being used in laboratories that responded to a survey by Smith (2006) show serious disagreements with those evidence-based values published by CLSI (2010). These data suggest that there is an urgent need for the development and use of improved and validated cut-off values.

The most efficient way in which the necessary improvement can be achieved would be for laboratories to generate their own cut-off values (CO_{wt}), using NRI analysis of their own data. When these CO_{wt} values were generated for species where CLSI (2006a, b) have published the relevant QC requirements, they could be submitted for acceptance as laboratory-independent ECO. For species where the QC requirements have not been set these CO_{wt} would still be of great value. They could not, of course, be treated as having universal applicability. They would have to be treated as laboratory-specific and should be applied only to the susceptibility data generated in the laboratory that produced them. In that limited context they would, however, significantly reduce the frequency of error.

7.7.2   The issue of quality control requirements

Performing susceptibility tests according to the CLSI protocol does not guarantee that the results obtained for control strains will be in the acceptable range (Smith, 2008c). It is quite possible, even for experienced laboratories, to have some difficulty in achieving compliance with these quality control requirements. This is not an insurmountable difficulty as the guidelines also specify the work needed to achieve compliance. However, the need for such work places significant demands on the capacity of the laboratories that wish to use the guidelines. This should cause little difficulty for laboratories with extensive experience in susceptibility testing and general

Table 7.2 Comparison of disc diffusion sensitivity breakpoints currently applied to *Aeromonas salmonicida* in testing laboratories (Smith, 2006) with the cut-off values accepted by the Clinical and Laboratory Standards Institute (CLSI, 2010)

Zone size (mm)	Number of laboratories using breakpoints of various sizes		
	Oxolinic acid	Oxytetracycline	Florfenicol
10–11	1		
12–13	2	1	
14–15	2		
16–17	1	2	1
18–19	1	12	5
20–21		2	4
22–23		1	1
24–25	1		2
26–27	1		
28–29		1	
30–31		1	1
32–33		1	1
34–35			
36–37			
38–39		1	
Laboratories prone to major error	100%	82%	87%

Shaded areas indicate the cut-off values that have been accepted by CLSI (2010). For oxolinic acid and oxytetracycline these are clinical breakpoints and for florfenicol they are epidemiological cut-off values.

Major errors occur when the breakpoint being used by a laboratory would classify some isolates as WT or sensitive when the classification, based on CLSI criteria, would be resistant or NWT.

microbiology. However, surveys have shown that many of the laboratories involved in frontline susceptibility testing of clinical isolates are handling a relatively small number of strains (Smith, 2006) and may not have this expertise. Thus, it is entirely possible that many of these small laboratories would have difficulty in achieving or maintaining full compliance with quality control requirements. It should be noted that, unless they can achieve compliance with these requirements, they could not, with any legitimacy, apply any interpretive criteria associated with the guidelines. The use of local, laboratory-specific cut-off values, calculated using NRI, would not encounter these quality control problems. It may well prove to be the case that relatively small laboratories would find it more effective to use such local cut-off values to interpret their data than to engage in extensive quality control work.

Another approach, the Single Plate Protocol (available at http://www.nuigalway.ie/microbiology/prof__peter_smith.html) that might be suitable for laboratories that analyse only a small number of clinical isolates per year has been proposed. In this protocol the susceptibility of a clinical isolate is tested on the same plate as the control strain. It therefore avoids the requirements for both quality control procedures and extensive determinations of relevant values of CO_{wt} and has been shown to be very robust (Smith et al., 2008). As it would enable a laboratory to determine whether a particular clinical isolate was WT or NWT on the basis of the results obtained from a single agar plate, this protocol may well be attractive to small front-line laboratories.

7.7.3 The issue of precision

A third and most possibly the most serious question concerns the precision of any laboratory-independent interpretive criteria that are produced by the CLSI guidelines. Because intra-laboratory variation is always greater than inter-laboratory variation there is an inevitable trade-off between universality and precision. The setting of acceptable ranges for control strains is an attempt to maximise universality while at the same time minimising loss of precision. The acceptable ranges that have so far been set are quite large. Acceptable ranges for control strains of up to 14 mm have been set for disc diffusion (CLSI, 2006a) and up to four two-fold dilutions for MIC tests (CLSI, 2006b). These relatively loose QC requirements automatically place a limit on the precision that can be achieved in setting or in applying any ECO or CB.

The importance of precision depends on the distribution of susceptibility measures that the cut-off values are to be applied to. With respect to oxy-tetracycline and A. salmonicida, the distribution of susceptibility measures is not only bimodal but the two modal groups are well separated (Miller and Reimschuessel, 2006; Uhland and Higgins, 2006; Smith et al., 2007). In this case, the difference in the susceptibility measures of WT and NWT strains is large and a degree of imprecision in applying a cut-off value to separate them will cause little problems. However, in situations where small decreases in susceptibility are regularly encountered, for example A. salmonicida and the quinolones (Ruane et al., 2007) or the potentiated sulphonamide agents (Douglas et al., 2007), the difference between the measures for the WT and NWT can be relatively small. When the differences between the measures for the WT and NWT are smaller than the acceptable range of results for the control strain, the lack of precision in CB or ECO can lead to inappropriate classification of isolates. The extent to which small changes in susceptibility prove to be clinically significant will ultimately inform the decision as to whether the loss of precision, that is the automatic consequence of the attempt to generate universal interpretive criteria, is acceptable.

7.8 Future trends: the way forward for susceptibility testing

The demonstration by Smith (2008b) that many laboratories are currently failing to correctly identify resistant strains (Table 7.2) underlines the urgent need for the development and propagation of improved interpretive criteria that have a sound empirical basis.

As argued above the first step for any laboratory interested in improved susceptibility testing must be to use NRI analysis to generate CO_{wt} values for those species where they have sufficient data. It is, of course, possible for laboratories to generate these CO_{wt} values using any test protocols and under any conditions they might choose. However, it will be more efficient if they adhere to the CLSI protocols and, where possible, use the test conditions that have been suggested for the species they are interested in. The demonstration that CO_{wt} values can be produced under the suggested test conditions would, in fact, provide the data that would allow CLSI to move them from the suggested to the accepted category. Whether CO_{wt} values produced in individual laboratories can be treated as laboratory-specific or as laboratory-independent interpretive criteria will depend on the stage of development of QC requirements. Therefore, the second step must be to initiate the international collaborative work necessary to expand the range of test condition for which the data necessary for setting QC requirements are available (CLSI, 2007, 2008).

As noted above there will be an inevitable loss of precision associated with universal, laboratory-independent interpretive criteria. Ultimately, the decision as to whether the advantages of universality outweigh the disadvantage of this loss of precision will be made by those who use the interpretive criteria. It is probable that, in making that decision, they will be influenced by the frequency with which they encounter strains with low-level but clinically significant resistance. Any imprecision in interpretive criteria renders detection of such strains unreliable. Luckily, the decision as to whether universal or laboratory-specific values are the most appropriate goal can be postponed. Much of the work needed to achieve either is the same, and possession of the data produced by that work will enable making the correct choice much easier.

7.9   Conclusions

The ability of bacteria to become clinically resistant to antibiotics is a major threat to the continued therapeutic use of these agents in aquaculture. As the emergence of resistance is an automatic consequence of antibiotic use, it follows that we will be able to conserve this therapy only if we learn to use antibiotics prudently. The essence of prudent use is that we use them as little and as wisely as possible. In this chapter it has been argued that reductions in antibiotic use could be achieved in a number of ways. Improvements in husbandry and management will reduce both the incidence and severity of infectious diseases that give rise to need for therapy. Development of vaccines, in hosts with competent immune systems, will augment this reduction. Improvements in diagnosis will, if they include a correct evaluation of the role of non-bacterial factors in disease epizootics, help to reduce the number of disease outbreaks we attempt, inappropriately, to control with antibiotic therapy. Finally improvements in susceptibility testing will reduce the frequency with which antibiotic therapies are attempted that, because of clinical resistance in the target bacterium, have no chance of benefiting either the animal's health or the farmer's bank balance.

7.10   Acknowledgements

I wish to acknowledge the part played in the genesis of the chapter by that inspirational scientist, the late Tony Ellis. In 1990, in a bar in Santiago, he urged me to focus my mind on antibiotics in aquaculture and in another bar, on a Greek island in 2006, he asked what I thought the problems were and listened to my attempt to answer until long after the sun had set.

7.11   References

ALDERMAN D and SMITH P (2001), 'Development of draft protocols of standard reference methods for antimicrobial agent susceptibility testing of bacteria associated with fish disease', *Aquaculture*, **196**, 211–243.

AMBROSE PG, BHAVNANI SM, RUBINO CM, LOUIE A, GUMBO T, FORREST A and DRUSANO GL (2007), 'Pharmacokinetics-pharmacodynamics of antimicrobial therapy: it's not just for mice anymore', *Clin Infect Dis*, **44**, 79–86.

ANGULO F (1999), 'Use of antimicrobial agents in aquaculture: potential for public health impact'. Public Health Service. Department of Health and Human Services, CDC, 18 October; http://www.fda.gov/ohrms/dockets/dailys/00/apr00/041100/c000019.pdf.

ANGULO F (2000), 'Antimicrobial agents in aquaculture: potential impact on public health', *Alliance for the Prudent Use of Antibiotics (APUA) Newsletter*, **18**, 1, 4.

ANGULO FJ and GRIFFIN PM (2000), 'Changes in antimicrobial resistance in *Salmonella enterica* serovar Typhimurium', *Emerg Infect Dis*, **6**, 436–438.

AVENDAÑO-HERRERA R, MOLINA A, MAGARIÑOS B, TORANZO AE and SMITH P (2011), 'Estimation of epidemiological cut-off values for disc diffusion susceptibility test data for *Streptococcus phocae*', *Aquaculture*, **314**, 44–48.

BALABAN NQ, MERRIN J, CHAIT R, KOWALIK L and LEIBLER S (2004), 'Bacterial persistence as a phenotypic switch', *Science*, **305**, 1622–1625.

BARCIA-MACAY M, SERAL C, MINGEOT-LECLERCQ M-P, TULKENS PM and VAN BAMBEKE F (2006), 'Pharmacodynamic evaluation of the intracellular activities of antibiotics against *Staphylococcus aureus* in a model of THP-1 macrophages', *Antimicrob Agents Chemother*, **50**, 841–851.

BARKER G, PAGE D and KEHOE E (1995), 'Comparison of 4 methods to determine MIC's of amoxycillin against *Aeromonas salmonicida*', *Bull Eur Assoc Fish Pathol*, **15**, 100–104.

BARNES AC, LEWIN CS, HASTINGS TS and AMYES SGB (1990). 'Cross resistance between oxytetracycline and oxolinic acid in *Aeromonas salmonicida* associated with alterations in outer membrane proteins', *FEMS Microbiol Lett*, **72**, 337–339.

BRAVO S and MIDTLYNG PJ (2007), 'The use of fish vaccines in the Chilean salmon industry 1999–2003', *Aquaculture*, **270**, 36–42.

BRYAN LE (1989), 'Two forms of antimicrobial resistance: bacterial persistence and positive function resistance', *J Antimicrob Chemother*, **23**, 817–823.

CABELLO FC (2006), 'Heavy use of prophylactic antibiotics in aquaculture: a growing problem for human and animal health and for the environment', *Environ Microbiol*, **8**, 1137–1144.

CARRYN S, CHANTEUX H, SERAL C, MINGEOT-LECLERCQ M-P, VAN BAMBEKE F and TULKEN PM (2003), 'Intracellular pharmacodynamics of antibiotics', *Infect Dis Clin N Am*, **17**, 615–634.

CLSI (2006a), 'Methods for antimicrobial disk susceptibility testing of bacteria isolated from aquatic animals', Approved guideline M42-A, Clinical and Laboratory Standards Institute, Wayne, Pennsylvania.

CLSI (2006b), 'Methods for broth dilution susceptibility testing of bacteria isolated from aquatic animals', Approved guideline M49-A, Clinical and Laboratory Standards Institute, Wayne, Pennsylvania.

CLSI (2007), 'Development of *in vitro* susceptibility testing criteria and quality control parameters for veterinary antimicrobial agents: approved guide, third edition', CLSI document M37-A3, Clinical and Laboratory Standards Institute, Wayne, Pennsylvania.

CLSI (2008), 'Performance standards for antimicrobial disc and dilution susceptibility tests for bacteria isolated from animals; approved standard, third edition', CLSI document M31-A3, Clinical and Laboratory Standards Institute, Wayne, Pennsylvania.

CLSI (2010), 'Performance standards for antimicrobial susceptibility testing of bacteria isolated from aquatic animals; first informational supplement'. CLSI document M42/49-S1, Clinical and Laboratory Standards Institute, Wayne, Pennsylvania.

COYNE R, BERGH Ø, SAMUELSEN O, ANDERSEN K, LUNESTAD BT, NILSEN H, DALSGAARD I and SMITH P (2004a), 'Attempt to validate breakpoint MIC values estimated from pharmacokinetic data obtained during oxolinic acid therapy of winter ulcer disease in Atlantic salmon (*Salmo salar*)', *Aquaculture*, **238**, 51–66.

COYNE R, SAMUELSEN O, BERGH Ø, ANDERSEN K, PURSELL L, DALSGAARD I and SMITH P (2004b). 'On the validity of setting breakpoint minimum inhibition concentrations at one quarter of the plasma concentration achieved following oral administration of oxytetracycline', *Aquaculture*, **239**, 23–35.

COYNE R, SMITH P, DALSGAARD I, NILSEN H, KONGSHAUG H, BERGH Ø and SAMUELSEN O (2006), 'Winter ulcer disease of post-smolt Atlantic salmon: an unsuitable case for treatment?', *Aquaculture*, **253**, 171–178.

CUÉLLAR-ANJEL J, CORTEEL M, GALLI L, ALDAY-SANZ V and HASSON KW (2010), 'Principal shrimp infectious diseases, diagnosis and management', in Alday-Sanz V, *The Shrimp Book*, Nottingham University Press, UK.

DARWISH AM, RAWLES D and GRIFFIN BR (2002), 'Laboratory efficacy of oxytetracycline for the control of *Streptococcus iniae* in blue tillapia', *J. Aquat Anim Health*, **13**, 184–190.

DOUGLAS I, RUANE NM, GEARY M, CARROLL C, FLEMING GTA, MCMURRAY J and SMITH P (2007), 'The advantages of the use of discs containing single agents in disc diffusion testing of the susceptibility of *Aeromonas salmonicida* to potentiated sulphonamides', *Aquaculture*, **272**, 118–125.

DRINAN, EM (1985). Studies on the pathogenisis of furunculosis in salmonids. PhD Thesis, National University of Ireland, Dublin.

DRUSANO GL (2004). 'Antimicrobial pharmacodynamics: critical interactions of "bug and drug"', *Nat Rev Microbiol*, **2**, 289–300.

EUCAST (2000), 'Terminology relating to methods for the determination of susceptibility of bacteria to antimicrobial agents', European Committee on Antimicrobial Susceptibility Testing Definitive document E.Def 1.2 May 2000; http://www.srga.org/Eucastwt/eucastdefinitions.htm.

FAO (2005), 'Responsible use of antibiotics in aquaculture', FAO Fisheries Technical Paper No. 469, Rome.

GIRAUD E, BLANC G, BOUJU-ALBERT A, WEILL F-X and DONNAY-MORENO C (2004), 'Mechanisms of quinolone resistance and clonal relationship among *Aeromonas salmonicida* strains isolated from reared fish with furunculosis', *J Med Microbiol*, **53**, 895–901.

GRÄSLUND S and BENGTSSON B-E (2001), 'Chemicals and biological products used in south east Asian shrimp farming, and their potential impact on the environment – a review', *Sci Total Environ*, **280**, 93–131.

GRÄSLUND S, KARLSON K and WONGTAVATCHAI J (2002), 'Responsible use of antibiotics in shrimp farming', *Aquaculture Asia*, **7**, 17–18.

GRAVE K, MARKESTAD A and BANGEN M (1996), 'Comparison in prescribing patterns of antibacterial drugs in salmonid farming in Norway during the periods 1980–1988 and 1989–1994', *J Vet Pharmacol Ther*, **19**, 184–191.

GRAVE K, LINGAAS E, BANGEN M and RØNNING M (1999), 'Surveillance of the overall consumption of antibacterial drugs in humans, domestic animals and farmed fish in Norway in 1992 and 1996', *J Antimicrob Chemother*, **42**, 243–252.

HASSON KW, WYLD ME, FAN Y, LINGSWEILLER SW, WEAVER SJ, CHENG J and VARNER PW (2009), 'Streptococcosis in farmed *Litopenaeus vannamei*: a new emerging bacterial disease of penaeid shrimp', *Dis Aquat Org*, **86**, 93–106.

HEUER OE, KRUSE H, GRAVE K, COLLIGNON P, KARUNASAGAR I and ANGULO FJ (2009), 'Human health consequences of use of antimicrobial agents in aquaculture', *Clin Infect Dis*, **49**, 1248–53.

HINEY MP and SMITH P (1998), 'Validation of polymerase chain reaction-based techniques for proxy detection of bacterial fish pathogens: framework, problems and possible solution for environmental applications', *Aquaculture*, **162**, 41–68.

HINEY MP and SMITH P (2000), 'Oil-adjuvanted furunculosis vaccines in commercial fish farms: a preliminary epizootiological investigation', *Aquaculture*, **200**, 1–9.

HONG G-E, KIM D-G, BAE J-Y, AHN S-H, BAI SC and KONG I-S (2007), 'Species-specific PCR detection of the fish pathogen, *Vibrio anguillarum*, using the *amiB* gene, which encodes *N*-acetylmuramoyl-l-alanine amidase', *FEMS Microbiol Lett*, **269**, 201–206.

JONES RN (1992), 'Recent trends in the college of American pathologists proficiency results for antimicrobial susceptibility testing: preparing for CLIA '88. *Clin Microbiol Newsletter*, **14**, 33–37.

KAHLMETER G, BROWN DFJ, GOLDSTEIN FW, MACGOWAN AP, MOUTON JW, ÖSTERLUND A, RODLOFF A, STEINBAKK M, URBASKOVA P and VATOPOULOS A (2003), 'European harmonization of MIC breakpoints for antimicrobial susceptibility testing of bacteria', *J Antimicrob Chemother*, **52**, 145–148.

KJAERGARD LL and ALS-NIELSEN B (2002), 'Association between competing interests and authors' conclusions: epidemiological study of randomised clinical trials published in the BMJ', *BMJ*, **325**, 249–252.

KRONVALL G (2003), 'Determination of the real standard distribution of susceptible strains in zone histograms', *Int J Antimicrob Agents*, **22**, 7–13.

KRONVALL G (2010), 'Normalized resistance interpretation as a tool for establishing epidemiological MIC susceptibility breakpoints', *J Clin Microbiol*, **48**, 4445–4452.

LAMBERT HP (1999), 'Don't keep taking the tablets', *Lancet*, **354**, 943–945.

LEES P, CONCORDET D, ALLIABADI FS and TOUTAIN P-L (2006), 'Drug selection and optimization of dosage schedules to minimise antimicrobial resistance', in Aarestrup FM, *Antimicrobial Resistance in Bacteria of Animal Origin*, Washington, ASM Press.

LI X-Z and NIKAIDO H (2009), 'Efflux-mediated drug resistance in bacteria: an update', *Drugs*, **69**, 1555–1623.

LILLEHAUG A, LUNESTAD BT and GRAVE K (2003), 'Epidemiological description of bacterial diseases in Norwegian aquaculture – a description based on antibiotic prescription data for the ten-year period 1991 to 2000', *Dis Aquat Org*, **53**, 115–125.

MARKESTAD A and GRAVE K (1997), 'Reduction of antibacterial drug use in Norwegian fish farming due to vaccination', *Fish Vaccinol*, **90**, 365–369.

MCCABE WR and TREADWELL TL (1986), '*In vitro* susceptibility tests: correlations between sensitivity testing and clinical outcome in infected patients', in: Lorian V, *Antibiotic in Laboratory Medicine*, 2nd edition, Baltimore Williams and Wilkins.

MCKEOWN T (1976), *The Modern Rise of Population*, London, Academic Press.

MILLER R and REIMSCHUESSEL R (2006), 'Epidemiological cutoff values for antimicrobial agents against *Aeromonas salmonicida* isolates determined by frequency distributions of minimal inhibitory concentration and diameter of zone of inhibition data', *Am J Vet Res*, **67**, 1837–1843.

NICGABHAINN S, AMEDEO M, BERGH O, DIXON B, DONACHIE L, CARSON J, COYNE R, CURTIN J, DALSGAARD I, MAXWELL G and SMITH P (2003), 'The precision and robustness of published protocols for disc diffusion assays of antimicrobial agent susceptibility: an inter-laboratory study', *Aquaculture*, **240**, 1–18.

NIKAIDO H (1989), 'Outer membrane barrier as a mechanism of antimicrobial resistance', *Antimicrob Agents Chemother*, **33**, 1831–1836.

OTTEM KE, NYLUND A, ISAKSEN TE, KARLSBAKK E and BERGH Ø (2008), 'Occurrence of *Francisella piscicida* in farmed and wild Atlantic cod, *Gadus morhua* L., in Norway', *J Fish Dis*, **31**, 525–534.

QIAN R-H, XIAO Z-H, ZHANG C-H, CHU W-Y, WANG L-S, ZHOU H-H, WEI Y-W and YU L (2008), 'A conserved outer membrane protein as an effective vaccine candidate from. *Vibrio alginolyticus*', *Aquaculture*, **278**, 5–9.

POOLE K (2005), 'Efflux-mediated antimicrobial resistance', *J Antimicrob Chemother*, **56**, 20–51.

REIMSCHUESSEL R, STEWART L, SQUIBB E, HIROKAWA K, BRADY T, BROOKS D, SHAIKH B and HODSDON C (2008), 'Phish-Pharm Database of Pharmacokinetics Data in Fish'; http://www.fda.gov/AnimalVeterinary/ScienceResearch/ToolsResources/Phish-Pharm/default.htm.

RODGERS CJ (2001), 'Resistance of *Yersinia ruckeri* to antimicrobial agents *in vitro*', *Aquaculture*, **196**, 325–345.

RUANE NM, DOUGLAS I, GEARY M, CARROLL C, FLEMING GTA and SMITH P (2007), 'Application of normalised resistance interpretation to disc diffusion data on the susceptibility of *Aeromonas salmonicida* to three quinolone agents', *Aquaculture*, **272**, 156–167.

SAMUELSEN OB (2007), 'Pharmacokinetics of quinolones in fish: a review', *Aquaculture*, **255**, 55–75.

SANDBERG A, HESSLER JH, SKOV RL, BLOM J and FRIMODT-MØLLER N (2009), 'Intracellular activity of antibiotics against *Staphylococcus aureus* in a mouse peritonitis model', *Antimicrob Agents Chemother*, **53**,1874–1883.

SCHMIDT S, BARBOUR A, SAHRE M, RAND KH and DERENDORF H (2008), 'PK/PD: new insights for antibacterial and antiviral applications', *Curr Opin Pharmacol*, **8**, 549–556.

SMITH P (1997), 'The epizootiology of Furunculosis: the present state of our ignorance', in: Ellis AE, Bernoth E-M, Midtlyng P, Olivier O and Smith P, *Furunculosis in Fish – A Multidisciplinary Review*, London, Academic Press.

SMITH P (2006), 'Breakpoints for disc diffusion susceptibility testing of bacteria associated with fish diseases: a review of current practice', *Aquaculture*, **261**, 1113–1121.

SMITH P (2007), 'Antimicrobial use in shrimp farming in Ecuador and emerging multi-resistance during the cholera epidemic of 1991: a re-examination of the data', *Aquaculture*, **271**, 1–7.

SMITH P (2008a), 'A cost–benefit analysis of the application of pharmacokinetic/pharmacodynamic-based approaches to setting disc diffusion breakpoints in aquaculture: a case study of oxolinic acid and *Aeromonas salmonicida*', *Aquaculture*, **284**, 2–18.

SMITH P (2008b), 'Antimicrobial resistance in aquaculture', *Rev Sci Tech*, **27**, 243–264.

SMITH P (2008c), 'How difficult is it to achieve compliance with the quality control requirements of the Clinical and Laboratory Standards Institute's guideline M42-A?', *Aquaculture*, **276**, 1–4.

SMITH P (2008d), 'The role of aquaculture in the emergence of florfenicol resistance in *Salmonella enterica* serovar Typhimurium DT 104; a re-evaluation of the data', *Emerg Infect Dis*, **14**, 1327–1328.

SMITH P (2010), 'An economic framework for discussing antimicrobial agent use in shrimp farming', in Alday-Sanz V, *The Shrimp Book*, Nottingham University Press, UK.

SMITH P, HINEY MP and SAMUELSEN OB (1994), 'Bacterial resistance to antimicrobial agents used in fish farming: a critical evaluation of method and meaning', *Ann Rev Fish Dis* **4**, 273–313.

SMITH P, RUANE NM, DOUGLAS I, CARROLL C, KRONVALL G and FLEMING GTA (2007). 'Impact of inter-lab variation on the estimation of epidemiological cut-off values for disc diffusion susceptibility test data for *Aeromonas salmonicida*', *Aquaculture*, **272**, 168–179.

SMITH P, FLEMING GTA and CARROLL C (2008), 'Reducing inter-operator variation in disc diffusion assays by the inclusion of internal controls in a standard susceptibility test protocol', *Aquaculture*, **285**, 273–276.

SMITH P, DOUGLAS I, MCMURRAY J and CARROLL C (2009), 'A rapid method of improving the criteria being used to interpret disc diffusion antimicrobial susceptibility test data for bacteria associated with fish diseases', *Aquaculture*, **290**, 172–178.

SØRUM H (2006), 'Antimicrobial drug resistance in fish pathogens', in Aarestrup FM, *Antimicrobial Resistance in Bacteria of Animal Origin*. Washington, DC, ASM Press.

SOTO E, ENDRIS RG and HAWKE JP (2010), '*In vitro* and *in vivo* efficacy of florfenicol for treatment of *Francisella asiatica* infection in tilapia', *Antimicrob Agents Chemother*, **54**, 4664–4670.

THRUSFIELD M (1986), *Veterinary Epidemiology*, London, Butterworth and Co.

TSOUMAS AT, ALDERMAN DJ and RODGERS CJ (1989), '*Aeromonas salmonicida*: development of resistance to 4-quinolone antimicrobials', *J Fish Dis*, **12**, 493–507.

TURNIDGE J and PATERSON DL (2007), 'Setting and revising antibacterial susceptibility breakpoints', *Clin Microbiol Rev*, **20**, 391–408.

UHLAND FC and HIGGINS R (2006), 'Evaluation of the susceptibility of *Aeromonas salmonicida* to oxytetracycline and tetracycline using antimicrobial disk diffusion and dilution susceptibility tests', *Aquaculture*, **257**, 111–117.

WOOD SC, MCCASHION RN and LYNCH WH (1986), 'Multiple low-level antibiotic resistance in *Aeromonas salmonicida*', *Antimicrob Agents Chemother*, **29**, 992–996.

WHO (2006), 'Report of a Joint FAO/OIE/WHO Expert Consultation on Antimicrobial Use in Aquaculture and Antimicrobial Resistance', Seoul, Republic of Korea, 13–16 June 2006. http://www.who.int/topics/foodborne_diseases/aquaculture_rep_13_16june2006%20.pdf (accessed 10 June 2009).

8

Considerations for the use of anti-parasitic drugs in aquaculture

A. P. Shinn and J. E. Bron, University of Stirling, UK

Abstract: The management and control of parasitic infections in aquaculture are a constant challenge, highly complicated by the current limited availability of efficacious licensed products; a situation exacerbated by the development of resistance to anti-parasitic drugs in parasite populations. In addition, parasite control in aquaculture requires a keen awareness of environmental, water quality and host parameters, and is subject to the constraints of economics and the requirement for aquaculture sustainability and environmental protection. Here, we look at some of the factors underlying general decision-making as it applies to drug treatment, including parasite identification, choice of chemical, strategy and dosage. Trigger-points, and the rationale underpinning these, are also considered.

Key words: treatment decisions, strategy, dosage, look-up table, identification.

8.1 Introduction

Despite continuing ambitions to expand the use of integrated pest management strategies for the control of parasites in aquaculture (e.g. Brooks, 2009), treatment with anti-parasitic drugs remains the most widely used tool for managing parasite infections in cultured aquatic species. While there has been considerable investment into the development of novel anti-parasitic drugs, the range of efficacious treatments currently available for use worldwide is extremely limited (Table 8.1), with use of many of the most successful treatment options now restricted owing to concerns over potential adverse effects on the environment, on treated aquaculture species, on farm staff or on consumers. Increasingly, the aquaculture industry is also observing the development of resistance to anti-parasitic drug in target parasite species. This chapter seeks to describe the main classes of anti-parasitic treatments employed for the management of parasites of aquatic species and to highlight some of the key issues that currently constrain or otherwise affect the use of anti-parasitic treatments in an aquaculture context. Also, it outlines strategies to maximise the efficacy and longevity of current and

Table 8.1 Some chemotherapeutants commonly used in aquaculture and ornamental fish husbandry with, in each case, examples of a target species/group against which partial efficacy has been demonstrated. These drugs however, may not be approved for food fish and/or the ornamental industry in all countries

Compound	Dose	Target species (for example)	Reference
Acetic acid*	1000–2000 mg l⁻¹ for 1–10 min	Ectoparasitic protozoa	see Noga (2010)
Acriflavin(e)**	10 mg l⁻¹ for 10 d	Piscinoodinium spp.	Jacobs (1946), van Duijn (1973)
Albendazole	1+ μg ml⁻¹ for 24 h	Spironucleus vortens	Sangmaneedet & Smith (1999)
	1.5 mg kg⁻¹/kg fish once per wk for 4 wks	Loma salmonae	Speare et al. (1999)
Aminosidine	40 g kg⁻¹ feed for 2 d	Hexamita salmonis	Tojo & Santamarina (1998)
	40 g kg⁻¹ feed for 2 d	Hexamita salmonis	Tojo & Santamarina (1998)
Amprolium hydrochloride³	63–104 mg kg⁻¹ for 10 d	Ichthyophthirius multifiliis	Shinn et al. (2003b)
Azadirachtin + camphor + curcumin (1:1:1)	400–700 mg l⁻¹ for 3 d	Aphanomyces invadans	Harikrishnan et al. (2009)
Azamethiphos (e.g. Salmosan)§	0.1–0.2 mg l⁻¹ bath for 1 h	Lepeophtheirus salmonis; Caligus elongatus	Roth et al. (1996)
Benzalkonium chloride§	various	General ectoparasiticide	see Noga (2010)
Benznidazole	2 g kg⁻¹ feed for 2 d	Hexamita salmonis	Tojo & Santamarina (1998)
Beta-glucans	10 mg kg⁻¹ fish i.p.	Loma salmonae	Guselle et al. (2010)
	200 g kg⁻¹ feed	Loma salmonae	Guselle et al. (2010)
Biothionol	1 mg l⁻¹ for 1 h	Neoparamoeba spp.	Florent et al. (2007a)
	25 mg kg⁻¹ feed for 2 wks	Neoparamoeba spp.	Florent et al. (2007b)
	4, 20 & 60 mg l⁻¹ for 3 h	Gyrodactylus [sic salaris] teuchis	Santamarina et al. (1991); see Schelkle et al. (2009)
Bronopol§	5 mg l⁻¹ continuous	Ichthyophthirius multifiliis	Picón-Camacho et al. (2011b)
	100 mg l⁻¹ for 30 min for 10 d	Ichthyophthirius multifiliis	Shinn et al. (2003a)
	20 mg l⁻¹ for 60 min for 14 d	Saprolegnia parasitica	Branson (2002)

Table 8.1 *Continued*

Compound	Dose	Target species (for example)	Reference
Calcium oxide*	2000 mg l⁻¹ for 5 s	Ectoparasitic protozoa	see Noga (2010)
Chloramine-T or B§§	100 mg l⁻¹ 30 min, 4 times over 10 d	*Ichthyophthirius multifiliis*	Shinn *et al.* (2001)
Chlorine	10 mg l⁻¹ for 1 h	*Neoparamoeba pemaquidensis*	Harris *et al.* (2004)
	120 mg l⁻¹ for 3 h	*Haliotrema* sp., *Euryhaliotrema* sp. eggs	Fajer-Ávila *et al.* (2007)
	1,500 mg l⁻¹ chlorine	*Glugea anomala*	Ferguson *et al.* (2007)
	1,600 mg l⁻¹ for 24 h	*Myxobolus cerebralis*	Wagner (2002)
	5,000 mg l⁻¹ for 10 min	*Myxobolus cerebralis*	Wagner (2002)
Chloroquine	10 mg l⁻¹ single treatment	*Amyloodinium ocellatum*	see Noga (2010)
	10 mg l⁻¹ every 5 d (4 doses)	*Cryptocaryon irritans*	see Noga (2010)
Copper sulphate§§	0.15 mg l⁻¹	*Amyloodinium ocellatum*	Bower (1983)
	0.15–0.25 mg l⁻¹ for 3–10 d	*Cryptocaryon irritans*	Herwig (1978)
	0.4–1.0 mg l⁻¹ for 8–10 d	*Ichthyophthirius multifiliis*	Schlenk *et al.* (1998)
Cypermethrin§§	5 µg l⁻¹ for 60 min	*Lepeophtheirus salmonis; Caligus elongatus*	Hart *et al.* (1997)
Deltamethrin§§	2–3 µg l⁻¹ for 40 min	*Lepeophtheirus salmonis; Caligus elongatus*	see Noga (2010)
Dichlorvos (e.g. Nuvan)[1]	0.5–2.0 mg l⁻¹ for 30–60 min	*Lepeophtheirus salmonis; Caligus elongatus*	Pike (1989)
Diethylcarbamacine	40 g kg⁻¹ feed for 2 d	*Hexamita salmonis*	Tojo & Santamarina (1998)
Diflubenzuron (Lepsidon)§§	75 mg kg⁻¹ bdy wt for 14 d	*Lepeophtheirus salmonis; Caligus elongatus*	Alderman (2002)
	0.03 mg l⁻¹	*Lernaea* sp.	Burtle & Morrison (1987)
Dimetridazole (Emtryl)**[1]	4+ µg ml⁻¹ for 48 h	*Spironucleus vortens*	Sangmaneedet & Smith (1999)
Dipterex (also see trichlorphon)[1]	2 mg l⁻¹ for 24 h	*Ergasilus intermedius*	Ingram & Philbey (1999)
	100 mg l⁻¹ for 1 h	*Argulus* spp.	Neubert (1984)
Doramectin[3]	1 mg kg⁻¹ bdy wt fish for 10 d	*Lernaea cyprinacea*	Hemaprasanth *et al.* (2008)

Treatment	Dose	Target	Reference
Dylox	various doses	Piscicola geometra, hirudinids	see review of Burreson (2006)
Emamectin benzoate[§§]	50 mg kg⁻¹ feed d⁻¹	Argulus coregoni	Hakalahti et al. (2004)
	10 kg ton⁻¹ fed at 0.5% bdy wt d⁻¹	Lepeophtheirus salmonis, Caligus elongatus	Stone et al. (2000)
Eugenol	1–10 g l⁻¹	Saprolegnia parasitica	Bouchard et al. (2001)
Fenbendazole[3]	1+ μg ml⁻¹ for 24 h	Spironucleus vortens	Sangmaneedet & Smith (1999)
	4–8 mg kg⁻¹ fish on d 1 & 8	Eubothrium crassum	R.H. Richards (pers. comm.)
Flumequine	10 mg kg⁻¹ bdy wt i.m.	Anguillicoloides crassus	van der Heijden et al. (1996)
Formaldehyde (e.g. Paracide-F)[†§]	15–25 mg l⁻¹ permanent	Protozoa e.g. Apiosoma, Epistylis, Trichodina	R.H. Richards (pers. comm.)
	20 mg l⁻¹ for 18 h	Gyrodactylus derjavinoides	Buchmann & Kristensson (2003)
	100–200 mg l⁻¹ for 6–9 h	Amyloodinium ocellatum	see Noga (2010)
	160–250 mg l⁻¹ for up to 60 min	Protozoa e.g. Apiosoma, Epistylis, Trichodina	R.H. Richards (pers. comm.)
Freshwater	200 mg l⁻¹ for 15 min	Dermophthirius nigrellii	Cheung et al. (1982)
	5 min bath	Amyloodinium ocellatum	Kingsford (1975); Lawler (1977)
Fumagillin[2]	3 h bath	Neoparamoeba spp.	Florent et al. (2007a,b)
	3 × 2 μg ml⁻¹ for 24 h at 2 d intervals	Glugea anomala	Schmahl et al. (1990)
i) Analogue TNP-470	15 mg kg⁻¹ fish for 3 weeks	Loma salmonae	Speare et al. (1999)
	1.0 mg kg⁻¹ fish d⁻¹ for 6 days	Tetracapsuloides bryosalmonae (PKD)	Hedrick et al. (1988), Morris et al. (2003)
Furanace (nitrofuran)[1]	2.5 mg l⁻¹	Amyloodinium ocellatum	see Noga & Levy (2006)
Furazolidone	50 mg kg⁻¹ fish d⁻¹ for 10 d	Hexamita	R.H. Richards (pers. comm.)
Garlic (whole)*[3]	200 mg l⁻¹ bath	Capillaria sp.	Pena et al. (1988)
Garlic (powder)[3]	0.5% inclusion in feed for 10 d	Ichthyophthirius multifiliis	Shinn (unpublished data)
Gracillin	0.18 mg l⁻¹ (EC50)	Dactylogyrus intermedius	Wang et al. (2010)

Table 8.1 Continued

Compound	Dose	Target species (for example)	Reference
Hydrogen peroxide[†]	75 mg l⁻¹ for 30 min, d 1 & 6	*Amyloodinium ocellatum*	Montgomery-Brock et al. (2001)
	100 mg l⁻¹ for 1 h	*Aphanomyces astaci*	Lilley & Inglis (1997)
	500 mg l⁻¹ for 1 h	*Aphanomyces invaderis*	Lilley & Inglis (1997)
	1500 mg l⁻¹ for 20 min	*Lepeophtheirus salmonis, Caligus elongatus*	Treasurer et al. (2000)
	500 mg l⁻¹ for 1 h	*Saprolegnia parasitica*	Lilley & Inglis (1997)
Ivermectin[3]	consecutive or alternative days 50 µg kg⁻¹ bdy wt once on d 1 & 8	*Lepeophtheirus salmonis, Caligus elongatus*	Pike & Wadsworth (1999)
Jenoclean (97% *Atacama* extract (Zeolites) + 3% citric acid)	*In vitro* 1–500 µg ml⁻¹ 50 mg l⁻¹ for 30 min	*Pseudoterranova decipiens Philasterides dicentrarchi*	Manley & Embil (1989) Harikrishnan et al. (2010)
Levamisole	Bath 1 mg l⁻¹ for 24 h Bath 2–5 mg l⁻¹	*Anguillicoloides crassus Anguillicoloides crassus*	Taraschewski et al. (1988) Hartmann (1989)
Lime i) unslaked	380 g/m²	*Myxobolus cerebralis*	Hoffman & Hoffman (1972)
ii) quick lime	10 kg acre⁻¹ (2.6 mg l⁻¹) 3 times over 10 d	*Argulus indicus, A. japonicus*	Jafri & Ahmed (1994)
Loperamid	50 mg kg⁻¹ over 3 consecutive d	*Echinorhynchus truttae*	Taraschewski et al. (1990)
Magnesium chloride	25 mM MgCl₂ for 24 h	*Aphanomyces astaci*	Rantimäki et al. (1992)
Magnesium sulphate*	70 mg ml⁻¹ for 24 h	*Spironucleus vortens*	Sangmaneedet & Smith (1999)
	30 g MgSO₄ + 7 g NaCl 5–10 min	Monogenean and crustacean infections	see Noga (2010)
Malachite green**[1]	0.5 mg l⁻¹ for 1 h	*Aphanomyces astaci, A. invaderis*	Lilley & Inglis (1997)
	5 mg l⁻¹ for 1 h	*Saprolegnia parasitica*	Lilley & Inglis (1997)
	1.2 g kg⁻¹ feed for 10 d	*Ichthyophthirius multifiliis*	Ruider et al. (1997)

Treatment	Dose	Target	Reference
Mebendazole[3]	0.5+ μg ml⁻¹ for 78 h	Spironucleus vortens	Sangmaneedet & Smith (1999)
	100 mg l⁻¹ for 10 min	Pseudodactylogyrus anguillae, P. bini	Székely & Molnár (1987)
	100 mg kg⁻¹ fish bdy wt d⁻¹	Proteocephalus ambloplitis	Boonyaratpalin & Rogers (1984)
Methylene blue**	3 mg l⁻¹ for 10 d	Piscinoodinium spp.	Jacobs (1946), van Duijn (1973)
Metriphonate	Bath 1 mg l⁻¹ for 24 h	Anguillicoloides crassus	Taraschewski et al. (1988)
Metronidazole**[,1]	6+ μg ml⁻¹ for 48 h	Spironucleus vortens	Sangmaneedet & Smith (1999)
	5 g kg⁻¹ feed for 2 d	Hexamita salmonis	Tojo & Santamarina (1998)
Monensin[3]	100 mg kg⁻¹ bdy wt d⁻¹	Calyptospora sp.	see Noga (2010)
	1 g kg⁻¹ feed for 9 weeks	Loma salmonae	Becker et al. (2002)
Niclosamide	1–20 g kg⁻¹ fish	Bothriocephalus, Caryophyllaeus, Khawia spp.	see Woo (2006)
Nitrofurazone (+ freshwater)	freshwater 10–15 min dip + 30 mg l⁻¹ dip for 5 min	Uronema marinum/ciliophorans	Basson & Van As (2006)
Nitroscanate	40 g kg⁻¹ feed for 2 d	Hexamita salmonis	Tojo & Santamarina (1998)
Peracetic acid (40% PAA)	1 mg l⁻¹ for 4 d	Ichthyophthirius multifiliis	Sudová et al. (2010)
Peracetic acid + H_2O_2 + acetic + peroctanoic acid	8–15 mg l⁻¹ for 60 min	Ichthyophthirius multifiliis	Picón-Camacho et al. (2011a)
Potassium ferrate	19.2 mg l⁻¹ for 3 d	Ichthyophthirius multifiliis	Ling et al. (2010)
Potassium permanganate[††§§]	3–5 mg l⁻¹	Ergasilus ceylonensis	Wijeyaratne & Gunawardene (1988)
Praziquantel (Droncit)[4]	5 kg/acre (1.3 mg l⁻¹) over 3 d	Argulus indicus, A. japonicus	Jafri & Ahmed (1994)
	Bath 0.75 mg l⁻¹ for 24 h	Bothriocephalus acheilognathi	Mitchell & Darwish (2009)
	Bath 1 mg l⁻¹ for 90 h	Diplostomum spathaceum	Székely & Molnár (1991)
	Bath 10 mg l⁻¹ for 3 h	Gyrodactylus [aculeati]	Schmahl & Taraschewski (1987)
	50–75 mg kg⁻¹ bdy wt d⁻¹ for 6 d	Benedenia seriolae/Zeuxapta seriolae	Williams et al. (2007)
	330 mg kg⁻¹ (single oral dose)	Diplostomum spathaceum	Székely & Molnár (1991)
	5–300 mg kg⁻¹ fish	Cestodes e.g. Eubothrium	R.H. Richards (pers. comm.)

Table 8.1 *Continued*

Compound	Dose	Target species (for example)	Reference
Proxitane 0510 (5% peracetic acid in H_2O_2)	100 mg l⁻¹ peracetic acid for 5 min	*Aphanomyces* spp./*Saprolegnia* spp.	Lilley & Inglis (1997)
Quinine hydrochloride	61 mg kg⁻¹ fish d⁻¹	*Loma salmonae*	Speare *et al.* (1998)
Ronidazole	2 g kg⁻¹ feed for 2 d	*Hexamita salmonis*	Tojo & Santamarina (1998)
Salinomycin sodium[1]	47 mg kg⁻¹ feed for 10 d p.i.	*Ichthyophthirius multifiliis*	Shinn *et al.* (2003b)
Secnidazole	5 g kg⁻¹ feed for 2 d	*Hexamita salmonis*	Tojo & Santamarina (1998)
Sodium chloride*§	6 g l⁻¹ for 96 h	*Piscinoodinium* sp.	Carneiro *et al.* (2002)
	10 g l⁻¹ for 3 days	*Argulus* spp.	R.H. Richards (pers. comm.)
	20 g l⁻¹ for 1 h	*Aphanomyces astaci, A. invaderis*	Lilley & Inglis (1997)
Sodium percarbonate	30 g l⁻¹ for 10–30 min	General parasiticide	see Noga (2010)
	80 mg l⁻¹ for 18 h	*Gyrodactylus derjavinoides*	Buchmann & Kristensson (2003)
Teflubenzuron§§	10 mg kg⁻¹ biomass d⁻¹ for 7 d	*Lepeophtheirus salmonis*; *Caligus elongatus*	Branson *et al.* (2000)
Toltrazuril[3]	Bath	*Glugea anomala*	Schmahl *et al.* (1990)
	10 mg l⁻¹ 4 h/d for 3 d	*Ichthyophthirius multifiliis*	Melhorn *et al.* (1988)
Trichlorphon[1] (e.g. Dipterex, Neguvon)	various doses	*Lernaea cyprinacea*	see Lester & Hayward (2006)
Trillin	26.48 mg l⁻¹ (EC50)	*Dactylogyrus intermedius*	Wang *et al.* (2010)

§ Approved for treating food fish in the UK; §§ Used in the UK food fish aquaculture industry; † Approved for treating food fish in the US; †† Used in the US food fish aquaculture industry; *Not formally approved for treating food fish in the US but are of low regulatory priority; **Unapproved for treating food fish in the US and considered high regulatory priority by the US FDA; ‡ Used in the non-food industry in either the UK or US; [1]Banned for the use in food producing fish; [2]Licensed in Canada for the treatment of microsporidiosis of honey bees; [3]Authorised for use in animals in the UK but food producing fish not specifically stated; [4]Authorised for use in the UK pet industry.

future treatment compounds. This chapter will also consider general decision-making as it applies to drug treatments that are commonly used rather than addressing the particular merits and drawbacks of individual compounds. The use of plant extracts, nutraceuticals, immunostimulants, probiotics and/or vaccines in the aquaculture and ornamental industries are not commented upon here but are discussed elsewhere (see Anderson, 1992; Gatesoupe, 1999; Sakai, 1999; Verschuere *et al.*, 2000; Irianto & Austin, 2002; Lorenzen & LaPatra, 2005; Balasubramanian *et al.*, 2007; Kesarcodi-Watson *et al.*, 2008; Wang, *et al.*, 2008; Benkendorff, 2009; Mohamed, 2010; *inter alia*).

In the UK, where control of parasites in aquaculture through drug treatments is widely employed, disease accounts for an estimated €90–115 million loss in production (Shinn unpublished data), of which parasites account for approximately 50% of the losses. Two parasite species, the caligid copepods *Caligus elongatus* and *Lepeophtheirus salmonis*, account for €34 million p.a. of this total (Costello, 2009). As global aquaculture continues to expand and intensify, the impact of parasitic disease is also likely to grow. The increased prevalence of parasitic disease in aquaculture is further exacerbated by greater globalisation of the trade in aquatic animals and their products (reviewed in Bondad-Reantaso *et al.*, 2005) and the attendant risks of parasite translocation and establishment into new areas. The consequences of introduced/invading host species and their parasites for native fish communities were the subject of a study by Torchin *et al.* (2003), who suggested that invaders not only introduce new pathogens but can also act as host reservoirs for endemic parasites.

While parasite impact in aquaculture is considerable, efficacious options for intervention other than anti-parasitic drugs remain limited. Aside from wider management strategies such as fallowing and use of single year-classes, interventions include the use of mechanical filtration (Larsen *et al.*, 2005; Heinecke & Buchmann, 2009), dislodgement (Anonymous, 1996), electrical grids (Schäperclaus, 1992), suction devices (Shinn *et al.*, 2009) and the use of removable substrates on which parasites lay their eggs (Bauer, 1970; Hoffman, 1977). Anti-parasitic drugs, on the other hand, show an extremely high diversity (see Table 8.1), although the number of compounds that are efficacious, consistent, safe and may be used across a wide range of environmental conditions is very much more restricted. Of the compounds listed in Table 8.1, very few are routinely employed and the scale of use of anti-parasitic drugs in aquaculture represents only a small proportion of the total animal health market for veterinary medicines (see the review of Alderman, 2009). Although aquaculture is a growth industry (8.9%) when compared to capture fisheries (1.2%) and terrestrial meat production (2.8%) (Bondad-Reantaso *et al.*, 2005), the regulatory demands that must be met and the potential profits in comparison to other animal health sectors remain key constraints to drug development and registration (Alderman, 2009). The list of effective anti-parasitic treatments commonly used in aquaculture and ornamental fish husbandry is therefore extremely short and has changed very little over the past 20 years (Table 8.1), with many compounds listed having been banned for use in aquaculture since their initial development or being of scientific interest only, with no current licensing for aquaculture use.

8.2   Factors in successful treatment with anti-parasitic drugs

8.2.1   The importance of correct parasite identification

'Know thy enemy and know yourself; in a hundred battles, you will never be defeated' (Sun Tzu, 496 BCE)

Before embarking on a course of treatment, it is important to have a proper identification of the causative disease agent and a clear understanding of its life cycle. This process involves a thorough examination of a sample of infected hosts as well as consideration of data concerning the fish's environment, in order to collect all the information necessary to make a clear diagnosis. An inaccurate diagnosis can lead either to the true problem going undiagnosed and becoming worse or to an inappropriate course of treatment being used. Most parasites, even those with direct life cycles, such as protists, fungi and monogeneans, have different stages (i.e. eggs, cysts, free-living and feeding, etc.) in their life cycles. For each event that requires treatment, understanding of the development of each stage and the timing between them is essential in devising an appropriate treatment strategy.

Having made an identification of the parasite involved, the next question is whether or not to employ anti-parasitic drugs. The finding of a parasite does not necessarily mean that it requires treatment. Mobile (e.g. *Trichodina* spp.) and sessile peritrichs (e.g. *Apiosoma* spp., *Epistylis* spp.), for example, are commonly encountered, and when present in low numbers cause negligible damage to their hosts. If the organic loading of the water increases, however, then this creates conditions that are ideal for these peritrichs to proliferate. Large numbers of *Trichodina* can result in severe irritation, injury and disintegration of the epithelium; likewise, marked changes in the number of sessile peritrichs can cause inflammation, hyperplasia, necrosis and ultimately ulceration of the epithelia. In this situation, the first course of action would be to improve water quality, to reduce the level of organic loading and then, following an evaluation of the host, decide upon whether a concurrent treatment should be applied or whether a period of monitoring, to see if parasite numbers decrease, is appropriate. The finding of highly pathogenic species such as the protist *Ichthyophthirius multifiliis*, which has the capacity to rapidly increase in number, should be considered a real threat to the fish population and an immediate course of remediation is recommended (Shinn *et al.*, 2009, 2011; Picon-Camacho *et al.*, 2011b).

Treatments may be administered for a variety of reasons other than to manage or attempt to eradicate a parasitic infection. Within a biosecurity framework, prophylactic treatments may be applied at identified hazard points in production as part of standard hygiene practices to protect the health of animal populations from risk organisms. The increasing record of *Gyrodactylus salaris* from sites across Europe, for instance, appears to be linked to the movement of cultured salmonids, notably rainbow trout, *Oncoryhnchus mykiss*, across national borders (Paladini *et al.*, 2009; Rubio-Godoy *et al.*, 2011). Likewise, the studies of García-Vásquez *et al.* (2010, 2011) demonstrate that *Gyrodactylus cichlidarum* has been spread globally with the movement of its Nile tilapia, *Oreochromis niloticus*, host. Gibson (1993) suggested that the introduction and establishment of the caryophyllidean cestode *Monobothrium wageneri* into Britain may have been through the importation of infected ornamental varieties of tench, *Tinca tinca*, for sport angling and fisheries, although he also discusses the possibility of parasite introduction having been via infected oligochaete worms, the suggested intermediate host, which are also imported from the Continent for bait. Therefore, prophylactic treatments may be advised when moving hosts, both to ensure disease-free stocks and to combat downstream effects of immuno-supression resulting from the stress of translocation.

Treatments may also be administered to remove parasite species that impact on their hosts in ways other than causing pathology, for example, to increase the aesthetic appearance and therefore the price of fish being sold to processors/consumers or to improve an aspect of growth and production. Decisions to treat the pseudophyllidean cestode *Eubothrium crassum* in the past, for instance, have been undertaken to remove unsightly worms which would otherwise lower the price of fish being sold. In addition, Mitchell (1993) suggested that low chronic infections of *E. crassum* could account for a potential 10–20% loss in growth and so the decision to treat also serves to improve the food conversion ratio. For food fish species, treatment must be performed in a responsible manner to ensure sustainability of the respective aquaculture industries and the problem-free trade of aquaculture products. For aquatic species destined for human consumption, food safety must be upheld, ensuring for example, that products are free from zoonotic agents or from harmful drug residues (Anonymous, 2009 a,b). Likewise, only licensed treatments that do not exert deleterious effects on biodiversity or the environment should be used (see the comprehensive review of Boxall *et al.*, 2004).

8.2.2   Choice of treatment strategy

'... let your methods be regulated by the infinite variety of circumstances'
(Sun Tzu, 496 BCE)

Experience has shown that while there are some rules of thumb that can be applied to the treatment of certain parasitic infections, it is vital that the particulars of each situation are carefully considered before a course of treatment is recommended. Such 'particulars' might include a diagnostic evaluation of the current health status of the population to be treated, the severity of the infection(s) and the pathology induced, water quality, and the size of the fish population and the environment to be treated. The attachment organ or opisthaptor of gill monogeneans like *Diplectanum aequans* for example, penetrate deep into the host's tissues, lying close to the basement membrane of the primary lamellae and subsequently inducing a range of pathological impacts including inflammation, hyperplasia, fusion of lamellae, erosion and necrosis with consequential respiratory distress (Dezfuli *et al.*, 2007). Heavy monogenean infections such as this and others, e.g. *Dactylogyrus*, are irritating to the host, causing gill hyperplasia, swelling, a reduction in surface area for respiration and excess mucus production, the latter potentially affording the parasite some protection from chemical treatments. While there is a temptation to use formaldehyde, which is a broad spectrum disinfectant, to curb the infection, it removes oxygen from the water column (each 5 mg l^{-1} of formaldehyde that is added removes 1 mg l^{-1} oxygen; Allison, 1962) and its use has also been documented to induce inflammatory changes including lamellar oedema, hyperplasia of the secondary lamellae resulting in an obliteration of the interlamellar space and fusion of the secondary lamellae (Dureza, 1988; Gregori *et al.*, 2007). Fish treated with formaldehyde are reported to ventilate heavily for several hours following treatment (see Thoney & Hargis, 1991) and so the use of formaldehyde to treat chronic gill infections may induce further oedema, respiratory distress and complications which may result in the loss of fish. If no treatment is given, however, bacterial gill disease may occur. While Ferguson (2006) comments on the gill's capacity to repair and recover from damage, this is a process that can take from days to weeks, and this recovery time must be taken into consideration when preparing a treatment schedule. For fish bearing heavy parasite infections with a degree of gill damage, therefore, it may be appropriate to consider a shuttle programme of treatments, using an appropriate treatment to first improve the general health condition of the gills before attempting to use more aggressive compounds, like formaldehyde, for the removal of gill flukes.

As well as the decision of which chemical treatment and dose to apply, there are other treatment regime considerations to be made, including which method of treatment delivery to use (bath versus in-feed), when to treat and whether multiple treatments are required. As with other farm activities the choice of treatment product and regime are further subject to the day-to-day constraints of economics. Different strategies present different pros and cons, which must be carefully considered for each situation. Treating a sea cage of Atlantic salmon to remove sea lice, for example, lends itself to both bath and in-feed treatments. Bath treatments, while highly labour intensive, have the benefit that all the fish nominally receive equal exposure to the treatment, assuming homogeneous mixing within the bath, while in-feed treatments are subject to factors such as dominance and appetence, differences between individuals that make treatments less uniform. Bath treatments, however, require considerable work to shallow cages, deploy tarpaulins, treat fish and monitor the treatment and, in addition, estimation of the bath volume can be problematic, especially if it is deformed by the prevailing water currents (Lader *et al.*, 2008), which can lead to fish being over- or under-dosed. The use of well-boats for the treatment of Atlantic salmon circumvents some of these problems in that conditions are standardised, the volume of chemical needed is reduced and the environmental concerns are lower (ACFFA, 2011). The transportation of fish in and their transfer to and from the well-boat, however, can be a stressful process (Iversen *et al.*, 2005). In-feed treatments are both easier to administer and provide longer-term protection (up to 9 weeks), with the drugs being incorporated into feed at a dose that is dependent on the pharmacokinetics of it reaching the parasite and exerting its effects. Any unmetabolised compound is then released from the body at a rate that, theoretically, poses no environmental concerns. However, the amount of the compound that is ingested can be variable, resulting either from different feeding rates shown by fish or the non-standard preparation of medicated diets (see Bravo *et al.*, 2008) and in addition, slow depletion of the product in flesh over time leads to reduced, sub-optimal concentrations, which favours the development of resistance in treated parasite populations.

For the effective treatment of many parasite infections, the timing of treatment is critical and should be specifically targeted to break key links in the parasite's life cycle. For oviparous monogenean species such as *Dactylogyrus*, the initial treatment should be administered to remove the juvenile and egg-laying adult population. Since the eggs of many monogeneans are resistant to most licensed treatments, a second, or if required, a third treatment should be given at specific times later, once all the eggs have hatched and before the parasite population reaches an egg-laying age. The implementation of these treatments is dependent on the time required by each parasite species to move between developmental stages and upon local water temperatures. Likewise it is also important to be aware that different strains/genotypes of a given parasite can behave very differently, not only in terms of their infectivity and host specificity but also in their susceptibility to treatment. This was successfully demonstrated by Straus *et al.* (2009) who found significantly different LC_{50} values for two strains of *I. multifiliis* being treated with copper sulphate (e.g. 3 h LC_{50} for strain 1 = 0.056 Cu mg l^{-1} versus 0.025 Cu mg l^{-1} for strain 2). Genetic diversity, therefore, is an important, though largely ignored, consideration in mitigating against the development of resistance in parasite populations.

8.2.3   Delivering the right dose

'The right dose of the right drug to the right patient at the right time'
Sidney Taurel, 2005

Having made an appropriate identification and decided upon a treatment, it is imperative that the pathogen is exposed to the correct target dose.

Overdosing, in addition to imposing unwarranted additional expense and increasing potential environmental impacts, can lead to the loss of stock, particularly when using chemicals such as peracetic acid which have narrow safety margins (i.e. 24 h LC_{50} 22 mg l^{-1}; 96 h LC_{50} 1.6 mg l^{-1}; Anonymous, 2007, 2011). Bath treatments may also be affected by environmental conditions such as temperature. Bruno & Raynard (1994) found that a 20 min, 1.23% hydrogen peroxide treatment for sea lice on Atlantic salmon smolts resulted in no mortalities when conducted in 10 °C seawater, however, a 35% mortality of stock within 2 hours of treatment was observed when conducted in 13.5 °C seawater. Further trials by Johnson *et al.* (1993a) and Kiemer & Black (1997) found that timing of treatments was critical, with there being a significant correlation between the level of exposure and the degree of observed gill damage (i.e. 0–10% mortality following a 1.5 g l^{-1} treatment at 11 °C for 20 min rising to 10–26.7% mortality when the treatment period was extended to 40 min). The toxicity of compounds such as chloramine-T (*N*-sodium-*N*-chloro-*q*-toluenesulphonamide) has been shown to be greater in soft, acidic waters than in hard, alkaline waters (Bills *et al.*, 1988).

There are, equally, concerns regarding the repercussions of under-dosing and/or the use of inappropriate strategies for the treatment of aquaculture fish species that have led to the development of resistance. Table 8.2, for example, includes some of the main compounds that have been used for the control and management of sea lice (*Caligus* and *Lepeophtheirus* spp.) and the resistance that has been subsequently reported. While the development of resistance within a parasite population can be attributed to many factors, under-dosing with the target drug in the feed and differential feeding rates by individual fish are possible causes. The analysis of medicated feed from four Chilean suppliers, for example, showed that diets had emamectin benzoate contents varying between 85% and 117% of the declared content (Bravo *et al.*, 2008). Despite this large variation in product preparation, the reduced sensitivity of the local sea lice population (*Caligus rogercresseyi*) to emamectin benzoate was not due to the incorrectly prepared diets but due to other factors including the feeding rate of individual salmon and/or the incorrect dosages being applied by the farms. Long-acting drugs such as emamectin benzoate may also show extended tailing off periods where the concentration of active compound in fish flesh falls below optimal concentrations. The efficacy of certain compounds like hydrogen peroxide, an oxidising agent which creates emboli within the haemolymph of sea lice (Bruno & Raynard, 1994), may be reduced by the organic loading of the water or heavily fouled nets (Johnson *et al.*, 1993b). The degree of biofouling on nets is a concern, not only because it can result in reduced flow rates passing through cages, lower dissolved oxygen and increased ammonia levels, and in extreme events, the asphyxiation of stock (Douglas-Helders *et al.*, 2003; Madin *et al.*, 2010), but also because it can also serve as a reservoir for pathogenic agents such as *Neoparamoeba*

Table 8.2 Chemotherapeutants that have been used in the treatment of salmon lice, *Lepeophtheirus salmonis* and *Caligus elongatus*, with comments on their use and, where reported, reduction in susceptibility

Compound	Recommended dose	Reported doses used	Reference
Azamethiphos [(S-6-chloro-2,3-dihydro-2-oxo-1,3-oxazolo[4,5-b]pyridin-3-ylmethyl O,O-dimethyl phosphorothioate] (Salmosan)	0.01 mg l^{-1} for 1 h 0.05 mg l^{-1} for 1 h (85%+ effective)	0.2 mg l^{-1} for 1 h (7.1–82.1% eff.) residual AChE activity ($E0$) 0.13–6.21 Ug^{-1}	Roth & Richards (1992) Roth et al. (1996) Fallang et al. (2004)
Cypermethrin [(RS)-α-cyano-3-phenoxybenzyl (1RS,3RS;1RS,3SR)-3-(2,2-dichlorovinyl)-2,2-dimethylcyclopropanecarboxylate] (Excis)	5 µg l^{-1} for 1 h	0.02–5.98 µg l^{-1} for 1 h (EC50)	Hart et al. (1997) Sevatdal et al. (2005a,b)
Deltamethrin [[(S)-cyano-(3-phenoxyphenyl)-methyl] (1R,3R)-3-(2,2-dibromoethenyl)-2,2-dimethyl-cyclopropane-1-carboxylate] (AlphaMax; AMX)	2–3 µg l^{-1} for 40 min 0.04–0.06 µg l^{-1} (24 h EC50)	0.25 µg l^{-1} (24 h EC50) 0.02–1.82 µg l^{-1} for 0.5 h (EC50)	see Noga (2010) Sevatdal & Horsberg (2003) Sevatdal et al. (2005a,b)
Dichlorvos [2,2-dichlorovinyl dimethyl phosphate] (Aquagard, Nogos 50EC, Nuvan EC500)	1 mg l^{-1} for 0.5–1 h (4–6 ppb 24 h LC50)	up to 40 ppb (24 h LC50) 60–202 µg l^{-1} (5 h LC50) No longer used	Jones et al. (1992) Tully & McFadden (2000)
Diflubenzuron [1-(4-chlorophenyl)-3-(2,6-difluorobenzoyl) urea] (Lepsidon)	75 mg kg^{-1} biomass for 14 d	Resistance suggested	Høy & Horsberg (1991) Sevatdal & Horsberg (2000)
Emamectin benzoate [avermectin B1, 4″-deoxy-4″-(methylamino)-,(4″R)-,benzoate (salt)] (Slice)	50 µg kg^{-1} biomass d^{-1} for 7 d	57–203 µg l^{-1} (24 h EC50) in summer 202–870 µg l^{-1} (24 h EC50) in winter Scottish treatments in 2006 11× less effective than those given in 2003. Winter treatments 11× more likely to fail than those in spring.	Stone et al. (1999) Bravo et al. (2008) Bravo et al. (2008) Lees et al. (2008a,b)

Table 8.2 *Continued*

Compound	Recommended dose	Reported doses used	Reference
Hydrogen peroxide [H₂O₂] (Paramove, Salartect 350) 1500 mg l⁻¹ for 20 min gave a 87–90% reduction in gravid female and a 97–99% reduction in other mobile stages		2000 mg l⁻¹ for 20 min gave a 7.5 % decrease in lice; 2500 mg l⁻¹ for 23 min to effect same reduction as initial 1500 mg l⁻¹ dose.	Treasurer *et al.* (2000)
Ivermectin [22,23 dihydro-avermectin] (Ivomec) 0.05 mg kg⁻¹ biomass twice a week Single in-feed treatment of 0.2 mg kg⁻¹			Johnson & Margolis (1993) see Roth *et al.* (1993)
Malathion [[5-1,2-bis(ethoxycarbonyl)ethyl 0,0-dimethyl phosphorodithioate] 5 mg l⁻¹ for 1 h		Experimental trial only	Høy & Horsberg (1991)
Metriphonate [1-napthyl methylcarbamate] (Carbaryl) 0.5 mg l⁻¹ for 40 min		Experimental trial only	Høy & Horsberg (1991)
4% Pyrethrin (Py-Sal-25 = 1% pyrethrum) 3 s dip		No longer used	Anon (1991)
4% Pyrethrin (Py-Sal-25 = 1% pyrethrum) + 4% piperonyl butoxide Deployed as a layer of oil for 5 d		No longer used	Jakobsen & Holm (1990)
Teflubenzuron [1-(3,5-dichloro-2,4-difluorophenyl)-3-(2,6-difluorobenzoyl) urea] (Calicide) 10 mg kg⁻¹ biomass d⁻¹ for 7 d			Branson *et al.* (2000)
Trichlorfon [O,O-dimethyl 2,2,2-trichloro-1-hydroxyethyl phosphonate] (DTHP, Dylox, D50, Dipterex, Masoten, Neguvon) 15–300 mg l⁻¹ for 15–60 min (temp dep.)		No longer used	Brandal & Egidius (1979)

pemaquidensis, a causative agent of amoebic gill disease of Atlantic salmon (Tan *et al.*, 2002). Additionally, if the tarpaulin on a cage needs to be dropped in an emergency to end a treatment prematurely, a heavily bio-fouled net can lead to a slower flush rate and longer than calculated treatment times with potentially serious consequences for the health of the stock.

Although delivering the correct bath dose is possible for experimental and small-scale systems where the precise dimensions of the fish's environment is known and where calculated doses can be administered through precision equipment, e.g. peristaltic pumps (Picón-Camacho *et al.*, 2011b), this becomes an almost impossible task when treating large water bodies or treatment tarpaulins enclosing cages that are deformed by the local water currents (Lader *et al.*, 2008). This is a recognised problem and analytical procedures for on-site monitoring and dose confirmation are now being developed. These include, for example, the use of DPD (*N*, *N*-diethyl-*p*-phenylenediamine)-based colorimetric methods to determine the concentration of chloramine-T in the water during the course of fish treatments (Dawson *et al.*, 2003) to ensure the target dose is maintained. Again, the use of well-boats for the treatment of large numbers of cage-reared stock can circumvent some of these problems by creating near-standardised conditions, although even in this case concentration hot-spots may develop within treatment chambers, requiring careful mixing/monitoring.

8.3   Trigger points for treatment with anti-parasitic drugs in aquaculture

While in most cases treatments are administered in order to conform with welfare requirements and the duty of care to farm animals and to remove or manage parasites to below levels that might otherwise harm or result in the loss of cultured stock, there are a number of other situations or trigger points under which treatments are advised or enforced. These include area management agreements for the coordinated, synchronised treatment of sea lice as part of national codes of practice for salmon farmers (see Andersen & Kvenseth, 2000; Eithun, 2000; Rae, 2000). These include agreements on monitoring infections, defining parasite number thresholds and the timing of treatments to maximise the potential benefits to all sites. In Scotland over the period of 1993 to 1995, for example, it was estimated that the implementation of these measures effected a 62% reduction in the mobile (adults and pre-adult) lice population with a resultant 83% reduction in sea lice-related mortalities, an 80% reduction in the number of fish being downgraded at harvest and, a 46% reduction in treatments (Rae, 2000). While for some, the appearance of sea lice on wild-caught salmon alludes to its recent return from the sea and its 'quality', for most vendors and consumers, their presence is undesirable and treatments prior to harvest (allowing for the appropriate withdrawal period) are required to ensure that the quality of the flesh and the aesthetic appeal of the fish are upheld.

8.4   Future trends

8.4.1   Minimising the speed of resistance development

One of the key properties of an integrated pest management strategy is the collection and dissemination of knowledge relevant to parasite control. This involves structured monitoring of hosts, parasites and treatments as well as the exchange of information within a specified management area. Given that integrated pest management seeks both to maintain the size of the parasite population below a predetermined threshold and to minimise the speed of development of resistance to drugs in parasite populations, modelling of resistance development has become an increasingly important tool in parasite control. In sheep husbandry for instance, computer simulation models are being used to look at the effects of combined treatment with a number of effective, unrelated drugs and to predict the frequency of anthelminthic resistance genes in gastrointestinal nematodes (e.g. *Trichostrongylus colubriformis, Haemonchus contortus* and *Teladorsagia (Ostertagia) circumcincta*) (Dobson *et al.*, 2011). While the use of computer models in aquaculture, for example, can assist in predicting the best times to treat sea lice (see for example Tucker *et al.*, 2002), models for the efficient use of veterinary treatments are only now being developed.

Given concerns regarding the development of resistance to anti-parasitic drugs in sea lice, Robbins *et al.* (2010), developed a simulation model to investigate how compounds such as cypermethrin (Excis; the second most commonly used treatment in Scotland after emamectin benzoate) could be most effectively administered. Forty Scottish salmon farms over the period 1998–2001, administered an average of eight cypermethrin treatments over the 2-year marine Atlantic salmon production cycle to control a mean infection burden of 331 lice/fish (sum of the monthly chalimus and mobile sea lice per fish over the 2-year period of production). The model showed that using only six treatments per annum, on weeks 39, 45, 59, 65, 78 and 84, could have reduced the average infection burden to 143 sea lice/fish. This regime would have effected a 57% reduction in lice numbers over the eight treatment regime that was actually used. Management agreements or legislation that require the number of parasites to be kept below a given threshold regardless of other considerations such as host size, pathogenicity, etc may, however, be detrimental to fish and environmental welfare in the long term despite good intentions. For sea lice, the development of resistance on a given site has been directly correlated to the number of treatments undertaken per annum (Jones *et al.*, 1992), such that imposition of unnecessary treatments in a year serves to accelerate the development of resistance and will thus potentially increase sea louse numbers in the long term. Reduction of the number of treatments required per annum and improvement of the efficacy of each treatment is therefore a priority. In particular, deliberate cycling of different classes of anti-parasitic drugs during the farm cycle can help prevent development of resistance and maintain therapeutic effectiveness. Other elements of control that serve to reduce parasite numbers, even by small margins, such as selective breeding of stock for disease resistance, have the potential to substantially reduce the number of treatments required in a season, thereby reducing development of resistance, treatment costs and environmental impact of treatments.

8.4.2 Identification of novel anti-parasitic treatments

In many situations, the choice of chemical and treatment strategy is not only dictated by the scale of the system to be treated and the efficacy of the regime but also by economics. While formaldehyde, as a broad spectrum disinfectant, is applied to treat a wide range of parasite groups, the costs of its application in large-scale systems (e.g. ponds and lakes) can be prohibitive, and its potential impacts on fish, users and the environment are becoming increasingly unacceptable. Likewise the high personnel costs of administering certain treatments can make them unfavourable options. The need for new compounds therefore mandates the identification of alternative treatments. High-throughput drug screening is being increasingly employed within the wider pharmaceutical industry and more recently by those developing novel compounds for treatment of parasites in aquaculture. Bioassays are traditionally used in sensitivity assays to determine the safe concentration of a particular chemical to use by pre-trialling it on a small number of fish prior to larger-scale trials, and they are also used to ensure target doses are achieved and as part of resistance monitoring (SEARCH, 2006; Westcott et al., 2008; Brooks, 2009). Bioassays are also, however, seeing increasing use in high-throughput screening protocols, aided by techniques such as image analysis, to accelerate the identification of novel anti-parasitic treatments for use in aquaculture (e.g. Toovey & Lyndon, 2000). While there are benefits to be had through the use of bioassays (rapid screening, reduced costs, etc), there are also limitations which include non-specific binding to the test vessels and therefore a potential loss of activity. For compounds that would ordinarily be incorporated into a commercial diet, the assay depends on the direct penetration and absorption of the compound (e.g. emamectin benzoate) through the cuticle, rather than through the normal route of exposure, potentially inaccurately reflecting the true sensitivity of the lice population under test.

8.4.3 Improving strategies for the use of traditional anti-parasitic drugs

The limited availability of licensed efficacious treatments remains one of the key barriers to effective parasite control in aquaculture, with this situation set to become worse following the removal from use of broad spectrum disinfectants such as formaldehyde. While screening for new, alternative compounds for the treatment of aquatic parasites continues, better strategies need to be developed in the interim to work with those currently available. Such approaches include the development of shuttle programmes of treatments using particular compounds in a set sequence to effect greater reductions than those currently achieved and structured cycling of treatments to minimise the development of resistance in populations and to provide a reduction in the volume of chemicals currently used.

8.5   References

ACFFA (ATLANTIC CANADA FISH FARMERS ASSOCIATION) (2011) Evaluation of well boat technology for the treatment of sea lice. Project Final Report. http://0101.nccdn. net/1_5/358/258/2b2/Final_Report_Mar_19_11_-_Evaluation_of_Well_Boat_ Technology_for_the_Treatmen.pdf

ALDERMAN, D.J. (2002) Trends in therapy and prophylaxis. *Bulletin of the European Association of Fish Pathologists*, **22**, 117–125.

ALDERMAN, D.J. (2009) Control of the use of veterinary drugs and vaccines in aquaculture in the European Union. In: Rodgers, C. & Basurco, B. (Eds.) *The use of veterinary drugs and vaccines in Mediterranean aquaculture*. Options Méditerranéennes, Série A (Séminaires Méditerranéens, No. 86. CIHEAM (Centre International de Hautes Etudes Agronomiques Méditerranéennes) / FAO (Food and Agriculture Organization of the United Nations), pp. 13–28.

ALLISON, R. (1962) The effects of formalin and other parasiticides upon oxygen concentrations in ponds. *Proceedings of the Annual Conference of the Southeast Association of Game and Fish Commissions*, **16**, 446–449.

ANDERSEN, P. & KVENSETH, P.G. (2000) Integrated lice management in mid-Norway. *Caligus*, **6** (March), 6–7.

ANDERSON, D.P. (1992) Immunostimulants, adjuvants, and vaccine carriers in fish: applications to aquaculture. *Annual Review of Fish Diseases*, **2**, 281–307.

ANONYMOUS (1991) Norwegians assess alternative delouser. *Fish Farmer*, **14**, 23–24.

ANONYMOUS (1996) Pumping with silkstream removes lice. *Fish Farming International*, **23** (9), 21.

ANONYMOUS (2007) Peracid. www.aeb-group.com/imgs/INGLESE/Schede/Food/ PERA CID_stsENG.pdf. Accessed 01/10/2011.

ANONYMOUS (2009a) Dangers of imported shrimp. *CBS Evening News*, 11 February, 2009.   http://www.cbsnews.com/stories/2004/09/17/eveningnews/consumer/main 644203.shtml.

ANONYMOUS (2009b) Nitrofuran a curse for shrimp export. *The Financial Express* (Dhaka), 28 April, 2009. http://www.thefinancialexpress-bd.com/2009/04/28/ 65003.html.

ANONYMOUS (2011) FMC Peracetic acid: General safety review of 15/10 PAA and other formulations as used for Poultry (Spectrum®), Direct Meat (Blitz®), F&V applications and VigorOx®. http://www.microbialcontrol.fmc.com/Portals/ Microbial/Content/Docs/15%20paa_sfty.pdf. Accessed 01/10/2011.

BALASUBRAMANIAN, G., SARATHI, M., RAJESH KUMAR, S. & SAHUL HAMEED, A.S. (2007) Screening the antiviral activity of Indian medicinal plants against white spot syndrome virus in shrimp. *Aquaculture*, **263** (1–4), 15–19.

BASSON, L. & VAN AS, J. (2006) Trichodinidae and other ciliophorans (Phylum Ciliophora). In: Woo, P.T.K. (Ed.) *Fish Diseases and Disorders*. Volume 1. CAB International. pp. 154–182.

BAUER, O.N. (1970) The ecology of parasites of freshwater fish. *Parasites of Freshwater Fish and the Biological Basis of their Control*, vol. XLIX Bulletin of the State Scientific Research Institute of Lake and Fisheries, pp. 3–207, Translated from Russian: Israel Program for Scientific Translations Jerusalem, 1962.

BECKER, J.A., SPEARE, D.J., DALEY, J. & DICK, P. (2002) Effects of monensin dose and treatment on xenoma reduction in microsporidial gill disease in rainbow trout, *Oncorhynchus mykiss* (Walbaum). *Journal of Fish Diseases*, **25**, 673–680.

BENKENDORFF, K. (2009) Aquaculture and the production of pharmaceuticals and nutraceuticals. In: Burnell, G. & Allen, G. (Eds.) *New Technologies in Aquaculture: Improving Production Efficiency, Quality and Environmental Management*. Woodhead Publishing, Cambridge, Ch 28, pp. 866–891.

BILLS, T.D., MARKING, L.L., DAWSON, V.K. & RACH, J.J. (1988) Effects of environmental factors on the toxicity of chloramine-T to fish. *U.S. Fish and Wildlife Service, Investigations in Fish Control*, **96**.

BONDAD-REANTASO, M.G., SUBASINGHE, R.P., ARTHUR, J.R., OGAWA, K., CHINABUT, S., ADLARD, R., TAN, Z. & SHARIFF, M. (2005) Disease and health management in Asian aquaculture. *Veterinary Parasitology*, **132**, 249–272.

BOONYARATPALIN, S. & ROGERS, W.A. (1984) Control of the bass tapeworm, *Proteocephalus ambloplitis* (Leidy), with mebendazole. *Journal of Fish Diseases*, **7**, 449–456.

BOUCHARD, L., PATEL, J. & LAHEY, L. (2001) The effect of clove oil on fungal infections of salmonid eggs. In: Hendry, C.I. & McGladdery, S.E. (Eds.) *Aquaculture Canada 2000*. Aquaculture Association of Canada Special Publication, Issue 4, pp. 110–112.

BOWER, C.E. (1983) *The Basic Marine Aquarium*. Charles C. Thomas, Springfield, IL.

BOXALL, A.B.A., FOGG, L.A., BLACKWELL, P.A., KAY, P., PEMBERTON, E.J. & CROXFORD, A. (2004) Veterinary medicines in the environment. *Reviews of Environmental Contamination and Toxicology*, **180**, 1–91.

BRANDAL, P.O. & EGIDIUS, E. (1979) Treatment of salmon lice (*Lepeophtheirus salmonis* Krøyer, 1838) with Neguvon® – Description of method and equipment. *Aquaculture*, **18** (2), 183–188.

BRANSON, E.J. (2002) Efficacy of bronopol against infection of rainbow trout (*Oncorhynchus mykiss*) with the fungus *Saprolegnia* species. *Veterinary Record*, **151**, 539–541.

BRANSON, E.J., RØNSBERG, S. & RITCHIE, G. (2000) Efficacy of teflubenzuron (Calcide^R) for the treatment of sea lice, *Lepeophtheirus salmonis* (Krøyer 1838), infestations of farmed Atlantic salmon (*Salmo salar* L.). *Aquaculture Research*, **31**, 861–867.

BRAVO, S., SEVATDAL, S. & HORSBERG, T.E. (2008) Sensitivity assessment of *Caligus rogercresseyi* to emamectin benzoate in Chile. *Aquaculture*, **282**, 7–12.

BROOKS, K.M. (2009) Considerations in developing an integrated pest management programme for control of sea lice on farmed salmon in Pacific Canada. *Journal of Fish Diseases*, **32** (1), 59–73.

BRUNO, D.W. & RAYNARD, R.S. (1994) Studies on the use of hydrogen peroxide as a method for the control of sea lice on Atlantic salmon. *Aquaculture International*, **2**, 10–18.

BUCHMANN, K. & KRISTENSSON, R.T. (2003) Efficacy of sodium percarbonate and formaldehyde bath treatments against *Gyrodactylus derjavini* infestations of rainbow trout. *North American Journal of Aquaculture*, **65**, 25–27.

BURRESON, E.M. (2006) Chapter 15. Phylum Annelida: Hirudinea as vectors and disease agents. In: Woo, P.T.K. (2006) *Fish Diseases and Disorders. Volume 1. Protozoan and Metazoan Infections*. Second Edition. CAB International, pp 566–591.

BURTLE, G. & MORRISON, J. (1987) Dimilin for control of *Lernaea* in golden shiner ponds. *Proceedings of the Arkansas Academy of Sciences*, **41**, 17–19.

CARNEIRO, P.C.F., MARTINS, M.L. & URBINATI, E.B. (2002) Effect of sodium chloride on physiological responses and the gill parasite, *Piscinoodinium* sp., in matrinxa, *Brycon cephalus*, (Telostei: Characidae) subjected to transport stress. *Journal of Aquaculture in the Tropics*, **17**, 337–348.

CHEUNG, R., NIGRELLI, R.F., RUGGIERI, G.D. & CILIA, A. (1982) Treatment of skin lesions in captive lemon sharks, *Negaprion brevirostris* (Poey), caused by monogeneans (*Dermophthirius* sp.). *Journal of Fish Diseases*, **5**, 167–170.

COSTELLO, M.J. (2009) The global economic cost of sea lice to the salmonid farming industry. *Journal of Fish Diseases*, **32** (1), 115–118.

DAWSON, V.K., MEINERTZ, J.R., SCHMIDT, L.J. & GINGERICH, W.H. (2003) A simple analytical procedure to replace HPLC for monitoring treatment concentrations of chloramine-T on fish culture facilities. *Aquaculture*, **217**, 61–72.

DEZFULI, B.S., GIARI, L., SIMONI, E., MENEGATTI, R., SHINN, A.P. & MANERA, M. (2007) Gill histopathology of cultured European sea bass, *Dicentrarchus labrax* (L.), infected with *Diplectanum aequans* (Wagener 1857) Diesing 1958 (Diplectanidae: Monogenea). *Parasitology Research*, **100**, 707–713.

DOBSON, R.J., HOSKING, B.C., BESIER, R.B., LOVE, S., LARSEN, J.W.A., ROLFE, P.F. & BAILEY, J.N. (2011) Minimising the development of anthelmintic resistance, and optimising the use of the novel anthelmintic monepantel, for the sustainable control of nematode parasites in Australian sheep grazing systems. *Australian Veterinary Journal*, **89**, 160–166.

DOUGLAS-HELDERS, G.M., TAN, C., CARSON, J. & NOWAK, B.F. (2003) Effects of copper-based antifouling treatment on the presence of *Neoparamoeba pemaquidensis* Page, 1987 on nets and gills of reared Atlantic salmon (*Salmo salar*). *Aquaculture*, **221**, 13–22.

DUREZA, L.A. (1988) *Toxicity and lesions in the gills of* Tilapia nilotica *fry and fingerlings exposed to formalin, furanace, potassium permanganate and malachite green.* Ph.D Thesis. Auburn University, Alabama, USA.

EITHUN, I. (2000) Measures to control sea lice in Norwegian fish farms. *Caligus*, **6** (March), 4–5.

FAJER-ÁVILA, E.J., VELÁSQUEZ-MEDINA, S.P. & BETANCOURT-LOZANO, M. (2007) Effectiveness of treatments against eggs, and adults of *Haliotrema* sp. and *Euryhaliotrema* sp. (Monogenea: Ancyrocephalinae) infecting red snapper, *Lutjanus guttatus*. *Aquaculture*, **264** (1–4), 66–72.

FALLANG, A., RAMSAY, J.M., SEVATDAL, S., BURKA, J.F., JEWESS, P., HAMMELL, K.L. & HORSBERG, T.E. (2004) Evidence for occurrence of an organophosphate-resistant type of acetylcholinesterase in strains of sea lice (*Lepeophtheirus salmonis* Krøyer). *Pest Management Science*, **60**, 1163–1170.

FERGUSON, H.W. (2006) *Systemic Pathology of Fish: A Text and Atlas of Normal Tissues in Teleosts and their Responses in Disease*, second edition. Scotian Press.

FERGUSON, J.A., WATRAL, V., SCHWINDT, A.R. & KENT, M.L. (2007) Spores of two fish microsporidia (*Pseudoloma neurophilia* and *Glugea anomala*) are highly resistant to chlorine. *Diseases of Aquatic Organisms*, **76**, 205–214.

FLORENT, R.L., BECKER, J.A. & POWELL, M.D. (2007a) Evaluation of bithionol as a bath treatment for amoebic gill disease caused by *Neoparamoeba* spp. *Veterinary Parasitology*, **144** (3–4), 197–207.

FLORENT, R.L., BECKER, J.A. & POWELL, M.D. (2007b) Efficacy of bithionol as an oral treatment for amoebic gill disease in Atlantic salmon *Salmo salar* L. *Aquaculture*, **270** (1–4), 15–22.

GARCÍA-VÁSQUEZ, A., HANSEN, H., CHRISTISON, K.W., RUBIO-GODOY, M., BRON, J.E. & SHINN, A.P. (2010) Gyrodactylids (Gyrodactylidae, Monogenea) infecting *Oreochromis niloticus niloticus* (L.) and *O. mossambicus* (Peters) (Cichlidae): a pan-global survey. *Acta Parasitologica*, **55** (3), 215–229.

GARCÍA-VÁSQUEZ, A., HANSEN, H., CHRISTISON, K.W., BRON, J.E. & SHINN, A.P. (2011) Description of three new species of *Gyrodactylus* Nordmann, 1832 (Monogenea) from oreochromids (*Oreochromis*, Cichlidae). *Acta Parasitologica*, **56** (1), 20–33.

GATESOUPE, F.J. (1999) The use of probiotics in aquaculture. *Aquaculture*, **180** (1–2), 147–165.

GIBSON, D.I. (1993) *Monobothrium wageneri*: another imported tapeworm established in wild British freshwater fishes? *Journal of Fish Biology*, **43**, 281–285.

GREGORI, M., PRETTI, C., INTORRE, L., BRACA, G. & ABRAMO, F. (2007) Gill histopathology in zebrafish model following exposure to aquacultural disinfectants. *Bulletin of the European Association of Fish Pathologists*, **27** (5), 185–191.

GUSELLE, N.J., SPEARE, D.J., MARKHAM, R.J.F. & PATELAKIS, S. (2010) Efficacy of intraperitoneally and orally administered ProVale, a yeast beta-(1,3)/(1,6)-D-glucan product, in inhibiting xenoma formation by the microsporidian *Loma salmonae* on rainbow trout gills. *North American Journal of Aquaculture*, **72** (1), 65–72.

HAKALAHTI, T., LANKINEN, Y. & VALTONEN, E.T. (2004) Efficacy of emamectin benzoate in the control of *Argulus coregoni* (Crustacea: Branchiura) on rainbow trout *Oncorhynchus mykiss*. *Diseases of Aquatic Organisms*, **60**, 197–204.

HARIKRISHNAN, R., BALASUNDARAM, C., DHARANEEDHARAN, S., MOON, Y.-G., KIM, M.-C., KIM, J.-S. & HEO, M.-S. (2009) Effect of plant active compounds on immune response and disease resistance in *Cirrhina mrigala* infected with fungal fish pathogen, *Aphanomyces invadans*. *Aquaculture Research*, **40**, 1170–1181.

HARIKRISHNAN, R., JIN, C.-N., KIM, M.-C., KIM, J.-S., BALASUNDARAM, C. & HEO, M.-S. (2010) Effectiveness and immunomodulation of chemotherapeutants against scuticociliate *Philasterides dicentrarchi* in olive flounder. *Experimental Parasitology*, **124**, 306–314.

HARRIS, J.O., POWELL, M.D., ATTARD, M. & GREEN, T.J. (2004) Efficacy of chloramine-T as a treatment for amoebic gill disease (AGD) in marine Atlantic salmon (*Salmo salar* L.). *Aquaculture Research*, **35**, 1448–1456.

HART, J.L., THACKER, J.R.M., BRAIDWOOD, J.C., FRASER, N.R. & MATTHEWS, J.E. (1997) Novel cypermethrin formulation for the control of sea lice on salmon (*Salmo salar*). *Veterinary Record*, **140**, 179–181.

HARTMANN, F. (1989) Investigations on the effectiveness of levamisol as a medication against the eel parasite *Anguillicola crassus* (Nematoda). *Diseases of Aquatic Organisms*, **7**, 185–190.

HEDRICK, R.P., GROFF, P.F. & MCDOWELL, T. (1988) Oral administration of Fumagilin DCH protects chinook salmon *Oncorhynchus tshawytscha* from experimentally induced proliferative kidney disease. *Diseases of Aquatic Organisms*, **4**, 165–168.

HEINECKE, R.D. & BUCHMANN, K. (2009) Control of *Ichthyophthirius multifiliis* using a combination of water filtration and sodium percarbonate: dose–response studies. *Aquaculture*, **288**, 32–35.

HEMAPRASANTH, K.P., RAGHAVENDRA, A., SINGH, R., SRIDHAR, N. & RAGHUNATH, M.R. (2008) Efficacy of doramectin against natural and experimental infections of *Lernaea cyprinacea* in carps. *Veterinary Parasitology*, **156** (3–4), 261–269.

HERWIG, N. (1978) Notes on the treatment of *Cryptocaryon*. *Drum and Croaker*, **18**, 6–12.

HOFFMAN, G.L. (1977) *Argulus*: a branchiuran parasite of freshwater fishes. *U.S. Fish and Wildlife Service, Fish Disease Leaflet*, **49**, 9.

HOFFMAN, G.L. & HOFFMAN, G.L. JR. (1972) Studies on the control of whirling disease (*Myxosoma cerebralis*). I. The effect of chemicals on spores *in vitro*, and of calcium oxide as a disinfectant in simulated ponds. *Journal of Wildlife Diseases*, **8**, 49–53.

HØY, T. & HORSBERG, T.E. (1991) *Chemotherapy of sea lice infestations in salmonids: pharmacological, toxicological and therapeutic properties of established and potential agents.* Ph.D. Thesis, Norwegian College of Veterinary Medicine, Oslo.

INGRAM, B.A. & PHILBEY, A.W. (1999) Occurrence of the parasite *Ergasilus intermedius* (Copepoda: Ergasilidae) on the gills of Macquarie perch, *Macquaria australasica* (Percichthyidae). *Proceedings of the Linnean Society of New South Wales*, **121**, 39–44.

IRIANTO, A. & AUSTIN, B. (2002) Probiotics in aquaculture. *Journal of Fish Diseases*, **25**, 633–642.

IVERSEN, M., FINSTAD, B., MCKINLEY, R.S., ELIASSEN, R.A., CARLSEN, K.T. & EVJEN, T. (2005) Stress responses in Atlantic salmon (*Salmo salar* L.) smolts during commercial well boat transports, and effects on survival after transfer to sea. *Aquaculture*, **243**, 372–382.

JACOBS, D.L. (1946) A new parasitic dinoflagellate from freshwater fish. *Transactions of the American Microscopic Society*, **65**, 1–17.

JAFRI, S.I.H. & AHMED, S.S. (1994) Some observations on mortality in major carp due to fish lice and their chemical control. *Pakistan Journal of Zoology*, **26**, 274–276.

JAKOBSEN, P.J. & HOLM, J.C. (1990) Promising test with new compound against salmon lice. *Norsk Fiskeoppdrett*, January, 16–18.

JOHNSON, S.C. & MARGOLIS, L. (1993) Efficacy of ivermectin for control of the salmon louse *Lepeophtheirus salmonis* on Atlantic salmon. *Diseases of Aquatic Organisms*, **17**, 101–105.

JOHNSON, S.C., CONSTIBLE, J.M. & RICHARD, J. (1993a) Laboratory investigations on the efficacy of hydrogen peroxide against the salmon louse *Lepeophtheirus salmonis* and its toxicological and histopathological effects on Atlantic salmon *Salmo salar* and chinook salmon *Oncorhynchus tshawytscha*. *Diseases of Aquatic Organisms*, **17**, 197–204.

JOHNSON, S.C., WHYTE, J.N.C. & MARGOLIS, L. (1993b) The efficacy of hydrogen peroxide against the salmon louse *Lepeophtheirus salmonis*, its toxicological effects on Atlantic and Chinook salmon, its stability in sea water, and its toxic effects on some non-target marine species. *Aquaculture Update*, **63** (December). http://www.dfo-mpo.gc.ca/Library/338131.pdf.

JONES, M.W., SOMMERVILLE, C. & WOOTTEN, R. (1992) Reduced sensitivity of the salmon louse, *Lepeophtheirus salmonis*, to the organophosphate dichlorvos. *Journal of Fish Diseases*, **15**, 197–202.

KESARCODI-WATSON, A., KASPAR, H., LATEGAN, M.J. & GIBSON, L. (2008) Probiotics in aquaculture: the need, principles and mechanisms of action and screening processes. *Aquaculture*, **274** (1), 1–14.

KIEMER, M.C.B. & BLACK, K.D. (1997) The effects of hydrogen peroxide on the gill tissues of Atlantic salmon, *Salmo salar* L. *Aquaculture*, **153**, 181–189.

KINGSFORD, E. (1975) *Treatment of Exotic Marine Fish Diseases*. Palmetto Publishing Co., St. Petersburg, Florida.

LADER, P., DEMPSTER, T., FREDHEIM, A. & JENSEN, Ø. (2008) Current induced net deformations in full-scale sea-cages for Atlantic salmon (*Salmo salar*). *Aquacultural Engineering*, **38**, 52–65.

LARSEN, A.H., BRESCIANI, J. & BUCHMANN, K. (2005) Pathogenicity of *Diplostomum* cercariae in rainbow trout, and alternative measures to prevent diplostomosis in fish farms. *Bulletin of the European Association of Fish Pathologists*, **25** (1), 20–27.

LAWLER, A.R. (1977) Dinoflagellate (*Amyloodinium*) infestation of pompano. In: Sindermann, C.J. (Ed.) *Disease Diagnosis and Control in North American Marine Aquaculture*, Elsevier, Amsterdam, pp 257–264.

LEES, F., BAILLIE, M., GETTINBY, G. & REVIE, C.W. (2008a) The efficacy of emamectin benzoate against infestations of *Lepeophtheirus salmonis* on farmed Atlantic salmon (*Salmo salar* L.) in Scotland, 2002–2006. *PLoS ONE*, **3**, e1549.

LEES, F., BAILLIE, M., GETTINBY, G. & REVIE, C.W. (2008b) Factors associated with changing efficacy of emamectin benzoate against infestations of *Lepeophtheirus salmonis* on Scottish salmon farms. *Journal of Fish Diseases*, **31** (12), 947–951.

LESTER, R.J.G. & HAYWARD, C.J. (2006) Phylum Arthropoda. In: Woo, P.T.K. (Ed.) *Fish Diseases and Disorders. Volume 1.* CAB International, pp 466–565.

LILLEY, J.H. & INGLIS, V. (1997) Comparative effects of various antibiotics, fungicides and disinfectants on *Aphanomyces invaderis* and other saprolegniaceous fungi. *Aquaculture Research*, **28** (6), 461–469.

LING, F., WANG, J. G., LIU, Q.F., LI, M., YE, L.T. & GONG, X.N. (2010) Prevention of *Ichthyophthirius multifiliis* infestation in goldfish (*Carassius auratus*) by postassium ferrate (VI) treatment. *Veterinary Parasitology*, **168**, 212–216.

LORENZEN, N. & LAPATRA, S.E. (2005) DNA vaccines for aquacultured fish. *Revue Scientifique et Technique Office International des Epizooties*, **24** (1), 201–213.

MADIN, J., CHONG, V.C. & HARTSTEIN, N.D. (2010) Effects of water flow velocity and fish culture on net biofouling in fish cages. *Aquaculture Research*, **41**, e602–e617.

MANLEY, K.M. & EMBIL, J.A. (1989) *In vitro* effect of ivermectin on *Pseudoterranova decipiens* survival. *Journal of Helminthology*, **63**, 72–74.

MELHORN, H., SCHMAHL, G. & HABERKORN, A. (1988) Toltrazuril effective against a broad spectrum of protozoan parasites. *Parasitology Research*, **75**, 64–66.

MITCHELL, A. & DARWISH, A. (2009) Efficacy of 6-, 12-, and 24-h praziquantel bath treatments against Asian tapeworms *Bothriocephalus acheilognathi* in grass carp. *North American Journal of Aquaculture*, **71** (1), 30–34.

MITCHELL, C.G. (1993) *Eubothrium. Aquaculture Information Series*, **14**, 1–4.

MOHAMED, K.S. (2010) Probiotics in aquaculture. *Coastal Fisheries and Aquaculture Management*, **2**, 303–322.

MONTGOMERY-BROCK, D., SATO, V.T., BROCK, J.A. & TAMARU, C.S. (2001) The application of hydrogen peroxide as a treatment for the ectoparasite *Amyloodinium ocellatum* (Brown 1931) on the Pacific threadfin *Polydactylus sexfilis*. *Journal of the World Aquaculture Society*, **32**, 250–254.

MORRIS, D.J., ADAMS, A., SMITH, P. & RICHARDS, R.H. (2003) Effects of oral treatment with TNP-470 on rainbow trout (*Oncorhynchus mykiss*) infected with *Tetracapsuloides bryosalmonae* (Malacosporea), the causative agent of proliferative kidney disease. *Aquaculture*, **221**, 51–64.

NEUBERT, J. (1984) Investigations on the application of trichlorfon in control of parasites and food organisms in inland fisheries. *Zeitschrift für die Binnenfischerei der DDR*, **31**, 334–336.

NOGA, E.J. (2010) *Fish Disease: Diagnosis and Treatment*. Second edition. Wiley-Blackwell, Hoboken, NJ.

NOGA, E.J. & LEVY, M.G. (2006) Phylum Dinoflagellata. In: Woo, P.T.K. (Ed.) *Fish Diseases and Disorders. Volume 1*. CAB International, pp 16–45.

PALADINI, G., GUSTINELLI, A., FIORAVANTI, M.L., HANSEN, H. & SHINN, A.P. (2009) The first report of *Gyrodactylus salaris* Malmberg, 1957 (Platyhelminthes, Monogenea) on Italian cultured stocks of rainbow trout (*Oncorhynchus mykiss*). *Veterinary Parasitology*, **165**, 290–297.

PENA, N., AURO, A. & SUMANO, H. (1988) A comparative trial of garlic, its extract, and ammonium-potassium tartrate as anthelmintics in carp. *Journal of Ethnopharmacology*, **24**, 199–203.

PICÓN-CAMACHO, S.M., MARCOS-LOPEZ, M., BELJEAN, A., DEBEAUME, S. & SHINN, A.P. (2011a) *In vitro* assessment of the chemotherapeutic action of a specific hydrogen peroxide, peracetic, acetic, and peroctanoic acid-based formulation against the free-living stages of *Ichthyophthirius multifiliis* (Ciliophora). *Parasitology Research*, DOI 10.1007/s00436-011-2575-1.

PICÓN-CAMACHO, S.M., TAYLOR, N.G.H., BRON, J.E., GUO, F.C. & SHINN, A.P. (2011b) Effects of continuous exposure to low doses of bronopol on the infection dynamics of

Ichthyophthirius multifiliis (Ciliophora), parasitising rainbow trout (*Oncorhynchus mykiss* Walbaum). *Veterinary Parasitology.*

PIKE, A.W. (1989) Sea lice – major pathogens of farmed Atlantic salmon. *Parasitology Today*, **5**, 291–297.

PIKE, A.W. & WADSWORTH, S.L. (1999) Sea lice on salmonids: their biology and control. *Advances in Parasitology*, **44**, 233–337.

RAE, G.H. (2000) A national treatment strategy for control of sea lice on Scottish salmon farms. *Caligus*, **6** (March), 2–4.

RANTIMÄKI, J., CERENIUS, L. & SÖDERHÄLL, K. (1992) Prevention of transmission of the crayfish plague fungus (*Aphanomyces astaci*) to the fresh-water crayfish *Astacus astacus* by treatment with MgCl₂. *Aquaculture*, **104** (1–2), 11–18.

ROBBINS, C., GETTINBY, G., LEES, F., BAILLIE, M., WALLACE, C. & REVIE, C.W. (2010) Assessing topical treatment interventions on Scottish salmon farms using a sea lice (*Lepeophtheirus salmonis*) population model. *Aquaculture*, **306**, 191–197.

ROTH, M. & RICHARDS, R.H. (1992) Trials on the efficacy of azamethiphos and its safety to salmon for the control of sea lice. In: Michel, C. & Alderman, D.J. (Eds.) *Chemotherapy in Aquaculture: From Theory to Reality.* Office International Des Epizooties, Paris, pp 212–218.

ROTH, M., RICHARDS, R.H. & SOMMERVILLE, C. (1993) Current practices in the chemotherapeutic control of sea lice infestations in aquaculture: a review. *Journal of Fish Diseases*, **16**, 1–26.

ROTH, M., RICHARDS, R.H., DOBSON, D.P. & RAE, G.H. (1996) Field trials on the efficacy of the organophosphorus compound azamethiphos for the control of sea lice (Copepoda: Caligidae) infestations of farmed Atlantic salmon (*Salmo salar*). *Aquaculture*, **140** (3), 217–239.

RUBIO-GODOY, M., PALADINI, G., FREEMAN, M.A., GARCÍA-VÁSQUEZ, A. & SHINN, A.P. (2011) Morphological and molecular characterisation of *Gyrodactylus salmonis* (Platyhelminthes, Monogenea) isolates collected in Mexico from rainbow trout (*Oncorhynchus mykiss* Walbaum). *Veterinary Parasitology.*

RUIDER, S., SCHMAHL, G., MEHLHORN, H., SCHMIDT, H. & RITTER, G. (1997) Effects of different malachite green derivatives and metabolites on the fish ectoparasite, *Ichthyophthirius multifiliis*, Fouquet 1876 (Hymenostomatida, Ciliophora). *European Journal of Protistology*, **33**, 375–388.

SAKAI, M. (1999) Current research status of fish immunostimulants. *Aquaculture*, **172** (1–2), 63–92.

SANGMANEEDET, S. & SMITH, S.A. (1999) Efficacy of various chemotherapeutic agents on the growth of *Spironucleus vortens*, an intestinal parasite of the freshwater angelfish. *Diseases of Aquatic Organisms*, **38** (1), 47–52.

SANTAMARINA, M.T., TOJO, J., UBEIRA, F.M., QUINTEIRO, P. & SANMARTÍN, M.L. (1991) Anthelmintic treatment against *Gyrodactylus* sp. infecting rainbow trout *Oncorhynchus mykiss*. *Diseases of Aquatic Organisms*, **10**, 39–44.

SCHÄPERCLAUS, W. (1992) *Fish Diseases.* Vol. 2. A.A. Balkema, Rotterdam.

SCHELKLE, B., SHINN, A.P., PEELER, E. & CABLE, J. (2009) Treatment of gyrodactylid infections in fish. *Diseases of Aquatic Organisms*, **86**, 65–75.

SCHLENK, D., GOLLON, J.L. & GRIFFIN, B.R. (1998) Efficacy of copper sulfate for the treatment of ichthyophthiriasis in channel catfish. *Journal of Aquatic Animal Health*, **10**, 390–396.

SCHMAHL, G., EL TOUKHY, A. & GHAFFAR, F.A. (1990) Transmission electron microscopic studies on the effects of toltrazuril on *Glugea anomala*, Moniez, 1887 (Microsporidia) infecting the three-spined stickleback *Gasterosteus aculeatus*. *Parasitology Research*, **76**, 700–706.

SCHMAHL, G. & TARASCHEWSKI, H. (1987) Treatment of fish parasites. 2. Effects of praziquantel, niclosamide, levamisole-HCl, and metrifonate on Monogenea

(*Gyrodactylus aculeati*, *Diplozoon paradoxum*). *Parasitology Research*, **73** (4), 341–351.

SEARCH (2006) *Sea Lice Resistance to Chemotherapeutants: A Handbook in Resistance Management*. Second edition. Available: http://www.rothamsted.bbsrc.ac.uk/pie/search-EU [accessed September 2011].

SEVATDAL, S. & HORSBERG, T.E. (2000) Kartlegging av pyretroidresistens hos lakselus. *Norsk Fiskeoppdrett*, **12**, 34–35 [in Norwegian].

SEVATDAHL, S. & HORSBERG, T.E. (2003) Determination of reduced sensitivity in sea lice (*Lepeophtheirus salmonis* Krøyer) against the pyrethroid deltamethrin using bioassays and probit modelling. *Aquaculture*, **218**, 21–31.

SEVATDAL, S., COPLEY, L., WALLACE, C., JACKSON, D. & HORSBERG, T.E. (2005a) Monitoring of the sensitivity of sea lice (*Lepeophtheirus salmonis*) to pyrethroids in Norway, Ireland and Scotland using bioassays and probit modelling. *Aquaculture*, **244**, 19–27.

SEVATDAL, S., FALLANG, A., INGEBRIGTSEN, K. & HORSBERG, T.E. (2005b) Monooxygenase mediated pyrethroid detoxification in sea lice (*Lepeophtheirus salmonis*). *Pest Management Science*, **61**, 772–778.

SHINN, A., WOOTTEN, R., SOMMERVILLE, C. & CONWAY, D. (2001) Putting the squeeze on whitespot. *Trout News*, **32**, 20–25.

SHINN, A.P., WOOTTEN, R. & SOMMERVILLE, C. (2003a) Alternative compounds for the treatment of *Ichthyophthirius multifiliis* infecting rainbow trout. *Trout News*, **35**, 38–41.

SHINN, A.P., WOOTTEN, R., CÔTE, I. & SOMMERVILLE, C. (2003b) Efficacy of selected oral chemotherapeutants against *Ichthyophthirius multifiliis* (Ciliophora: Ophyroglenidae) infecting rainbow trout *Oncorhynchus mykiss*. *Diseases of Aquatic Organisms*, **55**, 17–22.

SHINN, A.P., PICÓN-CAMACHO, S.M., BAWDEN, R. & TAYLOR, N.G.H. (2009) Efficacy of a mechanical device for the control of *Ichthyophthirius multifiliis* Fouquet, 1876 (Ciliophora) infections in a commercial rainbow trout, *Oncorhynchus mykiss* (Walbaum), hatchery system. *Aquacultural Engineering*, **41** (3), 152–157.

SHINN, A.P., PICÓN-CAMACHO, S., BRON, J.E., CONWAY, D., YOON, G.H., HUNTER, R., GUO, F.C. & TAYLOR, N.G.H. (2011) The anti-protozoal activity of bronopol on the key life-stages of *Ichthyophthirius multifiliis* Fouquet, 1876 (Ciliophora). *Veterinary Parasitology*.

SPEARE, D.J., RITTER, G. & SCHMIDT, H. (1998) Quinine hydrochloride treatment delays xenoma formation and dissolution in rainbow trout challenged with *Loma salmonae*. *Journal of Comparative Pathology*, **119** (4), 459–465.

SPEARE, D.J., ATHANASSOPOULOU, F., DALEY, J. & SANCHEZ, J.G. (1999) A preliminary investigation of alternatives to fumagillin for the treatment of *Loma salmonae* infection in rainbow trout. *Journal of Comparative Pathology*, **121** (3), 241–248.

STONE, J., SUTHERLAND, I.H., SOMMERVILLE, C., RICHARDS, R.H. & VARMA, K.J. (1999) The efficacy of emamectin benzoate as an oral treatment of sea lice, *Lepeophtheirus salmonis* (Krøyer), infestations in Atlantic salmon, *Salmo salar* L. *Journal of Fish Diseases*, **22**, 261–270.

STONE, J., SUTHERLAND, I., SOMMERVILLE, C., RICHARDS, R.H. & ENDRIS, R.G. (2000) The duration and efficacy following oral treatment with emamectin benzoate against infestation of sea lice, *Lepeophtheirus salmonis* (Krøyer), in Atlantic salmon *Salmo salar* L. *Journal of Fish Diseases*, **23**, 185–192.

STRAUS, D.L., HOSSAIN, M.M. & CLARK, T.G. (2009) Strain differences in *Ichthyophthirius multifiliis* to copper toxicity. *Diseases of Aquatic Organisms*, **83**, 31–36.

SUDOVÁ, E., STRAUS, D. L., WIENKE, A. & MEINELT, T. (2010) Evaluation of continuous 4-day exposure to peracetic acid as a treatment for *Ichthyophthirius multifiliis*. *Parasitology Research*, **106**, 539–542.

SZÉKELY, C. & MOLNÁR, K. (1987) Mebendazole is an efficacious drug against pseu-dodactylogyrosis in the European eel (*Anguilla anguilla*). *Journal of Applied Ichthyology*, **3** (4), 183–186.

SZÉKELY, C. & MOLNÁR, K. (1991) Praziquantel (Droncit) is effective against diplos-tomosis of grass carp (*Ctenophayngodon idella*) and silver carp (*Hypophthalmichthys molitrix*). *Diseases of Aquatic Organisms*, **11**, 147–150.

TAN, C.K.F., NOWAK, B.F. & HODSON, S.L. (2002) Biofouling as a reservoir of *Neoparamoeba pemaquidensis* (Page, 1970), the causative agent of amoebic gill disease in Atlantic salmon. *Aquaculture*, **210**, 49–58.

TARASCHEWSKI, H., RENNER, C. & MEHLHORN, H. (1988) Treatment of fish parasites. 3. Effects of levamisole HCl, metrifonate, fenbendazole, mebendazole, and ivermectin on *Anguillicola crassus* (nematodes) pathogenic in the air bladder of eels. *Parasitology Research*, **74** (3), 281–289.

TARASCHEWSKI, H., MEHLHORN, H. & RAETHER, W. (1990) Loperamid, an efficacious drug against fish-pathogenic acanthocephalans. *Parasitology Research*, **76**, 619–623.

THONEY, D.A. & HARGIS, W.J. JR. (1991) Monogenea (Platyhelminthes) as hazards for fish in confinement. *Annual Review of Fish Diseases*, **2**, 133–153.

TOJO, J.L. & SANTAMARINA, M.T. (1998) Oral pharmacological treatments for parasitic diseases of rainbow trout *Oncorhynchus mykiss*. I: *Hexamita salmonis*. *Diseases of Aquatic Organisms*, **33**, 51–56.

TOOVEY, J.P.G. & LYNDON, A.R. (2000) Cell culture bio-assays for potential anti-sea louse chemotherapeutants. *Caligus*, **6** (March), 11.

TORCHIN, M.E., LAFFERTY, K.D., DOBSON, A.P., MCKENZIE, V.J. & KURIS, A.M. (2003) Introduced species and their missing parasites. *Nature*, **421**, 628–630.

TREASURER, J., WADSWORTH, S. & GRANT, A. (2000) Resistance of sea lice, *Lepeophtheirus salmonis* (Krøyer), to hydrogen peroxide on Atlantic farmed salmon, *Salmo salar* L. *Aquaculture Research*, **31**, 855–860.

TUCKER, C.S., NORMAN, R., SHINN, A.P., BRON, J.E., SOMMERVILLE, C. & WOOTTEN, R. (2002) A single cohort, time delay model of the life cycle and control of sea lice (*Lepeophtheirus salmonis*) on Atlantic salmon. *Fish Pathology*, **37** (3), 107–118.

TULLY, O. & MCFADDEN, Y. (2000) Variation in sensitivity of sea lice [*Lepeophtheirus salmonis* (Kroyer)] to dichlorvos on Irish salmon farms in 1991–92. *Aquaculture Research*, **31**, 849–854.

VAN DER HEIJDEN, M.H.T., HELDERS, G.M., BOOMS, G.H.R., HUISMAN, E.A., ROMBOUT, J.H.W.M. & BOON, J.H. (1996) Influence of flumequine and oxytetracycline on the resistance of the European eel against the parasitic swimbladder nematode *Anguillicola crassus*. *Veterinary Immunology and Immunopathology*, **52** (1–2), 127–134.

VAN DUIJIN, C., JR. (1973) *Diseases of Fishes*. 3rd edition. Thomas, Springfield, IL.

VERSCHUERE, L., ROMBAUT, G., SORGELOOS, P. & VERSTRAETE, W. (2000) Probiotic bacteria as biological control agents in aquaculture. *Microbiology and Molecular Biology Reviews*, **64** (4), 655–671.

WAGNER, E.J. (2002) Whirling disease prevention, control and management. A review. In: Bartholomew, J.L. & Wilson, J.C. (Eds.), *Whirling Disease: Reviews and Current Topics*. American Fisheries Society Symposium No. 29, American Fisheries Society, Bethesda, MD, pp 217–225.

WANG, G-X., JIANG, D-X., LI, J., HAN, J., LIU, Y.-T. & LIU, X.-L. (2010) Anthelmintic activity of steroidal saponins from *Dioscorea zingiberensis* C. H. Wright against *Dactylogyrus intermedius* (Monogenea) in goldfish (*Carassius auratus*). *Parasitology Research*, **107** (6), 1365–1371.

WANG, Y.B., LI, J.R. & LIN, J. (2008) Probiotics in aquaculture: challenges and outlook. *Aquaculture*, **281** (1–4), 1–4.

WESTCOTT, J.D., STRYHN, H., BURKA, J.F. & HAMMELL, K.L. (2008) Optimization and field use of a bioassay to monitor sea lice *Lepeophtheirus salmonis* sensitivity to emamectin benzoate. *Diseases of Aquatic Organisms*, **79**, 119–131.

WIJEYARATNE, M.J.S. & GUNAWARDENE, R.S. (1988) Chemotherapy of ectoparasite, *Ergasilus ceylonensis* of Asian cichlid, *Etroplus suratensis. Journal of Applied Ichthyology*, **4**, 97–100.

WILLIAMS, R.E., ERNST, I., CHAMBERS, C.B. & WHITTINGTON, I.D. (2007) Efficacy of orally administered praziquantel against *Zeuxapta seriolae* and *Benedenia seriolae* (Monogenea) in yellowtail kingfish *Seriola lalandi. Diseases of Aquatic Organisms*, **77** (3), 199–205.

WOO, P.T.K. (2006) *Fish Diseases and Disorders. Volume 1. Protozoan and Metazoan Infections*. Second edition. CAB International.

9

Developments in vaccination against fish bacterial disease

B. Austin, University of Stirling, UK

Abstract: From initial work on the use of chemically inactivated whole-cell suspensions, the development of fish vaccines has included purified and often inactivated subcellular components, subunit, live attenuated, DNA and compound products containing antigens for multiple pathogens. Application is by injection, immersion in a dilute suspension and/or orally. The latter, which originally was least successful, has improved due to the use of new oralisers and micro-encapsulation techniques which protect antigens during passage through the digestive tract. There is uncertainty over the precise mode of action. Nevertheless despite so much research, only a comparatively few commercial products are available.

Key words: vaccine, antigen, immune response, adjuvant, immersion, injection, oral uptake.

9.1   Introduction

With terrestrial animals, vaccination is regarded as the primary means of disease control where application is often by injection and protection reflects antibody levels, but the numbers of individuals needed to be vaccinated is comparatively low compared with the populations in intensive aquaculture facilities. The first bacterial fish vaccine was described by Duff (1942), who used chloroform-inactivated cells of *Aeromonas salmonicida* to protect cutthroat trout (*Salmo clarki*) against furunculosis. There is evidence that interest in vaccines declined with the availability and effectiveness of antibiotics, but the subsequent development and spread of antibiotic-resistance rekindled research in vaccinology. Nevertheless despite the increasing number of serious fish pathogens, only a comparatively few commercial products have entered the market place. Moreover, interest has focused on less than half of all the bacteria taxa described as fish pathogens.

The development of vaccines for fish diseases has used most of the approaches considered for human and terrestrial animal pathogens, with research progressing from the simplistic and often effective approach of using inactivated (usually chemically inactivated involving formalin) whole cell preparations to recombinant products. Application to fish may be via injection, immersion/bath or oral; the latter is easiest but early attempts were least successful although the availability of modern, effective oralisers and micro-encapsulation techniques has considerably boosted the effectiveness of products.

9.1.1   Composition of vaccines

- *Chemically or heat-inactivated whole cell preparations.* Typically, inactivation is with 0.3–0.5% (v/v) formalin. This approach has been used successfully with *Vibrio anguillarum* and *Yersinia ruckeri*, which are the causal agents of vibriosis and enteric redmouth, respectively. If carried out carefully and with sufficient safety checks, the vaccines are inevitably safe, and from the commercial prospect, there is negligible opportunity of a competitor acquiring and cloning the bacterial strain(s).
- *Attenuated, live vaccines.* Certainly, a modern approach, and the scientific argument in their favour is that their use will more closely mimic an infection, thereby a natural and hopefully efficacious immune response will be mounted by the host.
- *Subcellular components.* These may be highly immunogenic, and have been included in other vaccine preparations, such as formalized whole cells. Some components, such as lipopolysaccharide (LPS), have broad spectrum immunostimulatory properties.
- *Subunit vaccines.*
- *DNA vaccines.* A modern and exciting concept, which possibly has greater value to viral than bacterial pathogens.
- *Compound products*, i.e. comprising two or more different components, such as formalized mixtures of pathogens with or without purified subcellular components.

9.2   Methods of administration of vaccines for fish

Essentially, three different methods are used regularly, namely injection with or without the presence of adjuvant, immersion in diluted vaccine (or bathing for prolonged periods of 2 h or more in an even more dilute preparation) or oral application. In terms of ease of application, injection is comparatively slow and requires prior anaesthesia, immersion is quick and easy although there are issues with the disposal of spent vaccine, and oral administration which has tended to be least successful (Midtlyng *et al.*, 1996) due to possible problems with the degradation of antigens during passage through the digestive tract (Johnson and Amend, 1983a). Generally, injection is most effective in terms of protection, followed by immersion and lastly oral uptake. Fortunately with the availability of micro-encapsulation, e.g. alginate microparticles (Joosten *et al.*, 1997) and new and improved oralizers, which protect antigens during passage through the digestive tract, the ability of oral vaccination to protect against disease is improving. For example, liposome-entrapped antigens of atypical *A. salmonicida* were fed to carp leading to greater protection, which was measured in terms of a reduction in ulceration compared to controls (Irie *et al.*, 2005). Furthermore, Yasumoto *et al.* (2006) entrapped *A. hydrophila* antigens in liposomes, which were fed to carp over 3 days, and led to demonstrable protection at 22 days (relative percent protection (after Amend, 1981); RPS = 55% or 63.6% depending on the severity of the challenge dose) (Yasumoto *et al.*, 2006). A polysaccharide matrix, termed MicroMatrix™, has demonstrated promise with *Piscirickettsia salmonis* (Tobar *et al.*, 2011). However, there are issues about dose, i.e. the amount of antigen to be administered to the fish, and the duration of feeding (the number of days). In a detailed examination of the effects of oral administration of formalized *V. anguillarum* vaccines in chinook salmon, Fryer *et al.* (1978) recorded that maximal protection followed the feeding of 2 mg of dried vaccine/g of food for 15 days at temperatures even as low as 3.9 °C. Longer feeding regimes did not result in better protection. Incorporation of vaccine in natural food, i.e. plankton, has also shown promise with ayu (Kawai *et al.*, 1989).

Other possible means of application include:

- The showering or low-pressure spraying of vaccine onto fish (Gould *et al.*, 1978), a process which may be automated and carried out during grading.
- Hyperosmotic infiltration whereby fish are dipped into a strong saline solution followed by immersion in the vaccine suspension (Croy and Amend, 1977; Giorgetti *et al.*, 1981). Although effective, the technique is stressful to fish (Busch *et al.*, 1978) and the level of protection is only comparable to immersion (Antipa *et al.*, 1980). Consequently, hyperosmotic infiltration is no longer used.
- Anal/oral intubation; the former allows the bypassing of the possible adverse effects of antigen passage in the stomach and intestine, but the antigen has to be introduced directly into the mouth or anus, and is slow and cumbersome.
- Ultrasonics/ultrasound; this is a new approach that requires further development (Zhou *et al.*, 2002; Navot *et al.*, 2011). In one example, ultrasound has been evaluated to administer vaccines for the control of goldfish ulcer disease, which is caused by atypical *A. salmonicida*. Thus, soluble surface layer protein (= A-protein) of *A. salmonicida* was applied by immersion after ultrasound (1 MHz frequency of ultrasound/1 min) pretreatment, and led to promising results after challenge (Navot *et al.*, 2011).

Although the use of adjuvants has often boosted the effectiveness of injectable vaccines, there is concern over the use of mineral oils, notably

Freund's complete adjuvant (FCA) and to a lesser extent Freund's incomplete adjuvant (FIA), with intra-abdominal adhesions occurring in Atlantic salmon following intraperitoneal (i.p.) injection of FCA-containing products (Gudmundsdóttir *et al.*, 2003). Side effects of FIA included the development of black/brown pigment, most likely melanin, in the stomach with inflammation in the form of granulomas and cysts (van Gelderen *et al.*, 2009). In addition, there have been indications of temporary immunosuppression (Inglis *et al.*, 1996). One solution to this problem has been the inclusion of antibiotics, namely amoxicillin, which are administered by injection with the vaccine and which counteract the initial effects of any opportunistic bacterial invaders (Inglis *et al.*, 1996). Alternatives to oil adjuvants include β-1,3 glucan (Vita-Stim-Taito), lentinan and formalin-killed cells of *R. salmoninarum*, which have enhanced the effectiveness of formalized whole cell furunculosis vaccines (Nikl *et al.*, 1991), and aluminium hydroxide and aluminium phosphate, with data pointing to increased RPS (Jiao *et al.*, 2010). Immunostimulants, e.g. levamisole, also enhance efficacy of vaccine preparations (Kajita *et al.*, 1990). Nevertheless, oil-based adjuvants continue to be favoured in some countries, including Norway. Indeed in one Norwegian field study, Midtlyng (1996) determined that i.p. administered furunculosis vaccine in a mineral oil adjuvant gave the best protection in Atlantic salmon.

9.3   Determination of effectiveness of the vaccines

Often, vaccine potency is determined by challenge methods and the calculation of RPS (after Amend, 1981).

$$RPS = 1 - \left(\frac{\text{vaccinated fish mortality \%}}{\text{non-vaccinated mortality \%}} \right) \times 100\%$$

Clearly, there are ethical issues with this approach, which involves infecting and deliberately killing experimental animals. However, to date, there is no successful and reliable alternative that could be used routinely by industry, as the precise nature of the protective immune response of fish resulting from vaccination is unclear.

9.4   Vaccine composition

From the initial reliance on conventional bacterial cultures, research has progressed to a consideration of growth conditions (medium composition, incubation temperature/time) and the precise antigenic composition of the bacterial cells. A topical example concerns *A. salmonicida*, which, when cultured in iron-depleted conditions, produces highly immunogenic iron regulated outer membrane proteins (IROMP) (e.g. Durbin *et al.*, 1999).

This forms the basis of production of some current commercial furunculosis vaccines. Complex preparations, such as those containing inactivated whole cells supplemented with toxoids and/or purified subcellular components, have been evaluated. Also, the relative benefits of polyvalent versus monovalent preparations have been considered.

9.4.1   Inactivated whole cell preparations

Inactivated whole-cell vaccines are the earliest form of bacterial fish vaccines, and have been used successfully for controlling many diseases, with examples including *A. hydrophila* (Ruangpan *et al.*, 1986), *Aliivibrio salmonicida* (Holm and Jørgensen, 1987), *Flavobacterium columnare* (Mano *et al.*, 1996), *F. psychrophilum* (Kondo *et al.*, 2003), *Photobacterium damselae* subsp. *piscicida* (Afonso *et al.*, 2005), *Piscirickettsia salmonis* (Smith *et al.*, 1995), *Pseudomonas anguilliseptica* (Nakai *et al.*, 1982), β-haemolytic *Streptococcus* (Sakai *et al.*, 1987, 1989), *Streptococcus difficilis* (Eldar *et al.*, 1995), *V. anguillarum* (e.g. Horne *et al.*, 1982) and *Y. ruckeri* (Tebbit *et al.*, 1981) (Table 9.1). A comparison of different methods of inactivation of *Y. ruckeri* convinced Anderson and Ross (1972) that 3% chloroform was better than sonication, 1% formalin, or 0.5% or 3% phenol. An alternative method of inactivation has included pressure (600 kgf/cm^2 for 5 s using a French press), which was used to inactivate *Edwardsiella tarda* cells, the use of which led to an RPS of >85% in Japanese eels, 6 months after vaccination (Hossain and Kawai, 2009).

Although many of these whole cell vaccines are produced in standard batch culture conditions, some ingenuity has been demonstrated. For examples, the ability of *A. hydrophila* to develop biofilms on surfaces has been exploited, and a study with walking catfish (*Clarias batrachus*) demonstrated that attached cells on chitin flakes led to higher RPS (= 91–100%) and serum antibody titre when administered orally for 20 days compared with preparations derived from suspensions in tryptone soya broth (RPS = 29–42%) (Nayak *et al.*, 2004). With *A. salmonicida*, the value of including A-protein producing strains has been highlighted in experiments whereby non-oily Montanide-adjuvanted injectable whole cell inactivated vaccines prepared from A-protein$^+$ cultures led to RPS values of 51–78% in Atlantic salmon whereby use of A-layer$^-$ components lacked efficacy (Lund *et al.*, 2003a). Furthermore, Lund *et al.* (2003b) confirmed the need for A-protein, but highlighted the necessity of incorporating atypical rather than typical cultures of *A. salmonicida* in vaccine preparations (RPS = 82–95% in wolffish). The explanation given was that atypical *A. salmonicida* had genetically and serological different A-protein from their typical counterparts (Lund *et al.*, 2003b). Subsequently, Lund *et al.* (2008) reinforced the importance of A-layer in vaccine preparations designed to protect Atlantic cod against atypical isolates by using oil adjuvanted preparations administered by i.p. injection. The vaccines contained formalized cultures with different cell

Table 9.1 Examples of formalin-inactivated bacterial fish vaccines

Pathogen	Fish species	Application	RPS	Reference(s)
Aeromonas salmonicida	Atlantic salmon	Injection	51–79%	Lund et al. (2003a)
Edwardsiella ictaluri	Channel catfish	Immersion and oral booster	96.9%	Plumb and Vinitnantharat (1993)
Edwardsiella ictaluri	Channel catfish	Immersion	93%	Plumb and Vinitnantharat (1993)
Edwardsiella tarda	Japanese eels	i.m. injection and booster	~25%	Gutierrez and Miyazaki (1994)
Flavobacterium psychrophilum	Rainbow trout	i.p. injection (with FCA)	83%	LaFrentz et al. (2002)
Flavobacterium psychrophilum	Ayu	i.p. injection (in Montanide IMS1312)	33% (in 2000) 39.6% (in 2001)	Nagai et al. (2003)
Moriella viscosa	Atlantic salmon	i.p. injection	97%	Greger and Goodrich (1999)
Photobacterium damselae subsp. *piscicida*	Ayu	i.p. injection	96%	Hanif et al. (2005)
Piscirickettsia salmonis	Atlantic salmon	i.p. injection	50%	Birkbeck et al. (2004)
Pseudomonas plecoglossicida	Ayu	i.p. injection – in saline	65–86%	Ninomiya and Yamamoto (2001)
		– in Montanide-ISA711	17–58%	Ninomiya and Yamamoto (2001)
		– in Montanide-ISA763A	57–92%	Ninomiya and Yamamoto (2001)
Pseudomonas plecoglossicida	Ayu	Oral (acetone killed)	40–79%	Kintsuji et al. (2006)
Renibacterium salmoninarum	Rainbow trout	Injection	10–23.8%	Sakai et al. (1993)
β-haemolytic streptococci	Rainbow trout	Immersion/injection	70%	Sakai et al. (1987; 1989)
Streptococcus iniae	Nile tilapia	i.p. injection	79–100%	Shoemaker et al. (2010)
Tenacibaculum maritimum	Atlantic salmon	i.p. injection (with FIA)	79.6%	van Gelderen et al. (2009)
Tenacibaculum maritimum	Atlantic salmon	i.p. injection (without FIA)	27.7%	van Gelderen et al. (2009)
Vibrio anguillarum	Pacific salmon	i.p. injection – formalized vaccine	55.7%	Antipa (1976)
		heat killed vaccine	73.8%	Antipa (1976)
Vibrio anguillarum	Coho salmon; rainbow trout	Spray (showering)	100%	Gould et al. (1978)

Table 9.1 Continued

Pathogen	Fish species	Application	RPS	Reference(s)
Vibrio anguillarum	Salmonids	Oral	6.2%	Baudin-Laurençin and Tangtrongpiros (1980)
Vibrio anguillarum	Salmonids	Immersion	93.7%	Baudin-Laurençin and Tangtrongpiros (1980)
Vibrio anguillarum	Salmonids	Injection	95.8%	Baudin-Laurençin and Tangtrongpiros (1980)
Vibrio anguillarum	Salmonids	Oral	48%	Amend and Johnson (1981)
Vibrio anguillarum	Salmonids	Immersion	92.3%	Amend and Johnson (1981)
Vibrio anguillarum	Salmonids	Spray	98%	Amend and Johnson (1981)
Vibrio anguillarum	Salmonids	i.p. injection	100%	Amend and Johnson (1981)
Vibrio anguillarum	Rainbow trout	i.p. injection	93%	Horne *et al.* (1982)
		Immersion	47%	Horne *et al.* (1982)
		Oral	6%	Horne *et al.* (1982)
Vibrio anguillarum	Ayu	Oral (incorporated in plankton)	78.7%	Kawai *et al.* (1989)
Vibrio anguillarum	Sea bass	Oral	72.3%	Dec *et al.* (1990)
Vibrio anguillarum	Turbot	Oral	70.6%	Dec *et al.* (1990)
Vibrio anguillarum serogroups O2a and O2b	Cod	Immersion	100%	Mikkelsen *et al.* (2007)
Vibrio anguillarum	Atlantic halibut, African catfish, sea bass	Immersion	100%	Bricknell *et al.* (2000), Bowden *et al.* (2002), Vervarcke *et al.* (2004), Angelidis *et al.* (2006)
Yersinia ruckeri	Sockeye salmon	Oral intubation		Johnson and Amend (1983b)
		– (vaccine in gelatin)	28.8%	Johnson and Amend (1983b)
		– (no gelatin) anal intubation	63.9%	Johnson and Amend (1983b)
		– (vaccine in gelatin)	92.7%	Johnson and Amend (1983b)
		– (no gelatin)	61.8%	Johnson and Amend (1983b)

surface components, specifically A-layer (including an A-protein⁻ isolate with re-attached A-protein) and LPS. The outcome was that whole cell preparations with A-protein elicited better protection than those without.

The salinity of the growth medium composition for *Photobacterium damselae* subsp. *piscicida* (Afonso *et al.*, 2005) affected the immune response, with 2.5% rather than 0.5% (w/v) NaCl being more effective (Nitzan *et al.*, 2004). Moreover, media containing peptones, yeast extract and salt led to the synthesis of a wider range of cellular components, including novel compounds of ~14 and ~21.3 kDa, than others. Indeed, these compounds were recognised by post-disease sea bass serum (Bakopoulos *et al.*, 2003). Amend *et al.* (1983), examining factors affecting the potency of *Y. ruckeri* vaccine preparations, reported that potency was not affected by medium pH values of 6.5 to 7.7, or by cultivation for up to 96 h in tryptic soy broth (TSB) at room temperature. However, protection was enhanced by culturing the cells for 48 h at pH 7.2, lysing them at pH 9.8 for 1–2 h, and then adding 0.3% (w/v) formalin.

The incubation temperature used to culture *Aliivibrio salmonicida* is an important aspect of vaccine production, with 10 °C (this coincides with the upper range of water temperatures at which the disease coldwater vibriosis is most likely to occur) rather than 15 °C giving a higher yield of cells in broth media (Colquhoun *et al.*, 2002).

Johnson *et al.* (1982a) examined the interval necessary from vaccination for the onset and duration of immunity to vibriosis to develop, and reported that a 5 s immersion in *Vibrio anguillarum* and *V. anguillarum* vaccines was sufficient to induce protection within 5 days at 18 °C, or 10 days at 10 °C. The minimum size of salmonids necessary for maximal protection was determined to be ~1.0–2.5 g. Of relevance, this team concluded that protection was correlated with size rather than age of the fish. In the case of 1.0, 2.0 and 4.0 g salmonids, immunity lasted for approximately 4, 6 and 12 months, respectively (Johnson *et al.*, 1982b). The water temperature was also regarded as important, with 15 °C being completely effective but 5 or 25 °C less so (Raida and Buchmann, 2008). Also, there was variation in efficacy of vaccination between species, with coho salmon and sockeye salmon retaining immunity for longer than pink salmon. Immunity of 7 and >14 months has been reported by Lamers and Muiswinkel (1984) and Cossarini-Dunier (1986), respectively.

Despite the success of vaccines, their use in the absence of a comprehensive disease control strategy may allow the entry of other pathogens. This may well have happened with enteric redmouth (ERM) vaccines in rainbow trout whereby the apparent failure of a previously effective product in the UK heralded the arrival of what transpired to be a new biogroup of *Y. ruckeri*, biogroup 2 (Austin *et al.*, 2005). A parallel situation occurred with Atlantic salmon culture in Tasmania whereby the standard ERM vaccine lost effectiveness, and led to an improved product involving the trypsinization of the component culture to expose the O-antigen and thereby improving antigenicity. The end result was improved protection (the RPS increased from 37% to 55.6%) (Costa *et al.*, 2011).

9.4.2   Subcellular component vaccines

Attention has focused on a range of potentially immunogenic molecules, including S/A-layer proteins, extracellular products (ECP), outer membrane proteins (OMP) and LPS. In the case of ECPs, the enzyme activity inevitably has to be inactivated such as by the addition of formalin before use.

With *A. salmonicida*, there is controversy over the wisdom of using ECP, with indications of immunosuppression (e.g. Sövényi *et al.*, 1990), in addition to immunogenicity (Kawahara *et al.*, 1990). Severe side effects have certainly resulted from injection of oil-adjuvanted vaccines, with the ECP component contributing to inflammation (Mutoloki *et al.*, 2006).

Purified 43 kDa OMP of *A. hydrophila* in Freund's complete adjuvant (FCA) with a booster after 3 weeks (without FCA) administered by injection led to protection and a demonstrable immune response in blue gourami (*Trichogaster richopterus*) (Fang *et al.*, 2000). *F. psychrophilum* OMP administered i.p. led to protection and a demonstrable immune response in ayu (RPS = 64 and 71%) and rainbow trout (RPS = 93 and 95%) (Rahman *et al.*, 2002). A surface protein, coined P18, was purified and the responsible gene identified which encoded a 166 amino acid OmpH-like protein. In vaccine trials using rainbow trout and i.p. administration with FCA, protection and high antibody levels ensued (RPS = 88%) (Dumetz *et al.*, 2006). 18–28, 41–49 and 70–100 kDa fractions were identified by western blotting in rainbow trout immune serum, and adjuvanted in FCA. Good protection was reported after i.p. injection of rainbow trout fry for the 41–49 (RPS = 58%) and 70–100 kDa (O-proteins and O-polysaccharide) fractions with an RPS (for the latter fraction) of 94% (LaFrentz *et al.*, 2004). Pang *et al.* (2010) used *V. harveyi* OmpN mixed with FCA to vaccinate estuary cod (*Epinephelus coioides*) via the i.p. route, leading to RPS = 60 and 70% depending on the nature of the challenge strain. Serine protease has been reported as a protective antigen, and enabled fish to resist challenge (Zhang *et al.*, 2008).

The putative hydrophobic cytoplasmic membrane protein, MtsB, of the ATP-binding cassette transporter system of *S. iniae*, was also protective (RPS = 69.9%) following i.p. injection in FCA into tilapia (Zou *et al.*, 2011).

Saeed (1983) and Saeed and Plumb (1987) used a LPS extract of *E. ictaluri*, administered intraperitoneally to channel catfish, which led to ≥80% survival after challenge (RPS = ~70%) and a demonstrable humoral immune response. Furthermore, intramuscular (i.m.) injection of eels and red sea bream with *E. tarda* LPS led to protection after challenge and a demonstrable humoral immune response (titre = 1:2048) and phagocytosis by T-lymphocytes (Salati *et al.*, 1987a,b). Indeed, the evidence was that LPS was more effective as an immunogen than a formalized culture (Salati *et al.*, 1987a,b). This was substantiated when injection of LPS into Japanese eels led to an RPS = ~38–66% after challenge; higher than a whole cell vaccine (Gutierrez and Miyazaki, 1994). The benefit of LPS from *Photobacterium damselae* subsp. *piscicida* has also been mooted (Fukuda and Kusuda, 1982).

9.4.3 Live and recombinant vaccines

There has certainly been a rush to use new technologies for the development of bacterial fish vaccines.

Aeromonas hydrophila

A recombinant S-layer protein of *A. hydrophila* was immunoprotective (RPS = 56–87%) when injected i.p. in adjuvant into common carp (Poobalane *et al.*, 2010). Separately, Zhao *et al.* (2011) used the glyceraldehyde-3-phosphate dehydrogenase (GAPDH) gene *gapA* to express *A. hydrophila* GAPDH in attenuated *V. anguillarum*, which was injected i.p. into turbot, with the outcome that cytoplasm GAPDH expressing strain AV/pUC-gapA-vaccinated fish challenged with *A. hydrophila* and *V. anguillarum* gave RPS of 42% and 92%, respectively. Tu *et al.* (2010) used ghost cells (these were produced with the lysis plasmid, pElysis, with lysing proceeding after incubation at 27–42 °C), which were fed to carp and led to an RPS of 76.8% after challenge with *A. hydrophila*. This compared with RPS of 58.9% in the group administered with formalized whole cells (Tu *et al.*, 2010).

Aeromonas salmonicida

A live aromatic-dependent *A. salmonicida* vaccine, *aroA*, was administered i.p., and resulted in a 253-fold increase in LD_{50} (Vaughan *et al.*, 1993). Similarly, i.p. administration of a live auxotrophic *aroA* mutant of *A. hydrophila* protected rainbow trout against furunculosis after 30 days (RPS = >60%).

Edwardsiella ictaluri

Novobiocin-resistant attenuated *E. ictaluri* cells have been administered i.p. and immersion, to channel catfish leading to RPS after challenge of >90% and 100%, respectively (Pridgeon and Klesius, 2011a).

Edwardsiella tarda

By multiple passaging through laboratory media, Sun *et al.* (2010c) isolated an attenuated strain of *E. tarda*, i.e. TX5RM, which was immunoprotective to Japanese flounder by i.p., immersion, oral and oral plus immersion. With the exception of injection, the other groups were administered boosters after 3 weeks, with challenge after 5 or 8 weeks. Oral plus immersion gave the highest RPS of 80.6% at 5 weeks after vaccination, decreasing to 69.4% at 8 weeks (Sun *et al.*, 2010c). A natural avirulent isolate of *E. tarda*

ATCC 15947 was used by i.p. injection with a booster at 3 weeks, leading to complete protection (RPS = 100%) in Japanese flounder. Some success resulted from oral application for 5 days (RPS = 56%) when the antigens were incorporated into alginate microspheres (Cheng *et al.*, 2010b).

Hu *et al.* (2011) developed a recombinant vaccine that expressed *V. harveyi* DegQ as a soluble antigen that elicited significant protection against both *E. tarda* and *V. harveyi* in laboratory experiments with turbot. Igarashi and Iida (2002) evaluated live attenuated and formalin-inactivated cells of an *E. tarda* mutant, SPM31, constructed with transposon Tn5 with reduced siderophore producing capability. Tilapia were vaccinated intraperitoneally, and protection was recorded after challenge for the live (0% mortality) but not the formalin-inactivated (mortality = 80–100%) preparation (Igarashi and Iida, 2002). Jiao *et al.* (2009) described the use of two antigens, Eta6 (= ecotin precursor) and FliC, (= FliCflagellin), the former of which was moderately protective in Japanese flounder (RPS = 53%) when administered intraperitoneally in *Bacillus* sp. B187 as an adjuvant. DNA vaccines, based on Eta6 and FliC, i.e. *pEta6*, and *pFliC* respectively, led to RPS of 50% and 33% respectively, when administered intramuscularly (Jiao *et al.*, 2009). A chimeric DNA vaccine, i.e. *eta6* covalently linked to *FliC*, led to superior protection, i.e. RPS = 72% (Jiao *et al.*, 2009).

Sun *et al.* (2010a) proposed use of a recombinant surface protein, Esa1, of 795 amino acid residues, which led to protection of Japanese flounder when administered orally in alginate microspheres (RPS = 52%) and by i.p. injection (RPS = 79%). Furthermore, Kwon *et al.* (2006) used ghost cells, which were generated by gene *E* mediated lysis, in tilapia, and demonstrated high levels of protection. Eta2 is a protein from *E. tarda* with sequence identity to OMP, and was prepared as a recombinant protein in *Escherichia coli*. Eta2 was administered intraperitoneally in aluminium hydroxide adjuvant as a subunit vaccine to Japanese flounder and led to RPS of 83% after challenge (Sun *et al.*, 2011). However, Eta2 was used also as DNA vaccine (plasmid pCEta2) which was administered i.m., and was protective (RPS = 67%) (Sun *et al.*, 2011).

Flavobacterium columnare

A modified live *Flavobacterium columnare* vaccine and a 1:1 bivalent product with a commercial vaccine for *E. ictaluri* were evaluated in channel catfish eggs by immersion with booster after 34 days, and led to RPS values of 50–76.8% after challenge (Shoemaker *et al.*, 2007).

Flavobacterium psychrophilum

An *aroA* mutant of *F. psychrophilum* has been suggested as a vaccine candidate (Thune *et al.*, 2003). An attenuated vaccine, generated by mutagenesis in which a mutant FP1033 was obtained with an inability to grow in iron-depleted medium, was reported as protective of rainbow trout after challenge (Álvarez *et al.*, 2008).

Francisella asiatica

An attenuated mutant of *Francisella asiatica*, Δ*iglc*, was protective in tilapia after vaccination by immersion for 30 min (RPS = 68.75%) or 180 min (RPS = 87.5%) (Soto *et al.*, 2011).

Mycobacterium marinum

A DNA vaccine involving the Ag85A gene which encodes a major secreted fibronectin-binding protein of *M. marinum* and cloned in a eukaryotic expression vector stimulated a protective humoral immune response in hybrid striped bass when administered i.m. (RPS = 80% and 90% depending on dose) (Pasnik and Smith, 2005, 2006).

Nocardia

Live, low virulent cells of *Nocardia seriolae* and *N. soli*, *N. fluminea* and *N. uniformis* were injected intraperitoneally into yellowtail with particular success with the first (Itano *et al.*, 2006).

Renibacterium salmoninarum

Two 'strains' were isolated which, unlike the normal expectation of nutritionally fastidious renibacteria, could grow on tryptone soya agar (TSA) and brain heart infusion agar, and were non-pathogenic in Atlantic salmon (Daly *et al.*, 2001). When evaluated as live vaccines, the culture which grew on TSA (= Rs TSA1) led to an RPS of 50 and 74% at 74 and 60 days after challenge, respectively (Daly *et al.*, 2001).

Streptococcus iniae

A recombinant subunit *S. iniae* vaccine centring on the putative iron-binding protein, Sip11, was expressed in *E. coli* and led to RPS = 69.7% when linked to an inert carrier protein and administered as a live vaccine by i.p. injection into Japanese flounder (Cheng *et al.*, 2010a). A DNA vaccine involving a putative secretory antigen, Sia10, of *S. iniae* was used in the form of a plasmid, pSia10, which led to an RPS of 73–92% in turbot (Sun *et al.*, 2010b). In a comparison of a formalized whole cell vaccine of *S. iniae* with live attenuated products in hybrid striped bass by bath and i.p. injection, the outcome was that the live vaccine lacking M-like protein gave complete protection (RPS = 100%) (Locke *et al.*, 2010). An attenuated novobiocin-resistant strain has been proposed as a vaccine for use in Nile tilapia with administration by i.p. injection leading to a RPS of up to 100% (Pridgeon and Klesius, 2011b).

Vibrio alginolyticus

A DNA vaccine was constructed containing the flagellin *flaA* gene from a culture of *V. alginolyticus* and injected intramuscularly into red snapper (*Lutjanus sanguineus*) and led to a RPS of 88% after challenge (Liang *et al.*, 2011). Use of recombinant *flaC* was a worthy vaccine candidate with a resulting RPS of 84% (Liang *et al.*, 2010).

Vibrio anguillarum

Using a DNA vaccine comprising a mutated zinc metalloprotease gene (m-*EmpA*) applied i.m., Japanese flounder were well protected after challenge (RPS = 85.7%) (Yang *et al.*, 2009). A field trial with an attenuated live *V. anguillarum* vaccine (VAN1000) involved bathing rainbow trout for 60 min at 9 °C in brackish water, and after a natural challenge, a RPS of 79.5% was recorded (Norquist *et al.*, 1994).

Vibrio harveyi

The *HL*1 gene, which encodes the haemolysin from *V. harveyi*, was inserted in yeast (*Saccharomyces cerevisiae*), and the protein (= haemolysin) was expressed on the cell surface. The live yeast was injected i.p. into flounder and turbot, which were subsequently protected against challenge (Zhu *et al.*, 2006). A live recombinant vaccine, which is centred on the OMP VhhP2, was administered i.p. to Japanese flounder (RPS = 92.3%) and orally (RPS = 61.2%) (Sun *et al.*, 2009). Zhang *et al.* (2011) studied two *V. harveyi* OMP genes, *OmpK* and *GADPH*, and expressed the recombinant proteins in the prokaryotic expression vector pET-30a(+), which was purified and used to vaccinate i.p. large yellow croaker (*Pseudosciaena crocea*) with booster doses after 3 weeks. The outcome was that r-OmpK and r-GADPH enabled RPS values of 37.7% and 40%, respectively, after challenge. OmpK was expressed in yeast, *Pichia pastoris*, and fed for 5 days to sea bass (*Lateolabrax japonicus*) on alginate microspheres with protection against challenge (RPS = 61.5%) (Mao *et al.*, 2011). A denatured inactive cytotoxic recombinant secreted protease, Vhp1, recovered from a pathogenic isolate, was an effective subunit vaccine (RPS = 70%) with improved performance when expressed in *E. coli* as a live vaccine (RPS = 90%) (Cheng *et al.*, 2010c). Furthermore, Wang *et al.* (2011) used the purified OMPOmpU and a DNA vaccine involving the insertion of the *ompU* gene into pEGFP-N1 plasmid injected i.m. into turbot. Use of the purified OmpU led to complete protection after challenge 5 weeks later (RPS = 100%) whereas lesser protection resulted with the DNA vaccine (RPS = 51.4%). Two potentially protective immunogens, DegQ and Vhp1, have been accommodated in DNA vaccines and used to vaccine Japanese flounder with promising results for pDV (RPS = 84.6%) (Hu and Sun, 2011). Hu *et al.* (2011) continued the work by developing a recombinant product that expressed *V. harveyi* DegQ as a soluble antigen that elicited significant protection against both *E. tarda* and *V. harveyi* in experiments with turbot when administered by i.p. (RPS = 90.9%) orally (RPS = 60.5%) or immersion (RPS = 47.1%) or a combination of oral plus immersion (RPS = 77.8% after one month in a mock field trial; RPS = 81.8% after two months).

Yersini ruckeri

A live auxotrophic *aroA* mutant of *Y. ruckeri* was evaluated by i.p. injection in rainbow trout, with a resulting RPS of 90% (Temprano *et al.*, 2005).

9.4.4   Compound vaccines

There has been much commercial interest in polyvalent vaccines which have resulted in numerous products. The benefit of this approach to controlling furunculosis may be illustrated by the observation that vibrio antigens, particularly *Aliivibrio salmonicida*, appear to enhance the humoral immune response to *A. salmonicida* (Hoel *et al.*, 1997). Moreover, vaccination with *Aliivibrio salmonicida* antigens led to protection against *A. salmonicida* following challenge by cohabitation (Hoel *et al.*, 1998). This approach could well overcome the perceived problem that *A. salmonicida* is a weak antigen (Tatner, 1989). Also, this ability to achieve cross-protection may explain the often superior protection afforded by polyvalent vaccines (Hoel *et al.*, 1998). However, some products appear to be unusually specific. Thus, a commercial polyvalent salmon vaccine containing a component to protect against *A. salmonicida* subsp. *salmonicida* failed to protect turbot from experimental challenge with *A. salmonicida* subsp. *achromogenes* (Björnsdóttir *et al.*, 2005). In contrast, Santos *et al.* (2005) experienced better success with turbot, although the specific pathogen in this case was not equated with subsp. *achromogenes*.

A bivalent *P. damselae* subsp. *piscicida* (RPS = ~88%) and *V. harveyi* (RPS = ~82%) vaccine based on formalized cells and ECP was administered by i.p. injection or immersion with booster dose to sole, and led to commendable protection for 4 months after which there was a decline in effectiveness (Arijo *et al.*, 2005). Toxoid enriched inactivated whole cells of *P. damselae* subsp. *piscicida* applied by immersion led to a RPS of 37–41% and a low antibody response in sea bream (Magariños *et al.*, 1994). An improved RPS of >60% after 35 days resulted from use of an LPS mixed chloroform-killed whole cell vaccine (Kawakami *et al.*, 1997).

A *V. vulnificus* serovar E vaccine, named Vulnivaccine, which contains capsular antigens and toxoids, was administered to eels by immersion for 1 h in three doses at 12 day intervals, and led to protection after challenge (RPS = 60–90%). In field trials with 9.5 million glass eels in Spain and parallel experiments in Denmark, Vulnivaccine was administered by prolonged immersion and boosting after 14 and 24–28 days, and led to RPS of 62–86% (Fouz *et al.*, 2001). With the appearance of *V. vulnificus* serotype A, a bivalent vaccine was developed, and determined to be efficacious in terms of protection and immunity following application orally, by anal and oral intubation and by i.p. injection (RPS = 80–100%) (Esteve-Gassent *et al.*, 2004).

9.4.5   Ribosomal and 'naked cell' vaccines

A ribosomal vaccine for *P. damselae* subsp. *piscicida* has been evaluated in yellowtail following administration by i.p. injection, with the initial evidence pointing to success with ribosomal antigen P (Ninomiya *et al.*, 1989). In a further development, this group experimented with a potassium thiocyanate extract and acetic acid treated 'naked cells' obtained from a virulent culture (Muraoka *et al.*, 1991). Yellowtail were vaccinated twice i.p., at one week intervals with the extract – with or without the naked cells – and were challenged two weeks after the second injection. Results indicated partial success for the extract when used alone. However, the extract used in conjunction with naked cells was better (RPS = 36.5%).

9.4.6 Nature of the protective antigens

If the structure of protective antigens was known, it should be possible to develop effective vaccines for every pathogen, but this is certainly not the case, and information is patchy. In the case of many Gram-negative bacterial pathogens, there is evidence that the host responds to LPS, ECPs, OMPs and other surface proteins. Specific examples of protective antigens include:

- *Aeromonas hydrophila* (LPS – Baba *et al.*, 1988; 43 kDa OMP – Fang *et al.*, 2000);
- *A. salmonicida* (A-protein, LPS O-antigen, ECP (including proteases, Ellis *et al.*, 1988; Hastings and Ellis, 1988; Lund *et al.*, 2003a), a 28 kDa outer membrane pore forming protein (= porin) – Lutwyche *et al.*, 1995);
- *Aliivibrio salmonicida* (20 kDa peptidoglycan-associated lipoprotein, Pal; Karlsen *et al.*, 2011);
- *F. psychrophilum* (LPS, ~20 kDa surface protein, a low molecular weight fraction of 25–33 kDa; Crump *et al.*, 2001, 2005; Högfors *et al.*, 2008);
- *M. viscosa* (lipooligosaccharides, ~17–19 kDa outer membrane antigen – Heidarsdóttir *et al.*, 2008; ~20 kDa OMP, termed MvOMP1 – Björnsson *et al.*, 2011);
- *P. damselae* subsp. *piscicida* (7 and 45 kDa proteins – Hirono *et al.*, 1997; ECPs, OMP, outer (extremely immunogenic) and cytoplasmic membranes, LPS and O-antigen – Arijo *et al.*, 2004);
- *V. anguillarum* (LPS – Evelyn and Ketcheson, 1980; Salati *et al.*, 1989; Kawai and Kusuda, 1995; ~40 and 49–51 kDa OMP – Chart and Trust, 1984);
- *V. vulnificus* (LPS, 70–80 kDa OMP, protease – Esteve-Gassent and Amaro, 2004; OMP – Boesen *et al.*, 1997).

9.5   Mode of action of vaccines

The question that needs to be resolved is whether or not humoral antibodies are responsible for protection in the host. There is much evidence that fish mount an antibody response to vaccination. Examples include *Aliivibrio salmonicida* (antibodies to LPS; Steine *et al.*, 2001), *F. columnare* (Fujihara and Nakatani, 1971; Grabowski *et al.*, 2004), *P. damselae* subsp. *piscicida* (Kusuda and Fukuda, 1980; Arijo *et al.*, 2004) and *V. anguillarum* (Groberg, 1982; Vervarcke *et al.*, 2005). Many studies link the presence of humoral antibodies to bacterial vaccines directly with protection, with examples including *A. hydrophila* (Azad *et al.*, 1999), atypical *A. salmoni-cida* LPS-specific antibodies (Lund *et al.*, 2008), *F. psychrophilum* (Crump *et al.*, 2001, 2005; Madetoja *et al.*, 2006), *M. viscosa* (Heidarsdöttir *et al.*, 2008), *Streptococcus difficilis* (Eldar *et al.*, 1995) and *V. vulnificus* (Esteve-Gassent *et al.*, 2003). Other work concludes that there is no correlation between protection and antibody production, e.g. *A. salmonicida* (Lund *et al.*, 2003a). This may reflect the apparent absence of antibody formation, such as has been reported for products to *F. columnare* (Mano *et al.*, 1996). Other work has reported only low antibody titres, e.g. in the case of the ribosomal vaccine and 'naked' cells of *P. damselae* subsp. *piscicida*, suggesting to the researchers that humoral antibodies did not play an important role in protection (Muraoka *et al.*, 1991).

Other possible modes of action may centre on stimulation of T cells (this occurred with the live *aroA* mutant of *A. salmonicida*; Marsden *et al.*, 1996), which introduces the role of cellular and innate rather than humoral immunity as the mode of action. For this, examples include *A. hydrophila* LPS (Baba *et al.*, 1988) and *E. tarda* ECPs (Lee *et al.*, 2010). Of course, there could be involvement of humoral, cell-mediated and innate immune parameters as stated for the i.p. administration of a live auxotrophic *aroA* mutant of *A. hydrophila* with effectiveness against furunculosis in rainbow trout (Vivas *et al.*, 2004). Other possibilities include the evidence that one commercial formalized whole cell *V. anguillarum* vaccine induces Mx gene (these are inducible by Type I interferons and have a role in antiviral activity) expression in Atlantic salmon after administration intraperitoneally (Acosta *et al.*, 2004). In another example, vaccination with *P. damselae* subsp. *piscicida* cells were found to enhance the nitric oxide response, i.e. the production of reactive nitrogen intermediates with their antimicrobial activities, to infection with the pathogen, and is correlated with the level of protection (Acosta *et al.*, 2005). There was inhibition of *F. columnare* adhesion to the skin of immersion vaccinated eel (Mano *et al.*, 1996). Finally, mention will be made of a possible mechanism of protection of *V. anguillarum* vaccines that may well involve the inhibition of bacterial attachment by unknown factors in the skin mucus (Kawai and Kusuda, 1995).

9.6   Conclusions

Clearly, some bacterial vaccines work exceedingly well at controlling disease. However, basic questions remain about the dose, duration of protection and whether or not boosters are needed. There have been questions about the precise mode of action insofar as there is not always a correlation with antibody titre. Instead, it is appropriate to look towards cell-mediated and innate immunity. The current dominance of challenge methods to determine the potency of vaccines is emotive. Consideration needs to be given to the use of non-lethal methods although reliable alternatives are awaited. The calculation of RPS has certainly helped with the comparison between vaccines, but a result of <100% indicates that some individuals have not responded and become protected against challenge. This raises an interesting question – why?

9.7   References

ACOSTA, F., LOCKHART, K., GAHLAWAT, S.K., REAL, F. and ELLIS, A.E. (2004) Mx expression in Atlantic salmon (*Salmo salar* L.) parr in response to *Listonella anguillarum* bacterin, lipopolysaccharide and chromosomal DNA. *Fish & Shellfish Immunology* **17**, 255–263.

ACOSTA, F., REAL, F., ELLIS, A.E., TABRAUE, C., PADILLA, D. and RUIZ DE GALARRETA, C.M. (2005) Influence of vaccination on the nitric oxide response of gilthead seabream following infection with *Photobacterium damselae* subsp. *piscicida*. *Fish & Shellfish Immunology* **18**, 31–38.

AFONSO, A., GOMES, S., DA SILVA, J., MARQUES, F. and HENRIQUE, M. (2005) Side effects in sea bass (*Dicentrarchus labrax* L.) due to intraperitoneal vaccination against vibriosis and pasteurellosis. *Fish & Shellfish Immunology* **19**, 1–16.

ÁLVAREZ, B., ÁLVAREZ, J., MENÉNDEZ, A. and GUIJARRO, J.A. (2008) A mutant in one of two *exbD* loci of a TonB system in *Flavobacterium psychrophilum* shows attenuated virulence and confers protection against cold water disease. *Microbiology* **154**, 1144–1151.

AMEND, D.F. (1981) Potency testing of fish vaccines. *Developments in Biological Standardization* **49**, 447–454.

AMEND, D.F. and JOHNSON, K.A. (1981) Current status and future needs of *Vibrio anguillarum* bacterins. *Developments in Biological Standardization* **49**, 403–417.

AMEND, D.F., JOHNSON, K.A., CROY, T.R. and McCARTHY, D.H. (1983) Some factors affecting the potency of *Yersinia ruckeri* bacterins. *Journal of Fish Diseases* **6**, 337–344.

ANDERSON, D.P. and ROSS, A.J. (1972) Comparative study of Hagerman redmouth disease oral bacterins. *Progressive Fish Culturist* **34**, 226–228.

ANGELIDIS, P., KARAGIANNIS, D. and CRUMP, E.M. (2006) Efficacy of a *Listonella anguillarum* (syn. *Vibrio anguillarum*) vaccine for juvenile sea bass *Dicentrarchus labrax*. *Diseases of Aquatic Organisms* **71**, 19–24.

ANTIPA, R. (1976) Field testing of injected *Vibrio anguillarum* bacterins in pen-reared Pacific salmon. *Journal of the Fisheries Research Board of Canada* **33**, 1291–1296.

ANTIPA, R., GOULD, R. and AMEND, D.P. (1980) *Vibrio anguillarum* vaccination of sockeye salmon (*Oncorhynchus nerka*) by direct immersion and hyperosmotic immersion. *Journal of Fish Diseases* **3**, 161–165.

ARIJO, S., BALEBONA, C., MARTINEZ-MANZANARES, E. and MORIÑIGO, M.A. (2004) Immune response of gilt-head seabream (*Sparus aurata*) to antigens from *Photobacterium damselae* subsp. *piscicida*. *Fish & Shellfish Immunology* **16**, 65–70.

ARIJO, S., RICO, R., CHABRILLON, M., DIAZ-ROSALES, P., MARTÍNEZ-MANZANARES, E., BALEBONA, M.C., TORANZO, A.E. and MORIÑIGO, M.A. (2005) Effectiveness of a divalent vaccine for sole, *Solea senegalensis* (Kaup), against *Vibrio harveyi* and *Photobacterium damselae* subsp. *piscicida*. *Journal of Fish Diseases* **28**, 33–38.

AUSTIN, D.A., ROBERTSON, P.A.W. and AUSTIN, B. (2005) Recovery of a new biogroup of *Yersinia ruckeri* from diseased rainbow trout (*Oncorhynchus mykiss*, Walbaum). *Systematic and Applied Microbiology* **26**, 127–131.

AZAD, I.S., SHANKAR, K.M., MOHAN, C.V. and KALITA, B. (1999) Biofilm vaccine of *Aeromonas hydrophila* – standardization of dose and duration for oral vaccination of carps. *Fish & Shellfish Immunology* **9**, 519–528.

BABA, T., IMAMURA, J., IZAWA, K. and IKEDA, K. (1988) Cell-mediated protection in carp, *Cyprinus carpio* L., against *Aeromonas hydrophila. Journal of Fish Diseases* **11**, 171–178.

BAKOPOULOS, V., PEARSON, M., VOLPATTI, D., GOUSMANI, L., ADAMS, A., GALEOTTI, M. and DIMITRIADIS, G.J. (2003) Investigation of media formulations promoting differential antigen expression by *Photobacterium damsela* ssp. *piscicida* and recognition by sea bass, *Dicentrarchus labrax* (L.), immune sera. *Journal of Fish Diseases* **26**, 1–13.

BAUDIN-LAURENÇIN, F. and TANGTRONGPIROS, J. (1980) Some results of vaccination against vibriosis in Brittany. In: Ahne, W. (ed.), *Fish Diseases, Third COPRAQ-Session*. Berlin, Springer-Verlag, p.60–68.

BIRKBECK, T.H., RENNIE, S., HUNTER, D., LAIDLER, L.A. and WADSWORTH, S. (2004) Infectivity of a Scottish isolate of *Piscirickettsia salmonis* for Atlantic salmon *Salmo salar* and immune response of salmon to this agent. *Diseases of Aquatic Organisms* **60**, 97–103.

BJÖRNSDÓTTIR, B., GUDMUNDSDÓTTIR, S., BAMBIR S.H. and GUDMUNDSDÓTTIR, B.K. (2005) Experimental infection of turbot, *Scophthalmus maximus* (L.), by *Aeromonas salmonicida* subsp. *achromogenes* and evaluation of cross protection induced by a furunculosis vaccine. *Journal of Fish Diseases* **28**, 181–188.

BJÖRNSSON, H., MARTEINSSON, V.P., FRIÐJÓNSSON, Ó.H., LINKE, D. and BENEDIKTSDÓTTIR, E. (2011) Isolation and characterization of an antigen from the fish pathogen *Moritella viscosa. Journal of Applied Microbiology* **111**, 17–25.

BOESEN, H.T., PEDERSEN, K., KOCH, C. and LARSEN, J.L. (1997) Immune response of rainbow trout (*Oncorhynchus mykiss*) to antigenic preparations from *Vibrio anguillarum* serogroup O1. *Fish & Shellfish Immunology* **7**, 543–553.

BOWDEN, T.J., MENOYO-LUQUE, D., BRICKNELL, I.R. and WEGELAND, H. (2002) Efficacy of different administration routes for vaccination against *Vibrio anguillarum* in Atlantic halibut (*Hippoglossus hippoglossus* L.). *Fish & Shellfish Immunology* **12**, 283–285.

BRICKNELL, I.R., BOWDEN, T.J., VERNER-JEFFREYS, D.W., BRUNO, D.W., SHIELDS, R.J. and ELLIS, A.E. (2000) Susceptibility of juvenile and sub-adult Atlantic halibut (*Hippoglossus hippoglossus* L.) to infection by *Vibrio anguillarum* and efficacy of protection induced by vaccination. *Fish & Shellfish Immunology* **10**, 319–327.

BUSCH, R.A., BURMEISTER, N.E. and SCOTT, A.L. (1978) Field and laboratory evaluation of a commercial enteric redmouth disease vaccine for rainbow trout. *Proceedings of the Joint 3rd Biennial Fish Health Section and 9th Annual Midwest Fish Disease Workshop*, p.67.

CHART, H. and TRUST, T.J. (1984) Characterization of the surface antigens of the marine fish pathogens, *Vibrio anguillarum* and *Vibrio ordalii. Canadian Journal of Microbiology* **30**, 703–710.

CHENG, S., HUA, Y.-H., JIAO, X.-D. and SUN, L. (2010a) Identification and immunoprotective analysis of a *Streptococcus iniae* subunit vaccine candidate. *Vaccine* **28**, 2636–2641.

CHENG, S., HU, Y.-H., ZHANG, M. and SUN, L. (2010b) Analysis of the vaccine potential of a natural avirulent *Edwardsiella tarda* isolate. *Vaccine* **28**, 2716–2721.

CHENG, S., ZHANG, W.-W., ZHANG, M. and SUN, L. (2010c) Evaluation of the vaccine potential of a cytotoxic protease and a protective immunogen from a pathogenic *Vibrio harveyi* strain. *Vaccine* **28**, 1041–1047.

COLQUHOUN, D.J., ALVHEIM, K., DOMMARSNES, K., SYVERTSEN, C. and SØRUM, H. (2002) Relevance of incubation temperature for *Vibrio salmonicida* vaccine production. *Journal of Applied Microbiology* **92**, 1087–1096.

COSSARINI-DUNIER, M. (1986) Protection against enteric redmouth disease in rainbow trout, *Salmo gairdneri* Richardson, after vaccination with *Yersinia ruckeri* bacterin. *Journal of Fish Diseases* **9**, 27–33.

COSTA, A.A., LEEF, M.J., BRIDLE, A.R., CARSON, J. and NOWAK, B.F. (2011) Effect of vaccination against yersiniosis on the relative percent survival, bactericidal and lysozyme response of Atlantic salmon, *Salmo salar*. *Aquaculture* **315**, 201–206.

CROY, T.R. and AMEND, D.F. (1977) Immunization of sockeye salmon (*Oncorhynchus nerka*) against vibriosis using the hyperosmotic infiltration technique. *Aquaculture* **12**, 317–325.

CRUMP, E.M., PERRY, M.B., CLOUTHIER, S.C. and KAY, W.W. (2001) Antigenic characterization of the fish pathogen *Flavobacterium psychrophilum*. *Applied and Environmental Microbiology* **67**, 750–759.

CRUMP, E.M., BURIAN, J., ALLEN, P.D. and KAY, W.W. (2005) Identification and expression of a host-recognized antigen, FspA, from *Flavobacterium psychrophilum*. *Microbiology* **151**, 3127–3135.

DALY, J.G., GRIFFITHS, S.G., KEW, A.K., MOORE, A.R. and OLIVIER, G. (2001) Characterization of attenuated *Renibacterium salmoninarum* strains and their use as live vaccines. *Diseases of Aquatic Organisms* **44**, 121–126.

DEC, C., ANGELIDUS, P. and BAUDIN-LAURENÇIN, F. (1990) Effects of oral vaccination against vibriosis in turbot, *Scophthalmus maximus* (L.), and sea bass, *Dicentrarchus labrax* (L.). *Journal of Fish Diseases* **13**, 369–376.

DUFF, D.C.B. (1942) The oral immunization of trout against *Bacterium salmonicida*. *Journal of Immunology* **44**, 87–94.

DUMETZ, F., DUCHAUD, E., LAPATRA, S.E., LE MARREC, C., CLAVEROL, S., URDACI, M.-C. and LE HÉNAFF, M. (2006) A protective immune response is generated in rainbow trout by an OmpH-like surface antigen (P18) of *Flavobacterium psychrophilum*. *Applied and Environmental Microbiology* **72**, 4845–4852.

DURBIN, M., MCINTOSH, D., SMITH, P.D., WARDLE, R. and AUSTIN, B. (1999) Immunization against furunculosis in rainbow trout with iron-regulated outer membrane protein vaccines: relative efficacy of immersion, oral and injection delivery. *Journal of Aquatic Animal Health* **11**, 68–75.

ELDAR, A., SHAPIRO, O., BEJERANO, Y. and BERCOVIER, H. (1995) Vaccination with whole-cell vaccine and bacterial protein extract protects tilapia against *Streptococcus difficile* meningoencephalitis. *Vaccine* **13**, 867–870.

ELLIS, A.E., BURROWS, A.S., HASTINGS, T.S. and STAPLETON, K.J. (1988) Identification of *Aeromonas salmonicida* extracellular proteases as a protective antigen against furunculosis by passive immunization. *Aquaculture* **70**, 207–218.

ESTEVE-GASSENT, M.D. and AMARO, C. (2004) Immunogenic antigens of the eel pathogen *Vibrio vulnificus* serovar E. *Fish & Shellfish Immunology* **17**, 277–291.

ESTEVE-GASSENT, M.D., NIELSEN, M.E. and AMARO, C. (2003) The kinetics of antibody production in mucus and serum of European eel (*Anguilla anguilla* L.) after vaccination against *Vibrio vulnificus:* development of a new method for antibody quantification in skin mucus. *Fish & Shellfish Immunology* **15**, 51–61.

ESTEVE-GASSENT, M.D., FOUZ, B. and AMARO, C. (2004) Efficacy of a bivalent vaccine against eel diseases caused by *Vibrio vulnificus* after its administration by four different routes. *Fish & Shellfish Immunology* **16**, 93–105.

EVELYN, T.P.T. and KETCHESON, J.E. (1980) Laboratory and field observations on anti-vibriosis vaccines. In: Ahne, W. (ed.), *Fish Diseases, Third-COPRAQ Session*. Berlin, Springer-Verlag, p.45–54.

FANG, H.M., LING, K.C., GE, R. and SIN, Y.M. (2000) Enhancement of protective immunity in blue gourami, *Trichogaster trichopterus* (Pallas), against *Aeromonas hydrophila* and *Vibrio anguillarum* by A. hydrophila major adhesin. *Journal of Fish Diseases* **23**, 137–145.

FOUZ, B., ESTEVE-GASSENT, M.D., BARRERA, R., LARSEN, J.L., NIELSEN, M.E. and AMARO, C. (2001) Field testing of a vaccine against eel diseases caused by *Vibrio vulnificus*. *Diseases of Aquatic Organisms* **45**, 183–189.

FRYER, J.L., ROHOVEC, J.S. and GARRISON, R.L. (1978) Immunization of salmonids for control of vibriosis. *Marine Fisheries Review* **40**, 20–23.

FUJIHARA, M.P. and NAKATANI, R.E. (1971) Antibody production and immune responses of rainbow trout and coho salmon to *Chondrococcus columnaris*. *Journal of the Fisheries Research Board of Canada* **28**, 1253–1258.

FUKUDA, Y. and KUSUDA, R. (1982) Detection and characterization of precipitating antibody in the serum of immature yellowtail immunized with *Pasteurella piscicida* cells. *Fish Pathology* **17**, 125–127.

GIORGETTI, G., TOMASIN, A.B. and CESCHIA, G. (1981) First Italian anti-vibriosis vaccination experiments of freshwater farmed rainbow trout. *Developments in Biological Standardization* **49**, 455–459.

GOULD, R.W., O'LEARY, P.J., GARRISON, R.L., ROHOVEC, J.S. and FRYER, J.L. (1978) Spray vaccination: a method for the immunization of fish. *Fish Pathology* **13**, 63–68.

GRABOWSKI, L.D., LAPATRA, S.E. and CAIN, K.D. (2004) Systemic and mucosal antibody response in tilapia, *Oreochromis niloticus* (L.), following immunization with *Flavobacterium columnare*. *Journal of Fish Diseases* **27**, 573–581.

GREGER, E. and GOODRICH, T. (1999) Vaccine development for winter ulcer disease, *Vibrio viscosus*, in Atlantic salmon, *Salmo salar* L. *Journal of Fish Diseases* **22**, 193–199.

GROBERG, W.J. (1982) Infection and the immune response induced by *Vibrio anguillarum* in juvenile coho salmon (*Oncorhynchus kisutch*). Ph.D. thesis, Oregon State University, Corvallis.

GUDMUNSDÓTTIR, S., LANGE, S., MAGNADÓTTIR, B. and GUDMUNDSDÓTTIR, B.K. (2003) Protection against atypical furunculosis in Atlantic halibut, *Hippoglossus hippoglossus* (L.); a comparison of a commercial furunculosis vaccine and an autogenous vaccine. *Journal of Fish Diseases* **26**, 331–338.

GUTIERREZ, M.A. and MIYAZAKI, T. (1994) Responses of Japanese eels to oral challenge with *Edwardsiella tarda* after vaccination with formalin-killed cells or lipopolysaccharide of the bacterium. *Journal of Aquatic Animal Health* **6**, 110–117.

HANIF, A., BAKOPOULOS, V., LEONARDOS, I. and DIMITRIADIS, G.J. (2005) The effect of sea bream (*Sparus aurata*) broodstock and larval vaccination on the susceptibility by *Photobacterium damsela* subsp. *piscicida* and on the humoral immune parameters. *Fish & Shellfish Immunology* **19**, 345–361.

HASTINGS, T.S. and ELLIS, A.E. (1988) The humoral immune response of rainbow trout, *Salmo gairdneri* Richardson, and rabbits to *Aeromonas salmonicida* extracellular products. *Journal of Fish Diseases* **11**, 147–160.

HEIDARSDÓTTIR, K.J., GRAVNINGEN, K. and BENEDIKTSDÓTTIR, E. (2008) Antigen profiles of the fish pathogen *Moritella viscosa* and protection in fish. *Journal of Applied Microbiology* **104**, 944–951.

HIRONO, I., KATO, M. and AOKI, T. (1997) Identification of major antigenic proteins of *Pasteurella piscicida*. *Microbial Pathogenesis* **23**, 371–380.

HOEL, K., SALONIUS, K. and LILLEHAUG, A. (1997) *Vibrio* antigens of polyvalent vaccines enhance the humoral immune response to *Aeromonas salmonicida* antigens in Atlantic salmon (*Salmo salar* L.). *Fish & Shellfish Immunology* **7**, 71–80.

HOEL, K., REITAN, L.J. and LILLEHAUG, A. (1998) Immunological cross reactions between *Aeromonas salmonicida* and *Vibrio salmonicida* in Atlantic salmon (*Salmo salar* L.) and rabbits. *Fish & Shellfish Immunology* **8**, 171–182.

HÖGFORS, E., PULLINEN, K.-R., MADETOJA, J. and WIKLUND, T. (2008) Immunization of rainbow trout, *Oncorhynchus mykiss* (Walbaum), with a low molecular mass fraction isolated from *Flavobacterium psychrophilum*. *Journal of Fish Diseases* **31**, 899–911.

HOLM, K.O. and JØRGENSEN, T. (1987) A successful vaccination of Atlantic salmon, *Salmo salar* L., against 'Hitra disease' or coldwater vibriosis. *Journal of Fish Diseases* **10**, 85–90.

HORNE, M.T., TATNER, M., MCDERMENT, S. and AGIUS, C. (1982) Vaccination of rainbow trout, *Salmo gairdneri* Richardson, at low temperatures and the long-term persistence of protection. *Journal of Fish Diseases* **5**, 343–345.

HOSSAIN, M.M.M. and KAWAI, K. (2009) Stability of effective *Edwardsiella tarda* vaccine developed for Japanese eel (*Anguilla japonica*). *Journal of Fisheries and Aquatic Science* **4**, 296–305.

HU, Y.-H. and SUN, L. (2011) A bivalent *Vibrio harveyi* DNA vaccine induces strong protection in Japanese flounder (*Paralichthys olivaceus*). *Vaccine* **29**, 4328–4333.

HU, Y.-H., CHENG, S., ZHANG, M. and SUN, L. (2011) Construction and evaluation of a live vaccine against *Edwardsiella tarda* and *Vibrio harveyi*: laboratory vs. mock field trial. *Vaccine* **29**, 4081–4085.

IGARISHI, A. and IIDA, T. (2002) A vaccination trial using live cells of *Edwardsiella tarda* in tilapia. *Fish Pathology* **37**, 145–148.

INGLIS, V., ROBERTSON, D., MILLER, K., THOMPSON, K.D. and RICHARDS, R.H. (1996) Antibiotic protection against recrudescence of latent *Aeromonas salmonicida* during furunculosis vaccination. *Journal of Fish Diseases* **19**, 341–348.

IRIE, T., WATARAI, S., IWASAKI, T. and KODAMA, H. (2005) Protection against experimental *Aeromonas salmonicida* infection in carp by oral immunisation with bacterial antigen entrapped liposomes. *Fish & Shellfish Immunology* **18**, 235–242.

ITANO, T., KAWAKAMI, H., KONO, T. and SAKAI, M. (2006) Live vaccine trials against nocardiosis in yellowtail *Seriola quinqueradiata*. *Aquaculture* **261**, 1175–1180.

JIAO, X.-D., ZHANG, M., HU, Y.-H. and SUN, L. (2009) Construction and evaluation of DNA vaccines encoding *Edwardsiella tarda* antigens. *Vaccine* **27**, 5195–5202.

JIAO, X.-D., CHENG, S., HU, Y.-H. and SUN, L. (2010) Comparative study of the effects of aluminium adjuvants and Freund's incomplete adjuvant on the immune response to an *Edwardsiella tarda* major antigen. *Vaccine* **28**, 1832–1837.

JOHNSON, K.A. and AMEND, D.F. (1983a) Comparison of efficacy of several delivery methods using *Yersinia ruckeri* bacterin on rainbow trout, *Salmo gairdneri* Richardson. *Journal of Fish Diseases* **6**, 331–336.

JOHNSON, K.A. and AMEND, D.F. (1983b) Efficacy of *Vibrio anguillarum* and *Yersinia ruckeri* bacterins applied by oral and anal intubation of salmonids. *Journal of Fish Diseases* **6**, 473–476.

JOHNSON, K.A., FLYNN, J.K. and AMEND, D.F. (1982a) Onset of immunity in salmonid fry vaccinated by direct immersion in *Vibrio anguillarum* and *Yersinia ruckeri* bacterins. *Journal of Fish Diseases* **5**, 197–205.

JOHNSON, K.A., FLYNN, J.K. and AMEND, D.F. (1982b) Duration of immunity in salmonid fry vaccinated by direct immersion with *Yersinia ruckeri* and *Vibrio anguillarum* bacterins. *Journal of Fish Diseases* **5**, 207–213.

JOOSTEN, P.H.M., TIEMERSMA, E., THREELS, A., CAUMARTIN-DHIEUX, C. and ROMBOUT, J.H.W.M. (1997) Oral vaccination of fish against *Vibrio anguillarum* using alginate microparticles. *Fish & Shellfish Immunology* **7**, 471–485.

KAJITA, Y., SAKAI, M., ATSUTA, S. and KOBAYASHI, M. (1990) The immunomodulatory effects of levamisole on rainbow trout, *Oncorhynchus mykiss*. *Fish Pathology* **25**, 93–98.

KARLSEN, C., ESPELID, S., WILLASSEN, N.-P. and PAULSEN, S.M. (2011) Identification and cloning of immunoprotective *Aliivibrio salmonicida* Pallike protein present in profiled outer membrane and secreted subproteome. *Diseases of Aquatic Organisms* **93**, 215–223.

KAWAHARA, E., OSHIMA, S. and NOMURA, S. (1990) Toxicity and immunogenicity of *Aeromonas salmonicida* extracellular products to salmonids. *Journal of Fish Diseases* **13**, 495–503.

KAWAI, K. and KUSUDA, R. (1995) A review: *Listonella anguillarum* infection in ayu, *Plecoglossus altivelis*, and its prevention by vaccination. *Israeli Journal of Aquaculture Bamidgeh* **47**, 173–177.

KAWAI, K., YAMAMOTO, S. and KUSUDA, R. (1989) Plankton-mediated oral delivery of *Vibrio anguillarum* vaccine to juvenile ayu. *Nippon Suisan Gakkaishi* **55**, 35–40.

KAWAKAMI, H., SHINOHARA, N., FUKUDA, Y., YAMASHITA, H., KIHARA, H. and SAKAI, M. (1997) The efficacy of lipopolysaccharide mixed chloroform-killed cell (LPS-CKC) bacterin of *Pasteurella piscicida* on yellowtail, *Seriola quinqueradiata*. *Aquaculture* **154**, 95–105.

KINTSUJI, H., NINOMIYA, K. and YAMAMOTO, M. (2006) Vaccination with acetone-dry bacterin against bacterial hemorrhagic ascites of ayu *Plecoglossus altivelis*. *Fish Pathology* **41**, 121–122.

KONDO, M., KAWAI, K., OKABE, M., NAKANO, N. and OSHIMA, S.-I. (2003) Efficacy of oral vaccine against bacterial coldwater disease in ayu *Plecoglossus altivelis*. *Diseases of Aquatic Organisms* **55**, 261–264.

KUSUDA, R. and FUKUDA, Y. (1980) Agglutinating antibody titers and serum protein changes of yellowtail after immunization with *Pasteurella piscicida* cells. *Bulletin of the Japanese Society of Scientific Fisheries* **46**, 801–807.

KWON, S.R., NAM, Y.K., KIM, S.K. and KIM, K.H. (2006) Protection of tilapia (*Oreochromis mossambicus*) from edwardsiellosis by vaccination with *Edwardsiella tarda* ghosts. *Fish & Shellfish Immunology* **20**, 621–626.

LAFRENTZ, B.R., LAPATRA, S.E., JONES, G.R., CONGLETON, J.L., SUN, B. and CAIN, K.D. (2002) Characterization of the serum and mucosal antibody responses and relative per cent survival in rainbow trout, *Oncorhynchus mykiss* (Walbaum), following immunization and challenge with *Flavobacterium psychrophilum*. *Journal of Fish Diseases* **25**, 703–713.

LAFRENTZ, B.R., LAPATRA, S.E., JONES, G.R. and CAIN, K.D. (2004) Protective immunity in rainbow trout *Oncorhynchus mykiss* following immunization with distinct molecular mass fractions isolated from *Flavobacterium psychrophilum*. *Diseases of Aquatic Organisms* **59**, 17–26.

LAMERS, C.H.J. and MUISWINKEL, W.B. (1984) Primary and secondary immune responses in carp (*Cyprinus carpio*) after administration of *Yersinia ruckeri* O-antigen. In: Acuigrup (ed.) *Fish Diseases*, Madrid, Editora ATP, p.119–127.

LEE, D.C., KIM, D.H. and PARK, S. (2010) Effects of extracellular products on the innate immunity in olive flounder *Paralichthys olivaceus*. *Fish Pathology* **45**, 17–23.

LIANG, H., XIA, L., WU, Z., JIAN, J. and LU, Y. (2010) Expression, characterization and immunogenicity of flagellin FlaC from *Vibrio alginolyticus* strain HY9901. *Fish & Shellfish Immunology* **29**, 343–348.

LIANG, H.Y., WU, Z.-H., JIAN, J.-C. and HUANG, Y.C. (2011) Protection of red snapper (*Lutjanus sanguineus*) against *Vibrio alginolyticus* with a DNA vaccine containing flagellin *flaA* gene. *Lettters in Applied Microbiology* **52**, 165–161.

LOCKE, J.B., VICKNAIR, M.R., OSTLAND, V.E., NIZET, V. and BUCHANAN, J.T. (2010) Evaluation of *Streptococcus iniae* killed bacterin and live attenuated vaccines in hybrid striped bass through injection and bath immersion. *Diseases of Aquatic Organisms* **89**, 117–123.

LUND, V., ARNESEN, J.A., COUCHERON, D., MODALSLI, K. and SYVERTSEN, C. (2003a) The *Aeromonas salmonicida* A-layer protein is an important protective antigen in oil-adjuvanted vaccines. *Fish & Shellfish Immunology* **15**, 367–372.

LUND, V., ESPELID, S. and MIKKELSEN, H. (2003b) Vaccine efficacy in spotted wolffish *Anarhichas minor*: relationship to molecular variation in A-layer protein of atypical *Aeromonas salmonicida*. *Diseases of Aquatic Organisms* **56**, 31–42.

LUND, V., ARNESEN, J.A., MIKKELSEN, H., GRAVNINGEN, K., BROWN, L. and SCHRØDER, M.B. (2008) Atypical furunculosis vaccines for Atlantic cod (*Gadus morhua*); vaccine efficacy and antibody responses. *Vaccine* **26**, 6791–6799.

LUTWYCHE, P., EXNER, M.M., HANCOCK, R.E.W. and TRUST, T.J. (1995) A conserved *Aeromonas salmonicida* porin provides protective immunity to rainbow trout. *Infection and Immunity* **63**, 3137–3142.

MADETOJA, J., LÖNNSTRÖM, L.-G., BJÖRKBLOM, C., ULUKÖY, G., BYLUND, G., SYVERTSEN, C., GRAVNINGEN, K., NORDERHUS, E.-A. and WIKLUND, T. (2006) Efficacy of injection vaccines against *Flavobacterium psychrophilum* in rainbow trout, *Oncorhynchus mykiss* (Walbaum). *Journal of Fish Diseases* **29**, 9–20.

MAGARIÑOS, B., ROMALDE, J.L., SANTOS, Y., CASAL, J.F., BARJA, J.L. and TORANZO, A.E. (1994) Vaccination trials on gilthead sea bream (*Sparus aurata*) against *Pasteurella piscicida*. *Aquaculture* **120**, 201–208.

MANO, N., INUI, T., ARAI, D., HIROSE, H. and DEGUCHI, Y. (1996) Immune response in the skin of eel against *Cytophaga columnaris*. *Fish Pathology* **31**, 65–70.

MAO, Z., HE, C., QIU, Y. and CHEN, J. (2011) Expression of *Vibrio harveyi* ompK in the yeast *Pichia pastoris*: the first step in developing an oral vaccine against vibriosis? *Aquaculture* **318**, 268–272.

MARSDEN, M.J., VAUGHAN, L.M., FOSTER, T.J. and SECOMBES, C.J. (1996) A live (Δ*aroA*) *Aeromonas salmonicida* vaccine for furunculosis preferentially stimulates T-cell responses relative to B-cell responses in rainbow trout (*Oncorhynchus mykiss*). *Infection and Immunity* **64**, 3863–3869.

MIDTLYNG, P.J. (1996) A field study on intraperitoneal vaccination of Atlantic salmon (*Salmo salar* L.) against furunculosis. *Fish & Shellfish Immunology* **6**, 553–565.

MIDTLYNG, P.J., REITAN, L.J., LILLEHAUG, A. and RAMSTAD, A. (1996) Protection, immune responses and side effects in Atlantic salmon (*Salmo salar* L.) vaccinated against furunculosis by different routes. *Fish & Shellfish Immunology* **6**, 599–600.

MIKKELSEN, H., LUND, V., MARTINSEN, L.-C., GRAVNINGEN, K. and SCHRØDER, M.B. (2007) Variability among *Vibrio anguillarum* O2 isolates from Atlantic cod (*Gadus morhua* L.): characterisation and vaccination studies. *Aquaculture* **266**, 16–25.

MURAOKA, A., OGAWA, K., HASHIMOTO, S. and KUSUDA, R. (1991) Protection of yellowtail against pseudotuberculosis by vaccination with a potassium thiocyanate extract of *Pasteurella piscicida* and co-operating protective effect of acid-treated, naked bacteria. *Nippon Suisan Gakkaishi* **57**, 249–253.

MUTOLOKI, S., BRUDESETH, B., REITE, O.B. and EVENSEN, Ø. (2006) The contribution of *Aeromonas salmonicida* extracellular products to the induction of inflammation in Atlantic salmon (*Salmo salar* L.) following vaccination with oil-based vaccines. *Fish & Shellfish Immunology* **20**, 1–11.

NAGAI, T., IIDA, Y. and YONEJI, T. (2003) Field trials of a vaccine with water-soluble adjuvant for bacterial coldwater disease in ayu *Plecoglossus altivelis*. *Fish Pathology* **38**, 63–65.

NAKAI, T., MUROGA, K. and WAKABAYASHI, H. (1982) An immuno-electrophoretic analysis of *Pseudomonas anguilliseptica*. *Bulletin of the Japanese Society of Scientific Fisheries* **48**, 363–367.

NAVOT, N., SINYAKOV, S. and AVTALION, R.R. (2011) Application of ultrasound in vaccination against goldfish ulcer disease: a pilot study. *Vaccine* **29**, 1382–1389.

NAYAK, D.K., ASHA, A., SHANKAR, K.M. and MOHAN, C.V. (2004) Evaluation of biofilm of *Aeromonas hydrophila* for oral vaccination of *Clarias batrachus* – a carnivore model. *Fish & Shellfish Immunology* **16**, 613–619.

NIKL, L., ALBRIGHT, L.J. and EVELYN, T.P.T. (1991) Influence of seven immunostimulants on the immune response of coho salmon to *Aeromonas salmonicida*. *Diseases of Aquatic Organisms* **12**, 7–12.

NINOMIYA, M., MURAOKA, A. and KUSUDA, R. (1989) Effect of immersion vaccination of cultured yellowtail with a ribosomal vaccine prepared from *Pasteurella piscicida*. *Nippon Suissan Gakkaishi* **55**, 1773–1776.

NINOMIYA, K. and YAMAMOTO, M. (2001) Efficacy of oil-adjuvanted vaccines for bacterial hemorrhagic ascites in ayu *Plecoglossus altivelis*. *Fish Pathology* **36**, 183–185.

NITZAN, S., SHWARTSBURD, B. and HELLER, E.D. (2004) The effect of growth medium salinity of *Photobacterium damselae* subsp. *piscicida* on the immune response of hybrid bass (*Morone saxatilis* × *M. chrysops*). *Fish & Shellfish Immunology* **16**, 107–116.

NORQUIST, A., BERGMAN, A., SKOGMAN, G. and WOLF-WATZ, H. (1994) A field trial with the live attenuated fish vaccine strain *Vibrio anguillarum* VAN1000. *Bulletin of the European Association of Fish Pathologists* **14**, 156–158.

PANG, H.-Y., LI, Y., WU, Z.-H., JIAN, J.-C., LU, Y.-S. and CAI, S.-H. (2010) Immunoproteomic analysis and identification of novel immunogenic proteins from *Vibrio harveyi*. *Journal of Applied Microbiology* **109**, 1800–1809.

PASNIK, D.J. and SMITH, S.A. (2005) Immunogenic and protective effects of a DNA vaccine for *Mycobacterium marinum* in fish. *Veterinary Immunology and Immunopathology* **103**, 195–206.

PASNIK, D.J. and SMITH, S.A. (2006) Immune and histopathologic responses of DNA-vaccinated hybrid striped bass *Morone saxatilis* × *M. chrysops* after acute *Mycobacterium marinum* infection. *Diseases of Aquatic Organisms* **73**, 33–41.

PLUMB, J.A. and VINITNANTHARAT, S. (1993) Vaccination of channel catfish, *Ictalurus punctatus* (Rafinesque), by immersion and oral booster against *Edwardsiella ictaluri*. *Journal of Fish Diseases* **16**, 65–71.

POOBALANE, S., THOMPSON, K.D., ARDÓ, L., VERJAN, N., HAN, H.-J., JENEY, G., HIRONO, I., AOKI, T. and ADAMS, A. (2010) Production and efficacy of an *Aeromonas hydrophila* recombinant S-layer protein vaccine for fish. *Vaccine* **28**, 3540–3547.

PRIDGEON, J.W. and KLESIUS, P.H. (2011a) Development of a novobiocin-resistant *Edwardsiella ictaluri* as a novel vaccine in channel catfish (*Ictalurus punctatus*). *Vaccine* **29**, 5631–5637.

PRIDGEON, J.W. and KLESIUS, P.H. (2011b) Development and efficacy of a novobiocin-resistant *Streptococcus iniae* as a novel vaccine in Nile tilapia (*Oreochromis niloticus*). *Vaccine* **29**, 5986–5993.

RAHMAN, M.H., KURODA, A., DIJKSTRA, J.M., KIRYU, I., NAKANISHI, T. and OTOTAKE, M. (2002) The outer membrane fraction of *Flavobacterium psychrophilum* induces protective immunity in rainbow trout and ayu. *Fish & Shellfish Immunology* **12**, 169–179.

RAIDA, M.K. and BUCHMANN, K. (2008) Bath vaccination of rainbow trout (*Oncorhynchus mykiss* Walbaum) against *Yersinia ruckeri*: effects of temperature on protection and gene expression. *Vaccine* **26**, 1050–1062.

RUANGPAN, L., KITAO, T. and YOSHIDA, T. (1986) Protective efficacy of *Aeromonas hydrophila* vaccines in Nile tilapia. *Veterinary Immunology and Immunopathology* **12**, 345–350.

SAEED, M. (1983) Chemical characterization of the lipopolysaccharides of *Edwardsiella ictaluri* and the immune response of channel catfish to this function and to whole cell antigen with histopathological comparisons. Ph.D. dissertation, Auburn University.

SAEED, M.O. and PLUMB, J.A. (1987) Serological detection of *Edwardsiella ictaluri* Hawke lipopolysaccharide antibody in serum of channel catfish *Ictalurus punctatus* Rafinesque. *Journal of Fish Diseases* **10**, 205–209.

SAKAI, M., KUBOTA, R., ATSUTA, S. and KOBAYASHI, M. (1987) Vaccination of rainbow trout *Salmo gairdneri* against β-haemolytic streptococcal disease. *Nippon Suisan Gakkaishi* **53**, 1373–1376.

SAKAI, M., ATSUTA, S. and KOBAYASHI, M. (1989) Protective immune response in rainbow trout *Oncorhynchus mykiss*, vaccinated with β-haemolytic streptococcal bacterin. *Fish Pathology* **24**, 169–173.

SAKAI, M., ATSUTA, S. and KOBAYASHI, M. (1993) The immune response of rainbow trout (*Oncorhynchus mykiss*) injected with five *Renibacterium salmoninarum* bacterins. *Aquaculture* **113**, 11–18.

SALATI, F., IKEDA, Y. and KUSUDA, R. (1987a) Effect of *Edwardsiella tarda* lipopolysaccharide immunization on phagocytosis in the eel. *Nippon Suisan Gakkaishi* **53**, 201–204.

SALATI, F., HAMGUCHI, M. and KUSUDA, R. (1987b) Immune response of red sea bream to *Edwardsiella tarda* antigens. *Fish Pathology* **22**, 93–98.

SALATI, F., KAWAI, S. and KUSUDA, R. (1989) Characteristics of the lipopolysaccharide from *Pasteurella piscicida*. *Fish Pathology* **24**, 143–147.

SANTOS, Y., GARCÍA-MARQUEZ, S., PEREIRA, P.G., PAZOS, F., RIAZA, A., SILVA, R., EL MORABIT, A. and UBEIRA, F.M. (2005) Efficacy of furunculosis vaccines in turbot, *Scophthalmus maximus* (L.): evaluation of immersion, oral and injection delivery. *Journal of Fish Diseases* **28**, 165–172.

SHOEMAKER, C.A., KLESIUS, P.H. and EVANS, J.J. (2007) Immunization of eyed channel catfish, *Ictalurus punctatus*, eggs with monovalent *Flavobacterium columnare* vaccine and bivalent *F. columnare* and *Edwardsiella ictaluri* vaccine. *Vaccine* **25**, 1126–1131.

SHOEMAKER, C.A., LAFRENTZ, B.R., KLESIUS, P.H. and EVANS, J.J. (2010) Protection against heterologous *Streptococcus iniae* isolates using a modified bacterin vaccine in Nile tilapia, *Oreochromis niloticus* (L.). *Journal of Fish Diseases* **33**, 537–544.

SMITH, P.A., LANNAN, C.N., GARCES, L.H., JARPA, M., LARENAS, J., CASWELL-RENO, P., WHIPPLE, M. and FRYER, J.L. (1995) Piscirickettsiosis: a bacterin field trial in coho salmon (*Oncorhynchus kisutch*). *Bulletin of the European Association of Fish Pathologists* **14**, 137–141.

SOTO, E., WILES, J., ELZER, P., MACALUSO, K. and HAWKE, J.P. (2011) Attenuated *Francisella asiatic aiglc* mutant induces protective immunity to francisellosis in tilapia. *Vaccine* **29**, 593–598.

SÖVÉNYI, J.F., YAMAMOTO, H., FUJIMOTO, S. and KUSUDA, R. (1990) Lymphomyeloid cells, susceptibility to erythrodermatitis of carp and bacterial antigens. *Developmental and Comparative Immunology* **14**, 185–200.

STEINE, N.O., MELINGEN, G.O. and WERGELAND, H.I. (2001) Antibodies against *Vibrio salmonicida* lipopolysaccharide (LPS) and whole bacteria in sera from Atlantic salmon (*Salmo salar* L.) vaccinated during the smolting and early post-smolt period. *Fish & Shellfish Immunology* **11**, 39–52.

SUN, K., ZHANG, W.-W., HOU, J.-H. and SUN, L. (2009) Immunoprotective analysis of VhhP2, a *Vibrio harveyi* vaccine candidate. *Vaccine* **27**, 2733–2740.

SUN, Y., LIU, C.-S. and SUN, L. (2010a) Identification of an *Edwardsiella tarda* surface antigen and analysis of its immunoprotective potential as a purified recombinant subunit vaccine and a surface-anchored subunit vaccine expressed by a fish commensal strain. *Vaccine* **28**, 6603–6608.

SUN, Y., HU, Y.-H., LIU, C.-S. and SUN, L. (2010b) Construction and analysis of an experimental *Streptococcus iniae* DNA vaccine. *Vaccine* **28**, 3905–3912.

SUN, Y., LIU, C.-S. and LI, S. (2010c) Isolation and analysis of the vaccine potential of an attenuated *Edwardsiella tarda* strain. *Vaccine* **28**, 6344–6350.

SUN, Y., LIU, C.-S. and SUN, L. (2011) Comparative study of the immune effect of an *Edwardsiella tarda* antigen in two forms: subunit vaccine vs DNA vaccine. *Vaccine* **29**, 2051–2057.

TATNER, M.F. (1989) The antibody response of intact and short term thymectomised rainbow trout (*Salmo gairdneri*) to *Aeromonas salmonicida*. *Developmental and Comparative Immunology* **13**, 387.

TEBBIT, G.L., ERICKSON, J.D. and VANDE WATER, R.B. (1981) Development and use of *Yersinia ruckeri* bacterins to control enteric redmouth disease. *Developments in Biological Standardization* **49**, 395–401.

TEMPRANO, A., RIAÑO, J., YUGUEROS, J., GONZÁLEZ, P., DE CASTRO, L., VILLENA, A., LUENGO, J.M. and NAHARRO, G. (2005) Potential use of a *Yersinia ruckeri* O1 auxotrophic *aroA* mutant as a live attenuated vaccine. *Journal of Fish Diseases* **28**, 419–428.

THUNE, R.L., FERNANDEZ, D.H., HAWKE, J.P. and MILLER, R. (2003) Construction of a safe, stable, efficacious vaccine against *Photobacterium damselae* ssp. *piscicida*. *Diseases of Aquatic Organisms* **57**, 51–58.

TOBAR, J.A., JEREZ, S., CARUFFO, M., BRAVO, C., CONTRERAS, F., BUCAREY, S.A. and HAREL, M. (2011) Oral vaccination of Atlantic salmon (*Salmo salar*) against salmonid rickettsial disease. *Vaccine* **29**, 2336–2340.

TU, F.P., CHU, W.H., ZHUANG, X.Y. and LU, C.P. (2010) Effect of oral immunization with *Aeromonas hydrophila* ghosts on protection against experimental fish infection. *Letters in Applied Microbiology* **50**, 13–17.

VAN GELDEREN, R., CARSON, J. and NOWAK, B. (2009) Effect of extracellular products of *Tenacibaculum maritimum* in Atlantic salmon, *Salmo salar* L. *Journal of Fish Diseases* **32**, 727–731.

VAUGHAN, L.M., SMITH, P.R. and FOSTER, T.J. (1993) An aromatic-dependent mutant of the fish pathogen *Aeromonas salmonicida* is attenuated in fish and is effective as a live vaccine against the salmonid disease furunculosis. *Infection and Immunity* **61**, 2172–2181.

VERVARCKE, S., LESCROART, O., OLLEVIER, F., KINGET, R. and MICHOEL, A. (2004) Vaccination of African catfish with *Vibrio anguillarum* O2: 1. ELISA development and response to IP and immersion vaccination. *Journal of Applied Ichthyology* **20**, 128–133.

VERVARCKE, S., OLLEVIER, F., KINGET, R. and MICHOEL, A. (2005) Mucosal response in African catfish after administration of *Vibrio anguillarum* O2 antigens via different routes. *Fish & Shellfish Immunology* **18**, 125–133.

VIVAS, J., RIAÑO, J., CARRACEDO, B., RAZQUIN, B.E., LÓPEZ-FIERRO, P., NAHARRO, G. and VILLENA, A.J. (2004) The auxotrophic *aroA* mutant of *Aeromonas hydrophila* as a live attenuated vaccine against *A. salmonicida* infections in rainbow trout (*Oncorhynchus mykiss*). *Fish & Shellfish Immunology* **16**, 193–206.

WANG, Q., CHEN, J., LIU, R. and JIA, J. (2011) Identification and evaluation of an outer membrane protein OmpU from a pathogenic *Vibrio harveyi* isolate as vaccine candidate in turbot (*Scophthalmus maximus*). *Letters in Applied Microbiology* **53**, 22–29.

YANG, H., CHEN, J., YANG, G., ZHANG, X.H., LIU, R. and XUE, X. (2009) Protection of Japanese flounder (*Paralichthys olivaceus*) against *Vibrio anguillarum* with a DNA vaccine containing the mutated zinc-metalloprotease gene. *Vaccine* **27**, 2150–2155.

YASUMOTO, S., YOSHIMURA, T. and MIYAZAKI, T. (2006) Oral immunization of common carp with a liposome vaccine containing *Aeromonas hydrophila* antigens. *Fish Pathology* **41**, 45–49.

ZHANG, C., YU, L. and QIAN, R. (2011) Cloning and expression of *Vibrio harveyi* *OmpK** and *GADPH** genes and their potential application as vaccines in large yellow croakers *Pseudosciaena crocea*. *Journal of Aquatic Animal Health* **20**, 1–11.

ZHANG, W.-W., SUN, K., CHENG, S. and SUN, L. (2008) Characterization of DegQ$_{vh}$, a serine protease and a protective immunogen from a pathogenic *Vibrio harveyi* strain. *Applied and Environmental Microbiology* **74**, 6254–6262.

ZHAO, Y., LIU, Q., WANG, X., ZHOU, L., WANG, Q. and ZHANG, Y. (2011) Surface display of *Aeromonas hydrophila* GAPDH in attenuated *Vibrio anguillarum* to develop a novel multivalent vector vaccine. *Marine Biotechnology* DOI 10.1007/s10126-010-9359-y.

ZHOU, Y.-C., WANG, J., ZHANG, B. and SU, Y.-Q. (2002) Ultrasonic immunization of sea bream, *Pagrus major* (Temminck & Schlegel), with a mixed vaccine against *Vibrio alginolyticus* and *V. anguillarum*. *Journal of Fish Diseases* **25**, 325–331.

ZHU, K., CHI, Z., LI, J., ZHANG, F., LI, M., YASODA, H.N. and WU, L. (2006) The surface display of haemolysin from *Vibrio harveyi* on yeast cells and their potential applications as live vaccines in marine fish. *Vaccine* **24**, 6046–6052.

ZOU, L., WANG, J., HUANG, B., XIA, M. and LI, A. (2011) MtsB, a hydrophobic membrane protein of *Streptococcus iniae*, is an effective subunit vaccine candidate. *Vaccine* **29**, 391–394.

10

Developments in adjuvants for fish vaccines

J. Bøgwald and R. A. Dalmo, University of Tromsø, Norway

Abstract: Use of injection vaccines plays a key role in controlling infectious diseases in the aquaculture industry. Since 1990 oil-adjuvanted vaccines have been the most widespread and most efficacious vaccines available. However, side effects such as retardation in growth, pigmentation and adherence of intestines have initiated a search for new adjuvants, even though considerable advances with respect to adverse side effects have been achieved during the latest years. Experimental vaccines have included immunostimulants such as beta-glucans, polyinosinic polycytidylic acid (poly I:C), CpG, flagellin, and cytokines free or encapsulated in polymers/particles. Encapsulation of vaccines in biodegradable poly(lactic-co-glycolic acid (PLGA) is especially intriguing.

Key words: adjuvants, fish vaccines, vaccinology.

10.1   Introduction

In the past, fish vaccines were made using a trial-and-error approach (conventional vaccine design) that included pathogen identification, pathogen cultivation and formulation of vaccines that contained whole-cell preparations and oils. In parallel, alternative approaches were being used to develop recombinant vaccines. The vaccines that were based on whole cells were quite efficient and resulted in dramatic reductions in mortality rates and antibiotic usage. Many of the most economically important diseases of today are caused by intracellular pathogens. Alternative approaches for the development of vaccines must therefore be sought since intracellular pathogens may be protected from, for example, antibody attacks. These include, rational vaccine design, which uses a tailored adjuvant system combined with the most appropriate antigen to create vaccines that may provide a more effective immune response against a specific pathogen. Highly specific adjuvants may be selected to induce responses against extra- or intracellularly localised pathogens.

Combination of different adjuvants may also be useful in cases where specific protection is conferred by mixed responses – such as both antibodies (opsonisation) and cell-mediated defence mechanisms (e.g. pathogen killing by molecules secreted by cells). In fish vaccinology, rational vaccine design is still in its infancy, but seems to be gaining in importance due to the abundance of virus diseases and the fact that intracellular bacteria may resist both antibody attack and antibiotics. This vaccine design method, operating at a post-genomic level, consists of genomic and proteomic analysis of pathogens to identify the most promising immunogenic epitopes using bioinformatic processing of data. Candidate vaccines are clinically tested *in vitro* with continuous assessment of immune responses through transcriptome/proteome/vaccinome/immunome profiling, and potency and efficacy testing. In the fish vaccinology field, elements of the post-genomic 'reverse' vaccine design have been described. However, because the tools required to analyse the multifaceted immune responses at a functional level are lacking, reverse vaccine design in fish vaccinology lags far behind the state of the art in human and mammalian vaccinology. Despite this, many successful studies have been reported that describe the development of vaccines against several fish pathogens – including, for example, DNA vaccine strategies (Lorenzen *et al.* 1998; Kurath 2008).

10.2   Fish immune responses: implications for the development of vaccines and adjuvants

10.2.1   Innate immune defences

Innate immunity refers to natural or native immunity that is always present in order to rapidly defend the host against pathogenic microorganisms. Adaptive immunity refers to acquired immunity that is activated by microorganisms that invade host tissues. The adaptive response is specific and slow. There is no doubt that the innate defence system of fish is strongly developed and can cope with many infectious agents. It enables the fish to eradicate viruses, bacteria and even parasites. However, many infectious agents may resist innate defence molecules. A robust adaptive immune response must then come into play to fight these pathogens (as described in Section 10.2.4). Numerous review articles have been published on the fish immune defence system, including the innate defence system (Plouffe *et al.* 2005; Chistiakov *et al.* 2007; Peatman & Liu 2007; Whyte 2007; Dalmo & Bøgwald 2008; Martin *et al.* 2008; Secombes 2008; Randelli *et al.* 2009; Rebl *et al.* 2010). Only a selection of innate immune defence molecules will therefore be described further here.

There are many similarities between 'man and mouse' and between 'man/mouse' and fish with respect to the immune system and its responses. However, many striking differences between mammals and fish also exist. To complicate matters further, there may be significant inherited differences from one fish species to another, and many fish species have undergone a genome-wide duplication event that has resulted in the 'doubling' of the number of immune genes. Currently, there are close to 22 000 different fish species, and most of them have their 'immune peculiarities'. Pattern recognition receptors (PRRs) are a class of receptors that recognise molecules with repeating units – such as DNA, RNA, certain carbohydrates, fatty acid derivatives, and proteins with attached sugars. Over the last decade these receptors have been the focus of much research attention since they are involved in the host immune response against pathogens. Generally, receptors bring about intracellular signalling that leads to gene activation and the production of functionally active proteins such as cytokines, growth factors, etc. There are many types of PRRs, one of which is the Toll-like receptor (TLR) family; interestingly, the number of TLR genes has expanded during evolution at a greater rate in many fish species compared with the rate in mammals. The increased number of TLRs may result in a more diverse and efficient immune response in fish compared with higher vertebrates. TLRs are discussed in more detail in the following section.

10.2.2   The role of Toll-like receptors (TLRs)

As discussed briefly in Section 10.2.1, there are a number of TLR ligands (agonists) that may induce strong innate responses that may be decisive for the outcome of T-cell responses. These signaling molecules include differentiation factors for T-helper (Th) 1, Th2 and Th17 responses. An excellent detailed overview of fish TLRs has recently been written by Rebl *et al.* (2010).

It has been elucidated that there are many more TLRs in fish than in mammals, presumably due to ancient gene duplication. Whereas it is clear that many mammalian species have up to 11 different TLRs, fish (e.g. fugu and zebrafish) also possess TLR21 and -22. Many similarities between mammals and fish exist with respect to intracellular and downstream signalling events, but there are dissimilarities that warrant focus. An example is TLR4 that is discussed at the end of this section.

Broadly speaking, different TLRs can be categorised as 'antiviral' or 'antibacterial' receptors, since upon ligand binding unique intracellular signalling occurs. TLR3, TLR7, TLR9 and TLR21 are regarded as antiviral for their induction of, for example, alpha and beta interferons (IFNs) and IFN-inducible factors; TLR3, -4, -7 and -9 may induce type I IFN responses via interferon regulating factors; other TLRs may induce cytokines which help accessory cells and lymphocytes to fight and eliminate bacteria. In general, those TLRs, after ligand binding, that induce the production of interleukin 12 (IL-12) favour a Th1 response (TLR3, -4, -5, -7, -8, -9 and -11). In addition, activation of these TLRs may induce cross-presentation of antigens facilitating a cytotoxic T-cell response under certain conditions

(Manicassamy and Pulendran 2009). Molecules that can bind TLRs may act as adjuvants (helper substances which increase antimicrobial responses) in fish vaccines; as such, different ligands of aquatic microbes have been shown to induce immune responses, probably through the TLR pathways. This paves the way for a more rational vaccine design – to produce vaccines that are capable of targeting highly problematic pathogens such as viruses.

As previously mentioned, a higher number of TLRs exist in fish compared with mammals, and there may also be several TLR isotypes (gene subtypes) present, as has been shown for TLRs 7, 8, 9 and 22 in Atlantic cod (Star *et al.* 2011). Some of the TLR isotypes may be highly functionally relevant – for example soluble TLR5. Membrane-bound TLR5 (mTLR5) binds bacterial monomeric flagellins that are present as 'polymers' in the flagella, helping bacteria to move around. When bacteria are being degraded by, for example, phagocytic cells, the release of monomeric flagellin may occur and this binds to mTLR5 and induces cell activation. It is suggested that soluble and circular TLR5 (sTLR5) either 'competes' for the flagellin and therefore amplifies the flagellin-induced immune response (Tsujita *et al.* 2004) or 'cools down' the response by 'catching' the flagellin before mTLR5 does.

Toll-like receptor 4 (TLR4)
As mentioned above, TLR4 is a TLR that has been discovered in a small number of fish species and is an example of how fish TLRs can differ from mammalian TLRs. One hypothesis to explain the action of fish TLR is that lipopolysaccharide (LPS) (a TLR4 agonist) down-regulates MyD88-dependent signalling which, in turn, induces a higher level of LPS resistance in fish. Together with this down-regulation, the LPS co-receptors MD-2 and CD14 may be absent from fish and this may also explain the relatively high LPS non-responsiveness in fish (Sepulcre *et al.* 2009). However, LPS may induce activation of protein kinase C enzymes (Olavarria *et al.* 2010) that may phosphorylate several transcription factors, e.g. IkB (nuclear factor of kappa light polypeptide gene enhancer in B-cells inhibitor), modulating gene expression in antigen-stimulated T-cells. Following on, the degree of activation of rainbow trout macrophages by ultra-pure LPS was much lower than that by 'contaminated' LPS – as shown in a recent report by MacKenzie *et al.* (2010). Obviously, commercially obtained and laboratory-made LPS may induce cellular activation in fish, but the strength of response may be dependent on the purity of LPS. More research on the intracellular signalling cascade is needed to confirm regulatory activities by cells during LPS stimulation.

10.2.3   Innate immune response and acquired immunity
The innate immune defence system has been described in the previous two sections. Interestingly, there is also a close link between the innate and the

adaptive systems (Section 10.2.4) that has not received much attention in fish immunology studies. The link is governed by several innate receptors and signalling molecules such as cytokines and transcription factors. After infection by, for example, bacteria, the host may start to produce cytokines. It is acknowledged that the first cytokines that may appear are IL-1β, , IL-8 (CXCL8) and tumour necrosis factor (TNF)-α (Koj 1996). It has been proposed that IL-6 family cytokines, in the presence of endogenous IL-4, act on Th differentiation – especially favouring Th2 differentiation (Sokol *et al.* 2008). IL-6 inhibits Th1 differentiation by up-regulating suppressor of cytokine signalling (SOCS)-1 expression to interfere with IFN-γ signalling and the development of Th1 cells (Diehl & Rincón 2002). It has also been suggested that IL-6, together with IL-23, induces the Th17 T-cell lineage (Weaver *et al.* 2006). Th17 is often referred to in autoimmune disorders, and many members of IL-17 have been found in fish (Kumari *et al.* 2009; Korenaga *et al.* 2010; Wang *et al.* 2010). Cloning and functional characterisation of IL-6 in rainbow trout have been reported (Bird *et al.* 2005; Iliev *et al.* 2007; Raida and Buchmann 2008a,b, 2009). Similarly, the expression profile of rainbow trout fry IL-6 has been described following bath treatments of plasmid DNA and lactoferrin (Zhang *et al.* 2009).

10.2.4   The adaptive immune defence system in fish and current vaccination strategies

The immune response to different categories of pathogens follows a general pattern in higher vertebrates, i.e. immunity to extracellular pathogens is generally mediated by humoral immune responses (for example, production of antibodies), while immunity to intracellular pathogens (including viruses) often relies on cytotoxic immune responses (e.g. killing of viral infected cells). However, there are many exceptions. Among the lower vertebrates and cartilaginous and bony fish (teleosts), even less information is available. It has been shown that all jawed vertebrates have developed an adaptive immune system with both humoral and cell-mediated immune responses. Genes for both immunoglobulin (Ig; i.e. B-cell receptor (BCR), humoral responses) and T-cell receptors (TCRs; i.e. cell-mediated responses) have been identified in all lineages of gnathostomes (jawed vertebrates). Fish Igs are expressed as three isotypes (IgM, IgD and IgT). The functional importance of the last two types of Ig is still very poorly understood but the neutralisation capacity of the IgM against viruses and bacteria is very well established, and appears to be a key parameter in the protection mechanism.

Fish TCR genes were first described in rainbow trout (Partula *et al.* 1995) and later TCR-α and -β genes were identified in many fish species. The presence of TCR genes and also the presence of myosin heavy chain (MHC) I and II genes with typical patterns of polymorphism are suggestive of T-cell responses being an integrated part of the fish immune response. Recent studies have confirmed the capacity of fish lymphocytes to mount cytotoxic responses (Somamoto *et al.* 2006). Other studies employing spectratyping of the complementarity determining region 3 (CDR-3) length of TCR-β chains (Boudinot *et al.* 2001) have shown that T-cell responses are elicited following viral infection (rhabdovirus) with skewed CDR-3 profiles for many VβJβ combinations, indicating strongly clonal T-cell expansion during both primary and secondary responses to infection. Antigen-specific CD8+ cytotoxic T-lymphocytes recognise viral peptides in the context of MHC I. The limitations in the number of expressed MHC class I and class II loci and their physical separation have influenced their evolution. Interestingly, MHC II has not been found in gadoid fish species that may suggest that vaccination of, for example, Atlantic cod results in suboptimal protection (Star *et al.* 2011). This is not, however, the case since specific disease protection occurs after vaccination of cod (Mikkelsen *et al.* 2011). Following on, it seems that the achieved protection may be due to an increased antibody response against lipopolysaccharide which is a B-cell mitogen. The understanding of specific T-cell subsets in fish is still under development with the characterisation of marker molecules such as CD4, CD3 and CD8 (reviewed by Randelli *et al.* 2009). In addition, several key cytokines and chemokines have been cloned and sequenced in salmonids, for example: IL1β, TNFα, several IFNs – including IFNγ, IFNα, IL-4,IL-10, IL-12, IL-8, IL-21, lymphotoxin, IL-2, and IL-17. Their expression patterns and regulatory functions should soon provide more insights in T-helper cell sub-populations that may constitute functional counterparts of mammalian CD4+ Th1, Th2 and Th17. In addition, key T-cell co-stimulatory receptors like CD28, cytotoxic T-lymphocyte antigen 4 (CTLA4) and B- and T-lymphocyte attenuator (BTLA) have been characterised (Bernard *et al.* 2007), and this paves the way for the phenotypic analysis of T-cell differentiation during the immune response. However, few of these cytokines are yet available as recombinant proteins, and few well-characterised monoclonal antibodies are available as specific T-cell markers. With limited knowledge and limited tools available, rational assessment of the importance of the T-cells in responses to vaccines remains complicated. Thus, the design of vaccines is also a difficult undertaking and there are currently few, if any, vaccines commercially available that confer strong immunity to many viruses.

The use of injected vaccines plays a key role in controlling infectious diseases in aquaculture. Oil-adjuvanted vaccines were introduced in the early 1990s, and they are still the dominant mode of delivery although they elicit sub-optimal immunity against viruses (Sommerset *et al.* 2005) and result in unwanted injection site reactions (Evensen *et al.* 2005; Koppang *et al.* 2009). Several other research efforts, such as the use of vaccines and immunostimulants (adjuvants) (Sakai 1999; Sommerset *et al.* 2005) especially in combination, have also been reported (Anderson 1997), and the concept of adding immunostimulants has gradually become popular since it represents a higher degree of food safety and is more environmentally friendly. The first-generation vaccines were developed more or less based on empirical knowledge, not a rational design basis (Schijns and Tangeras 2005). However, a rational design process for vaccines has attracted much attention over the last decade – especially for the design of vaccines for domesticated animals and humans. In recent years, similar design approaches have been used to develop fish vaccines.

10.2.5   Adjuvants and the polarisation of T-cell responses

Adjuvants have traditionally been defined as helper substances that increase the magnitude of an adaptive response to a vaccine (potency), or its ability to prevent infection and death (efficacy). However, nowadays scientists have acknowledged that adjuvants may become more important by guiding the type of adaptive response against a specific pathogen. It is acknowledged that different adjuvants may polarise immune responses into Th1, Th2, Th17 or Tcyt, but often mixed responses occur as a result of the chemical features of the adjuvants or as a result of the antigen–adjuvant combination (Coffman et al. 2010). In general, and in humans, alum (with antigen) induces a high antibody response (signature of a Th2 response) together with a Th1 response in humans. Other licensed adjuvants include MF59 and AS03, based on squalene-in-water emulsions and these also induce a mixed Th1 and Th2 response and antibody response. Alum plus MPL (AS04) with antigen induces an antibody response and a Th1 response.

There are several experimental adjuvants that include ligands for TLRs. Polarised Th1 responses normally occur when, for example, agonists of TLR3, TLR4, TLR7, TLR8 and TLR9 are included in the vaccines. Thus, by combining a certain adjuvant with an antigen one may predict the resulting CD4+ T-cell response (Coffman et al. 2010). The task is then to decide which adjuvants to choose when the preferable T-cell response should be a cytotoxic one. Exogenous vaccine antigens need to escape from the endosomal compartment to enter the MHC class I pathway that may induce a robust CD8+ T-cell response. Several vaccines based on live or attenuated pathogens that induce protection have been developed. The most promising future vaccines that induce protection against viruses are DNA vaccines. The plasmid DNA constructs may encode for an immunogenic peptide, made by the host cells themselves, that could be presented on MHC class I, whereas expressed and secreted immunogens may be endocytosed by antigen presenting cells (APC) and as such, be presented by MHC class II molecules. Thus, both MHC class I and II presentations may occur. DNA vaccines may be used in combination with prime-boost strategies. Following on, nanoparticle-based vaccines hold great promise in the search for vaccines that elicit CD8+ T-cell responses. These particles may be made of synthetic polymers such as poly(lactic-co-glycolic acid) (PLGA) and should contain at least one adjuvant to induce a robust reponse (Hamndy et al. 2008; Coffman et al. 2010; Yewdell 2010).

10.3   Oil-adjuvanted fish vaccines

10.3.1   Freund's adjuvants

The most widely used and most effective adjuvant for experimental purposes has been Freund's complete adjuvant (FCA) and specific examples of its use are given in this section. FCA is composed of heat-killed mycobacteria combined with mineral oil and a surfactant (Freund *et al.* 1937; Stills, 2005). Before injection, the antigen in an aqueous solution is mixed with the FCA producing a stable water-in-oil emulsion. Unfortunately, the use of FCA has been associated with a variety of side effects including injection-site granulomas. Immunisation with FCA with antigens results in strong Th1 and Th17 responses mostly via the MyD88 pathway. The use of FCA has therefore been limited to research on animals, including fish, to establish an effective immune response.

There are many studies in the literature that discuss the use of Freund's adjuvants in fish vaccines. A range of fish species and diseases have been investigated and the results reported to date have been mixed. The examples below describe fish diseases caused by a variety of pathogens.

Pasteurellosis, caused by *Pasteurella piscicida*, also named *Photobacterium damsela* subsp. *piscicida* is one of the major diseases in many species of wild and farmed fish in Asia, USA and Europe. The pathogen is causing disease in yellowtail (*Seriola quinqueradiata*) aquaculture in Japan (Kawakami *et al.* 1998). Vaccination with an LPS-mixed chloroform-killed bacterin resulted in protection against challenge with the virulent bacterium. Inclusion of FCA in the vaccine did not significantly enhance the protective effect (Kawakami *et al.* 1998).

Edwardsiella tarda is a Gram-negative bacterium that can infect both marine and freshwater fish. The Japanese flounder (*Paralichthys olivaceus*) is a susceptible species and experiments have been initiated to develop effective vaccines against the pathogen (Jiao *et al.* 2010a,b). Fish were intraperitoneally (i.p.) injected with a vaccine containing a major antigenic protein of *E. tarda* in the absence or presence of Freund's incomplete adjuvant (FIA). Protection against experimental challenge in the absence of adjuvant resulted in a relative percent survival (RPS) of 34%, and in the presence of FIA an RPS of 81% was achieved. Vaccination with the oil-adjuvanted antigen stimulated the expression of a series of genes, for example complement component 3 (C3), MHC class I and MHC class II, CD8α, CD40, Mx, IFNγ, TNF-α and IL-6. Vaccination with the antigen alone resulted in increased expression of just IgM, MHC class I and class II, and Mx (Jiao *et al.* 2010a).

Aeromonas hydrophila is a Gram-negative, facultative anaerobic bacterium known to cause motile aeromonad septicaemia (MAS) in freshwater fish farming. It is also considered a human pathogen. The major adhesin of *A. hydrophila*, a 43 kDa outer membrane protein, was cloned and expressed by Fang *et al.* (2004). The molecule was used in a vaccine for the blue gourami (*Trichogaster trichopterus*) emulsified in FCA. The vaccine was i.p. injected and after 3 weeks a booster was given without FCA. Two weeks after the booster the fish were challenged with two strains of *A. hydrophila*. The recombinant adhesin protected against challenge with the homologous strain of *A. hydrophila*, and also against a heterologous strain. The recombinant adhesin provided the same immune protection as the native adhesin (Fang *et al.* 2004). Recently, LaPatra and colleagues (2010) reported on a vaccine containing bacterial lysate emulsified in FCA to protect rainbow trout against *A. hydrophila*. These authors developed a new challenge model for *A. hydrophila* in rainbow trout by injection of the bacteria into the dorsal sinus. The reason why it has been difficult to develop effective vaccines against *A. hydrophila* is most probably a high degree of antigenic variation. Some success has, however been achieved. A bacterial lysate was shown to give protection after i.p administration, and this protection could be potentiated in the presence of FCA (LaPatra *et al.* 2010). In addition, fish that survived an *A. hydrophila* challenge were very resistant to reinfection.

Streptococcus iniae is a Gram-positive bacterium associated with disease in several commercial species including tilapia (*Oreochromis aureus* and *O. niloticus*), yellowtail (*S. quinqueradita*), hybrid striped bass (*Morone saxatilis*), turbot (*Scophthalmus maximus*) and rainbow trout (*Oncorhynchus mykiss*). Vaccination of rainbow trout with a formalin-killed culture of *S. iniae* resulted in good protection against experimental challenge which was not significantly potentiated in the presence of FCA (Soltani *et al.* 2007).

Nocardia seriolae is a Gram-positive acid-fast bacterium that causes nocardiosis in cultured marine and freshwater fish in Taiwan, Japan and China. Although the disease results in considerable economic loss, there is no vaccine against *N. seriolae*. Very recently, an oil-adjuvanted vaccine was developed and tested for its protection against challenge with a virulent strain of *N. seriolae* (Shimahara *et al.* 2010). Formalin-inactivated whole-cell antigen was used as a vaccine with FIA. Even though antibody levels increased when the two were used together, no protective effects were found.

Another Gram-positive bacterium that causes disease (lactococcosis) and mortality in rainbow trout is *Lactococcus garvieae*. Recently a vaccine was prepared based on formalin-inactivated bacterin, bacterin together with FIA, bacterin combined with β-glucan, and phosphate-buffered saline (PBS) as control. Fish were given i.p. injections and challenged by exposure to virulent bacteria 30, 75 and 125 days after vaccination (Kubilay *et al.* 2008). At 125 days after vaccination the RPS in fish vaccinated with bacterin only was 54%; for fish vaccinated with bacterin + FIA, the RPS was 85%.

Flavobacterium psychrophilum is a widespread Gram-negative pathogen in freshwater causing rainbow trout fry syndrome (RTFS) and bacterial cold water disease (BCWD) (Högfors *et al.* 2008). In addition to rainbow trout, coho salmon (*Oncorhynchus kisutch*) Walbaum is also very a susceptible species, but non-salmonids are also affected. Injection of a low molecular weight fraction emulsified in FCA resulted in an enhanced level of protection (Högfors *et al.* 2008).

Flavobacterium columnare is a Gram-negative bacterium responsible for columnaris disease. The disease was first described in 1917 in several warmwater fish species from the Mississippi River, and has since been isolated from freshwater fish species worldwide (Grabowski *et al.* 2004). Specific antibodies were found in tilapia (*Oreochromis niloticus*) plasma and mucus following i.p. injection of formalin-killed sonicated or whole cells of *F. columnare* in FCA within 2 weeks. After a secondary immunisation the antibody response increased. At 10 weeks post-immunisation the titre remained elevated. In addition, antibodies were observed in the cutaneous mucus of fish that had been i.p. immunised with formalin-killed sonicated cells in FCA 6 and 8 weeks post-immunisation (Grabowski *et al.* 2004).

The major bacterial disease of farmed Atlantic cod is classical vibriosis (Samuelsen *et al.* 2006). Cod vaccinated by injection with mineral oil-adjuvanted vaccines against both ***Vibrio anguillarum*** and atypical ***Aeromonas salmonicida*** were very well protected against homologous challenges (Mikkelsen *et al.* 2004). Even without adjuvant the fish were protected against *V. anguillarum*, but not against atypical *A. salmonicida* challenge. In a study on coho salmon (*Oncorhynchus kisutch*), formalin-killed *A. salmonicida* were i.p. injected in the absence or presence of modified Freund's complete and incomplete adjuvant (MFCA and MFIA) (a mixture of saline, Tween 80, light mineral oil, Arlacel C and killed *Mycobacterium butyricum* cells). The mixture of MFCA and *A. salmonicida* gave better protection compared with the antigen in saline or in modified Freund's incomplete adjuvant (MFIA). Interestingly, fish injected with MFCA (without antigen) showed some protection even 90 days after the challenge (Olivier *et al.* 1985).

Tenacibaculum maritimum is a marine bacterium that causes flexibacteriosis worldwide. In Australia (Tasmania), Atlantic salmon and rainbow trout are the most heavily affected species, and so far the disease has been treated with trimethoprim and oxytetracycline with negative impacts on the environment as a result (van Gelderen *et al.* 2009). Salmon injected with formalin-inactivated bacteria mixed with FIA showed protection against challenge with *T. maritimum*. A vaccine without the adjuvant could not provide sufficient protection against moderate challenge of *T. maritimum*.

10.3.2 Other mineral oil adjuvants

The mode of action of mineral oil adjuvant vaccines is complex and includes the initial antibody stimulus resulting from dispersal of antigen, the slow release of antigen and the inflammatory response (McKinney and Davenport 1961; Jansen *et al.* 2005). As with FCA, the use of other mineral oil adjuvants has also been associated with side effects such as abdominal adhesions and growth retardation; further research is therefore needed. A few examples of mineral oil adjuvants and their side effects are given below.

Moritella viscosa is the causative agent of winter ulcers in farmed fish such as Atlantic salmon (*Salmo salar*) and Atlantic cod (*Gadus morhua*). Vaccination of Atlantic salmon against *M. viscosa* is performed with oil-adjuvanted polyvalent injection vaccines based on formalin-inactivated bacterial cultures (Gudmundsdottir and Björnsdottir 2007). However, in spite of the availability and extensive use of vaccines, winter ulcer is the main bacterial disease in the aquaculture of Atlantic salmon in Norway (Lillehaug *et al.* 2003). Marine species such as turbot (*Scophthalmus maximus*) and halibut (*Hippoglossus hippoglossus*) have been shown to be sensitive to *M. viscosa* experimental challenges. A multivalent commercial salmon vaccine containing *M. viscosa* as one of five bacteria and a mineral oil adjuvant, did not protect turbot against the challenge (Björnsdottir *et al.* 2004); moderate intra-abdominal adhesions were detected in vaccinated fish.

Effective vaccines against several Gram-negative pathogens of salmonids – such as *Aliivibrio salmonicida, Vibrio anguillarum, Yersinia ruckerii* and *Aeromonas salmonicida* – have been developed for many years (Kajita *et al.* 1992; Adams *et al.* 1988; Olivier *et al.* 1985). Oil-adjuvanted vaccines have for many years been shown to give protection in Atlantic salmon against bacterial diseases such as vibriosis, coldwater vibriosis and furunculosis. However, side effects and retardation in growth have also been clearly demonstrated (Midtlyng *et al.* 1996; Midtlyng and Lillehaug 1998). Mutoloki and coworkers investigated the intraperitoneal lesions induced by an oil-adjuvanted vaccine against infection with *A. salmonicida* and *M. viscosa* in Atlantic salmon (Mutoloki *et al.* 2010). The cellular composition of the lesions was typical of granulomas containing large macrophages, eosinophilic granular cells, lymphocytes and multinucleate cells. The expression of TGF-β, IL-17A and up-regulation of arginase correlate well with strong infiltration of neutrophils and macrophages. Oil-adjuvanted vaccines are also used to protect sea bass (*Dicentrarchus labrax*) from bacterial diseases such as vibriosis and pasteurellosis. Sea bass is one of the most common aquacultured fish species in the Mediterranean area, and the fish suffer from infection by *V. anguillarum* and *Photobacterium damsela* subsp. *piscicida*. Oil-adjuvanted vaccines against these diseases have been prepared and injected i.p. A granulomatous peritonitis was also recognised in these fish (Afonso *et al.* 2005).

Montanide
Montanide is a metabolizable oil adjuvant that has been used in both mammalian and fish vaccines (Lawrence *et al.* 1997; Ravelo *et al.* 2006).

Non-metabolisable oils may also be included in different montanide preparations; surfactants may also be added.

Philasterides dicentrarchi is a scuticocilate parasite that causes mortality and significant economic losses in cultured turbot (Lamas *et al.* 2008). Optimisation of a vaccine was performed on the basis of antigenic dose, concentration of inactivating agent (formalin) and proportion of the adjuvant (Montanide ISA711 and Montanide ISA763A) in the emulsion. The results of the study showed that a high concentration of antigen, 0.2% formalin and 50% adjuvant generated the longest time of survival after challenge (at 30 days after the second injection), and the highest levels of antibodies in the vaccinated fish (Lamas *et al.* 2008).

Pseudomonas plecoglossicida is a bacterium causing bacterial haemorrhagic ascites of cultured ayu (*Plecoglossus altivelis*). To develop a vaccine against the disease formalin-killed *P. plecoglossicida* bacterin were emulsified with montanide, and injected i.p. As a control, saline was i.p injected and the fish were challenged with i.p injection of virulent *P. plecoglossicida* 22 and 52 days after vaccination (Ninomiya and Yamamoto 2001). The RPS rates of vaccinated fish were 17–58% without adjuvant, 57–92% with Montanide ISA711 and 65–86% with Montanide ISA763A. Another study on the same disease and adjuvant (Montanide ISA 763A) concluded that there is a good correlation between antibody levels and protection against disease in a challenge test (Sitja-Bobadilla *et al.* 2008).

Atlantic halibut were injected i.p. with an experimental vaccine of human gamma globulin with either FCA or Montanide ISA711 as adjuvants (Bowden *et al.* 2003). Antibody responses and intraperitoneal adhesions were examined every month for up to 12 months. FCA produced the highest and fastest antibody response and the fastest growth of intraperitoneal adhesions. In the group injected with the montanide adjuvant, only 4 of 47 fish reached a titre of 1:1000 (month 6) compared to 27 of 48 fish in the FCA group (after 2 months) (Bowden *et al.* 2003).

Several studies have investigated the use of montanide in vaccines against *A. hydrophila* infection in fish. The bacterium is very heterogeneous and no good common antigen(s) has been identified; as mentioned previously, there is still no commercial vaccine available against *A. hydrophila*. In a very recent study in carp, a recombinant S-layer protein of *A. hydrophila* was used, and the ability of this protein to protect fish against six virulent isolates of *A. hydrophila* was assessed. A recombinant S-layer protein of *A. hydrophila* was produced, diluted in PBS and mixed with a montanide adjuvant at a ratio of 30:70. Common carp *Cyprinus carpio* L. were i.p. injected with the emulsion, and after 35 days the fish were challenged with six different isolates of *A. hydrophila* (Poobalane *et al.* 2010). The RPS values varied between the different challenge isolates (40–75%). This indicates that the S-layer protein is a good candidate for obtaining efficacous vaccines against this bacterium.

10.4   Vaccines adjuvanted with substances other than oil

10.4.1   Alum

The adjuvant property of aluminium salts was discovered in 1926 (Glenny *et al.* 1926). Aluminium compounds, especially aluminium phosphate and aluminium hydroxide, are some of the few adjuvants that have been authorised as safe to use in human vaccines. Aluminium adjuvants have been shown to induce only Th2 responses (Jiao *et al.* 2010a), and also a certain level of Th1 response in humans (Coffman *et al.* 2010). It is believed that alum activates 'NACHT, LRR and PYD domains-containing protein 3' (NLRP3) inflammasome and induces necrotic cell deaths that release the danger signal uric acid (Coffman *et al.* 2010). These insults may both induce Th1 and Th2 T-cell responses. Few studies have been performed with aluminium adjuvants for the development of vaccines for farmed fish. A vaccine against *A. salmonicida*, mixed with potassium aluminium sulphate (alum) as an adjuvant, was tested in Atlantic salmon over a decade ago (Mulvey *et al.* 1995). Alum appeared to enhance protection against challenge, but the results were not statistically significant. In another study at about the same time (Tyler and Klesius 1994), an *Escherichia coli* mutant was used for vaccination against *Edwardsiella ictaluri*-induced enteric septicaemia of catfish (*Ictalurus punctatus*). Killed *E. coli* bacteria, with or without aluminium hydroxide (alum), were administered i.p to catfish and the fish were challenged with virulent *E. ictaluri* bacteria (Tyler and Klesius 1994). Fish given *E. coli* in alum had enhanced survival rates (92%) compared with those given *E. coli* alone (54%) or fish given saline (56%).

Recently, an aluminium hydroxide-adjuvanted *E. tarda* vaccine was prepared and injected i.p in Japanese flounder. The resultant protection was found give an RPS of 69% (Jiao *et al.* 2010a). Immunisation with the antigen alone gave an RPS of 34%, so in this system alum seemed to be a promising adjuvant. However, when FIA was used an RPS of 81% was seen, thus showing that the oil adjuvant would be a more promising line of research.

10.4.2   β-Glucans – ligands for dectin-1

β-Glucans are known to stimulate the non-specific immune response of both mammals and fish and the receptor dectin-1 is thought to be involved (Robertsen *et al.* 1999; Dalmo and Bøgwald 2008). To obtain protective effects against diseases, the glucan is injected i.p. and there seems to be a dose-dependent, short-lived protection. In addition, there have been numerous reports on the adjuvant effect of β-glucans (Nikl *et al.* 1991; Chen and Ainsworth 1992; Rørstad *et al.* 1993; DeBaulney *et al.* 1996; Midtlyng *et al.* 1996; Figueras *et al.* 1998; Midtlyng and Lillehaug 1998; Ashida *et al.* 1999; Kamilya *et al.* 2006; Selvaraj *et al.* 2006).

Vibriosis is one of the most common and serious diseases of turbot (*Scophthalmus maximus*). DeBaulney and coworkers (1996) prepared an oral vaccine against vibriosis; the fish were fed the vaccine for 5 days and were then challenged 28 days later. Fish given the vaccine alone had an RPS of 52%, while a combination of the vaccine and the β-glucan gave an RPS of 61%; however, this result was not statistically significant (DeBaulney *et al.* 1996). In Spain, *Vibrio damsela* has been identified as the cause of high mortality in cultured turbot (Figueras *et al.* 1998). These authors tried to develop vaccines and immunisation protocols to obtain the highest immune response against *V. damsela*. The turbot were i.p. injected with the O-antigen of *V. damsela* in combination with i.p. injection of β-glucan. A phagocytic index of head kidney macrophages was used to evaluate the effects of the vaccine. Compared with the fish that were injected with β-glucan before being injected with the antigen, the enhancement of the phagocytic index lasted longer in fish that were injected with β-glucan at the same time as the antigen or in those where β-glucan was injected after the antigen. Similar results were obtained with regard to antibody titers (Figueras *et al.* 1998).

In the early 1990s, furunculosis was a great threat to the aquaculture of Atlantic salmon. Yeast glucan (mainly a β-1,3-D glucan) was included in a vaccine that consisted of a formalin-killed culture of *A. salmonicida* and *V. salmonicida* (Rørstad *et al.* 1993). The vaccine with or without β-glucan was injected i.p., and salmon were challenged 3–46 weeks after vaccination. Vaccines supplemented with glucan induced significantly higher protection against furunculosis than vaccines without (Rørstad *et al.* 1993). The i.p. injection of glucan alone did not result in protection (after 11 weeks). In another study, glucan-adjuvanted vaccines against furunculosis seemed to give protection at an early timepoint after vaccination (6 weeks), but no protection was seen after 3 and 6 months (Midtlyng *et al.* 1996). The average weight of the glucan-adjuvanted group was significantly lower then that of the controls, and average weight of the oil-adjuvanted group was significantly lower than that of the glucan-adjuvanted group (Midtlyng and Lillehaug 1998).

In a study on furunculosis in coho salmon, Nikl *et al.* (1991) evaluated the potentiating effect of seven substances on the protection after vaccination with formalin-killed *A. salmonicida* bacterin. Compared with the group receiving bacterin alone, a statistically significant improvement in survival was noted in groups receiving bacterin combined with β-glucans (i.e. Vitastim-Taito and lentinan). Agglutinin levels were significantly elevated in all cases where the bacterin was injected, and no significant elevation in agglutinin titre occurred as a result of combining an immunostimulant with the bacterin (Nikl *et al.* 1991).

Catla (*Catla catla* Hamilton) is one of the major carp species in India and *A. hydrophila* is a bacterial pathogen associated with disease in this species. A formalin-inactivated *A. hydrophila* vaccine was developed and protection was studied in the absence and presence of a β-glucan adjuvant (Kamilya *et al.* 2006). A reduction in mortality was found when glucan was added to the vaccine, compared with the vaccine alone, but this effect was not statistically significant (RPS of 67.7% with adjuvant and 58.0 without).

In another study on carp (*Cyprinus carpio*) a vaccine against *A. hydrophila* showed a higher antibody titer when glucan was i.p. injected prior to vaccination. Bath and oral administration of glucan before vaccination did not result in enhanced induction of antibodies (Selvaraj *et al.* 2005). In another study by Selvaraj and coworkers, carp (*C. carpio*) were vaccinated against *A. hydrophila* with LPS from a virulent strain of the bacterium in the presence of different concentrations of β-glucan administered through various routes: i.p, oral and bathing (Selvaraj *et al.* 2006). The RPS was significantly higher in i.p.-injected groups, even at the lowest concentration of glucan (10 mg LPS + 100 mg glucan). Fish given the LPS–glucan mixture orally (1% glucan and 0.25% LPS) obtained a higher RPS compared with the controls. However, administration of the LPS–glucan by the bathing route did not result in increased survival. An obvious control group was missing in this study, namely the protective effect of LPS without adjuvant. The antibody levels of fish injected with 10 mg LPS + 100 mg glucan were increased, but no increased levels of antibodies were seen after bathing and oral administration (Selvaraj *et al.* 2006).

Edwardsiellosis of Japanese flounder may be a serious problem in high-temperature seasons. Formalin-killed *E. tarda* cells were administered to fish by feeding in the absence or presence of curdlan or curdlan + quillaja saponin. The *E. tarda* + curdlan-containing diet gave higher survival rates, but the group administered *E. tarda* + curdlan + quillaja (Quill A) saponin was the only group with significantly better survival (Ashida *et al.* 1999).

In another study, i.p. injection of β-glucan in channel catfish (*Ictalurus punctatus*) and subsequent i.p. immunization with *E. ictaluri* resulted in higher serum antibody levels relative to control catfish receiving PBS before administration of *E. ictaluri* (Chen and Ainsworth 1992). The fish received glucan or PBS on days 1 and 3 followed by two immunizations of *E. ictaluri* on days 7 and 14. Serum antibody levels were determined on day 7 (day 21) after the last immunisation. The antibody levels were typically two-fold higher than those in the control fish.

Blue gourami (*Trichogaster trichopterus*) is a freshwater tropical fish species often used as an ornamental fish, and *A. hydrophila* is a common pathogen of the fish. In a study to investigate possible treatments against the disease, laminaran, a β-1,3-D-glucan was injected i.p. in the absence and presence of formalin-killed *A. hydrophila* bacteria (Samuel *et al.* 1996). A single i.p. injection of 20 mg kg^{-1} laminaran was sufficient to protect the fish against infection by a virulent strain of *A. hydrophila* up until 29 days after injection. Fish injected with a mixture of 20 mg kg^{-1} laminaran and formalin-killed *A. hydrophila* did not show significantly improved protection. The protective immunity was closely correlated with the increased phagocytic activity of head kidney phagocytes (Samuel *et al.* 1996).

10.4.3 Cytokines

Cytokines, such as interleucins and interferons, are small cell-signalling protein molecules used in intercellular communication. Interferon regulatory factors (IRFs) form a large family of transcription factors. IRF-1 has been shown to be involved in host defence against pathogens and in cytokine signalling. In addition, IRF-1 has been shown to be up regulated upon virus infection and to induce an antiviral state in fish cells (Caipang *et al.* 2005). In a recent study, the role of IRF-1 in the regulation of the immune system of Japanese flounder was investigated (Caipang *et al.* 2009). Co-injection of IRF-1 with a DNA vaccine encoding the major capsid protein (MCP) gene of red seabream iridovirus (RSIV) resulted in elevated serum neutralisation antibodies, but the level was not significantly different from that in fish vaccinated with the DNA vaccine alone. IRF-1 is also responsible for the up regulation of antiviral substances like nitric oxide (NO), interferon β (IFN-β) and interferon inducible genes like Mx (Caipang *et al.* 2009).

Interleukin 8 (IL-8) is an α-chemokine produced by many cell types in mammals – e.g.macrophages, monocytes, epithelial cells, neutrophils and fibroblasts – upon infection or when stimulated by cytokines like IL-1β and TNF-α. In fish, IL-8 has been characterised in rainbow trout and Japanese flounder. In mammals, chemokines have been used as adjuvants in vaccines against viral infections. In rainbow trout, co-injection with IL-8 was studied as an adjuvant in vaccines against viral haemorrhagic septicaemia virus (VHSV) (Sanchez *et al.* 2007; Jimenez *et al.* 2006). A pMCV1.4-G plasmid (coding for the glycoprotein gene of VHSV) was used as a DNA vaccine. This plasmid was intra-muscularly (i.m.) injected in rainbow trout in the absence or presence of pIL8+ encoding IL-8. When pIL-8+ was administered together with pMCV1.4-G, an increase of IL-1β in the spleen was found 3 and 7 days after injection. When pIL8+ was injected, a greater cellular infiltration was produced at the site of inoculation. Furthermore, fish injected with pIL8+ alone showed a significantly higher expression of TNF-α, IL-11, TGF-β and IL-18 in the spleen at day 3 and higher expression of TGF-β and IL-18 at day 7 (Jimenez *et al.* 2006). This shows that IL-8 is able to modulate the early immune response and could be a potent vaccine adjuvant in fish vaccines designed to protect against viral infection. In another study in trout, Sanchez *et al.* (2007) studied the expression of inducible CC chemokines in response to a VHSV DNA vaccine and IL-8. The study showed that inducible CC chemokines were expressed in response to the DNA vaccine, mainly in the head kidney. In addition, when IL-8 was used as an adjuvant, the expression of the chemokines CK5A, CK6, CK7 and CK5B was modulated.

The administration of IL-1β-derived peptides to rainbow trout by i.p. injection induced reduced mortality of fish when exposed to VHSV 2 days later (Peddie *et al.* 2003). The peptide also induced leucocyte migration into the peritoneal cavity 1–3 days post-injection. The role of IL-1β as an adjuvant was also investigated in carp after i.p. injection of killed *A. hydrophila* in the absence and presence of recombinant C-terminal peptide of carp IL-1β. It was found that the agglutinating antibody titre 3 weeks after vaccination was significantly higher in the fish injected with killed bacteria + recombinant IL-1β peptide compared with the titre in fish that were vaccinated with killed bacteria alone (Yin and Kwang 2000).

10.4.4   Poly(lactide-co-glycolide) (PLGA) particles

Nano- and microparticles can be used to improve the antigenicity of weak antigens and thereby act as adjuvants (Morris *et al.* 1994). The encapsulation of vaccines in biocompatible and biodegradable PLGA polymers has been studied for over 20 years. Antigen is released from the microspheres by diffusion through matrix pores and by matrix degradation. Biodegradation rates can be regulated by alterations in polymer composition and molecular weights.

To date, only a few studies have been carried out on fish with regard to uptake and degradation of PLGA particles and the immune response obtained. For the most part these studies have focused on oral administration (O'Donnell *et al.* 1996; Lavelle *et al.* 1997; Tian *et al.* 2008; Altun *et al.*, 2010; Tian and Yu 2011). Recently, an article has been published which reported on parenteral immunisation of a major Indian carp species, rohu (*Labeo rohita*), with a PLGA-encapsulated antigen (Behera *et al.* 2010). Outer membrane proteins (OMPs) of *A. hydrophila* were encapsulated in PLGA microparticles. OMPs were mixed with FIA in an emulsion, and OMPs alone were i.p. injected in rohu. The antibody titres 21 and 42 days post-immunisation were significantly higher in the PLGA-encapsulated antigen group and in the FIA group compared with the OMP group (Behera *et al.* 2010). No significant differences in antibody titres were found between the FIA and PLGA groups.

Oral vaccines encapsulated in PLGA have been used in Japanese flounder (Tian *et al.* 2008; Tian and Yu 2011) and salmonids like rainbow trout (Lavelle *et al.* 1997; Altun *et al.* 2010) and Atlantic salmon (O'Donnell *et al.* 1996). The Japanese flounder, an important aquacultured species, is suffering from infection with lymphocystis disease virus (LCDV). A plasmid encoding the MCP of LCDV was constructed and encapsulated in PLGA. Controls were naked plasmid vaccine and blank PLGA particles (Tian and Yu 2011). The fish were orally intubated, and 28 days post-vaccination the fish were challenged by i.m. injection with LCDV. Vaccine effects were evaluated by observing the presence of lymphocystis nodules. The cumulative percentage of Japanese flounder with nodules after challenge was greatly reduced in the group receiving the plasmid coding for the LCDV protein in PLGA particles in the period between 15 and 120 days post-immunisation (Tian and Yu 2011). The MCP was expressed in tissues of fish vaccinated with plasmid DNA (pDNA) particles. In addition, the levels of antibody in the sera of fish vaccinated with PLGA microcapsules increased until 9 weeks post-immunisation, and then started to decrease (Tian *et al.* 2008).

In rainbow trout, human gamma globulin (HGG) was microencapsulated in PLGA (Lavelle *et al.* 1997). Specific antibodies were detected in the intestinal mucus of fish that had been administered with the microencapsulated antigen after boosting with soluble HGG, but not in fish that had been primed with the soluble antigen. The fate of orally administered free and encapsulated HGG has also been determined in Atlantic salmon. At 15 min after administration, the HGG–PLGA was found in the intestine, resembling the observation for free HGG (O'Donnell *et al.* 1996). The results from this study indicate that orally delivered HGG–PLGA had higher levels and greater persistence of HGG systemically than free HGG. In rainbow trout, oral vaccination (as a feed additive) against lactococcosis was attempted with antigens encapsulated in PLGA particles (Altun *et al.* 2010). The RPS of the PLGA-vaccine group was 63% compared with non-vaccinated fish. Booster vaccination with oral administration of the PLGA-vaccine gave an RPS of more than 60% 120 days after the first vaccination.

10.4.5   Polyinosinic polycytidylic acid (Poly I:C) – a TLR3 agonist

Poly I:C, a double-stranded polyribonucleotide, has been used to induce a type I IFN in many species, including fish (Eaton 1990; Jensen *et al.* 2002; Plant *et al.* 2005). IFNs are cytokines with a major role in the early defence against virus infections. Poly I:C induces a non-specific antiviral state after binding to TLR3 and activation of intracellular signaling events. A novel immunisation method using Poly I:C was tested recently in rainbow trout infected with infectious haematopoietic necrosis virus (IHNV) (Kim *et al.* 2009). When exposed to a virus while in the antiviral state, the fish acquire a specific and protective immunity against the corresponding viral disease. Fish pre-injected with Poly I:C were protected against IHNV challenge 2 days later, and IHNV-specific antibodies were detected in survivors. The survivors showed a 100% survival rate following rechallenge with IHNV at both 21 and 49 days after the primary IHNV challenge (Kim *et al.* 2009). A similar study was performed in the seven band grouper *Epinephelus septemfasciatus* (Nishizawa *et al.* 2009) after immunisation against the nodavirus red-spotted grouper nervous necrosis virus (RGNNV). Fish injected i.m. with 50 mg or more Poly I:C per fish, and then challenged i.m. with RGNNV 2 days post-injection, showed more than 90% survival rate. Surviving fish were re-challenged with RGNNV 3 weeks after the primary challenge. No mortality was observed in the re-challenged Poly I:C–RGNNV group. Antibodies against RGNNV were produced after the primary challenge Poly I:C group. Survivors that were re-challenged with RGNNV showed even higher levels of specific antibodies. In addition, the RGNNV titres in brain tissues of the survivors in the Poly I:C–RGNNV–RGNNV group were all under the detection limit (Nishizawa *et al.* 2009).

The most recent paper on this immunisation procedure is by Takami and coworkers (2010). In this study, Japanese flounder were experimentally infected with viral haemorrhagic septicaemia virus (VHSV). The survival rate after VHSV challenge following Poly I:C administration was 100%. In the group that did not receive Poly I:C injections, all fish died within 9 days. Survival rates of the fish given a secondary VHSV challenge were 100% in the Poly I:C–VHSV group (Poly I:C–VHSV–VHSV group). The control fish (naive–VHSV group) showed 0% survival.

10.4.6   DNA vaccines with immunostimulatory CpG motifs – TLR9 agonists

One of the unique features of DNA vaccines is the ability to stimulate both cellular and humoral immune responses (Weiner and Kennedy 1999). DNA vaccines are administered in the form of pDNA carrying a promoter and the gene of interest. They are used to obtain rapid and long-lasting protection against a variety of diseases caused by intracellular pathogens. pDNA possesses intrinsic immunostimulatory capacity due to the presence of CpG motifs. The CpGs may bind TLR9 which induces production of antiviral molecules. The plasmid is propagated in bacteria, purified and administered by i.m. injection to activate protein expression and induce an immune response against the antigen and protection against the disease.

DNA vaccines encoding the viral glycoproteins of VHSV and IHNV have proved highly efficient in rainbow trout (*O. mykiss*) (Lorenzen *et al.* 1998, 2009; Corbeil *et al.* 1999). A recent study by Skinner and coworkers found a reduced weight of Atlantic salmon vaccinated with an oil-adjuvanted commercial vaccine, but no weight reduction in the group receiving a DNA vaccine (a DNA vaccine encoding the G-protein of IHNV) 106 degree days after administration (Skinner *et al.* 2008). The same authors examined the effects of concurrent vaccine injection (a rhabdovirus-specific DNA vaccine and a polyvalent oil-adjuvanted vaccine against bacterial diseases) in Atlantic salmon (Skinner *et al.* 2010a). The production of anti-*A. Salmonicida* antibodies was significantly greater in the combined vaccine group at 296 degree days post-injection, while production of anti-*Listonella anguillarum* antibodies was significantly greater at 106 degree days post-injection. However, the production of IHNV-specific neutralising antibodies appeared to be delayed or eliminated when the DNA vaccine was injected concurrently with the polyvalent oil-adjuvanted vaccine (Skinner *et al.* 2010a). The same authors found a significant increase in routine metabolic rate following injection of a DNA vaccine concurrently with a polyvalent, oil-adjuvanted vaccine in rainbow trout (Skinner *et al.* 2010b). The lysozyme activity of the combined vaccine group was significantly higher than that of the control group and the DNA vaccine group. Regardless of the increased immune response and transient increase in routine metabolic rate, the overall growth performance was not significantly affected. In yet another study, Atlantic salmon were injected with a DNA vaccine against IHNV alone or concurrently with a polyvalent, oil-adjuvanted bacterial vaccine (Skinner *et al.* 2010c). The influence of supra-physiological levels of cortisol on lysozyme activity and specific antibody titres was examined. The data obtained indicated that chronically supra-physiological levels of plasma cortisol suppressed the innate immune response (lysozyme), but did not affect the antibody production (Skinner *et al.* 2010c).

DNA vaccines encoding *E. tarda* antigens Eta6 and FliC (flagellin) were constructed and injected in Japanese flounders (Jiao *et al.* 2009). A chimeric DNA vaccine including Eta6 covalently linked to FliC, which encodes Eta6 fused in-frame to the C-terminus of FliC, was also made and these vaccines were i.m. injected. FliC–Eta6 induced a higher level of protection than Eta6 alone.

10.4.7   Synthetic CpG oligonucelotides – TLR9 agonists
Bacterial DNA and synthetic oligodeoxynucleotides (ODNs) expressing unmethylated CpG motifs trigger an immunostimulatory cascade that culminates in the maturation, differentiation and proliferation of multiple immune cells, including B and T lymphocytes, natural killer (NK) cells, monocytes, macrophages and dendritic cells. CpG motifs are approximately 20 times less common in mammalian DNA than in microbial DNA, and they stimulate cells that express TLR9. Synthetic CpG oligonucleotides (ODNs) function as adjuvants when co-administered with protein–antigen-based vaccines, and may both accelerate and magnify the immune response. In fish, many studies have been carried out on the immunomodulatory effect of CpGs, but few studies have investigated the adjuvant effect of these molecules. Recently, studies have been performed in salmonids, turbot and the Japanese flounder (Rhodes *et al.* 2004; Carrington and Secombes 2007; Liu *et al.* 2010a,b) and these will be outlined in the following paragraphs.

Chinook salmon (*O. tshawytscha*) reared in the Pacific Northwest of the United States suffer from infection with *Renibacterium salmoninarum*, the causative agent of bacterial kidney disease (BKD) (Rhodes *et al.* 2004). The conclusion from this study was that whole-cell vaccines with or without CpG adjuvants provided limited protection against i.p. challenge by *R. salmoninarum*. However, a combined vaccine of a commercial vaccine (Renogen) with *R. salmoninarum* with or without CpG adjuvant significantly reduced the level of bacterial antigens in the kidneys of naturally infected fish (Rhodes *et al.* 2004).

In a study in rainbow trout, four groups were i.m. injected with a commercially available, unadjuvanted aqueous vaccine against furunculosis containing inactivated cultures of *A. salmonicida* (Aquavac Furovac 5) alone, or containing non-CpG ODN 1982, or CpG ODN 2133 or ODN 2143.

The fish were challenged with i.p. injection of a pathogenic strain of *A. salmonicida* 7 weeks after injection. The only group that showed a significantly lower mortality compared with those injected with Furovac alone (mortality of 52%), or Furovac + non-CpG ODN 1982 (mortality 58%), was the group injected with Furovac + CpG ODN 2143 where only 21% of the fish died (Carrington and Secombes 2007).

The adjuvant effect of CpG motifs was also studied by Liu and coworkers in turbot (*S. maximus*) and Japanese flounder (*P. olivaceus*) (Liu *et al.* 2010a,b). In this work, 16 CpG ODNs were synthesised and examined for the ability to inhibit bacterial dissemination in Japanese flounder blood. Four ODNs with the strongest inhibitory effects were selected and a plasmid pCN6 was constructed that contained the sequences of the four selected ODNs. Japanese flounders were injected i.m. with plasmids pCN6 and pCN3 (control) and PBS. Four weeks post-vaccination the fish were challenged with *A. hydrophila* and mortality was monitored over a period of 20 days. Accumulated mortalities were 30%, 66.7% and 63.3% in pCN6-, pCN3- and PBS-immunised fish respectively (Liu *et al.* 2010b). In addition, fish were vaccinated as above and challenged with *E. tarda* 4 weeks after vaccination. Mortalities were 53.3%, 90%, and 93.3% respectively. The immunoprotection elicited by pCN6 seems to be non-specific, since comparable levels of protection were seen against both *A. hydrophila* and *E. tarda* infections.

In order to analyse the adjuvant effect of CpGs in turbot, fish were vaccinated with a *Vibrio harveyi* recombinant subunit vaccine, DegQ, in different formulations. pCN5 is a CpG that was shown to exibit anti-bacterial effects after i.p. injection. The DegQ was mixed with pCN5 (DegQ–pCN5), and fish were vaccinated by i.p. injection. Appropriate controls were included. At 28 days after vaccination, the fish were challenged by a virulent strain of *V. harveyi*, and accumulated mortalities were recorded. The only vaccine formulation that induced a significant protection was DegQ–pCN5. The adjuvant effect was found to last at least 50 days after vaccination (Liu *et al.* 2010a).

10.4.8   Lipopeptides

Lipoproteins and lipopeptides have been found in a large number of microorganisms, the most prominent being mycobacteria and mycoplasms. These molecules have been found to exibit both a strong innate (inflammatory) response and a long-lasting adaptive immune response in mammals. Very few studies on lipopeptides have been performed in fish. The adjuvant effect of polar glycopeptidolipids in experimental vaccines against *A. salmonicida* was investigated by Hoel and Lillehaug (1997). Polar glycopeptidolipids (pGPL-*Mc*) were extracted from *Mycobacterium chelonae*, which is one of three fish-pathogenic mycobacteria. At 12 weeks post-i.p. vaccination, the antibody response of fish given 0.25 mg kg^{-1} pGPL-*Mc* in combination with

A. salmonicida bacterin was significantly higher than that induced by a non-adjuvanted bacterin. Increased doses of pGPL-*Mc* suppressed the antibody response. No significant side effects were observed in the peritoneal cavity after use of this adjuvant (Hoel and Lillehaug 1997).

10.4.9 Flagellin – a TLR5 agonist

The structural protein of Gram-negative flagella is called flagellin. Flagellin is a potent activator of a broad range of cell types in innate and adaptive immunity. Several studies have demonstrated the ability of flagellin to act as an adjuvant and to promote cytokine production (Mizel and Bates 2010). Flagellin is known to induce immune responses via TLR5 signalling, resulting in a mixed Th1 and Th2 response (Coffman *et al.* 2010), although it is reported that inflammasomes containing Nod-like receptor family CARD domain-containing protein 4 (NLRC4, also known as IPAF) may bind cytosolically located flagellin. Over the last decade, the adjuvant effect of flagellin has been studied in vertebrates, and in the last couple of years it has also been investigated in fish (Wilhelm *et al.* 2006; Jiao *et al.* 2009, 2010b). These studies are detailed in the following paragraphs.

Piscirickettsiosis is a severe disease that has been reported in salmonids. The bacterium *Piscirickettsia salmonis* was isolated in 1989 from a moribund coho salmon, and was found to be the aetiological agent. The pathogen is a Gram-negative obligate intracellular bacterium. The disease has also been reported to affect Atlantic salmon and rainbow trout and other farmed salmonid species (Wilhelm *et al.* 2006). The disease has caused particularly severe problems for the Chilean aquaculture industry. A recombinant subunit vaccine was developed in order to control the disease due to poor responses after treatment with antibiotics. Three experimental formulations were prepared containing two or three recombinant proteins of the bacterium. The formulations were emulsified with one volume of FIA (Wilhelm *et al.* 2006). The highest protective response was obtained with a vaccine formulation containing the subunit of the flagellum and chaperonins Hsp60 and Hsp70 of *P. salmonis*. The results indicated that the use of more than one recombinant protein antigen would lead to a good protective effect against the disease.

Another Gram-negative bacterium *E. tarda*, is a pathogen with a broad host range. Jiao *et al.* have been studying different vaccine concepts in the Japanese flounder to obtain effective protective formulations, based on both recombinant proteins and DNA vaccine constructs (Jiao *et al.* 2009, 2010b). The most promising vaccine concept was one consisting of a chimeric DNA vaccine coding for the *E. tarda* proteins Eta6 fused in-frame to FliC (pCE6). pCE6 was found to induce a significantly higher level of protection than pEta6. Fish immunised with pEta6 and pCE6 produced specific serum antibodies and exhibited enhanced expression of the genes that are involved in both innate and adaptive immune responses (IL-1β, IFN, Mx,

CD8α, MHC Ia, MHC IIa, IgM). The expression levels of most of the genes were significantly higher in pCE6-vaccinated fish than in pEta6-vaccinated fish (Jiao *et al.* 2009).

10.5   Future trends and conclusions

The immune system is unresponsive to most foreign proteins that are injected in a soluble, deaggregated form, but when injected together with an immunostimulatory agent, these foreign proteins can generate a robust immunity and long-lived memory to the antigen (Pulendran 2004). The mechanisms by which adjuvants achieve these effects include the generation of antigen depots, enhancement of presentation of vaccine antigens by activated antigen presenting cells (APCs), and induction of appropriate co-stimulatory molecules to help direct the immune response (Duthie *et al.* 2011). Antigen processing and presentation play a significant role in the adjuvant-mediated increase in vaccine–antigen immunogenicity (Sun *et al.* 2003), and dendritic cells (DCs) are the most important cells for activating naive T cells. In addition, DCs need to be activated in order to mature into potent APCs, as has been shown in mammals.

A decade ago, an *in vitro* study was performed on murine macrophages stimulating both the TLR4 (LPS) and TLR9 (CpG ODN) pathways giving rise to a synergistic TNF production (Gao *et al.* 2001). In another study, Poly I:C (TLR3 agonist) + CpG ODN synergistically stimulated murine macrophages to produce IL-12, TNF, IL-6 and NO and up-regulated MHC class I in splenic DCs (Whitmore *et al.* 2004). Future vaccines may well consist of several targeting adjuvants that may induce more robust responses that result in increased vaccine efficacy.

Nanoparticles can enhance antigen uptake and/or stimulate APCs. PLGA-based nanoparticles loaded with oligonucleotides induced greater cytokine production and T-cell proliferation than oligonucleotide alone (Diwan *et al.* 2004). Also, the amount of antigen required to achieve a high antibody response was an order of magnitude lower than for immunisation with Freund's adjuvant (Dobrovolskaia and McNeil 2007). This fact may be applied in developing more potent vaccines, especially against virus infections, since the amount of antigen is often a limiting factor in such vaccines. Development of vaccine adjuvants has been mostly empirical. Examples are aluminium salts and oil emulsions (Guy 2007). At the current time, the focus is on the induction of well-defined cell-mediated responses in addition to antibodies. For this approach new adjuvants/immunostimulants may be required or combinations thereof (Kornbluth and Stone 2006). However, as discussed by Guy (2007), the induction of an immune response is not a black and white process. As stated in a recent review by Secombes (2008) very few studies can be found on the stimulatory potential of co-stimulatory molecules such as cytokines, either as recombinant molecules or as DNA vaccines encoding both the protective antigen and co-stimulatory molecule(s). This is therefore an area for future investigation.

10.6   Acknowledgements

The support of the Research Council of Norway (contract no. 183204) is greatly appreciated. The authors would also like to acknowledge the funding given by the University of Tromsø ('Tromsø forskningsstiftelse').

10.7   References

ADAMS A, AUCHINACHIE N, BUNDY A, TATNER M, HORNE MT. The potency of adjuvanted injected vaccines in rainbow trout (*Salmo gairdneri* Richardson) and bath vaccines in Atlantic salmon (*Salmo salar* L.) against furunculosis. *Aquaculture* 1988, **69**, 15–26.

AFONSO A, GOMES S, DA SILVA J, MARQUES F, HENRIQUE M. Side effects in sea bass (*Dicentrarchus labrax* L.) due to intraperitoneal vaccination against vibriosis and pasteurellosis. *Fish Shellfish Immunol* 2005, **19**, 1–16.

ALTUN S, KUBILAY A, EKICI S, DIDINEN BI, DILER O. Oral vaccination against lactococcosis in rainbow trout (*Oncorhynchus mykiss*) using sodium alginate and poly (lactide-co-glycolide) carrier. *Kafkas Universitesi Veteriner Fakultesi Dergisi* 2010, **16**, S211-S217.

ANDERSON DP. Adjuvants and immunostimulants for enhancing vaccine potency in fish. *Dev Biol Stand* 1997, **90**, 257–265.

ASHIDA T, OKIMASU E, UI M, HEGURI M, OYAMA Y, AMEMURA A. Protection of Japanese flounder *Paralichthys olivaceus* against experimental edwardsiellosis by formalin-killed *Edwardsiella tarda* in combination with oral administration of immunostimulants. *Fisheries Science* 1999, **65**, 527–530.

BEHERA T, NANDA PK, MOHANTY C, MOHAPATRA D, SWAIN P, DAS BK, ROUTRAY P, MISHRA BK, SAHOO SK. Parenteral immunization of fish, *Labeo rohita* with poly D,L-lactide-co-glycolic acid (PLGA) encapsulated antigen microparticles promotes innate and adaptive immune responses. *Fish Shellfish Immunol* 2010, **28**, 320–325.

BERNARD D, SIX A, RIGOTTIER-GOIS L, MESSIAEN S, CHILMONCZYK S, QUILLET E, BOUDINOT P, BENMANSOUR A. Costimulatory receptors in jawed vertebrates: conserved CD28, odd CTLA4 and multiple BTLAs. *Dev Comp Immunol* 2007, **31**, 255–271.

BIRD S, ZOU J, SAVAN R, KONO T, SAKAI M, WOO J, SECOMBES C. Characterisation and expression analysis of an interleukin 6 homologue in the Japanese pufferfish, *Fugu rubripes*. *Dev Comp Immunol* 2005, **29**, 775–789.

BJÖRNSDOTTIR B, GUDMUNDSDOTTIR S, BAMBIR SH, MAGNADOTTIR B, GUDMUNDSDOTTIR BK. Experimental infection of turbot, *Scophthalmus maximus* (L.), by *Moritella viscosa*, vaccination effort and vaccine-induced side-effects. *J Fish Dis* 2004, **27**, 645–655.

BOUDINOT P, BOUBEKEUR S, BENMANSOUR A. Rhabdovirus infection induces public and private T cell responses in teleost fish. *J Immunol* 2001, **167**, 6202–6209.

BOWDEN TJ, ADAMSON K, MACLACHLAN P, PERT CC, BRICKNELL IR. Long-term study of antibody response and injection-site effects of oil adjuvants in Atlantic halibut (*Hippoglossus hippoglossus* L.). *Fish Shellfish Immunol* 2003, **14**, 363–369.

CAIPANG CM, HIRONO I, AOKI T. Induction of antiviral state in fish cells by Japanese flounder, *Paralichthys olivaceus*, interferon regulatory factor-1. *Fish Shellfish Immunol* 2005, **19**, 79–91.

CAIPANG CMA, HIRONO I, AOKI T. Modulation of the early immune response against viruses by a teleostean interferon regulatory factor-1 (IRF-1). *Comp Biochem Physiol A* 2009, **152**, 440–446.

CARRINGTON AC, SECOMBES CJ. CpG oligodeoxynucleotides up-regulate antibacterial systems and induce protection against bacterial challenge in rainbow trout (*Oncorhynchus mykiss*). *Fish Shellfish Immunol* 2007, **23**, 781–792.

CHEN D, AINSWORTH AJ. Glucan administration potentiates immune defence mechanisms of channel catfish, *Ictalurus punctatus* Rafinesque. *J Fish Dis* 1992, **15**, 295–304.

CHISTIAKOV DA, HELLEMANS B, VOLCKAERT FAM. Review on the immunology of European sea bass *Dicentrarchus labrax*. *Vet Immunol Immunopathol* 2007, **117**, 1–16.

COFFMAN RL, SHER A, SEDER RA. Vaccine adjuvants: putting innate immunity to work. *Immunity* 2010, **33**, 492–503.

CORBEIL S, LAPATRA SE, ANDERSON ED, JONES J, VINCENT B, HSU Y-L, KURATH G. Evaluation of the protective immunogenicity of the N, P, M, NV and G proteins of infectious hematopoietic necrosis virus in rainbow trout *Oncorhynchus mykiss* using DNA vaccines. *Dis Aquat Org* 1999, **39**, 29–36.

DALMO RA, BØGWALD J. β-Glucans as conductors of immune symphonies. *Fish Shellfish Immunol* 2008, **25**, 384–396.

DEBAULNEY MO, QUENTEL C, FOURNIER V, LAMOUR F, LE GOUVELLO R. Effect of long-term oral administration of β-glucan as an immunostimulant or an adjuvant on some non-specific parameters of the immune response of turbot *Scophthalmus maximus*. *Dis Aquatic Org* 1996, **26**, 139–147.

DIEHL S, RINCÓN M. The two faces of IL-6 on Th1/Th2 differentiation. *Mol Immunol* 2002, **39**, 531–536.

DIWAN M, ELAMANCHILI P, CAO M, SAMUEL J. Dose sparing of CpG oligodeoxynucleotide vaccine by nanoparticle delivery. *Curr Drug Deliv* 2004, **1**, 405–412.

DOBROVOLSKAIA MA, MCNEIL SE. Immunological properties of engineered nanomaterials. *Nature Nanotechnol* 2007, **2**, 469–478.

DUTHIE MS, WINDISH HP, FOX CB, REED SG. Use of defined TLR ligands as adjuvants within human vaaccines. *Immunological Rev* 2011, **239**, 178–196.

EATON WD. Antiviral activity in four species of salmonids following exposure to poly inosinic:cytidylic acid. *Dis Aquat Org* 1990, **9**, 193–198.

EVENSEN Ø, BRUDESETH B, MUTOLOKI S. The vaccine formulation and its role in inflammatory processes in fish. *Dev Biol Stand* 2005, **121**, 117–125.

FANG H-M, GE R, SIN YM. Cloning, characterisation and expression of *Aeromonas hydrophila* major adhesin. *Fish Shellfish Immunol* 2004, **16**, 645–658.

FIGUERAS A, SANTAREM MM, NOVOA B. Influence of the sequence of administration of β-glucans and a *Vibrio damsela* vaccine on the immune response of turbot (*Scophthalmus maximus* L.). *Vet Immunol Immunopathol* 1998, **64**, 59–68.

FREUND J, CASALS J, HOSMER EP. Sensitization and antibody formation after injection of tubercle bacilli and paraffin oil. *Proc Soc Exp Biol Med* 1937, **37**, 509–513.

GAO JJ, XUE Q, PAPASIAN CJ, MORRISON DC. Bacterial DNA and lipopolysaccharide induce synergistic production of TNF-a through a post-transcriptional mechanism. *J Immunol* 2001, **166**, 6855–6860.

GLENNY AT, POPE CG, WADDINGTON H, WALLACE U. Immunological notes. XXIII. The antigenic value of toxoid precipitated by potassium alum. *J Pathol Bacteriol* 1926, **29**, 31–40.

GRABOWSKI LD, LAPATRA SE, CAIN KD. Systemic and mucosal antibody response in tilapia, *Oreochromis niloticus* (L.), following immunization with *Flavobacterium columnare*. J Fish Dis 2004, **27**, 573–581.

GUDMUNDSDOTTIR BK, BJÖRNSDOTTIR B. Vaccination against atypical furunculosis and winter ulcer disease of fish. *Vaccine* 2007, **25**, 5512–5523.

GUY B. The perfect mix: recent progress in adjuvant research. *Nature Rev Microbiol* 2007, **5**, 505–517.

HAMNDY S, MOLAVI O, MA ZS, HADDADI A, ALSHAMSAN A, GOBTL Z, ELHASI S, SAMUEL J. LAVASANIFAR A. Co-delivery of cancer-associated antigen and Toll-like receptor 4 ligand in PLGA nanoparticles induces potent CD8(+) T cell-mediated antitumor immunity. *Vaccine* 2008, **26**, 5046–5057.

HOEL K, LILLEHAUG A. Adjuvant activity of polar glycopeptidolipids from *Mycobacterium chelonae* in experimental vaccines against *Aeromonas salmonicida* in salmonid fish. *Fish Shellfish Immunol* 1997, **7**, 365–376.

HÖGFORS E, PULLINEN K-R, MADETOJA J, WIKLUND T. Immunization of rainbow trout, *Oncorhynchus mykiss* (Walbaum), with a low molecular mass fraction isolated from *Flavobacterium psychrophilum*. *J Fish Dis* 2008, **31**, 899–911.

ILIEV DB, CASTELLANA B., MACKENZIE S, PLANAS JV, GOETZ FW. Cloning and expression analysis of an IL-6 homolog in rainbow trout (*Oncorhynchus mykiss*), *Mol Immunol* 2007, **44**, 1803–1807.

JANSEN T, HOFMANS MPM, THEELEN MJG, SCHIJNS VEJC. Structure-activity relations of water-in-oil vaccine formulations and induced antigen-specific antibody responses. *Vaccine* 2005, **23**, 1053–1060.

JENSEN I, ALBUQUERQUE A, SOMMER AI, ROBERTSEN B. Effect of poly I:C on the expression of Mx proteins and resistance against infection by infectious salmon anaemia virus in Atlantic salmon. *Fish Shellfish Immunol* 2002, **13**, 311–326.

JIAO X-D, ZHANG M, HU Y-H, SUN L. Construction and evaluation of DNA vaccines encoding *Edwardsiella tarda* antigens. *Vaccine* 2009, **27**, 5195–5202.

JIAO X-D, CHENG S, HU Y-H, SUN L. Comparative study of the effects of aluminium adjuvants and Freund's incomplete adjuvant on the immune response to an *Edwardsiella tarda* major antigen. *Vaccine* 2010a, **28**, 1832–1837.

JIAO X-D, HU Y-H, SUN L. Dissection and localization of the immunostimulating domain of *Edwardsiella tarda* FliC. *Vaccine* 2010b, **28**, 5635–5640.

JIMENEZ N, COLL J, SALGUERO BFJ, TAFALLA C. Co-injection of interleukin 8 with the glycoprotein gene from viral haemorrhagic septicaemia virus (VHSV) modulates the cytokine response in rainbow trout (*Oncorhynchus mykiss*). *Vaccine* 2006, **24**, 5615–5626.

KAJITA Y, SAKAI M, ATSUTA S, KOBAYASHI M. Immunopotentiation activity of Freund's complete adjuvant in rainbow trout *Oncorhynchus mykiss*. *Nippon Suisan Gakkaishi* 1992, **58**, 433–437.

KAMILYA D, MAITI TK, JOARDAR SN, MAL BC. Adjuvant effect of mushroom glucan and bovine lactoferrin upon *Aeromonas hydrophila* vaccination in catla, *Catla catla* (Hamilton). *J Fish Dis* 2006, **29**, 331–337.

KAWAKAMI H, SHINOHARA N, SAKAI M. The non-specific stimulation and adjuvant effects of *Vibrio anguillarum* bacterin, M-glucan, chitin and Freunds complete adjuvant against *Pasteurella piscicida* infection in yellowtail. *Fish Pathol* 1998, **33**, 287–292.

KIM HJ, OSEKO N, NISHIZAWA T, YOSHIMIZU M. Protection of rainbow trout from infectious hematopoietic necrosis (IHN) by injection of infectious pancreatic necrosis virus (IPNV) or Poly(I:C). *Dis Aquat Org* 2009, **83**, 105–113.

KOJ A. Initiation of acute phase response and synthesis of cytokines. *Biochim Biophys Acta* 1996, **1317**, 84–94.

KOPPANG EO, BJERKAS I, HAUGARVOLL E, CHAN EKL, SZABO NJ, ONO N, AKIKUSA B, JIRILLO E, POPPE TT, SVEIER H, TORUD B, SATOH M. Vaccination-induced systemic autoimmunity in farmed Atlantic salmon. *J Immunol* 2009, **181**, 4807–4814

KORENAGA H, KONO T, SAKAI M. Isolation of seven IL-17 family genes from the Japanese pufferfish *Takifugu rubripes*. *Fish Shellfish Immunol* 2010, **28**, 809–818.

KORNBLUTH RS, STONE GW. Immunostimulatory combinations: designing the next generation of vaccine adjuvants. *J. Leukocyte Biol* 2006, **80**, 1084–1102.

KUBILAY A, ALTUN S, ULUKOY G, EKICI S, DILER O. Immunization of rainbow trout (*Oncorhynchus mykiss*) against *Lactococcus garvieae* using vaccine mixtures. *Israeli J Aquaculture Bamidgeh* 2008, **60**, 268–273.

KUMARI J, LARSEN AN, BØGWALD J, DALMO RA. Interleukin-17D in Atlantic salmon (*Salmo salar*): Molecular characterization, 3D modelling and promoter analysis. *Fish Shellfish Immunol* 2009, **27**, 647–659.

KURATH G. Biotechnology and DNA vaccines for aquatic animals. *Rev Sci Techn Int Epizoot* 2008, **27**, 175–196

LAMAS J, SANMARTIN ML, PARAMA AI, CASTRO R, CABALEIRO S, DE OCENDA MVR, BARJA JL, LEIRO J. Optimization of an inactivated vaccine against a scuticociliate parasite of turbot: effect of antigen, formalin and adjuvant concentration on antibody response and protection against the pathogen. *Aquaculture* 2008, **278**, 22–26.

LAPATRA SE, PLANT KP, ALCORN S, OSTLAND V, WINTON J. An experimental vaccine against *Aeromonas hydrophila* can induce protection in rainbow trout, *Oncorhynchus mykiss* (Walbaum). *J Fish Dis* 2010, **33**, 143–151.

LAVELLE EC, JENKINS PG, HARRIS JE. Oral immunization of rainbow trout with antigen microencapsulated in poly(DL-lactide-co-glycolide) microparticles. *Vaccine* 1997, **15**, 1070–1078.

LAWRENCE GW, SAUL A, GIDDY AJ, KEMP R, PYE D. Phase I trial in humans of an oil-based adjuvant seppic montanide isa 720. *Vaccine* 1997, **15**, 176–178.

LILLEHAUG A, LUNESTAD BT, GRAVE K. Epidemiology of bacterial diseases in Norwegian aquaculture – a description based on antibiotic prescription data for the ten-year period 1991 to 2000. *Dis Aquat Org* 2003, **53**, 115–125.

LIU C-S, SUN Y, HU Y-H, SUN L. Identification and analysis of the immune effects of CpG motifs that protect Japanese flounder (*Paralichthys olivaceus*) against bacterial infection. *Fish Shellfish Immunol* 2010a, **29**, 279–285.

LIU C-S, SUN Y, HU Y-H, SUN L. Identification and analysis of a CpG motif that protects turbot (*Scophthalmus maximus*) against bacterial challenge and enhances vaccine-induced specific immunity. *Vaccine* 2010b, **28**, 4153–4161.

LORENZEN N, LORENZEN E, EINER-JENSEN K, HEPPELL J, WU T, DAVIS H. Protective immunity to VHS in rainbow trout (*Oncorhynchus mykiss*, Walbaum) following DNA vaccination. *Fish Shellfish Immunol* 1998, **8**, 261–270.

LORENZEN E, EINER-JENSEN K, RASMUSSEN JS, KJÆR TE, COLLET B, SECOMBES CJ, LORENZEN N. The protective mechanisms induced by a fish rhabdovirus DNA vaccine depend on temperature. *Vaccine* 2009, **27**, 3870–3880.

MACKENZIE SA, ROHER N, BOLTANA S, GOETZ FW. Peptidoglycan, not endotoxin, is the key mediator of cytokine gene expression induced in rainbow trout macrophages by crude LPS. *Mol Immunol* 2010, **47**, 1450–1457.

MANICASSAMY S, PULENDRAN B. Modulation of adaptive immunity with Toll-like receptors. *Sem Immunol* 2009, **21**, 185–193.

MARTIN SAM, COLLET B, MACKENZIE S, EVENSEN Ø, SECOMBES CJ. Genomic tools for examining immune function in salmonid fish. *Rev Fish Sci* 2008, **16**, 112–118.

MCKINNEY RW, DAVENPORT FM. Studies on the mechanism of action of emulsified vaccines. *J Immunol* 1961, **86**, 91–100.

MIDTLYNG PJ, LILLEHAUG A. Growth of Atlantic salmon *Salmo salar* after intraperitoneal administration of vaccines containing adjuvants. *Dis Aquat Org* 1998, **32**, 91–97.

MIDTLYNG PJ, REITAN LJ, SPEILBERG L. Experimental studies on the efficacy and side-effects of intraperitoneal vaccination of Atlantic salmon (*Salmo salar* L.) against furunculosis. *Fish Shellfish Immunol* 1996, **6**, 335–350.

MIKKELSEN H, SCHRØDER MB, LUND V. Vibriosis and atypical furunculosis vaccines; efficacy, specificity and side effects in Atlantic cod, *Gadus morhua* L. *Aquaculture* 2004, **242**, 81–91.

MIKKELSEN H, LUND V, LARSEN R, SEPPOLA M. Vibriosis vaccines based on various sero-subgroups of *Vibrio anguillarum* O2 induce specific protection in Atlantic cod. *Fish Shellfish Immunol* 2011, **30**, 330–339.

MIZEL SB, BATES JT. Flagellin as an adjuvant: cellular mechanisms and potential. *J Immunol* 2010, **185**, 5677–5682.

MORRIS W, STEINHOFF MC, RUSSELL PK. Potential of polymer microencapsulation technology for vaccine innovation. *Vaccine* 1994, **12**, 5–11.

MULVEY B, LANDOLT ML, BUSCH RA. Effects of potassium aluminium sulphate (alum) used in an *Aeromonas salmonicida* bacterin in Atlantic salmon, *Salmo salar* L. *J Fish Dis* 1995, **18**, 495–506.

MUTOLOKI S, COOPER GA, MARJARA IS, KOOP BF, EVENSEN Ø. High gene expression of inflammatory markers and IL-17A correlates with severity of injection site reactions of Atlantic salmon vaccinated with oil-adjuvanted vaccines. *BMC Genomics* 2010, **11**, 336

NIKL L, ALBRIGHT LJ, EVELYN TPT. Influence of seven immunostimulants on the immune response of coho salmon to *Aeromonas salmonicida*. *Dis Aquat Org* 1991, **12**, 7–12.

NINOMIYA K, YAMAMOTO M. Efficacy of oil-adjuvanted vaccines for bacterial hemorrhagic ascites in ayu *Plecoglossus altivelis*. *Fish Pathol* 2001, **36**, 183–185.

NISHIZAWA T, TAKAMI I, KOKAWA Y, YOSHIMIZU M. Fish immunization using a synthetic double-stranded RNA Poly(I:C), an interferon inducer, offers protection against RGNNV, a fish nodavirus. *Dis Aquat Org* 2009, **83**, 115–122.

O'DONNELL GB, REILLY P, DAVIDSON GA, ELLIS AE. The uptake of human gamma globulin incorporated into poly (D,L-lactide-co-glycolide) microparticles following oral intubation in Atlantic salmon, *Salmo salar* L. *Fish Shellfish Immunol* 1996, **6**, 507–520.

OLAVARRIA VH, GALLARDO L, FIGUEROA JE, MULERO V. Lipopolysaccharide primes the respiratory burst of Atlantic salmon SHK-1 cells through protein kinase C-mediated phosphorylation of p47phox. *Dev Comp Immunol* 2010, **34**, 1242–1253.

OLIVIER G, EVELYN TPT, LALLIER R. Immunity to *Aeromonas salmonicida* in Coho salmon (*Oncorhynchus kisutch*) induced by modified Freund's complete adjuvant: its non-specific nature and the probable role of macrophages in the phenomenon. *Dev Comp Immunol* 1985, **9**, 419–432.

PARTULA S, GUERRA A DE, FELLAH JS, CHARLEMAGNE J. Structure and diversity of the T-cell antigen receptor beta-chain in a teleost fish. *J Immunol* 1995, **155**, 699–706.

PEATMAN E, LIU ZJ. Evolution of chemokines in teleost fish: a case study in gene duplication and implications for immune diversity. *Immunogenetics* 2007, **59**, 613–623.

PEDDIE S, MCLAUCHLAN PE, ELLIS AE, SECOMBES CJ. Effect of intraperitoneally administered IL-1β-derived peptides on resistance to viral haemorrhagic septicaemia in rainbow trout *Oncorhynchus mykiss*. *Dis Aquat Org* 2003, **56**, 195–200.

PLANT KP, HARBOTTLE H, THUNE RL. Poly I:C induces an antiviral state against Ictalurid Herpesvirus 1 and Mx1 transcription in the channel catfish (*Ictalurus punctatus*). *Develop Comp Immunol* 2005, **29**, 627–635.

PLOUFFE DA, HANINGTON PC, WALSH JG, WILSON EC, BELOSEVIC M. Comparison of select innate immune mechanisms of fish and mammals. *Xenotransplantation* 2005, **4**, 266–277.

POOBALANE S, THOMPSON KD, ARDO L, VERJAN N, HAN H-J, JENEY G, HIRONO I, AOKI T, ADAMS A. Production and efficacy of an *Aeromonas hydrophila* recombinant S-layer protein vaccine for fish. *Vaccine* 2010, **28**, 3540–3547.

PULENDRAN B. Modulating vaccine responses with dendritic cells and Toll-like receptors. *Immunol Rev* 2004, **199**, 227–250.

RAIDA MK, BUCHMANN K. Bath vaccination of rainbow trout (*Oncorhynchus mykiss* Walbaum) against *Yersinia ruckeri*: effects of temperature on protection and gene expression. *Vaccine* 2008a, **26**, 1050–1062.

RAIDA MK, BUCHMANN K. Development of adaptive immunity in rainbow trout, *Oncorhynchus mykiss* (Walbaum) surviving an infection with *Yersinia ruckeri*. *Fish Shellfish Immunol* 2008b, **25**, 533–541.

RAIDA MK, BUCHMANN K. Innate immune response in rainbow trout (*Oncorhynchus mykiss*) against primary and secondary infections with *Yersinia ruckeri* O1. *Dev Comp Immunol* 2009, **33**, 35–45.

RANDELLI E, BUONOCORE F, CASANI D, FAUSTO AM, SCAPIGLIATI G. An immunome gene panel for transcriptomic analysis of immune defence activities in the teleost sea bass (*Dicentrarchus labrax* L.): a review. *Italian J Zool* 2009, **76**, 146–157.

RAVELO C, MAGARINOS B, HERRERO MC, COSTA L, TORANZO AE, ROMALDE JL. Use of adjuvanted vaccines to lengthen the protection against lactococcosis in rainbow trout (*Oncorhynchus mykiss*). *Aquaculture* 2006, **251**, 153–158.

REBL A, GOLDAMMER T, SEYFERT HM. Toll-like receptor signaling in bony fish. *Vet Immunol Immunopathol* 2010, **134**, 139–150.

RHODES LD, RATHBONE CK, CORBETT SC, HARRELL LW, STROM MS. Efficacy of cellular vaccines and genetic adjuvants against bacterial kidney disease in chinook salmon (*Oncorhynchus tshawytscha*). *Fish Shellfish Immunol* 2004, **16**, 461–474.

ROBERTSEN B. Modulation of the non-specific defence of fish by structurally conserved microbial polymers. *Fish Shellfish Immunol* 1999, **9**, 269–290.

RØRSTAD G, AASJORD PM, ROBERTSEN B. Adjuvant effect of a yeast glucan in vaccines against furunculosis in Atlantic salmon (*Salmo salar* L.). *Fish Shellfish Immunol* 1993, **3**, 179–190.

SAKAI M. Current research status of fish immunostimulants. *Aquaculture* 1999, **172**, 63–92.

SAMUEL M, LAM TJ, SIN YM. Effect of laminaran [β(1,3)-D-glucan] on the protective immunity of blue gourami, *Trichogaster trichopterus* against *Aeromonas hydrophila*. *Fish Shellfish Immunol* 1996, **6**, 443–454.

SAMUELSEN OB, NERLAND AH, JØRGENSEN T, SCHRØDER MB, SVÅSAND T, BERGH Ø. Viral and bacterial diseases of Atlantic cod, *Gadus morhua*, their prophylaxis and treatment: a review. *Dis Aquat Org* 2006, **71**, 239–254.

SANCHEZ E, COLL J, TAFALLA C. Expression of inducible CC chemokines in rainbow trout (*Oncorhynchus mykiss*) in response to a viral haemorrhagic septicaemia virus (VHSV) DNA vaccine and interleukin 8. *Develop Comp Immunol* 2007, **31**, 916–926.

SCHIJNS VE, TANGERAS A. Vaccine adjuvant technology: from theoretical mechanisms to practical approaches. *Dev Biol (Basel)* 2005, **121**, 127–134.

SECOMBES C. Will advances in fish immunology change vaccination strategies? *Fish Shellfish Immunol* 2008, **25**, 409–416.

SELVARAJ V, SAMPATH K, SEKAR V. Administration of yeast glucan enhances survival and some non-specific and specific immune parameters in carp (*Cyprinus carpio*) infected with *Aeromonas hydrophila*. *Fish Shellfish Immunol* 2005, **19**, 293–306.

SELVARAJ V, SAMPATH K, SEKAR V. Adjuvant and immunostimulatory effects of β-glucan administration in combination with lipopolysaccharide enhances survival and some immune parameters in carp challenged with *Aeromonas hydrophila*. *Vet Immunol Immunopathol* 2006, **114**, 15–24.

SEPULCRE MP, ALCARAZ-PÉREZ F, LÓPEZ-MUNOZ A, ROCA FJ, MESEGUER J, CAYUELA ML, MULERO V. Evolution of lipopolycaccharide recognition and signaling: fish TLR4 does not recognize LPS and negatively regulates NF-κB activation. *J Immunol* 2009, **182**, 1836–1845.

SHIMAHARA Y, HUANG Y-F, TSAI M-A, WANG P-C, CHEN S-C. Immune response of large-mouth bass, *Micropterus salmoides*, to whole cells of different *Nocardia seriolae* strains. *Fish Sci* 2010, **76**, 489–494.

SITJA-BOBADILLA A, PALENZUELA O, ALVAREZ-PELLITERO P. Immune response of turbot, *Psetta maxima* (L.) (Pisces: Teleostei), to formalin-killed scuticociliates (*Ciliophora*) and adjuvanted formulations. *Fish Shellfish Immunol* 2008, **24**, 1–10

SKINNER LA, SCHULTE PM, LAPATRA SE, MCKINLEY RS. Growth and performance of Atlantic salmon, *Salmo salar* L., following administration of a rhabdovirus DNA vaccine alone or concurrently with an oil-adjuvanted, polyvalent vaccine. *J Fish Dis* 2008, **31**, 687–697.

SKINNER LA, LAPATRA SE, ADAMS A, THOMPSON KD, BALFRY SK, MCKINLEY RS, SCHULTE PM. Concurrent injection of a rhabdovirus-specific DNA vaccine with a polyvalent, oil-adjuvanted vaccine delays the specific anti-viral immune response in Atlantic salmon, *Salmo salar* L. *Fish Shellfish Immunol* 2010a, **28**, 579–586.

SKINNER LA, SCHULTE PM, BALFRY SK, MCKINLEY RS, LAPATRA SE. The association between metabolic rate, immune parameters, and growth performance of rainbow trout, *Oncorhynchus mykiss* (Walbaum), following the injection of a DNA vaccine alone and concurrently with a polyvalent, oil-adjuvanted vaccine. *Fish Shellfish Immunol* 2010b, **28**, 387–393.

SKINNER LA, LAPATRA SE, ADAMS A, THOMPSON KD, BALFRY SK, MCKINLEY RS, SCHULTE PM. Supra-physiological levels of cortisol suppress lysozyme but not antibody response in Atlantic salmon, *Salmo salar* L., following vaccine injection. *Aquaculture* 2010c, **300**, 223–230.

SOKOL CL, BARTON GM, FARR AG, MEDZHITOV R. A mechanism for the initiation of allergen-induced T helper type 2 responses. *Nature Immunol* 2008, **9**, 310–318.

SOLTANI M, ALISHAHI M, MIRZARGAR S, NIKBAKHT GH. Vaccination of rainbow trout against *Streptococcus iniae* infection: comparison of different routes of administration and different vaccines. *Iranian J Fish Sci* 2007, **7**, 129–140.

SOMAMOTO T, YOSHIURA Y, SATO A, NAKAO M, NAKAMSHI T, OKAMOTO N, OTOTAKE M. Expression profiles of TCR beta and CD8 alpha mRNA correlate with virus-specific cell-mediated cytotoxic activity in ginbuna crucian carp. *Virology* 2006, **348**, 370–377.

SOMMERSET I, KROSSØY B, BIERING E, FROST P. Vaccines for fish in aquaculture. *Expert Rev Vaccines* 2005, **4**, 89–101.

STAR B, NEDERBRAGT AJ, JENTOFT S, GRIMHOLT U, MALMSTRØM M, GREGERS TF, ROUNGE TB, PAULSEN J, SOLBAKKEN MH, SHARMA A, WETTEN OF, LANZÉN A, WINER R, KNIGHT J, VOGEL J-H, AKEN B, ANDERSEN Ø, LAGESEN K, TOOMING-KLUNDERUD A, EDVARDSEN RB, TINA KG, ESPELUND M, NEPAL C, PREVITI C, KARLSEN BO, MOUM T, SKAGE M, BERG PR, GJØEN T, KUHL H, THORSEN J, MALDE K, REINHARDT R, DU L, JOHANSEN SD, SEARLE S, LIEN S, NILSEN F, JONASSEN I, OMHOLT SW, STENSETH NS, JAKOBSEN KS. The genome sequence of Atlantic cod reveals a unique immune system. *Nature* 2011, **477**, 207–210.

STILLS JR HF. Adjuvants and antibody production: dispelling the myths associated with Freund's complete and other adjuvants. *ILAR J* 2005, **46**, 280–293.

SUN H, POLLOCK KGJ, BREWER JM. Analysis of the role of vaccine adjuvants in modulating dendritic cell activation and antigen presentation *in vitro*. *Vaccine* 2003, **21**, 849–855.

TAKAMI I, KWON SR, NISHIZAWA T, YOSHIMIZU M. Protection of Japanese flounder *Paralichthys olivaceus* from viral hemorrhagic septicaemia (VHS) by Poly(I:C) immunization. *Dis Aquat Org* 2010, **89**, 109–115.

TIAN J, YU J. Poly(lactic-co-glycolic acid) nanoparticles as candidate DNA vaccine carrier for oral immunization of Japanese flounder (*Paralichthys olivaceus*) against lymphocystis disease virus. *Fish Shellfish Immunol* 2011, **30**, 109–117.

TIAN J, SUN X, CHEN X, YU J, QU L, WANG L. The formulation and immunisation of oral poly(DL-lactide-co-glycolide) microcapsules containing a plasmid vaccine against lymphocystis disease virus in Japanese flounder (*Paralichthys olivaceus*). *Int Immunopharmacol* 2008, **8**, 900–908.

TSUJITA T, TSUKADA H, NAKAO M, OSHIUMI H, MATSUMOTO M, SEYA J. Sensing bacterial falgellin by membrane and soluble ortologs of Toll-like receptor 5 in rainbow trout (*Onchorhynchus mykiss*). *J Biol Chem* 2004, **19**, 485888–485897.

TYLER JW, KLESIUS PH. Protection against enteric septicemia of catfish (*Ictalurus punctatus*) by immunization with the R-mutant, *Escherichia coli* (J5). *American J Vet Res* 1994, **55**, 1256–1260.

VAN GELDEREN R, CARSON J, NOWAK B. Experimental vaccination of Atlantic salmon (*Salmo salar* L.) against marine flexibacteriosis. *Aquaculture* 2009, **288**, 7–13.

WANG TH, MARTIN SAM, SECOMBES CJ. Two interleukin-17C-like genes exist in rainbow trout *Oncorhynchus mykiss* that are differentially expressed and modulated. *Dev Comp Immunol* 2010, **34**, 491–500.

WEAVER CT, HARRINGTON LE, GAVRIELI M, MURPHY KM. Th17: An effector CD4 T cell lineage with regulatory T cell ties. *Immunity* 2006, **24**, 677–688.

WEINER DB, KENNEDY RC. Genetic vaccines. *Scientific American* 1999, **281**, 50–57.

WILHELM V, MIQUEL A, BURZIO LO, ROSEMBLATT M, ENGEL E, VALENZUELA S, PARADA G, VALENZUELA PDT. A vaccine against the salmonid pathogen *Piscirickettsia salmonis* based on recombinant proteins. *Vaccine* 2006, **24**, 5083–5091.

WHITMORE MM, DEVEER MJ, EDLING A, OATES RK, SIMONS B, LINDNER B, WILLIAMS BR. Synergistic activation of innate immunity by double-stranded RNA and CpG DNA promotes enhanced antitumor activity. *Cancer Res* 2004, **64**, 5850–5860.

WHYTE SK. The innate immune responses of finfish – a review of current knowledge. *Fish Shellfish Immunol* 2007, **23**, 1127–1151.

YEWDELL JW. Designing CD8+ T cell vaccines: it's not rocket science (yet). *Curr Opin Immunol* 2010, **22**, 402–410.

YIN Z, KWANG J. Carp interleukin-1β in the role of an immuno-adjuvant. *Fish Shellfish Immunol* 2000, **10**, 375–378.

ZHANG Z, SWAIN T, BØGWALD J, DALMO, RA, KUMARI, J. Bath immunostimulation of rainbow trout (*Oncorhynchus mykiss*) fry induces enhancement of inflammatory cytokine transcripts, while repeated bath induce no changes. *Fish Shellfish Immunol* 2009, **26**, 677–684.

Part III

Development of specific pathogen-free populations and novel approaches for disease control

11

Development of specific pathogen-free (SPF) shrimp stocks and their application to sustainable shrimp farming

D. V. Lightner and R. M. Redman, University of Arizona, USA

Abstract: According to the FAO, global production of marine penaeid shrimp from farms reached nearly 3.5 million tonnes in 2009, accounting for nearly half of the world's total shrimp supply. With most of the world's shrimp fisheries at maximum sustainable yield, the ratio of farmed to fished shrimp appears likely to continue to increase. This production is from a very young food producing industry that began to emerge in the mid-1970s. The remarkable growth of sustainable shrimp farming has been accomplished in part through the successful development of domesticated shrimp stocks, many of which are free of specific diseases, and the development of the necessary infrastructure, in terms of biosecurity, diagnostic methods and trained personnel, to successfully prevent disease or to manage disease outbreaks when they occur.

Key words: specific pathogen-free (SPF), shrimp diseases, biosecurity, shrimp farming, domesticated stocks.

11.1 Introduction

Disease has had a major impact on shrimp aquaculture since it became a significant commercial entity in the 1970s. Diseases due to viruses, rickettsial-like bacteria, true bacteria, protozoa, and fungi have emerged as major diseases of farmed shrimp (see reviews by Brock and Lightner 1990; Lightner 1993, 1996a; Brock and Main 1994; Flegel 1997, 2006; Flegel and Alday-Sanz 1998; Lightner *et al.* 2009; Walker and Mohan 2009). Many of the bacterial, fungal and protozoan-caused diseases are now managed using improved culture practices, routine sanitation, and the use of chemotherapeutics. However, the virus diseases have been far more problematic to manage and they have been responsible for the most costly epizootics (Moss 2002; Lightner 2005; Flegel 2006; Lightner *et al.* 2009; Walker and

Mohan 2009). Notorious viral epizootics include the Taura syndrome pandemic that began in 1991–92 when the disease emerged in Ecuador, and the subsequent white spot disease pandemics that emerged at about the same time in SE Asia. Because of their socioeconomic significance to shrimp farming, five of the nine crustacean diseases currently listed by the World Organisation for Animal Health Office International des Epizooties, (OIE) are virus diseases of shrimp (OIE 2009a, 2009b).

In the wake of the extraordinary losses that occurred as a result of the viral pandemics, the industry began to mature into a much more sustainable, technology-based industry. The industry has largely recovered from the major viral pandemics and its production has begun a new phase of rapid growth (FAO 2006). The adoption of new shrimp farming technologies and the abandonment of practices which posed high disease risks have contributed to the industry's recovery and current expansion (FAO/NACA/UNEP/WB/WWF 2006). Among the most notable changes in culture practices has been the shift of the industry away from using wild stocks for seed production to the use of domesticated stocks (Moss 2002; Lightner 2003a, 2003b, 2005; FAO 2006; Lightner *et al.* 2009; Moss and Moss 2009). This has been a consequence of the ever-increasing incidence of diseases such as white spot syndrome virus (WSSV) and infectious hypodermal and hematopoietic necrosis virus (IHHNV) in wild shrimp stocks, which has made the collection of wild postlarvae (PLs) and adult broodstock, for the production of PLs for use as seed stock, a risky practice (Lightner 2005; Flegel 2006; Lightner *et al.* 2009; Walker and Mohan 2009). With the declining dependence of the industry on wild stocks in Asia and in the Americas, the use of domesticated lines of specific pathogen-free (SPF) *Litopenaeus vannamei* (the Pacific white shrimp) recently surpassed *Penaeus monodon* (the giant black tiger shrimp) as the dominant farmed shrimp species in Asia (FAO 2006). This paradigm switch in the species being farmed occurred within 5 years after SPF *L. vannamei* stocks were introduced in quantity to Asia. The use of SPF *L. vannamei* has led to improved production and predictable crops virtually everywhere that was once dominated by the culture of *Fenneropenaeus chinensis* (the Chinese white shrimp) or *P. monodon*.

The terms 'SPF' and 'biosecurity' were not in widespread use in the shrimp farmer's vocabulary a decade ago, but today most farms incorporate some form of these practices for disease exclusion and management (Lightner and Pantoja 2001; Moss 2002; Lee and O'Bryen 2003; Lightner 2005; Scarfe *et al.* 2006). Despite the significant challenges posed by disease, the shrimp farming industry has responded to the challenges posed by disease and it has developed methods to manage its diseases and mature into a sustainable industry. Adoption of the SPF concept in the domestication of *L. vannamei* and development of the species for aquaculture were among the milestones that led to the industry's current explosive growth and apparent sustainability (FAO 2006). How the industry developed and adopted the SPF concept is the topic of this review.

11.2 A historical perspective on the concept of domesticated specific pathogen-free (SPF) shrimp

The term SPF was in widespread use in large number of terrestrial animal, aquatic animal and plant agriculture industries prior to its being applied to shrimp aquaculture (Pruder *et al.* 1995). Among the more successful SPF culture practices are those developed and used in the poultry, swine and trout producing industries. SPF culture practices for other animals were commonplace for many years before they were adapted for use with shrimp (Wyban *et al.* 1992; Carr *et al.* 1994, 1996; Pruder *et al.* 1995; Lotz 1997a, 1997b; Zavala 1999). However, the application of the SPF concept to shrimp farming is a relatively recent event and it occurred well after the technologies had been developed that were necessary to close the life cycle of the penaeid shrimp in the laboratory and begin the process of producing domesticated breeding lines of penaeid shrimp. Because the term SPF is poorly understood and often misused, the term 'high health' has also been borrowed from other animal-producing industries for use with shrimp to designate shrimp stocks that were developed as SPF, and which may be free of infection by specific disease agents, but which are no longer contained within a designated biosecure SPF facility (Pruder *et al.* 1995; Moss *et al.* 2003; Lightner *et al.* 2009).

As early as the mid-1970s a number of penaeid shrimp research programs were developing culture systems and methods to close the life cycle of several penaeid shrimp species in captivity. Some early research groups and institutions were successful in growing, maturing, mating, spawning, and producing progeny from founder shrimp stocks that had been reared for a full generation in captivity. Successes by some of the early pioneers in the quest to close the penaeid shrimp life cycle in captivity were documented in the literature of the period by several groups. Particularly noteworthy milestones in closing the cycle of penaeid shrimp in the laboratory were accomplished and reported for *P. monodon* in England (Forester and Beard 1974; Wickins and Beard 1978), *Farfantepenaeus californiensis*, *L. stylirostris*, and *L. vannamei* in Mexico (Salser *et al.* 1978; Moore and Brand 1993), and *Fenneropenaeus merguiensis*, *Fe. indicus*, *L. stylirostris*, *L. vannamei*, and *P. monodon* in French Polynesia (Aquacop 1983). Despite the early successes in developing captive breeding populations of penaeid shrimp at these various facilities, most of the shrimp farming industry remained dependent on the direct or indirect use of wild shrimp stocks for the postlarvae (PLs or 'seed') used to stock its farms (Argue and Alcivar-Warren 1999; Lightner 2005; Lightner *et al.* 2009). Nonetheless, during this period when the industry remained largely dependent upon wild stocks for its seed (~1980 to ~2000), the industry was experiencing much of its initial rapid growth.

The reasons for the dependence of the shrimp farming industry on wild shrimp stocks for seed were partially technical, but mostly economic. For example, in most large shrimp farming regions of the Americas, the PL requirements were highly seasonal. Hatcheries (called 'laboratories' in most of Latin America) were expensive to build, staff, and run, and the seasonal requirements for PLs left them operating at below capacity for lengthy periods each year. Further, wild PLs ('wild seed') could be obtained in large numbers seasonally (and often when needed most for seasonal stocking plans) and for less cost than hatchery produced PLs ('lab seed'). Another reason, with both economic and technical implications, was that the prevalence of IHHNV (Table 11.1) in captive-wild *L. vannamei* broodstock typically increased the longer the captive-wild stocks were held in maturation and/or hatchery facilities. This made persistently IHHNV-infected captive-wild broodstock essentially worthless within 2–3 months of use as broodstock due to their declining performance (Motte *et al.* 2003). The use

Table 11.1 OIE listed crustacean diseases as of 2009–2010, recently de-listed diseases and those being considered for listing (OIE 2009a, 2010)

Disease name	Pathogen type	Pathogen name & acronym	Principal host group
Taura syndrome	ssRNA virus	Taura syndrome virus (TSV)	Penaeid shrimp
White spot disease	dsDNA virus	White spot syndrome virus (WSSV)	Penaeid shrimp
Yellowhead disease	ssRNA virus	Yellow head virus (YHV) & gill-associated virus (GAV)	Penaeid shrimp
Tetrahedral baculovirosis*	dsDNA virus	*Baculovirus penaei*, BP	Penaeid shrimp
Spherical baculovirosis*	dsDNA virus	Monodon baculovirus, MBV	Penaeid shrimp
Infectious hypodermal and hematopoietic necrosis (IHHN)	ssDNA virus	IHHN virus, IHHNV	Penaeid shrimp
Crayfish plague	Fungus	*Aphanomyces astaci*	Freshwater crayfish
Infectious myonecrosis (IMN)	dsRNA virus	IMN virus (IMNV)	Penaeid shrimp
Necrotizing hepatopancreatitis (NHP)**	Bacteria	NHP-bacterium (NHP-B)	Penaeid shrimp
White tail disease	ssRNA virus	*Macrobrachium nodavirus* (MrNV)	*Macrobrachium rosenbergii*
Milky hemolymph disease of spiny lobsters***	Bacteria	Rickettsia-like bacteria	*Panulirus* spp.

Notes: * Removed from OIE list in 2009.
   ** Listed by OIE in 2010 (OIE 2010).
   *** Listing of this disease was under study by the OIE in 2009 (OIE 2009a).

of *L. vannamei* broodstock with high IHHNV prevalence resulted in poor survival of infected larvae and the production of poor quality PLs. The surviving PLs ('lab seed' or 'maturation seed') had a very high IHHNV prevalence relative to wild PLs (Motte *et al.* 2003). Ponds stocked with such PLs typically had poorer production levels due to the development of IHHNV-caused runt-deformity syndrome (RDS) than did ponds stocked with wild PLs (Kalagayan *et al.* 1991; Browdy *et al.* 1993; Bray *et al.* 1994; Brock and Main 1994; Lightner 1996a; Motte *et al.* 2003; OIE 2009b). Long-term and increasing problems with IHHNV, and subsequently with Taura syndrome (caused by TSV), and the arrival of white spot disease (caused by WSSV) (Table 11.1) in 1999 to Central America, Mexico and Ecuador resulted in rapid changes in shrimp farming strategies in the Americas (Lightner 2005; Lightner *et al.* 2009).

In retrospect it may seem ironic that until as recently as 2000 most of the world's multibillion-dollar penaeid shrimp farming industry was still dependent on the capture of wild postlarvae or broodstock to provide the 'seed stock' used to stock farms (Argue and Alcivar-Warren 1999; Lightner 2005; FAO 2006). For example, before WSSV was introduced into Ecuador in 1999, more than 100 000 people were involved in the collection of wild post-larvae from the littoral zone for use in stocking Ecuador's more than 175 000 ha of shrimp ponds (Rosenberry 2001). The white spot disease and Taura syndrome pandemics in Asia and the Americas that began about 1992 forced the world's shrimp farming industries to change how shrimp are farmed.

FAO (2006) credits the development and export (from producers in the USA) of SPF *L. vannamei* and *L. stylirostris* for this paradigm shift in shrimp farming. The FAO report goes on to comment that while the export of SPF shrimp stocks from the USA to Asia and elsewhere in the world may not have been significant in terms of quantity or total value, but that their impact has been considerable on both the total quantity of shrimp produced and on global shrimp pricing. The FAO report concludes that without the import and use of USA-produced SPF shrimp stocks it is argu-able whether Asia's major shrimp producing countries could have recov-ered from disease outbreaks and the severe shortage of healthy wild-caught broodstock of native penaeids, much less grown to achieve the record levels of production that currently characterize the Asian shrimp farming sector (FAO 2006).

11.3 The development of *Litopenaeus vannamei* as the dominant species in the Americas

11.3.1 Ralston Purina and its selection of *L. vannamei*

Among the first commercial shrimp farming development projects in the western hemisphere were those run by Ralston Purina. In about 1971

Ralston began operating a shrimp farming research facility near Crystal River, Florida. The company began construction in 1974 of its first commercial farm in the Republic of Panama. By 1973 the company was evaluating potential penaeid species as candidates for commercial development at its Crystal River research facility. Local species from the Gulf of Mexico were evaluated, as were stocks of *Litopenaeus stylirostris* and *L. vannamei* imported from Panama (R. Staha, in Rosenberry 2006). Anecdotal information from company employees at the time tell a story of the imported stocks of *L. stylirostris* and *L. vannamei* performing very well at the Crystal River research facility until a stock of *P. monodon* from the Philippines was imported and tested at the research facility. At that time, the Southeast Asian Fisheries Development Center (SEAFDC), the Philippine Fisheries Commission, and a commercial hatchery (on Iloilo Island, Philippines) were exporting *P. monodon* for research and commercial purposes (Forester and Beard 1974; Lightner *et al.* 1992c). After the introduction of *P. monodon* at its Crystal River facility, the company had problems with poor survival in culture systems stocked with *L. stylirostris*, and the company justifiably shifted its emphasis to *L. vannamei* for commercial development at its farm in Panama. The reason for the sudden change in culture performance of the *L. stylirostris* stocks was not determined at that time. However, when methods became available to genotype isolates of IHHNV from wild and farmed penaeid shrimp from the Americas (Hawaii, North, Central, and South America) and from East and Southeast Asia it was found that all of the isolates from the Americas shared >99% sequence homology to an isolate of IHHNV from *P. monodon* collected from the Philippines (Tang and Lightner 2002; Tang *et al.* 2003). Other IHHNV isolates from elsewhere in Asia and from East Africa showed much less homology (~96 and 86%, respectively) to isolates from the Americas (Tang *et al.* 2003; Tang and Lightner 2004). The latter findings support the hypothesis suggested by Lightner (1999) that what became the American strain of IHHNV was introduced with one or more introductions of Asian *P. monodon* in the mid-1970s and early 1980s from the central Philippines. Subsequent transmission of the virus to *L. vannamei* that were being co-cultured with imported *P. monodon* led to the establishment of the disease in infected, but disease resistant, *L. vannamei* stocks. Transfer of the progeny of these stocks after that time eventually led to IHHNV being disseminated into virtually all shrimp farming regions of the Americas long before its discovery in 1981 in Hawaii (Lightner *et al.* 1992c, 1997; Lightner 1996b, 1999; Tang *et al.* 2003; Tang and Lightner 2004).

11.3.2   Pond studies at Texas A&M University

Other research from the same period (the 1970s and early 1980s), showed *L. stylirostris* to perform as well or better in pond, tank, and raceway culture than *L. vannamei*. Among the available examples was a study in which the length–frequency distribution was compared using pooled data from 1972 to 1979 for five species of penaeids (*Farfantepenaeus duorarum, L. occidentalis, L. setiferus, L. stylirostris*, and *L. vannamei*) cultured in ponds at the Texas A&M University research facility near Corpus Christi, Texas. That study showed *L. stylirostris* and *L. vannamei* to reach similar sizes, which were much larger than the other three species studied, but the data presented in the paper also showed that after the 1975 season, only *L. vannamei* was cultured (Hutchins *et al.* 1979), possibly because IHHNV had become established in one or more of the shrimp stocks being tested at the facility.

11.3.3 Taura syndrome virus (TSV) and IHHNV resistant *L. stylirostris*: Super Shrimp™ and SPR-43

A notable exception to the industry's general preference for *L. vannamei* occurred after the Taura syndrome virus (TSV) pandemic had reached most of the major shrimp farming countries (Brock 1997; Hasson *et al.* 1999). Soon after the TSV pandemic began in Ecuador in 1992 (Jimenez 1992; Lightner *et al.* 1995, 1997; Brock 1997), it was noted that *L. stylirostris* being co-cultured in ponds or farms with *L. vannamei* had better survival rates during TSV outbreaks (Lightner 1996a; Lightner and Redman 1998b; Fegan and Clifford 2001). This innate resistance of *L. stylirostris* to Taura syndrome disease, despite infection by TSV, provided an opportunity for significant commercial development of *L. stylirostris*, if domesticated breeding lines or stocks could be developed that were also resistant to IHHNV. After the TSV pandemic reached Mexico in 1994 and caused severe losses the highly TSV susceptible lines of *L. vannamei* then in use throughout the country, two companies, with interests primarily in Mexico, were formed to produce and market domesticated lines of *L. stylirostris* that were resistant to both IHHNV and TSV (Lightner 1996a; Fegan and Clifford 2001; Zarin-Herzberg and Ascencio-Valle 2001; Erickson *et al.* 2002). The companies, Nova Aquaculture in Nayarit and Super Shrimp in Sonora and Sinaloa, began to market SPR-43 and Super Shrimp™, respectively, mostly to clients in Mexico (Lightner and Redman 1998b).

The SPR-43 and Super Shrimp™ lines of *L. stylirostris* were developed over time by breeding survivors of captive stocks of *L. stylirostris* that were persistently infected with IHHNV. Breeding survivors to survivors eventually resulted in continuous domesticated lines of *L. stylirostris* with a high degree of resistance to IHHN disease, despite being persistently infected with the virus (Weppe *et al.* 1992). Some lines of Super Shrimp™ were found in laboratory challenge studies with IHHNV to be resistant even to infection and to quickly clear the virus after challenge (Tang *et al.* 2000). SPR-43 was developed from survivors of IHHNV-infected stocks in Tahiti and New Caledonia and this stock was successfully used to develop the shrimp culture industries of Tahiti and New Caledonia (Weppe *et al.* 1992). The Super Shrimp™ line was likewise developed using the same strategy of breeding survivors of stocks initially imported into Venezuela from Panama (Fegan and Clifford 2001). When the stocks were introduced into Mexico in ~1995, the SPR-43 stock was at about its 18th generation in captivity and the Super Shrimp lines were at about their 16th generation (Weppe *et al.* 1992; Fegan and Clifford 2001; Lightner and Redman 1998b; Lightner 2003a, 2011).

For several years the economic impact of the use of SPR-43, and especially Super Shrimp™, in Mexican farms was significant, and these stocks accounted for nearly 80% of the farmed shrimp produced in Mexico in 1998 (Clifford, in Rosenberry 1998; Zarin-Herzberg and Ascencio-Valle 2001; Moss 2002). Unfortunately, for the Super Shrimp company and for Mexican shrimp farmers, a new strain of TSV emerged in 1998–1999, which was highly pathogenic to both the Super Shrimp™ line of *L. stylirostris* and to *L. vannamei* (Robles-Sikisaka *et al.* 2001; Erickson *et al.* 2002). The emergence of the new strain of TSV and the arrival of the WSSV pandemic into Mexico in 1999 (GAA 1999a, 1999b) effectively negated any advantage of culturing Super Shrimp™ and other selected lines of IHHNV and TSV-resistant *L. stylirostris* in the Americas.

11.3.4 Experiences in super-intensive culture systems with *L. stylirostris* and *L. vannamei*

In another section of this review, the history of the Marine Culture Enterprises (MCE) project is reviewed as it related to its decision to develop IHHNV-free (SPF) *L. stylirostris* for commercial development in its super-intensive raceway culture system. This decision was made because SPF *L. stylirostris* performed better than *L. vannamei* in terms of growth rate, production per square meter, average size at harvest, and food conversion efficiency (Moore and Brand 1993). Following the outbreak of and discovery of IHHNV at the MCE research facility in 1981 (Brock *et al.* 1983; Lightner *et al.* 1983a, 1983b) and the discovery that *L. vannamei* was relatively tolerant to the virus (Bell and Lightner 1984), *L. vannamei* was evaluated again at MCE, but was found, for a variety of reasons, including especially their lower growth rates and their 'panic' tail-flip behavior (that resulted in a high prevalence of unsightly melanized wounds which made much of the product unmarketable), to have less potential for profitable culture in the MCE super-intensive raceway system than did *L. stylirostris* (Lightner *et al.* 2009).

11.3.5 Selection of *L. vannamei* for development in the Americas

Despite the apparent advantages of *L. stylirostris* over *L. vannamei* in terms of size at harvest (Fig. 11.1), growth rate, and ease of culture, its extreme susceptibility to IHHNV severely limited its culture potential in the Americas and this ultimately led the early pioneers of shrimp farming in the Americas to develop *L. vannamei* as the dominant farmed species.

Fig. 11.1   Photograph of harvest-size pacific white (*Litopenaeus vannamei* – middle) and Pacific blue shrimp (*L. stylirostris* – top and bottom). The shrimp are the same age. They were stocked as PLs into the same pond at the same time. Note that *L. vannamei* weighs approximately 30% less than *L. stylirostris*.

The commercial culture of most genetic lines of *L. stylirostris* was simply not possible in locations where IHHNV was present in wild or in other farmed shrimp stocks (Lightner and Redman 1998b; Lightner *et al.* 2009). By 1990 IHHNV was enzootic and highly prevalent in wild populations of *L. vannamei*, *L. stylirostris*, and other penaeids along the Pacific coast of the Americas and in cultured stocks of these species throughout the major shrimp farming countries of the Americas (Lotz 1992; Lightner 1996a, 1996b; Pantoja *et al.* 1999; Morales-Covarrubias and Chavez-Sanchez 1999; Morales-Covarrubias *et al.* 1999; Fegan and Clifford 2001; Nunan *et al.* 2001). In summary, the high susceptibility of *L. stylirostris* to IHHN disease and the nature of IHHNV in terms of its modes of transmission (i.e. vertical by the transovarian routes as well as by typical horizontal routes – Motte *et al.* 2003), caused the industry to develop *L. vannamei* as the dominant species cultured in the Americas.

11.4   The adaptation of the specific pathogen-free (SPF) concept to domesticated shrimp stocks

11.4.1   The concept of specific pathogen-free (SPF)

Although marketers commonly use the term 'disease-free' to describe the live shrimp products being sold in commerce, they are in reality marketing shrimp that are free of specific disease causing agents. Because nothing that is living is completely free of some sort of disease, such 'disease-free shrimp' are more correctly referred to as having 'freedom from disease' or being free of certain specific pathogens (OIE 2009a). This is the underlying principle that defines a shrimp stock as being SPF.

Disease management through exclusion of specific pathogens is commonplace in modern agriculture (Bullis and Pruder 1999; Zavala 1999; Moss 2002). The concept of developing shrimp stocks that are SPF, and rearing of these stocks in regions where the specific pathogens of concern are excluded, has been used in the western hemisphere with mixed success. The successful application of the SPF concept is dependent upon the documented absence of the pathogen(s) of concern in the stocks being reared (or that are present) for a predefined period of time, on the availability of sensitive and accurate detection and diagnostic methods for the pathogen(s), and the presence of an effective barrier (i.e., facility design and geographic location, government mandated import restrictions, etc.) to prevent the introduction of the specific pathogen(s) intended to be excluded. These are essentially the same principles outlined by the OIE for a country, zone or compartment to make a self-declaration of freedom from a specific OIE-listed disease (OIE 2009a; Tables 11.1 and 11.2). In the western hemisphere, SPF stocks of L. stylirostris and L. vannamei have been developed and these are being cultured successfully in many locations (Wyban 1992; Wyban et al. 1992; Carr et al. 1994; Pruder et al. 1995; FAO 2006; Lightner et al. 2009; Moss and Moss 2009).

In actual use by the industry, the term SPF implies that the stock of interest is free of one or more specific pathogens (Lotz et al. 1995; Lotz and Lightner 1999; Fegan and Clifford 2001). To the United States Marine Shrimp Farming Program (USMSFP), SPF means the stock of interest has at least two years of documented historical freedom of the disease agents listed on its working list of specific pathogens, that the stock has been cultured in biosecure facilities, and that either the stock was cultured under conditions where the listed disease agents would have produced recognizable disease if any were present and/or the stock has been subjected to routine surveillance and testing for the listed pathogens (Lightner 2005; Lightner et al. 2009). Those pathogens on the USMSFP SPF list have also met certain criteria including: (a) the pathogen(s) must be excludable; (b) adequate diagnostic and pathogen detection methods are available; and (c) the pathogen(s) poses significant threat of disease and production losses (Lotz et al. 1995; Lightner 2003a, 2003b), which are also among the criteria required for disease listing by the World Animal Health Organization (OIE 2009a).

11.4.2   The Marine Culture Enterprises (MCE) experiment

Among the early projects that closed the life cycle of certain penaeid shrimp species, the commercial development project run jointly by the University

Table 11.2   Current U.S. Marine Shrimp Farming Consortium (USMSFC) working list of 'specific' and excludable pathogens of American penaeids and Asian penaeids for 2010–2011 (adapted from Lightner 2005; Lightner et al. 2009)

Pathogen type	Pathogen/pathogen group	Pathogen category[a]
Viruses	* WSSV – white spot syndrome virus (Nimaviridae)[b]	C-1
	* YHV, GAV, LOV – the Oka viruses (Roniviridae)[b]	C-1
	* TSV – Taura syndrome virus (Dicistroviridae)[b]	C-1
	* IHHNV – a systemic parvovirus (Parvoviridae)[b]	C-1
	*** BP – *Baculovirus penaei* (occluded; Baculoviridae)[c]	C-2
	*** MBV – Monodon baculovirus (occluded; Bacuolviridae)[c]	C-2
	*** BMN – Baculoviral midgut gland baculovirus (non-occluded; Baculoviridae)[c]	C-2
	*** SMV – Spawner mortality virus (Parvoviridae)[b]	C-1
	HPV – Hepatopancreatic parvovirus (Parvoviridae)[b]	C-2
	* IMNV – Infectious myonecrosis virus (Totiviridae)[b]	C-1
	PvNV – *Penaeus vannamei* nodavirus (Nodaviridae)[b]	
Procaryotes	** NHP – bacterium – Alpha proteobacteria	C-2
	MHD – milky hemolymph disease (rickettsia-like bacteria)	C-2
Protozoa	Microsporidians	C-2
	Haplosporidians	C-2
	Gregarines (Apicomplexa)	C-3

Notes: * OIE listed pathogen (OIE 2009a, 2009b).
　　** Listed by OIE as of May 2010 (OIE 2010).
　　*** Formerly listed by OIE but de-listed prior to 2009.
　　[a] Pathogen category (modified from Lotz et al. 1995), with C-1 pathogens defined as excludable pathogens that can potentially cause catastrophic losses in one or more American penaeid species; C-2 pathogens are serious, potentially excludable; and C-3 pathogens have minimal effects, but may be excluded from breeding centers, hatcheries, and some types of farms.
　　[b] Faquet et al. 2005.
　　[c] The 1995 Committee report on virus taxonomy (Murphy et al. 1995) removed crustacean baculoviruses from the Baculoviridae and assigned them to a position of unknown taxonomic position. Nonetheless, BP, MBV, and BMN are most like members of the Baculoviridae (Faquet et al. 2005) and, for practicality, they are listed here as baculoviruses.

of Arizona (UAZ) and Marine Culture Enterprises (MCE) in Hawaii was unique in its efforts to develop domesticated shrimp stocks that were specifically free of IHHNV (SPF for IHHNV). The MCE project in Hawaii had two phases. The first was the 'species selection phase' which began in 1980 and ended in 1983. The second phase, which ran from 1984 to 1987, focused on the commercial development of IHHNV-free domesticated stocks of *L. stylirostris* for use in the commercial prototype super-intensive growout facility. The MCE project began in 1980 with the construction of an experimental super-intensive recirculating raceway shrimp farm near the village of Kahuku on the Island of Oahu, Hawaii, and this was followed in

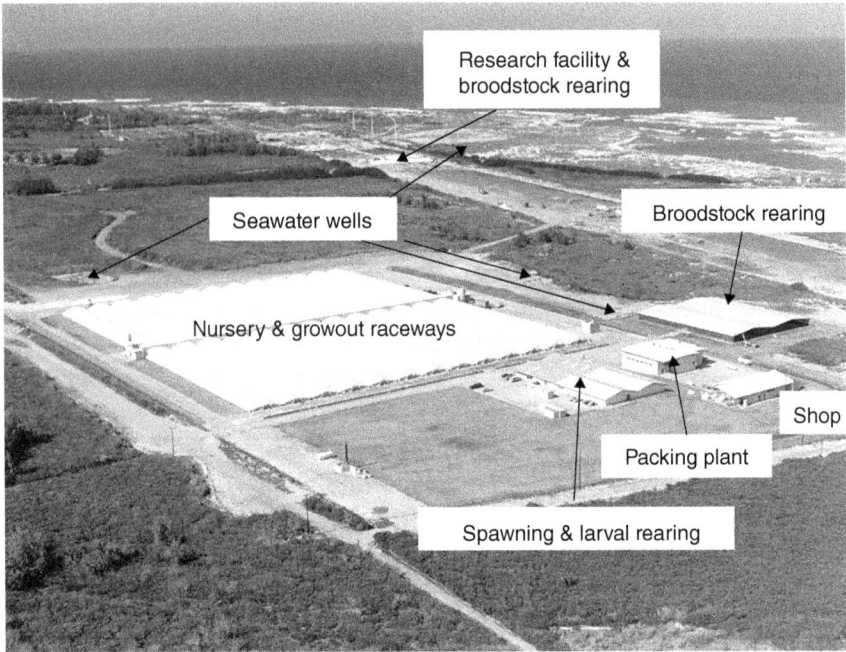

Fig. 11.2   Aerial view of the Marine Culture Enterprises (MCE) super-intensive shrimp culture facility in 1985. The main components of the MCE facility referred to in the text are indicated on the photograph.

1984 with the construction of a 2 ha commercial prototype farm at the same location (Moore and Brand 1993). The experimental MCE farm was developed and operated by the University of Arizona (Fig. 11.2). Because none of the species initially cultured at either the MCE experimental farm, and later at the commercial prototype farm, were indigenous to Hawaii, all of the farmed stocks had to be imported. The first, or research, phase of the MCE project was also seen as an opportunity by the project's management to acquire and test a variety of non-indigenous (to Hawaii) penaeids from the Americas and Asia as candidates for commercial culture in a super-intensive raceway growout system (Figs 11.3 and 11.4). The initial outbreaks of IHHN at the experimental facility in late 1980 through mid-1981 caused severe losses in the populations of *L. stylirostris* being cultured there, but seemed to cause little or no mortality in populations of *L. vannamei*, *P. monodon*, and *Fenneropenaeus indicus* being cultured in adjacent raceways (Brock *et al.* 1983; Lightner *et al.* 1983b). Except for using only filtered seawater (from a ~30 m deep seawater well located behind the beach in coastal deposits of sand and coral reef litter) and the use of raceways within air-inflated plastic greenhouse structures, no other biosecurity measures were in effect (Moore and Brand 1993).

Fig. 11.3   Photograph taken inside of an MCE 'aquacell' containing two growout raceways.

The MCE project may have been the first to intentionally develop domesticated SPF lines of penaeid shrimp. As indicated above, IHHNV was first discovered at the MCE experimental prototype farm in Hawaii in 1981 where it caused in excess of 90% mortality in the nursery phase and juvenile phases of the culture of *L. stylirostris* (Brock *et al.* 1983; Lightner 1983, 1988; Lightner *et al.* 1983a, 1983b; Brock and Lightner 1990). The most severely affected stocks of *L. stylirostris* were naive to the disease and they had been recently imported from the northern Gulf of California, Mexico, where IHHNV had not previously been observed (Lightner *et al.* 1983b, 1992a, 1992b; Pantoja *et al.* 1999). Research performed at the time using histopathological and live shrimp bioassay methods showed that IHHNV had been imported into the MCE facility in Hawaii with several populations of *L. vannamei*, *L. stylirostris*, and *P. monodon* from Central America, Micronesia, or southeast Asia (Lightner *et al.* 1983b). Because of the extreme virulence of IHHNV in Mexican lines of *L. stylirostris*, and because Mexican lines of *L. stylirostris* were found to be the species most suitable for culture in the MCE super-intensive raceway culture system, it was essential to the future development of the MCE facility to eradicate IHHNV.

To eradicate IHHNV from the MCE research facility in 1983, it was depopulated, disinfected, dried, and fallowed for several months. To restock the facility, only PLs of *L. stylirostris* were used which had been produced in a designated quarantine area at the MCE facility from new batches of

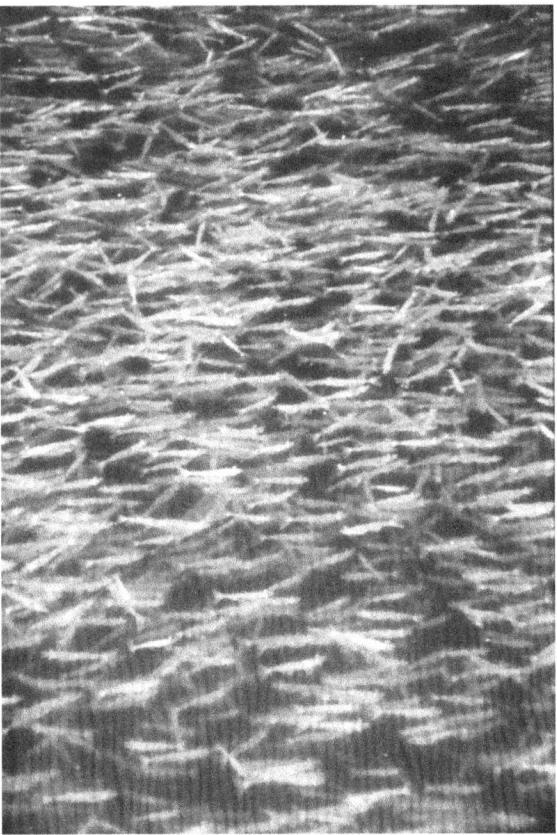

Fig. 11.4 Photograph of ~18 g average weight *Litopenaeus stylirostris* in a super-intensive MCE raceway at a harvest density of approximately 7.9 kg/m^2 of bottom area.

imported wild shrimp broodstock that had been fished from the Gulf of California in Sonora and Sinaloa, Mexico. In the quarantine area, imported broodstock from Mexico were received, matured, mated, spawned, and PLs were produced. Based on previous experimental findings on the biology of IHHN disease in PLs produced from infected *L. stylirostris* broodstock, criteria were defined for assessing freedom from IHHNV in the imported broodstock and the PLs produced from them for the purposes of restocking the facility with IHHNV-free stocks. Because IHHNV in vertically infected PLs was found to cause disease in *L. stylirostris* during the nursery phase of culture (between 20 and 50 days of culture or at about PL30–60), the criteria for defining freedom from IHHNV infection was defined as having at least four batches of PLs produced from at least four separate females from each population of broodstock in quarantine that showed no signs of IHHN disease either grossly or by histology when sampled as young juveniles after 40–60 days of culture (= ~PL50–70) (Lightner *et al*. 1983b; Moore and Brand 1993).

This scheme worked well and the MCE facility operated IHHNV-free until 1987 when sudden high mortality episodes were noted in several of the growout raceways at the MCE commercial prototype farm. IHHNV was diagnosed as the cause of the disease outbreak. The source of IHHNV that caused the 1987 outbreak at MCE was never determined. However, the MCE growout facility was located in an aquaculture park and its seawater supply was from three vertical seawater wells located within the farm's perimeter fence (Fig. 11. 2). Several neighboring farms were culturing *L. vannamei* at the time and discharging their pond effluents into a dispersion canal which ran parallel to the MCE perimeter fence and which returned all effluent water by gravity into the same aquifer from which MCE drew its seawater. The close proximity of neighboring shrimp farms and the location of their effluent dispersion canal relative to the MCE wells may have facilitated the spread of the virus among the farms through the aquifer (Moore and Brand 1993).

11.4.3   International principles for responsible shrimp farming

This document was published jointly by five major international agencies that made up the Consortium on Shrimp Farming and the Environment (FAO/NACA/UNEP/WB/WWF 2006). The five agencies that made up this Consortium were the Food and Agricultural Organization of the United Nations (FAO), the Network of Aquaculture Centres in Asia-Pacific (NACA), United Nations Environmental Programme (UNEP), the World Bank Group (WB), and the World Wildlife Fund (WWF). The purpose of the document was to list guiding principles for responsible shrimp farming and recommended means to achieve that goal. The underlying need for the development of the international principles was that while shrimp farming is recognized as one of the fastest growing aquaculture sectors in many parts of the world, it is also one of the most controversial. Although the rapid expansion of shrimp farming has resulted in new employment, trade and income opportunities for many developing counties, this change has been accompanied by rising concerns over environmental and social impacts. The Consortium's document addresses those concerns. Among the eight principles for responsible shrimp farming outlined in the Consortium's document were two that are of special relevance to the present review. These were 'Principle 4 – Broodstock and Postlarvae' and 'Principle 6 – Health Management.' Principle 4 states that shrimp farmers should use domesticated selected stocks of disease-free and/or resistant shrimp broodstock and post-larvae to enhance biosecurity, reduce disease incidence and increase production, whilst reducing demand on wild stocks. Principle 6 states that health management plans should be adopted to reduce stress and minimize the risks of disease that might affect cultured and wild stocks as well as food safety. The guidance provided for the implementation of this Principle specifically identifies the need to maintain biosecurity to minimize disease transmission between broodstock, hatchery, and growout systems both within the same farms and between farms (FAO/NACA/UNEP/WB/ WWF 2006). The level of biosecurity needed to accomplish this Principle is consistent with the level needed to protect and manage the culture of SPF shrimp stocks (Lee and O'Bryen 2003; Lightner 2005; Lightner *et al.* 2009).

The SPF Program of the US Marine Shrimp Farming Program (USMSFP)
Like the MCE project that preceded it, the USMSFP found soon after its inception that domesticated shrimp stocks that were free of diseases like IHHN produced progeny that performed better than infected stocks. Early in the USMSFP's efforts to domesticate *L. vannamei*, side-by-side comparisons of IHHNV-infected and IHHNV-free stocks from the same genetic lines clearly demonstrated that IHHNV was the cause of runt-deformity syndrome (Kalagayan *et al.* 1991). Hence, the initial efforts to develop domesticated SPF stocks by the USMSFP were directed at producing IHHNV-free *L. vannamei* (Lotz 1992; Wyban 1992; Wyban *et al.* 1992). Because *Baculovirus penaei* (BP) and hepatopancreatic parvovirus (HPV) were also of concern at the time, these viruses and certain excludable, but potentially common, parasites were also added to the list of specific diseases/pathogens targeted by the initial USMSFP SPF stock development plan (Table 11.3) (Wyban *et al.* 1992).

SPF stocks developed by the USMSFP were developed in the spirit of the ICES Code (the International Council for the Exploration of the Sea; Code of Practice to Reduce the Risks of Adverse Effects Arising from the Introduction on Non-indigenous Marine Species, 1973, as reviewed in Sindermann 1988, 1990; Turner 1988; Bartley *et al.* 1996; ICES 1995, 2004) for the development of domesticated stocks of *L. vannamei* (Table 11.4; Fig. 11.5). The determination of which specific pathogens the selected stocks were to be free of was based on working lists of pathogenic, diagnosable, and excludable pathogens (Wyban 1992; Lotz *et al.* 1995), which necessarily changed over time as new diseases such as those due to WSSV and TSV emerged and caused serious pandemics (Tables 11.1, 11.2 and 11.3). The most current working list for the US Marine Shrimp Farming Program (USMSFP; Table 11.2) includes 11 viruses or virus groups (WSSV, the YHV group, TSV, IHHNV, hepatopancreatic parvovirus (HPV), *Baculovirus penaei* (BP), monodon baculovirus (MBV), baculoviral mid-gut gland necrosis (BMN), and spawner-isolated mortality virus (SMV), infectious myonecrosis (IMN), *Penaeus vannamei* nodavirus (*Pv*NV), certain classes of parasitic protozoa (microsporidians, haplospordians, and gregarines), and the bacterial agent of necrotizing hepatopancreatitis, or NHP (Lightner *et al.* 2009 and unpublished).

To begin the process of developing an SPF stock (Fig. 11.5), each 'SPF candidate population' of wild or cultured shrimp stocks of interest was

Table 11.3 Chronological list of specific diseases targeted by shrimp early domestication and breeding programs that developed specific pathogen-free (SPF) stocks of *Litopenaeus vannamei* and *L. stylirostris*

Program & SPF species cultured	Listed diseases by pathogen name	Time period	Key reference(s)
Marine Culture Enterprises (MCE), Hawaii *L. stylirostris* *L. vannamei*	IHHNV	1981–1986	Lightner *et al.* (1983b), Moore and Brand (1993)
USMSFP, Oceanic Institute, Hawaii *L. vannamei*	IHHNV BP HPV Microsporidians Haplosporidians Gregarines Metazoan parasites	1991–1994	Wyban (1992), Wyban *et al.* (1992)
USMSFP, Oceanic Institute, Hawaii *L. vannamei*	IHHNV HPV BP MBV BMN YHV WSBV (WSSV) Microsporidians Haplosporidians Gregarines Metazoan parasite larvae	1995–2000	Pruder (1995)
USMSFP, Oceanic Institute, Hawaii *L. vannamei*	IHHNV HPV BP MBV BMN YHV WSBV (WSSV) SMV Microsporidians Haplosporidians Gregarines Metazoan parasite larvae	2001–2004	Lightner & Pantoja (2001), Lightner (2003b), Moss *et al.* (2003)

identified. If available, samples of the stock were taken and tested using appropriate diagnostic and pathogen detection methods for the specific pathogens of concern. If none was found, a founder population (F_0) of the 'candidate SPF' stock was acquired and reared in primary quarantine. During primary quarantine, the F_0 stock was monitored for signs of disease, sampled, and tested periodically for specific pathogens. If any pathogens of concern were detected, the stock was destroyed. Those stocks that tested

Table 11.4 Recommended steps in the ICES Code for risk reduction in aquatic species introductions (modified from Sindermann 1988, 1990; Lightner 2005)

Original ICES code	Adapted to SPF shrimp development
1. Conduct comprehensive disease study in native habitat.	1. Identify stock of interest (i.e., cultured or wild).
2. Transfer (founder stock) system in recipient area.	2. Evaluate stock's health/disease history.
3. Maintain and study closed system population.	3. Acquire and test samples for specific listed pathogens (SLPs) and pests.
4. Develop broodstock in closed system.	4. Import and quarantine founder (F_0) population; monitor F_0 stock.
5. Grow isolated F_1 individuals; destroy original introductions.	5. Produce F_1 generation from F_0 stock.
6. Introduce small lots to natural waters – continue disease study.	6. Culture F1 stock through critical stage(s); monitor general health and test for SLPs.
	7. If SLPs, pests, other significant pathologies are not detected, F_1 stock may be defined as SPF and released from quarantine.

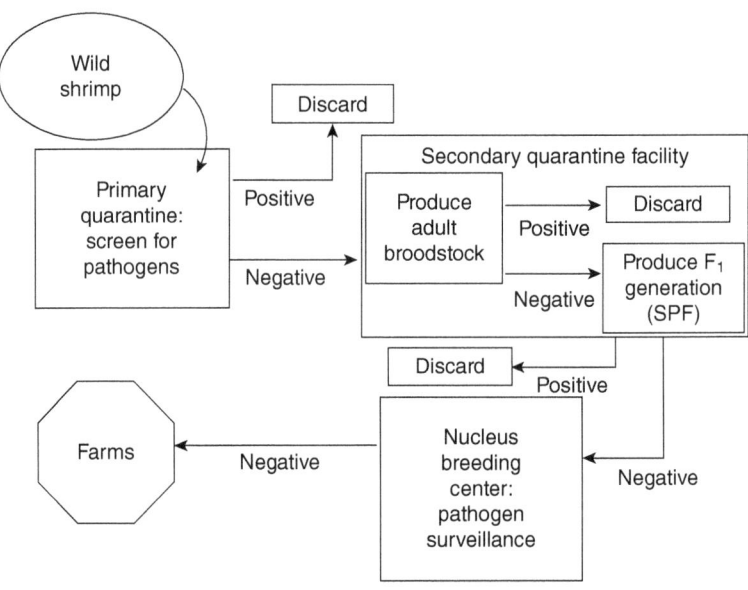

Fig. 11.5 Schematic of the steps followed in developing domesticated SPF shrimp stocks by the U.S. Marine Shrimp Farming Program.

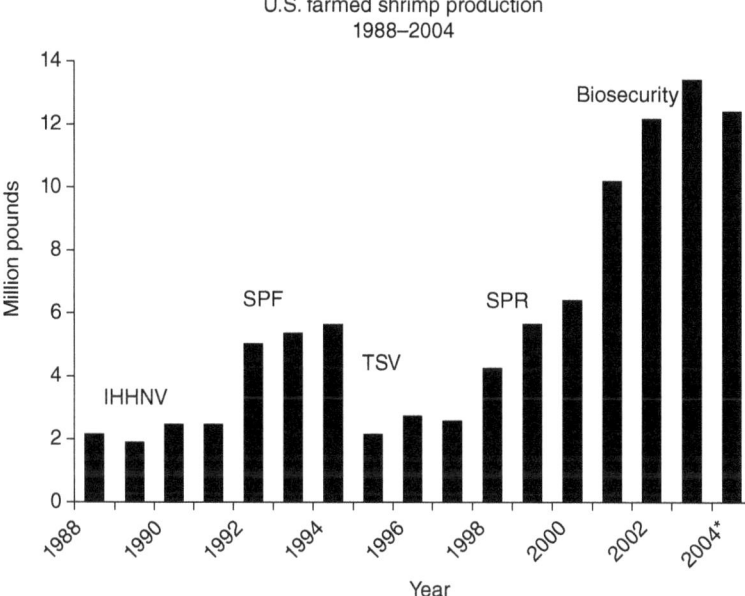

Fig. 11.6 Effect of the introductions and application of SPF and SPR stocks of *Litopenaeus vannamei* and improved biosecurity in USA shrimp farming.

negative for pathogens of concern through primary quarantine (which ran from 30 days to as much as one year for some stocks) were moved to a separate secondary quarantine facility for maturation, selection, mating, and production of a second (F_1) generation. The F_1 stocks were maintained in quarantine for further testing for specific pathogens of concern. Those that tested negative were designated as SPF, and used to produce domesticated lines of SPF and 'high health' shrimp (Wyban *et al.* 1992; Lotz *et al.* 1995; Pruder *et al.* 1995; Lightner *et al.* 2009). SPF and high health stocks of *L. vannamei* were introduced and used successfully in US shrimp farms in 1993 and 1994 (Fig. 11.6). This resulted in nearly doubling the production per crop that had been previously obtained at the same farms in previous years when the farms cultured non-selected lines of *L. vannamei*, which in previous crops, had been persistently affected by RDS due to chronic infection by IHHNV (Brock and Main 1994; Pruder *et al.* 1995; Lightner 1996a, 1996b; Moss *et al.* 2002, 2003).

11.4.4   Lessons learned in developing specific pathogen-free (SPF) penaeid shrimp stocks

The first domesticated IHHNV-free stock of *L. vannamei* developed by the USMSFP was developed from a relatively small founder population that consisted of about six mated wild adult female *L. vannamei* that had been collected by trawling near Mazatlan, Sinaloa, Mexico in April or May of 1989 (Wyban 1992). These were the first wild broodstock to be introduced into and spawned successfully in a newly completed commercial shrimp hatchery located near Mazatlan. The larvae and PL produced from the pooled spawns from the six females were reared together in one larval rearing tank.

Approximately 2000 PLs were acquired from the hatchery in June in exchange for diagnostic services and transferred to the University of Arizona in Tucson where they were sampled for histology and examined for signs of infection by IHHNV or other significant diseases. No signs of IHHNV infection, nor of other significant diseases, were found. Hence, in June, the stock was transferred to the Oceanic Institute (Oahu, Hawaii) for secondary quarantine. After two rounds of additional sampling and testing (using histological methods at ~30 day intervals), no signs of IHHNV infection or of other significant diseases were found. With these findings, the stock was considered to be an F_0 'candidate SPF' population and it was reared to produce mature adults. In the spirit of the ICES Code (Table 11.3; Lightner 2005), the F_1 PLs and juveniles produced from the F_0 stock were tested three times at ~30 day intervals from ~PL25 and examined (using histological methods) for signs of infection by IHHNV or by other significant diseases. None was found, and this 'Mazatlan' stock of *L. vannamei* became the first domesticated stock of this species to be declared as SPF through use of the ICES Code (Wyban 1992; Lotz 1992; Pruder *et al.* 1995; Moss and Moss 2009).

To obtain breeding stocks of SPF *L. vannamei* from sites throughout the geographic distribution of the species (from central Mexico to Peru), the USMSFP began to collect potential founder stock populations from wild stocks and from broodstock facilities and hatcheries at regional locations that ranged from Peru in the extreme south end of the range for *L. vannamei* to additional sites in southwestern Mexico, the northern limit for the species. The development of subsequent SPF domesticated stocks of *L. vannamei* from the various adult and PL stocks that were acquired and tested in primary quarantine was not as easily accomplished as it was with the Mazatlan SPF line. It was soon found that IHHNV had a relatively high prevalence throughout the geographic range of *L. vannamei* (Lotz 1992; Nunan *et al.* 2001; Motte *et al.* 2003). Hence, many of the attempts failed because in primary quarantine the prevalence of IHHNV increased rapidly, often reaching nearly 100% within 30 days of quarantine. To effectively deal with IHHNV in geographic regions where its prevalence in wild stocks was higher than 5–10%, a strategy for the collection, transport, and primary quarantine of such stocks had to be developed.

The development of DNA probes and PCR tests for IHHNV in 1992–1993 (Mari *et al.* 1993; Lightner *et al.* 1994) added a valuable tool needed for SPF stock development. With DNA-based molecular test methods (e.g.

polymerase chain reaction, PCR) replacing histology as the screening method for infections due to IHHNV, it became routine to screen quarantine founder stocks and candidate SPF stocks for the virus with test methods that were several orders of magnitude more sensitive than histological methods. A further advantage of PCR for IHHNV was that it could be used to test larger numbers of specimens for the same cost as histology, and it could be applied to non-lethal testing of larger juvenile and adult animals (Lightner and Redman 1998a). DNA-based molecular methods soon became routine for screening founder and candidate SPF stocks for IHHNV, HPV, WSSV, and most other OIE or USMSFP listed diseases.

Molecular tests were combined with the collection of a limited number of individual shrimp to constitute the founder stock populations used to develop the SPF stocks of *L. vannamei* from Panama (Nunan *et al.* 2001) and *Fenneropenaeus chinensis* from the Yellow Sea in China (Hennig *et al.* 2005; Pantoja *et al.* 2005). This was considered necessary because in these regions certain pathogens of concern were known to be present at high prevalence in wild stocks (Lightner and Redman 1998a, 1998b; Lightner 2003a, b). In the example cited here by Nunan *et al.* (2001), 104 wild *L. vannamei* were captured by trawl off the Pacific coast of Panama. On-shore, the collected shrimp were held in a large (~15 Mt) tank, tagged with an eye-tag for individual identification, and hemolymph was collected from each shrimp to test for IHHNV and WSSV using a relatively rapid on-site dot-blot test with DIG-labeled DNA probes to the two viruses. Some 27% (28 of 104) were found to test positive for IHHNV and 2% tested positive for WSSV (2 of 104). The virus positive shrimp were removed from the holding tanks, and those that tested negative were divided into smaller groups of ~5 shrimp each and held in smaller tanks until being shipped to the University of Arizona for primary quarantine. During subsequent testing, additional IHHNV positive shrimp were found which resulted in the elimination and destruction of several of the ~5 shrimp subpopulations. Eventually, no additional IHHNV positive shrimp were found and the remaining shrimp were transferred to the Oceanic Institute for secondary quarantine, and production of F_0 and F_1 candidate SPF lines (Nunan *et al.* 2001).

A similar strategy was used to develop SPF stocks of *Fe. chinensis* from the Yellow Sea region of northeast China. In this region, HPV (Lightner and Redman 1985; Lightner *et al.* 1992b, 1992c, 1993; Lightner 1993) and WSSV are known to occur at high prevalence (Lo *et al.* 1997; OIE 2009b). As was done in Panama, the adult wild *Fe. chinensis* were transferred to individual tanks as soon as possible after capture and held until PCR testing could be completed on hemolymph or fecal strands collected as a non-lethal samples for screening. In the 36 adult female *Fe. chinensis* collected, seven (19%) were WSSV positive and two (5%) were positive for HPV (Pantoja *et al.* 2005). After additional screening, spawns from eight of the remaining shrimp were transferred to the University of Arizona in Tucson for primary quarantine (as the F_0 stock). After three rounds of additional screening at ~30 day intervals for USMSFP and OIE listed diseases, six of the eight original F_0 stocks were eventually transferred to the Oceanic Institute for secondary quarantine and development of SPF breeding lines of *Fe. chinensis* (Hennig *et al.* 2005; Pantoja *et al.* 2005).

The key to the success of these examples of SPF stock development was dividing the founder population into small groups or to one individual shrimp or one spawn from a single female as soon as possible after capture in order to minimize the possibility of shrimp-to-shrimp transmission of IHHNV, HPV, WSSV, or other infectious agents that might have been present.

11.5   Maintenance of specific pathogen-free (SPF) status: disease surveillance and control programs

While the SPF concept has been endorsed as a disease prevention and control strategy, the process does not end with the initial declaration that a particular domesticated shrimp stock is free of one or more specific diseases. The maintenance of SPF status requires that the stock(s) be reared in biosecure facilities and that the stock(s) are subjected to routine surveillance (Lotz *et al.* 1995; Lotz and Lightner 1999; Moss and Pruder 1999; Fegan and Clifford 2001; Lightner 2005; OIE 2009a). In most applications this means that the SPF stock(s) is the subject of an active disease surveillance and disease control program (Lightner 2003b, 2005; OIE 2009a).

The OIE (2009a, 2009b; Corsin *et al.* 2009) suggests that after a shrimp production facility (and the shrimp stocks reared therein) has been recognized to be free of certain pathogens (i.e., declared to be SPF, a process which may require from as little as 6 months but usually to two years or longer of surveillance with laboratory tests and in the absence of any suspect clinical signs) twice-yearly inspections should continue in order to document that SPF status has been maintained. Random collection of specimens (targeted surveillance) for testing, however, may be reduced to 30 specimens, including (especially) broodstock (OIE 2009b). This inspection and sampling scheme assumes a prevalence of ~30%. Moribund shrimp observed during inspection visits, however, must be collected for further laboratory examination (passive surveillance – Corsin *et al.* 2009). Hence, in an SPF shrimp program, domesticated stocks are subjected to both targeted surveillance for specific diseases/pathogens and to general or passive surveillance for any other significant disease that may occur (Lightner *et al.* 2009).

11.5.1   Listed shrimp diseases and pathogens

What should be on a list of pathogens in an SPF program? Because not all potential causes of disease can be excluded, the development of a list of specific pathogens to be excluded is among the essential elements of a workable list of specifically listed diseases (SLDs) and the biosecurity plan that is required to prevent the SPF shrimp stocks from coming into contact with SLDs. Such lists of SLDs, and the corresponding biosecurity plan, may be designed for a single culture facility or 'compartment', a group of farms, a country, or a region comprising several countries (OIE 2009a; Corsin *et al.* 2009). Some disease agents share common characteristics that make them excludable. Among these are a limited geographic distribution, a limited host range, and often being an obligate parasite/pathogen that requires a suitable host for replication (Lightner 2005; OIE 2009a). Not all potential causes of disease in shrimp aquaculture can be excluded by the application of a biosecurity program. Shrimp have as part of their natural microbial flora and in their aquatic environment, a large and diverse population of microorganisms, some of which are facultative pathogens ready to strike when the shrimp become compromised by any number of stressors. Certain *Vibrio* spp. provide a good example of a group of organisms that live in the shrimp's environment, often as part of their normal microflora inhabiting the surface of their cuticle or colonizing areas of the gut or hepatopancreas (Brock and Lightner 1990; Fulks and Main 1992), but which can become serious pathogens in a stressed individual shrimp or in a stressed population. Hence, a listed pathogen in an SPF development and maintenance plan should not be ubiquitous in the culture environment, and it should not be a commensal organism that can become an opportunistic pathogen in a compromised host population. Robust, sensitive and reliable pathogen detection methods must be readily available for listed pathogens (OIE 2009a, 2009b; Corsin *et al.* 2009). Another criterion to consider for listing a specific pathogen is whether it poses a significant threat of clinical disease and production losses (OIE 2009a). Excluding pathogens, where the costs far exceed the benefits, may not be justified in some aquaculture situations (Fegan and Clifford 2001; Lightner 2003b, 2005).

Examples of specific lists of excludable pathogens are available, and may be helpful if referred to when formulating a pathogen list for a facility, a particular region or zone, or a country. The USMSFP publishes in its annual publications a list of pathogens (Table 11.2) that it strives to exclude from its facilities and lines of domesticated shrimp (Lightner 2003a; Lightner *et al.* 2009). Also indicated in Table 11.2 are those disease agents listed by the OIE (2009a, 2009b, 2010). The OIE maintains at its website (www.oie.int) and publishes regularly an International *Aquatic Animal Health Code and Diagnostic Manual*. The 2010 OIE *Aquatic Animal Health Code* (OIE 2010) currently lists eight crustacean diseases. Six of the eight diseases are diseases of penaeid shrimp; five are viral diseases and one is a bacterial disease of penaeid shrimp. Two of the eight diseases, crayfish plague of European freshwater crayfish and white tail disease of *Macrobrachium rosenbergii*, are not diseases of farmed penaeid shrimp. The diseases listed by OIE pose a threat to international commerce, fisheries, and aquaculture of crustaceans (especially shrimp). Before a disease may be included on the OIE list of notifiable and listed diseases, OIE has set criteria that must be met: (a) the etiological agent must be known; (b) reliable diagnostic(s) methods must be available; and (c) the disease must be a significant disease of local, regional, or international importance (OIE 2009a). The OIE criteria for listing diseases are very similar to those used by the USMSFC. Pathogen and disease lists, such as the OIE and USMSFP lists (Tables 11.1 and 11.2), are useful models for setting up a biosecurity program that is based on exclusion of a list of specific pathogens and the diagnostic methods for surveillance and diagnosis.

While biosecurity has as its goal the exclusion of known pathogens for which epizootiological data is available and for which there are adequate diagnostic and detection methods, the application of biosecure practices can also reduce the likelihood of an unknown or poorly understood pathogen being introduced. To be most effective, however, the epizootiology of a pathogen (i.e., its hosts, biology, and methods of transmission) that is the etiological agent of a particular disease must be sufficiently known to permit managers to understand how the pathogen is transmitted and how to prevent its entry and spread. It is impractical, if not impossible, to expect biosecurity to lead to the development of 'disease-free' or 'pathogen-free' shrimp stocks. It is equally impractical to expect to farm such stocks in an environment where every potential pathogen is excluded (Lightner *et al.* 2009).

11.5.2   Disease diagnosis and surveillance

The application of biosecurity by any component of the penaeid shrimp culture industry (i.e., a facility, compartment, a geographic region, or a country – OIE 2009a) is dependent upon the availability of sensitive, accurate, cost-effective disease diagnosis, and pathogen detection methods (Table 11.5). Highly sophisticated methods for pathogen detection in various sorts of samples are of little value to the shrimp farming industry if those methods exist only in the laboratories that developed them, and are of even less value if they are not readily available to an industry that could benefit from their application and use. Likewise, of little value to biosecurity programs are other diagnostic and pathogen detection methods that may be generally available, but which are not sufficiently sensitive or accurate to meet the industry's requirements (Lightner 2003b; Walker and Mohan 2009).

Various manuals can serve as guides to the available methods for disease screening and diagnosis. The Food and Agriculture Organization of the United Nation (FAO) *Diagnostic Guide to Aquatic Animal Diseases* (Bondad-Reantaso *et al.* 2001) and the OIE *Diagnostic Manual for Aquatic Animal Diseases* (OIE 2009b) are examples of manuals that contain methods that have national or international approval for disease screening and

Table 11.5 Methods available to diagnosticians for shrimp disease diagnosis and pathogen detection (adapted from Lightner 2005; Lightner *et al.* 2009)

Method	Tests and data obtained
History	History of disease at facility or in region, facility design, source of seed stock (e.g., wild or domestic specific pathogen-free, SPF, or resistant, SPR), type of feed used, environmental conditions, etc.
Gross, clinical signs	Lesions visible, behavior, abnormal growth, feeding or food conversion efficiency, etc.
Direct microscopy	Bright-field, phase contrast or dark-field microscopic examination of stained or unstained tissue smears, whole-mounts, wet-mounts, etc. of diseased or abnormal specimens.
Histopathology	Routine histological or histochemical (with special stains) analysis of tissue sections.
Electron microscopy	Ultrastructural examination of tissue sections, negatively stained virus preparations, or sample surfaces.
Culture and biochemical identification	Routine culture and isolation of bacterial isolates on artificial media and identification using biochemical reactions on unique substrates.
Enhancement	Rearing samples of the appropriate life stages of shrimp under controlled, stressful conditions to 'enhance' expression of latent or low grade infections.
Bioassay	Exposure of susceptible, indicator shrimp to presumed carriers of a pathogenic agent.
Antibody-based methods	Use of specific antibodies as diagnostic reagents in immunoblot, immunohistochemistry, agglutination, IFAT, ELISA, lateral flow immunoassay, or other tests.
Hematology and clinical chemistry	Determination of hemocyte differential count, hemolymph clotting time, glucose, lactic acid, fatty acids, certain enzymes, etc.
Toxicology/ analysis	Detection of toxicants by analysis and verification of toxicity by bioassay
DNA probes	Detection of unique portions of a pathogen's nucleic acid using a labeled DNA probe.
PCR/RT-PCR	Amplification of unique sections of a pathogen's genome to readily detectable concentrations using specific primer pairs.
Tissue culture	*In vitro* culture of shrimp pathogens in non-shrimp tissue culture systems or in primary cell cultures derived from shrimp.

diagnosis. Tables 11.1, 11.6 and 11.7 list the OIE notifiable and listed shrimp viruses and the available diagnostic and detection methods for each.

Modern penaeid shrimp diagnostic and research laboratories employ methods that are based on traditional methods of disease diagnosis and pathogen detection that have been adapted from methods used in fish, veterinary, and human diagnostic laboratories. Methods for the detection of pathogens and the diagnosis of diseases that are currently in use by shrimp pathologists and by diagnostic labs (Table 11.5) have been reviewed

Table 11.6 Diagnostic and pathogen detection methods for OIE listed (and recently de-listed) viral diseases of penaeid shrimp (modified from Lightner 2005; Lightner *et al.* 2009; OIE 2009a)

Method	WSSV	TSV	IHHNV	YHV-group	BP	MBV	IMNV
Direct BF / LM / PH / DF	+	+	–	++	+++	+++	–
Histopathology	++	+++	++	+++	++	++	++
Bioassay	+	+	+	+	+	–	+
TEM	+	+	+	+	+	+	+
Antibody-based tests with PAb / MAb	++	++	–	–	+	–	–
DNA probes DBH / ISH	+++	+++	+++	+++	++	++	++
PCR / RT-PCR	+++	+++	+++	+++	+++	+++	+++

Definitions for each virus:
– = no known or published application of technique.
+ = application of technique known or published, but not commonly practiced or readily available.
++ = application of technique considered by authors of present paper to provide sufficient diagnostic accuracy or pathogen detection sensitivity for most applications.
+++ = technique provides a high degree of sensitivity in pathogen detection.
Methods: BF = bright field LM of tissue impression smears, wet-mounts, stained whole mounts; LM = light microscopy; PH = phase microscopy; DF = dark-field microscopy; TEM/SEM = transmission/scanning electron microscopy of sections or of purified or semi-purified virus; ELISA = enzyme-linked immunosorbent assay; PAb = polyclonal antibodies; MAb = monoclonal antibodies; DBH = dot blot hybridization; ISH = *in situ* hybridization; PCR = polymerase chain reaction; RT-PCR = reverse transcription PCR.

many times in the past decade (Lightner 1988, 1992, 1993, 1996a, 1999, 2005, 2011; Liu 1989; Brock and Lightner 1990; Johnson 1990, 1995; Brock 1991, 1992; Lightner and Redman 1991, 1992, 1998a; Brock and LeaMaster 1992; Flegel *et al.* 1992; Fulks and Main 1992; Lightner *et al.* 1992a, 1992b, 1994; Limsuwan 1993; Brock and Main 1994; Chanratchakool *et al.* 1998; Flegel 1997; Bondad-Reantaso *et al.* 2001; OIE 2009b). In penaeid shrimp pathology, diagnosticians rely heavily on case history, gross signs and behavior, morphological pathology (direct bright-field or phase contrast light microscopy and electron microscopy), and classical microbiology (bacteriology and mycology) (Table 11.5). Among the most important of these are gross and clinical signs, with the most commonly applied laboratory tests being direct examination and microscopy using the light microscope, classical microbiology with isolation and culture of the agent, and routine histology and histochemistry (Bell and Lightner 1988; Lightner 1996a). Virtually every functional shrimp pathology/diagnostic laboratory today is equipped to use direct light microscopic methods and routine procedures in histology and bacteriology (Vanpatten and Lightner 2001). Paradoxically, important

Table 11.7 Summary of methods for surveillance and confirmatory diagnosis of OIE listed viral diseases of penaeid shrimp (adapted from Lightner 2005; Lightner *et al.* 2009; OIE 2009b)

Agent	Surveillance	Confirmatory diagnosis
TSV	RT-PCR	RT-PCR, ISH with DNA probes, AB, histology
WSSV	PCR, AB	PCR, ISH or DBH with DNA probes, AB, histology, bioassay
YHV/GAV	RT-PCR	RT-PCR, ISH with DNA probes, AB, histology, bioassay
MBV	PCR, direct microscopy, histology	Direct microscopy, histology, PCR
BP	PCR, direct microscopy, histology	Direct microscopy, histology, PCR
IHHNV	PCR, DNA probes	PCR, ISH or DBH with DNA probes, histology
IMNV	RT-PCR	ISH with DNA probes, histology

Method abbreviations: AB = antibody-based method; DBH = dot blot hybridization; ISH = *in situ* hybridization; PCR = polymerase chain reaction; RT-PCR = reverse transcription PCR.

techniques involving tissue and cell culture, hematology, and clinical chemistry, which are virtual cornerstones of vertebrate biomedical research, diagnostics, and pathology, have either not been successfully applied as routine diagnostic tools in penaeid shrimp pathology (in the case of cell and tissue culture), or have not provided routinely practical diagnostic data (in the case of hematology and clinical chemistry of hemolymph samples) (Crane and Benzie 1999). Likewise, the development of antibody-based diagnostic methods for penaeid shrimp diseases has not been remarkable (Lightner and Redman 1998a; Lightner 1999, 2003a, 2003b, 2005). Some methods based on pathogen detection using monoclonal antibodies (Poulos *et al.* 1999, 2001; Takahashi *et al.* 2003; Bradley-Dunlop *et al.* 2004; Houghton *et al.* 2009) have been reported and some are commercially available. In marked contrast, molecular methods (using gene probes and gene amplification methods that employ PCR) have been found to provide accurate and standardizable methods for disease diagnosis and pathogen detection to the penaeid shrimp culture industries, especially for certain penaeid viruses (Lightner 1996a, 1999; Lightner *et al.* 2006; Walker and Subasinghe 2000; OIE 2009b; Tables 11.6 and 11.7).

Molecular diagnostic methods have become as important as classical methods (such as routine histopathology and microbiology) to the shrimp culture industry in recent years (Bondad-Reantaso *et al.* 2001; Lightner 1999; Walker and Subasinghe 2000; OIE 2009b). The first diagnostic tests using DNA-based technologies were reported less than 20 years ago. Vickers and co-workers (1992) reported on the application of PCR to detect MBV, and, in the same year, the first report of a non-radioactively labeled

DNA probe was reported for IHHNV (Lightner *et al.* 1992d; Mari *et al.* 1993). Today, DNA-based diagnostic tests are commonly employed for the detection of most of the major shrimp viruses and for several bacterial and parasitic diseases. For many of the penaeid shrimp virus diseases, molecular tests have become the 'gold standard' for disease diagnosis and for detection of their etiological agents (Bondad-Reantaso *et al.* 2001; Walker and Subasinghe 2000; OIE 2009b). For the OIE listed crustacean diseases, gene amplification methods (PCR and reverse transcription PCR, RT-PCR) are recommended for surveillance (screening) for all of the five currently listed shrimp viruses (Table 11.7).

There is a continuing need to standardize and validate the DNA-based diagnostic methods and the laboratories that use them (Walker and Subasinghe 2000). Standardization of DNA-based diagnostic methods is almost inherent in the nature of the tests. That is, a specific DNA probe, or a specific set of primers, is used to demonstrate the presence or absence of a unique DNA or RNA sequence that does not vary from batch to batch. Hence, with proper controls, these DNA-based methods are readily standardized (Walker and Subasinghe 2000). However, despite the growing dependence of the shrimp culture industry on DNA-based diagnostic methods, most of the tests that are available from commercial sources and from the technical published literature on the topic have not been validated using controlled field tests. Likewise, there are few formal accreditation or certification programs in place to assure that test results from technicians and laboratories running the tests are indeed accurate and properly controlled (Lightner and Redman 1998a; Lotz and Lightner 1999; Lightner 1999, 2003b, 2005; Lightner *et al.* 2009). The implementation of a formal program by appropriate international agencies or professional societies is needed to validate new diagnostic methods and to periodically review the accreditation and certification of diagnosticians and diagnostic laboratories. The establishment of regional reference laboratories for DNA-based diagnostic methods of penaeid shrimp/prawn pathogens would fit well into such a program with the goal of making these methods uniform, reliable, and readily applicable to disease control and management strategies for viral diseases of cultured penaeids.

Recognizing the need to validate commercially available diagnostic kits, the International Committee of the OIE passed a resolution in 2003 that established the legal basis for the OIE (with the direct assistance of OIE Reference Laboratories) to establish a registry of assays with levels of validation specified. Commercial kits registered by the OIE must be found by independent evaluation to be 'fit for purpose,' which means that the kit has to be validated to such a level to show that the kit's results can be interpreted to have a defined meaning in terms of diagnosis or another biological property being examined (see Certification of Diagnostic Assays at www. oie.int). A commercial PCR kit for detection of WSSV was among the first commercial kits certified by OIE.

11.5.3   Diagnosis versus surveillance and screening

Disease diagnosis and disease surveillance of shrimp stocks being reared in biosecure facilities are essential components of a functional biosecurity program in shrimp aquaculture. While the same diagnostic methods may be used for both surveillance (or screening) and disease diagnosis, their purpose and application in shrimp farming are different. The distinction between the two concepts is best illustrated by reviewing the pertinent OIE and FAO definitions for each and then adapting the concepts they represent to the context of disease control in biosecure production using SPF shrimp stocks. According to OIE (2009a), 'diagnosis' is the *determination of the nature of disease*, and 'disease' means *clinical or nonclinical infection with one or more etiological agents.* 'Surveillance' means *a systematic series of investigations of a given population of aquatic animals to detect the occurrence of disease for control purposes, and which may involve testing of samples of a population.*

The FAO (Subasinghe *et al.* 2004) further divides the term 'surveillance' into 'general (or passive) surveillance' and 'targeted surveillance.' 'General (passive) surveillance' is *an ongoing observation of the endemic disease profile of a susceptible population, so that unexpected and/or abnormal changes can be detected and acted upon as rapidly as possible.* 'Targeted surveillance' *collects information on a specific disease or condition so that its presence within a defined population can be measured, or its absence can be substantiated.*

For the purposes of disease diagnosis in biosecure culture systems, in most instances the intent is to determine the cause of an outbreak of clinical disease. The expression of disease can range from poor culture performance (i.e., poor growth, elevated food conversion ratio (FCR) values, reduced spawning success, etc.) to major mortality episodes. In clinical disease episodes, carefully selected quality specimens with representative lesions should be obtained. These specimens should be moribund or presenting clear signs of clinical disease and be representative of the disease(s) that is (are) affecting the cultured stock of interest. The samples should be preserved using methods appropriate for the intended or anticipated diagnostic test(s) (i.e., fixation for histology, preservation for PCR/RT-PCR testing, preservation for bacterial isolation and culture, etc.). Collection of dead specimens should be avoided, as post-mortem change and bacterial growth typically make such samples essentially useless. Hence, for diagnosis of clinical disease, the necessary sample size is usually small. For the diagnosis of clinical disease, it is suggested in the General Information chapter for crustacean diseases of OIE *Manual of Diagnostic Tests for Aquatic Animals* (OIE, 2009b) that the following guidelines for sample size be considered: shrimp larval stages, 100 specimens; PLs, 50 specimens, and 10 for juveniles and adults. Random sampling of affected populations during disease episodes should be avoided whenever specimens presenting clear signs of clinical disease(s) are apparent and available.

Surveillance and screening have a different purpose in biosecure shrimp farming. One major use of surveillance in biosecure shrimp production is for the development and maintenance of SPF stocks. For this application, targeted and general (passive) surveillance is performed. Because surveillance is typically carried out on 'healthy' populations when signs of clinical disease are not apparent, the sampling process must be random (Corsin *et al.* 2009). Although the prevalence of disease (asymptomatic infection in this context) may range from 0 to 100%, it is typically low in otherwise healthy populations (Lightner 2003a, 2005; Corsin *et al.* 2009; OIE 2009b). To provide a guideline for the required sample size for detection (using one or more specific tests for specific disease agents) of asymptomatic infections in this application of targeted surveillance, the OIE (see Corsin *et al.* 2009) recommends the use of computer programs such as FreeCalc Version 2 to determine the sample size required for 95% confidence that a disease agent, if present, will be detected at an assumed prevalence (usually 2, 5 or 10%). Because of their high sensitivity and specificity, samples taken for molecular (PCR/RT-PCR) or antibody-based tests may be combined as pooled samples of no more than five individuals for the actual testing (OIE 2009b). As a reminder, it may be necessary to pool larger samples of eggs, larvae or very small juveniles to obtain adequate nucleic acid for such tests. For expensive assays such as PCR/RT-PCR, the pooling of samples prior to testing may constitute very significant savings in diagnostic costs. The program FreeCalc Version 2 will calculate the sample size required for different levels of statistical confidence at assumed prevalence rates in populations. FreeCalc can also adjust sample size when the sensitivity and specificity of the screening test to be used are known to be less than 100%. FreeCalc Version 2 and the supporting documentation is available from the Australian Centre for International Agricultural Research (ACIAR) at http://www.aciar.gov.au/web.nsf/doc/JFRN-5J46ZY.

11.5.4   Biosecurity and the culture of wild seed and broodstock

While some application of biosecurity principles are possible in an industry that uses wild stocks for seed production, consistency in preventing the introduction of diseases and pathogens is problematic because of a variety of problems inherent in having laboratory testing performed. Such problems may include limitations to the accuracy and sensitivity of the test(s) used, representative sampling and sample sizes needed for statistical confidence, and problems with getting the required samples to diagnostic laboratories, tested, and reported within what is often a relatively short period of time between the time the wild seed stock is collected or spawned and the time by which transport and/or stocking must occur. Furthermore, the prevalence and severity of infection of significant pathogens in wild populations may be quite low, making their detection a difficult task. These factors lead frequently to false negative results when wild stocks (nauplii, larvae, postlarvae, or broodstock) are sampled and screened using even the most sensitive molecular methods available. Hence, while more sensitive and accurate diagnostic tests are becoming available each year, no test is likely to ever be 100% accurate (Corsin *et al.* 2009; OIE 2009b). The best way to be sure of the pathogen status of any given shrimp stock is have control of the stock (e.g. in quarantine or in culture) and to monitor it for specific pathogens over time, thus building a documented history of that particular stock as being free of specific pathogens. This is the concept of programs that develop domesticated lines of SPF shrimp, and this is the principal reason why domesticated lines of the Pacific white shrimp, *Litopenaeus vannamei*, so rapidly became the dominant shrimp species farmed in the world within 5 years after their introduction to East and SE Asia (FAO 2006).

11.5.5 Biosecurity through environmental control and best management practices

A variety of environmental and best management practice strategies have been adopted for the control of viral and other significant excludable diseases in penaeid shrimp aquaculture (Lee and O'Bryen 2003; Lightner 2005; Scarfe *et al.* 2006). These strategies range from the use of improved culture practices (i.e., where sources of virus contamination are reduced or eliminated, source water is treated, filtered, and aged to remove potential vectors, culture ponds are cleaned, plowed, and fallowed and treated between crops, routine sanitation practices are improved, stocking densities are reduced, etc.) to stocking domesticated SPF or SPR shrimp stocks. Some opportunistic disease agents (e.g., certain *Vibrio* spp.) are part of the shrimp's normal microflora, but can become deadly pathogens in 'stressed' shrimp. 'Stress' in shrimp is a poorly defined condition that is difficult to measure, and it has multiple causes that are not well understood. Its causes can range from the shrimp being subjected to environmental extremes to over- or under-feeding. Most penaeid shrimp have the best culture performance (i.e., growth and food conversion efficiency) at water temperatures near their upper tolerance limit for a particular life stage of the species (Villalon 1991; Chanratchakool *et al.* 1998; Fegan and Clifford 2001). Farms and management practices must recognize this factor, be tailored to benefit from the effect, and mitigate it when 'stress' and disease could result from water temperatures becoming too high for too long. Hence, the farm site, culture system design, the quality of feed used, stocking density, the farm's routine management practices, and other factors can have a profound effect on the amount of 'stress' to which farmed shrimp stocks are subjected. Therefore, diseases due to 'abiotic' agents (i.e., 'stress', toxicants, environmental extremes, nutritional imbalances, etc.), or those due to opportunistic 'biotic' agents that are either commonly present in the culture environment or part of the shrimp's normal microflora, are not excludable and should not be among the listed disease agents to be excluded in a biosecurity plan for a facility, compartment, zone, etc., or in an SPF stock domestication and development program. The management of such diseases, however, through farm design, the use of appropriate feeds and feed application, and the quality of overall management are nonetheless essential components of successful shrimp farming. Because these topics are beyond the scope of the present review, the authors refer the reader to reviews published elsewhere in which this topic has been thoroughly reviewed (Browdy and Jory 2001; Lee and O'Bryen 2003; Scarfe *et al.* 2006).

11.6   Conclusions

In the wake of the global shrimp disease pandemics, due principally to the shrimp-viruses TSV, WSSV and IMNV (OIE 2009b), that swept through the main penaeid shrimp-growing regions of both Asia and the Americas, there has been a paradigm change in what the industry farms and how it is done. 'SPF,' 'SPR' and 'biosecurity' were terms seldom heard in shrimp farming establishments a decade ago, but today these terms, and the concepts and practices they represent, are increasingly being applied by the global shrimp farming industry. The application of SPF, SPR and biosecurity concepts to many of the existing types of shrimp farming, as they have been applied to poultry for example, is not something that can be accomplished easily or in the short term. The industry has thousands of hectares of farms and hundreds of hatcheries (Rosenberry 2001) which were not designed to afford managers with much of an opportunity to totally prevent particular pathogens from being introduced and becoming established, or to exclude them during normal farming activities even if SPF shrimp are stocked. Nonetheless, with the use of applicable elements of the concepts of SPF, SPR and biosecurity much can be done to reduce losses due to particular pathogens by utilizing 'seed stocks' that are free of the major pathogens of concern and by modifying existing farms and their management routines to apply biosecure practices. The progress made by the shrimp farming industry by changing from farming mostly shrimp stocks obtained directly or indirectly from wild stocks to culturing domesticated stocks of *L. vannamei*, which are SPF for the major shrimp diseases and which have been improved through selective breeding for commercially valuable traits and improved performance characteristics, including disease resistance. This change has contributed significantly to making the industry more sustainable and environmentally responsible. These remarkable changes, and the trend away from farming wild shrimp to culturing only domesticated SPF shrimp, are likely to continue well into future.

11.7   Acknowledgements

Grant support for the author of this review was provided by the United States Marine Shrimp Farming Consortium under Grant No. 2010-38808-

21115, Hatch Project ARZT-136860-H-02-135 (both through the National Institute of Food and Agriculture, U.S. Department of Agriculture), and special grants from Darden Restaurants (Orlando, FL), Morrison Enterprises (Hastings, NE) and the National Fisheries Institute (McLean, VA).

11.8   References

AQUACOP. 1983. Construction of broodstock, maturation, spawning, and hatching systems for penaeid shrimps in the Centre Oceanologique du Pacifique. Pages 105–121 in: J.P. McVey (editor) *CRC Handbook of Mariculture*. Volume 1 *Crustacean Aquaculture*. CRC Press, Boca Raton, FL.

ARGUE, B., and A. ALCIVAR-WARREN. 1999. Genetics and breeding applied to the penaeid shrimp farming industry. Pages 29–53 in: R.A. Bullis and G.D. Pruder (editors) *Controlled and Biosecure Production Systems. Evolution and Integration of Shrimp and Chicken Models*. Proceedings of a Special Session, Sydney, Australia, 27–30 April 1999. World Aquaculture Society, Baton Rouge, LA.

BARTLEY, D.M., R. SUBASINGHE, and D. COATES. 1996. *Draft Framework for The Responsible Use of Introduced Species*. European Inland Fisheries Advisory Commission. EIFAC/XIX/96/Inf.8.

BELL, T.A., and D.V. LIGHTNER. 1984. IHHN virus: infectivity and pathogenicity studies in *Penaeus stylirostris* and *Penaeus vannamei*. *Aquaculture* **38**: 185–194.

BELL, T.A., and D.V. LIGHTNER. 1988. *A Handbook of Normal Shrimp Histology*. Special Publication No. 1, World Aquaculture Society, Baton Rouge, LA.

BONDAD-REANTASO, M.G., S.E. MCGLADDERY, I. EAST, and R.P. SUBASINGHE. 2001. *Asia Diagnostic Guide to Aquatic Animal Diseases*. FAO Fisheries Technical Paper 402/2, Food and Agriculture Organization of the United Nations, Rome, Italy.

BRADLEY-DUNLOP, D., C.R. PANTOJA, and D.V. LIGHTNER. 2004. Development of monoclonal antibodies for detecting necrotizing hepatopancreatitis in penaeid shrimp. *Dis. Aquatic Organisms* **60**: 233–240.

BRAY, W.A., A.L. LAWRENCE, and J.R. LEUNG-TRUJILLO. 1994. The effect of salinity on growth and survival of *Penaeus vannamei*, with observations on the interaction of IHHN virus and salinity. *Aquaculture* **122**: 133–146.

BROCK, J.A. 1991. An overview of diseases of cultured crustaceans in the Asia Pacific region. Pages 347–395 in: *Fish Health Management in Asia-Pacific*. Report on a Regional Study and Workshop on Fish Disease and Fish Health Management. ADB Agriculture Department Report Series No. 1. Network of Aquaculture Centres in Asia-Pacific. Bangkok, Thailand.

BROCK, J.A. 1992. Current diagnostic methods for agents and diseases of farmed marine shrimp. Pages 209–231 in: W. Fulks and K. Main (editors) *Proceedings of the Asian Interchange Program Workshop on the Diseases of Cultured Penaeid Shrimp in Asia and the United States*, Honolulu, Hawaii, 27–30 April 1992. The Oceanic Institute, Honolulu, HI.

BROCK, J.A. 1997. Special topic review: Taura syndrome, a disease important to shrimp farms in the Americas. *World Journal of Microbiology & Biotechnology* **13**: 415–418.

BROCK, J.A., and D.V. LIGHTNER. 1990. Diseases of crustacea. Diseases caused by microorganisms. Pages 245–349 in: O. Kinne (editor) *Diseases of Marine Animals*, Vol. III, Biologische Anstalt Helgoland, Hamburg, Germany.

BROCK, J.A., and B. LEAMASTER. 1992. A look at the principal bacterial, fungal and parasitic diseases of farmed shrimp. Pages 212–226 in: J. Wyban (editor) *Proceedings of the Special Session on Shrimp Farming*, Orlando, Florida, 22–25 May 1992, Orlando, Florida. World Aquaculture Society, Baton Rouge, LA.

BROCK, J.A., and K. MAIN. 1994. *A Guide to the Common Problems and Diseases of Cultured* Penaeus vannamei. Oceanic Institute, Hawaii, HI.

BROCK, J.A., D.V. LIGHTNER, and T.A. BELL. 1983. A review of four virus (BP, MBV, BMN, and IHHNV) diseases of penaeid shrimp with particular reference to clinical significance, diagnosis and control in shrimp aquaculture. *Proceedings of the 71st International Council for the Exploration of the Sea*, C.M. 1983/Gen:10/1-18.

BROWDY, C.L., and D.E. JORY (editors). 2001. *The New Wave, Proceedings of the Special Session on Sustainable Shrimp Culture*. The World Aquaculture Society, Baton Rouge, LA.

BROWDY, C.L., J.D. HOLLOWAY, C.O. KING, A.D. STOKES, J.S. HOPKINS, and P.A. SANDIFER. 1993. IHHN virus and intensive culture of *Penaeus vannamei*: effects of stocking density and water exchange rates. *J. Crustacean Biol.* **13**: 87–94.

BULLIS, R.A., and G.D. PRUDER (editors). 1999. *Shrimp Biosecurity: Pathogens and Pathogen Exclusion. Controlled and Biosecure Production Systems. Evolution and Integration of Shrimp and Chicken Models.* Proceeding of a Special Session, World Aquaculture Society, Sydney, Australia, Oceanic Institute, Honolulu, HI.

CARR, W.H., J.M. SWEENEY, and J.S. SWINGLE. 1994. The Oceanic Institute's specific pathogen free (SPF) shrimp breeding program: preparation and infrastructure. In: U.S. Marine Shrimp Farming Program 10th Anniversary Review, Gulf Coast Research Laboratory Special Publication. Ocean Springs, MS. *Gulf Research Reports* **1**: 47–54.

CARR, W.H., J.N. SWEENEY, L.M. NUNAN, D.V. LIGHTNER, H. H. HIRSCH, and J.J. REDDINGTON. 1996. The use of an infectious hypodermal and hematopoietic necrosis virus gene probe serodiagnostic field kit for the screening of candidate specific pathogen-free *Penaeus vannamei* broodstock. *Aquaculture* **147**: 1–8.

CHANRATCHAKOOL, P., J.F. TURNBULL, S.J. FUNGE-SMITH, I.H. MACRAE, and C. LIMSUWAN. 1998. *Health Management in Shrimp Ponds.* Aquatic Animal Health Research Institute, Dept. of Fisheries, Kasetsart University, Bangkok, Thailand.

CORSIN, F., M. GEORGIADIS, K.L. HAMMELL, and B. HILL. 2009. *Guide for Aquatic Animal Health Surveillance.* The World Animal Organisation for Animal Health (OIE), Paris.

CRANE, M. ST. J., and J.A.H. BENZIE (editors). 1999. Proceedings of the Aquaculture CRC International Workshop on Invertebrate Cell Culture, 2–4 November 1997, University of Technology, Sydney, Australia. *Methods of Cell Science* **21**(4): 171–272.

ERICKSON, H.S., M. ZARIN-HERZBERG, and D.V. LIGHTNER. 2002. Detection of Taura syndrome virus (TSV) strain differences using selected diagnostic methods: diagnostic implications in penaeid shrimp. *Diseases of Aquatic Organisms* **52**: 1–10.

FAO. 2006. *State of World Aquaculture.* FAO Fisheries Technical Paper 500, Food and Agriculture Organization of the United Nations, Rome.

FAO/NACA/UNEP/WB/WWF (Food and Agricultural Organization of the United Nations, United Nations Environmental Programme, World Bank Group, World Wildlife Fund). 2006. *International Principles for Responsible Shrimp Farming*. Network of Aquaculture Centres in Asia-Pacific (NACA). Bangkok, Thailand.

FAUQUET C.M., M.A. MAYO, J. MANILOFF, U. DESSELBERGER, and L.A. BALL. 2005. *Virus Taxonomy. Classification and Nomenclature of Viruses. Eighth Report of the International Committee on Taxonomy of Viruses.* Elsevier Academic Press.

FEGAN, D.F., and H.C. CLIFFORD III. 2001. Health management for viral diseases in shrimp farms. Pages 168–198 in: C.L. Browdy and D.E. Jory (editors) *The New Wave, Proceedings of the Special Session on Sustainable Shrimp Culture. Aquaculture 2001.* The World Aquaculture Society, Baton Rouge, LA.

FLEGEL, T.W. 1997. Major viral diseases of the black tiger prawn (*Penaeus monodon*) in Thailand. *World J. Microbiol. Biotechnol.* **13**: 433–442.

FLEGEL, T.W. 2006. Detection of major penaeid shrimp viruses in Asia, a historical perspective with emphasis on Thailand. *Aquaculture* **258**: 1–33.

FLEGEL, T.W., and V. ALDAY-SANZ. 1998. The crisis in Asian shrimp aquaculture: current status and future needs. *J. Appl. Ichthyol* **14**: 269–273.

FLEGEL, T.W., D.F. FEGAN, S. KONGSOM, S. VUTHIKOMUDOMKIT, S. SRIURAIRTANA, S. BOONYARATPALIN, C. CHANTANACHOOKIN, J.E. VICKERS, and O.D. MACDONALD. 1992. Occurrence, diagnosis and treatment of shrimp diseases in Thailand. Pages 57–112 in: W. Fulks and K.L. Main (editors) *Diseases of Cultured Penaeid Shrimp in Asia and the United States*. The Oceanic Institute, Honolulu, HI.

FORESTER, J.R.M., and T.W. BEARD. 1974. Experiments to assess the suitability on nine species of prawns for intensive cultivation. *Aquaculture* **3**: 355–368.

FULKS, W., and K.L. MAIN (editors). 1992. *Diseases of Cultured Penaeid Shrimp in Asia and the United States*. The Oceanic Institute, Honolulu, HI.

GAA (Global Aquaculture Alliance). 1999a. Shrimp white spot virus confirmed in Central America. *GAA Newsletter*, vol. 2 issue 2.

GAA (Global Aquaculture Alliance). 1999b. Shrimp white spot disease in Latin America: an update. *GAA Newsletter*, vol. 2 issue 3.

HASSON K.W., D.V. LIGHTNER, L.L. MOHNEY, R.M. REDMAN, B.T. POULOS, J. MARI, and J.R. BONAMI. 1999. The geographic distribution of Taura Syndrome Virus (TSV) in the Americas: determination by histology and *in situ* hybridization using TSV-specific cDNA probes. *Aquaculture* **171**: 13–26.

HENNIG O.L., C.R. PANTOJA, S.M. ARCE, B. WHITE-NOBLE, S.M. MOSS, and D.V. LIGHTNER. 2005. Development of a specific pathogen free population of the Chinese fleshy prawn *Fenneropenaeus chinensis*. Part 2: Secondary quarantine. *Aquaculture* **250**: 579–585.

HOUGHTON, R.L., J. CHEN, S. MORKOWSKI, S. RAYCHAUDHURI, C. PANTOJA, B.T. POULOS, and D.V. LIGHTNER. 2009. Rapid test detects NPH in penaeid shrimp. *Global Aquaculture Advocate*, **July/August**: 64.

HUTCHINS, D.L., G.W. CHAMBERLAIN, and J.C. PARKER. 1979. Length-weight relations for several species of penaeid shrimp cultured in ponds near Corpus Christi, Texas. *Proc. World Maricult. Soc.* **10**: 565–570.

ICES. 1995. *ICES Code of Practice on the Introductions and Transfers of Marine Organisms 1994 / Code de Conduite du CIEM pour les Introductions et Transferts d'Organismes Marins 1994*. ICES Cooperative Research Report No. 204. Copenhagen, Denmark.

ICES. 2004. ICES code of practice on the introduction and transfers of marine organisms 2004. www.ices.dk/reports/general/2004/ICESCOP2004.pdf

JIMENEZ, R. 1992. Sindrome de Taura (Resumen). Pages 1–16 in: *Acuacultura del Ecuador*. Camara Nacional de Acuacultura, Guayaquil, Ecuador.

JOHNSON, S.K. 1990. *Handbook of Shrimp Diseases*. Sea Grant Publ. No. TAMU-SG-90-601. Texas A&M University, College Station, TX.

JOHNSON, S.K. 1995. *Handbook of Shrimp Diseases*. Sea Grant Publ. No. TAMU-SG-95-601. Texas A&M University, College Station, TX.

KALAGAYAN, G., D. GODIN, R. KANNA, G. HAGINO, J. SWEENEY, J. WYBAN, and J. BROCK. 1991. IHHN virus as an etiological factor in runt-deformity syndrome of juvenile *Penaeus vannamei* cultured in Hawaii. *J. World Aquaculture Soc.* **22**: 235–243.

LEE, C.S., and P.J. O'BRYEN (editors). 2003. *Biosecurity in Aquaculture Production Systems: Exclusion of Pathogens and Other Undesirables*. The World Aquaculture Society, Baton Rouge, Louisiana, USA.

LIGHTNER, D.V. 1983. Diseases of Cultured Penaeid Shrimp. Page 289–320 in: J.P. McVey (editor) *CRC Handbook of Mariculture*. Vol. 1. *Crustacean Aquaculture*, CRC Press, Boca Raton, FL.

LIGHTNER, D.V. 1988. Diseases of cultured penaeid shrimp and prawns. Pages 8–127 in: C.J. Sindermann and D.V. Lightner (editors) *Disease Diagnosis and Control in North American Marine Aquaculture*, Elsevier, Amsterdam, The Netherlands.

LIGHTNER, D.V. 1992. Shrimp viruses: diagnosis, distribution and management. Pages 238–253 in: J. Wyban (editor) *Proceedings of the Special Session on Shrimp Farming*, Orlando, Florida, 22–25 May 1992. World Aquaculture Society, Baton Rouge, LA.

LIGHTNER, D.V. 1993. Diseases of penaeid shrimp. Pages 393–486 in: J.P. McVey (editor) *CRC Handbook of Mariculture: Crustacean Aquaculture*, 2nd Edition. CRC Press, Boca Raton, FL.

LIGHTNER D.V. (editor). 1996a. *A Handbook of Shrimp Pathology and Diagnostic Procedures for Diseases of Cultured Penaeid Shrimp.* World Aquaculture Society, Baton Rouge, LA.

LIGHTNER, D.V. 1996b. Epizootiology, distribution and the impact on international trade of two penaeid shrimp viruses in the Americas. *Rev. Sci. Techn. Office Int. Epizooties* **15**: 579–601.

LIGHTNER, D.V. 1999. The penaeid shrimp viruses TSV, IHHNV, WSSV, and YHV: current status in the Americas, available diagnostic methods and management strategies. *J. Appl. Aquacult.* **9**: 27–52.

LIGHTNER, D.V. 2003a. The penaeid shrimp viral pandemics due to IHHNV, WSSV, TSV and YHV: history in the Americas and current status. Pages 1–24 in: Y. Sakai, J.P. McVey, D. Jang, E. McVey and M. Caesar (editors) *Proceedings of the Thirty-second US Japan Symposium on Aquaculture.* US–Japan Cooperative Program in Natural Resources (UJNR). US Department of Commerce, N.O.A.A., Silver Spring, MD. http://www.lib.noaa.gov/japan/aquaculture/aquaculture_panel.htm

LIGHTNER, D.V. 2003b. Exclusion of specific pathogens for disease prevention in a penaeid shrimp biosecurity program. Pages 81–116 in: C.-S, Lee and P.J. O'Bryen (editors) *Biosecurity in Aquaculture Production Systems: Exclusion of Pathogens and Other Undesirables.* The World Aquaculture Society, Baton Rouge, LA.

LIGHTNER, D.V. 2005. Biosecurity in shrimp farming: pathogen exclusion through use of SPF stock and routine surveillance. *J. World Aquaculture Soc.* **36**: 229–248.

LIGHTNER, D.V. 2011. Virus diseases of farmed shrimp in the Western Hemisphere. *J. Invertebrate Pathology* **106**: 110–130. (G. Stentiford, editor special issue on diseases of edible crustaceans).

LIGHTNER D.V., and C.R. PANTOJA. 2001. Biosecurity in shrimp farming. Pages 123–165 in M.C. Haws and C.E. Boyd (editors) *Methods for Improving Shrimp Farming in Central America.* USDA, Hurricane Mitch Reconstruction Project, UCA Press, Managua (in English and Spanish).

LIGHTNER, D.V., and R.M. REDMAN. 1985. A parvo-like virus disease of penaeid shrimp. *J. Invertebr. Pathol.* **45**: 47–53.

LIGHTNER, D.V., and R.M. REDMAN. 1991. Hosts, geographic range and diagnostic procedures for the penaeid virus diseases of concern to shrimp culturists in the Americas. Pages 173–196 in: P. DeLoach, W.J. Dougherty, and M.A. Davidson (editors) *Frontiers of Shrimp Research.* Elsevier, Amsterdam, The Netherlands.

LIGHTNER, D.V., and R.M. REDMAN. 1992. Penaeid virus diseases of the shrimp culture industry of the Americas. Pages 569–588 in: A.W. Fast and L.J. Lester (editors) *Culture of Marine Shrimp: Principles and Practices.* Elsevier, Amsterdam, The Netherlands.

LIGHTNER, D.V., and R.M. REDMAN.1998a. Shrimp diseases and current diagnostic methods. *Aquaculture* **164**: 201–220.

LIGHTNER, D.V., and R.M. REDMAN. 1998b. Strategies for the control of viral diseases of shrimp in the Americas. *Fish Pathology* **33**: 165–180.

LIGHTNER, D.V., R.M. REDMAN, and T.A. BELL. 1983a. Infectious hypodermal and hematopoietic necrosis a newly recognized virus disease of penaeid shrimp. *J. Invertebr. Pathol.* **42**: 62–70.

LIGHTNER, D.V., R.M. REDMAN, T.A. BELL, and J.A. BROCK. 1983b. Detection of IHHN virus in *Penaeus stylirostris* and *P. vannamei* imported into Hawaii. *J. World Mariculture Soc.* **14**: 212–225.

LIGHTNER, D.V., R.R. WILLIAMS, T.A. BELL, R.M. REDMAN, and L.A. PEREZ 1992a. A collection of case histories documenting the introduction and spread of the virus disease IHHN in penaeid shrimp culture facilities in northwestern Mexico. *ICES Marine Science Symposia* **194**: 97–105.

LIGHTNER, D.V., R.M. REDMAN, T.A. BELL, and R.B. THURMAN. 1992b. Geographic dispersion of the viruses IHHNV, MBV and HPV as a consequence of transfers and introductions of penaeid shrimp to new regions for aquaculture purposes. Pages 155–173 in: A. Rosenfield and R. Mann (editors) *Dispersal of Living Organisms into Aquatic Ecosystems*. University of Maryland System, College Park, MD, UM-SG-TS-92-04.

LIGHTNER, D.V., T.A. BELL, R.M. REDMAN, L.L. MOHNEY, J.M. NATIVIDAD, A. RUKYANI, and A. POERNOMO. 1992c. A review of some major diseases of economic significance in penaeid prawns/shrimps of the Americas and Indopacific. Pages 57–80 in: I.M Shariff, R.P. Subasinghe, and J.R. Arthur (editors) *Diseases in Asian Aquaculture*. Fish Health Section, Asian Fisheries Society, Manila, Philippines.

LIGHTNER, D.V., B.T. POULOS, R.M. REDMAN, J. MARI, and J.R. BONAMI. 1992d. New developments in penaeid virology: application of biotechnology in research and disease diagnosis for shrimp viruses of concern in the Americas. Pages 233–253 in: W. Fulks and K.L. Main (editors) *Diseases of Cultured Penaeid Shrimp*. The Oceanic Institute, Honolulu, HI.

LIGHTNER, D.V., R.M. REDMAN, D.W. MOORE, and M.A. PARK. 1993. Development and application of a simple and rapid diagnostic method to studies on hepatopancreatic parvovirus of penaeid shrimp. *Aquaculture* **116**: 15–23.

LIGHTNER, D.V., B.T. POULOS, L. BRUCE, R.M. REDMAN, L. NUNAN, C. PANTOJA, J. MARI, and J.R. BONAMI. 1994. Development and application of genomic probes for use as diagnostic and research reagents for the penaeid shrimp parvoviruses IHHNV and HPV, and the baculoviruses MBV and BP. U.S. Marine Shrimp Farming Program 10th Anniversary Review, Gulf Coast Research Laboratory Special Publication, Ocean Springs, MS, USA. *Gulf Research Reports* **1**: 59–85.

LIGHTNER, D.V., R.M. REDMAN, K.W. HASSON, and C.R. PANTOJA. 1995. Taura syndrome in *Penaeus vannamei* (Crustacea: Decapoda): gross signs, histopathology and ultrastructure. *Diseases of Aquatic Organisms* **21**: 53–59.

LIGHTNER, D.V., R.M. REDMAN, B.T. POULOS, L.M. NUNAN, J.L. MARI, and K.W. HASSON. 1997. Risk of spread of penaeid shrimp viruses in the Americas by the international movement of live and frozen shrimp. *Rev. Sci. Tech. Office Int. Epizooties* **16**: 146–160.

LIGHTNER, D.V., B.T. POULOS, K.F.J. TANK-NELSON, C.R. PANTOJA, L.M. NUNAN, S.A. NAVARRO, R.M. REDMAN, and L.L. MOHNEY. 2006. Application of molecular diagnostic methods to penaeid shrimp diseases: advances of the past 10 years for control of viral diseases in farmed shrimp. Pages 117–122 in: P. Vannier and D. Espeseth (editors) *New Diagnostic Technology: Application in Animal Health and Biologics Controls*. Dev Biol (Basel). Basel, Karger, Vol. 126.

LIGHTNER, D.V., R.M. REDMAN, S. ARCE, and S.M. MOSS. 2009. Chapter 16: Specific pathogen-free shrimp stocks in shrimp farming facilities as a novel methods for disease control in crustaceans. Pages 384–424 in: S.E. Shumway and G.E. Rodrick (editors) *Shellfish Safety and Quality*. Woodhead Publishing Limited, CRC Press, Boca Raton, FL.

LIMSUWAN, C. 1993. *Diseases of Black Tiger Shrimp* Penaeus monodon *Fabrics in Thailand*. Technical Bulletin, American Soybean Association, M.I.T.A.(P) No. 518/12/92, Vol. AQ39, Singapore.

LIU, C.I. 1989. Shrimp disease, prevention and treatment. Pages 64–74 in: D.M. Akiyama (editor) *Proceedings of the Southeast Asia Shrimp Farm Management Workshop*. American Soybean Association, Singapore.

LO, C.F., C.H. HO, S.E. PENG, C.H. CHEN, H.C. HSU, Y.L. CHIU, C.F. CHANG, K.F. LIU, M.S. SU, C.H. WANG, and G.H. KOU. 1997. White spot syndrome baculovirus (WSBV) detected in cultured and captured shrimp, crabs and other arthropods. *Diseases of Aquatic Organisms* **27**: 215–225.

LOTZ, J.M. 1992. Developing specific pathogen-free (SPF) animal populations for aquaculture: a case study for IHHN virus of penaeid shrimp. Pages 269–284 in: W. Fulks and K.L. Main (editors) *Diseases of Cultured Penaeid Shrimp in Asia and the United States*. The Oceanic Institute, Honolulu.

LOTZ, J.M. 1997a. Disease control and pathogen status assurance in an SPF-based shrimp aquaculture industry, with particular reference to the United States. Pages 243–254 in: Flegel T.W. and MacRae I.H. (editors) *Diseases in Asian Aquaculture III*, Fish Health Section, Asian Fisheries Society, Manila, The Philippines.

LOTZ, J.M. 1997b. Special topic review: viruses, biosecurity and specific pathogen-free stocks in shrimp aquaculture. *World J. Microbiol. Biotechnol.* **13**: 405–413.

LOTZ, J.M., and D.V. LIGHTNER. 1999. Shrimp biosecurity: pathogens and pathogen exclusion. Pages 67–74 in: R.A. Bullis and G.D. Pruder (editors) *Controlled and Biosecure Production Systems. Evolution and Integration of Shrimp and Chicken Models*. Proceedings of a Special Session, 27–30 April 1999, Sydney, Australia. World Aquaculture Society, Baton Rouge, LA.

LOTZ, J.M., C.L. BROWDY, W.H. CARR, P.F. FRELIER, and D.V. LIGHTNER. 1995. USMSFP suggested procedures and guidelines for assuring the specific pathogen status of shrimp broodstock and seed. Pages 66–75 in: C.L. Browdy and J.S. Hopkins (editors) *Swimming through Troubled Water*, Proceedings of the Special Session on Shrimp Farming, Aquaculture '95. San Diego, California, 1–4 February 1995. World Aquaculture Society, Baton Rouge, LA.

MARI, J., J.R. BONAMI, and D. LIGHTNER. 1993. Partial cloning of the genome of infectious hypodermal and hematopoietic necrosis virus, an unusual parvovirus pathogenic for penaeid shrimps; diagnosis of the disease using a specific probe. *J. General Virology* **74**: 2637–2643.

MOORE, D.W., and C.W. BRAND. 1993. The culture of marine shrimp in controlled environment superintensive systems. In: J.P. McVey (editor) *CRC Handbook of Mariculture*. Volume 1 *Crustacean Aquaculture*, 2nd Edition. CRC Press, Boca Raton, FL. pp. 315–348.

MORALES-COVARRUBIAS, M.S., and M.C. CHAVEZ-SANCHEZ. 1999. Histopathological studies on wild broodstock of white shrimp *Penaeus vannamei* in the Platanitos area, adjacent to San Blas, Nayarit, Mexico. *J. World Aquaculture Soc.* **30**: 192–200.

MORALES-COVARRUBIAS, M.S., L.M. NUNAN, D.V. LIGHTNER, J.C. MOTA-URBINA, M.C. GARZA-AGUIRRE, and M.C. CHAVEZ-SANCHEZ. 1999. Prevalence of IHHNV in wild broodstock of *Penaeus stylirostris* from the upper Gulf of California, Mexico. *J. Aquat. Anim. Health* **11**: 296–301.

MOSS, S.M. 2002. Marine shrimp farming in the Western Hemisphere: past problems, present solutions, and future visions. *Revi. Fisheries Sci.* **10**(3 & 4): 601–620.

MOSS, S.M., and D.R. MOSS. 2009. Selective breeding of penaeid shrimp. Pages 425–452 in: S.E. Shumway and G.E. Rodrick (editors) *Shellfish Safety and Quality*. Woodhead Publishing Limited, CRC Press, Boca Raton, FL.

MOSS, S.A., and G.D. PRUDER. 1999. Shrimp biosecurity: environmental management and control. Pages 79–83 in: R.A. Bullis and G.D. Pruder (editors) *Controlled and Biosecure Production Systems. Evolution and Integration of Shrimp and Chicken Models*. Proceeding of a Special Session, World Aquaculture Society, 27–30 April 1999, Sydney, Australia. The Oceanic Institute, Honolulu, HI, USA.

MOSS, S.M., S.M. ARCE, B.J. ARGUE, C.A. OTOSHI, F.R.O. CALDERON, and A.G.J. TACON. 2002. Greening of the blue revolution: efforts toward environmentally responsible shrimp culture. Pages 1–19 in: C.L. Browdy and D.E. Jory (editors) *The New Wave, Proceedings of the Special Session on Sustainable Shrimp Culture, Aquaculture 2001*. The World Aquaculture Society, Baton Rouge, LA.

MOSS, S.M., D.R. MOSS, S.M. ARCE, and C.A. OTOSHI. 2003. Disease prevention strategies for penaeid shrimp culture. Pages 34–46 in: Y. Sakai, J.P. McVey, D. Jang, E. McVey and M. Caesar (editors) *Proceedings of the Thirty-second US Japan Symposium on Aquaculture*. US–Japan Cooperative Program in Natural Resources (UJNR). US Department of Commerce, N.O.A.A., Silver Spring, MD. http://www.lib.noaa.gov/japan/aquaculture/aquaculture_panel.htm

MOTTE, E., E. YUGCHA, J. LUZARDO, F. CASTRO, G. LECLERCQ, J. RODRÍGUEZ, P. MIRANDA, O. BORJA, J. SERRANO, M. TERREROS, K. MONTALVO, A. NARVÁEZ, N. TENORIO, V. CEDEÑO, E. MIALHE, and V. BOULO. 2003. Prevention of IHHNV vertical transmission in the white shrimp *Litopenaeus vannamei. Aquaculture* 219: 57–70.

MURPHY, F.A., C.M. FAQUET, D.H.L. BISHOP, S.A. GHABRIAL, A.W. JARVIS, G.P. MARTELLI, M.A. MAYO, and M.D. SUMMERS. 1995. *Virus Taxonomy. Classification and Nomenclature of Viruses. Sixth Report of the International Committee on Taxonomy of Viruses. Archives of Virology Supplement 10*, Springer-Verlag, Wein, NY.

NUNAN, L.M., S.M. ARCE, R.J. STAHA, and D.V. LIGHTNER. 2001. Prevalence of infectious hypodermal and hematopoietic necrosis virus (IHHNV) and white spot syndrome virus (WSSV) in *Litopenaeus vannamei* in the Pacific Ocean off the coast of Panama. *J. World Aquaculture Soc.* 32: 330–334.

OIE (Office International des Epizooties). 2009a. *Aquatic Animal Health Code*, 12th edition. Office International des Epizooties, Paris, France.

OIE (Office International des Epizooties). 2009b. *Manual of Diagnostic Tests for Aquatic Animal Diseases*, 6th edition. Office International des Epizooties, Paris, France.

OIE (Office International des Epizooties). 2010. *Aquatic Animal Health Code*, 13th edition. Office International des Epizooties, Paris, France.

PANTOJA, C.R., D.V. LIGHTNER, and K.H. HOLTSCHMIT. 1999. Prevalence and geographic distribution of IHHN parvovirus in wild penaeid shrimp (Crustacea: Decapoda) from the Gulf of California, Mexico. *J. Aquatic Animal Health* 11: 23–34.

PANTOJA, C.R., X. SONG, L. XIA, H. GONG, J. WILKENFELD, B. NOBLE, and D.V. LIGHTNER. 2005. Development of a specific pathogen free (SPF) population of the Chinese fleshy prawn *Fenneropenaeus chinensis*. Part 1: Disease pre-screening and primary quarantine. *Aquaculture* 250: 573–578.

POULOS, B.T., R. KIBLER, D. BRADLEY-DUNLOP, L.L. MOHNEY, and D.V. LIGHTNER. 1999. Production and use of antibodies for the detection of the Taura syndrome virus in penaeid shrimp. *Diseases of Aquatic Organisms* 37: 99–106.

POULOS, B.T., C.R. PANTOJA, D. BRADLEY-DUNLOP, J. AGUILAR, and D.V. LIGHTNER. 2001. Development and application of monoclonal antibodies for the detection of white spot syndrome virus of penaeid shrimp. *Diseases of Aquatic Organisms* 47: 13–23.

PRUDER, G.D. 1995. *U.S. Marine Shrimp Farming Program Progress Report*. Volume I, page 18, Institution Reports. Oceanic Institute, Honolulu, HI.

PRUDER G.D., C.L. BROWN, J.N. SWEENEY, and W.H. CARR. 1995. High health shrimp systems: seed supply – theory and practice. Pages 40–52 in: C.L. Browdy and J.S. Hopkins (editors) *Swimming through Troubled Water, Proceedings of the Special Session on Shrimp Farming, Aquaculture '95*. San Diego, California, 1–4 February 1995. World Aquaculture Society, Baton Rouge, LA.

ROBLES-SIKISAKA, R., D.K. GARCIA, K.R. KLIMPEL, and A.K. DHAR. 2001. Nucleotide sequence of 3′-end of the genome of Taura syndrome virus of shrimp suggests that it is related to insect picornaviruses. *Arch. Virol.* 146: 941–952.

ROSENBERRY, B. (editor). 1998. *World Shrimp Farming 1998*. Number 11, Shrimp News International, San Diego, CA.

ROSENBERRY, B. (editor). 2001. *World Shrimp Farming 2000*. Number 13, Shrimp News International, San Diego, CA.

ROSENBERRY, B. (editor). 2006. *World Shrimp Farming 2006*. Number 19, Shrimp News International, San Diego, CA.

SALSER B., L. MAHLER, D.V. LIGHTNER, J. URE, D. DANALD, C. BRAND, N. STAMP, D. MOORE, and B. COLVIN. 1978. Controlled-environment aquaculture of penaeids. Pages 345–355 in: P.N. Kaul and C.J. Sindermann (editors) *Drugs and Food from the Sea, Myth or Reality?* The University of Oklahoma Press, Norman, OK.

SCARFE A.D., C.S. LEE, and P.J. O'BRYEN (editors). 2006. *Aquaculture Biosecurity. Prevention, Control, and Eradication of Aquatic Animal Disease*. Blackwell, Ames, IA.

SINDERMANN, C.J. 1988. Disease problems created by introduced species. Pages 394–398 in: C.J. Sindermann and D.V. Lightner (editors) *Disease Diagnosis and Control in North American Marine Aquaculture. Developments in Aquaculture and Fisheries Science*, Vol. 17. Elsevier, Amsterdam, The Netherlands.

SINDERMANN, C.J. 1990. *Principal Diseases of Marine Fish and Shellfish*, Volume 2, 2nd Edition. Academic Press, NY.

SUBASINGHE, R.P., S.E. MCGLADDERY, and B.J. HILL. 2004. *Surveillance and Zoning for Aquatic Animal Diseases*. FAO Fisheries Technical Paper 451. FAO, Rome, Italy.

TAKAHASHI, Y., K. FUKUDA, M. KONDO, A. CHONGTHALEONG, K. NISHI, M. NISHIMURA, K. OGATA, I. SHINYA, K. TAKISE, Y. FUJISHIMA, and M. MATSUMURA. 2003. Detection and prevention of WSSV infection in cultured shrimp. *Asian Aquaculture Magazine*, **November/December**: 25–27.

TANG, K.F.J., and D.V. LIGHTNER. 2002. Low sequence variation among isolates of infectious hypodermal and hematopoietic necrosis virus (IHHNV) originating from Hawaii and the Americas. *Diseases of Aquatic Organisms* **49**: 93–97.

TANG, K., and D.V. LIGHTNER. 2004. Geographic variation among IHHNV isolates provides insights into their pathology, spread. *Global Aquaculture Advocate* **7**(1): 91–92.

TANG, K.F.J., S.V. DURAND, B.L. WHITE, R.M. REDMAN, C.R. PANTOJA and D.V. LIGHTNER. 2000. Postlarvae and juveniles of a selected line of *Penaeus stylirostris* are resistant to infectious hypodermal and hematopoietic necrosis virus infection. *Aquaculture* **190**: 203–210.

TANG, K.F.J., B.T. POULOS, J. WANG, R.M. REDMAN, H. SHIH, and D.V. LIGHTNER. 2003. Geographic variations among infectious hypodermal and hematopoietic necrosis virus (IHHNV) isolates and characteristics of their infection. *Diseases of Aquatic Organisms* **53**: 91–99.

TURNER, G. E. 1988. *Codes of Practice and Manual of Procedures for Consideration of Introductions and Transfers of Marine and Freshwater Organisms*. EIFAC Occasional Paper No. 23. European Inland Fisheries Advisory Commission. Food and Agriculture Organization of the United Nations.

VANPATTEN, K.A., and D.V. LIGHTNER. 2001. Molecular diagnostic methods used by 92% of labs; respondents support 'ring test' for greater standardization. *Global Aquaculture Advocate* **4**(4): 50–51.

VICKERS, J.E., R.J.G. LESTER, P.B. SPRADBROW, and J.M. PEMBERTON. 1992. Detection of *Penaeus monodon*-type baculovirus (MBV) in digestive glands of postlarval prawns using polymerase chain reaction. Pages 127–133 in: M. Shariff, R.P. Subasinghe, and J.R. Arthur (editors) *Diseases in Asian Aquaculture I. Proceeding of the First Symposium on Diseases in Asian Aquaculture*, Bali, Indonesia, 26–29 November 1990. Fish Health Section, Asian Fisheries Society, Manila, Philippines.

VILLALON, J.R. 1991. *Practical Manual for Semi-intensive Commercial Production of Marine Shrimp*. Texas A&M University Sea Grant College Program, TAMU-SG-91-501.

WALKER, P., and R. SUBASINGHE, editors. 2000. *DNA-based Molecular Diagnostic Techniques. Research Needs for Standardization and Validation of the Detection of Aquatic Animal Pathogens and Diseases*. FAO Fisheries Technical Paper 395. Report and Proceedings, Bangkok, Thailand, February 7–9, 1999.

WALKER, P.J., and C.V. MOHAN. 2009. Viral disease emergence in shrimp aquaculture: origins, impact and the effectiveness of health management strategies. *Rev. Aquaculture* 1: 125–154.

WEPPE, M., AQUACOP, J.R. BONAMI and D.V. LIGHTNER. 1992. Demonstracion de las altas cualidades de la cepa de *P. stylirostris* (AQUACOP SPR 43) resitente al virus IHHN. Pages 229–232 in: J. Calderon and L. Shartz (editors) *Proceedings of the Primer Congresso Ecuatoriano de Acuiculture*. 19–23 October 1992, Guayaquil Ecuador.

WICKINS, J.F., and T.W. BEARD. 1978. *Prawn Culture Research*. Ministry of Agriculture Fisheries and Food, Directorate of Fisheries Research. Laboratory Leaflet No. 42, Lowestoft.

WYBAN J.A. 1992. Selective breeding of specific pathogen-free (SPF) shrimp for high health and increased growth. Pages 257–268 in: W. Fulks and K.L. Main (editors) *Diseases of Cultured Penaeid Shrimp in Asia and the United States*. The Oceanic Institute, Honolulu, HI.

WYBAN J.A., J. SWINGLE, J.N. SWEENEY, and G.D. PRUDER. 1992. Development and commercial performance of high health shrimp from SPF broodstock *Penaeus vannamei*. Pages 254–260 in: J. Wyban (editor) *Proceedings of the Special Session on Shrimp Farming*, Orlando, Florida, 22–25 May 1992. World Aquaculture Society, Baton Rouge, LA.

ZARIN-HERZBERG, M., and F. ASCENSIO-VALLE. 2001. Taura syndrome in Mexico: follow-up study in shrimp farms of Sinaloa. *Aquaculture* **193**: 1–9.

ZAVALA, G. 1999. Biosecurity in the poultry industry. Pages 75–78 in: R.A. Bullis and G.D. Pruder (editors) *Controlled and Biosecure Production Systems. Evolution and Integration of Shrimp and Chicken Models. Proceedings of a Special Session*, Sydney, Australia, 27–30 April 1999. The Oceanic Institute, Honolulu, HI.

12

The role of risk analysis in the development of biosecurity programmes for the maintenance of specific pathogen-free populations

C. J. Rodgers, IRTA-Sant Carles de la Ràpita, Spain and
E. J. Peeler, Centre for Environment, Fisheries and Aquaculture
Science (CEFAS), UK

Abstract: Risk analysis has an increasingly important role to play in developing surveillance and biosecurity at a national level, as aquaculture and trade in aquatic animals grows. A key aim of health management in aquaculture is to maintain disease freedom at the level of the farm, compartment, geographic zone or country. International recognition of disease freedom allows appropriate biosecurity measures to be put in place. Risk analysis is an important tool for assessing routes of exotic disease entry and establishment and thus allows a country to develop and defend its biosecurity strategy. Under international guidelines a risk assessment may be required to justify trade restrictive measures. Disease surveillance is needed to both establish disease-free areas and maintain them by ensuring early detection of exotic pathogen incursions. Applying risk approaches to the design of surveillance programmes ensures the most efficient use of the available resources. The constraints to the current application of risk assessment (e.g. lack of data) and the development of improved methods, notably in the area of consequence assessment and the assessment of emerging diseases, are discussed. Despite these constraints, risk analysis has to date been a useful tool in decision making related to managing the spread of established diseases of farmed aquatic animals.

Key words: aquaculture health management, risk analysis, biosecurity, surveillance.

12.1 Introduction

Health management in aquaculture relies to a great extent on maintaining the freedom from specified pathogens status of discrete populations of animals. A population may be a single farm, a collection of farms under the same biosecurity system (i.e. a compartment) or in the same region (a zone) or country. Tools and practices have been developed to assist the health manager maintain the disease freedom of populations under his or her care.

Risk analysis is essentially a tool to aid decision making in the face of uncertainty. It uses a structured approach to identify potential hazards and the likelihood that they occur, as well as the subsequent consequences of their presence. The results of this process can be used to determine ways of reducing the likelihood and impact of the hazards. In this chapter the hazards considered are pathogens.

The recent growth of aquaculture has been associated with the use of the term 'blue revolution' to describe its major potential for supplying high quality protein in answer to the demand for food from an increasing global population. For instance, the FAO (2008a) predicts that 'aquaculture is for the first time set to contribute half of the fish consumed by the human population worldwide', with an annual average growth rate of 7% from 1970 to 2006. This increase has put pressure on many aquatic ecosystems that have been increasingly poorly managed and are in danger of collapse through overexploitation. The emergence and intensification of aquaculture have therefore led to the development of a greater diversity of aquatic animal species and the systems used for their sustainable production. However, this growth has not been without its problems, since significantly increased trade of live aquatic animals and, to a lesser extent, their products has led to the emergence of pathogens or the appearance of certain diseases in new geographical areas (Rodgers *et al.*, 2011).

Effective surveillance and monitoring systems have been developed to generate epidemiological data about the presence or absence of diseases and their prevalence. Such systems provide vital information about the spread of diseases that can be used in the control and containment of hazards (e.g. pathogens) or the demonstration of freedom from disease status. In fact, the analysis of surveillance data can produce early warning indicators and trends in disease patterns that can support the need for biosecurity contingency plans, such as import controls or the establishment of buffer zones.

In this chapter, the primary role risk analysis can play in developing surveillance and biosecurity at a national level is discussed. The application of the same principles at other population levels is also considered within the context of aquaculture.

12.2   Aquaculture disease risk analysis

Risk analysis has a number of applications in aquatic health management, including supporting the development of sanitary measures, broader biosecurity policy and surveillance (Peeler *et al.*, 2007), as well as understanding disease emergence in aquaculture (Murray and Peeler, 2005) and the effects due to climate change (Marcos-López *et al.*, 2010). The structured approach of risk analysis can be applied qualitatively or quantitatively for evaluating the likelihood and severity of hazards (e.g. pathogen introduction). Identifying all the necessary steps needed for the hazard to occur allows a systematic evaluation of the data currently available and a synthesis of the complex interrelationships of disease processes. Outputs need to capture the likelihood of establishment, the potential economic and environmental costs and the uncertainties in these estimates, so that they inform the decision making process. Risk communication should involve all interested parties at critical stages of the risk analysis process, to ensure both transparency and wide acceptance of the results.

The development of aquaculture disease risk analysis can be traced back to the World Trade Organization (WTO) Agreement on the Application of Sanitary and Phytosanitary Measures (SPS Agreement) in the context of international trade, and movements of live animals and their products (WTO, 1994). Since then there has been an increase in trade that has to be considered in terms of the harmonization of health control measures, and the fact that globalization should not jeopardize the health status of the animal populations of trading nations. The science-based standard setting organization for international trade in animals and their products, and the promotion of aquatic animal health within this scenario, is the World Organisation for Animal Health (Office International des Epizooies; OIE, http://www.oie.int/en/).

The basic step-wise procedure generally used in aquatic animal risk analysis, as recommended by the OIE, comprises hazard identification, risk assessment, risk management and risk communication:

- Hazard identification – in the context of trade this is the process of identifying the pathogenic agent(s) that could potentially be introduced in the commodity considered for importation (OIE, 2010a) (i.e. which specific diseases are of concern to an epidemiological unit; Håstein *et al.*, 2008).
- Risk assessment – the evaluation of the likelihood and the biological and economic consequences of entry, establishment and spread of a hazard within the territory of an importing country (OIE, 2010a). Thus, risk assessment can be broken down into three parts: (i) release assessment (entry into the importing country), (ii) exposure assessment (contact of the hazard with a susceptible population), and, (iii) consequence assessment (establishment of infection and spread, ecological and economic impact) (see Fig. 12.1).
- Risk management – the process of identifying, selecting and implementing measures that can be applied to reduce the level of risk (OIE, 2010a).
- Risk communication – the interactive exchange of information and opinions throughout the risk analysis process concerning risk, risk-related factors and risk perceptions among risk assessors, risk managers, risk communicators, the general public and other interested parties

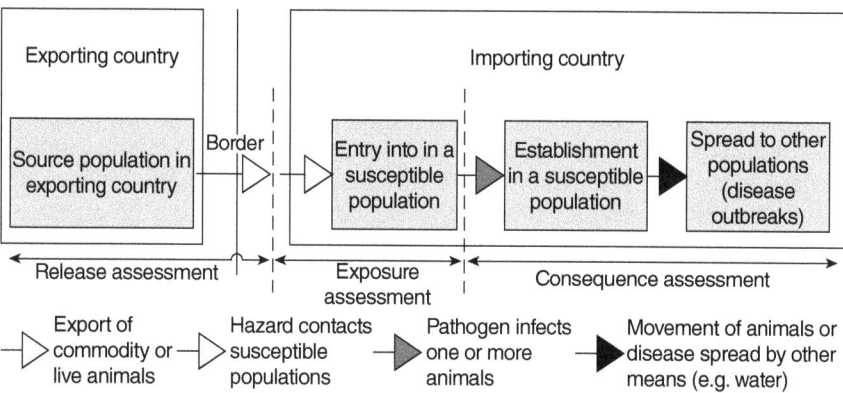

Fig. 12.1   The steps of risk assessment, pathways of disease introduction and spread.

(OIE, 2010a), as well as how the identified risks will be assessed for effectiveness on an ongoing basis (Håstein *et al.*, 2008).

The growth of aquaculture in recent decades has led to an enhanced awareness of biosecurity related to the complexity of the different hazards that represent risks to the industry sector (FAO, 2008b). This, in turn, has put pressure on decision makers that need to be able to use a reliable technique to manage the growing number of risks associated with current trade and production practices.

The risk evaluation level can vary depending on the area of application (e.g. financial services, biological ecosystems, environmental pollution, disease processes, food safety, trade logistics). However, essentially, it is a good cost-effective technique for targeting finite resources to manage adverse repercussions before any possible hazard can gain entry and establish itself in a balanced system. This approach has been developed in recent decades largely following a change in perspective from insistence on impractical risk avoidance (the zero risk position) to acceptable risk management for reducing the effects of any identified undesirable hazard. However, the issues leading to the development of risk analysis-based techniques have become more complex, particularly with increasing globalization of trade. Consequently, the methodology to deal with new problems is continuously evolving in order to adapt to the complicated interdependencies of different biological compartments exposed to constant and changing pressures.

12.3   Application of risk analysis to surveillance and biosecurity strategies

Application of risk analysis is critical for providing the evidence base which supports transparent and consistent decision making for the level of acceptable risk associated with the commercial import of aquatic animals, disease spread within countries, and disease interaction between aquaculture and wild fish populations. Risk-based methods have effectively been used to improve the efficiency of surveillance for terrestrial livestock and shift analysis from inputs towards output standards (e.g. confidence that disease is absent). The same approaches are now being applied in surveillance systems for aquatic animal diseases (stimulated by the requirement of the EC Fish Health Directive 2006/88 (EC, 2006).

12.3.1   Risk-based surveillance (RBS)

Essentially, aquatic animal disease surveillance provides 'scientifically accurate, cost-effective, information for assessing and managing risks of disease transfer associated with trade (intra and international)' (FAO, 2004). This approach towards management and control therefore helps to prevent the spread of diseases through the proactive identification of outbreaks, as well as providing baseline prevalence data and other important information in support of risk assessments.

In general, the objectives of veterinary risk-based surveillance (RBS) have been defined as the identification of surveillance needs to protect the health of livestock, to set priorities, and to allocate resources effectively and efficiently (Stärk et al., 2006). In addition, the purpose of RBS is to ensure that limited resources are used to best effect (Stärk et al., 2006).

The design of an RBS programme includes the selection of the most important disease or pathological agent (e.g. the hazard with the highest consequences), as well as the selection of strata and units (sampling) which are at highest risk of infection (Stärk et al., 2006). Thus, based on existing epidemiological information, high risk rivers, bays or farms will be sampled more frequently than other epidemiological units. This will produce a surveillance system which generates a higher level of confidence in disease freedom, compared with a system which uses the same level of resources but in which all animals have an equal chance of being sampled. Risk factors may include, among others, frequency of live fish movements, proximity to processing plants, density of farms, and contact with wild populations.

It is recognized that disease surveillance is a highly important component of a competent authority's monitoring strategy designed to provide data concerning the presence (or absence) of certain key pathogens, especially exotic pathogens. Nevertheless, such strategies are nearly always faced with resource constraints. This means that effectively most surveillance plans are targeted towards notifiable diseases rather than also include screening programmes for emerging (or re-emerging) diseases that are, arguably, as potentially important in situations where they might become increasingly problematic, such as those related to changing production or trade practices. The data generated by surveillance is vital for disease emergency preparedness (Baldock, 2004) because early detection is particularly critical to minimize the impact of exotic disease incursions and new diseases. Nevertheless, ideally, given adequate (usually greater) resources, surveillance programmes should also include a monitoring component for identifying the patterns and trends of existing diseases in any given country or region, since the possibility of transmission to new areas or naïve susceptible populations is high.

12.3.2   Freedom from disease and containment

Establishing and demonstrating freedom from disease is of fundamental importance in protecting the health of farmed (and wild) aquatic animal populations. Under international guidelines (e.g. the SPS Agreement), a declaration of disease freedom may be recognized. The OIE publishes guidance on how international recognition of disease freedom may be achieved at the level of the member state (MS), zone or compartment (OIE, 2010a).

The currently accepted definitions of a compartment and zone are as follows (OIE, 2010a):

- Compartment – means one or more aquaculture establishments under a common biosecurity management system containing an aquatic animal population with a distinct health status with respect to a specific disease or diseases for which required surveillance and control measures are applied and basic biosecurity conditions are met for the purpose of international trade.
- Zone – means a portion of one or more countries comprising:
  - an entire water catchment from the source of a waterway to the estuary or lake, or more than one water catchment, or
  - part of a water catchment from the source of a waterway to a barrier that prevents the introduction of a specific disease or diseases, or
  - part of a coastal area with a precise geographical delimitation, or
  - an estuary with a precise geographical delimitation, that consists of a contiguous hydrological system with a distinct health status with respect to a specific disease or diseases.

Disease surveillance is required to generate evidence that the country (or zone or compartment) has been able to maintain disease freedom. Appropriate biosecurity measures to maintain disease freedom need to be applied at the same level as the surveillance system. Under international agreements, disease freedom allows countries to restrict the introduction of animals (and products not considered safe) from areas of a lower disease status. A country may consider restrictions of trade not explicitly allowed for under international agreement, for instance, in the case where a proposed import of an aquatic animal product presents an unacceptable disease risk. In these circumstances, risk analysis has a clearly defined role within the SPS Agreement, and the importing country must support its decision to restrict trade with a scientifically based risk analysis.

Thus, international recognition of disease freedom has a role to play in facilitating safe trade, at international, regional and national levels. Furthermore, risk analysis has a more general role, outside the narrow confines of international trade agreements, in designing biosecurity tailored to the most important routes of disease introduction and surveillance systems for detecting any exotic disease incursion.

12.3.3   Import risk analysis (IRA) and biosecurity

The types of import risk analysis (IRA) and their uses are summarized in Table 12.1. IRA for aquatic animals has largely been applied to gain greater understanding of disease introduction caused by international trade (Peeler et al., 2007). The SPS Agreement requires the WTO member countries to use the risk analysis procedure to justify scientifically any possible restrictions on international trade applied through excessive or additional health measures related to a health risk, over and above those measures sanctioned by international agreement (Rodgers, 2004; Peeler et al., 2007). It is therefore important that a consistent and systematic approach to an IRA is used. The OIE provides an overview of the process (OIE, 2010b). In addition, the OIE Handbook on Import Risk Analysis for Animals and Animal Products (Murray et al., 2004) contains a reasonably comprehensive breakdown of the IRA process, which is also encapsulated in the handbooks published by Biosecurity Australia (2009a) and the New Zealand Ministry of Agriculture (Murray, 2002). The main stages of IRA are summarized in Table 12.2. The outputs of an IRA need to be compared with the acceptable level of risk that a country must apply equally across commodities.

Aquatic IRAs have been reviewed by Peeler et al. (2007). One of the first and best known aquatic IRAs was undertaken by the Australian Government to assess the disease risk associated with the import of wild caught salmon from Canada (Kahn et al., 1999a). This work was completed with the explicit purpose of implementing appropriate risk mitigation measures. Other similar work includes assessment of disease risks associated with existing trade: import of ornamental fish (Kahn et al., 1999b) and penaeid shrimp to Australia (Biosecurity Australia, 2009b). Peeler et al. (2006) analysed the likely risk of Gyrodactylus salaris disease spread associated with a potential new route of introduction (live smolts imported from seawater sites). Within Norway, risk analysis has been used to inform decision making around the spread of the parasite between river catchments (Høgåsen and Brun, 2003) and from a farm to a wild population (Paisley et al., 1999).

More generally, IRA supports biosecurity measures to minimize the likelihood of disease entry and establishment. For example, mechanical introduction of pathogens through the movement of vehicles was assessed by Peeler and Thrush (2009).

Table 12.1 Types of import risk analysis for aquatic animal health

Reason for an IRA	Example	Approach
Proposed new commodity import	Import of wild caught salmon carcasses from Canada to Australia (Kahn *et al.*, 1999a)	Hazard identification (HI), followed by pathogen-specific IRA for most important hazard(s)
Proposed change in legislation allowing new patterns of trade	Change in EU legislation allowing import of salmon smolts to areas declared free of *Gyrodactylus salaris* (Peeler *et al.*, 2006)	Comparison of existing and proposed pathways of hazard introduction
Review of current mitigation measures for: A specific pathogen	Review of pathways of introduction of *Gyrodactylus salaris* to the UK (Peeler and Thrush, 2004)	Comparison of pathways of introduction and impact of current risk mitigation.
Current trade	Import of penaeid shrimp to Australia (Biosecurity Australia, 2009b) Import of ornamental fish into Australia (Kahn *et al.*, 1999b)	HI, followed by IRA for most important hazards.
Introduction of a non-native species	No available examples	HI, e.g. via the disease module of the FISK risk protocol (Copp *et al.*, 2009), followed by pathogen-specific IRA for most important hazard(s)
Review of biosecurity related to movement of vehicles or other mechanical routes of introduction	Movement of live fish transporters between the UK and continental Europe (Peeler and Thrush, 2009)	Comparison of likelihood of the introduction of different pathogens via the same pathway.

12.3.4 Application of risk analysis to support health management for a compartment or zone

Farms and businesses are not under the same legal obligations as national governments to use risk analysis to support measures which are trade restrictive. However, the principles of risk analysis can be applied equally to a single farm or group of farms to ensure measures are in place to minimize disease introduction and establishment. However, since risk analyses are generally undertaken by governments for legal or trade reasons, the incentives to use them at geographic levels below the level of a country are

Table 12.2 The import risk analysis (IRA) process (based on the Australian and New Zealand published IRA procedures)

Action

1. Import risk analysis initiated
2. Establish a project team
3. Develop risk communication strategy (identify stakeholders)
4. Scope the study
5. Initial consultation with stakeholders
6. Assemble relevant technical documents
7. Undertake the IRA
8. Internal peer review of draft IRA
9. External peer-review
10. Submit draft report for stakeholder consultations
11. Compile a report of stakeholder comments and documented responses
12. Modify the IRA on the basis of technical submissions by the stakeholders
13. Release of final report

less. Nevertheless, a few have been conducted. La Patra *et al.* (2001) used risk analysis to investigate the risk of spread of infectious haematopoietic necrosis virus with carcasses from an endemic to a free area. The spread of *G. salaris* from a single salmon farm (Paisley *et al.*, 1999) and from a single river have also been assessed (Høgåsen and Brun, 2003). These studies show that a risk analysis which is spatially explicit and discrete allows a more accurate assessment to be made.

12.4   Constraints and future trends

Despite the usefulness of risk analysis applied to the various aquaculture sectors, there can be problems with its application to aquatic animal diseases when the quantity or quality of the available data that needs to be assessed is variable. This is often related to a lack of understanding of pathogen transmission processes or pathogen survival parameters in fish products. In addition, there are data gaps that lead to interpretation difficulties or uncertainties, which vary greatly between aquacultured fish, shellfish and crustaceans. Methodological challenges also remain, since satisfactory methods for combining qualitative likelihood estimates have not been achieved. Similarly, approaches that combine epidemiological and economic modelling are needed to improve ad hoc assessments of the consequences of disease spread.

The expansion of aquatic animal risk analysis into other areas is a distinct possibility in the future. For instance, the current European legislation (e.g. Directive 2006/88) is driving a risk-based approach for characterizing aquaculture farms through risk ranking exercises, with improved identification of risk factors related to aquatic animal health and disease spread. Risk-based surveillance could also benefit from the use of network analysis to interpret national or international fish movements using information from contact chasing. Current risk approaches are not well suited to assessing the threat of new and emerging diseases (because the hazard has not been clearly defined); thus, there is a need for better data on generic pathways for introduction and establishment. To this end, more databases, validated data sources and supportive epidemiological studies are required.

Many published risk assessments have not attempted to assess consequences or have only superficially estimated likely impacts (Peeler *et al.*, 2007). In part, this may be explained by the lack of available methodology to assess the environmental and economic costs of disease incursion. This is an important area for further development, which should include the integration of disease modelling to estimate the likely extent of outbreaks (Thrush *et al.*, 2011). For some commodities (e.g. ornamental aquatic animals and bait fish) there is a need for greater delineation and understanding of the risk pathways involved in pathogen transmission.

Although the current health standards are designed to reduce the disease risk associated with transboundary movements of live aquatic animals, it is clear that transmission of disease between aquacultured populations still occurs. Consequently, risk analysis has an important role to play in reinforcing the understanding of the available epidemiological data in the face of the predicted further increase of trade. In this respect, there has been a suggestion that a more holistic approach is needed to deal with risk-based threats at source, which would involve the modification of existing standards to improve their efficacy in the face of increasing pressure from the expansion of aquaculture (Rodgers *et al.*, 2011).

The future increase in the application of risk analysis to aquatic health management and disease processes will require improved training and capacity building, as well as supportive targeted research for closing any identified data gaps, which, at the moment, can be provided only through the use of solicited expert opinion. Nevertheless, improved use of expert opinion is also needed and this could be in conjunction with risk-based studies through the application of techniques such as the Delphi method.

Although risk analysis has wide-ranging application to many areas of aquaculture, there is still potential for its development in areas where it is less well used (e.g. environmental, biological, operational, financial and social risks) (Arthur, 2008).

12.5   Conclusions

There is an inevitable tension between free trade and maintaining disease freedom. In the past, countries have used health standards for trade protectionist reasons. Consequently, international agreements and standards have evolved that allow countries to use agreed measures to maintain disease freedom (in a country or zone). Additional measures must be scientifically justified and risk analysis is the tool that is used to systematically assess the supportive evidence. The advantages of this approach are that a solid evidence base is generated, and assumptions and judgements are documented. However, there are limitations and, most importantly, risk analysis does not assess the hazards which have not been identified, for example, pathogens which emerge only when introduced to new hosts (Peeler *et al.*, 2010). Risk analysis has been applied extensively for purposes other than to inform the management of international trade in aquatic animals and products. It has proven to be a useful tool in decision making related to managing the spread of established diseases (e.g. *G. salaris* in Norway). Moreover, the principles of risk analysis are now starting to be applied to the design of surveillance systems and greater uptake of the methods will allow its use in other areas of aquaculture health management.

12.6   References

ARTHUR J.R. (2008), General principles of the risk analysis process and its application to aquaculture. In: M.G. Bondad-Reantaso, J.R. Arthur and R.P. Subasinghe (eds). *Understanding and Applying Risk Analysis in Aquaculture*. FAO Fisheries and Aquaculture Technical Paper 519, 3–8.

BALDOCK C. (2004), Disease surveillance. In: J.R. Arthur and M.G. Bondad-Reantaso. (eds.) *Capacity and awareness building on import risk analysis for aquatic animals*. Proceedings of the Workshops held 1–6 April 2002 in Bangkok, Thailand and 12–17 August 2002 in Mazatlan, Mexico. APEC FWG 01/2002, NACA, Bangkok, 37–42.

BIOSECURITY AUSTRALIA. (2009a), Australian Government Department of Agriculture, Fisheries and Forestry 2009, *Import Risk Analysis Handbook 2007* (update 2009), Canberra. Available at: http://www.daff.gov.au/__data/assets/pdf_file/0003/1177833/IRA_handbook_2009_FINAL_FOR_WEB.pdf (accessed on 30th March 2011).

BIOSECURITY AUSTRALIA. (2009b), *Generic Import Risk Analysis Report for Prawns and Prawn Products*. Biosecurity Australia, Canberra, Australia. Available at: http://www.daff.gov.au/__data/assets/pdf_file/0012/1302411/Final_prawn_IRA_report_7_Oct_09.pdf (accessed on 30th March 2011).

COPP G.H. VILIZZI L. MUMFORD J. FENWICK G.V. GODARD M.J. and GOZLAN R.E. (2009), Calibration of FISK, an invasiveness screening tool for non-native freshwater fishes. *Risk Analysis*, **29**, 457–467.

EC (EUROPEAN COMMISSION). (2006), Council Directive 2006/88/EC of 24 October 2006 on animal health requirements for aquaculture animals and products thereof, and on the prevention and control of certain diseases in aquatic animals. *OJEU*, 24.11.2006, L 328/14–56.

FAO (FOOD AND AGRICULTURE ORGANIZATION OF THE UNITED NATIONS). (2004), *Surveillance and zoning for aquatic animal diseases*. R.P. Subasinghe, S.E. McGladdery and B.J. Hill (eds), FAO Fisheries and Aquaculture Technical Paper 451.

FAO (FOOD AND AGRICULTURE ORGANIZATION OF THE UNITED NATIONS). (2008a), *The State of World Fisheries and Aquaculture*. Available at: http://www.fao.org/docrep/011/i0250e/i0250e00.htm (accessed on 30th March 2011).

FAO (FOOD AND AGRICULTURE ORGANIZATION OF THE UNITED NATIONS). (2008b), *Understanding and applying risk analysis in aquaculture.* M.G. Bondad-Reantaso, J.R. Arthur and R.P. Subasinghe (eds), FAO Fisheries and Aquaculture Technical Paper 519.

HÅSTEIN T. BINDE M. HINE M. JOHNSEN S. LILLEHAUG A. OLESEN N.J. PURVIS N. SCARFE A.D. and WRIGHT B. (2008), National biosecurity approaches, plans and programmes in response to diseases in farmed aquatic animals: evolution, effectiveness and the way forward. *Rev Sci Tech Off Int Epiz*, **27**, 125–145.

HØGÅSEN H.R. and BRUN E. (2003), Risk of inter-river transmission of *Gyrodactylus salaris* by migrating Atlantic salmon smolts, estimated by Monte Carlo simulation. *Dis Aquat Org*, **57**, 247–254.

KAHN S.A. BEERS P.T. FINDLAY V.L. PEEBLES I.R. DURHAM P.J. WILSON D.W. and GERRITY S.E. (1999a), *Import Risk Analysis on Non-viable Salmonids and Non-salmonids Marine Finfish.* Australian Quarantine and Inspection Service, Canberra, p. 409. Available at: http://www.daff.gov.au/__data/assets/pdf_file/0016/16081/finalfinfish. pdf (accessed on 30th March 2011).

KAHN S.A. WILSON D.W. PERERA R.P. HAYDER H. and GERRITY S.E. (1999b), *Import Risk Analysis on Live Ornamental Finfish.* Australian Quarantine and Inspection Service, Canberra, p. 172. Available at: http://www.daff.gov.au/__data/assets/pdf_ file/0018/16362/finalornamental.pdf (accessed on 30th March 2011).

LAPATRA S.E. BATTS W.N. OVERTURF K. JONES G.N. SHEWMAKER W.D. and WINTON J.R. (2001), Negligible risk associated with the movement of processed rainbow trout, *Oncorhynchus mykiss* (Walbaum), from an infectious haematopoietic necrosis virus (IHNV) endemic area. *J Fish Dis*, **24**, 399–408.

MARCOS-LÓPEZ M. GALE P. OIDTMANN B.C. and PEELER E.J. (2010), Assessing the impact of climate change on disease emergence in freshwater fish in the United Kingdom. *Transbound Emerg Dis*, **57**, 293–304.

MURRAY A.G. and PEELER E.J. (2005), A framework for understanding the potential for emerging diseases in aquaculture. *Prev Vet Med*, **67**, 223–235.

MURRAY N. (2002), *Handbook on Animal Import Risk Analysis.* Ministry of Agriculture and Forestry, Biosecurity New Zealand. Available at: http://www.biosecurity. govt.nz/pests-diseases/animals/risk/import-risk-analysis-handbook.htm (accessed on 30th March 2011).

MURRAY N. MACDIARMID S.C. WOOLDRIDGE M. GUMMOW B. MORLEY R.S. WEBER S.E. GIOVANNINI A. and WILSON D. (2004), *OIE Handbook on Import Risk Analysis for Animals and Animal Products – Introduction and qualitative risk analysis.* Paris, France: OIE.

OIE (WORLD ORGANISATION FOR ANIMAL HEALTH). (2010a), *Aquatic Animal Health Code.* 13th Edition. Available at: http://www.oie.int/en/international-standard-setting/aquatic-code/access-online/ (accessed on 30th March 2011).

OIE (WORLD ORGANISATION FOR ANIMAL HEALTH). (2010b), *Aquatic Animal Health Code.* 13th Edition. Section 2. Risk analysis. Available at: http://www.oie. int/index.php?id=171&L=0&htmfile=titre_1.2.htm (accessed on 30th March 2011).

PAISLEY L.G. KARLSEN E. JARP J. and MO T.A. (1999), A Monte Carlo simulation model for assessing the risk of introduction of *Gyrodactylus salaris* to the Tana River, Norway. *Dis Aquat Org*, **37**, 145–152.

PEELER E.J. and THRUSH M.A. (2004), Qualitative analysis of the risk of introducing *Gyrodactylus salaris* into the United Kingdom. *Dis Aquat Org*, **62**, 103–113.

PEELER E.J. and THRUSH M.A. (2009), Assessment of exotic fish disease introduction and establishment in the United Kingdom via live fish transporters. *Dis Aquat Org*, **83**, 85–95.

PEELER E. THRUSH M. PAISLEY L. and RODGERS C. (2006), An assessment of the risk of spreading the fish parasite *Gyrodactylus salaris* to uninfected territories in the

European Union with the movement of live Atlantic salmon (*Salmo salar*) from coastal waters. *Aquaculture*, **258**, 187–197.

PEELER E.J. MURRAY A.G. THEBAULT A. BRUN E. GIOVANINNI A. and THRUSH M.A. (2007), The application of risk analysis in aquatic animal health management. *Prev Vet Med*, **81**, 3–20.

PEELER E.J. OIDTMANN B. MIDTLYNG P.J. MIOSSEC L. and GOZLAN R.E. (2010), Non-native aquatic animals introductions have driven disease emergence in Europe. *Biol Invasions*, **13**, 1291–1303.

RODGERS C.J. (2004), Risk analysis in aquaculture and aquatic animal health. In: J.R. Arthur and M.G. Bondad-Reantaso (eds.) *Capacity and Awareness Building on Import Risk Analysis for Aquatic Animals*. Proceedings of the Workshops held 1–6 April 2002 in Bangkok, Thailand and 12–17 August 2002 in Mazatlan, Mexico. APEC FWG 01/2002, NACA, Bangkok, 59–64.

RODGERS C.J. MOHAN C.V. and PEELER E.J. (2011), The spread of pathogens through trade in aquatic animals and their products. *OIE Sci Tech Rev*, **30** (1), 241–256.

STÄRK K.D. REGULA G. HERNANDEZ J. KNOPF L. FUCHS K. MORRIS R.S. and DAVIES P. (2006), Concepts for risk-based surveillance in the field of veterinary medicine and veterinary public health: review of current approaches. *BMC Health Services Research* **6**, 20.

THRUSH M.A. MURRAY A.G. BRUN E. WALLACE S. and PEELER E.J. (2011), The application of risk and disease modelling to emerging freshwater diseases in wild aquatic animals. *Freshwater Biol*, **56**, 658–675.

WTO (WORLD TRADE ORGANIZATION). (1994), Agreement on the application of sanitary and phytosanitary measures. In: *The Results of the Uruguay Round of Multilateral Trade Negotiations: The Legal Texts*. General Agreement on Tariffs and Trade (GATT), World Trade Organization, Geneva. Available at: http://www.wto.org/english/docs_e/legal_e/15sps_01_e.htm (accessed on 30th March 2011).

13

Developments in genomics relevant to disease control in aquaculture

S. MacKenzie and S. Boltaña, Universitat Autonoma de Barcelona, Spain, B. Novoa and A. Figueras, Instituto de Investigaciones Marinas (IIM), CSIC, Spain and F. W. Goetz, University of Wisconsin-Milwaukee, USA

Abstract: The use of genomic technologies in aquaculture has experienced a rapid expansion over the past decade where functional genomics exemplified by microarray technology has experienced a more comprehensive development than actual genome sequencing. The significant increase in genomic resources driven by large-scale expressed sequence tag (EST) sequencing projects and more recently high throughput sequencing technologies has driven this development. Health management and prevention of disease are major issues in aquaculture, and genomics has contributed to a number of key areas in disease research including basic research aimed toward mRNA transcript discovery relevant to the immune response in diverse phylogenetic groups, evaluation of vaccine efficacy and immunostimulant diets, identification of the underlying mechanisms of disease resistance and the development of prognostic markers.

Key words: microarrays, transcriptome, fish, shellfish, response, pathogen, gene discovery.

13.1 Introduction

In all organisms, vertebrate and invertebrate, the nonspecific or innate immune response is immediately active in response to pathogen challenge and tissue damage, and is not antigen-specific. Innate immunity maintains host integrity and is mainly based upon physiological and inflammatory responses although response diversity and interaction with other regulatory events in the organism reflect a complex network of reactions that in many cases is poorly understood. Furthermore, indirect effects of the highly regulated immune response such as excessive or inadequate responses can cause significant deleterious effects upon the individual.

Fish are the largest group of vertebrates and the first phylogenetic group to display a complete immune system formed by the innate and adaptive immune response in comparison to the invertebrates. The adaptive immune response is in principle similar across the vertebrates, including the mammals. However, there are some distinct differences specifically emphasizing the function and organization of the adaptive immune response, antibody repertoire and the existence of a specific memory response between the fish and other vertebrate groups, including mammals. Research in comparative immunology seeks to delineate these differences, both in innate and adaptive responses across significant phylogenetic distances and provide understanding that can be exploited to drive the development of disease management and tools therein. Owing to the diverse interacting factors from both biological and environmental sources found within the context of intensive aquaculture systems functional genomics approaches provide a 'holistic' approach toward understanding of the disease process. In this chapter, split into two major sections (firstly finfish followed by shellfish and crustaceans), we will review the use of functional genomics in disease control and provide selected examples.

13.2   The genomics toolbox in comparative immunology

13.2.1   Microarrays

Over the past decade microarray analysis, i.e. simultaneous measurement of thousands of mRNAs, has become increasingly more common and popular in comparative immunology. This technology has evolved from the generation of custom-built cDNA array platforms to the availability of commercial oligonucleotide array platforms for a number of fish species. Although microarrays are in principle able to reflect the diversity of mRNAs in an experimental sample, the information obtained is in effect only as good as the sequences on the array. Thus underlying the application of microarray technology is the representation of mRNAs and therefore the quality of the sequencing and annotation of different mRNA populations. This is of significant importance in all biological systems where the transcriptome is a dynamic and highly regulated entity. Thus if the strategy for gene discovery has accounted for the activation of a particular biological system in a time-dependent manner then it is more likely to obtain a wider representation of mRNAs relevant to that response (Goetz et al., 2004). This has a significant impact in particular in immunity as many sequencing projects have used 'healthy' individuals, therefore the possibility of identifying transcripts relevant to an immune response is clearly much lower. In the particular case of the immune system a specific response has been shown to mobilize more transcriptional remodelling than most physiological responses studied in vitro in established mammalian cell lines (Abbas et al., 2005). Thus we must consider the immune response as a highly complex network involving multiple cellular phenotypes/modules and a large range of related proteins that form specific regulatory networks. In terms of transcriptomics the most current array platforms available for both fish and invertebrates have a relatively poor coverage of transcripts related to immunity due to a lack of genomic resources, i.e. large EST collections, and this may be due to the origin of the cDNA libraries generated for sequencing.

In general the impetus for array development in fish biology has been for use in developmental biology, biomedicine or aquaculture and this is reflected in the arrays that are published in the public domain and available for study. High density cDNA or oligonucleotide arrays have been developed for fishes (review: Douglas, 2007) including for zebrafish, diverse salmonids including the rainbow trout and the Atlantic salmon, catfish, seabream and several flatfish including turbot, sole and the Japanese flounder. The availability of microarrays to investigators has been a bottleneck in developing this approach as over the past decade the majority of array designs have been based upon cDNA microarrays that require a labour-intensive approach to physically produce the arrays. Here the limiting factor is spotting cDNA onto the array and the resulting quality of the array. This requires a significant effort in the laboratory to produce a large number of individual cDNAs and polymerase chain reaction (PCR) products for spotting. However the public availability of on-line development tools for oligonucleotide array design, e.g. eArray by Agilent Technologies, coupled to a demand approach (one can create and buy a single array) has made this technology both easily manageable and affordable for most laboratories. Commercial arrays are also available for a number of species, e.g. zebrafish and salmon, with the advantage that the development and supporting bioinformatics are available.

In comparative immunology published studies are dominated by both *in vitro* and *in vivo* experimental models in salmonids due to the strong industrial emphasis upon these species. However the zebrafish, a classical experimental model species for developmental studies with a fully sequenced genome, is becoming more and more popular in fish immunology due to the ease of handling and husbandry and the available experimental tools. In this context in recent years a significant increase in papers addressing immunological questions in the Pleuronectiformes can be appreciated again due to the commercial value of these species and increasing production. As mentioned above the availability of online platforms and synthetic array technology coupled to increasing genomic resources will likely cause a significant change in this area over the next few years as more array platforms with increased transcript representation become available.

13.2.2 'Deep' sequencing (RNASeq)

Significant advances have been made in the throughput of sequencers allowing for an exhaustive description or 'deep' sequencing of specific pools of mRNAs. Simply put, in contrast to previous studies where potentially several thousands of cDNAs were sequenced using the Sanger methodology in a labour-intensive setting, NGS, based upon polymerase amplification, is capable of producing sequences at several orders of magnitude greater ($>10^6$/lane). Thus productivity is much higher at less cost. However, the bioinformatic strategies and computation needs that are required to unravel this huge amount of information have also grown in parallel. Without the support from genomes sequenced at high resolution a major problem encountered is in the annotation of the sequences obtained (a description of the putative biological function). Typically a putative function can only be assigned often to <50% of the sequences obtained and this is species-dependent (Goetz *et al.*, 2010; Goetz and MacKenzie, 2008). Thus a tremendous amount of relevant biological information is lost. Owing to significant sequence divergence across phylogenetic distance and a continued reliance upon functional annotations in mammals this represents a major task for biologists working with model organisms or organisms of significant commercial importance with no available genome.

RNASeq technology represents using NGS to provide a complete picture of the transcriptome under study (global annotation) and simultaneously provide quantitative data related to differential gene expression. Quantitative studies require biological replication and several libraries generated from the mRNA must be sequenced in parallel. In theory and in practice RNASeq can provide gene discovery and gene expression data (Goetz *et al.*, 2010). In the absence of a genome for annotation the success of these approaches are dependent upon the quality of existing genome resources and in the absence of a genome require the construction of a global annotated transcriptome from existent expressed sequence tag (EST) collections. Although RNASeq produces huge number of sequence reads these sequences are small in size. Contigs and isotigs must be assembled and annotated using the backbone and identified both using the backbone and by Blasting against the relevant databases. The sheer volume of data produced requires an informatics strategy that goes further than a desktop personal computer in order to carry out annotation studies. On the other hand once the annotation has been completed, analysis follows a similar methodology as that developed for array studies. A significant advantage for this approach is that it is direct, unlike nucleic acid hybridization, and in theory one day should be able to sequence all transcripts in the experimental pool thus significantly reducing variation.

13.3   Genomics in finfish disease control

Disease control in aquaculture aims to mitigate the effects of pathogen outbreaks throughout the production cycle and improve productivity. Disease is a major issue in aquaculture as in all intensive culture systems and accounts for significant mortalities and economic loss in industry. The availability of microarray technology has provided tools for researchers to describe viral, bacterial and parasitic disease processes in a number of host species with an emphasis upon commercial species. Many of these studies aimed to extend information from previous studies and have concentrated upon basic research. However in recent years an increasing number of studies have addressed areas of specific interest in aquaculture and disease or related areas such as product development. The major translational objectives are to use these technologies to: (1) improve disease resistance by understanding and selecting for underlying molecular mechanisms of resistance, (2) to develop comprehensive screening methodologies to identify compounds with immunomodulatory potential, i.e. additives to diet, and (3) to identify the mechanisms of vaccine-induced protection in fish therefore facilitating the design of more effective vaccines and evaluating efficacy.

Resistance to disease involves a complex interplay between the two arms of the immune system, the innate and adaptive immune responses, that result in the development of an effective and pathogen-specific response thus preventing the pathogen from colonizing the individual. The interactions between the host and pathogen vary depending upon the selective pressures present in the environment, host integrity and pathogen virulence. Culture conditions in aquaculture represent a modified environment, which as a general rule does not resemble a natural environment for intensively cultured species. One must also consider that selective pressure on pathogens also becomes modified under intensive culture conditions.

The identification of prognostic biomarkers for disease resistance is undoubtedly a major aim for aquaculture. Underlying this is the basic description of the underlying mechanisms involved in the regulation of the fish immune response to pathogens. Genomics has the potential to identify genetic markers related to resistance potentially at both the structural and functional levels (DNA–RNA) and here the integration of genomic resources is a key issue. Thus computation and organisation of bioinformatic resources are increasingly major requirements for research teams and companies involved in such efforts. However few finfish species outwith salmonid culture have a defined genetic background. This significantly increases the technical difficulties associated with development of genetic markers for resistance. Disease resistance is normally measured by assessing cumulative mortalities to a particular pathogen in a restricted laboratory setting. In a background of defined families, surviving fish, or non-challenged siblings from the same family, are then considered as 'resistant' and in some cases quantitative trait loci (QTLs) as markers have been derived and used (i.e. infectious pancreatic necrosis (IPN) virus resistance). This process is costly and limited therefore there is a need for molecular prognostics based upon identifiable genetic determinants of resistance that can be measured by alternative methods. Microarray analyses have been used primarily to describe the biological processes involved; however, the potential to identify markers when measuring many thousands of transcripts is clearly of significant value. The effects of a range of bacterial pathogens have been investigated, in *in vivo* challenge, using microarrays including: *Mycobacterium* sp. (Kato *et al.*, 2010), *Mycobacterium marinum* (Meijer *et al.*, 2005; van der Sar *et al.*, 2009), *Streptococcus iniae* (Dumrong-phol *et al.*, 2009), *Streptococcus suis* (Wu *et al.*, 2010), *Piscirickettsia salmonis* (Rise *et al.*, 2004), *Aeromonas salmonicida* (Ewart *et al.*, 2005; Martin *et al.*, 2006; Millán *et al.*, 2010; Skugor *et al.*, 2009; Tsoi *et al.*, 2003), *Vibrio anguillarum* (Ching *et al.*, 2009; Gerwick *et al.*, 2007), *Edwardsiella ictaluri* (Peatman *et al.*, 2007, 2008), *Edwardsiella tarda* (Matsuyama *et al.*, 2007; Yasuike *et al.*, 2010), and lipoplysaccharide (LPS) (Li and Waldbieser, 2006; Mackenzie *et al.*, 2008; Osuna-Jiménez *et al.*, 2009). Studies on viral disease more prevalent in salmonids and the Japanese flounder (*Paralichthys olivaceus*) have been published including; infectious salmon anemia (ISA) (Jørgensen *et al.*, 2008), rhabdovirus infections including infectious hematopoietic necrosis (IHN) (Mackenzie *et al.*, 2008; Purcell *et al.*, 2011), viral hemorrhagic septicemia (VHS) (Byon *et al.*, 2005, 2006), hirame rhabdovirus infection (HIRRV) (Yasuike *et al.*, 2007) and nodavirus in the turbot (*Scophthalmus maximus*, Park *et al.*, 2009).

One of the first microarray publications by Rise *et al.* (2004) used the GRASP 3.7K cDNA array to identify responses to *P. salmonis* infection. The authors proposed 19 highly regulated transcripts as potential biomarkers to evaluate the efficacy of vaccines against *P. salmonis*. Interestingly, C-type lectin 2-1, a gene whose product is involved in endocytosis and the C/EBP-driven inflammatory response (Matsumoto *et al.*, 1999) was identified. Since this study this transcript has been identified in almost all reports in which bacterial preparations have been used to challenge live fish; (Ewart *et al.*, 2005; Mackenzie *et al.*, 2008; Martin *et al.*, 2006; Rise *et al.*, 2004). A similar study was reported aiming to identify biomarkers at the transcriptional level between triploid and diploid Chinook salmon under live *Vibrio anguillarum* challenge (Ching *et al.*, 2009). Twelve annotated mRNAs were identified as showing significant differences between the diploid and triploid fish; however, the underlying molecular mechanisms that could contribute to the observed reduced immune function of the triploid salmon remain unclear.

Several studies have addressed *A. salmonicida* infection, a primary study using a salmon cohabitation model reported 16 up-regulated mRNAs in spleen, liver and head kidney (Ewart *et al.*, 2005). Owing to the abundance of acute phase response transcripts the authors suggested that infected fish underwent a typical acute phase response to infection. A more recent study using the same pathogen in the turbot, *S. maximus* (Millán *et al.*, 2010) corroborated this view as 48 differentially regulated mRNAs mostly related to the acute-phase and the stress/defence immune response were identified in the spleen of the challenged turbot. In both channel and blue catfish the effects of a Gram-negative bacterial infection highlighted an up-regulation of mRNA transcripts involved in iron homeostasis, transport proteins, complement components, and inflammatory and humoral immune response

(Peatman *et al.*, 2007, 2008). This data collectively indicates that a conserved acute phase response to Gram-negative bacterial infection occurs in most fish species. However more studies to elucidate expression patterns resulting from Gram-negative bacterial infection between both similar and phylogenetically distant fish are required in order to describe common and divergent responses. Again this may lead to the development of marker systems, a consensus upon the acute phase response (APR) in fish and treatments tailored to certain species all of which have significant applied interest.

In order to reduce individual variation resulting from live fish challenges, *in vitro* cell cultures have become popular tools to explore the mechanisms of disease resistance in fish. A major aim of these studies is to understand which molecular components of pathogens are the most potent inducers of immunological activity. These molecules are known as PAMPs (pathogen associated molecular patterns). The majority of microarray studies have been carried out under *in vitro* conditions. The salmonid fish array (SFA) microarray platform accounts for nine studies conducted with bacterial and viral PAMPs across a number of different cell types including erythrocytes (Morera and MacKenzie 2011), hepatocytes (Skugor *et al.*, 2009) and adherent differentiated macrophages (Iliev *et al.*, 2010; MacKenzie *et al.*, 2006). An emergent pattern in these studies is that certain PAMPs stimulate significantly different responses in salmonid fish and this trend is similar across species. Iliev *et al.* (2010) showed that synthetic unmethylated CpG motifs (representing single-stranded bacterial DNA and touted as strong inducers of the immune response via the dietary route) induced a more pronounced response in adherent salmon monocyte/macrophage cultures with higher numbers of genes up-regulated in contrast to that observed for LPS challenge in trout monocyte/macrophages (MacKenzie *et al.*, 2006). CpG-B stimulation induced a specific and divergent response that was suggested to activate cellular differentiation rather than pro-inflammatory responses (Iliev *et al.*, 2010). On the other hand peptidoglycans and CpGs have recently been shown to be contaminants in lipopolysaccharide preparations and responsible for a strong and specific activation of fish macrophages (MacKenzie *et al.*, 2010).

Boltaña *et al.* (2011) recently reported the response of adherent trout monocyte/macrophages stimulated with peptidoglycan (PGN) from two different strains of *Escherichia coli* (O111:B4 and K12 serotypes). PGNs induced a strong inflammatory response in macrophages characterized by increases in pro-inflammatory cytokines mRNA expression and the release of inflammatory products; prostaglandin E_2 and D_2. In this study, 819 transcripts were differentially expressed in both treatments and the shifts in mRNA expression were registered at early activation times (270 transcripts at 1 h). A dramatic difference observed was that PGN-B4 induced a low number of transcripts (285) with a high intensity (fold change (FC) > 2; 215) and in contrast PGN-K12 induced a response highlighting a higher mRNA diversity (456 transcripts) but a lower intensity (FC > 2; 190), suggesting structure-specific recognition.

PGNs stimulated mRNA expression for a large cohort of closely related proteins involved in PAMP recognition, and show a marked similitude with the differentially expressed genes induced by whole bacteria including the macrophage receptor MARCO, or cystein-rich protein. Other transcripts identified included the PGN-recognition or processing protein PGLYR6 and transcripts involved in the toll-like receptor (TLR) pathway such as the adaptor protein MyD88, TRAF6 or serine/threonine-protein kinase all of which are involved in signal transduction between the TLRs and the transcription factor, NF-κβ. Thus at least for bacterial recognition it appears that PAMP recognition in fish is structure and potentially bacterial strain-specific. However, the molecular basis for such differences at the structural level remains unresolved. Microarray studies have tentatively uncovered differential responses in salmonid adherent/macrophages that reflect PAMP diversity and pathogen recognition receptor (PRR) specificity highlighting PAMP purity as key consideration. The potential of this approach and its applicability in aquaculture for the development of health products such as vaccine adjuvant or diet additives holds future promise.

In direct relation to the above, i.e. development of compounds for immunomodulation via diet, a recent study evaluated how high (fed fish) and low (starved fish) individual energy reserves, using the liver as a model tissue, modulate the response to an infection with *A. salmonicida* in the Atlantic salmon (Martin *et al.*, 2010). Fifty-three genes showed a higher expression in challenged starved fish (CSF, low energy reserves), where complement C7, the acute phase protein transcripts serum amyloid (SAA), serotransferrin, and hepcidin indicated that the CSF responded differentially in comparison to fed fish under the same experimental infection. The authors suggested that this effect may be a compensatory response in starved fish due to the observed down-regulation of the immune system in non-challenged starved fish. Immunostimulants (IS) have been used as feed additives for many years in aquaculture, and yeast β-glucan may be the most utilized. IS diets are thought to activate the innate immune system by increasing non-specific defence mechanisms. A transcriptomic approach was reported by Doñate and colleagues (2010) to elucidate the effects of dietary administration of β-glucans in the processes involved in immune activation in the mucosal immune system of trout (gills and intestine). Immunostimulant diets significantly changed gene expression profiles in a tissue-specific manner after a 4 week period of diet. Biological processes that were modified at these portals of entry included tissue remodelling processes and antigen presentation. Of particular interest was that in the gills a significant increase in antigen presenting capacity coupled to a tissue remodelling and cell recruitment response was observed. This is suggestive of a reinforcement of mucosal defences directly related to diet. On the other hand central immune processes such as haematopoiesis were significantly depleted over this time period (C. Doñate, personel communication) highlighted that the duration for diet application may be a key factor in balancing the potential positive and negative effects of chronic stimulation of the immune system.

The study of viral disease using microarrays has enjoyed more success possibly due to the apparent specificity of the viral response in fish and the intensity and magnitude of the transcriptomic response to viral pathogens. As an example MacKenzie and collaborators compared the response of the rainbow trout head kidney to *in vivo* virulent IHNV, attenuated IHNV and bacterial LPS challenge. From selected differentially expressed genes ($p < 0.01$), the differential response was evaluated (Fischer's exact $p < 0.01$) to identify transcripts co-expressed in each or both experimental conditions. A total of 49 regulated transcripts, often directionally opposed, specific to both LPS and viral stimuli were identified. However responses to the LPS were significantly weaker than those to IHN infection. Further analysis restricted to viral-response specific transcripts gave a total of 28 genes that were highly specific to the anti-viral response when compared across multiple experiments (MacKenzie *et al.*, 2008). Thus even although in its infancy studies addressing the anti-viral response in fish using microarrays, both cDNA and more recently oligonucleotide-based arrays, have been able to begin to identify distinct gene expression profiles and specific cassettes of responsive genes. This can be highlighted in studies using the Rhabdoviridae where specific regulatory patterns are conserved across different fish species (Goetz and MacKenzie, 2008).

Recently, Purcell *et al.* evaluated the potential mechanisms responsible for host-specific virulence in rainbow trout infected with high and low virulence strains of IHNV (Purcell *et al.*, 2011). A marked down-regulation in biological processes including the immune response, lymphocyte activation, response to stress, transcription and translation, together with a greater viral load resulting from infection with the high virulence strain, suggest that virulence is due to the ability of the virus to suppress the immune response by sequestering the cellular machinery, a classical viral mechanism. MacKenzie *et al.* (2008) compared IHNV with attenuated IHNV challenges in rainbow trout, over a short time frame of 1 and 3 days post-challenge. At 3 days post-infection, a significant change in the transcriptional program of the head kidney revealed a IHNV-dependent shift in the host immune response orientated toward the activation of adaptive immunity. The rapid systemic spreading of IHNV and not the attenuated virus inhibited TNF-alpha, major histocompatibility complex (MHC) class I, and several macrophage and cell cycle/differentiation markers and favored an MHC class II, immunoglobulin and MMP/TBX4 enhanced immune response (MacKenzie *et al.*, 2008). Thus it appears that the effectivity of the host response may be dependent upon the shift between the two arms of the immune system and the degree of interference from the virus.

Parallel studies by Aoki and colleagues addressed the efficacy of DNA vaccines containing the viral G proteins of VHSV and HIRRV in the Japanese flounder (Byon *et al.*, 2005; Yasuike *et al.*, 2007). All DNA vaccines containing the viral G glycoprotein conferred specific protection to fish challenged 1 month after vaccination. The underlying mechanisms were suggested to occur via the type 1 Interferon (IFN) system due to the number of IFN-related transcripts up-regulated in both studies. Interestingly, Mx, an antiviral protein that is commonly used as a marker for antiviral activity in animal species, was consistently up-regulated across vaccinations (Yasuike *et al.*, 2007, 2010). Similar observations have been reported at the site of DNA vaccination against IHNV in trout (Purcell *et al.*, 2011) and in turbot challenged with nodavirus (Park *et al.*, 2009). These results suggest that both the host-expressed viral glycoprotein and the virulent rhadovirus induces a systemic anti-viral state indicative of the non-specific type 1 IFN innate immune response and that this canonical response may be conserved throughout the fishes. However, the mechanisms for the development of a specific cytotoxic T or B lymphocyte-mediated humoral response in fish vaccinated with plasmid DNA-IHNV G which confers protective immunity are yet to be identified and remains an area of uncertainty in fish immunology (Purcell *et al.*, 2011).

As a final point relevant to the development of prognostic markers for viral disease. Jorgensen *et al.* (2008) reported upon an extensive analysis of a highly virulent ISAV infection in Atlantic salmon aiming to characterize differences between early and late mortalities and identify molecular determinants of resistance. Using linear discriminant analysis based upon quantitative PCR (QPCR), they were able to identify a minimum set of genes (5-lipoxygenase activating protein, cytochrome P450 2K4, galectin-9 and annexin A1) chosen from an unbiased (only expression profiles, no inference of function) microarray data set (derived from individual liver samples) that could be used to predict which class, early or late mortality, an individual fish would belong to. This highly promising result suggests that prognostic markers based upon PCR assays may become a reality in the near future. However the utility of for example C-type lectin in salmon and other potential species-specific biomarkers for disease resistance will require a significant effort to further develop. The future publication of several fish genomes coupled to array platforms with a significantly increased transcript representation may provide an exciting route to further develop this strategy by combining both functional and structural genomics for species of commercial interest with a sequenced genome.

13.4   Genomics in shellfish and crustacean disease control

Invertebrates do not possess acquired/adaptive immunity and therefore their defence mechanisms rely solely upon the innate immune system.

Furthermore in comparison with the vertebrates and taking into account the diversity of these animal groups little is known about their immune responses in great depth. In an effort to increase knowledge and understanding of invertebrate immune systems and their relevance to global aquaculture over the past decade a significant effort has been made to uncover, identify and characterize the main facets of immunity in these organisms. Here genomics has played a significant role in providing the methodologies and tools to carry out gene discovery on a large scale. The results of these studies have led to the initiation of an innovative research area applying molecular biology to both shellfish and crustacean aquaculture. In this section we will concentrate upon reviewing the current status of genomic resources for these organisms and provide examples of where it has been applied to disease control.

13.4.1   Shellfish

Genomic resources; genome sequencing

Five bivalve species have been selected for genome sequencing: the scallop *Argopecten irradians*, the oyster *Crassostrea virginica,* the Atlantic surf clam *Spisula solidissima,* the pearl oyster *Pinctada maxima* (Project ID: 41527) and the mussel *Mytilus californianus* (http://www.jgi.doe.gov). Large-insert genomic bacterial artificial chromosome (BAC) libraries have been constructed from two economically important cultured oyster species, *C. virginica* and *C. gigas* (Cunningham *et al.*, 2006), representing a 11.8-fold genome coverage and a 9-fold coverage is available from two BAC libraries from nuclear DNA of the Zhikong scallop *Chlamys farreri* (Zhang *et al.*, 2008). In addition to these efforts several mitochondrial genomes are also available in GenBank.

Genomic resources; ESTs

EST sequencing from cDNA or subtractive hybridized libraries derived from numerous shellfish species have been carried out to identify specific mRNA transcripts that are expressed within a tissue or in the course of physiological disturbances. Many of these libraries were constructed from stimulated or infected animals, often using hemocytes elicited by injection of PAMPs or whole pathogens, to identify transcripts putatively related with the immune system. Consortia of several laboratories have helped to produce this data, such as the Marine Genomics project in America (www.marinegenomics.org) or the Marine Genomics Europe, Network of Excellence (MGE) (www.euromarineconsortium.eu) (Saavedra and Bachère, 2006).

The first libraries to be reported were on the eastern oyster, *C. virginica*, from hemocytes and embryos with the aim of identifying transcripts involved in the stress response for use as bioindicators of exposure to environmental pollutants and to toxic and infectious agents (Jenny *et al.*, 2002). This was rapidly followed by studies in *C. gigas* challenged with bacteria (Gueguen *et al.*, 2003). A study of genes expressed in response to *Perkinsus marinus* challenge was carried out in both species identifying 19 ESTs related to immunity and cell communication (Tanguy *et al.*, 2004). A similar low yield of potential immune-related transcripts genes was also found in libraries from *C. gigas* affected by summer mortalities. However, this technology paved the way for the identification of antimicrobial peptides belonging to the mussel defensin family in oysters (Gueguen *et al.*, 2006; Peatman *et al.*, 2004). Recently, Taris *et al.* (2008) suggested that most of the differences in transcription patterns of immune and metabolic genes between tolerant and sensitive families could be largely attributable to constitutive differences in gene transcription. Further advances in EST sequencing of the two oyster species were reported (Fleury *et al.*, 2009; Quilang *et al.*, 2007; Roberts *et al.*, 2009) and have led to the identification of interleukin 17, astacin, cystatin B, the EP4 receptor for prostaglandin E, the ectodysplasin receptor, c-jun, and the p100 subunit of nuclear factor-κβ as transcripts strongly up-regulated under bacterial challenge. All of these sequences, together with existing public sequence data, have been compiled into a publicly available website   (http://public-contigbrowser.sigenae.org:9090/Crassostrea_gigas/index.html). In addition to the transcript database a total of 208 *in silico* microsatellites with 173 having sufficient flanking sequence for primer design and a total of 7530 putative *in silico*, single-nucleotide polymorphisms using existing and newly generated EST resources for the Pacific oyster were identified.

The development of appropriate bioinformatics tools is essential to optimize the use of this information. A set of expressed sequence tag-simple sequence repeat (EST-SSR) markers of the Pacific oyster, *C. gigas*, was developed through bioinformatic mining of the GenBank public database (L. Wang *et al.*, 2008). Allele numbers ranged from 3 to 10, and the observed and expected heterozygosity values varied from 0.125 to 0.770 and from 0.113 to 0.732 respectively. Cross-species amplification was examined for five other *Crassostrea* species and reasonable results were obtained, promising usefulness of these markers in oyster genetics. In another study, a total of 147 microsatellite-containing ESTs (3.63%) were detected from 4053 ESTs of *C. gigas* in GenBank (Yu and Li, 2008). Twenty EST-SSRs were tested on three families of *C. gigas* for examination of inheritance mode of EST-SSRs. The results obtained in this study suggest that *C. gigas* EST-SSRs can complement the currently available genomic SSR markers and may be useful for comparative mapping, marker-assisted selection and evolutionary studies. In the Mediterranean mussel *Mytilus galloprovincialis*, transcripts encoding AMP myticin, methallothionein and heat shock proteins were among the 426 singletons identified from hemolymph, gills, digestive gland, foot, adductor muscle and mantle cDNA samples from unstressed *M. galloprovincialis* mussels (Venier *et al.*, 2003). These ESTs were enriched with cDNA and subtracted libraries from bacteria stimulated mussels (Pallavicini *et al.*, 2008) and, in 2009, the same research group constructed, sequenced and annotated 17 cDNA libraries from different *M. galloprovincialis* tissues (gills, digestive gland, foot, anterior and posterior adductor muscle, mantle and hemocytes) (Venier *et al.*, 2009). A total of 24 939 clones were sequenced from these libraries, generating 18 788 high-quality ESTs that were assembled into 2446 overlapping clusters and 4666 singletons, resulting in a total of 7112 non-redundant sequences. Bioinformatic screening of the non-redundant *M. galloprovincialis* sequences identified 159 microsatellite-containing ESTs. Clusters, consensuses and related similarity and gene ontology searches were organized in a searchable database (http://mussel.cribi.unipd.it).

The triangle sail mussel (*Hyriopsis cumingii*) is the most important mussel species commercially exploited for freshwater pearl production in China and a total of 5290 ESTs (481 contigs and 1165 singletons) have been reported. BLAST similarity analysis indicated almost half (46.5%) of these ESTs were homologs of known genes. A total of 201 microsatellites were identified from these ESTs, with 31 having sufficient flanking sequences for primer design (Bai *et al.*, 2009).

Several other EST projects have been reported for other bivalve species including; the adult bay scallop *A. irradians irradians* (Song *et al.*, 2006), whole body tissues of the zhikong scallop, *C. farreri*, challenged by *Listonella anguillarum* (Y. Wang *et al.*, 2008) and carpet shell clams (*Ruditapes decussatus*) after stimulation with dead bacteria (Gestal *et al.*, 2007) or with *Perkinsus olseni* (Prado-Alvarez *et al.*, 2009). Recently Tanguy *et al.* (2008) reported the construction of normalized cDNA libraries for four different species (*C. gigas, Mytilus edulis, R. decussatus* and *Bathymodiolus azoricus*), using numerous tissues and physiological conditions. Each EST library was independently assembled and 1300–3000 unique sequences were identified in each species. For the different species, functional categories could be assigned to only about 16–27% of ESTs using the GO annotation tool.

Although much progress has been made in characterizing transcriptomic resources obtained from bivalves, annotation of resources (obtaining a putative biological function) remains a significant bottleneck in order to develop tools for study in disease control. Significant phylogenetic distance and diversity coupled to a lack of annotated genomes represents an important hurdle. In addition, an article reported by Taris *et al.* (2008), highlighting the potential difficulties for using Q-PCR as a validation tool due to the presence of sequence polymorphism in oysters is worthy of mention in this context. The authors emphasized the need for extreme caution and thorough primer testing when assaying genetically diverse biological materials such as Pacific oysters. Their findings suggest that melt-curve analysis alone (a benchmark QC method in QPCR analysis) may not be sufficient as a means of identifying acceptable QPCR primers. Minimally, testing numerous primer pairs seems to be necessary to avoid false conclusions from flawed QPCR assays for which sequence variation among individuals produces artifactual and unreliable quantitative results.

Microarrays

Microarrays have been constructed for both mussel (*M. galloprovincialis*) and oysters (*C. gigas*, *C. virginica* and *Ostrea edulis*) from the EST resources available from the studies described in the above section. The mussel cDNA array includes 1714 mussel probes (76% singletons, approximately 50% putatively identified transcripts) plus unrelated controls (Venier *et al.*, 2006). An international group of collaborators have constructed a 27 496-feature cDNA microarray containing 4460 sequences derived from *C. virginica*, 2320 from *C. gigas*, and 16 non-oyster DNAs serving as positive and negative controls. The performance of the array was assessed by gene expression profiling using gill and digestive gland RNA derived from both *C. gigas* and *C. virginica*, and digestive gland RNA from *Crassostrea ariakensis*. The utility of the microarray for detection of homologous genes by cross-hybridization between species was also assessed and the correlation between hybridization intensity and sequence homology for selected genes determined. The oyster cDNA microarray is publicly available to the research community. This microarray has been used for the study of the transcriptome profiling of selectively bred Pacific oyster *C. gigas* families that differ in tolerance of heat shock (Jenny *et al.*, 2002; Lang *et al.*, 2009).

The recent production of these tools has led to a time lag in respect to studies using microarray in fish that over the past decade represent 1% of the total number of publications in fish biology. Clearly resources are significantly more developed for the oyster; however, as with fish this situation is likely to change rapidly over the next few years as several NGS projects come to fruition. It can be expected that the availability of genomic tools for studies in bivalves will increase as resources expand and further insights into the biology of bivalve disease resistance will be revealed.

13.4.2   Crustaceans

The crustaceans, a very diverse and ancient group of arthropods, have been largely studied in order to understand animal evolution and physiology, and for use as models in environmental studies. Advances in sequencing methodologies have resulted in a significant volume of information related to genomics. For an in-depth and comprehensive review see Stillman *et al.* (2008).

Genomic resources; ESTs

ESTs have been generated for many crustacean species, providing an invaluable resource for peptide discovery in members of this arthropod subphylum. Recently this data was mined for novel peptide-encoding transcripts where mature peptides derived from the ESTs predicted using a combination of online peptide prediction programs and homology to known arthropod sequences. In total, 70 mature full-length/partial peptides representing members of 16 protein families/subfamilies were predicted, the vast majority being novel (Christie *et al.*, 2010).

EST projects have been described for the black tiger shrimp *Penaeus monodon* in which gene discovery was the primary objective. Initially, 15 cDNA libraries were constructed from different tissues (eyestalk, hepato-pancreas, haematopoietic tissue, haemocyte, lymphoid organ, and ovary) of shrimp, reared under normal or stress conditions, to identify tissue-specific transcripts and those responding to infection and heat stress. A total of 10 100 clones were analyzed by single-pass sequencing from the 5′ end. In addition, bioinformatic mining of microsatellites from the *P. monodon* ESTs identified 997 unique microsatellite containing ESTs in which 74 loci resided within the genes of known functions. The EST sequence data and the BLAST results were stored and made available through a web-accessible database (http://pmonodon.biotec.or.th/) (Tassanakajon *et al.*, 2006). This work was further extended by the sequencing of two cDNA libraries constructed from normal and white spot syndrome virus (WSSV)-infected postlarvae producing a total of 15 981 high-quality ESTs. Compara-tive EST analyses suggested that, in postlarval shrimp, WSSV infection strongly modulates the gene expression patterns in several organs or tissues, including the hepatopancreas, muscle, eyestalk and cuticle. Several basic cellular metabolic processes are likely to be affected, including oxidative phosphorylation, protein synthesis, the glycolytic pathway and calcium ion balance. A group of immune-related chitin-binding protein genes is also likely to be strongly up-regulated after WSSV infection. A database con-taining all the sequence data and analysis results is accessible at http://xbio.lifescience.ntu.edu.tw/pm/ (Leu *et al.*, 2007).

Two cDNA libraries from normal and *Vibrio harveyi*-challenged *P. monodon* hemocytes resulted in the discovery of the antimicrobial peptide (AMP) homologues, antilipopolysaccharide factors (ALF), pen-aeidins and crustins. They predominated among immune-related genes, representing 29.2% and 64.0% of the normal and challenged libraries, respectively. Several types of each AMP homologue were found (Supungul *et al.*, 2004). ALFs were originally characterized in the horseshoe crab. In order to characterize the properties and biological activities of this immune effector in shrimp, ALFPm3, the most abundant isoform found in *P. monodon*, was expressed in the yeast *Pichia pastoris*. Antimicrobial assays demonstrated that rALFPm3 has a broad spectrum of antifungal properties against filamentous fungi, and antibacterial activities against both Gram-positive and Gram-negative bacteria, associated with a bacteri-cidal effect (Somboonwiwat *et al.*, 2005). Cationic AMPs in penaeid shrimps composed of penaeidins, crustins and anti-lipopolysaccharide factors comprise multiple classes or isoforms and possess antibacterial and anti-fungal activities against different strains of bacteria and fungi. Shrimp AMPs are primarily expressed in circulating hemocytes, which is the main site of the immune response, and hemocytes expressing AMPs. Subtractive suppression hybridization (SSH) has also been used to identify yellow head virus (YHV)-responsive transcripts in black tiger shrimp hemocytes. Novel YHV-responsive genes were uncovered from these SSH libraries, including caspases, histidine triad nucleotide-binding protein 2, Rab11, beta-integrin, tetraspanin, prostaglandin E synthase, transglutaminase, Kazal-type serine proteinase inhibitor and antimicrobial peptides.

The expression of four up-regulated immune-related genes – anti-lipopolysaccharide factor isoform 6 (ALFPm6), crustin isoform 1 (crust-inPm1), transglutaminase and Kazal-type serine proteinase inhibitor isoform 2 (SPIPm2) – was evaluated by real-time RT-PCR confirming their differential expression and up-regulation to challenge (Prapavorarat *et al.*, 2010). In contrast to the volume of information available for the Black tiger shrimp, the Pacific white shrimp *Litopenaeus vannamei*, which is one of the most economically important marine aquaculture species in the world, has little information available. Recently, Zhang *et al.* (2008) described a BAC genomic library for this species where 92 160 clones were spotted onto high-density nylon filters for hybridization screening. Such efforts should increase the number of studies addressing this important commercial species.

A cDNA library was constructed from hemocytes of the Chinese mitten crab *Eriocheir sinensis* challenged with a mixture of *Listonella anguillarum* and *Staphylococcus aureus*, and randomly sequenced to collect genomic information and identify transcripts involved in the immune defense response. BLAST analysis revealed that 1706 unigenes (58.0% of the total) or 4593 ESTs (61.0% of the total) were novel transcripts that had no significant match to any protein sequences in the public databases. The rest of the unigenes (1237; 42.0% of the total) could be matched to the known genes or sequences deposited in public databases (Gai *et al.*, 2009). These sequences further extend a previous study where a non-normalized cDNA library derived from the hepatopancreas of this crab was constructed, resulting in 3297 high-quality expressed sequence tags representing 1178 unigenes (Jiang *et al.*, 2009).

Microarrays

In terms of relevance to disease resistance little information is available for the crustaceans. The exceptions to the rule are two studies published for *P. monodon*. A cDNA microarray composed of 2028 different ESTs from two shrimp species, *P. monodon* and *Masupenaeus japonicus*, was employed to identify YHV-responsive genes in hemocytes of *P. monodon*. A total of 105 differentially expressed transcripts were identified and grouped into five different clusters according to their expression patterns (up, down, up-down, down-up, etc). One of these clusters, which comprised of five transcripts including cathepsin L-like cysteine peptidase, hypothetical proteins and unknowns, was of particular interest because the transcript abundance rapidly increased (≤0.25 hours) and reached high expression levels in response to YHV injection (Pongsomboon *et al.*, 2008). A second version cDNA microarray representing 9990 different ESTs obtained from the *P. monodon* EST project was employed to identify viral (white spot and yellow head viruses) and bacterial (*V. harveyi*) responsive transcripts in the hemocytes of *P. monodon* (Pongsomboon *et al.*, 2011). The number of differentially expressed transcripts found was highest in shrimps infected with white spot virus followed by yellow head virus and finally lower in *V. harveyi* infected individuals. Whether this reflected specificity of the response or virulence of the pathogens or the development of a specific immune response aimed toward these pathogen groups remains to be answered.

Other microarray platforms are available for different crustacean species, e.g. *Carcinus maenas*; however, no studies have been reported on the immune response or immune system for these organisms.

13.5   Future trends

The impact of next generation sequencing technologies upon genomics in commercially farmed finfish, shellfish and crustaceans has been apparent over the past few years. The most important factors have been accessibility to sequencing platforms, cost and development of bioinformatics to manage data. This latter part, however, will still require much more work in the coming years, as the need for specialized analytical 'pipelines' that can effectively manage these huge data sets becomes an imperative in this area of study. This coupled to a decreasing reliance upon mammalian data sets and genomes should provide a more fruitful approach when considering organisms separated by significant phylogenetic distances. The above technologies are driving development in host-pathogen genomics, the genetics of the immune response and the identification of virulence factors in pathogens, by whole genome sequencing, of commercial importance. The overall result of these efforts should be an increased understanding of the molecular mechanisms involved in host–pathogen interactions that in turn should have a translational impact upon industry. Important challenges such as selection for disease resistance within a breeding population, the development of more effective vaccines and the development of novel therapeutics for intensively farmed aquatic organisms will benefit from this and should provide the foundations for a strong aquatic biotechnology industry supporting global aquaculture.

13.6   References

ABBAS, A.R., BALDWIN, D., MA, Y., OUYANG, W., GURNEY, A., MARTIN, F., FONG, S., VAN LOOKEREN CAMPAGNE, M., GODOWSKI, P., WILLIAMS, P.M., CHAN, A.C. & CLARK, H.F. 2005. Immune response in silico (IRIS): immune-specific genes identified from a compendium of microarray expression data. *Genes Immun*, 6(4): 319–331.

BAI, Z., YIN, Y., HU, S., WANG, G., ZHANG, X. & LI, J. 2009. Identification of genes involved in immune response, microsatellite, and SNP markers from expressed sequence tags generated from hemocytes of freshwater pearl mussel (*Hyriopsis cumingii*). *Mar Biotechnol (NY)*, **11**(4): 520–530.

BOLTAÑA, S., REYES-LOPEZ, F., MORERA, D., GOETZ, F. & MACKENZIE, S.A. 2011. Divergent responses to peptidoglycans derived from different *E. coli* serotypes influence inflammatory outcome in trout, *Oncorhynchus mykiss*, macrophages. *BMC Genomics*, **12**: 34.

BYON, J.Y., OHIRA, T., HIRONO, I. & AOKI, T. 2005. Use of a cDNA microarray to study immunity against viral hemorrhagic septicemia (VHS) in Japanese flounder (*Paralichthys olivaceus*) following DNA vaccination. *Fish Shellfish Immunol*, **18**(2): 135–147.

BYON, J.Y., OHIRA, T., HIRONO, I. & AOKI, T. 2006. Comparative immune responses in Japanese flounder, *Paralichthys olivaceus* after vaccination with viral hemorrhagic septicemia virus (VHSV) recombinant glycoprotein and DNA vaccine using a microarray analysis. *Vaccine*, **24**(7): 921–930.

CHING, B., JAMIESON, S., HEATH, J.W., HEATH, D.D. & HUBBERSTEY, A. 2009. Transcriptional differences between triploid and diploid Chinook salmon (*Oncorhynchus tshawytscha*) during live *Vibrio anguillarum* challenge. *Heredity*, **104**(2): 224–234.

CHRISTIE, A.E., DURKIN, C.S., HARTLINE, N., OHNO, P. & LENZ, P.H. 2010. Bioinformatic analyses of the publicly accessible crustacean expressed sequence tags (ESTs) reveal numerous novel neuropeptide-encoding precursor proteins, including ones from members of several little studied taxa. *Gen Comp Endocrinol*, **167**(1): 164–178.

CUNNINGHAM, C., JENNY, M.J., CHAPMAN, R.W., FANG, G.C., SASKI, C., LUNDQVIST, M.L., WING, R.A., CUPIT, P.M., GROSS, P.S., WARR, G.W. & TOMKINS, J.P. 2006. New resources for marine genomics: bacterial artificial chromosome libraries for the Eastern and Pacific oysters (*Crassostrea virginica* and *C. gigas*). *Mar Biotechnol (NY)*, **8**(5): 521–533.

DOÑATE, C., BALASCH, J., CALLOL, A., BOBE, J., TORT, L. & MACKENZIE, S. 2010. The effects of immunostimulation through dietary manipulation in the rainbow trout; evaluation of mucosal immunity. *Mar Biotechnol*, **12**(1): 88–99.

DOUGLAS, S.E. 2007. Microarray studies of gene expression in fish. *OMICS*, **10**(4): 474–489.

DUMRONGPHOL, Y., HIROTA, T., KONDO, H., AOKI, T. & HIRONO, I. 2009. Identification of novel genes in Japanese flounder (*Paralichthys olivaceus*) head kidney up-regulated after vaccination with *Streptococcus iniae* formalin-killed cells. *Fish Shellfish Immunol*, **26**(1): 197–200.

EWART, K.V., BELANGER, J.C., WILLIAMS, J., KARAKACH, T., PENNY, S., TSOI, S.C.M., RICHARDS, R.C. & DOUGLAS, S.E. 2005. Identification of genes differentially expressed in Atlantic salmon (*Salmo salar*) in response to infection by *Aeromonas salmonicida* using cDNA microarray technology. *Dev Comp Immunol*, **29**(4): 333–347.

FLEURY, E., HUVET, A., LELONG, C., DE LORGERIL, J., BOULO, V., GUEGUEN, Y., BACHÈRE, E., TANGUY, A., MORAGA, D., FABIOUX, C., LINDEQUE, P., SHAW, J., REINHARDT, R., PRUNET, P., DAVEY, G., LAPÈGUE, S., SAUVAGE, C., CORPOREAU, C., MOAL, J., GAVORY, F., WINCKER, P. & MOREEWS, F. 2009. Generation and analysis of a 29 745 unique expressed sequence tags from the Pacific oyster (*Crassostrea gigas*) assembled into a publicly accessible database: the GigasDatabase. *BMC Genomics*, **10**: 341.

GAI, Y., WANG, L., ZHAO, J., QIU, L., SONG, L., LI, L., MU, C., WANG, W., WANG, M., ZHANG, Y., YAO, X. & YANG, J. 2009. The construction of a cDNA library enriched for immune genes and the analysis of 7535 ESTs from Chinese mitten crab *Eriocheir sinensis*. *Fish Shellfish Immunol*, **27**(6): 684–694.

GERWICK, L., CORLEY-SMITH, G. & BAYNE, C.J. 2007. Gene transcript changes in individual rainbow trout livers following an inflammatory stimulus. *Fish Shellfish Immunol,* **22**(3): 157–171.

GESTAL, C., COSTA, M.M., FIGUERAS, A. & NOVOA, B. 2007. Analysis of differentially expressed genes in response to bacterial stimulation in hemocytes of the carpet-shell clam *Ruditapes decussatus*: identification of new antimicrobial peptides. *Gene,* **406**(1–2): 134–143.

GOETZ, F.W. & MACKENZIE, S. 2008. Functional genomics with microarrays in fish biology and fisheries. *Fish and Fisheries,* **9**(4): 1467–2979.

GOETZ, F.W., ILIEV, D.B., MCCAULEY, L.A., LIARTE, C.Q., TORT, L.B., PLANAS, J.V. & MACKENZIE, S. 2004. Analysis of genes isolated from lipopolysaccharide-stimulated rainbow trout (*Oncorhynchus mykiss*) macrophages. *Mol Immunol,* **41**(12): 1199–1210.

GOETZ, F., ROSAUER, D., SITAR, S., GOETZ, G., SIMCHICK, C., ROBERTS, S., JOHNSON, R., MURPHY, C., BRONTE, C.R. & MACKENZIE, S. 2010. A genetic basis for the phenotypic differentiation between siscowet and lean lake trout (*Salvelinus namaycush*). *Mol Ecol,* **19**: 176–196.

GUEGUEN, Y., CADORET, J.P., FLAMENT, D., BARREAU-ROUMIGUIÈRE, C., GIRARDOT, A.L., GARNIER, J., HOAREAU, A., BACHÈRE, E. & ESCOUBAS, A. 2003. Immune gene discovery by expressed sequence tags generated from hemocytes of the bacteria-challenged oyster, *Crassostrea gigas. Gene,* **303**: 139–145.

GUEGUEN, Y., HERPIN, A., AUMELAS, A., GARNIER, J., FIEVET, J., ESCOUBAS, J.M., BULET, P., GONZALEZ, M., LELONG, C., FAVREL, P. & BACHÈRE, E. 2006. Characterization of a defensin from the oyster *Crassostrea gigas*. Recombinant production, folding, solution structure, antimicrobial activities, and gene expression. *J Biol Chem,* **281**(1): 313–323.

ILIEV, D.B., JØRGENSEN, S.M., RODE, M., KRASNOV, A., HARNESHAUG, I. & JØRGENSEN, J.B. 2010. CpG-induced secretion of MHCIIbeta and exosomes from salmon (*Salmo salar*) APCs. *Dev Comp Immunol,* **34**(1): 29–41.

JENNY, M.J., RINGWOOD, A.H., LACY, E.R., LEWITUS, A.J., KEMPTON, J.W., GROSS, P.S., WARR, G.W. & CHAPMAN, R.W. 2002. Potential indicators of stress response identified by expressed sequence tag analysis of hemocytes and embryos from the American oyster, *Crassostrea virginica. Mar Biotechnol (NY),* **4**(1): 81–93.

JIANG, H., CAI, Y.M., CHEN, L.Q., ZHANG, X.W., HU, S.N. & WANG, Q. 2009. Functional annotation and analysis of expressed sequence tags from the hepatopancreas of mitten crab (*Eriocheir sinensis*). *Mar Biotechnol (NY),* **11**(3): 317–326.

JØRGENSEN, S.M., AFANASYEV, S. & KRASNOV, A. 2008. Gene expression analyses in Atlantic salmon challenged with infectious salmon anemia virus reveal differences between individuals with early, intermediate and late mortality. *BMC Genomics,* **9**: 179.

KATO, G., KONDO, H., AOKI, T. & HIRONO, I. 2010. BCG vaccine confers adaptive immunity against *Mycobacterium* sp. infection in fish. *Dev Comp Immunol,* **34**, 133–140.

LANG, R.P., BAYNE, C.J., CAMARA, M.D., CUNNINGHAM, C., JENNY, M.J. & LANGDON, C.J. 2009. Transcriptome profiling of selectively bred Pacific oyster *Crassostrea gigas* families that differ in tolerance of heat shock. *Mar Biotechnol (NY),* **11**(5): 650–668.

LEU, J.H., CHANG, C.C., WU, J.L., HSU, C.W., HIRONO, I., AOKI, T., JUAN, H.F., LO, C.F., KOU, G.H. & HUANG, H.C. 2007. Comparative analysis of differentially expressed genes in normal and white spot syndrome virus infected *Penaeus monodon. BMC Genomics,* **8**: 210.

LI, R.W. & WALDBIESER, G.C. 2006. Production and utilization of a high-density oligonucleotide microarray in channel catfish, *Ictalurus punctatus. BMC Genomics,* **7**: 134.

MACKENZIE, S., ILIEV, D., LIARTE, C., KOSKINEN, H., PLANAS, J.V., GOETZ, F.W., MÖLSÄ, H., KRASNOV, A. & TORT, L. 2006. Transcriptional analysis of LPS-stimulated activation of trout (*Oncorhynchus mykiss*) monocyte/macrophage cells in primary culture treated with cortisol. *Mol Immunol,* **43**(9): 1340–1348.

MACKENZIE, S., BALASCH, J.C., NOVOA, B., RIBAS, L., ROHER, N., KRASNOV, A. & FIGUERAS, A. 2008. Comparative analysis of the acute response of the trout, *O. mykiss*, head kidney to *in vivo* challenge with virulent and attenuated infectious hematopoietic necrosis virus and LPS-induced inflammation. *BMC Genomics*, 9: 141.

MACKENZIE, S.A., ROHER, N., BOLTAÑA, S. & GOETZ, F.W. 2010. Peptidoglycan, not endotoxin, is the key mediator of cytokine gene expression induced in rainbow trout macrophages by crude LPS. *Mol Immunol*, 47(7–8): 1450–1457.

MARTIN, S.A., BLANEY, S.C., HOULIHAN, D.F. & SECOMBES, C.J. 2006. Transcriptome response following administration of a live bacterial vaccine in Atlantic salmon (*Salmo salar*). *Mol Immunol*, 43(11): 1900–1911.

MARTIN, S.A., DOUGLAS, A., HOULIHAN, D.F. & SECOMBES, C.J. 2010. Starvation alters the liver transcriptome of the innate immune response in Atlantic salmon (*Salmo salar*). *BMC Genomics*, 11: 418.

MATSUMOTO, M., TANAKA, T., KAISHO, T., SANJO, H., COPELAND, N.G., GILBERT, D.J., JENKINS, N.A. & AKIRA, S. 1999. A novel LPS-inducible C-type lectin is a transcriptional target of NF-IL6 in macrophages. *J Immunol*, 163: 5039–5048.

MATSUYAMA, T., FUJIWARA, A., NAKAYASU, C., KAMAISHI, T., OSEKO, N., HIRONO, I. & AOKI, T. 2007. Gene expression of leucocytes in vaccinated Japanese flounder (*Paralichthys olivaceus*) during the course of experimental infection with *Edwardsiella tarda*. *Fish Shellfish Immunol*, 22(6): 598–607.

MEIJER, A.H., VERBEEK, F.J., SALAS-VIDAL, E., CORREDOR-ADÁMEZ, M., BUSSMAN, J., VAN DER SAR, A.M., OTTO, G.W., GEISLER, R. & SPAINK, H.P. 2005. Transcriptome profiling of adult zebrafish at the late stage of chronic tuberculosis due to *Mycobacterium marinum* infection. *Mol Immunol*, 42(10): 1185–1203.

MILLÁN, A., GÓMEZ-TATO, A., FERNÁNDEZ, C., PARDO, B.G., ALVAREZ-DIOS, J.A., CALAZA, M., BOUZA, C., VÁZQUEZ, M., CABALEIRO, S. & MARTÍNEZ, P. 2010. Design and performance of a turbot (*Scophthalmus maximus*) oligo-microarray based on ESTs from immune tissues. *Mar Biotechnol*, 12(4): 452–465.

MORERA, D. & MACKENZIE, S. 2011. Is there a direct role for erythrocytes in the immune response? *Vet Res*, 42(1): 89.

OSUNA-JIMÉNEZ, I., WILLIAMS, T.D., PRIETO-ALAMO, M.J., ABRIL, N., CHIPMAN, J.K. & PUEYO, C. 2009. Immune- and stress-related transcriptomic responses of *Solea senegalensis* stimulated with lipopolysaccharide and copper sulphate using heterologous cDNA microarrays. *Fish Shellfish Immunol*, 26(5): 699–706.

PALLAVICINI, A., COSTA, M.M., GESTAL, C., DREOS, R., FIGUERAS, A., VENIER, P. & NOVOA, B. 2008. High sequence variability of myticin transcripts in hemocytes of immune-stimulated mussels suggests ancient host–pathogen interactions. *Dev Comp Immunol*, 32(3): 213–226.

PARK, K.C., OSBORNE, J.A., MONTES, A., DIOS, S., NERLAND, A.H., NOVOA, B., FIGUERAS, A., BROWN, L.L. & JOHNSON, S.C. 2009. Immunological responses of turbot (*Psetta maxima*) to nodavirus infection or polyriboinosinic polyribocytidylic acid (pIC) stimulation, using expressed sequence tags (ESTs) analysis and cDNA microarrays. *Fish Shellfish Immunol*, 26(1): 91–108.

PEATMAN, E.J., WEI, X., FENG, J., LIU, L., KUCUKTAS, H., LI, P., HE, C., ROUSE, D., WALLACE, R., DUNHAM, R. & LIU, Z. 2004. Development of expressed sequence tags from Eastern oyster (*Crassostrea virginica*): lessons learned from previous efforts. *Proc Mar Biotechnol*, 6: S491–S496.

PEATMAN, E., BAOPRASERTKUL, P., TERHUNE, J., XU, P., NANDI, S., KUCUKTAS, H., LI, P., WANG, S., SOMRIDHIVEJ, B., DUNHAM, R. & LIU, Z. 2007. Expression analysis of the acute phase response in channel catfish (*Ictalurus punctatus*) after infection with a Gram-negative bacterium. *Dev Comp Immunol*, 31(11): 1183–1196.

PEATMAN, E., TERHUNE, J., BAOPRASERTKUL, P., XU, P., NANDI, S., WANG, S., SOMRIDHIVEJ, B., KUCUKTAS, H., LI, P., DUNHAM, R. & LIU, Z. 2008. Microarray analysis of gene expression in the blue catfish liver reveals early activation of the MHC class I pathway after infection with *Edwardsiella ictaluri*. *Mol Immunol*, **45**(2): 553–566.

PONGSOMBOON, S., TANG, S., BOONDA, S., AOKI, T., HIRONO, I., YASUIKE, M. & TASSANAKA-JON, A. 2008. Differentially expressed genes in *Penaeus monodon* hemocytes following infection with yellow head virus. *BMB Rep*, **41**(9): 670–677.

PONGSOMBOON, S., TANG, S., BOONDA, S., AOKI, T., HIRONO, I. & TASSANAKAJON, A. 2011. A cDNA microarray approach for analyzing transcriptional changes in *Penaeus monodon* after infection by pathogens. *Fish Shellfish Immunol*, **30**(1): 439–446.

PRADO-ALVAREZ, M., GESTAL, C., NOVOA, B. & FIGUERAS, A. 2009. Differentially expressed genes of the carpet shell clam *Ruditapes decussatus* against *Perkinsus olseni*. *Fish Shellfish Immunol*, **26**(1): 72–83.

PRAPAVORARAT, A, VATANAVICHARN, T., SÖDERHÄLL, K. & TASSANAKAJON, A. 2010. A novel viral responsive protein is involved in hemocyte homeostasis in the black tiger shrimp, *Penaeus monodon*. *J Biol Chem*, **285**(28): 21467–21477.

PURCELL, M., MARJARA, K., MARJARA, I.S., BATTS, W., KURATH, G. & HANSEN, J.D. 2011. Transcriptome analysis of rainbow trout infected with high and low virulence strains of infectious hematopoietic necrosis virus. *Fish Shellfish Immunol*, **30**(1): 84–93.

QUILANG, J., WANG, S., LI, P., ABERNATHY, J., PEATMAN, E., WANG, Y., WANG, L., SHI, Y., WALLACE, R., GUO, X. & LIU, Z. 2007. Generation and analysis of ESTs from the eastern oyster, *Crassostrea virginica* Gmelin and identification of microsatellite and SNP markers. *BMC Genomics*, **8**: 157.

RISE, M.L., JONES, S.R., BROWN, G.D., VON SCHALBURG, K.R., DAVIDSON, W.S. & KOOP, B.F. 2004. Microarray analyses identify molecular biomarkers of Atlantic salmon macrophage and hematopoietic kidney response to *Piscirickettsia salmonis* infection. *Physiol Genomics*, **20**(1): 21–35.

ROBERTS S., GOETZ, G., WHITE S. & GOETZ F. 2009. Analysis of genes isolated from plated hemocytes of the Pacific oyster, *Crassostreas gigas*. *Mar Biotechnol (NY)*, **11**(1): 24–44.

SAAVEDRA, C. & BACHÈRE E. 2006. Bivalve genomics. *Aquaculture*, **256**: 1–14.

SKUGOR, S., JORGENSEN, S.M., GJERDE, B. & KRASNOV, A. 2009. Hepatic gene expression profiling reveals protective responses in Atlantic salmon vaccinated against furunculosis. *BMC Genomics*, **10**: 503.

SOMBOONWIWAT, K., MARCOS, M., TASSANAKAJON, A., KLINBUNGA, S., AUMELAS, A., ROME-STAND, B., GUEGUEN, Y., BOZE, H., MOULIN, G. & BACHÈRE, E. 2005. Recombinant expression and anti-microbial activity of anti-lipopolysaccharide factor (ALF) from the black tiger shrimp *Penaeus monodon*. *Dev Comp Immunol*, **29**(10): 841–851.

SONG, L., XU, W., LI, C., LI, H., WU, L., XIANG, J. & GUO, X. 2006. Development of expressed sequence tags from the bay scallop, *Argopecten irradians irradians*. *Mar Biotechnol (NY)*, **8**(2): 161–169.

STILLMAN, J.H., COLBOURNE, J.K., LEE, C.E., PATEL, N.H., PHILLIPS, M.R., TOWLE, D.W., EADS, B.D., GELEMBUIK, G.W., HENRY, R.P., JOHNSON, E.A., PFRENDER, M.E. & TERWILLIGER, N.B. 2008. Recent advances in crustacean genomics. *Integrative Comp Biol*, **48**(6): 852–868.

SUPUNGUL, P., KLINBUNGA, S., PICHYANGKURA, R., HIRONO, I., AOKI, T. & TASSANAKAJON, A. 2004. Antimicrobial peptides discovered in the black tiger shrimp *Penaeus monodon* using the EST approach. *Dis Aquat Organ*, **61**(1–2): 123–135.

TANGUY, A., GUO, X. & FORD, S.E. 2004. Discovery of genes expressed in response to *Perkinsus marinus* challenge in Eastern (*Crassostrea virginica*) and Pacific (*C. gigas*) oysters. *Gene*, **338**(1): 121–131.

TANGUY, A., BIERNE, N., SAAVEDRA, C., PINA, B., BACHÈRE, E., KUBE, M., BAZIN, E., BON-HOMME, F., BOUDRY, P., BOULO, V., BOUTET, I., CANCELA, L., DOSSAT, C., FAVREL, P., HUVET, A., JARQUE, S., JOLLIVET, D., KLAGES, S., LAPÈGUE, S., LEITE, R., MOAL, J., MORAGA, D. &

REINHARDT, R. 2008. Increasing genomic information in bivalves through new EST collections in four species: development of new genetic markers for environmental studies and genome evolution. *Gene*, **408**(1–2): 27–36.

TARIS, N., LANG, R.P. & CAMARA, M.D. 2008. Sequence polymorphism can produce serious artefacts in real-time PCR assays: hard lessons from Pacific oysters. *BMC Genomics*, **9**: 234.

TASSANAKAJON, A., KLINBUNGA, S., PAUNGLARP, N., RIMPHANITCHAYAKIT, V., UDOMKIT, A., JITRAPAKDEE, S., SRITUNYALUCKSANA, K., PHONGDARA, A., PONGSOMBOON, S., SUPUNGUL, P., TANG, S., KUPHANUMART, K., PICHYANGKURA, R. & LURSINSAP, C. 2006. *Penaeus monodon* gene discovery project: the generation of an EST collection and establishment of a database. *Gene*, **384**: 104–112.

TSOI, S.C., CALE, J.M., BIRD, I.M., EWART, V., BROWN, L.L. & DOUGLAS, S. 2003. Use of human cDNA microarrays for identification of differentially expressed genes in Atlantic salmon liver during *Aeromonas salmonicida* infection. *Mar Biotechnol*, **5**(6): 545–554.

VAN DER SAR, A.M., SPAINK, H.P., ZAKRZEWSKA, A., BITTER, W. & MEIJER, A.H. 2009. Specificity of the zebrafish host transcriptome response to acute and chronic mycobacterial infection and the role of innate and adaptive immune components. *Mol Immunol*, **46**(11–12): 2317–2332.

VENIER, P., PALLAVICINI, A., DE NARDI, B. & LANFRANCHI, G. 2003. Towards a catalogue of genes transcribed in multiple tissues of *Mytilus galloprovincialis*. *Gene*, **18**(314): 29–40.

VENIER, P., DE PITTÀ, C., PALLAVICINI, A., MARSANO, F., VAROTTO, L., ROMUALDI, C., DONDERO, F., VIARENGO, A. & LANFRANCHI, G. 2006. Development of mussel mRNA profiling: can gene expression trends reveal coastal water pollution? *Mut Res Fundamental and Molecular Mechanisms of Mutagenesis*, **602**(1–2): 121–134.

VENIER, P., DE PITTÀ, C., BERNANTE, F., VAROTTO, L., DE NARDI, B., BOVO, G., ROCH, P., NOVOA, B., FIGUERAS, A., PALLAVICINI, A. & LANFRANCHI, G. 2009. MytiBase: a knowledgebase of mussel (*M. galloprovincialis*) transcribed sequences. *BMC Genomics*, **9**(10): 72.

WANG, L., SONG, L., ZHAO, J., QIU, L., ZHANG, H., XU, W., LI, H., LI, C., WU, L. & GUO, X. 2008. Expressed sequence tags from the zhikong scallop (*Chlamys farreri*): discovery and annotation of host-defense genes. *Fish Shellfish Immunol*, **26**(5): 744–750.

WANG, Y., REN, R. & YU, Z. 2008. Bioinformatic mining of EST-SSR loci in the Pacific oyster, *Crassostrea gigas*. *Anim Genet*, **39**(3): 287–289.

WU, Z., ZHANG, W., LU, Y. & LU, C. 2010. Transcriptome profiling of zebrafish infected with *Streptococcus suis*. *Microbial Pathogenesis*, **48**(5): 178–187.

YASUIKE, M., KONDO, H., HIRONO, I. & AOKI, T. 2007. Difference in Japanese flounder, *Paralichthys olivaceus* gene expression profile following hirame rhabdovirus (HIRRV) G and N protein DNA vaccination. *Fish Shellfish Immunol*, **23**(3): 531–541.

YASUIKE, M., TAKANO, T., KONDO, H., HIRONO, I. & AOKI, T. 2010. Differential gene expression profiles in Japanese flounder (*Paralichthys olivaceus*) with different susceptibilities to edwardsiellosis. *Fish Shellfish Immunol*, **29**(5): 747–752.

YU, H. & LI, Q. 2008. Exploiting EST databases for the development and characterization of EST-SSRs in the Pacific oyster (*Crassostrea gigas*). *J Hered*, **99**(2): 208–214.

ZHANG, Y., ZANG, X., SCHEURING, C.F., ZHANG, H.B., HUAN, P., LI, F. & XIANG, J. 2008. Construction and characterization of two bacterial artificial chromosome libraries of Zhikong scallop, *Chlamys farreri* Jones et Preston, and identification of BAC clones containing the genes involved in its innate immune system. *Mar Biotechnol (NY)*, **10**: 358–365.

14

Bacteria and bacteriophages as biological agents for disease control in aquaculture

A. Carrias, C. Ran, J. S. Terhune and M. R. Liles, Auburn University, USA

Abstract: Diseases afflicting aquaculture are a worldwide problem, and are in need of cost-effective and sustainable therapeutants. This chapter reviews the literature on both bacterial and bacteriophage agents that have been applied as biological control agents in aquaculture to reduce the severity and mortality associated with various infectious diseases. The specific interactions between pathogen, host, and biological control agent, and how these interactions are influenced by environmental factors, can be unique for each biological control application. Each biological control agent should be evaluated independently for its efficacy and potential for application in aquaculture. Taken together, these studies demonstrate both the successes and challenges of these approaches for biological control of disease, and point toward increasing adoption of efficacious biological control agents worldwide in aquaculture.

Key words: aquaculture, biological control, bacteria, bacteriophage, pathogen, probiotic.

14.1 Introduction

Diseases are extremely detrimental to the global farming of aquatic organisms, and are a direct consequence of the high-density practices typical of aquaculture. The monetary value of losses attributed to diseases in world aquaculture is difficult to quantify; however, there is no doubt that diseases exact a significant toll on aquaculture producers worldwide. Infectious diseases in aquaculture are caused by bacterial pathogens, viruses, fungi, protozoa, and parasites. There are at least 30 bacterial pathogens associated with disease in cultured aquatic organisms, with some of them affecting multiple species (Stickney, 2009). Vaccines have been developed for at least 19 of the major bacterial pathogens. Compiling a list of the antibiotics used in world aquaculture is more difficult owing to the lack of a system for monitoring antibiotic usage in some countries that are major contributors to world aquaculture production totals. Both vaccines and antibiotics have improved the well-being of cultured organisms and have played a major role in reducing disease-related losses. However, vaccines and antibiotics are not available for all diseases or in some cases may not be the best options to use. For example, vaccines have limited usefulness in early larval stages, may not be cost-effective for all aquaculture animals or production systems, and may not be available for all pathogens, or all hosts (e.g. cultured invertebrates).

The indiscriminate use of antibiotics, particularly in some countries with insufficient regulations and limited enforcement for controlling antimicrobial agent use, has contributed to antibiotic resistance problems that may pose potential public health hazards. Antibiotic resistance in aquaculture can impact human health either by the development of acquired resistance in bacteria in aquatic environments that can infect humans or by development of acquired resistance in bacteria in aquatic environments that may serve as carriers of resistant genes that may be transferred to bacteria that can infect humans. Furthermore, residues from the use of antibiotics in aquaculture may remain in fish tissue and without proper withdraw periods before harvest/slaughter may cause several problems including the exertion of selective pressure on the dominant intestinal flora of humans. This has led to limited approval of drugs for use in fish and difficulty in disease control strategies. For these reasons there has been a growing interest in the development of effective biological control (herein, 'biocontrol') strategies that are based on naturally occurring compounds and strategies with limited human impacts through consumption of aquaculture products to reduce the onset or severity of disease.

There have been many changes and adaptations to the definition of biocontrol over the years. One of these (Wilson, 1997) simply defines biocontrol as: 'the control of disease with a natural biological process or the product of a natural process'. Fuller (1989) defined a probiotic as 'a live microbial feed supplement which beneficially affects the host animal by improving its intestinal microbial balance'. Verschuere *et al.* (2000) gave a more specific definition of probiotics as 'a live microbial adjunct which has a beneficial effect on the host by modifying the host-associated or ambient microbial community, by ensuring improved use of the feed or enhancing its nutritional value, by enhancing the host response towards disease, or by improving the quality of its ambient environment'. This chapter will focus specifically on the biocontrol activity of bacteria or bacteriophages, and not other probiotic benefits that may derive from their application. It should be also appreciated that many probiotic benefits to a host can result in improved resistance or survival after exposure to a pathogen through improved immune function/stimulation or overall well-being of the animal; however, this chapter will only describe those probiotic studies in which there are experiments related to biocontrol. In aquaculture farmed fish, the skin, gill, and intestinal microbiota can all contribute to disease prevention, with complex interactions occurring between the host, its associated micro-organisms, and the surrounding environment that can contribute to biocontrol efficacy.

In aquaculture, bacteria or lytic bacteriophages possessing inhibitory activity against other microorganisms have received increasing consideration for use as biocontrol agents for the control of diseases caused by bacterial pathogens. However, the realization of biocontrol strategies to control diseases in aquaculture will be possible only if agents used are safe, effective, and economically viable. Hence, consideration of biocontrol therapy for treating diseases in aquaculture should include as the primary criteria: (1) the efficacy of the biological agent in controlling the target disease(s), (2) the safety to the aquatic host, the environment, and to human consumers, (3) the cost and ease in producing large-scale quantities of the biocontrol agent, (4) the stability and shelf-life capacity, (5) route and cost of administration, and (6) the regulatory and approval process. Additionally, consideration should be given to the type of killing spectrum that the biocontrol agent possesses. Antimicrobial agents can be either broad-spectrum, which can inhibit the growth or kill a wide range of pathogens, or narrow-spectrum, which are more specific for treatment of certain genera or species (Schwarz *et al.*, 2001). Bacteria (generally broad-spectrum antimicrobial, Table 14.1) and bacteriophage (generally narrow-spectrum antimicrobial, Table 14.2) are the primary examples and the subject of this chapter. Given the increased bacterial resistance to antibiotics and the interest in consumers for ecologically or organically grown food products the use of biocontrol agents may find increasing use in sustainable aquaculture practices.

14.1.1 Bacteria as biocontrol agents for aquatic animal larvae

During initial feeding, it is possible to induce an artificially high dose of a bacterium into the fish rearing water (Strøm & Ringø, 1993) or to the culture medium of the live food (Gatesoupe, 1994). Gatesoupe introduced a lactic acid bacteria (LAB) strain daily to the enrichment medium of roti-fers used as live food for turbot *Psetta maxima* larvae (Gatesoupe, 1994). The added LAB could be retrieved in large amounts from the turbot larvae. While the addition of LAB had no significant effect on growth and survival rates of turbot reared in normal conditions, a significant reduction of larval mortality was observed when the larvae were challenged with a pathogenic *Vibrio* sp. on day nine post-hatch. The hypothesis was that the LAB acted as a microbial antagonist against the pathogenic *Vibrio* and might curb the invasion by the pathogen into the host fish. Similarly, rotifer enrichment with a siderophore-producing strain of *Vibrio* type E improved the survival of larval turbot after a 48 h challenge with the pathogenic *Vibrio* type P (Gatesoupe, 1997).

In another study, *Bacillus* strain IP5832 spores were introduced into the culture medium of rotifers, which were fed to turbot larvae (Gatesoupe,

Table 14.1 Overview of literature evaluating bacteria as biological control agents in aquaculture

Bacteria	Origin	Method of administration	Observations	Suggested mode of action	Reference
Bacillus	DMS series products	Addition to pond water	Increase of survival of penacid shrimps, decrease of luminous *Vibrio* in pond water and sediment	Antagonism	Moriarty (1998)
Bacillus S11	Black tiger shrimp habitats	Feed supplement	Increase of mean weight and survival of *Penaeus monodon* larvae and postlarvae; decrease of mortality after challenge with the pathogen *Vibrio harveyi*	Antagonism	Rengpipat et al. (1998)
Carnobacterium strain K1	Atlantic salmon intestines	Feed supplement	Growth inhibition of *Vibrio anguillarum* and *Aeromonas salmonicida* in fish intestinal mucus and fecal extracts *in vitro*	Antagonism	Jöborn et al. (1997)
Fluorescent pseudomonad F19/3	Mucus of brown trout	Bathing	Decrease of mortality of Atlantic salmon presmolts with stress-inducible *A. salmonicida* infection	Competition for iron	Smith & Davey (1993)
Pseudomonas fluorescens AH2	Iced freshwater fish (*Lates niloticus*)	Addition to tank water	Decrease of mortality of rainbow trout juveniles challenged with a pathogenic *V. anguillarum*	Competition for iron	Gram et al. (1999)
Vibrio alginolyticus	Shrimp hatchery in Ecuador	Bathing	Decrease of mortality of Atlantic salmon juveniles challenged with a pathogenic *A. salmonicida, V. anguillarum,* and *Vibrio ordalii*	Antagonism	Austin et al. (1995)
Carnobacterium divergens	Intestine of Atlantic cod	Feed supplement	Improvement of disease resistance of Atlantic cod challenged with *V. anguillarum*	?	Gildberg et al. (1997)

Probiotic	Source	Application	Effect	Mechanism	Reference
Aeromonas hydrophila, Vibrio fluvialis, Carnobacterium sp. and an unidentified Gram-positive coccus	Intestine of Atlantic salmon, rainbow trout and turbot	Feed supplement	Decrease of mortality of rainbow trout fingerlings and fry challenged by A. salmonicida	Stimulation of cellular immunity	Irianto & Austin (2002)
Lactobacillus rhamnosus	Probiotic for human use	Feed supplement	Decrease of mortality of rainbow trout challenged by A. salmonicida	?	Nikoskelainen et al. (2001)
Enterococcus faecium SF68 and Bacillus toyoi	Commercial products	Feed supplement	Decrease or delay of mortality of European eel, Anguilla anguilla L. challenged by Edwardsiella tarda	Antagonism?	Chang (2002)
Lactic acid bacteria	Rotifers	Applied to enrichment medium of rotifers	Decrease of the mortality of turbot larvae challenged by pathogenic vibrio	?	Gatesoupe (1994)
Aeromonas media strain A199	?	Addition to tank water	Recovery of eel Anguilla australis Richardson infected with Saprolegnia parasitica	Antagonism	Lategan et al. (2004a)
Aeromonas sobria GC2	Digestive tract of rainbow trout and carp	Feed supplement	Decrease of mortality of rainbow trout challenged by Lactococcus garvieae and Streptococcus iniae	Stimulation of innate immunity	Brunt & Austin (2005)
Carnobacterium sp.	Intestine of Atlantic salmon	Feed supplement	Improvement in survival of rainbow trout and Atlantic salmon following challenge with A. salmonicida, V. ordalii, and Yersinia ruckeri	Antagonism?	Robertson et al. (2000)
Clostridium butyricum	Probiotic for human use	Oral administration of bacterin by intubation	Decreased mortality of rainbow trout challenged by V. anguillarum	Increased phagocytic activity of leucocytes	Sakai et al. (1995)
Aeromonas media A 199	?	Addition to tank water	Decrease of mortality and suppression of the pathogen of Pacific oyster larvae when challenged with a pathogenic Vibrio tubiashii	Antagonism	Gibson et al. (1998)

Table 14.2 Overview of literature evaluating phage as biological control agents in aquaculture

Phage	Host bacteria	Host organism	Disease	Reference
ET-1	*Edwardsiella tarda*	Loaches (*Misgurnus anguillcaudatus*)	Edwardsiellosis	Wu and Chao (1982)
PLgY-16	*Lactococcus garvieae*	Yellowtail (*Seriola quinqueradiata*)	Lactococcosis	Nakai et al., (1999)
PPpW-3	*Pseudomonas plecoglossicida*	Ayu (*Plecoglossus altivelis*)	Bacterial hemorrhagic ascites	Park et al., (2000)
PPpW-4				Park and Nakai (2003)
	Vibrio harveyi	Black tiger shrimp (*Penaeus monodon*)	Luminous vibriosis	Vinod et al. (2006)
				Karunasagar et al. (2005)
				Karunasagar et al. (2007)
	Streptococcus iniae	Japanese flounder (*Paralichthys olivaceus*)	Streptococcosis	Masuoka et al. (2007)
HER 110	*Aeromonas salmonicida*	Atlantic salmon (*Salmo salar*) and Brook trout (*Salvelinus fontinalis*)	Furanculosis	Imbeault et al. (2006)
O, R and B				Verner-Jeffreys et al. (2007)
PFpW-3	*Flavobacterium psychrophilum*	Ayu (*Plecoglossus altivelis*)	Bacterial cold water disease	Nakai et al. (2010)

1991). A decrease in the proportion of members of the Vibrionaceae in the rotifers was observed, and the mean weight of the turbot larvae fed with the spore-fed rotifers was significantly improved on day ten. Survival of turbot larvae fed with the spore-fed rotifers was significantly higher than that of the control group (31 and 10%, respectively) ten days after challenge with an opportunistic *Vibrio* sp. pathogen. Although the author suggested that the likeliest mode of action was the production of antibiotics, it is not clear whether an improvement of the nutritional status of the larvae could have also contributed to the increased resistance to infection (Verschuere *et al.*, 2000).

14.1.2   Bacteria as biocontrol agents for juvenile and adult fish
Furunculosis
Irianto and Austin (2002) isolated four bacterial strains from intestinal contents of Atlantic salmon *Salmo salar* and rainbow trout *Oncorhynchus mykiss* with antagonistic activity against *Aeromonas salmonicida*. These isolates were identified as *Aeromonas hydrophila*, *Vibrio fluvialis*, *Carnobacterium* sp. and an unidentified Gram-positive coccus. The bacterial strains were applied to dry rainbow trout feed separately or as an equal mixture at a dose of 10^7 to 10^8 colony forming units (CFU)/g of feed. Groups of rainbow trout fingerlings and fry were fed with amended feed for 14 days and then challenged with *A. salmonicida* by cohabitation, intraperitoneal (i.p.) injection (for fingerlings), or immersion (for fry). In the i.p. injection challenge experiment with fingerlings, no mortality was recorded in the four groups fed amended feed seven days after challenge, while the control group suffered a 48% mortality. In the cohabitation challenge experiment, administration of each of the bacterial strains with the exception of the *Vibrio* strain resulted in a significant reduction in mortality. Further experiments with rainbow trout fry confirmed the potential benefit of the biocontrol strains used separately or as an equal mixture. In terms of the mode of action, no serum or mucous antibodies to *A. salmonicida* were detected in fish fed with biocontrol strains; however, an increased number of erythrocytes, macrophages, lymphocytes, and leucocytes, and enhanced lysozyme activity were found, indicating a stimulation of cellular rather than humoral immunity.

Nikoskelainen *et al.* (2001) chose a LAB strain *Lactobacillus rhamnosus* ATCC 53103 for evaluation of furunculosis biocontrol in rainbow trout. Amended feed was given to rainbow trout for 51 days. Sixteen days after the start of the *Lactobacillus* feeding, the fish were challenged with *A. salmonicida* by cohabitation. The group receiving *L. rhamnosus*-amended feed had an average mortality rate of 19% and 46% for the 10^9 CFU/g feed and 10^{12} CFU/g feed groups, respectively, whereas the control group had a 53% mortality. The *in vitro* antagonistic activity of the biocontrol strain against *A. salmonicida* was not characterized, nor was the mode of action studied.

In a study by Smith and Davey (1993), a *Pseudomonas fluorescens* strain F19/3 was tested for its activity to exclude *A. salmonicida* from salmon pre-smolts with stress-inducible furunculosis. Fish were removed from a hatchery population where stress-inducible infections were prevalent and transported to the laboratory. Half of the fish were bathed in water containing strain F19/3 for 24 h at 16 °C on the day after their arrival, with the rest as the control. This was followed by administration of the stressors and each fish was held in an individual tank for 14 days to avoid cross infections. A significant reduction in the frequency of stress-inducible infections was observed in the group of fish bathed in F19/3 compared to the control group. Since it was demonstrated that strain F19/3 did not colonize the fish internally, this indicated an exclusionary effect of strain F19/3 by bacterial colonization of fish external surfaces.

Saprolegniosis
The Saprolegniales are a group of oomycetes encountered in fish and in aquatic environments. *Saprolegnia*, a genus of this order, is generally recognized as a saprophytic opportunist (Dick, 1990) with a few species parasitic on higher plants and animals (Alexopoulos, 1962). *Saprolegnia* has been known to affect a variety of freshwater fish (Copland & Willoughby, 1982; Rowland & Ingram, 1991), particularly under intensive farming conditions where trauma, stress, and poor or rapidly changing water quality provide the ideal conditions for the aquatic mould to proliferate. Outbreaks result from the attachment of the zoospore to the skin of the fish. This is followed by germination and hyphal invasion of the epidermis (Gosper, 1996), usually when immune function is weakened.

Lategan and Gibson (2003) reported that an *Aeromonas media* strain A199 can inhibit the growth of *Saprolegnia* sp. *in vitro*. The antagonism was suggested to be derived from a bacteriocin-like inhibitory substance (BLIS) produced by strain A199. In four independent *in vivo* tank observations of fish affected with saprolegniosis, the daily addition of *A. media* strain A199 to tank water contributed to the rapid recovery of affected hosts from pathogen invasion. In the four observations, there was no control and only one fish was observed, which was too low to warrant statistical analysis.

In a study by Lategan *et al.* (2004a) the potential of *A. media* strain A199 as a candidate for biocontrol of winter saprolegniosis was tested during a winter outbreak on a farm and in a laboratory challenge trial on silver perch, *Bidyanus bidyanus*. Fish showing early symptoms of the disease were sampled from the *Saprolegnia*-affected pond and randomly distributed to tanks in a laboratory. Biocontrol treatments were conducted by adding 10^4–10^5 strain A199 cells/ml of tank water, resulting in a significant difference in the mortality rate between fish exposed to A199 ($n = 3$ mortalities) and those in the control tanks ($n = 7$ mortalities) by 35 days. In the challenge trial, silver perch were treated with a *Saprolegnia* sp. for three days with and without inoculation of treatment tanks with and without 10^4–10^5 strain A199 cells/ml of tank water.

To address the effect of seasonality on biocontrol efficacy, a study on eel *Anguilla australis* was performed (Lategan *et al.*, 2004b). Healthy elvers were selected and distributed randomly into experimental tanks. A physiological stress followed by a physical stress was used to emulate conditions that precede saprolegniosis. A decrease in water temperature (21 °C to 10 °C over a period of seven days) provided the physiological stress, while the physical stress of handling was given on day eight. On day six, cyst suspensions of a *Saprolegnia* sp. were added to all treatment tanks, and strain A199 was added to all the experimental test tanks at the first appearance of visual saprolegniosis symptoms (day 12). The incidence of saprolegniosis in this laboratory challenge followed the same trend typically observed at the eel farm. Saprolegniosis-affected fish appeared simultaneously in both the A199-treated tanks and non-treated control tanks. The fish treated with A199 had an overall morbidity of 27%, compared to 44% for control fish. At day 18 the temperature was increased to 14 °C, after which the fish in the treatment tanks recovered rapidly yet the fish in the control tanks remained affected with the disease after the temperature increase. With the recovery of fish in the treatment tanks on day 22 of the investigation, the nontreated control tanks were subjected to a daily treatment of A199 at concentrations of 10^5 CFU/ml for a period of four days, resulting in a recovery of the control fish affected with saprolegniosis. Temperature was demonstrated to be an important factor regulating the therapeutic benefit of strain A199.

Edwardsiellosis
Edwardsiellosis, a bacterial septicaemia caused by the Gram-negative bacterium *Edwardsiella tarda*, is a common but serious bacterial disease in cultured warm-water animals, including eels. Two bacterial strains, *Enterococcus faecium* SF68 and *Bacillus toyoi*, were isolated from a commercial product based on their inhibitory effects against *E. tarda* (Chang, 2002). Firstly, the colonization potential for the two strains in European eels *Anguilla anguilla* L. was tested by feeding eels with amended feed for two weeks with three eels randomly selected from each group for intestinal microbial flora test every second day. This part of the experiment showed that *E. faecium* began to colonize eel intestines on day four post-inoculation and numbers reached 1.6×10^5 CFU/g on day 14, constituting 73% of the cultured intestinal microbiota, whereas in the group of eels fed on *B. toyoi* supplement, total viable counts in intestines continuously decreased during the feeding period to 3×10^4 CFU/g, approximately 10% of the number at the beginning of the experiment. A challenge experiment followed, where the eels were fed with the amended feed for two weeks and then challenged with *E. tarda* 981210L1 suspensions at a concentration of 7×10^8 CFU/ml by anal injection. Two weeks after challenge, the survival

rates of eels fed on *E. faecium* SF68 supplement was significantly higher than those of control eels and those fed on *B. toyoi* supplement, whereas no significant difference was observed between the survival rates of the *B. toyoi* and control groups which was consistent with the results of the colonization test.

Enteric septicemia of catfish
Enteric septicemia of catfish (ESC), caused by the Gram-negative bacterium *Edwardsiella ictaluri*, is the most important endemic infectious disease in catfish aquaculture industry in the United States (Hawke *et al.*, 1981, 1998; Plumb, 1999; Hawke & Khoo, 2004) and has recently become a serious emerging disease in cultured *Pangasius* ssp. Losses resulting from infections with ESC in the US were reported in over 78% of all operations with outbreaks being reported in 42% of catfish production ponds, with an economic loss between $20 and $30 million yearly (Wagner *et al.*, 2002; US Department of Agriculture, 2003a, 2003b).

Queiroz and Boyd (1998) applied a commercial product Biostart containing various *Bacillus* sp. strains to channel catfish *Ictalurus punctatus* ponds three times a week and demonstrated that survival and net production of fish treated with *Bacillus* spp. was significantly greater than the control. During the feeding regime, some fish in all ponds were infected by proliferative gill disease and enteric septicemia of catfish. The authors suggested the greater survival of fish in ponds treated with *Bacillus* spp. suggested the possible protective effect against ESC.

Chao *et al.* (unpublished data) selected *Bacillus* spp. for biocontrol of ESC from a collection of *Bacillus* spp. strains isolated from the soil as well as the intestine of channel catfish. During *in vitro* screening, the antimicrobial activity of the *Bacillus* strains was evaluated against a panel of fish pathogens commonly encountered on aquaculture operations including *E. ictaluri*, *E. tarda*, *Streptococcus iniae*, *Yersinia ruckeri*, *Flavobacterium columnare* and *Saprolegnia ferax*. The *Bacillus* strains that showed good antimicrobial activity *in vitro* were tested for survival and growth within the catfish intestine by feeding channel catfish with *Bacillus* spores supplemented feed for seven days followed by normal feed for three days, and then determining *Bacillus* CFU/g of intestinal tissue of catfish. *Bacillus* strains with good pathogen antagonism that attained high numbers within the intestine are being evaluated for their respective ability to reduce mortality due to *E. ictaluri*. The specific *Bacillus* strains are predominantly isolates of *B. subtilis* and *B. amyloliquefaciens*, and vary in terms of their specific antibiotics produced that are inhibitory to *E. ictaluri*, the colonization of the fish gastrointestinal tract, and their degree of biological control of ESC. Future research will evaluate the dose-responsiveness of *Bacillus*-mediated biological control of ESC and the safety of each respective strain within channel catfish, and the effect of *Bacillus* strain(s) on the gut microbiota.

Penaeid shrimp

In contrast to the already broad application of probiotics in commercial penaeid shrimp hatcheries, relatively few in-depth studies have been published on biocontrol in shrimp. The use of a soil bacterial strain, PM-4 (a Gram-negative motile strain, no species designation was given), was reported to promote the growth of the black tiger shrimp *Penaeus monodon* (Maeda & Liao, 1992). This strain also showed an *in vitro* inhibitory effect against a *V. anguillarum* strain. When added to tanks inoculated with diatoms and rotifers, the strain resulted in 57% survival of the larvae after 13 days, while without the bacterium all the larvae had died after 5 days.

The use of *Bacillus* strain S11 for prevention of disease due to *Vibrio harveyi* when administered in enriched *Artemia* to larvae of *P. monodon* in laboratory aquaria was studied (Rengpipat *et al.*, 1998, 2003). The *P. monodon* larvae fed the *Bacillus*-fortified *Artemia* had significantly shorter development times and fewer disease problems than did larvae reared without the *Bacillus* strain. After being fed for 100 days with the *Bacillus* strain S11-supplemented feed, *P. monodon* were challenged with *V. harveyi* by immersion. Ten days later all of the groups treated with *Bacillus* strain S11 showed 100% survival whereas the control group had only 26% survival.

In another experiment, *Bacillus* strain BS11 was used as a supplement in feed for *P. monodon* in two earthen pond field-trials carried out for 100 days during two different seasons (Rengpipat *et al.*, 1998, 2003). Growth and survival were compared with those of shrimp receiving control feed, and shrimp fed BS11-amended feed grew significantly larger and had significantly higher survival than shrimp fed control feed. Projected yields on an annual basis were 49% greater with BS11-fed shrimp. After being fed for 100 days in pond cage trials, *P. monodon* were transferred to aquaria and challenged by bath exposure to *V. harveyi* in aquarium water at 10^7 CFU/ml. In the two aquaria challenge experiments shrimp fed control feed all died within six days while survival for shrimp fed BS11-feed was 5% and 9% for hot and cool seasons, respectively. The survival rates in the challenge test would not be acceptable in a farm situation and were also inconsistent with the earlier biocontrol effect demonstrated in laboratory studies. The authors suggest that the disparity between the two studies could have been the consequence of the stress caused by acclimatization of shrimp from pond cages to aquaria and/or the higher ammonia and nitrite concentrations encountered in the challenge trials. The lack of a significant protective effect could also have been the consequence of lower levels of BS11 in hepatopancreas and intestines of BS11-fed shrimp in the latter study, when BS11 levels were 10^4–10^6 CFU/g before challenge and decreased by nearly 2 logs after *V. harveyi* challenge. However, in the laboratory studies, the BS11 level in BS11-fed shrimp intestines was 10^6–10^8 CFU/g before challenge and stayed constant during the challenge. These disparate results achieved with BS11 illustrate the diversity of factors (i.e., host, pathogen, biocontrol agent, and environment) that can influence the ultimate disease outcome.

Moriarty (1998) investigated the value of adding selected strains of *Bacillus* to control disease due to *Vibrio* by comparing farms in Indonesia using the same water sources, which contain luminous *Vibrio* strains. The farms that did not use the *Bacillus* cultures experienced almost complete failure in all ponds, with luminescent *Vibrio* disease killing the shrimp before 80 days of culture was reached, whereas the addition of several *Bacillus* cultures in penaeid culture ponds allowed the culture of the shrimps for over 160 days without significant mortality. The bacterial species composition was different in the pond water on the two farms, with low *Vibrio* numbers in ponds where a large abundance of selected *Bacillus* species was maintained in the water column. While many questions are left unanswered from this study (e.g., bacterial community composition in different ponds), the results of this study do suggest that it is possible to change bacterial species composition and improve prawn production in large water bodies.

14.2   Isolation of bacteria for biocontrol

Some biocontrol bacteria used in aquaculture are derived from commercial products for human or livestock use. One of the advantages of choosing biocontrol bacteria that have already been approved for use as a 'probiotic' for human use is their existing safety data and ease of regulatory approval (Nikoskelainen *et al.*, 2001). The first trial for incorporation of a probiotic bacteria into aquaculture feeds used commercial preparations designed for land animals. Spores of *B. toyoi* reduced the mortality of Japanese eel *Anguilla japonica* experimentally infected by *Edwardsiella* sp. (Kozasa, 1986). In some cases, bacteria from the product were directly used to control disease without characterizing their activity *in vitro* (Nikoskelainen *et al.*, 2001; Queiroz & Boyd, 1998). As a more rational way to select strains that may be more likely to express antagonistic activity, bacteria from probiotic products were screened for their antagonistic activity against target pathogens first and strains that inhibit the growth of pathogens *in vitro* were tested further *in vivo* (Moriarty, 1998; Chang, 2002; Chao *et al.*, in preparation).

Beneficial bacteria can also be isolated from the intestine and skin mucus of the host. In juvenile fish and shellfish, the autochthonous microbes may be isolated from the digestive tract after dissection. The microbes adherent to epithelial cells can be separated from those adherent to mucus and from those transient in the lumen (Westerdahl *et al.*, 1991; Gatesoupe, 1999). These methods are not applicable to larvae and live food organisms, but the external surface of larval fish may be washed with a 0.1% benzalkonium chloride saline solution to differentiate the microbes adherent to the external surface from those present in the gut (Blanch *et al.*, 1997; Gatesoupe, 1999). Some bacterial strains isolated from the animals and culture units also showed potential biocontrol activity (Rengpipat *et al.*, 1998). It should also be understood that the characterization of potential biocontrol agents should be for specific bacterial strains, and that the typing of a specific strain as a member of a certain bacterial genus or species does not satisfy the need for carefully controlled efficacy and safety studies on individual bacterial strains.

14.3   Antagonistic activity of bacterial agents

In most cases, *in vitro* antagonism against target pathogens was used as the primary criterion for screening of candidate biocontrol bacteria. The screening methods include a cross-streaking method (Robertson *et al.*, 2000), and a disk diffusion double-layer method (Jack *et al.*, 1996). Different methods may give different results and the composition of the growth medium may affect the amount of active metabolites produced or the amount released into the medium (Olsson *et al.*, 1992). So it is more reasonable to try different methods and media during screening. Antagonism may be mediated not only by antibiotics, but also by many other inhibitory substances, for example organic acids, hydrogen peroxide (Ring & Gatesoupe, 1998; Gatesoupe, 1999), and siderophores (Gram & Melchiorsen, 1996). The inhibition due to such compounds is highly dependent on the experimental conditions, which are potentially different *in vitro* and *in vivo*. Also, the most important advantage of biocontrol agents over antibiotics is that multiple mechanisms are potentially involved in the biocontrol process, making it difficult for the pathogens to evolve all of the necessary resistant genes together (Moriarty, 1998). For example, the ability of some bacteria to adhere to intestinal mucus may block the intestinal infection route common to many pathogens (Evelyn, 1996), and some bacteria may stimulate the innate immunity of fish (Brunt & Austin, 2005). Therefore, the expression of antagonism *in vitro* is not a sufficient criterion to select candidate biocontrol agents (Riquelme *et al.*, 1997), nor is the absence of antagonism sufficient to rule the strains out (Rico-Mora *et al.*, 1998). Other potentially beneficial properties should also be considered during screening of candidate bacteria.

14.4   Colonization and persistence within the host

The colonization potential is another important criterion to characterize biocontrol agents. The process of colonization is characterized by attraction of bacteria to the mucosal surface, followed by association within the mucous gel or attachment to epithelial cells (Balcázar *et al.*, 2006). Adhesion and colonization of the mucosal surfaces are possible protective mechanisms against pathogens through competition for binding sites and nutrients (Westerdahl *et al.*, 1991), or immune modulation (Salminen *et al.*, 1998). Some bacteria can persist in the intestine for a long time if they have colonized the mucus of intestine, which is advantageous economically as repeated application of the biocontrol agent may be avoided.

In some cases, the bacteria do not truly colonize the gastrointestinal tract but rather achieve a sustained transient state (Fuller, 1992; Irianto & Austin, 2002). Jöborn *et al.* (1997) studied the ability of *Carnobacterium* strain K1 to colonize the intestinal tract of rainbow trout. *In vitro* experiments demonstrated that the strain was able to grow in extracts of intestinal mucus and feces of rainbow trout. Furthermore, 10^5 CFU/g of *Carnobacterium* strain K1 were recovered from the feces of rainbow trout after oral administration of K1-amended feed for six days followed by four days of normal feeding, which indicated that the strain can survive and persist in the intestine of rainbow trout and suggested the potential colonization ability of the strain.

A potential biocontrol bacterium also showed an inability to permanently colonize the intestine of a host in a study where rainbow trout were fed for 28 days with diets containing a strain of *Carnobacterium* sp. isolated from the intestine of Atlantic salmon (Robertson *et al.*, 2000). An increase in the number of *Carnobacterium* cells in the digestive tract during amended feeding was observed with a maximum population (7×10^6 CFU/g of intestine) achieved after feeding for 28 days. However, these levels declined rapidly after cessation of feeding with the amended diet, such that *Carnobacterium* could not be isolated from intestinal samples six days later. Although some bacteria may not be effective at colonizing their host, transient bacteria may also be efficient at biocontrol if the cells are introduced at a relatively high dose either continuously or semi-continuously (Gournier-Chateau *et al.*, 1994; Gatesoupe, 1999). In practice, it is therefore essential to evaluate the persistence of biocontrol agents in the gut (or other tissue of the host) which can guide the frequency of application to aquatic animals.

Colonization of the intestine of aquatic animals by bacteria was also demonstrated in other studies. Gildberg fed Atlantic salmon fry for five weeks with diets supplemented with a *Carnobacterium divergens* strain isolated from salmon intestines and then challenged them with *A. salmonicida* (Gildberg *et al.*, 1995). Fish were sampled for intestinal microbiota before the start of challenge and four weeks after challenge. For the samples taken before challenge, high numbers of *C. divergens* were recovered from the intestine of fish given supplemented feed, whereas many total bacterial counts and no *C. divergens* were detected from the control fish, indicating the colonization of this bacterium within the intestine of salmon fry. Colonization of *C. divergens* in the internal mucous layer of pyloric caeca of

Atlantic cod fry was further verified by immunostaining and light micros-copy (Gildberg & Mikkelsen, 1998). This study demonstrated bacterial colo-nization by culturing from the intestine of fish fed amended feed. However, the further persistence of the bacteria in the intestine of fish after the ces-sation of amended feed was not investigated.

Similarly, Atlantic cod fry were fed with feed containing a LAB isolated from Atlantic cod *Gadus morhua* (Gildberg *et al.*, 1997). After three weeks of feeding the fry were exposed to a virulent strain of *Vibrio anguillarum*. Three weeks after challenge, surviving fish were taken for microbial analysis by the same method mentioned above except that the intestinal contents were analyzed for microbial culturable counts. The intestine from fish fed with LAB had a virtual monoculture of LAB, whereas the intestine of control fish had a mixed microbiota. As the intestinal content from control fish also contained *Pseudomonas*-like bacteria, it was concluded that the LAB had displaced other potential colonizers.

In another study, a *Vibrio alginolyticus* strain used in a commercial shrimp hatchery in Ecuador was applied to Atlantic salmon by immersion for 10 min (Austin *et al.*, 1995). Challenge experiments seven days after application of *V. alginolyticus* revealed a reduction in mortality after expo-sure to *A. salmonicida* and to a lesser extent after exposure to *V. anguilla-rum* and *Vibrio ordalii*. *V. alginolyticus* was cultured from the intestine 21 days after the initial application.

14.5   Considerations for the design of pathogen challenge tests

The improvement in the survival response of aquatic animals to pathogens has been shown in many challenges that followed biocontrol treatments. There are, however, many factors that may influence the sensitivity of the animal to pathogens, and the biocontrol efficacy. These effects were some-times difficult to observe repeatedly (Gildberg & Mikkelsen, 1998; Har-zevili *et al.*, 1998; Gatesoupe, 1999). In some studies mortality was delayed only in comparison with the control without treatment (Gatesoupe, 1994; Gildberg & Mikkelsen, 1998; Gatesoupe, 1999).

The methods of challenge with pathogens include immersion, injection, and cohabitation. Immersion and cohabitation may emulate the natural infection process better than injection. The protection achieved by coloniz-ing the mucus of the host and competitively excluding pathogens may not be assessed by injection challenge, as the pathogen is administered directly to the inner cavity of the fish. Bacteria can be applied to the culture water of fish directly or incorporated in the feed and then fed to fish. In some cases, bacteria were applied to the enrichment medium of live food organ-isms and then fed to larvae (Gatesoupe, 1994, 1997).

14.5.1   Important parameters to consider during challenge

Dosage of protective bacteria

In a study evaluating rotifer enrichment using LAB, bacterial dosages greater than 10^7 CFU/ml of enrichment medium were demonstrated to be necessary to obtain the maximum protection effect after challenge by *Vibrio* spp. (Gatesoupe, 1991). However, the survival rate of the non-challenged groups was significantly decreased with LAB concentrations as high as 5×10^7 CFU/ml in comparison with the control group. The optimal LAB concentration corresponded therefore to a level of approximately 10^7 CFU/ml of enrichment medium.

In a study evaluating the biocontrol efficacy of *Aeromonas sobria* strain GC2 supplemented feed, bacteria were incorporated into the feed at a level between 10^3 and 10^{10} CFU/g feed (Brunt & Austin, 2005). Challenge with *S. iniae* and *Lactococcus garvieae* indicated that the best protection was given by dosage levels of 10^7 and 10^8 CFU/g, with the dosages of 10^3 CFU/g or 10^{10} CFU/g leading to less protection.

Similar results were achieved in the control of furunculosis by *L. rhamnosus*, where treatments fed 10^{12} CFU/g feed suffered a higher mortality than groups fed 10^9 CFU/g feed in a challenge trial of rainbow trout by *A. salmonicida* through cohabitation (Nikoskelainen *et al.*, 2001). It is also worth noting that the mortality of the cohabitants, challenged by i.p. injection, was significantly lower in the 10^{12} CFU/g feed group than in the 10^9 CFU/g group. The authors noted that the opposite result observed here might be because that i.p. injection presented more harsh challenge conditions than cohabitation and higher bacterial dosage was therefore necessary to observe a reduction in mortality. Thus, the optimal dosage is dependent upon the challenge conditions. This dependency upon the route of pathogen administration is important in terms of the dosage determination for pond trials, since typically there are relatively moderate levels of pathogens in culture units as compared to the challenge conditions in laboratory studies, suggesting a lower dosage of biocontrol agent may be necessary in commercial conditions. Also, the duration of feeding with biocontrol agents before challenge may also be an important factor influencing dosage. As a whole, the dosage of biocontrol agents should be determined in different situations that ultimately emmulate commercial conditions to avoid a resultant lower efficacy and unnecessary costs.

Timing of challenge

The timing for the application of biocontrol bacteria is another important parameter in challenges. Chang (2002) fed *E. faecium*-amended feed to fish 14 days before challenge by *E. tarda*. The author demonstrated that *E. faecium* strain SF68 suppressed the growth of *E. tarda in vitro* only if its initial inoculum was much higher (ca. 10^3) than that of *E. tarda*, suggesting that the oral application of *E. faecium* is more effective in disease prevention than treatment.

The duration of feeding with biocontrol bacteria before challenge in different studies varied from 5 to 100 days, with seven to 14 days mostly used. Robertson *et al.* (2000) also demonstrated that the duration of time that *Carnobacterium* sp. was fed before challenge was an important parameter for the reduction in mortality in Atlantic salmon. In contrast, the application of a *Roseobacter* sp. to the culture water of scallop larvae together with the pathogen *Vibrio pectenicida* did not result in any protective effect (Ruiz-Ponte *et al.*, 1999). The authors suggested that a protective effect may have been achieved if the *Roseobacter* sp. had been applied for a certain amount of time prior to challenge. In some cases a reduction in mortality may be achieved by applying biocontrol agents after challenge as demonstrated by Lategan *et al.* (2004b).

14.6   Safety of bacterial biocontrol agents and perspectives on future development

Tests of pathogen antagonism, survival within a host, and disease challenge are essential to select the candidate biocontrol agents, and once a collection of candidate bacterial strains is available it is important to conduct long-term feeding trials to conclude that the strains are harmless. Among these long-term feeding trials, there are few reports of improvement for survival (Ringø & Vadstein, 1998). This paucity of information is not surprising, since the protective effect of biocontrol agents may only be observed in the presence of a pathogen. The practical evaluation of prospective biocontrol strains will require long-term surveys (Gatesoupe, 1999).

The potential advantages of biocontrol agents compared to antibiotics were discussed by Moriarty (1998), with most attention directed towards the production of inhibitory substances by the beneficial bacterial strains. The possibility of selecting for biocontrol-resistant pathogens must not be underestimated, and it is particularly important to search for strains that express diverse antagonistic properties that may lower the risk of resistance. For example, the ability of some bacteria to adhere to the intestinal epithelium may inhibit the intestinal infection route common to many pathogens (Evelyn, 1996; Gatesoupe, 1999).

With the application of biocontrol bacteria in aquaculture, the fate of the bacteria in the rearing medium or in the gastrointestinal tract remains undetermined in many cases. Immunological and molecular probes will be useful tools to track the presence and relative abundance of biocontrol bacteria (Austin *et al.*, 1995; Ringø *et al.*, 1995, 1996; Gatesoupe, 1999). Also, the influence of inoculated biocontrol bacteria on the gastrointestinal microbiota remains poorly described, and it is anticipated that in the near future molecular approaches (e.g., next-generation sequencing of rRNA gene amplicons) will be applied in many systems to examine bacterial communities during biocontrol administration (Gatesoupe, 1999; Raskin *et al.*, 1997; Wallner *et al.*, 1997; Hugenholtz *et al.*, 1998).

14.7   Biocontrol using bacteriophages

The use of lytic bacteriophages to control pathogenic bacteria may be another option to control diseases in aquaculture. Bacteriophages are narrow-spectrum, in many cases species-specific, thereby causing minimal impact to the indigenous microbiota (Greer, 2005). This specificity differentiates the use of bacteriophages from the use of antibiotics to treat bacterial diseases. Additionally, bacteriophages are natural components of the environment (Goyal *et al.*, 1987), and bacteriophage preparations are relatively easy and inexpensive to prepare (Greer, 2005).

This section will review numerous reports in the literature on the therapeutic use of bacteriophages (herein, 'phage') to control bacterial pathogens of aquaculture importance. While reviewing the literature on the topic of phage therapy it is important to recognize the fervor that exists amongst both the proponents and the critics for phage therapy; therefore, as a reader one should be careful not to make conclusions without the availability of objective data. In aquaculture, only a few of the publications proposing phages as therapeutic agents are peer-reviewed, and this section will confine discussion only to these peer-reviewed reports (Table 14.2). The peer-reviewed studies on phage therapy include a study of the control of *E. tarda* (Wu & Chao, 1982), *E. ictaluri* (Walakira *et al.*, 2008; Carrias *et al.*, 2011), *L. garvieae* (Park *et al.*, 1997, 1998; Nakai *et al.*, 1999) *Pseudomonas plecoglossicida* (Park *et al.*, 2000; Park & Nakai, 2003), *V. harveyi* (Vinod *et al.*, 2006; Karunasagar *et al.*, 2007; Shivu *et al.*, 2007; Karunasagar *et al.*, 2005), *S. iniae* (Matsuoka *et al.*, 2007), and *A. salmonicida* (Imbeault *et al.*, 2006; Verner-Jeffreys *et al.*, 2007). Additionally, phages have been isolated against *Flavobacterium psychrophilum* (Stenholm *et al.*, 2008; Kim *et al.*, 2010) and a brief report was published on their therapeutic potential to protect ayu *Plecogloccus altivelis* from *F. psychrophilum* infection (Nakai *et al.*, 2010).

14.7.1   Phage therapy of *Edwardsiella tarda* infection in loaches

The first reported investigation of phage as biocontrol agents in aquaculture were phages isolated against *E. tarda* and tested for control of edwardsiellosis in loaches (Wu & Chao, 1982). The study demonstrated that a single phage (ΦET-1) was capable of lysing 25 out of 27 strains of *E. tarda*. In water, the phage reduced bacterial counts to 0.15% of the initial count and phage titers increased by one $\log_{10}$. The effect of phage-infected *E. tarda* on loach mortality was investigated, with no difference in mortality (100%) observed after 48 h compared to a control group in which loaches were inoculated with bacteria only. Loaches that were exposed to water containing phage-infected *E. tarda* after two hours post-mixing resulted in 5% survival for four days, fish exposed after 8 h post-mixing resulted in 90% survival.

14.7.2   Phage therapy of *Lactococcus garvieae* infection in yellowtail

The Gram-positive cocci *L. garvieae* is the etiological agent of lactococcal infection in yellowtail *Seriola quinqueradiata*, resulting in a systemic fish infection that affects the intestine, liver, spleen, and kidney (Kusuda *et al.*, 1991). Experimentally, mortality occurs within a short time (2–3 days) post-infection (Itami *et al.*, 1996). *L. garvieae* is an opportunistic pathogen that is ubiquitous in fish and is transmitted through the fecal-oral route (Kitao *et al.*, 1979). *L. garvieae* has great phenotypic heterogeneity and genetic diversity among isolates (Vela *et al.*, 2000). The high mortality caused by the disease in yellowtail and development of resistance to antibiotics used warranted development of additional treatment agents for its control (Kawanishi *et al.*, 2005).

Lytic phages specific to *L. garvieae* have been isolated (Park *et al.*, 1997, 1998) and three of these phages were evaluated for their stability in various conditions, their fate in yellowtail, and for their capacity to protect the fish from experimentally induced *L. garvieae* infection (Nakai *et al.*, 1999). Each phage was more stable in sterilized water than in unsterilized natural sea-water, and phage titers decreased with increasing temperature, after application onto feed, after incubation with yellowtail serum, or when the pH was lower than 3.5 (Nakai *et al.*, 1999). When a broad host-range *L. garvieae* phage (PLgY-16) was administered to yellowtail by i.p. injection it appeared in the spleens of fish 3 h post-administration and was undetectable after 48 h post-administration (Nakai *et al.*, 1999). In the intestines phages were recovered up to 3 h post-administration when given orally in phage-impregnated feed. Simultaneous administration of phage and *L. garvieae* resulted in the recovery of phages from the spleen up to five days post-i.p. administration, and recovery from the intestines up to 24 h post-oral administration. These results suggest that these phages are capable of migrating into the sites of infection within the fish.

The capacity of a *L. garvieae* phage (PLgY-16) to control experimentally induced *L. garvieae* infection in yellowtail was evaluated (Nakai *et al.*, 1999). A single administration of phage by i.p. injection resulted in higher survival of fish groups receiving phage (90%) compared with the control group that did not receive phage (45%). The i.p administration of phage at different times (0, 1, and 24 hours) post-bacterial challenge (also by i.p. administration) resulted in higher survival rates for fish that received phage treatment at the earlier time. Survival percentages over ten days were 100%, 80%, 50%, for fish groups that received phage at 0, 1, 24 h post-challenge, and 10% for the control group receiving no phage, respectively. A delay in mortality was also observed in fish groups that received phage compared with control groups that did not receive phage. In another experiment, fish that received phage through oral administration of phage-impregnated feed and were challenged 30 minutes later by anal intubation had survival rates of 90%, 80%, and 35% for fish groups that received phage, phage and *L. garvieae* together, and no phage control, respectively. The authors suggested that these results demonstrate a protective effect of these phages against *L. garvieae* infection over the tested time frame, and indicate that they may be useful as a therapeutant at early stages of systemic infection or with low burdens of bacterial infection in yellowtail (Nakai *et al.*, 1999). Additionally, it is important to point out that better protection was observed when phages were administered immediately after challenge or at earlier times compared to administering phage at 24 h post-challenge, possibly indicating that phage treatment for *L. garvieae* infection may be more effective as a prophylactic treatment rather than a therapeutic treatment. No phage-resistant strains were isolated during challenge studies and no antibody production was detected, although that is not surprising considering the short time frame of these experiments. The authors point out the need for experimentation with larger fish, but to date no report on the efficacy of these phages to treat infection of yellowtail at the commercial/farm scale level has been reported.

14.7.3   Phage therapy of *Pseudomonas plecoglossicida* in ayu

The Gram-negative, aerobic, rod-shaped bacteria *P. plecoglossicida* is the etiological agent of bacterial hemorrhagic ascites (BHA), a systemic infection of ayu (Nishimori *et al.*, 2000). The main route of *P. plecoglossicida* infection is unclear, although there are reports that injuries to the skin and fin are the main sites of bacterial adherence (Sukenda & Wakabayashi, 2001). Ayu have been experimentally infected by immersion (Sukenda & Wakabayashi, 2000), intramuscular injection and oral administration (Park & Nakai, 2003). The disease mainly affects the kidney and spleen (Kobayashi *et al.*, 2004). *P. plecoglossicida* isolates are biochemically homogeneous (Nakatsugawa & Iida, 1996; Nishimori *et al.*, 2000) and consists of a single serotype (Park *et al.*, 2000). Outbreaks of BHA are common shortly after fish are introduced into culture ponds and after chemotherapy of cold water disease caused by *F. psychrophilum* (Park & Nakai, 2003). The disease affects juvenile and market-size fish with mortalities ranging from 30 to 50% in pond cultured ayu (Kobayashi *et al.*, 2004). Presently, there are no licensed chemotherapeutants effective against BHA (Kobayashi *et al.*, 2004).

Isolation of *P. plecoglossicida* phages PPpW-3 (Myoviridae) and PPpW-4 (Podoviridae) from diseased ayu and from rearing pond water was reported (Park *et al.*, 2000). Both phages were reported to be lytic and specific for *P. plecoglossicida*. The co-inoculation of *P. plecoglossicida* and phage in pond water resulted in a decrease of bacterial titers and an increase in phage titers over time. Phages administered orally through feed only were unde-tectable in the kidneys of ayu at 24 h, but were detected at 24 h in the kidneys when phages were administered shortly after a bacterial challenge, suggesting the bacteria were infected with the phage and moved them into the fish via the infected bacterial cell. No phage-resistant strains were recov-ered from dead challenged fish; however, studies of *P. plecoglossicida* sus-ceptibility to phages resulted in variant strains of *P. plecoglossicida* that were phage-resistant. Furthermore, these phage-resistant strains were non-pathogenic to ayu at doses that cause high mortality using the wild-type isolate. Importantly, this indicates that while phage-resistant pathogen strains may evolve under strong selective pressure during phage administra-tion, the genetic loci that are required for phage infectivity may similarly be required for a fully virulent phenotype. This will of course be unique to each phage–host interaction, and will not be generally true for all phage-resistant strains of bacterial pathogens.

There are two published reports regarding the evaluation of phages PPpW-3 and PPpW-4 as therapeutic agents against *P. plecoglossicida* infection. In the first study (Park *et al.*, 2000), fish that received oral admin-istration of phage-impregnated feed 15 min post-challenge with oral admin-istration of bacteria-impregnated feed resulted in significantly lower mor-tality (23%) compared with a control group (65%) that did not receive phage treatment. In another experiment where smaller fish were used, mortality from fish that were challenged by oral administration of bacteria-impregnated feed and received phage 1 h post-challenge resulted in 78% lower mortality than the comparative control group, and fish that received phage 24 h post-challenge resulted in 68% lower mortality than control fish. Additionally, in both experiments there was a delay in mortality in the groups that received phage. The authors also reported that *P. plecoglossi-cida* was typically isolated from the kidneys of dead fish from the control group that received no phage but not from the group that received phage.

A second study proceeded to further evaluate phages PPpW-3 and PPpW-4 for their therapeutic potential (Park & Nakai, 2003). In this study it was demonstrated that phage suppressed *P. plecoglossicida* growth espe-cially when high titers of phage (10^6 plaque forming units (PFU)/ml) were co-inoculated with bacteria (5×10^2 CFU/ml) compared to lower multiplic-ity of infections. Bacterial growth was not completely suppressed even though a large increase in phage titers from 10^6 PFU/ml to 10^{10} PFU/ml was observed in some cases. Therapeutic effects of these phages were evaluated for both oral and water borne infection models, as well as in the field in a commercial pond. When fish groups received a single oral administration of phage-impregnated feed immediately after challenge with oral adminis-tration of *P. plecoglossicida* also through feed, mortalities of 93%, 53%, 40%, and 20%, were observed for fish groups that did not receive phage (control), that received phage PPpW-3, that received phage PPpW-4, and that received a phage mixture (PPpW-3 and PPpW-4), respectively. In the water-borne infection model, fish that were injected intramuscularly with *P. plecoglossicida* served as a source of infection for groups that received phage-impregnated feed at 24 h and 72 h post-injection. In two trials, mortality in the treated group was 63% and 73% lower than in the controls that did not received phages. In the field trial, three administrations of feed containing a phage mixture (PPpW-3 and PPpW-4) was administered through feed to fish in a pond with an ongoing *P. plecoglossicida* infection, and the bacterial and phage titers present in dead fish, live fish, and pond water were quantified at various time points before and after phage administration. *P. plecoglossicida* was isolated from 98% of dead fish and from 40% of apparently healthy fish sampled two days prior to phage administration. In the field trial a reduction in bacterial levels and an increase in phage levels were observed in dead fish, live fish, and pond water after phage administration. The authors use this observation in conjunction with the observation that mortality in fish consistently decreased by 5% daily after phage administration to suggest protective effects. However, because no control pond was available it is difficult to determine if the decrease in daily mortality was a consequence of phage administration. Further complicating the results as noted by the authors was the observation of mixed infection with *F. psychrophilum* during the trial.

14.7.4   Phage therapy of *Vibrio harveyi* infection in shrimp

The Gram-negative luminous *V. harveyi* is the etiological agent of luminous vibriosis, an infection that causes losses in penaeid shrimp *P. monodon* cultures in hatcheries (Prayitno & Latchford, 1995) and in growout ponds (Ruangpan & Kitao, 1991; Nithimathachoke *et al.*, 1995). The pathogenesis of the disease is not fully understood; however, challenge studies suggest that the haemolymph and the hepatopancreas may be major sites of infection. *V. harveyi* is the most abundant aerobic flora in shrimp hatcheries and has the capability to produce one of the widest spectrums of disease in aquaculture (Shivu *et al.*, 2007). Additionally, genetic fingerprinting and protein profiling suggest that *V. harveyi* is a very diverse species (Pizzutto & Hirst, 1995), suggesting a need for broad-spectrum phages as candidates for therapy. Persistence and survival of *V. harveyi* in shrimp hatcheries may be attributed to its ability to form biofilms that exhibit resistance to disinfectants and antibiotics (Karunasagar *et al.*, 1994). The ban imposed by seafood importers on the use of unregulated antibiotics (Shivu *et al.*, 2007) further complicates control efforts of the disease, hence the need for alternative therapeutic agents. However, consideration should be given when selecting candidates for phage therapy of *V. harveyi* as there is evidence that phages may mediate its toxicity (Munro *et al.*, 2003; Austin *et al.*, 2003).

    The first reported use of phage to control *V. harveyi* infections used phages isolated from water from shrimp hatcheries or shrimp grow-out ponds (Karunasagar *et al.*, 2005). One of the phage types had the capacity to effectively lyse 46 *V. harveyi* pathogenic isolates from a variety of sources. In a laboratory microcosm two applications of phage at 24 h intervals was able to reduce *V. harveyi* levels by three $\log_{10}$ and resulted in 80% shrimp survival compared with 10% survival in the control. In a field trial phages were tested for biocontrol at a hatchery experiencing mortality due to *V. harveyi*. A reduction of *V. harveyi* counts from 10^6 CFU/ml to 10^3 CFU/ml was observed by 48 h and *V. harveyi* was undetectable after 72 h. The survival of larvae was 78% in the tank that received phage but no survival is reported on the control tank although one was included.

In another study, a *V. harveyi*-specific phage belonging to the Siphoviridae family of phages was isolated from water at a shrimp farm in India and was evaluated for its potential to reduce mortality due to *V. harveyi* (Vinod *et al.*, 2006). Similar to the first study, the capacity of phage to reduce *V. harveyi* in a water sample demonstrated that two administrations of phage, at time 0 and 24 h post-bacterial inoculation, resulted in a three $\log_{10}$ reduction in *V. harveyi* counts (10^6 CFU/ml to 10^3 CFU/ml) and 80% shrimp survival, compared with 25% shrimp survival in the control which also had a 10-fold increase in *V. harveyi* counts. In a hatchery trial, triplicate tanks each containing 500 L of salt water and stocked with 35000 nauplii of *P. monodon* were used to evaluate the impact of phage addition on shrimp survival. Daily phage administration for 17 days at a final concentration of 2×10^5 PFU/ml in tank water resulted in 86% larvae survival and no *V. harveyi* counts were detected during the trial. This is in comparison to an antibiotic treatment (kanamycin at 10 ppm and oxytetracycline at 5 ppm) that resulted in 40% larvae survival and *V. harveyi* counts of 10^5 CFU/ml, as well as a control treatment that received neither phage or antibiotics and resulted in larvae survival of 17% with *V. harveyi* counts of 10^6 CFU/ml at the end of the study.

In another study, four additional phages lytic to *V. harveyi* were isolated from oyster tissue and from hatchery water and tested for their therapeutic potential in shrimp (Karunasagar *et al.*, 2007). The four phages had the ability to form clear plaques on 55–70% of 100 *V. harveyi* pathogenic isolates, and using a combination of all four phages lysed 94% of these *V. harveyi* strains. Because *V. harveyi* is known to survive in hatchery environments within biofilms (Karunasagar *et al.*, 1996), an experiment was performed to test the capacity of phage to control *V. harveyi* within biofilms. It was demonstrated that treatment of biofilm on a polyethylene (HDPE) surface with one phage type at 10^8 PFU/ml reduced bacterial counts by three $\log_{10}$ after 18 h. The authors noted a dose-dependent response of phage titer on the reduction of bacterial counts. In a hatchery exhibiting mortality due to *V. harveyi*, two phage types were inoculated for four consecutive days into duplicate 10 ton larval tanks each containing 0.5 million *P. monodon* shrimp post-larvae, and another two tanks received antibiotic (oxytetracycline at 5 mg/L and kanamycin at 10 mg/L). The shrimp that received phage had a 87% mean survival, whereas shrimp that received antibiotics had a 67% mean survival. However, in the hatchery trial only two replicate tanks were used for each treatment and a control tank receiving no treatment was not included, precluding any statistical analysis of these results.

14.7.5   Phage therapy of *Aeromonas salmonicida* infection in salmonids

As previously described, *A. salmonicida* is an economically important disease to the Atlantic salmon industry (Hirväla-Koski *et al.*, 1994). The bacterium is highly pathogenic and has many virulence factors, including haemolysins, cytotoxins, enterotoxins, endotoxins, and adhesions, that allows it to evade the host defenses and cause disease (Ellis, 1991). There is evidence to suggest that the pathogen obtains entry into fish through the intestines by damaging the intestinal wall (Ringø *et al.*, 2004). *A. salmonicida* is composed of four sub-species, one of which (*A. salmonicida* subsp. *salmonicida*) is the causative agent of furunculosis in salmonids. As a species *A. salmonicida* is genetically heterogeneous; however, *A. salmonicida* subsp. *salmonicida* consists of a single type strain within the species (O'Hici *et al.*, 2000). In infected fish the pathogen is readily isolated from the kidney and from surface lesions of fish (O'Hici *et al.*, 2000). Presently, antibiotics and commercial vaccines are available for treatment of *A. salmonicida* infections.

A phage (HER 110, also known as phage 65) that was isolated from the La Petite Mouge River in France was evaluated for its interactions with its host *A. salmonicida* and its efficacy in biocontrol in aquarium water (Imbeault *et al.*, 2006). For the host interaction experiments, aquaria either contained phage alone (10^9–10^{10} PFU/ml), *A. salmonicida* (10^8 CFU/ml) alone, or both bacteria and phage were added to the aquaria on days 1 and 2, and were incubated for 21 days. When bacteria alone were added to the water, *A. salmonicida* in open water increased by two logs at day 8 and then decreased to undetectable levels thereafter. In interstitial water *A. salmonicida* increased by four logs by day 8 and only slightly decreased thereafter. When phage only was present in the water, the titers of phage decreased over time until they were undetectable in open water at day 11. However, phage levels remained constant up to 14 days in interstitial water. When both bacteria and phage were added to the water, phage titers remained stable at high titers both in open and interstitial water, while bacterial titers decreased rapidly to undetectable levels three days after addition of phage. In the same study, the phage HER 110 was evaluated for its biocontrol potential in brook trout *Salvelinus fontinalis* yearlings. Fish were challenged with *A. salmonicida* (10^8 CFU/ml) added to aquaria with other aquaria only receiving sterile culture media. A high titer of phage (10^9–10^{10} PFU/ml) was added to some aquaria, and the fish receiving phage treatment (or no bacteria) survived and initially showed no signs of infection. Fish that were exposed to bacteria and did not receive phage rapidly became sick, and by day 45 all were dead. In contrast, the group that received phage remained in good health for a longer period and it was not until day 35 that any fish were seriously affected by furanculosis, and only 10% of the fish died at the end of the study. From infected fish, 17 *A. salmonicida* isolates were obtained that were resistant to phage lysis, and these isolates were divided into six groups based on colony morphology. Cellular lysis tests showed that 8 of these 17 phage-resistant strains recovered sensitivity to phage HER 110 after repeated passaging.

In a second study, three previously isolated *A. salmonicida* lytic phages (Rodgers *et al.*, 1981) were evaluated for their retention within fish, their safety, and their capacity to control furunculosis in Atlantic salmon caused by *A. salmonicida* subsp. *salmonicida* (Verner-Jeffreys *et al.*, 2007). First, the safety of phage was tested *in vivo* using rainbow trout, a salmonid that responds better to feeding and stress conditions in tank confinement compared to Atlantic salmon. In one experiment, administration of a single phage type (phage O) by i.p. injection (2×10^6 PFU) to a group of 18 rainbow trout resulted in the recovery of low phage concentrations (10^2 PFU/g) in the stomach of fish at 2 h post-injection. The phage titers present in the stomach reached a level of 1×10^5 PFU/g and then titers declined to undetectable levels at 48 h post-phage administration. In some fish, plaques were observed in samples from the lower intestine up to 72 h post-administration, while phages were still detected in spleen and kidney samples up to 96 h post-administration. In another group of fish, a single administration of phage given through feed containing phage O ($10^{6.5}$ PFU/g) had higher phage titers (10^5 PFU/g) present in the stomach of fish after 1 h after which titers declined to undetectable levels at 24 h post-administration. However, phages were still present in the upper and lower intestines 96 h post-phage administration. Phages were also transiently detected in the spleen and kidney up to 4 h post-feeding. In both experiments no adverse effects attributable to the administration of phage were observed.

In a second experiment, a group of fish received an i.p. injection on days 0, 14, and 28 of a suspension containing a combination of three phages (O, R, and B, 6×10^7 PFU/fish). No adverse effect attributed to the repeated administration of phage was observed. Similarly, when a group of fish was fed a combination of three phages (8×10^7 PFU/g) at 1.5% body weight each day for three and seven day intervals, no adverse effect was observed. Similarly to the first set of experiments, phages were also rapidly eliminated. This study also included two experiments that evaluated the efficacy of phage to control furunculosis. In one experiment, when a combination of three phages (O, R, and B) were administered through i.p. injection to Atlantic salmon at two different time points post-challenge (0 h and 24 h) with *A. salmonicida* (also by i.p. injection) a delay in onset of mortality and a delay in the time to reach 60% mortality was observed in the group that received phage at time zero but not at 24 h post-challenge. However, at the end of the study no significant difference in mortality was observed among the three treatments.

In another experiment by the same authors, as part of the same study, a cohabitation method in which infected fish were used as a source of infection was used to evaluate the efficacy of a daily bath, daily oral administration, or single injection of phages (O, R, and B) on the control of furunculosis. In addition, administration of oxolinic acid for 10 days was also included as a treatment. Administration of phage and oxolinic acid was performed four days post-injection of the challenge source fish. No protection was provided by the control and the two treatments that received phage, while protection was provided by the group that received oxolinic acid treatment. During the challenge experiments a more translucent phenotype of *A. salmonicida* was recovered from fish that had succumbed to furunculosis. This mutant was resistant to all three phages used.

14.7.6 Phage therapy of *Flavobacterium psychrophilum* infection in trout

The Gram-negative rod *F. psychrophilum* is the etiological agent of bacterial coldwater disease (CWD) and rainbow trout fry syndrome (RTFS). This disease causes severe economic losses in salmonids worldwide (Nematollahi *et al.*, 2003). The disease causes high mortality in adult fish and in fry may cause mortalities of 50 to 60% (Lorenzen *et al.*, 1991). Studies have shown that *F. psychrophilum* isolates from different locales are phenotypically and genetically similar (Lorenzen *et al.*, 1997). The transmission and portal of entry of the bacterium are not fully understood; however, there is data that suggests that the gastrointestinal tract is a route of entry (Liu *et al.*, 2001). There are suggestions that the bacterium may be part of the normal aerobic flora present in the skin and gills of salmonids (Madetoja *et al.*, 2002). Several antibiotics have been used to treat *F. psychrophilum* and presently florfenicol is the antibiotic of choice (Stenholm *et al.*, 2008). Currently, no commercial vaccine is available for use on an industry wide level (Stenholm *et al.*, 2008).

The use of phages in therapy trials for the control of *F. psychrophilum* infections has only briefly been reported. The isolation of 22 *F. psychrophilum*-specific phages from rainbow trout farms in Denmark was reported in 2008 (Stenholm *et al.*, 2008). Infectivity studies showed that these phages had pronounced differences in their ability to infect different *F. psychrophilum* strains, with 24 out of 28 *F. psychrophilum* strains showing susceptibility to one or more of these phages, while four strains were resistant. In another study, five phages that infect *F. psychrophilum* were collected in Japan from ayu farms (Kim *et al.*, 2010). These phages also had highly variable patterns of infectivity for 128 *F. psychrophilum* isolates tested, with four of the phages able to infect the majority of 72 isolates obtained from ayu and 17 of these *F. psychrophilum* isolates showing complete resistance to phage infectivity. Phages tested were stable in pond water for 14 days and were stable in pH 4 to 9, but not at pH 2, 3, or 11. One of the phages (PFpW-3) was highly infective of *F. psychrophilum* isolates obtained both from ayu and other fish and was effective in reducing *in vitro* bacterial growth.

The only report on the use of phages to control *F. psychrophilum* infection was reported in a review on the use of phages in aquaculture (Nakai *et al.*, 2010). This review reports previously unpublished preliminary results in which ayu fish were challenged with *F. psychrophilum* by intramuscular injection or by immersion in a bacterial suspension, and then dipped in a phage (PFpW-3) suspension. The authors report the observation of gross signs (hemorrhage and ulceration on the skin) on the untreated group of fish while the in the phage treated group the signs were seen but were less severe. They reported that in both the injection and immersion challenge the phage-treated group had significantly lower mortality compared to the control.

14.7.7   Phage therapy of *Streptococcus iniae* infection in yellowtail

The hemolytic Gram-positive bacteria *Streptococcus iniae* causes meningo-encephalitis in several fish including tilapia (Perera *et al.*, 1994), yellowtail (Teixeira *et al.*, 1996), rainbow trout, and coho salmon (Kitao *et al.*, 1981), and may result in farm outbreaks with mortality rates of up to 50% in trout (Eldar *et al.*, 1995). There is evidence to suggest that *S. iniae* has an intracellular lifestyle, and that its ability to survive and multiply within macrophages and induce apoptosis is important in its pathogenesis (Zlotkin *et al.*, 2003). Two distinct serotypes have been recognized (Bachrach *et al.*, 2001; Barnes *et al.*, 2003). The main route of *S. iniae* infection is unclear, and fish have been experimentally infected orally, by bath exposure, and by i.p. injection (Bromage & Owens, 2002; Shoemaker *et al.*, 2000). It is also suggested that infection may occur through the olfactory route (Bromage & Owens, 2002). The disease affects 27 fish species affecting mainly adult and sub-adult fish; however, juveniles may also be susceptible (Agnew & Barnes, 2007). In Japan, *S. iniae* infections cause mortality of up to 8% in Japanese flounder *Paralichthys olivaceus* (Nguyen & Kanai, 1999). A commercial vaccine and chemotherapeutics are available for use in *S. iniae* infections; however, pathogen control has been difficult and this has encouraged development of other therapeutic options.

Six lytic *streptococcus iniae* – specific phages previously isolated between 2003 and 2005 from fish culture environments in Japan were evaluated for their therapeutic effect against experimental streptococcicosis of Japanese flounder *P. olivaceus* (Matsuoka *et al.*, 2007). All six phages (PSiJ31, PSiJ32, PSiJ41, PSiJ42, PSiJ51, PSiJ52) were assigned to the Siphoviridae family based on morphology. Therapeutic effects of the use of these phages were evaluated in five experiments. In the first four experiments, fish were injected i.p. with *S. iniae* strain Psi402 and then received i.p. injections of phage 1 h post-administration of bacteria. Fish received $10^{7.7}$ CFU/fish and $10^{8.2}$ PFU/fish in experiment 1, $10^{6.3}$ CFU/fish and $10^{8.2}$ PFU/fish in experiment 2, $10^{6.4}$ CFU/fish and 10^{8} PFU/fish in experiment 3, and $10^{5.4}$ CFU/fish and $10^{8.4}$ PFU/fish in experiment 4. Experiments 1 and 2 received a mixture of two phages (PSiJ31and PSiJ32), and experiments 3 and 4 received a mixture of four phages (PSiJ31, PSiJ32, PSiJ41, and PSiJ42). Survival rates of fish after 15 days were 28%, 33%, 48%, and 80% in experiments 1, 2, 3, and 4, respectively, while the percent survival of the control in all four experiments was zero. Although the authors indicate significant differences in all four experiments, limited replication was included (only one group containing 25 fish and one group containing 30 fish were included in experiments 1 and 2, respectively, and 2 groups each containing 25 and 20 fish were included in experiments 3 and 4, respectively). In experiment 5, four groups of 20 fish were i.p. injected with bacteria ($10^{5.4}$ CFU/fish) and then two of those groups received either at 12 h or 24 h post-bacteria administration a mixture of phages (PSiJ31, PSiJ32, PSiJ41, and PSiJ42) at a dose of $10^{8.7}$ PFU/fish. Mean survival rates were 45%, 33%, and 0% for the groups of fish that received phage at 12 h, 24 h, and for the control that did not receive phage, respectively. Similarly in this experiment only two replicates were included in each treatment.

During these challenge studies *S. iniae* strains re-isolated from dead fish were resistant to infection by the phages used. These resistant strains remained virulent to Japanese flounder when tested by i.p. injection at 10^5 CFU/fish. The authors indicate that there is a need for new phages that can kill these resistant strains, and that further investigations are required to establish phage therapy of *S. iniae* infections in Japanese flounder.

14.8   Strengths and challenges to phage biocontrol

Phage therapy has garnered interest because of the need for alternative therapies, and due to some of the unique attributes of phage that could provide a comparative advantage over chemotherapy. For example, in some cases where antibiotics do not work well because of the deep location of the infection, or when bacteria are resistant to available antibiotics, phage therapy has shown some degree of efficacy compared to available chemotherapeutants (Kochetkova *et al.*, 1989; Meladze *et al.*, 1982; Sakandelidze, 1991). Antibiotics and other chemotherapeutants may cause several side effects including intestinal disorders and secondary infections due to their broad-spectrum killing of beneficial microbiota, while the specificity of phages in targeting specific pathogens avoids harm to beneficial microorganisms (Chernomordik, 1989). The use of phages to control bacterial diseases like any therapeutic strategy has its difficulties and concerns. Some of these concerns are legitimate and some are perceived notions that derive from poorly designed initial studies involving phage therapy. However, it should be recognized that phage therapy, unlike the use of antibiotics or bacterial biocontrol agents, has a unique set of difficulties inherent to phage that restrict their efficacy and adoption as biocontrol agents.

The specificity of phages is a dual-edged sword, on one side providing an advantage in targeting specific bacterial pathogens, while also in some cases (1) requiring testing of the infecting strain for phage sensitivity prior to treatment, (2) not covering the diversity of pathogenic bacterial strains that may be encountered in aquaculture, and (3) selecting for resistant bacterial pathogens that may retain their virulence and evade phage infection (Levin & Bull, 2004). Additionally, companies may see it as too risky to invest in phage therapy since their return on their investment may be limited to a narrow range of pathogenic bacterial species. These limitations may be minimized by using rapid methods for bacterial identification, by using 'multivalent' phages that can attach to multiple receptors, or by using phage cocktails that collectively possess a broader host-range (Levin & Bull, 2004; Sagor et al., 2005).

Bacterial resistance to phages is one of the most significant potential limitations for phage therapy. There are multiple mechanisms by which bacteria may become resistant to phage, including an alteration or loss of a receptor leading to decreased (or absent) phage adsorption, and restriction modification immunity in which bacteria or plasmids they carry encode restriction endonucleases that can degrade phage genomes (Levin & Bull, 2004). Phage-resistant bacterial mutants may be susceptible to phage infection by other phages possessing a similar target range (Sagor et al., 2005). In contrast, resistance development to an antibiotic may occur for many bacterial species and not just for the target bacteria (Salyers & Amabile-Cuevas, 1997). However, as opposed to antibiotics, phages do have the capacity to co-evolve, counteracting the mutations of bacteria (Carlton, 1999). Additionally, selecting new phages to target phage-resistant bacterial mutants can be performed relatively easily and quickly within days or weeks (Sulakvelidze et al., 2001) as opposed to developing new antibiotics for antibiotic-resistant bacteria which may take years (Silver & Bostian, 1993; Chopra et al., 1997). It has been proposed that similarly to the multi-drug treatment strategy used against tuberculosis and HIV the use of a phage 'cocktail' to target several receptors on bacterial cells can be used to reduce the incidence of bacterial resistance (Leverentz et al., 2003; Levin & Bull, 2004); additionally, it has been reported that phage-resistant strains isolated in several studies, some in aquaculture (Loc Carrillo et al., 2005; Park et al., 2000), have reduced virulence or are completely avirulent. and in one case, it was reported that phage-resistant Salmonella enterica isolates resulted in good vaccines that provided protection from S. enterica infections (Capparelli et al., 2009).

Because of the self-replicating nature of phages, they may only need to be applied once which may result in a cheaper dose application cost than chemotherapeutants (Sulakvelidze et al., 2001). While there is potential for only a single administration to treat an infection, for this to occur the target bacterial population should be sufficiently dense, and physiologically and genetically susceptible to phage infection (Levin & Bull, 2004). Additionally, self-amplification will not occur if bacterial densities are too low for the phage to replicate faster that they are lost, or if the phage does not propagate in the bacteria even if it kills it (Levin & Bull, 2004).

It is well recognized that some phages (lysogenic phages) have the capacity to carry virulence factors or toxic genes (Brussow et al., 2004; McGrath et al., 2004; Mesyanzhinov et al., 2002) that they may transfer to their bacterial host (i.e., lysogenic conversion). This problem may be avoided by selecting virulent phages that lack the capacity for lysogenic conversion, via establishing the phage–host life cycle and sequencing the genome of phages that are candidates for therapy (Carlton, 1999).

Another possible limitation of phage therapy is the rapid elimination of phages by the reticuloendothelial system (Merril et al., 1996; Geier et al., 1973). If a phage does not rapidly find a bacterial host in which to multiply they will rapidly be eliminated before they reach their target, resulting in reduced effectiveness of phage therapy efforts especially in systemic applications. To mediate this problem it may be possible to serially passage phage to select phage variants that are better at evading the reticuloendothelial system (Merril et al., 1996).

Neutralizing anti-phage antibodies have been reported in phage therapy trials in humans (Kucharewicz-Krukowska & Slopek, 1987) and animals (Smith et al., 1987) and is considered a possible challenge to the success of phage therapies. It has been proposed that treating with a higher dose of phage may compensate for the portion of bacteria that may be eliminated by the antibody response (Carlton, 1999). Recent studies showed low immunogenicity of phage in mice (Capparelli et al., 2007) and in ayu (Park & Nakai, 2003), suggesting that antibody production may be minimal in some animals. Another consideration is that antibody production does not occur immediately, hence allowing some time for control of a pathogen prior to any interference by the fish immune system.

Apart from the specific problems that can potentially affect phage therapy there are other problems that are common to both antibiotics and phage therapy. For success, both antibiotics and phages have to be maintained in sufficient concentrations and densities, both agents must be able to reach the sites of infection or entry into the animal, and both must have access to bacteria when they are susceptible (Levin & Bull, 2004). Even if all of these practical issues associated with phage therapy are addressed, realization of phage therapy faces other hurdles regarding intellectual property and approval for use by regulatory agencies. Because the concept of using phages is over 100 years old it is unpatentable, although specific phages and unique applications may yield intellectual property (Thiel, 2004). Investors can secure a patent for individual phages; however, there is nothing to stop competing parties from finding a different but yet equally effective phage, making profitability questionable. It is a general consensus that using phage cocktails will make phage therapy more effective; however, to secure regulatory approval in the US, the US Food and Drug Administration (FDA) requires proven efficacy and safety at both the individual and collective level, a process that is very expensive and may impede investment. Additionally, the pharmacokinetics of self-replicating phages is different from that of normal drugs and demonstrating an effective dose for regulatory approval purposes may prove complicated. The treatment dose depends on density-dependent thresholds, and the ability to predict those thresholds may be difficult but is necessary for phage therapy to be practical (Payne & Jansen, 2000). It has been proposed that instead of having to prove efficacy and safety for phage cocktails, the broadening of a single phage type that can infect the majority or all of its targeting pathogenic strains may be an effective strategy (Thiel, 2004).

The approval of phage therapy in humans is certainly complicated from a regulatory perspective. However, approval for the use of phages in agriculture may prove less stringent as some approvals have already been given for the use of phages for some applications, and may serve as a route for approval of phage therapy use in other areas of agriculture. For example, the company Omnilytics (Salt Lake City, UT; www.omnilytics.com) has cleared regulatory hurdles for the use of phage against *Salmonella* on live animals prior to slaughtering, produces phage products to combat tomato and pepper spot, and has received FDA approval for an anti-*Escherichia coli* hide wash spray-mist for use on live animals prior to slaughter. The FDA has also granted approval to Intralytix (www.intralytix.com), for the use of *LMP 102*, a cocktail of six different phages to be used as an additive against *L. monocytogenes* during packaging of poultry and ready-to-eat meat products. In order to understand the potential of phage therapy, additional studies that help us understand the biology, ecology, and genetics of phages, as well as further studies evaluating the efficacy of phage therapy in various animal models require further exploration.

14.9   Future trends

The future of biological agents for the control of diseases in aquaculture will be shaped by many factors, including the demand for its use by consumers and by regulatory agents, the willingness of companies to invest in biological products, and most important the demonstration of the safety and effectiveness of these biological agents to control disease in various aquaculture systems. It should be noted that the awareness of the potential harm to the environment and to human consumers posed by the misuse of chemoterapeutic agents has awoken a wave of interest in consumers for products that have been organically or ecologically grown. This in turn is shaping the way in which products for food consumption are produced. As an example, regulatory agencies in the European Union and the United States now demand that exporters of fish products abide to stricter rules on the use of antibiotics. Countries that want to export have no choice but to abide by these regulations. This has resulted in an opportunity for alternative products such as biocontrol agents for use in improving health particularly in the shrimp industry, which is now evident in countries such as Mexico, Latin America, and South East Asia. This opportunity provides incentives for companies to invest in developing and marketing these alternative products. Presently, several companies (e.g., Biomin, San Antonio, TX, http://www.biomin.net; Novus International, St. Louis, MO, http://www. novusint.com) now carry probiotic products for aquaculture use.

The commercial use of biocontrol agents in aquaculture may be adopted first in countries that have had historically poor regulatory oversight but are now required to have more stringent control of chemotherapeutic use. With regard to regulatory approval, probiotics are already widely available for aquaculture use, and phages have been approved for use in agricultural products. However, before the commercial use of biocontrol agents becomes a common practice more data is required to establish the efficacy of these agents for the control of disease in various aquaculture systems and in various species. Presently, available data is restricted to a few well-designed studies that have reported variable results on the success of phage and bacterial-based therapy in aquaculture. Overall, there is generally greater enthusiasm for the application of bacterial biocontrol agents in aquaculture compared to phage therapy, given the more extensive literature concerning the efficacy of bacterial strains in preventing disease, the ability of some bacterial strains to control multiple aquaculture pathogens, and the greater familiarity of scientists and investors with bacteria (and their associated antibiotic products) for agricultural and industrial applications. Lastly, with any biocontrol strategy there is a necessity for a greater understanding of host–pathogen–biocontrol agent interactions and the impact of environmental factors and the microbial community on these interactions. It is our sincere hope that advancing our collective knowledge of these factors will contribute to the efficacy of biocontrol strategies and advance the environmental and economic sustainability of aquaculture practices.

14.10   References

AGNEW, W. & BARNES, A. C. (2007) *Streptococcus iniae*: an aquatic pathogen of global veterinary significance and a challenging candidate for reliable vaccination. *Veterinary Microbiology*, **122**, 1–15.

ALEXOPOULOS, C. J. (1962) *Introductory Mycology*. John Wiley & Sons, New York.

AUSTIN, B., STUCKEY, L. F., ROBERTSON, P. A. W., EFFENDI, I. & GRIFFITH, D. R. W. (1995) A probiotic strain of *Vibrio alginolyticus* effective in reducing diseases caused by *Aeromonas salmonicida*, *Vibrio anguillarum* and *Vibrio ordalii*. *Journal of Fish Diseases*, **18**, 93–96.

AUSTIN, B., PRIDE, A. C. & RHODIE, G. A. (2003) Association of a bacteriophage with virulence in *Vibrio harveyi*. *Journal of Fish Diseases*, **26**, 55–58.

BACHRACH, G., ZLOTKIN, A., HURVITZ, A., EVANS, D. L. & ELDAR, A. (2001) Recovery of *Streptococcus iniae* from diseased fish previously vaccinated with a *Streptococcus* vaccine. *Applied and Environmental Microbiology*, **67**, 3756.

BALCÁZAR, J. L., BLAS, I., RUIZ-ZARZUELA, I., CUNNINGHAM, D., VENDRELL, D. & MÚZQUIZ, J. L. (2006) The role of probiotics in aquaculture. *Veterinary Microbiology*, **114**, 173–186.

BARNES, A. C., YOUNG, F. M., HORNE, M. T. & ELLIS, A. E. (2003) *Streptococcus iniae*: serological differences, presence of capsule and resistance to immune serum killing. *Diseases of Aquatic Organisms*, **53**, 241–247.

BLANCH, A. R., ALSINA, M., SIMON, M. & JOFRE, J. (1997) Determination of bacteria associated with reared turbot (*Scophthalmus maximus*) larvae. *Journal of Applied Microbiology*, **82**, 729–734.

BROMAGE, E. S. & OWENS, L. (2002) Infection of barramundi *Lates calcarifer* with *Streptococcus iniae*: effects of different routes of exposure. *Diseases of Aquatic Organisms*, **52**, 199–205.

BRUNT, J. & AUSTIN, B. (2005) Use of a probiotic to control lactococcosis and streptococcosis in rainbow trout, *Oncorhynchus mykiss* (Walbaum). *Journal of Fish Diseases*, **28**, 693–701.

BRUSSOW, H., CANCHAYA, C. & HARDT, W. D. (2004) Phages and the evolution of bacterial pathogens: from genomic rearrangements to lysogenic conversion. *Microbiology and Molecular Biology Reviews*, **68**, 560.

CAPPARELLI, R., PARLATO, M., BORRIELLO, G., SALVATORE, P. & IANNELLI, D. (2007) Experimental phage therapy against *Staphylococcus aureus* in mice. *Antimicrobial Agents and Chemotherapy*, **51**, 2765.

CAPPARELLI, R., NOCERINO, N., IANNACCONE, M., ERCOLINI, D., PARLATO, M., CHIARA, M. & IANNELLI, D. (2009) Bacteriophage therapy of *Salmonella enterica*: A fresh appraisal of bacteriophage therapy. *Journal of Infectious Diseases*, **201**, 52–61.

CARLTON, R. M. (1999) Phage therapy: past history and future prospects. *Archivum Immunologiae et Therapiael Experimentalis*, **47**, 267–274.

CARRIAS, A., WELCH, T. J., WALDBIESER, G. C., MEAD, D. A., TERHUNE, J. S., & LILES, M. R. (2011) Comparative genomic analysis of bacteriophages specific to the channel catfish pathogen *Edwardsiella ictaluri*. *Virology Journal*, **8**, 6.

CHANG, C. (2002) An evaluation of two probiotic bacterial strains, *Enterococcus faecium* SF68 and *Bacillus toyoi*, for reducing edwardsiellosis in cultured European eel, *Anguilla anguilla* L. *Journal of Fish Diseases*, **25**, 311–315.

CHAO, R., CARRIAS, A., WILLIAMS, M. A., CAPPS, N., NEWTON, J. C., KLOEPPER, J. W., TERHUNE, J. S., & LILES, M. R. (in preparation) Identification of *Bacillus* strains for biological control of disease in catfish.

CHERNOMORDIK, A. B. (1989) Bacteriophages and their therapeutic-prophylactic use. *Medit sinskai a sestra*, **48**, 44.

CHOPRA, I., HODGSON, J., METCALF, B. & POSTE, G. (1997) The search for antimicrobial agents effective against bacteria resistant to multiple antibiotics. *Antimicrobial Agents and Chemotherapy*, **41**, 497.

COPLAND, J. W. & WILLOUGHBY, L. G. (1982) The pathology of *Saprolegnia* infections of *Anguilla anguilla* L. elvers. *Journal of Fish Diseases*, **5**, 421–428.

DICK, M. W. (1990) Phylum Oomycota. In Margulis, L., Corliss, J., Melkonian, M., & Chapman, D.J. (Ed.) *Handbook of Protoctista*. Jones and Bartlett, Boston.

ELDAR, A., BEJERANO, Y., LIVOFF, A., HOROVITCZ, A. & BERCOVIER, H. (1995) Experimental streptococcal meningo-encephalitis in cultured fish. *Veterinary Microbiology*, **43**, 33–40.

ELLIS, A. E. (1991) An appraisal of the extracellular toxins of *Aeromonas salmonicida* ssp. *salmonicida*. *Journal of Fish Diseases*, **14**, 265–277.

EVELYN, T. P. T. (1996) Infection and disease. In Iwama, G. & Nakanishi, T. (Ed.) *The Fish Immune System, Organism, Pathogen, and Environment.* Academic Press, San Diego.

FULLER, R. (1989) Probiotics in man and animals. *Journal of Applied Bacteriology*, **66**, 365.

FULLER, R. (1992) *Probiotics: The scientific basis*, Chapman & Hall, London.

GATESOUPE, F. J. (1991) The effect of three strains of lactic bacteria on the production rate of rotifers, *Brachionus plicatilis*, and their dietary value for larval turbot, *Scophthalmus maximus. Aquaculture*, **96**, 335–342.

GATESOUPE, F. J. (1994) Lactic acid bacteria increase the resistance of turbot larvae, *Scophthalmus maximus*, against pathogenic *Vibrio. Aquatic Living Resources*, **7**, 277–282.

GATESOUPE, F. J. (1997) Siderophore production and probiotic effect of *Vibrio* sp. associated with turbot larvae, *Scophthalmus maximus. Aquatic Living Resources*, **10**, 239–246.

GATESOUPE, F. J. (1999) The use of probiotics in aquaculture. *Aquaculture*, **180**, 147–165.

GEIER, M. R., TRIGG, M. E. & MERRIL, C. R. (1973) Fate of bacteriophage lambda in non-immune germ-free mice. *Nature*, **246**, 221–223.

GIBSON, L. F., WOODWORTH, J. & GEORGE, A. M. (1998). Probiotic activity of *Aeromonas media* on the Pacific oyster, *Crassostrea gigas*, when challenged with *Vibrio tubiashii. Aquaculture*, **169**, 111–120.

GILDBERG, A. & MIKKELSEN, H. (1998) Effects of supplementing the feed to Atlantic cod (*Gadus morhua*) fry with lactic acid bacteria and immuno-stimulating peptides during a challenge trial with *Vibrio anguillarum. Aquaculture*, **167**, 103–113.

GILDBERG, A., JOHANSEN, A. & BØGWALD, J. (1995) Growth and survival of Atlantic salmon (*Salmo salar*) fry given diets supplemented with fish protein hydrolysate and lactic acid bacteria during a challenge trial with *Aeromonas salmonicida. Aquaculture*, **138**, 23–34.

GILDBERG, A., MIKKELSEN, H., SANDAKER, E. & RINGØ, E. (1997) Probiotic effect of lactic acid bacteria in the feed on growth and survival of fry of Atlantic cod (*Gadus morhua*). *Hydrobiologia*, **352**, 279–285.

GOSPER, D. (1996) *Guide to the Diseases and Parasites Occurring in Eels.* Aquaculture Sourcebook. Department of Primary Industries, Queensland.

GOURNIER-CHATEAU, N., LARPENT, J. P., CASTELLANOS, M. I. & LARPENT, J. L. (1994) *Les probiotiques en alimentation animale et humaine.* Tec & Doc, Lavoisier, Paris, France, 192.

GOYAL, S. M., GERBA, C. P. & BITTON, G. (1987) *Phage Ecology.* Wiley series in ecological and applied microbiology, New York.

GRAM, L. & MELCHIORSEN, J. (1996) Interaction between fish spoilage bacteria *Pseudomonas* sp. and *Shewanella putrefaciens* in fish extracts and on fish tissue. *Journal of Applied Microbiology*, **80**, 589–595.

GRAM, L., MELCHIORSEN, J., SPANGGAARD, B., HUBER, I. & NIELSEN, T. F. (1999) Inhibition of *Vibrio anguillarum* by *Pseudomonas fluorescens* AH2, a possible probiotic treatment of fish. *Applied and Environmental Microbiology*, **65**, 969.

GREER, G. G. (2005) Bacteriophage control of foodborne bacteria. *Journal of Food Protection*, **68**, 1102–1111.

HARZEVILI, A. R. S., DUFFEL, H., DHERT, P., SWINGS, J. & SORGELOOS, P. (1998) Use of a potential probiotic *Lactococcus lactis* AR21 strain for the enhancement of growth in the rotifer *Brachionus plicatilis* (Müller). *Aquaculture Research*, **29**, 411–417.

HAWKE, J. P. & KHOO, L. H. (2004) Infectious diseases. In Tucker, C.S. & Hargreaves, J.A. (Ed.) *Biology and Culture of Channel Catfish.* Elsevier, Amsterdam, The Netherlands.

HAWKE, J. P., MCWHORTER, A., STEIGERWALT, A. G. & BRENNER, D. (1981) *Edwardsiella ictaluri* sp. nov., the causative agent of enteric septicemia of catfish. *International Journal of Systematic and Evolutionary Microbiology*, **31**, 396.

HAWKE, J. P., DURBOROW, R. M., THUNE, R. L. & CAMUS, A. C. (1998) *Enteric Septicemia of Catfish*. *SRAC Publication no. 477*. Texas A&M University, College Station, TX.

HIRVÄLA-KOSKI, V., KOSKI, P. & NIIRANEN, H. (1994) Biochemical properties and drug resistance of *Aeromonas salmonicida* in Finland. *Diseases of Aquatic Organisms*, **10**, 191–206.

HUGENHOLTZ, P., PITULLE, C., HERSHBERGER, K. L. & PACE, N. R. (1998) Novel division level bacterial diversity in a Yellowstone hot spring. *Journal of Bacteriology*, **180**, 366.

IMBEAULT, S., PARENT, S., LAGACÉ, M., UHLAND, C. F. & BLAIS, J. F. (2006) Using bacteriophages to prevent furunculosis caused by *Aeromonas salmonicida* in farmed brook trout. *Journal of Aquatic Animal Health*, **18**, 203–214.

IRIANTO, A. & AUSTIN, B. (2002) Use of probiotics to control furunculosis in rainbow trout, *Oncorhynchus mykiss* (Walbaum). *Journal of Fish Diseases*, **25**, 333–342.

ITAMI, T., KONDO, M., UOZU, M., SUGANUMA, A., ABE, T., NAKAGAWA, A., SUZUKI, N. & TAKAHASHI, Y. (1996) Enhancement of resistance against *Enterococcus seriolicida* infection in yellowtail, *Seriola quinqueradiata* (Temminck & Schlegel), by oral administration of peptidoglycan derived from *Bifidobacterium thermophilum*. *Journal of Fish Diseases*, **19**, 185–187.

JACK, R. W., WAN, J., GORDON, J., HARMARK, K., DAVIDSON, B. E., HILLIER, A. J., WETTENHALL, R. E., HICKEY, M. W. & COVENTRY, M. J. (1996) Characterization of the chemical and antimicrobial properties of piscicolin 126, a bacteriocin produced by *Carnobacterium piscicola* JG126. *Applied and Environmental Microbiology*, **62**, 2897–2903.

JÖBORN, A., OLSSON, J. C., WESTERDAHL, A., CONWAY, P. L. & KJELLEBERG, S. (1997) Colonization in the fish intestinal tract and production of inhibitory substances in intestinal mucus and faecal extracts by *Carnobacterium* sp. strain Kl. *Journal of Fish Diseases*, **20**, 383–392.

KARUNASAGAR, I., PAI, R., MALATHI, G. R. & KARUNASAGAR, I. (1994) Mass mortality of *Penaeus monodon* larvae due to antibiotic-resistant *Vibrio harveyi* infection. *Aquaculture*, **128**, 203–209.

KARUNASAGAR, I., OTTA, S. K. & KARUNASAGAR, I. (1996) Biofilm formation by *Vibrio harveyi* on surfaces. *Aquaculture*, **140**, 241–245.

KARUNASAGAR, I., VINOD, M. G., KENNEDY, B., VIJAY, A., DEEPANJALI, K. R., UMESHA, K. R. & KARUNASAGAR, I. (2005) Biocontrol of bacterial pathogens in aquaculture with emphasis on phage therapy. *Diseases in Asian Aquaculture V*, 535–542.

KARUNASAGAR, I., SHIVU, M. M., GIRISHA, S. K., KROHNE, G. & KARUNASAGAR, I. (2007) Biocontrol of pathogens in shrimp hatcheries using bacteriophages. *Aquaculture*, **268**, 288–292.

KAWANISHI, M., KOJIMA, A., ISHIHARA, K., ESAKI, H., KIJIMA, M., TAKAHASHI, T., SUZUKI, S. & TAMURA, Y. (2005) Drug resistance and pulsed field gel electrophoresis patterns of *Lactococcus garvieae* isolates from cultured *Seriola* (yellowtail, amberjack and kingfish) in Japan. *Letters in Applied Microbiology*, **40**, 322–328.

KIM, J. H., GOMEZ, D. K., NAKAI, T. & PARK, S. C. (2010) Isolation and identification of bacteriophages infecting ayu *Plecoglossus altivelis* specific *Flavobacterium psychrophilum*. *Veterinary Microbiology*, **140**, 109–115.

KITAO, T., AOKI, T. & IWATA, K. (1979) Epidemiological study on streptococcicosis of cultured yellowtail (*Seriola quinqueradiata*) – I. Distribution of *Streptococcus* sp. in sea water and muds around yellowtail farms. *Bulletin of the Japanese Society of Scientific Fisheries*, **45**, 567–572.

KITAO, T., AOKI, T. & SAKOH, R. (1981) Epizootic caused by beta-hemolytic *Streptococcus* species in cultured freshwater fish. *Fish Pathology*, **19**, 173–180.

KOBAYASHI, T., IMAI, M., ISHITAKA, Y. & KAWAGUCHI, Y. (2004) Histopathological studies of bacterial haemorrhagic ascites of ayu, *Plecoglossus altivelis* (Temminck & Schlegel). *Journal of Fish Diseases*, **27**, 451–457.

KOCHETKOVA, V. A., MAMONTOV, A. S., MOSKOVTSEVA, R. L., ERASTOVA, E. I., TROFIMOV, E. I., POPOV, M. I. & DZHUBALIEVA, S. K. (1989) Phagotherapy of postoperative suppurative-inflammatory complications in patients with neoplasms. *Sovetskaia meditsina*, 23.

KOZASA, M. (1986) Toyocerin (*Bacillus toyoi*) as growth promoter for animal feeding. *Microbiol. Aliment. Nutr*, **4**, 121–135.

KUCHAREWICZ-KRUKOWSKA, A. & SLOPEK, S. (1987) Immunogenic effect of bacteriophage in patients subjected to phage therapy. *Archivum Immunologiae et Therapiae Experimentalis*, **35**, 553.

KUSUDA, R., KAWAI, K., SALATI, F., BANNER, C. R. & FRYER, J. L. (1991) *Enterococcus seriolicida* sp. nov., a fish pathogen. *International Journal of Systematic and Evolutionary Microbiology*, **41**, 406.

LATEGAN, M. J. & GIBSON, L. F. (2003) Antagonistic activity of *Aeromonas media* strain A199 against *Saprolegnia* sp., an opportunistic pathogen of the eel, *Anguilla australis* Richardson. *Journal of Fish Diseases*, **26**, 147–153.

LATEGAN, M. J., TORPY, F. R. & GIBSON, L. F. (2004a) Biocontrol of saprolegniosis in silver perch *Bidyanus bidyanus* (Mitchell) by *Aeromonas media* strain A199. *Aquaculture*, **235**, 77–88.

LATEGAN, M. J., TORPY, F. R. & GIBSON, L. F. (2004b) Control of saprolegniosis in the eel *Anguilla australis* Richardson, by *Aeromonas media* strain A199. *Aquaculture*, **240**, 19–27.

LEVERENTZ, B., CONWAY, W. S., CAMP, M. J., JANISIEWICZ, W. J., ABULADZE, T., YANG, M., SAFTNER, R. & SULAKVELIDZE, A. (2003) Biocontrol of *Listeria monocytogenes* on fresh-cut produce by treatment with lytic bacteriophages and a bacteriocin. *Applied and Environmental Microbiology*, **69**, 4519.

LEVIN, B. R. & BULL, J. J. (2004) Population and evolutionary dynamics of phage therapy. *Nature Reviews Microbiology*, **2**, 166–173.

LIU, H., IZUMI, S. & WAKABAYASHI, H. (2001) Detection of *Flavobacterium psychrophilum* in various organs of ayu *Plecoglossus altivelis* by *in situ* hybridization. *Fish Pathology*, **36**, 7–12.

LOC CARRILLO, C., ATTERBURY, R. J., EL-SHIBINY, A., CONNERTON, P. L., DILLON, E., SCOTT, A. & CONNERTON, I. F. (2005) Bacteriophage therapy to reduce *Campylobacter jejuni* colonization of broiler chickens. *Applied and Environmental Microbiology*, **71**, 6554.

LORENZEN, E., DALSGAARD, I., FROM, J., HANSEN, E. M., HORLYCK, V., KORSHOLM, H., MELLERGUARD, S. & OLESEN, N. J. (1991) Preliminary investigations of fry mortality syndrome in rainbow trout. *Bulletin of the European Association of Fish Pathologists*, **11**, 77–79.

LORENZEN, E., DALSGAARD, I. & BERNARDET, J. F. (1997) Characterization of isolates of *Flavobacterium psychrophilum* associated with coldwater disease or rainbow trout fry syndrome I: phenotypic and genomic studies. *Diseases of Aquatic Organisms*, **31**, 197–208.

MADETOJA, J., DALSGAARD, I. & WIKLUND, T. (2002) Occurrence of *Flavobacterium psychrophilum* in fish-farming environments. *Diseases of Aquatic Organisms*, **52**, 109–118.

MAEDA, M. & LIAO, I. (1992) Effect of bacterial population on the growth of a prawn larva, *Penaeus monodon*. *Bulletin of the National Research Institute of Aquaculture*, **21**, 25–29.

MATSUOKA, S., HASHIZUME, T., KANZAKI, H., IWAMOTO, E., PARK, S. C., YOSHIDA, T. & NAKAI, T. (2007) Phage therapy against β-hemolytic streptococcicosis of Japanese flounder *Paralichthys olivaceus*. *Fish Pathology*, **42**, 181–189.

MCGRATH, S., FITZGERALD, G. F. & SINDEREN, D. (2004) The impact of bacteriophage genomics. *Current Opinion in Biotechnology*, **15**, 94–99.

MELADZE, G. D., MEBUKE, M. G., CHKHETIA, N. S., KIKNADZE, N. I., KOGUASHVILI, G. G., TIMOSHUK, I. I., LARIONOVA, N. G. & VASADZE, G. K. (1982) The efficacy of staphylococcal bacteriophage in treatment of purulent diseases of lungs and pleura. *Grudnaia Khirugiia*, **1**, 53–56.

MERRIL, C. R., BISWAS, B., CARLTON, R., JENSEN, N. C., CREED, G. J., ZULLO, S. & ADHYA, S. (1996) Long-circulating bacteriophage as antibacterial agents. *Proceedings of the National Academy of Science of the United States of America*, **93**, 3188–3192.

MESYANZHINOV, V. V., ROBBEN, J., GRYMONPREZ, B., KOSTYUCHENKO, V. A., BOURKALTSEVA, M. V., SYKILINDA, N. N., KRYLOV, V. N. & VOLCKAERT, G. (2002) The genome of bacteriophage [phi] KZ of *Pseudomonas aeruginosa* PAO1. *Journal of Molecular Biology*, **317**, 1–19.

MORIARTY, D. J. W. (1998) Control of luminous *Vibrio* species in penaeid aquaculture ponds. *Aquaculture*, **164**, 351–358.

MUNRO, J., OAKEY, J., BROMAGE, E. & OWENS, L. (2003) Experimental bacteriophage-mediated virulence in strains of *Vibrio harveyi*. *Diseases of Aquatic Organisms*, **54**, 187–194.

NAKAI, T., SUGIMOTO, R., PARK, K. H., MATSUOKA, S., MORI, K., NISHIOKA, T. & MARUYAMA, K. (1999) Protective effects of bacteriophage on experimental *Lactococcus garvieae* infection in yellowtail. *Diseases of Aquatic Organisms*, **37**, 33–41.

NAKAI, T., SABOUR, P. M. & GRIFFITHS, M. W. (2010) Application of bacteriophages for control of infectious diseases in aquaculture. *Bacteriophages in the Control of Food – And Waterborne Pathogens*, 257.

NAKATSUGAWA, T. & IIDA, Y. (1996) *Pseudomonas* sp. isolated from diseased ayu, *Plecoglossus altivelis*. *Fish Pathology*, **31**, 221–227.

NEMATOLLAHI, A., DECOSTERE, A., PASMANS, F. & HAESEBROUCK, F. (2003) *Flavobacterium psychrophilum* infections in salmonid fish. *Journal of Fish Diseases*, **26**, 563–574.

NGUYEN, H. T. & KANAI, K. (1999) Selective agars for the isolation of *Streptococcus iniae* from Japanese flounder, *Paralichthys olivaceus*, and its cultural environment. *Journal of Applied Microbiology*, **86**, 769–776.

NIKOSKELAINEN, S., OUWEHAND, A., SALMINEN, S. & BYLUND, G. (2001) Protection of rainbow trout (*Oncorhynchus mykiss*) from furunculosis by *Lactobacillus rhamnosus*. *Aquaculture*, **198**, 229–236.

NISHIMORI, E., KITA-TSUKAMOTO, K. & WAKABAYASHI, H. (2000) *Pseudomonas plecoglossicida* sp. nov., the causative agent of bacterial haemorrhagic ascites of ayu, *Plecoglossus altivelis*. *International Journal of Systematic and Evolutionary Microbiology*, **50**, 83.

NITHIMATHACHOKE, C., PRATANPIPAT, P., THONGDAENG, K., WITHYACHUMNARNKUL, B. & NASH, G. (1995) Luminous bacterial infection in pond reared *Penaeus monodon*. *Asian Shrimp Culture Council. Asian Shrimp News*, **3**, 1–3.

O'HICI, B., OLIVIER, G. & POWELL, R. (2000) Genetic diversity of the fish pathogen *Aeromonas salmonicida* demonstrated by random amplified polymorphic DNA and pulsed-field gel electrophoresis analyses. *Diseases of Aquatic Organisms*, **39**, 109–119.

OLSSON, J. C., WESTERDAHL, A., CONWAY, P. L. & KJELLEBERG, S. (1992) Intestinal colonization potential of turbot (*Scophthalmus maximus*) and dab (*Limanda limanda*)-associated bacteria with inhibitory effects against *Vibrio anguillarum*. *Applied and Environmental Microbiology*, **58**, 551–556.

PARK, K. H., MATSUOKA, S., NAKAI, T. & MUROGA, K. (1997) A virulent bacteriophage of *Lactococcus garvieae* (formerly *Enterococcus seriolicida*) isolated from yellowtail *Seriola quinqueradiata*. *Diseases of Aquatic Organisms*, **29**, 145–149.

PARK, K. H., KATO, H., NAKAI, T. & MUROGA, K. (1998) Phage typing of *Lactococcus garvieae* (formerly *Enterococcus seriolicida*) a pathogen of cultured yellowtail. *Fisheries Science – Tokyo*, **64**, 62–64.

PARK, S. C. & NAKAI, T. (2003) Bacteriophage control of *Pseudomonas plecoglossicida* infection in ayu *Plecoglossus altivelis*. *Diseases of Aquatic Organisms*, **53**, 33–39.

PARK, S. C., SHIMAMURA, I., FUKUNAGA, M., MORI, K. I. & NAKAI, T. (2000) Isolation of bacteriophages specific to a fish pathogen, *Pseudomonas plecoglossicida*, as a candidate for disease control. *Applied Environmental Microbiology*, **66**, 1416–1422.

PAYNE, R. J. H. & JANSEN, V. A. A. (2000) Phage therapy: the peculiar kinetics of self-replicating pharmaceuticals. *Clinical Pharmacology & Therapeutics*, **68**, 225–230.

PERERA, R. P., JOHNSON, S. K., COLLINS, M. D. & LEWIS, D. H. (1994) *Streptococcus iniae* associated with mortality of *Tilapia nilotica* × *T. aurea* hybrids. *Journal of Aquatic Animal Health*, **6**, 335–340.

PIZZUTTO, M. & HIRST, R. G. (1995) Classification of isolates of *Vibrio harveyi* virulent to Penaeus monodon larvae by protein profile analysis and M13 DNA fingerprinting. *Diseases of Aquatic Organisms*, **21**, 61–68.

PLUMB, J. A. (1999) Catfish bacterial diseases. In Plumb, J.A. (Ed) *Health Maintenance and Principal Microbial Diseases of Cultured Fish*. Iowa State University Press, Ames, IO.

PRAYITNO, S. B. & LATCHFORD, J. W. (1995) Experimental infections of crustaceans with luminous bacteria related to *Photobacterium* and *Vibrio*. Effect of salinity and pH on infectiosity. *Aquaculture*, **132**, 105–112.

QUEIROZ, J. F. & BOYD, C. E. (1998) Effects of a bacterial inoculum in channel catfish ponds. *Journal of the World Aquaculture Society*, **29**, 67–73.

RASKIN, L., CAPMAN, W. C., SHARP, R., POULSEN, L. K. & STAHL, D. A. (1997) Molecular ecology of gastrointestinal ecosystems. In Mackie, R.I., Withe, B.A., & Isaacson, R.E. (Ed.) *Gastrointestinal Microbiology*. International Thomson Publishing, New York.

RENGPIPAT, S., PHIANPHAK, W., PIYATIRATITIVORAKUL, S. & MENASVETA, P. (1998) Effects of a probiotic bacterium on black tiger shrimp *Penaeus monodon* survival and growth. *Aquaculture*, **167**, 301–313.

RENGPIPAT, S., TUNYANUN, A., FAST, A. W., PIYATIRATITIVORAKUL, S. & MENASVETA, P. (2003) Enhanced growth and resistance to *Vibrio* challenge in pond-reared black tiger shrimp *Penaeus monodon* fed a *Bacillus* probiotic. *Diseases of Aquatic Organisms*, **55**, 169–173.

RICO-MORA, R., VOLTOLINA, D. & VILLAESCUSA-CELAYA, J. A. (1998) Biological control of *Vibrio alginolyticus* in *Skeletonema costatum* (Bacillariophyceae) cultures. *Aquacultural Engineering*, **19**, 1–6.

RING, E. & GATESOUPE, F. J. (1998) Lactic acid bacteria in fish: a review. *Aquaculture*, **160**, 177–203.

RINGØ, E. & VADSTEIN, O. (1998) Colonization of *Vibrio pelagius* and *Aeromonas caviae* in early developing turbot (*Scophthalmus maximus* L.) larvae. *Journal of Applied Microbiology*, **84**, 227–233.

RINGØ, E., STRØM, E. & TABACHEK, J. (1995) Intestinal microflora of salmonids: a review. *Aquaculture Research*, **26**, 773–789.

RINGØ, E., BIRKBECK, T. H., MUNRO, P. D., VADSTEIN, O. & HJELMELAND, K. (1996) The effect of early exposure to *Vibrio pelagius* on the aerobic bacterial flora of turbot, *Scophthalmus maximus* (L) larvae. *Journal of Applied Bacteriology*, **81**, 207–211.

RINGØ, E., JUTFELT, F., KANAPATHIPPILLAI, P., BAKKEN, Y., SUNDELL, K., GLETTE, J., MAYHEW, T. M., MYKLEBUST, R. & OLSEN, R. E. (2004) Damaging effect of the fish pathogen *Aeromonas salmonicida* ssp. *salmonicida* on intestinal enterocytes of Atlantic salmon (*Salmo salar* L.). *Cell and Tissue Research*, **318**, 305–311.

RIQUELME, C., ARAYA, R., VERGARA, N., ROJAS, A., GUAITA, M. & CANDIA, M. (1997) Potential probiotic strains in the culture of the Chilean scallop *Argopecten purpuratus* (Lamarck, 1819). *Aquaculture*, **154**, 17–26.

ROBERTSON, P. A. W., O'DOWD, C., BURRELLS, C., WILLIAMS, P. & AUSTIN, B. (2000) Use of *Carnobacterium* sp. as a probiotic for Atlantic salmon (*Salmo salar* L.) and rainbow trout (*Oncorhynchus mykiss*, Walbaum). *Aquaculture*, **185**, 235–243.

RODGERS, C. J., PRINGLE, J. H., MCCARTHY, D. H. & AUSTIN, B. (1981) Quantitative and qualitative studies of *Aeromonas salmonicida* bacteriophage. *Microbiology*, **125**, 335.

ROWLAND, S. J. & INGRAM, B. A. (1991) *Diseases of Australian native freshwater fishes with particular emphasis on the ectoparasitic and fungal diseases of Murray cod* (Maccullochella peeli), *golden perch* (Macquaria ambigua) *and silver perch* (Bidyanus bidyanus), NSW Fisheries Bulletin Agriculture.

RUANGPAN, L. & KITAO, T. (1991) *Vibrio* bacteria isolated from black tiger shrimp, *Penaeus monodon* Fabricius. *Journal of Fish Diseases*, **14**, 383–388.

RUIZ-PONTE, C., SAMAIN, J. F., SANCHEZ, J. L. & NICOLAS, J. L. (1999) The benefit of a *Roseobacter* species on the survival of scallop larvae. *Marine Biotechnology*, **1**, 52–59.

SAGOR, M. M., ISLAM, K. K., ALI, M. R., ABDUL-AWAL, S. M., ADHIKARY, P. P., SARKER, P. K. & RAKIB, A. S. M. (2005) Bacteriophage: a potential therapeutic agent (a review). *Journal of Medical Sciences*, 5.

SAKAI, M., YOSHIDA, T., ASTUTA, S. & KOBAYASHI, M. (1995). Enhancement of resistance to vibriosis in rainbow trout, *Oncorhynchus mykiss* (Walbaum) by oral administration of *Clostridium butyricum* bacteria. *Journal of Fish Diseases*, **18**, 187–190.

SAKANDELIDZE, V. M. (1991) The combined use of specific phages and antibiotics in different infectious allergoses. *Vrachebnoe delo*, 60.

SALMINEN, S., BOULEY, C., BOUTRON, M. C., CUMMINGS, J. H., FRANCK, A., GIBSON, G. R., ISOLAURI, E., MOREAU, M. C., ROBERFROID, M. & ROWLAND, I. (1998) Functional food science and gastrointestinal physiology and function. *British Journal of Nutrition*, **80**, 147–171.

SALYERS, A. A. & AMABILE-CUEVAS, C. F. (1997) Why are antibiotic resistance genes so resistant to elimination? *Antimicrobial Agents and Chemotherapy*, **41**, 2321.

SCHWARZ, S., KEHRENBERG, C. & WALSH, T. R. (2001) Use of antimicrobial agents in veterinary medicine and food animal production. *International Journal of Antimicrobial Agents*, **17**, 431–437.

SHIVU, M. M., RAJEEVA, B. C., GIRISHA, S. K., KARUNASAGAR, I., KROHNE, G. & KARUNASAGAR, I. (2007) Molecular characterization of *Vibrio harveyi* bacteriophages isolated from aquaculture environments along the coast of India. *Environmental Microbiology*, **9**, 322–331.

SHOEMAKER, C. A., EVANS, J. J. & KLESIUS, P. H. (2000) Density and dose: factors affecting mortality of *Streptococcus iniae* infected tilapia (*Oreochromis niloticus*). *Aquaculture*, **188**, 229–235.

SILVER, L. L. & BOSTIAN, K. A. (1993) Discovery and development of new antibiotics: the problem of antibiotic resistance. *Antimicrobial Agents and Chemotherapy*, **37**, 377.

SMITH, H. W., HUGGINS, M. B. & SHAW, K. M. (1987) Factors influencing the survival and multiplication of bacteriophages in calves and in their environment. *Microbiology*, **133**, 1127.

SMITH, P. & DAVEY, S. (1993) Evidence for the competitive exclusion of *Aeromonas salmonicida* from fish with stress inducible furunculosis by a fluorescent pseudomonad. *Journal of Fish Diseases*, **16**, 521–524.

STENHOLM, A. R., DALSGAARD, I. & MIDDELBOE, M. (2008) Isolation and characterization of bacteriophages infecting the fish pathogen *Flavobacterium psychrophilum*. *Applied Environmental Microbiology*, **74**, 4070–4078.

STICKNEY, R. R. (2009) Diseases of aquaculture species. In Stickney, R.R. (Ed), *Aquaculture: an introductory text*, CABI, Wallingford, 148–173.

STRØM, E. & RINGØ, E. (1993) Changes in bacterial composition of early developing cod *Gadus morhua* (L.) larvae following inoculation of *Lactobacillus plantarum* into the water. In Walther, B.T. & Fyhn, H.J. (Ed.) *Physiological and Biochemical Aspects of Fish Development*. University of Bergen, Bergen, Norway.

SUKENDA & WAKABAYASHI, H. (2000) Tissue distribution of *Pseudomonas plecoglossicida* in experimentally infected ayu *Plecoglossus altivelis* studied by real-time quantitative PCR. *Gyobyo Kenkyu (Fish Pathology)*, **35**, 223–228.

SUKENDA & WAKABAYASHI, H. (2001) Adherence and infectivity of green fluorescent protein-labeled *Pseudomonas plecoglossicida* to ayu *Plecoglossus altivelis*. *Fish Pathology*, **36**, 161–167.

SULAKVELIDZE, A., ALAVIDZE, Z. & MORRIS JR, J. G. (2001) Bacteriophage therapy. *Antimicrobial Agents and Chemotherapy*, **45**, 649–659.

TEIXEIRA, L. M., MERQUIOR, V. L. C., VIANNI, M. C. E., CARVALHO, M. G. S., FRACALANZZA, S. E. L., STEIGERWALT, A. G., BRENNER, D. J. & FACKLAM, R. R. (1996) Phenotypic and genotypic characterization of atypical *Lactococcus garvieae* strains isolated from water buffalos with subclinical mastitis and confirmation of *L. garvieae* as a senior subjective synonym of *Enterococcus seriolicida*. *International Journal of Systematic and Evolutionary Microbiology*, **46**, 664.

THIEL, K. (2004) Old dogma, new tricks – 21st Century phage therapy. *Nature Biotechnology*, **22**, 31–36.

US DEPARTMENT OF AGRICULTURE. (2003a) *Part I: Reference of Fingerling Catfish Health and Production Practices in the United States*. USDA, Fort Collins, CO.

US DEPARTMENT OF AGRICULTURE. (2003b) *Part II: Reference of Foodsize Catfish Health and Production Practices in the United States*. National Health Monitoring System. USDA Fort Collins, CO.

VELA, A. I., VÁZQUEZ, J., GIBELLO, A., BLANCO, M. M., MORENO, M. A., LIÉBANA, P., ALBENDEA, C., ALCALÁ, B., MENDEZ, A. & DOMÍNGUEZ, L. (2000) Phenotypic and genetic characterization of *Lactococcus garvieae* isolated in Spain from lactococcosis outbreaks and comparison with isolates of other countries and sources. *Journal of Clinical Microbiology*, **38**, 3791.

VERNER-JEFFREYS, D. W., ALGOET, M., POND, M. J., VIRDEE, H. K., BAGWELL, N. J. & ROBERTS, E. G. (2007) Furunculosis in Atlantic salmon (*Salmo salar* L.) is not readily controllable by bacteriophage therapy. *Aquaculture*, **270**, 475–484.

VERSCHUERE, L., ROMBAUT, G., SORGELOOS, P. & VERSTRAETE, W. (2000) Probiotic bacteria as biological control agents in aquaculture. *Microbiology and Molecular Biology Reviews*, **64**, 655–671.

VINOD, M. G., SHIVU, M. M., UMESHA, K. R., RAJEEVA, B. C., KROHNE, G., KARUNASAGAR, I. & KARUNASAGAR, I. (2006) Isolation of *Vibrio harveyi* bacteriophage with a potential for biocontrol of luminous vibriosis in hatchery environments. *Aquaculture*, **255**, 117–124.

WAGNER, B. A., WISE, D. J., KHOO, L. H. & TERHUNE, J. S. (2002) The epidemiology of bacterial diseases in food-size channel catfish. *Journal of Aquatic Animal Health*, **14**, 263–272.

WALAKIRA, J. K., CARRIAS, A. A., HOSSAIN, M. J., JONES, E., TERHUNE, J. S. & LILES, M. R. (2008) Identification and characterization of bacteriophages specific to the catfish pathogen, *Edwardsiella ictaluri*. *Journal of Applied Microbiology*, **105**, 2133–2142.

WALLNER, G., FUCHS, B., SPRING, S., BEISKER, W. & AMANN, R. (1997) Flow sorting of microorganisms for molecular analysis. *Applied and Environmental Microbiology*, **63**, 4223–4231.

WESTERDAHL, A., OLSSON, J. C., KJELLEBERG, S. & CONWAY, P. L. (1991) Isolation and characterization of turbot (*Scophtalmus maximus*)-associated bacteria with inhibitory effects against *Vibrio anguillarum*. *Applied and Environmental Micro-biology*, **57**, 2223–2228.

WILSON, C. L. (1997) Biological control and plant diseases – a new paradigm. *Journal of Industrial Microbiology and Biotechnology*, **19**, 158–159.

WU, J. L. & CHAO, W. J. (1982) Isolation and application of a new bacteriophage, φ ET1, which infect *Edwardsiella tarda*, the pathogen of Edwardsiellosis. *Fish Disease Research*, **1**, 8–17.

ZLOTKIN, A., CHILMONCZYK, S., EYNGOR, M., HURVITZ, A., GHITTINO, C. & ELDAR, A. (2003) Trojan horse effect: phagocyte-mediated *Streptococcus iniae* infection of fish. *Infection and Immunity*, **71**, 2318.

15

Managing the microbiota in aquaculture systems for disease prevention and control

P. De Schryver, T. Defoirdt, N. Boon, W. Verstraete and P. Bossier, Ghent University, Belgium

Abstract: The growing human population necessitates fast expansion of the global aquaculture industry. To achieve this goal, disease management strategies are required, and antibiotics have mainly been relied on in the past. This unsustainable strategy is fortunately being replaced by environmentally integrated approaches. Several of these find their origin in the microbiological processes ubiquitous in nature that result in antagonistic activity towards pathogens. The goal is to use technologies to manage and control these microbiological processes in aquaculture systems to reduce losses resulting from infections with pathogenic microorganisms. In this chapter an overview of existing and novel biotechnologies is provided, and the need for further exploration is illustrated, together with potential methodology.

Key words: host/microbe interactions, microbial community organization, probiotics, prebiotics, sustainable.

15.1   Introduction

In the search for fast increases in production and return on investment, more than one farmer, often without realizing the impact of his or her choice, has made the mistake of increasing stocking densities beyond recommended biosecurity or ecosystem resilience levels with disastrous results (FAO, 2008). Decreased growth performance, reduced immune response and lowered disease resistance have all been observed (Ashley, 2007). The problem of disease in aquaculture systems is extremely relevant in regions with high aquaculture pond or cage densities, where metabolic waste and biological pollutants (pathogens) are very often recirculated among farms. This leads to an increased number of disease outbreaks over time, possibly ending finally in the collapse of the complete aquaculture industry in the area (Kautsky *et al.*, 2000). An example of this is Thailand, where the *Monodon* Bacillovirus in 1989–1990 resulted in the abandoning of 70–80% of the shrimp ponds in several regions and in some places even in the complete cessation of the shrimp farming business (Stevensson, 1997). As the occurrence of disease entails major socio-economic losses, the prevalence and control of pathogens in aquaculture systems is of major importance.

Over the last 10–15 years, studies have focused on how diseases proliferate and how pathogens can be controlled by using their microbiological counterparts. It seems that disease-causing organisms are ubiquitous in natural waters and in many circumstances the host and pathogen co-exist within the same systems with few adverse effects or none at all (Kautsky *et al.*, 2000; Olafsen, 2001). When the health status of the host becomes suboptimal, however, pathogenic activity can have detrimental results. In order for animals to be successfully raised in highly dense aquaculture systems, considerable amelioration of animal husbandry practices, water quality, and microbial management are therefore required (Subasinghe, 2005). Attempts are being made to limit negative pathogenic activity through the manipulation, control and steering of natural microbiological processes pertaining to aquaculture systems.

15.2   Control strategies against diseases

15.2.1   Antibiotics: the non-sustainable strategy

In an attempt to control the pathogens within aquaculture systems, farmers traditionally took refuge in the use of antibiotics. These are compounds of biological or chemical origin that interfere with important biochemical reactions in the pathogenic cell, but also influence microorganisms in general (Umbreit, 1955). Through their use, the growth and/or activity of unwanted microorganisms can be avoided (Serrano, 2005). Examples are chloramphenicol, tetracyclines, quinolones, sulfonamides, etc. (Serrano, 2005). Huge amounts have been and are still being used (Sapkota *et al.*, 2008). Estimations are that the use of antibiotics in fish farming in Norway reached a level of 32.6 tonnes in 1988 and that about 430 g of antibiotics were used for the production of 1 tonne of salmon (Wu, 1995). In Asian countries a variety of antibiotics equivalent to 500–600 tonnes was used on a yearly basis in the aquaculture industry (Moriarty, 1999).

Although antibiotics initially showed great promise in reducing death caused by infectious diseases, the goal of antibiotic use has too often shifted to mitigation of sanitary shortcomings within the aquaculture systems (Cabello, 2006) or improvement of the growth performance of the cultured animals (Serrano, 2005). Few thoughts were given to the detrimental effects that treatments could have in the long run. Prophylactic application, often at suboptimal doses, has resulted in multiple drug resistance in pathogens, rendering treatment ineffective and even in an increased risk of the transfer of resistant plasmids to human pathogens (Das *et al.*, 2009). Fortunately, awareness of this problem has resulted in legal frameworks throughout the world in an attempt to regulate the use of chemotherapeutics in aquaculture (Serrano, 2005).

15.2.2   Microbial management: a sustainable strategy

As the use of antibiotics in fish feed has become a debatable topic among farmers, researchers and planners, alternative treatments were and are in high demand. The focus lies on the search for methodologies that are low cost, relate to natural processes that have been shown to counteract pathogenic activity, and are ubiquitous in nature. The idea is that these strategies can be integrated with standard culturing practices and do not impose a burden on the environment.

In order to recognize the specific targets and to maximize the chances of success of applied strategies, one has to consider the different elements present in aquaculture systems. An aquaculture ecosystem comprises three compartments: (1) the physical environment in which the cultured animals live, (2) the cultured animals (also often called the host), and (3) the microbiota, consisting of both harmful as well as beneficial microorganisms, associated with either the environment, the animals, or both (Fig. 15.1). Over the past years, awareness has grown that the compartments stand in close relationship with the environment and each other. The best disease management strategy would thus be a holistic one, including the physicochemical environment, the animals and the microorganisms. In this way, the most sustainable outcome for the environment as well as for the farmer can be achieved. In Fig. 15.1, an overview of the close interaction between the different aquaculture ecosystem compartments is provided as well as of different types of biological control strategies currently proposed. Several of these strategies focus on the manipulation of the microbiological actors within the system to beneficially influence the host. These actors can be located either within the surrounding environment or within the gut of the host, while the manipulations can target the proliferation of the positive actors or the counteraction of the negative ones. The type of bacteria, the number of bacteria, the activity of the bacteria and the composition of the microbial community as a whole are all factors that can be adjusted.

It should be mentioned that the antagonistic modes of action of these microbiological control strategies are not a direct replacement of antibiotics in the event of an infection. Their prophylactic application is, however, said to lead to a considerable reduction in infection risk. Although their effectiveness has become generally accepted in the aquaculture world, the exact mode of action of several microbiological approaches has not yet been unequivocally established.

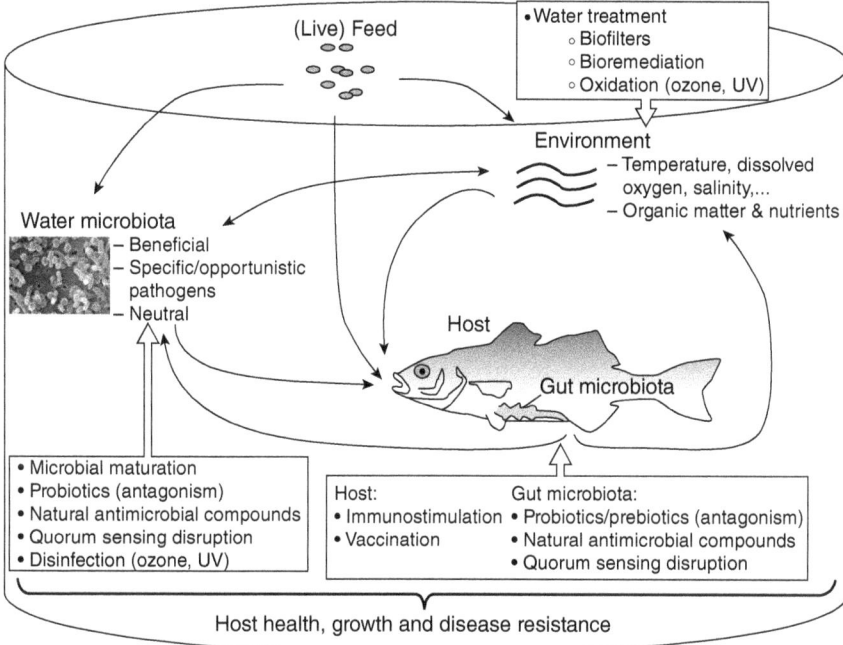

Fig. 15.1   Schematic representation of the interaction between the environment, the aquatic animals, and the microorganisms in aquaculture systems and overview of possible biocontrol strategies to prevent and control pathogens in aquaculture without using antibiotics (redrafted after Vadstein *et al.*, 2004; Defoirdt *et al.*, 2007a; Panigrahi and Azad, 2007; Gatesoupe, 2010).

15.3   The microbial maturation concept

The optimal scenario would be to prevent pathogens from invading and persisting in the aquatic environment. Diseases are thus prevented by taking away the causing agent. Techniques for continuous disinfection of the water flowing within the system have been developed in this respect. Ozonation and ultraviolet (UV) irradiation can be applied as they lead to the destruction of the cell membrane of microbiota and thus also of pathogens. A problem associated with these techniques, however, is that they do not specifically target the pathogens, but all bacteria present. Application thus disturbs the natural microbial balance in the water and renders the treated water more or less bacteria-free. The result is a low number of different bacterial species and a large reservoir of organic material, especially in periods of live feed provision. These are conditions that select for opportunistic microorganisms with a relatively high growth rate, the so-called *r strategists*, that normally do not have the opportunity to proliferate in the water (Vadstein *et al.*, 1993) and of which several can be classified as facultative pathogens (Salvesen *et al.*, 1999). Although the opportunists do not necessarily have to be pathogenic species, their interactions with fish and shrimp larvae (especially under stress conditions or due to the poorly developed immune system) can have detrimental results (Skjermo *et al.*, 1997). With these considerations in mind, the microbial maturation concept or the *r/K* concept was proposed by Vadstein *et al.* (1993).

The idea of the microbial maturation concept is the prevention of the negative effects of the opportunistic microorganisms by competition rather than by their elimination from the system. This can be achieved by making a selection for non-opportunists (*K strategists*) in the water. *K* strategists are microorganisms with a relatively slow growth rate that are able to grow in crowded populations and need only a low amount of substrate per bacterium. By proliferation of the non-opportunists, a highly diverse but also stable microbial community, relatively insensitive to perturbations in the organic matter concentration in the water, is obtained. If fish or shrimp larvae are reared in such a water, the colonization of the mucosal layers of the larvae by opportunistic (and potentially harmful) microorganisms can be largely avoided. As a result, the survival of the larvae is expected to be higher. To obtain matured water, the water can be filtered for the removal of organic material and bacteria and subsequently passed over a maturation biofilter containing a matured microbial community, thus mainly composed of *K* strategists, on the filter material. The water going to the culture unit and coming into contact with the animals is recolonized by a mature microbial community in this way. This approach was for the first time proven successful in experiments by increasing the survival of Atlantic halibut yolk sac larvae (Skjermo *et al.*, 1997), and the growth rate of turbot larvae (Salvesen *et al.*, 1999).

15.4   Manipulating type and number of bacteria: probiotics and prebiotics

15.4.1   Probiotics

Probiotics are the intervention most widely recognized for the manipulation of microorganisms in aquaculture systems. Probiotics are live beneficial microbial actors originally used as dietary supplements to increase the intestinal balance for improvement of the health status of aquatic animals (Verschuere *et al.*, 2000). However, the gastrointestinal microbiota stand in close relationship with the three-dimensional water matrix in which their aquatic hosts live. Due to the excretion of faeces in the water, manipulations of the dietary microbial community will also affect the water ecology, and vice versa because of the water flow passing through the digestive tract (Kesarcodi-Watson *et al.*, 2008). Therefore, the term probiotic can be broadened for aquaculture purposes to 'living microbial water or feed additives administered with the aim of improving health' (Gatesoupe, 1999).

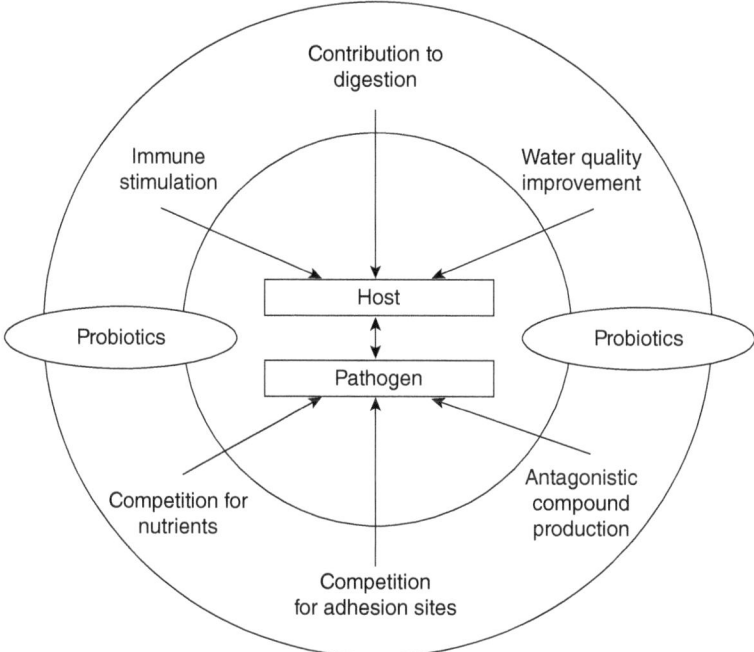

Fig. 15.2 Overview of the modes of action of probiotics.

Probiotics can induce beneficial effects in multiple ways (Fig. 15.2). Figure 15.2 gives an overview of the beneficial effects associated with the use of probiotics, which are described below.

Probiotic supplementation as water treatment: bioremediation
A first beneficial activity of probiotics can be ascribed to the improvement of the water quality. As the water quality is a key factor in the level of stress experienced by the cultured animals, it determines to a large extent the risk of infections within an aquaculture system. Several methodologies and (bio) technologies are available to remove the excess of organic material, nitrogen and phosphorus from the aquaculture systems as described in the reviews of Crab *et al.* (2007) and Timmons and Ebeling (2007). As an alternative, the supplementation to the water of bacteria able to convert or remove unwanted and toxic pollutants – bioaugmentation – or the supply of compounds to stimulate autochthonous bacteria with such capacities – biostimulation – has also been applied (Qi *et al.*, 2009). These two concepts are grouped under the name bioremediation. Very often, the microbial preparations that are added to the water contain nitrifying bacteria or *Bacillus* species (Gatesoupe, 1999). These two kinds of bacteria are given as an example here because they do not only differ in activity but their application also differs in concept according to Gatesoupe (1999). The

nitrifying bacteria have strict ecological niches, and they are not detected in the gastrointestinal tract of aquatic animals. Their activity thus merely focuses on the removal of nitrogen from the water. *Bacillus* species can on one hand be added for their abilities of enzymatic excretion and nutrient removal from the water and thus contribute to a better environment. On the other hand, several *Bacillus* species have been selected for their competitiveness with pathogens in the water. The fact that *Bacillus* species have been isolated from the intestine of several fish species makes it likely that they can also be active in the intestinal tract introducing some beneficial activity there. However, as the primary aim of bioremediation supplements is to improve water quality, this is most often not investigated in practice. For an overview of examples concerning the application of probiotics as a bioremediation strategy, the reader is referred to Gatesoupe (1999).

Probiotics as a dietary supplementation
Large variations in the intestinal microbiota have been observed in sibling larvae reared in the same tank (Fjellheim *et al.*, 2007). The types of microorganisms that persist inside the gastrointestinal tract depend on some deterministic factors (salinity, temperature, pH, etc.) but mainly on the stochastic inflow of bacteria associated with the water and the feed (Vine *et al.*, 2006). Upon hatching, larvae possess a sterile – and immature – digestive system. Gut colonization starts at mouth opening caused by the inflow of water loaded with bacteria. This colonization process is in general increased at the start of first feeding and results in a microbiota that normally shifts towards the microorganisms asociated with the feed. These early colonization mechanisms are thought to determine to a large degree the final microbial community composition in the gut (Fjellheim *et al.*, 2007). At later life stages, the community is still subject to changes, although to a lesser degree than at the larval stage.

Owing to the possibility of manipulating the intestinal microbiota at these earliest life stages, the larval development stage is the best at which to apply probiotic treatment. Microorganisms can be supplemented to the feed for the larvae, with the aim of including chosen bacterial species in the gut community.

The causes of the beneficial effects resulting from the dietary supplementation of probiotics are diverse. In the passage that follows, each of the biocontrol activities that have been associated with probiotics are briefly mentioned. For examples from both research and practice and more detailed information, some excellent reviews have been written (Gatesoupe, 1999, 2010; Kesarcodi-Watson *et al.*, 2008; Qi *et al.*, 2009; Verschuere, 2000; Vine *et al.*, 2006).

1. *Digestive enzyme excretion*: Changes to the microbiota in the gut may result in an alteration in intestinal food digestion patterns. Probiotics may provide their host with digestive enzymes and assist in assimilating

nutrients (Vine *et al.*, 2006). Some fermentation products may reinforce the intestinal barrier in fish and thus stimulate the defense of the host (Gatesoupe, 1999). More specifically, an increase in short chain fatty acid (SCFA) production is often targeted. SCFA, and especially butyrate, are the preferred energy sources of the colonocytes (Van Immerseel *et al.*, 2003). Higher SCFA production has been related to improved intestinal health status in the host. The animal is supplied with a higher amount of energy and thus is stronger and can better resist diseases. Moreover, the increased intestinal cell proliferation allows for increased uptake of nutrients originating from the feed so that an increase in the feed uptake ratio and a better growth can be expected (Vine *et al.*, 2006).

2. *Immune stimulation*: An immunostimulant is a naturally occurring compound that modulates the immune system and as a result increases the host's resistance against diseases that in most circumstances are caused by pathogens (Bricknell and Dalmo, 2005). Fish larvae have a poorly developed immune system, relying mainly on non-specific or innate defence mechanisms. The supplementation of probiotics to larvae has been shown to result in an immunomodulating effect and a higher response to a variety of antigens (Vadstein, 1997). The probiotics interact with the immune cells to introduce the immune respone (Nayak, 2010). In some cases the inactivated form of the probiotic can augment the systemic and mucosal immune respones similar to viable probiotics. This can be ascribed to the interaction of the preserved microbial components such as polysaccharides, peptidoglycans or lipoteichoic acids. At the systemic level, an increase in phagocytic activity, respiratory burst activity, lysozyme production, and peroxidase production and increased complement activity are examples of immunomodulation. At the gut level, an increase in the number of Ig^+ cells, acidophilic granulocytes, T cells and an increased production of lysozyme and increased phagocytic activity have been observed (Nayak, 2010). A downside to the use of immunostimulants can be that long-term oral administration may result in a decreased efficacy and that immunosuppression may occur at overdoses. In addition, some researchers are of the opinion that administering immunostimulants at larval stages can have detrimental impact on the development of the immune system (Sakai, 1999; Bricknell and Dalmo, 2005).

3. *Production of antagonistic compounds*: Several probiotics have been selected for their ability to produce compounds that are inhibitory towards the proliferation of (opportunistic) pathogens (Verschuere *et al.*, 2000). In that way, they can be considered to be natural antibiotics. The bactericidal or bacteriostatic activity can mostly be ascribed to the production of bacteriocins, lysozymes, hydrogen peroxide, carbon dioxide or siderophores and/or the alteration in pH by the production of organic acids (Verschuere *et al.*, 2000; Vine *et al.*, 2006). The production of antagonistic compounds is a common trait for both marine and

freshwater bacteria isolated from the intestinal tract of fish and has been shown to be inhibitory towards several fish pathogens (Gatesoupe, 1999).

4. *Competitive exclusion*: The competition between probiotics and pathogens for both adhesion sites and energy sources will determine how both will coexist in the same intestinal environment. The attachment of a bacterium to the gut tissue surface is a first prerequisite for establishment in the intestine. By attaching to the intestinal mucus layer, probiotics can prolong their residence time in the gut and proliferate without the need for high growth rates. The ability of pathogenic bacteria to attach to the mucus has been related to virulence and is considered the propagating step in a bacterial infection in the gut (Vine *et al.*, 2006). Both probiotic and pathogen thus profit from cell wall adhesion and if pre-emptive colonization by probiotics with superior attachment capacity as compared to pathogens can be induced, this can be considered a first line of defence against invasions. Alternatively, the probiotic attachment may be crucial for efficient antagonistic compound production as these metabolites are often only produced during the stationary growth phase of the probiotic (Vine *et al.*, 2004). A second type of competitive exclusion is related to the availability of chemicals and/or energy sources. However, there is a lack of understanding of the different nutritional factors in the piscine gut that determine relative bacterial growth rates (Verschuere *et al.*, 2000). Therefore, the application of this competitive exclusion principle remains a complex matter. Efforts should focus on developing this approach, as a considerable advantage of competitive exclusion compared with antibacterial activity is that the pathogen is not necessarily killed and therefore is less pressurized to develop resistance.

Just as in the case of antibiotics, it is possible that pathogenic microorganisms become resistant to the action of a probiotic. Therefore, the best probiotic is one that has different mechanisms to outcompete the pathogen and/or strengthen the host.

15.4.2 Prebiotics

An alternative approach focuses on the beneficial activity of microorganisms already present in the aquaculture system; more specifically, the stimulation of health-promoting microbial species already resident in the gastrointestinal tract of the aquatic animals (Ringø *et al.*, 2010). The desired results are a decrease in the presence or activity of intestinal pathogens and/ or the promotion of bacterial metabolite production to increase growth or improve gastrointestinal health (Gibson and Roberfroid, 1995). This can be accomplished by the application of prebiotics, which are non-digestible food ingredients that selectively stimulate health-promoting bacteria such as

Lactobacillus spp., *Carnobacterium* spp. and several other *Bacillus* spp. in the fish gut (Rurangwa *et al.*, 2009). In some cases, it appears more practical to manipulate the gastrointestinal tract microbial community by the application of prebiotics than probiotics. While probiotics have to adjust to environmental conditions in the gut, the conditions in the gut following prebiotic treatment are altered in such a way that natural selection favours certain bacterial species with the goal of enhancing fish growth and decreasing susceptibility of the host. The most common prebiotics that have been applied for aquatic species are inulin, fructooligosaccharides, short-chain fructooligosaccharides, mannanoligosaccharides, galactooligosaccharides, xylooligosaccharides, arabinoxylooligosaccarides and isomaltooligosaccharides. For a detailed overview of the effects of each of these prebiotics, the reader is referred to the review of Ringø *et al.* (2010).

15.5   Managing bacterial activity by means of quorum sensing (QS)

15.5.1   Quorum sensing (QS)

For many aquaculture pathogens, the expression of the virulence depends on the ability to communicate with other members of the population by means of quorum sensing (QS) (Bruhn *et al.*, 2005). This is a gene regulatory mechanism in which bacteria monitor one another's presence by the production, release and detection of small signalling molecules (Camilli and Bassler, 2006). A higher detection of signalling molecules indicates a higher presence of pathogens, which is the sign to become virulent. As such, the virulence of a pathogenic population can be regulated as a function of pathogen density making chances for successful infection considerably higher. Three types of QS systems exist depending on the type of signalling molecules: acylated homoserine lactones (AHLs), peptides, or autoinducer 1 (AI-1) and autoinducer 2 (AI-2) (Fig. 15.3).

Figure 15.3 (a) depicts AHL-mediated quorum sensing. The I protein is the AHL synthase enzyme. The AHL molecules diffuse freely through the plasma membrane. As population density increases, the AHL concentration increases as well and once a critical concentration has been reached, AHL binds to the R protein, a response regulator. The AHL-R protein complex activates or inactivates transcription of the target genes. Figure 15.3 (b) depicts peptide-mediated quorum sensing in Gram-positive bacteria. A peptide signal (PS) precursor protein is cleaved, releasing the actual signal molecule. The peptide signal is transported out of the cell by an ATP binding cassette (ABC) transporter. Once a critical extracellular peptide signal concentration is reached, a sensor kinase (SK) protein is activated to phosphorylate the response regulator (RR). The phosphorylated response regulator activates transcription of the target genes. Figure 15.3 (c) shows QS in *Vibrio harveyi*. In this system, there are two types of signal molecules.

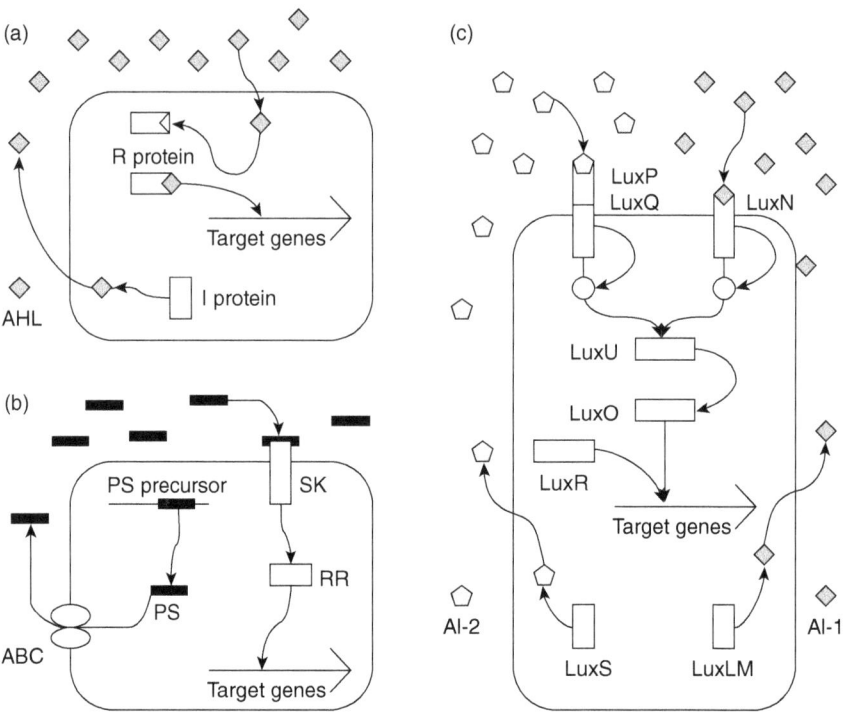

Fig. 15.3   The three major quorum sensing systems (from Defoirdt *et al.*, 2004).

AI-1 is an AHL and its biosynthesis is catalysed by the luxLM enzyme. AI-2 is a furanosyl borate diester; its biosynthesis is catalysed by the LuxS enzyme. AI-1 and AI-2 are detected at the cell surface by the LuxN and LuxP–LuxQ receptor proteins, respectively. At low cell density, LuxN and LuxQ autophosphorylate and transfer phosphate to LuxO via LuxU. The phosphorylated LuxO is an active repressor for the target genes. At high cell density, LuxN and LuxQ interact with their autoinducers and change from kinases to phosphatases that drain phosphate away from LuxO via LuxU. The dephosphorylated LuxO is inactive. Subsequently, transcription of the target genes is activated by LuxR.

Recently, the disruption of bacterial quorum sensing has been proposed as a new anti-infective strategy and several techniques that could be used to disrupt quorum sensing have been investigated. These techniques comprise (1) the inhibition of signal molecule biosynthesis, (2) the application of QS antagonists (including naturally occurring as well as synthetic halogenated furanones, antagonistic QS molecules and undefined exudates of higher plants and algae), (3) the chemical inactivation of QS signals by oxidized halogen antimicrobials, (4) signal molecule biodegradation by

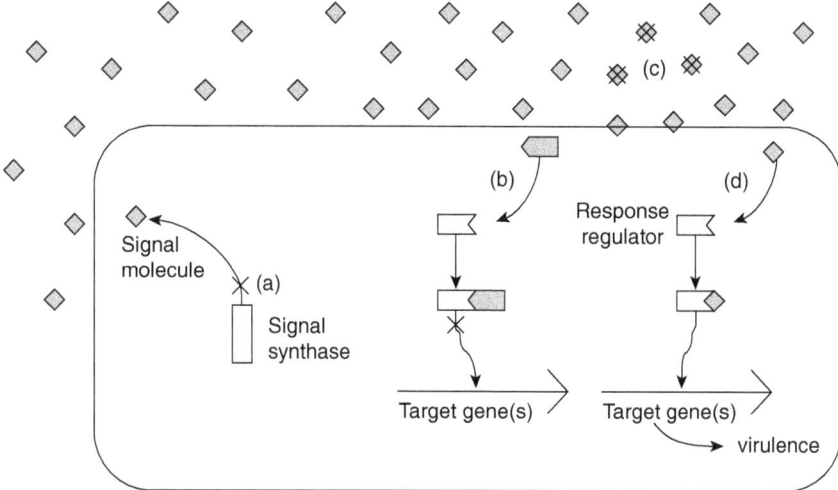

Fig. 15.4 Schematic overview of different strategies that have been developed to disrupt bacterial quorum sensing. (a) Inhibition of signal molecule biosynthesis by the application of substrate analogues. (b) Blocking signal transduction by the application of quorum sensing antagonists. (c) Chemical inactivation and biodegradation of signal molecules. (d) Application of quorum sensing agonists to evoke virulence factor expression at low population density (from Defoirdt *et al.*, 2004).

bacterial lactonases and by bacterial and eukaryotic acylases and (5) the application of QS agonists to trigger host responses (Fig. 15.4).

Although the technique seems very promising, a lot remains unknown about QS inhibition and it will take a while before practical application comes within reach (Tinh *et al.*, 2008). Up to now, the studies remain limited to laboratory experiments under controlled conditions like the ones performed and described by Rasch *et al.* (2004), Defoirdt *et al.* (2004), and Tinh *et al.* (2007).

15.5.2   Managing the microbial community organization

Aquaculture systems contain high numbers of opportunistic pathogens and the mutual relationships with the other microorganisms often determine whether these can become dangerous or not (Vadstein, 1997; Panigrahi and Azad, 2007). To be able to come to grips with randomized microbial interference in aquaculture, it is thus of prime importance to relate host health, and especially specific events regarding mortality and growth, to the state and changes of the microbial community in the intestinal tract and the environment. Such an approach is illustrated by a study of Reid *et al.* (2009), who followed the *Vibrio* community by a culture method in function of time in six tanks containing cod larvae. The larvae were colonized by several *Vibrio* spp. and every tank showed about 75–80% mortality after 80 days. In one tank, a sudden drop in survival occurred after 72 days, which coincided with an elevated level of *Vibrio anguillarum*. Unfortunately, analysis was performed on the complete microbial community structure in neither the larvae nor the water. This could have yielded information on the shifts in the microbial populations preceding this event. Sun *et al.* (2009) investigated the microbial communities in slow growing (SG) and fast growing (FG) grouper larvae by a culture-based method. They observed that the community in SG grouper contained four opportunistic *Vibrio* spp. representing about 12% of the total microbial abundance. The FG grouper contained only two of the *Vibrio* spp. at 3% of the total microbial abundance, in addition to *Bacillus pumilus*, *Bacillus clausii* and *Psychrobacter* sp. that dominated. These latter three showed an *in vitro* antagonistic activity against the *Vibrio* isolates, which may explain relative abundances and the higher growth performance. Bjornsdottir *et al.* (2009) adapted the same methodology of trying to relate larval survival to specific microbial community members. Also in these two studies, no analysis of the complete microbial community by means of molecular techniques was performed.

An important feature of studies on host/microbe interactions should be that they include a qualitative as well as a quantitative assessment of the total microbial community. In addition to looking at what is present, an interesting approach may be to investigate how management of the microbial community organization may contribute to the health status of the host. In relation to this, some hypotheses can be brought forward regarding the exploitation of the microbial community composition in the gut to fight infections:

- The organization of the intestinal microbial community relates to the relative abundances of the different species (Wittebolle *et al.*, 2009). It can be hypothesized that more equal abundances (= increased evenness) will contribute to the well-being of the fish as it is more difficult for a pathogen to invade a community with a fair number of equal players relative to a situation with a few dominant species (Ley *et al.*, 2006). Wittebolle *et al.* (2009) stated that species evenness can be an important element in managing invasions.
- The richness of a microbial community refers to the genetic diversity or the number of different species in an environment (Bell *et al.*, 2005). Bell *et al.* (2005) stated that microbial communities that are rich contribute to the better functioning of an ecosystem because overall more resources are used. Losure *et al.* (2007) described this as the fact that increased richness leads to increased niche complementarity. When a greater variety of different species are present, this implies a more efficient use of resources such as nutrients. The opportunities for an invader to find

a niche in a highly diverse ecosystem are thus considerably decreased. Alternatively, Bell *et al.* (2005) related ecosystem functioning to high microbial richness because in an ecosystem rich in microorganisms it is more likely that species are present which have a large effect on the function of the ecosystem. Translated to pathogenic invasions, in rich communities it is more likely that at least one of the community members has an antagonistic activity against the pathogen. The manipulation of the organization of the microbial community can be performed in several ways, the most evident being the application of probiotics and prebiotics. As far as the authors are aware, however, there are hardly any studies available that provide information on the effects of pro- and prebiotics on alteration of the microbial community, let alone relate the organization to the health of the host. This research methodology thus currently is a black box; efforts need to be dedicated to unravelling the complex host/microbe interactions. Some examples of how information can be obtained are given in Section 15.6.4.

15.6   Host–microbe interactions affecting host health in aquaculture: the need for knowledge

The microbiota associated with the host gut is generally accepted as being related to the occurrence of diseases in aquaculture (Vadstein *et al.*, 2004). The stochastic inflow of microorganisms at the larval life stages can result in major variations in the microbial community that colonizes the intestinal tract of the host (Fjellheim *et al.*, 2007). The randomness of the microbial ecology and the fact that numerous members are opportunistic pathogens can result in unpredictable and high mortality (Olafsen, 2001). One approach to counteract the mortality has involved identification of the microbial members that are likely to be most problematic (Bjornsdottir *et al.*, 2009; Reid *et al.*, 2009) and attempts to counteract them through the antagonistic activity of, for example, probiotics (Gatesoupe, 1999). The primary focus in such studies on microbial manipulation strategies in aquaculture has mainly been the observation of positive effects at the level of the host. Indeed, altered disease resistance, and also growth performance, associated with such a treatment is of prime concern to the farmer. It is, however, important to establish the mode of action of these methodologies so that these can be developed and applied at maximum efficiency. During recent years, several approaches have been developed that can contribute considerably to this field.

15.6.1   Gnotobiotic model systems as a research tool

A large number of studies have been performed on different kinds of dietary treatments with probiotic bacteria, but only in a limited number of cases have the persistence of the probiotic in the intestinal tract and its mode of action been investigated. In order to clarify the different contributing factors that make up the mode of action of probiotics and prebiotics, arbitrary influences during investigations need to be excluded as much as possible. In aquaculture environments, this is a rather complicated matter due to the concentration of bacteria in the culture water of larvae and the microorganisms associated with the feed. If mechanistic investigations need to be performed in such an environment, it is impossible to establish clear trigger–response relationships. The development of test systems in which the researcher has complete control over the intestinal microbial community structure was revolutionary in this respect. Such gnotobiotic model environments have been developed for mice (Sudo *et al.*, 2004), pigs (Splichal *et al.*, 2007) and birds (Waligora-Dupriet *et al.*, 2009). Relevant for aquaculture, gnotobiotic model systems have been developed for zebra fish, cod and sea bass with the purpose of studying host-microbe interactions (Rawls *et al.*, 2004; Dierckens *et al.*, 2009; Forberg *et al.*, 2011). The larvae of the fish are completely controlled at the bacterial level, as is the culture water and the food chain. Every factor inside the system is known to the researcher and thus this allows optimal interpretation of the results. By choosing combinations of probiotics and pathogens that are introduced to the sterile environment containing only the host, the interaction between the probiotic, pathogen and host can be perfectly monitored. In addition, the development of the gut can be measured in treated larvae by histological and stereological analysis (Rekecki *et al.*, 2009). The gut volume to body volume ratio, the intestinal epithelium height, the specific surface area of the villi, the villi number etc, are all factors that can be investigated.

15.6.2   Gene expression as a research tool

At the level of the host, the easiest factors to score are survival and growth. However, transcriptional patterns can also yield interesting information on the immunomodulatory effect of probiotics and prebiotics towards genes encoding for an innate or acquired immune response. Rawls *et al.* (2004) used RNA microarrays to study the influence of microbial presence and colonization on gene expression in zebra fish. Arrays allow monitoring of the expression of thousands of selected genes simultaneously. Darias *et al.* (2008) also made use of a similar approach to study gene expression in sea bass. Since no microarray had been developed for sea bass, a heterologous strategy was applied using the microarray developed for trout. This so-called cross-species hybridization also yields reliable estimates of gene expression (Renn *et al.*, 2004). Another option for transcriptomic analysis is the cDNA-AFLP (amplified fragment length polymorphic) technique (Breyne *et al.*, 2003; Wenne *et al.*, 2007). This offers the advantage that it gives an overview of the global gene expression, and the specificity and sensitivity allows the detection of poorly expressed genes. However, the data sets are not easily interpreted as no readily available information is obtained on the sequences of the genes expressed.

If a specific gene is targeted, the use of quantitative polymerase chain reaction (PCR) is most appropriate. Such an approach was used in the research on the expression of sea bass genes involved in interferon production (Casani *et al.*, 2009). Currently, a large diversity of genes have been identified that are involved in immune functioning in fish and can be selected for quantitative PCR analysis (Randelli *et al.*, 2009; Sarropoulou *et al.*, 2009). These encode for antimicrobial peptides, cytokines, cyclooxygenase-2, lectins, immunoglobulin, CD8 receptors, etc. The method with the most potential to extensively study gene expression patterns is the 454 pyrosequencing technology (Johanson *et al.*, 2009). It is a technique to accurately and quantitatively determine the complete scope of RNA sequences in a sample. It combines the quantitative aspects of microarrays but more sensitive with the completeness of cDNA-AFLP and with immediate sequencing of all genes. Although this technique is rather costly at the moment, is it the one that is proposed to be most important in future research.

15.6.3   Cell lines as a research tool

Parameswaran and coworkers (2006) developed cell lines from embryonic, kidney and spleen cells derived from fish. It thus seems possible to develop a cell line for the epithelial cells of the fish gut to study the invasive capacity of the pathogens after for example a probiotic treatment. Ideally, the results obtained would be combined with information on the gene activity in the pathogen. For example, hemolysin production by pathogenic *Vibrio*s may be interesting to study as it is an important factor in the virulence (San Luis and Hedreyda, 2006). The genes encoding for this exotoxin have been reported for *V. harveyi* and *V. anguillarum* (San Luis and Hedreyda, 2006), the latter being the pathogen in the gnotobiotic sea bass system (Dierckens *et al.*, 2009). Of course, a thorough screening of the literature should be performed to include as many genes as possible involved in the virulence expression. Fish cell lines can also be used to assess the effects of probiotics and prebiotics and their degradation products on the proliferation of gut epithelial cells (Blishchenko *et al.*, 2002). This may show whether or not the cell growth is indeed beneficially influenced.

15.6.4   Novel molecular approaches as a research tool

The way in which treatments modulate the intestinal microbial community is an important feature but is seldom investigated. Even when this is the case, the observations have often remained limited to increases or decreases in the viable count numbers or verification of the presence/absence of certain probiotic species after treatment. This empirical approach reflects the lack of knowledge on microbial relationships relating to the health of the host and how to steer them. As a result, the direct relationship between a probiotic or prebiotic treatment and a certain beneficial effect at the level of the host as observed in many studies has hardly ever been shown. In other words, the observed beneficial probiotic effects are almost always explained by empirical modes of action. This can be partly explained by the absence of effective tools to study the microbial community. Culture-dependent techniques for enumeration are time-consuming and less than 1% of the total microbial community can be cultivated. Moreover, it is not feasible to assess species richness merely based on colony morphology (Amann et al., 1995; Hovda et al., 2007; Navarrete et al., 2010). Moriarty stated already in 1997 that 'there is a need for efficient techniques to follow structure and abundance of bacterial populations in the water or host to gain insight and understanding in the microbial management of aquaculture systems'.

The availability of molecular ecology techniques based on sequence comparisons of nucleic acids (DNA and RNA) allows working unbiased by the limitations of culturability (Huber et al., 2004). They are valuable tools in the identification and quantification of community members and can be used for the monitoring of taxonomy, functioning and activity (Simpson et al., 2002). Examples of molecular tools applied in aquaculture research are clone libraries (Holben et al., 2002), terminal restriction fragment length polymorphism (T-RFLP) (Verner-Jeffreys et al., 2003), amplified ribosomal DNA restriction analysis (ARDRA) (Michel et al., 2007), and denaturing gradient gel electrophoresis (DGGE) (Li et al., 2007). The number of studies using these techniques has increased considerably during the past decade, though to date they have mainly lead to descriptive conclusions. By 16S rRNA gene analysis, assessments were made of the major populations in the gastrointestinal tract of for example salmon (Romero and Navarrete, 2006; Hovda et al., 2007), rainbow trout (Huber et al., 2004; Navarrete et al., 2010), cod (Brunvold et al., 2007; McIntosh et al., 2008), scallop (Sandaa et al., 2003) and halibut (Jensen et al., 2004). These studies focused solely on the identification of the species in the community. Although several reoccurring members, such as Vibrio spp. and Aeromonas spp., are known to be opportunistic pathogens, no information can be obtained from these studies on the beneficial or detrimental effects that these may bring to the cultured animals.

Other studies used the molecular techniques to look into the alterations that certain biotreatments enforce in the intestinal microbial community. Zhou et al. (2009) assessed the effect of the growth promotor potassium diformate on the growth and the intestinal microbial community of hybrid tilapia. Next to an increased growth performance of the fish, an increased intensity for some DGGE bands and a decreased intensity for others was described. In pacific white shrimp, Li *et al.* (2007) observed that short-chain fructooligosaccharides supplementation to the diet induced considerable changes in the intestinal microbial fingerprinting patterns. Like these, several other studies have been performed describing changes in the molecular intestinal microbiota patterns but lack an explanation for these changes.

A suggested approach is to describe the fingerprints obtained from intestinal samples in a quantitative way. As such, different patterns can be compared with each other in a defined reference system. The composition and structure of the microbial community can, for example, be expressed in terms of its range-weighted richness (Rr), dynamics of change (Dy), or community organization (Co) as proposed by Marzorati *et al.* (2008) and applied by De Schryver *et al.* (2009, 2010). In the latter studies, poly-β-hydroxybutyrate (PHB) was applied as a prebiotic compound for sea bass to illustrate how more information can be obtained from molecular analysis techniques. The bacterial storage polymer PHB is a compound that serves as an intracellular energy and carbon reserve for bacteria (Madison and Huisman, 1999; Tokiwa and Calabia, 2004). It is insoluble in water and has been shown to be biologically degradable into β-hydroxybutyric acid (Defoirdt *et al.*, 2007b). The latter can exhibit growth inhibition towards certain pathogens and protect *Artemia* like other SCFA (Defoirdt *et al.*, 2007b). As such, if PHB is supplemented through the feed and subsequently degraded in the gastrointestinal tract of aquaculture organisms, the locally released SCFA or PHB oligomers may induce their beneficial effects. In several experiments, this approach increased the survival of *Artemia fransiscana* up to 73% upon infection with the pathogen *Vibrio campbellii* (Defoirdt *et al.*, 2007b; Halet *et al.*, 2007).

In the investigation of De Schryver *et al.* (2009), juvenile sea bass were fed for six weeks with diets containing different levels of PHB. As part of the measured parameters, the growth performance was determined as well as the range-weighted richness (Rr) that was calculated from DGGE band patterns (Fig. 15.5). The Rr can be used to study a microbial community based on the base pair composition of the DNA sequence – or its content of (guanine + cytosine) more specifically – and the percentage of denaturing gradient in the DGGE gel needed to describe the total diversity of the sample analysed (Marzorati *et al.*, 2008). The higher the Rr is, the higher the probability the environment can host a larger number of different species with higher genetic variability. A Pearson product–moment correlation coefficient of 0.977 between the Rr and the growth performance was found, indicating that the condition at the level of the intestinal microbiota was closely related to the condition at the level of the host. It is highly likely that several such relationships exist between microbiota and host, some of which also relate to disease resistance. It is the task of science to determine which conditions/compositions of the microbiota result in higher resistance in the host.

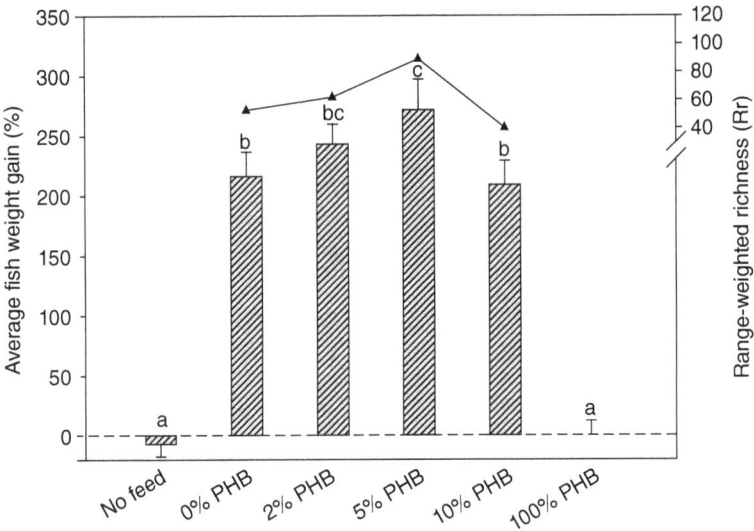

Fig. 15.5 Average fish weight gain of juvenile European sea bass (bars) and range-weighted richness of the gut microbial community (line) after a 6 weeks feeding trial with diets containing different levels of PHB (from De Schryver *et al.*, 2009).

15.7   Conclusions

The search for and development of microbial strategies to control the high mortality at the larval stages resulting from infections in aquaculture are proceeding at full speed. Until recently, science was looking into a black box in which the interactions between the cultured animal, the environment and the microorganisms were completely unclear. By global-scale efforts, this complex web is being unravelled little by little and is becoming better understood. But although progress is being made, we are still not able to control the (microbial) processes pertaining to aquaculture systems. More knowledge is required, and the future lies with the application of novel research technologies that allow the identification of the metabolic, microbial and genetic processes behind the macroscopic observations. If one is able to understand the complex host–microbe interactions, the step towards a high level of control over these interactions may not be that big.

15.8   Acknowledgements

Peter De Schryver and Tom Defoirdt are supported as postdoctoral fellows by the FWO-Vlaanderen (Fonds voor Wetenschappelijk Onderzoek – Vlaanderen / Fund for Scientific Research – Flanders).

15.9   References

AMANN, R.I., LUDWIG, W., and SCHLEIFER, K.H. (1995) Phylogenetic identification and *in-situ* detection of individual microbial-cells without cultivation. *Microbiological Reviews* **59**: 143–169.

ASHLEY, P.J. (2007) Fish welfare: current issues in aquaculture. *Applied Animal Behaviour Science* **104**: 199–235.

BELL, T., NEWMAN, J.A., SILVERMAN, B.W., TURNER, S.L., and LILLEY, A.K. (2005) The contribution of species richness and composition to bacterial services. *Nature* **436**: 1157–1160.

BJORNSDOTTIR, R., JOHANNSDOTTIR, J., COE, J., SMARADOTTIR, H., AGUSTSSON, T., SIGURGIS-LADOTTIR, S., and GUDMUNDSDOTTIR, B.K. (2009) Survival and quality of halibut larvae (*Hippoglossus hippoglossus* L.) in intensive farming: possible impact of the intestinal bacterial community. *Aquaculture* **286**: 53–63.

BLISHCHENKO, E., SAZONOVA, O., SUROVOY, A., KHAIDUKOV, S., SHEIKINE, Y., SOKOLOV, D. *et al.* (2002) Antiproliferative action of valorphin in cell cultures. *Journal of Peptide Science* **8**: 438–452.

BREYNE, P., DREESEN, R., CANNOOT, B., ROMBAUT, D., VANDEPOELE, K., ROMBAUTS, S. *et al.* (2003) Quantitative cDNA-AFLP analysis for genome-wide expression studies. *Molecular Genetics and Genomics* **269**: 173–179.

BRICKNELL, I., and DALMO, R.A. (2005) The use of immunostimulants in fish larval aquaculture. *Fish & Shellfish Immunology* **19**: 457–472.

BRUHN, J.B., DALSGAARD, I., NIELSEN, K.F., BUCHHOLTZ, C., LARSEN, J.L., and GRAM, L. (2005) Quorum sensing signal molecules (acylated homoserine lactones) in Gram-negative fish pathogenic bacteria. *Diseases of Aquatic Organisms* **65**: 43–52.

BRUNVOLD, L., SANDAA, R.A., MIKKELSEN, H., WELDE, E., BLEIE, H., and BERGH, Ø. (2007) Characterisation of bacterial communities associated with early stages of intensively reared cod (*Gadus morhua*) using denaturing gradient gel electrophoresis (DGGE). *Aquaculture* **272**: 319–327.

CABELLO, F.C. (2006) Heavy use of prophylactic antibiotics in aquaculture: a growing problem for human and animal health and for the environment. *Environmental Microbiology* **8**: 1137–1144.

CAMILLI, A., and BASSLER, B.L. (2006) Bacterial small-molecule signaling pathways. *Science* **311**: 1113–1116.

CASANI, D., RANDELLI, E., COSTANTINI, S., FACCHIANO, A.M., ZOU, J., MARTIN, S. *et al.* (2009) Molecular characterisation and structural analysis of an interferon homologue in sea bass (*Dicentrarchus labrax* L.). *Molecular Immunology* **46**: 943–952.

CRAB, R., AVNIMELECH, Y., DEFOIRDT, T., BOSSIER, P., and VERSTRAETE, W. (2007) Nitrogen removal in aquaculture towards sustainable production. *Aquaculture* **270**: 1–14.

DARIAS, M., ZAMBONINO-INFANTE, J., HUGOT, K., CAHU, C., and MAZURAIS, D. (2008) Gene expression patterns during the larval development of European sea bass (*Dicentrarchus labrax*) by microarray analysis. *Marine Biotechnology* **10**: 416–428.

DAS, A., SAHA, D., and PAL, J. (2009) Antimicrobial resistance and *in vitro* gene transfer in bacteria isolated from the ulcers of EUS-affected fish in India. *Letters in Applied Microbiology* **49**: 497–502.

DEFOIRDT, T., BOON, N., BOSSIER, P., and VERSTRAETE, W. (2004) Disruption of bacterial quorum sensing: an unexplored strategy to fight infections in aquaculture. *Aquaculture* **240**: 69–88.

DEFOIRDT, T., BOON, N., SORGELOOS, P., VERSTRAETE, W., and BOSSIER, P. (2007a) Alternatives to antibiotics to control bacterial infections: luminescent vibriosis in aquaculture as an example. *Trends in Biotechnology* **25**: 472–479.

DEFOIRDT, T., HALET, D., VERVAEREN, H., BOON, N., VAN DE WIELE, T., SORGELOOS, P. *et al.* (2007b) The bacterial storage compound poly-beta-hydroxybutyrate protects

Artemia franciscana from pathogenic *Vibrio campbellii. Environmental Microbiology* 9: 445–452.

DE SCHRYVER, P., SINHA, A.K., KUNWAR, P.S., BARUAH, K., VERSTRAETE, W., BOON, N., DE BOECK, G., and BOSSIER, P. (2009) The bacterial storage compound poly-β-hydroxybutyrate (PHB) increases growth performance and intestinal bacterial range-weighted richness in juvenile European sea bass. *Applied Microbiology and Biotechnology* 86: 1535–1541.

DE SCHRYVER, P., DIERCKENS, K., BAHN THI, Q.Q., AMALIA, R., MARZORATI, M., BOSSIER, P., BOON, N., and VERSTRAETE, W. (2010) Convergent dynamics of the juvenile European sea bass gut microbiota induced by poly-β-hydroxybutyrate. *Environmental Microbiology* 13: 1042–1051.

DIERCKENS, K., REKECKI, A., LAUREAU, S., SORGELOOS, P., BOON, N., VAN DEN BROECK, W., and BOSSIER, P. (2009) Development of a bacterial challenge test for gnotobiotic sea bass (*Dicentrarchus labrax*) larvae. *Environmental Microbiology* 11: 526–533.

FAO (2008) *The State of the World Fisheries and Aquaculture 2008.* FAO Fisheries Department, Food and Agriculture Organization of the United Nations, Rome, Italy, p. 176.

FJELLHEIM, A.J., PLAYFOOT, K.J., SKJERMO, J., and VADSTEIN, O. (2007) Fibrionaceae dominates the microflora antagonistic towards *Listonella anguillarum* in the intestine of cultured Atlantic cod (*Gadus morhua* L.) larvae. *Aquaculture* 269: 98–106.

FORBERG, T., ARUKWE, A., and VADSTEIN, O. (2011) A protocol and cultivation system for gnotobiotic Atlantic cod larvae (*Gadus morhua* L.) as a tool to study host microbe interactions. *Aquaculture* 315: 222–227.

GATESOUPE, F.J. (1999) The use of probiotics in aquaculture. *Aquaculture* 180: 147–165.

GATESOUPE, F.J. (2010) Probiotics and other microbial manipulations in fish feeds: prospective health benefits. In Watson, R.R., and Preedy, V.R. (eds): *Bioactive Foods in Promoting Health.* Oxford, Academic Press, pp. 541–552.

GIBSON, G.R., and ROBERFROID, M.B. (1995) Dietary modulation of the human colonic microbiota – introducing the concept of prebiotics. *Journal of Nutrition* 125: 1401–1412.

HALET, D., DEFOIRDT, T., VAN DAMME, P., VERVAEREN, H., FORREZ, I., VAN DE WIELE, T. *et al.* (2007) Poly-beta-hydroxybutyrate-accumulating bacteria protect gnotobiotic *Artemia franciscana* from pathogenic *Vibrio campbellii. FEMS Microbiology Ecology* 60: 363–369.

HOLBEN, W.E., WILLIAMS, P., SAARINEN, M., SARKILAHTI, L.K., and APAJALAHTI, J.H.A. (2002) Phylogenetic analysis of intestinal microflora indicates a novel mycoplasma phylotype in farmed and wild salmon. *Microbial Ecology* 44: 175–185.

HOVDA, M.B., LUNESTAD, B.T., FONTANILLAS, R., and ROSNES, J.T. (2007) Molecular characterisation of the intestinal microbiota of farmed Atlantic salmon (*Salmo salar* L.). *Aquaculture* 272: 581–588.

HUBER, I., SPANGGAARD, B., APPEL, K.F., ROSSEN, L., NIELSEN, T., and GRAM, L. (2004) Phylogenetic analysis and *in situ* identification of the intestinal microbial community of rainbow trout (*Oncorhynchus mykiss*, Walbaum). *Journal of Applied Microbiology* 96: 117–132.

JENSEN, S., ØVREAS, L., BERGH, Ø., and TORSVIK, V. (2004) Phylogenetic analysis of bacterial communities associated with larvae of the Atlantic halibut propose succession from a uniform normal flora. *Systematic and Applied Microbiology* 27: 728–736.

JOHANSON, S.D., COUCHERON, D.H., ANDREASSEN, M., KARLSEN, B.O., FURMANEK, T., JORGENSEN, T.E., EMBLEM, A., BREINES, R., NORDEIDE, J.T., MOUM, T., NEDERBRAGT, A.J., STENSETH, N.C., and JAKOBSEN, K.S. (2009) Large-scale sequence analyses of Atlantic cod. *New Biotechnology* 25: 263–271.

KAUTSKY, N., RÖNNBÄCK, P., TEDENGREN, M., and TROELL, M. (2000) Ecosystem perspectives on management of disease in shrimp pond farming. *Aquaculture* **191**: 145–161.

KESARCODI-WATSON, A., KASPAR, H., LATEGAN, M.J., and GIBSON, L. (2008) Probiotics in aquaculture: the needs, principles and mechanisms of action and screening processes. *Aquaculture* **274**: 1–14.

LEY, R.E., PETERSON, D.A., and GORDON, J.I. (2006) Ecological and evolutionary forces shaping microbial diversity in the human intestine. *Cell* **124**: 837–848.

LI, P., BURR, G.S., GATLIN, D.M., HUME, M.E., PATNAIK, S., CASTILLE, F.L., and LAWRENCE, A.L. (2007) Dietary supplementation of short-chain fructooligosaccharides influences gastrointestinal microbiota composition and immunity characteristics of pacific white shrimp, *Litopenaeus vannamei*, cultured in a recirculating system. *Journal of Nutrition* **137**: 2763–2768.

LOSURE, D.A., WILSEY, B.J., and MOLONEY, K.A. (2007) Evenness-invasibility relationships differ between two extinction scenarios in tallgrass prairie. *Oikos* **116**: 87–98.

MADISON, L.L., and HUISMAN, G.W. (1999) Metabolic engineering of poly(3-hydroxy-alkanoates): from DNA to plastic. *Microbiology and Molecular Biology Reviews* **63**: 21–53.

MARZORATI, M., WITTEBOLLE, L., BOON, N., DAFFONCHIO, D., and VERSTRAETE, W. (2008) How to get more out of molecular fingerprints: practical tools for microbial ecology. *Environmental Microbiology* **10**: 1571–1581.

MCINTOSH, D., JI, B., FORWARD, B.S., PUVANENDRAN, V., BOYCE, D., and RITCHIE, R. (2008) Culture-independent characterization of the bacterial populations associated with cod (*Gadus morhua* L.) and live feed at an experimental hatchery facility using denaturing gradient gel electrophoresis. *Aquaculture* **275**: 42–50.

MICHEL, C., PELLETIER, C., BOUSSAHA, M., DOUET, D.G., LAUTRAITE, A., and TAILLIEZ, P. (2007) Diversity of lactic acid bacteria associated with fish and the fish farm environment, established by amplified rRNA gene restriction analysis. *Applied and Environmental Microbiology* **73**: 2947–2955.

MORIARTY, D.J.W. (1997) The role of microorganisms in aquaculture ponds. *Aquaculture* **151**: 333–349.

MORIARTY, D.J.W. (1999) Disease control in shrimp aquaculture with probiotic bacteria. In Bell, C.R., Brylinsky, M., and Johnson-Green, P. (eds): *Microbial Biosystems: New Frontiers. Proceedings of the 8th International Symposium on Microbial Ecology.* Halifax, Canada: Atlantic Canada Society for Microbial Ecology.

NAVARRETE, P, MAGNE, F., MARDONES, P, RIVEROS, M., OPAZO, R., SUAU, A., POCHART, P., and ROMERO, J. (2010) Molecular analysis of intestinal microbiota of rainbow trout (Oncorhynchus mykiss). *FEMS Microbiology Ecology* **71**: 148–156.

NAYAK, S.K. (2010) Probiotics and immunity: a fish perspective. *Fish & Shellfish Immunology* **29**: 2–14.

OLAFSEN, J.A. (2001) Interactions between fish larvae and bacteria in marine aquaculture. *Aquaculture* **200**: 223–247.

PANIGRAHI, A., and AZAD, I.S. (2007) Microbial intervention for better fish health in aquaculture: the Indian scenario. *Fish Physiology and Biochemistry* **33**: 429–440.

PARAMESWARAN, V., SHUKLA, R., BHONDE, R., and HAMEED, A.S.S. (2006) Establishment of embryonic cell line from sea bass (*Lates calcarifer*) for virus isolation. *Journal of Virological Methods* **137**: 309–316.

QI, Z., ZHANG, X.-H., BOON, P., and BOSSIER, P (2009) Probiotics in aquaculture in China – Current state, problems and prospects. *Aquaculture* **290**: 15–21.

RANDELLI, E., BUONOCORE, F., CASANI, D., FAUSTO, A.M., and SCAPIGLIATI, G. (2009) An immunome gene panel for transcriptomic analysis of immune defence activities in the teleost sea bass (*Dicentrarchus labrax* L.): a review. *Italian Journal of Zoology* **76**: 146–157.

RASCH, M., BUCH, C., AUSTIN, B., SLIERENDRECHT, W.J., EKMANN, K.S., LARSEN, J.L. et al. (2004) An inhibitor of bacterial quorum sensing reduces mortalities caused by vibriosis in rainbow trout (Oncorhynchus mykiss, Walbaum). Systematic and Applied Microbiology 27: 350–359.

RAWLS, J.F., SAMUEL, B.S., and GORDON, J.I. (2004) Gnotobiotic zebrafish reveal evolutionarily conserved responses to the gut microbiota. Proceedings of the National Academy of Sciences of the United States of America 101: 4596–4601.

REID, H.I., TREASURER, J.W., ADAM, B., and BIRKBECK, T.H. (2009) Analysis of bacterial populations in the gut of developing cod larvae and identification of Vibrio logei, Vibrio anguillarum and Vibrio splendidus as pathogens of cod larvae. Aquaculture 288: 36–43.

REKECKI, A., DIERCKENS, K., LAUREAU, S., BOON, N., BOSSIER, P., and VAN DEN BROECK, W. (2009) Effect of germ-free rearing environment on gut development of larval sea bass (Dicentrarchus labrax L.). Aquaculture 293: 8–15.

RENN, S.C.P., AUBIN-HORTH, N., and HOFMANN, H.A. (2004) Biologically meaningful expression profiling across species using heterologous hybridization to a cDNA microarray. BMC Genomics 5: 42.

RINGØ, E., OLSEN, R.E., GIFSTAD, T.Ø., DALMO, R.A., AMLUND, H., HEMRE, G.I., and BAKKE, A.M. (2010) Prebiotics in aquaculture: a review. Aquaculture Nutrition 16: 117–136.

ROMERO, J., and NAVARRETE, P. (2006) 16S rDNA-based analysis of dominant bacterial populations associated with early life stages of coho salmon (Oncorhynchus kisutch). Microbial Ecology 51: 422–430.

RURANGWA, E., LARANJA, J.L., HOUDT, R.V., DELAEDT, Y., GERAYLOU, Z., VAN DE WIELE, T. et al. (2009) Selected nondigestible carbohydrates and prebiotics support the growth of probiotic fish bacteria mono-cultures in vitro. Journal of Applied Microbiology 106: 932–940.

SAKAI, M. (1999) Current research status of fish immunostimulants. Aquaculture 172: 63–92.

SALVESEN, I., SKJERMO, J., and VADSTEIN, O. (1999) Growth of turbot (Scophthalmus maximus L.) during first feeding in relation to the proportion of r/K-strategists in the bacterial community of the rearing water. Aquaculture 175: 337–350.

SANDAA, R.A., MAGNESEN, T., TORKILDSEN, L., and BERGH, Ø. (2003) Characterisation of the bacterial community associated with early stages of great scallop (Pecten maximus), using denaturing gradient gel electrophoresis (DGGE). Systematic and Applied Microbiology 26: 302–311.

SAN LUIS, B.B., and HEDREYDA, C.T. (2006) Analysis of a gene (vch) encoding hemolysin isolated and sequenced from Vibrio campbellii. Journal of General and Applied Microbiology 52: 303–313.

SAPKOTA, A., SAPKOTA, A.R., KUCHARSKI, M., BURKE, J., MCKENZIE, S., WALKER, P., and LAWRENCE, R. (2008) Aquaculture practices and potential human health risks: current knowledge and future priorities. Environment International 34: 1215–1226.

SARROPOULOU, E., SEPULCRE, P., POISA-BEIRO, L., MULERO, V., MESEGUER, J., FIGUERAS, A. et al. (2009) Profiling of infection specific mRNA transcripts of the European seabass Dicentrarchus labrax. BMC Genomics 10: 157.

SERRANO, P.H. (2005) Responsible Use of Antibiotics in Aquaculture. FAO fisheries Technical Paper 469, Rome, p. 97.

SIMPSON, J.M., KOCHERGINSKAYA, S.A., AMINOV, R.I., SKERLOS, L.T., BRADLEY, T.M., MACKIE, R.I., and WHITE, B.A. (2002) Comparative microbial diversity in the gastrointestinal tracts of food animal species. Integrative and Comparative Biology 42: 327–331.

SKJERMO, J., SAIVESEN, I., ØIE, G., OLSEN, Y., and VADSTEIN, O. (1997) Microbially matured water: a technique for selection of a non-opportunistic bacterial flora in water that may improve performance of marine larvae. Aquaculture International 5: 13–28.

SPLICHAL, I., RYCHLIK, I., GREGOROVA, D., SEBKOVA, A., TREBICHAVSKY, I., SPLICHALOVA, A. et al. (2007) Susceptibility of germ-free pigs to challenge with protease mutants of *Salmonella enterica* serovar Typhimurium. *Immunobiology* **212**: 577–582.

STEVENSSON, N.J. (1997) Disused shrimp ponds: option for redevelopment of mangroves. *Coastal Management* **25**: 425–435.

SUBASINGHE, R.P. (2005) Fisheries Topics: Governance. Fish health management in aquaculture. In FAO Fisheries and Aquaculture department (online), Rome, http://www.fao.org/fishery/topic/13545/en

SUDO, N., CHIDA, Y., AIBA, Y., SONODA, J., OYAMA, N., YU, X.N. et al. (2004) Postnatal microbial colonization programs the hypothalamic-pituitary-adrenal system for stress response in mice. *Journal of Physiology – London* **558**: 263–275.

SUN, Y.Z., YANG, H.L., LING, Z.C., CHANG, J.B., and YE, J.D. (2009) Gut microbiota of fast and slow growing grouper *Epinephelus coioides*. *African Journal of Microbiology Research* **3**: 713–720.

TIMMONS, M.B., and EBELING, J.M. (2007) *Recirculating Aquaculture*. New York: Cayuga Aqua ventures.

TINH, N., LINH, N.D., WOOD, T.K., DIERCKENS, K., SORGELOOS, P., and BOSSIER, P. (2007) Interference with the quorum sensing systems in a *Vibrio harveyi* strain alters the growth rate of gnotobiotically cultured rotifer *Brachionus plicatilis*. *Journal of Applied Microbiology* **103**: 194–203.

TINH, N.T.N., DIERCKENS, K., SORGELOOS, P., and BOLLIER, P. (2008) A review of the functionality of probiotics in the larviculture food chain. *Marine Biotechnology* **10**: 1–12.

TOKIWA, Y., and CALABIA, B.P. (2004) Review – Degradation of microbial polyesters. *Biotechnology Letters* **26**: 1181–1189.

UMBREIT, W.W. (1955) Mode of action of the antibiotics. *The American Journal of Medicine* **18**: 717–722.

VADSTEIN, O. (1997) The use of immunostimulation in marine larviculture: possibilities and challenges. *Aquaculture* **155**: 401–417.

VADSTEIN, O., ØIE, G., OLSEN, Y., SALVESEN, I., SKJERMO, J., and SKJÅK-BRÆK, G. (1993) A strategy to obtain microbial control during larval development of marine fish. In Reinertsen, H., Dahle, L.A., Jørgensen, L., and Tvinnereim, K. (eds): *Fish Farming Technology*, Rotterdam: A.A. Balkema Publishers, pp. 69–75.

VADSTEIN, O., MO, T.A., and BERGH, Ø. (2004) Microbial interactions, prophylaxis and diseases. In Moksness, E., Kjørsvik, E., and Olsen, Y. (eds): *Culture of Cold-Water Marine Fish*. Oxford: Blackwell Publishing, pp. 28–72.

VAN IMMERSEEL, F., DE BUCK, J., PASMANS, F., VELGE, P., BOTTREAU, E., FIEVEZ, V. et al. (2003) Invasion of *Salmonella enteritidis* in avian intestinal epithelial cells in vitro is influenced by short-chain fatty acids. *International Journal of Food Microbiology* **85**: 237–248.

VERNER JEFFREYS, D.W., SHIELDS, R.J., BRICKNELL, I.R., and BIRKBECK, T.H. (2003) Changes in the gut-associated microflora during the development of Atlantic halibut (*Hippoglossus hippoglossus* L.) larvae in three British hatcheries. *Aquaculture* **219**: 21–42.

VERSCHUERE, L., ROMBAUT, G., SORGELOOS, P., and VERSTRAETE, W. (2000) Probiotic bacteria as biological control agents in aquaculture. *Microbiology and Molecular Biology Reviews* **64**: 655–671.

VINE, N.G., LEUKES, W.D., KAISER, H., DAYA, S., BAXTER, J., and HECHT, T. (2004) Competition for attachment of aquaculture candidate probiotic and pathogenic bacteria on fish intestinal mucus. *Journal of Fish Diseases* **27**: 319–326.

VINE, N.G., LEUKES, W.D., and KAISER, H. (2006) Probiotics in marine larviculture. *FEMS Microbiology Reviews* **30**: 404–427.

WALIGORA-DUPRIET, A.J., DUGAY, A., AUZEIL, N., NICOLIS, I., RABOT, S., HUERRE, M.R., and BUTEL, M.J. (2009) Short-chain fatty acids and polyamines in the pathogenesis of necrotizing enterocolitis: kinetics aspects in gnotobiotic quails. *Anaerobe* **15**: 138–144.

WENNE, R., BOUDRY, P., HEMMER-HANSEN, J., LUBIENIECKI, K.P., WAS, A., and KAUSE, A. (2007) What role for genomics in fisheries management and aquaculture? *Aquatic Living Resources* **20**: 241–255.

WITTEBOLLE, L., MARZORATI, M., CLEMENT, L., BALLOI, A., DAFFONCHIO, D., HEYLEN, K. *et al.* (2009) Initial community evenness favours functionality under selective stress. *Nature* **458**: 623–626.

WU, R.S.S. (1995) The environmental impact of marine fish culture: towards a sustainable future. *Marine Pollution Bulletin* **31**: 159–166.

ZHOU, Z.G., LIU, Y.C., HE, S.X., SHI, P.I., GAO, X.H., YAO, B., and RINGØ, E. (2009) Effects of dietary potassium diformate (KDF) on growth performance, feed conversion and intestinal bacterial community of hybrid tilapia (*Oreochromis niloticus* female × *O. aureus* male). *Aquaculture* **291**: 89–94.

16

Natural antimicrobial compounds for use in aquaculture

T. Citarasu, **Manonmaniam Sundaranar University, India**

Abstract: Infectious diseases cause significant economic losses to the aquaculture industry. Current disease control protocols are often difficult to administer, not especially effective, costly and sometimes even environmentally hazardous. Alternative approaches could be developed by focusing on identifying antimicrobial compounds derived from natural resources. This chapter reviews the importance of natural antimicrobials, which include active principles from medicinal plants, marine invertebrates, micro- and macro-algae, bacteria, actinobacteria and fungi. Medicinal plants are the most significant of these, as a result of their versatile characteristics; they are effective, eco-friendly, easily available, economically attractive and non-biomagnifiable, and have no harmful side effects. This approach is regarded as an alternative to the use of synthetic drugs/antibiotics.

Key words: aquaculture, alternative medicines, natural antimicrobials, herbal treatments, marine-derived antimicrobials.

16.1 Introduction: microbial diseases in aquaculture

Infectious diseases in aquaculture are a major factor inhibiting expansion and socio-economic development. As aquaculture production becomes more intensive, the incidence of diseases has increased, leading to significant economic losses. Among the microbial diseases, pathogenic bacteria, viruses and fungi cause severe damage and economic losses in the hatchery and in grow-out ponds.

16.1.1 Bacterial diseases

Bacteria are the principal and most significant pathogens in cultured and wild finfish and shellfish worldwide. In aquaculture, bacteria occur in nurseries, during rearing and in grow-out ponds, causing heavy mortalities and serious economic losses to farmers. The important bacterial diseases in finfish include *Clostridium botulinum* (botulism produced in salmonids), *Vagococcus salmoninarum* (septicaemia in Atlantic salmon and rainbow trout), *Streptococcus difficilis* (meningo-encephalitis in carp, trout, silver pomfret and tilapia), *Renibacterium salmoninarum* (bacterial kidney disease of salmonids), pathogenic *Bacillus* sp. (septicaemia and bacillary necrosis in various fresh water fishes), *Micrococcus luteus* (micrococcosis in rainbow trout), *Mycobacterium* sp. (mycobacteriosis (fish tuberculosis) in most fish species), *Staphylococcus aureus* (eye diseases in carp, red sea bream and yellow tail), *Aeromonas hydrophila* (haemorrhagic septicaemia, motile *Aeromonas* septicaemia produced in many fresh water fish species including ornamental fishes), *Aeromonas salmonicida* (furunculosis in salmonids), *Edwardsiella ictaluri* (enteric septicaemia in catfish), *Yersinia ruckeri* (enteric redmouth in salmonids), *Flavobacterium* sp. (gill diseases in turbot, barramundi and many fresh water fishes), *Photobacterium damselae* (photobacteriosis in damsel fish, sea bream and yellow tail), *Pseudomonas* sp. (bacterial haemorrhages in most freshwater fish species), *Vibrio alginolyticus* (eye diseases and septicaemia in cobia and groupers) and *Vibrio anguillarum* (vibriosis in many marine fish species) (Austin and Austin, 2007). The important shellfish diseases are vibriosis or bacterial disease, penaeid bacterial septicaemia, penaeid vibriosis or luminescent vibriosis caused by *Vibrio harveyi*, *Vibrio fischeri* and *Vibrio parahaemolyticus* in shrimp species (Aguirre-Guzmán *et al.*, 2004)

16.1.2 Viral diseases

Viruses are small infectious agents in the aquatic environment which infect wild and cultivated fin and shellfish species, often causing devastating levels of mortality. Viruses are classified on the basis of structure, genomic features and protein properties. The DNA virus families of Iridoviridae, Adenoviridae and Herpesviridae widely infect teleost fish species such as rainbow trout, salmon, carp, catfish, sea bream, perch, turbot, eel, grouper, goldfish and sturgeon. The RNA virus families of the Reoviridae, Aquareoviridae, Picornaviridae, Nodoviridae, Togaviridae, Paramyxoviridae, Orthomyxoviridae and Rhabdoviridae cause infection in salmon, grass carp, cyprinids, channel catfish, turbot, halibut, flounder, milk fish striped bass, sea bass, grouper and ayu. The reverse transcriptase family, Retroviridae, infects chinook salmon, Atlantic salmon and damselfish (Roberts, 2003). More than 30 viral diseases are now known to occur in crustaceans, including penaeids. Major groups of viruses reported in crustaceans include the Reoviridae, Picornaviridae, Parvoviridae, Togaviridae, Baculoviridae, Paramyxoviridae, Rhabdoviridae and Iridoviridae (Bonami and Lightner, 1991). They include monodon baculovirus (MBV), baculovirus penaei (BP), baculoviral midgut gland necrosis virus (BMNV), type C baculo virus (TCBV), haemocycte infecting baculovirus (HB), infectious hypodermal and haematopoietic necrosis virus (IHHNV), Taura syndrome virus (TSV), hepatopancreatic parvo-like virus (HPV), lymphoidal parvo-like virus (LOPV), lymphoid organ vacuolization virus (LOVV), reo virus-3 and 4, yellow head virus (YHV) and white spot syndrome virus (WSSV).

16.1.3   Fungal diseases

Generally, fungus is the secondary invader in wounds, lesions or abrasions caused by bacterial pathogens, parasitic organisms, abusive handling or unfavourable environmental conditions. Fungal growths on the surface of eggs and larvae of fish and shellfish can cause extensive direct mortalities (Meyer, 1991). Fungi such as *Achlya, Aphanomyces, Calyptralegnia, Dictyuchus, Leptolegnia, Pythiopsis, Saprolegnia* and *Thraustotheca* spp. belong to the order of Saprolegniales, and are responsible for infections in freshwater fish and eggs. Branchiomycosis (*Branchiomysis demigrans* and *Branchiomysis sanguinis*) is a serious fungal disease affecting fishes almost all over the world, and especially farmed carp (Rehulka, 1991). Ulcerative mycosis (UM) and epizootic ulcerative syndrome (EUS) are caused by *Aphanomyces invadens* in fresh and brackish water fish (Noga, 1993). *Dermocystidium percae* is reported as an infectious agent in perch (Pekkarinen and Lotman, 2003). Ichthyophonus species including *Ichthyophonus hoferi* and *Ichthyophonus gasterophilum* have been the causative agents in more than 80 fishes including salmonoids (Rehulka, 1991). *Lagendium callinectes* and *Serolpidium parasitica* infect crustaceans at the larval stage and can cause a 100% mortality rate among shrimp, crabs and lobsters (Sindermann and Lightner, 1988; Karunasagar *et al.*, 2004). Fusariosis and black gill disease caused by *Fusarium solani* and *Fusarium moniliformae* in penaeid shrimp lead to high mortalities of 90% (Ramasamy *et al.*, 1996).

16.2   Current problems in microbial disease control

In aquaculture, hormones, antibiotics, vitamins and several other synthetic chemicals have been tested as growth promoters and antibacterial agents, and for other purposes (Jayaprakas and Sambhu, 1996). Even though they have positive effects on both fish and shrimps (Sambhu, 1996), antibiotics in particular cannot be recommended in commercial aquaculture operations for a variety of reasons. Their negative aspects in aquaculture include their relatively high cost, prohibited or uncertain regulatory status, unfeasible administration routes, poor absorption, toxicity, biomagnification, certain effects on the environment, possibility of consignment rejection and the fact that resistant bacteria may be transferred to humans through food handling and consumption (Committee on Drug Use in Food Animals, 1999). These resistant bacterial strains may have a negative impact on treatment for diseases in fish and humans and on the environment of the fish farms (Smith *et al.*, 1994). In general, large commercial ponds for fish are not drained after harvesting, so high levels of drugs may still remain, affecting newly growing fish, which are then exposed to the antibiotic residues and actively resistant bacteria (Committee on Drug Use in Food Animals, 1999). In marine fish hatcheries, the indiscriminate use of antibiotics as prophylactic treatments has led to the development of resistant strains and has necessitated a switch to other antibiotics (Brown, 1989). Antibiotics may also reduce larval growth and inhibit the defence mechanisms of fish larvae. Weak immune systems and responses may result in very high mortalities due to specific pathogens that cannot be combated with antibiotics (FAO/OIE/WHO, 2006).

Nowadays, most antibiotics are no longer effective in controlling diseases in aquaculture, especially fish systemic bacterial diseases, due to increasing antibiotic resistance among pathogenic bacteria. Furthermore, many countries have banned the use of antibiotics in aquaculture due to public health concerns and the environmental hazards that they pose (Lee *et al.*, 2009). Since 2006, the use of antibiotics as growth promoters in aquaculture (as well as any other domestic animal) has been completely banned in the EU. The Marine Product Export Development Authority of India (MPEDA) has also instructed hatchery operators and farmers not to use certain antibiotics due to their negative impact on health and the environment (Sanandakumar, 2002).

16.2.1   Antimicrobial resistance

Antimicrobial resistance is now a global public health problem in human as well as non-human antimicrobial usage. As in humans and terrestrial animals, bacterial resistance to antimicrobials has become widespread in aquaculture systems. The resistance of certain aquatic pathogens that cause high morbidity and mortality among infected fish and shellfish species is significant, as treatments for these diseases incur elevated costs. These increased production costs affect the livelihood of producers; moreover, consumers are affected by the availability and cost of animal products, and there is potential for international trade implications (Morley *et al.*, 2005). To avoid or reduce resistance problems, attention should be concentrated on the correct usage and concentration of antimicrobials, the principles of disease diagnosis, rational chemotherapeutic selection and mode of administration, empirical therapy and prophylactic use.

Origin of resistance

Generally resistance arises from both natural and spontaneous mutations, followed by multiplication of the resistant strains. Although some might argue that all resistance is acquired at some level, the acquisition of resistance genes requires that insertion occur either in the chromosome or plasmid, and a logical conclusion is that these insertion sites have been programmed, or have evolved, over time to accommodate molecular additions (Morley *et al.*, 2005). As stated by the report of a joint FAO/OIE/WHO consultation:

> 'It is generally acknowledged that any use of antimicrobial agents can select for the emergence of antimicrobial resistant microorganisms and can further promote the dissemination of resistant bacteria and resistance genes ... Antimicrobial resistance deriving from usage of antimicrobials in aquaculture presents a risk to public health owing to either 1. Development of acquired resistance in bacteria in aquatic environments that can infect humans and can be regarded as a direct spread of resistance from aquatic environments to humans; 2. Development of acquired resistance in bacteria in aquatic environments whereby such resistant bacteria can act as a reservoir of resistance genes from which the genes can be further disseminated and ultimately end up in human pathogens. This can be viewed as an indirect spread of resistance from aquatic environments to humans caused by horizontal gene transfer' (FAO/OIE/WHO, 2006).

Improper management such as false diagnosis, incorporation of high concentration of antimicrobials into the feeds, poor methods of administration, prolonged immersion therapy and improper discharge of effluent can also cause resistance to develop.

Mechanism of resistance
Bacteria can be resistant to the action of antimicrobial drugs because of their inherent structure or physiology (constitutive resistance), which includes a lack of cellular mechanisms required for antimicrobial action (i.e. penicillin resistance because of a lack of correct binding proteins) a growth rate that is too slow to allow effective action (β-lactam antimicrobials), and resistance in anaerobic bacteria to aminoglycosides from a lack of oxygen-dependent uptake of the antimicrobial into the bacterial cell. Acquired resistance, on the other hand, refers to the development of mechanisms to circumvent the action of drugs through genetic mutation or through acquisition of genetic elements. Several mechanisms are also associated with acquired resistance such as drug inactivation, drug modification, production of competitive metabolites, target mutation, target substitution, target modification, decreased cell wall permeability to drugs, active efflux of drugs, and failure to metabolize a drug to its active form (Morley *et al.*, 2005).

Transmission of resistance
Generally, resistance can be transmitted from one bacterium to another that has never been exposed to the antibiotic: this phenomenon is known as horizontal gene transfer (Serrano, 2005). Plasmids carry the resistant genes from one bacterium to another; thus, while an antimicrobial kills the sensitive bacteria, any resistant bacteria can survive and multiply. The more antimicrobials are used the greater will be the 'selective pressure' favouring resistant strains. This is an example of Darwin's theory of evolution, the 'survival of the fittest'. Antimicrobials create an environment

which favours the growth of resistant variants that already exist in nature or that arise by chance (Wiuff *et al.*, 2010). The molecular mechanisms involved include conjugation, whereby a plasmid is passed from one organism to another through a pilus. The spread of gene coding for antibiotic resistance is facilitated by mobile genetic elements called transposons, which can move from plasmids to the bacterial chromosome and vice versa, in a process known as transformation. Transduction is the third means by which genetic material can be acquired from an infecting bacteriophage (Serrano, 2005). Other methods of transmission are: carrier animals between herds (regionally, nationally or even internationally); exposure through feed and water or through the environment (e.g. contaminated soils and facilities); direct or indirect contact with infected humans, and vectors and vehicles such as wildlife, insects and birds (Morley *et al.*, 2005). The method of animal management adopted can also affect the likelihood of transmission.

16.3   Public health and ethical issues relating to the use of antibiotics for disease control in aquaculture

16.3.1   Effect on human intestinal flora

The resistant bacteria transmitted from aquatic fin and shellfishes to humans cause infections. Direct spread of resistance from aquatic environments to humans may occur as a result of consumption of aquaculture food products, through drinking water, through direct contact with water or aquatic organisms, or through the handling of aquaculture food products. Major pathogens including *Aeromonas* spp., *Edwardsiella tarda*, *Salmonella* spp., *Vibrio cholerae*, *V. parahaemolyticus*, and *V. vulnificus*, and to a lesser extent non-O1 *V. cholerae* and *Plesiomonas shigelloides*, can cause blood-borne infections, especially in individuals with risk factors such as chronic liver disease and iron overload, as well as in people with immune disorders. *V. vulnificus*, *V. parahaemolyticus* and some non-O1 *V. cholerae* serogroups are well-recognized causes of gastrointestinal illnesses, despite the incidence being low in some countries. Motile *Aeromonas* are common causes of self-limiting diarrhoeal illnesses (WHO, 1999).

16.3.2   Antimicrobial residues

Antimicrobial residues can cause serious illness, and therefore present a hazard to public health. Drug allergies are mediated by a number of different immunological mechanisms, including type 1 immediate immune response mediated through Immunoglobulin E (IgE). Symptoms include anaphylaxis, skin-rashes, urticaria and angioedema (FAO/OIE/WHO, 2006). The antibiotics used in aquaculture, either for prophylactic or therapeutic purposes, often accumulate in the tissue of aquatic animals.

The presence of antimicrobial drug residues in the edible tissues can cause allergies, toxic effects, changes in the intestinal microbial fauna and acquisition of drug resistance. Chloramphenicol residues in food consumed by humans can even result in aplastic anemia, which causes very serious bone marrow diseases. Nitrofuran antibiotics are known to cause cancer and many other diseases. It is for this reason that most countries that import fish products have banned the use of certain antibiotics (Sanandakumar, 2002).

16.3.3 Ethical issues

Ethical issues are some of the important restricting factors in the application of antibiotics. The Food and Agricultural Organization (FAO), the World Health Organization (WHO), the International Office of Epizootics (OIE) and a number of national governments have already raised the issue of the irresponsible use of antibiotics in all production sectors, with particular concern relating to the potential risks to public health. Many governments around the world have introduced, changed or tightened national regulations on the use of antibiotics, in general and within the aquaculture sector (FAO/OIE/WHO, 2006). The treatment protocols and the antibiotics used raise a number of ethical issues. The safety of the worker, consumer and the environment are important considerations in antimicrobial treatments carried out in aquaculture systems. The potential risk to the workers is due to contact with potentially carcinogenic chemical substances; while consumer safety is affected by aquaculture-derived microbes and biotoxins (Grigorakis, 2010). Uneaten antibiotic medicated feeds, unabsorbed drugs and residues are transferred in the water column and accumulated in the sediment, where they affect the non-target organisms including scavengers, secondary aquaculture species and beneficial microbes (Ma *et al.*, 2006).

Rigos and Troisi (2005) investigated the pharmacokinetics of antibacterial agents in aquaculture, and observed that withdrawal periods should be determined for each drug, each target species, and at different temperature conditions, in order to ensure that no residues above maximum residual level (MRL) exist in the tissues of farmed products; however, the current knowledge provided by research is not sufficient at present. In addition, as described above, the use of antibacterials in aquaculture has also been connected to the development of resistance in human pathogens (Sapkota *et al.*, 2008) and therefore to higher vulnerability to potential human disease outbreaks. Where disease and its treatment are concerned, the well-being of the environment is related to two aspects of biota conservation: transmission of microbial pathogens to the wild populations and pollution from chemotherapeutics (Grigorakis, 2010). From the ethical point of view, increased knowledge and awareness about the diseases and antimicrobials are required before the latter are used in aquaculture.

The following stages should be followed for an effective strategy:

1. Establish the cause of the disease condition.
2. Establish an antibiogram or sensitivity pattern for the pathogen.
3. Use the correct dosage for the recommended duration.
4. Adhere to careful storage of antibiotics.
5. Use as narrow a spectrum of antibiotic as possible and avoid indis-criminate use of drugs, especially with live feeds (rotifers and *Artemia*).
6. Avoid oral therapy if fish are inappetent.
7. Avoid repeated use of the same antibiotic and blanket treatment for prophylactic use.
8. Routinely monitor antibiotic resistance.
9. Avoid polypharmacy.
10. Whenever possible use products licensed for the species.
11. For all treatments, ensure that the prescribing clinician records the date of examination, the clients, the number of fish treated, the diagnosis, product prescribed, dosage, duration of treatment and withdrawal period recommended (Inglis, 1996).

16.4   Alternative antimicrobial compounds

A number of different alternative antimicrobial compounds are widely used today to control major aquatic pathogens. They are a good alternative source to synthetic drugs and antibiotics in the aquaculture industry, and reduce the side effects observed with synthetic compounds and antibiotics. Plants are rich in a wide variety of secondary metabolites with antimicrobial properties, and are thus a major source of antimicrobial compounds. Plants are the storehouses and sources of safer and cheaper chemicals (Prasad and Variyur Padhyoy, 1993); of the 80% of pharmaceuticals derived from plants, very few are currently used as antimicrobials (Perumalsamy and Gopalakrishnakone, 2008). Herbal compounds inhibit the pathogenic bacteria and exhibit greater selective toxicity towards the infecting microbes than conventional treatments. After plants, the other antimicrobial compounds employed to control pathogens in aquaculture are derived from animals, fishes, invertebrates, micro-algae, macro-algae, microbes such as antagonistic bacteria, and Actinomycetes.

16.4.1   Alternative antibacterial agents

Alternative antibacterials should not biomagnify, leave or cause resistant strain development. They should also be low cost. The most important low cost alternative antibacterials are derived from plants, seaweed, micro-algae and secondary metabolites of microbial origin; these are highly effective in controlling bacterial pathogens, and avoid the disadvantages associated with commercial antibiotics. Herbal compounds have proved to be particularly effective due to their versatile active principles. They provide not only antibacterial effects, but also offer a number of other benefits including anti-stress effects, immunostimulation, antioxidant activity and appetite stimulation, and are also good for liver function. Thanks to the broad range of activity offered by these plants, they are highly effective in controlling pathogens in aquaculture operations and are good alternatives to commercial antibiotics.

16.4.2   Alternative antiviral agents

As a result of the failure of synthetic chemicals to cure a wide range of viral diseases in aquaculture, the frequency of viral resistance has increased, and only a small number of antiviral drugs are currently used. In the search for alternative antivirals that can be used in place of synthetic drugs, recent interest has focused on medicinal herbs (McCutcheon *et al.*, 1995). The plant antiviral metabolites provide the prototypes for designing potentially superior new chemotherapeutic drugs to combat major aquatic viruses. Recently, various herbal extracts have been shown to successfully control shrimp viruses such as YHV and WSSV both *in vitro* and *in vivo* (Direkbusarakom *et al.*, 1998; Citarasu *et al.*, 2006; Balasubramanian *et al.*, 2007). The other alternative antivirals are bacterial exopolysaccharides (EPS) (Arena *et al.*, 2006) from extreme marine environments; sulphated EPS, polysaccharides from marine algae (Gerber *et al.*, 1958); Fucoidan, a sulfated polysaccharide from seaweed (Béress *et al.*, 1993); sulphoglycolipids from cyanobacteria (Reshef *et al.*, 1997); and sponge-derived secondary metabolites (Munro *et al.*, 1999).

16.4.3   Alternative antifungal agents

There is a real need to search for newer compounds with potential antifungal activities and worldwide spending on discovering new antifungal agents is expected to increase in the coming years. New sources, especially plant-derived antifungal compounds, have been investigated extensively in recent years. Antimicrobial peptides, secondary metabolites and other compounds from associated bacteria and actinobacteria are the new fungal agents that have been identified as suitable for use in aquaculture. New research should focus on finding antifungal compounds from bacteria, actinobacteria and fungi isolates of extremophilic origin, for example, those that flourish at high temperature, and high salinity, and those found in tannery effluent and highly polluted sewage and so on. Soil and terrestrial fungal-derived compounds are also novel sources of antifungal agents. These compounds are promising in pathogenic fungal control in aquaculture operations and replace synthetic antifungal agents such as antiseptic chemicals or antibiotics which are currently used.

16.5   Origin of alternative antimicrobials

Eighty per cent of the world population mainly depends on traditional medicines from plants, animals and microorganisms for their health care (Farnsworth *et al.*, 1985). Although this usage dates back more than 100 years, the usage of alternative antimicrobials in aquaculture is a very recent development.

16.5.1   Terrestrial origin

To date more than a million natural products derived from living organisms, including plants, animals and microbes, have been discovered (Berdy, 2005). About 50% of these are derived from terrestrial plants, and 20–25% possess various bioactive properties including antibacterial, antifungal, antiviral, antiprotozoal, antinematode, anticancer and anti-inflammatory activities.

Plants
Medicinal plants, including herbs, shrubs, grasses and higher plants, represent rich sources of antimicrobial active compounds. They are the oldest source of pharmacologically active compounds such as essential oils, alkaloids, flavonoids, sesquiterpene lactones, diterpenes, triterpenes, phenols, quinines and lectins. In most cases, phenolics, alkaloids, polysaccharides and flavonoids play a major role in preventing or controlling infectious microbes. The active extracts of several plants have been successfully used in the control of major aquatic pathogens *in vitro* and *in vivo*. *A. hydrophila* and Trichodinaiasis were successfully controlled at *in vivo* level in the Tilapia *Oreochromis niloticus* pond by the application of garlic (*Allium sativum*) extract. Palavesam *et al.* (2006) successfully controlled bacterial pathogens such as *Streptococcus pyogenes*, *Staph. aureus*, *Serratia salinaria*, *Alcaligen faecalis* and *V. parahaemolyticus*, isolated from the infected grouper *Epinephelus tauvina*, using extracts of *Withania somnifera*, *Tinospora cordifolia*, *Solanum xanthocarpum*, *Daemia extensa* and *Andrographis paniculata*. *Pseudomonas aeruginosa*, *Staph. aureus*, *Salmonella typhi* and *A. hydrophila* were controlled *in vitro* and *in vivo* in the *Penaeus monodon* larviculture tanks through the use of organic solvent extracts such as *Solanum trilobatum*, *A. paniculata*, *Psoralea corylifolia* and *Aegle marmelos* (Citarasu *et al.*, 2003). Immanuel *et al.* (2004) studied the bacterial load of shrimp *P. indicus* juveniles fed with seaweed and herbal extracts and reared in *V. parahaemolyticus* inoculated culture. Shangliang *et al.* (1990) reported the antimicrobial activity of five Chinese herbal extracts, *Stellaria aquatica*, *Impatiens biflora*, *Oenothera biennis*, *Artemisia vulgaris* and *Lonicera japonica*, against 13 bacterial fish pathogens. Chitmanat *et al.* (2005) demonstrated the inhibitory activity of Indian almond, *Terminalia catappa*, extract against *A. hydrophila*. Several herbs were also effective

in successfully controlling major aquatic viruses and fungi (Direkbusara-kom *et al.*, 1996; Khan *et al.*, 2004; Adigüzel *et al.*, 2005; Citarasu *et al.*, 2006; Balasubramanian *et al.*, 2007). Further information is given in Sections 16.6 to 16.8.

16.5.2   Fresh water/marine origin

The sea is an immense and practically unexploited source of new potentially useful biologically active substances. Fresh water and marine organisms are rich sources of biological active metabolites. Many bioactive and pharma-cologically important compounds such as alginate, carrageenan and algal fatty acids have potent antimicrobial activity. Invertebrates such as sponges, molluscs, bryozoans and tunicates produce a great number of natural marine products including alkaloids, peptides, terpenes and polyketides (Proksch *et al.*, 2003). Sponges have emerged as an effective means of combating invading microbes, inducing the production of secondary metabolites. Marine microorganisms such as bacteria and fungi have been reported to produce antibacterial (Rosenfeld and Zobell, 1947), antifungal, antiviral and antitumour substances (Bernan *et al.*, 1997). Several earlier studies have suggested that such marine bacteria can be used to combat epizootics in aquaculture systems (Maeda and Liao, 1992).

Fish and invertebrates

Varying amounts of natural antimicrobial compounds have been isolated from a number of different marine organisms such as sponges, coelenter-ates, echinoderms, tunicates, molluscs, bryozoans and crustaceans and they are highly effective in controlling bacterial, viral and fungal pathogens. Eukaryotic organisms, especially those from the marine environment, represent a rich hunting ground for the discovery of novel natural micro-bicidal agents. Accordingly, during the course of their evolution, marine eukaryotes have developed a plethora of anti-infective molecules and strat-egies by which they protect themselves against prokaryotic and viral attack (Smith *et al.*, 2010). Marine sponges have since been considered a 'gold mine'. More than 15000 marine products have been described from marine organisms. Sponges, in particular, are responsible for more than 5300 dif-ferent products, and every year hundreds of new compounds are being discovered. Various bioactive compounds such as terpenes, sterols, cyclic peptides, alkaloids, fatty acids, peroxides and amino acid derivatives are isolated from sponges (Tziveleka *et al.*, 2003; Sipkema *et al.*, 2005). Sponge-derived metabolites such as Bryostatin, ET-743, Dolastatin and Ziconotide, Halichondrin B derivatives are all important in aquaculture (Munro *et al.*, 1999).

Marine invertebrates express different types of defence proteins in the blood plasma, which have antimicrobial effects. A number of lectins from immune cells or haemolymph from marine invertebrates have been reported to be effective antibacterial agents, inhibiting the growth of the important aquatic pathogens *V. vulnificus* and *V. pelagicus* (Schröder *et al.*, 2003). Lectins from the hemolymph of commercially important mollusc species such as *Cerastoderma edule* (Cardiidae), *Ruditapes philippinarum* (Veneridae), *Ostrea edulis* (Ostreidae), *Crepidula fornicata* (Calyptraeidae), *Buccinum undatum* (Buccinidae) and colonial ascidians *Synoicum pulmonaria* have been shown to exert antibacterial activity against Gram-positive and Gram-negative bacteria that are significant in aquaculture (Defer *et al.*, 2009). A sialic-acid binding lectin from the horse-mussel, *Modiolus modiolus*, is active against *Vibrio* species (Chattopadhyay and Chatterjee, 1993). Lectins such as scyllin, (*Scylla serratau*), tachylectin 1 (*Tachypleus tridentatus*) and HSL (*Holothuria scabra*) were also shown to effectively suppress Gram-positive and Gram-negative bacteria (Morita *et al.*, 1985; Saito *et al.*, 1995; Gowda *et al.*, 2008).

Another option is provided by antimicrobial peptides (AMPs) recently discovered from shrimp, oysters and other vertebrates; these have effectively suppressed Gram-negative and Gram-positive bacteria, as well as viruses, yeasts and moulds. These molecules could be used to replace antibiotics, since they are less likely to cause resistance in the target microorganisms, due to their direct action on membranes, and to their fast degradability, both of which help to avoid the accumulation of residues (Nicolas *et al.*, 2007). Centrocins 1 (4.5 kDa), 2 (4.4 kDa) and SpStrongylocin purified from the green sea urchin *Strongylocentrotus droebachiensis* have effectively inhibited *V. anguillarum*, *Staph. aureus* and *Corynebacterium glutamicum* (Li *et al.*, 2010). AMPs such as arasin 1 and arasin 2, isolated from the hemocyte extracts of small spider crab, *Hyas araneus*, are effective against *V. anguillarum*, *Staph. aureus* and *C. glutamicum* (Stensvåg *et al.*, 2008). Tissue extracts and haemolymph/haemocyte of *Pandalus borealis* (northern shrimp), *Pagurus bernhardus* (hermit crab), and *Paralithodes camtschatica* (king crab) suppressed *Escherichia coli*, *V. anguillarum* and *C. glutamicum* (Haug *et al.*, 2002).

Macro-algae
Prior to the 1950s, the medicinal properties of seaweeds were restricted to traditional and folk medicines; their bioactivity was not discovered until after the 1980s (Mayer and Hamann, 2000). Seaweeds are considered a source of bioactive compounds as they are able to produce a great variety of secondary metabolites characterized by a broad spectrum of biological activities. Compounds with cytostatic, antiviral, anthelmintic, antifungal and antibacterial activities have been detected in green, brown and red algae (Newman *et al.*, 2003). The major bioactive compounds are brominates, aromatics, nitrogen-heterocyclics, nitrosulphuric-heterocyclics, sterols, dibutanoids, proteins, peptides, sulphated polysaccharides, amino acids, terpenoids, phlorotannins, acrylic acid, phenoliccompounds, steroids, halogenated ketones, alkanes, cyclic polysulphides and fatty acids (Mtolera and Semesi,

1996). Owing to the huge number of varieties with untapped bioactive compounds, seaweeds might be a potential new source of alternative anti-microbials in the future. Several bioactive compounds from seaweeds have recently been shown to be highly effective against pathogenic bacteria, viruses and fungi related to aquaculture. Methanolic extracts of kappa-phycus and padina species effectively suppressed major aquatic bacterial pathogens such as *Pseudomonas fluoresens*, *Staph. aureus* and *V. cholera* (Rajasulochana *et al.*, 2009). Dubber and Harder (2008) investigated the antibacterial effects of hexane and methanol extracts of the macro-algae *Mastocarpus stellatus*, *Laminaria digitata* and *Ceramium rubrum* on seven prominent fish pathogenic bacteria. *Egregia menziesii*, *Codium fragile*, *Sargassum muticum*, *Endarachne binghamiae*, *Centroceras clavulatum* and *Laurencia pacifica* collected from Todos Santos Bay, México and their organic solvent extracts showed good activities against *Staph. aureus*, *Klebsiella pneumoniae* and *P. aeruginosa* (Villarreal-Gómez *et al.*, 2010). Anti-bacterial activity has been detected in a number of seaweeds such as *Caulerpa racemosa*, *Ulva lactuca*, *Gracillaria folifera*, *Hypneme muciformis*, *Sargassum myricocystum*, *S. tenerrimum* and *Padina tetrastomatica* by Kolanjinathan *et al.* (2009) against the fish pathogenic bacteria *Streptococcus* sp., *Pseudomonas* sp., *Bacillus* sp., *Staphylococcus* sp., *Enterobacter* sp., *Acinetobacter* sp., *Moraxella* sp., *Alkaligens* sp., *Micrococcus* sp., *Serratia* sp., *Vibrio* sp. and *Bacillus* sp. Rhodophyceae of *Asparagopsis taxiformis*, *Laurencia ceylanica*, *Laurencia brandenii* and *Hypnea valentiae* were found to be highly active against shrimp pathogenic *Vibrio* sp. Organic solvents of *Gracilaria corticata*, *Ulva fasciata* and *Enteromorpha compressa* have been shown to inhibit the growth of six virulent strains of bacteria pathogenic to fish viz., *E. tarda*, *V. alginolyticus*, *P. fluorescens*, *P. aeruginosa* and *A. hydrophila* (Choudhury *et al.*, 2005).

Compounds from seaweeds have *in vitro* or *in vivo* activity against a wide range of viruses, including herpes viruses, togaviruses, paramyxoviruses, rhabdoviruses, and other important aquatic viruses. Fucoidan is a sulphated polysaccharide isolated from the brown seaweed *Fucus vesiculosus* (Béress *et al.*, 1993) and has an inhibitory effect on the replication of DNA viruses. Galactan sulphate (GS), a polysaccharide isolated from the red seaweed *Agardhiella tenera*, shows activity against enveloped viruses, including herpes viruses, togaviruses, arenaviruses and others (Witvrouw *et al.*, 1994). Brown algae such as *Cutleria cylindrica*, *Dictyopteris latiuscula*, *Dictyopteris polypodioides*, *Ectocarpus siliculosus*, *Padina arborescens*, *Sargassum polycystum*, *Sargassum polyporum*, *Sargassum fulvellum*, *Sargassum ringgoldianum* and *Zonaria stipitata* extracts were found to inhibit infectious hematopoietic necrosis virus (IHNV), *Paralichthys olivaceus* virus and *Oncorhynchus masou* virus, all pathogenic to salmonids, in cell lines (Kamei and Aoki, 2007). *Laurencia* sp. (Glombitza, 1979) and *Caulerpa* sp. were also reported to exert antifungal activity against several fungal species, including those affecting aquaculture.

Micro-algae
Micro-algae are useful bioactive agents, with anti-inflammatory, antiviral, antimicrobial, antihelmintic, cytotoxic, immunological and enzyme inhibition properties (Dufosse *et al.*, 2005), and since the second half of the 20th century they have been screened for their biological activities (Duff *et al.*, 1966). Because of their phototrophic nature and permanent exposure to high oxygen and radical stresses, they are able to produce numerous efficient protective chemicals (Tsao and Deng, 2004). Most of the micro-algal active principles belong to the groups of polyketides, amides, alkaloids and peptides (Ghasemi *et al.*, 2004). Das *et al.* (2005) investigated the effects of organic extracts of *Euglena viridis* against different strains of fish and shell-fish pathogens such as *A. hydrophila*, *Pseudomonas putida*, *P. aeruginosa*, *P. fluorescens*, *E. tarda*, *V. alginolyticus*, *V. anguillarum*, *V. harveyi*, *V. fluvialis*, and *V. parahaemolyticus*. Desbois *et al.* (2009) isolated an antibacterial polyunsaturated fatty acid, eicosapentaenoic acid (EPA) from the marine diatom *Phaeodactylum tricornutum* Bohlin, which showed activity against a range of both Gram-positive and Gram-negative bacteria. Blue–green algae *Anabaena wisconsinense* and *Oscillatoria curviceps* were isolated from fish farms and their antimicrobial effects were studied by El-Sheekh *et al.* (2008) against Gram-positive bacteria such as *Lactobacillus* sp. and *Bacillus firmus*; the Gram-negative bacteria *A. hydrophila*, *P. fluorescens* and *Pseudomonas anguilliseptica*; and the fungi *Aspergillus niger* and *Saprolegnia parasitica* which were isolated from diseased fish. *Nostoc muscorum* (DeCano *et al.*, 1990), *Chroococcus* sp., *Oscillatoria* sp., *Anabaena* sp., *Synechocystis aquatilis* and various micro-algae (Katircioglu *et al.*, 2006), as well as *Fischerella* sp. (Raveh and Carmeli, 2007) all possess antibacterial or/and antifungal activities.

Austin *et al.* (1992) indicated that extracts derived from *Tetraselmis suecica* inhibited *A. hydrophila*, *A. salmonicida*, *V. anguillarum* and *V. salmonicida in vitro*. Organic solvent of the marine diatome *Pleurosigma elongatem* showed inhibitory activity against *Staph. aureus*, *S. typhi* and *V. cholerae* (Manivasaham *et al.*, 1989). Cell extracts from the model marine diatom, *P. tricornutum* Bohlin, have been shown to be antibacterial against numerous bacterial species (Kellam and Walker, 1989). A mixture of fatty acids named chlorellin, derived from Chlorella, demonstrated inhibitory activity against Gram-positive and Gram-negative bacteria (Pratt *et al.*, 1951). Li and Tsai (2009) fed *Nannochloropsis oculata* transgenic algae to medaka fish to combat *V. parahaemolyticus* infection; the algae proved to offer a bactericidal defense mechanism. Ishida *et al.* (1997) observed that kawaguchipeptin B, isolated from *Microcystis aeruginosa*, inhibited the growth of *Staph. aureus*. It was observed that thermo-tolerant cyanobacterium *Phormidium* sp. inhibited the growth of *Staph. aureus* and *Candida albicans* (Fish and Codd, 1994). Özdemir *et al.* (2001) found that *Spirulina* extracts obtained from various solvents exhibited antimicrobial activity on both Gram-positive and Gram-negative organisms. Methanolic extracts of

Antarctic cyanobacterium Nostoc CCC 537 acts as an antibacterial against *S. typhi* MTCC 3216, *P. aeruginosa* ATCC 27853 and *Enterobacter aerogenes* MTCC 2822 (Asthana *et al.*, 2009). Extracellular sulphated polysaccharides A1 and A2 isolated and purified from *Cochlodinium polykrikoides* marine microalgae (Hasui *et al.*, 1995) inhibit the cytopathic effects of influenza virus types A and B grown on MDCK cells.

A number of cyanobacteria species have been found to produce sulph-oglycolipids with strong anti-HIV action (Reshef *et al.*, 1997). Calcium spirulan (Ca-SP), a sulphated polysaccharide, isolated from a marine blue–green alga, *Arthrospira platensis*, was shown to be a potent antiviral agent against HIV (Hayashi *et al.*, 1996). Cyanovirin-N (CV-N) isolated from an aqueous cellular extract of the cyanobacterium *Nostoc ellipsosporum* (Boyd *et al.*, 1996) prevents the *in vitro* replication and cytopathic effects of primate retroviruses, including SIV and diverse laboratory strains and clinical isolates of HIV-1 and HIV-2 (Bewley *et al.*, 1998; Gustafson *et al.*, 1997). Pereira *et al.* (2004) reported an extensive study on the mechanism of action of two diterpenes, Da-1 and AcDa-1, isolated from the marine alga *Dictyota menstrualis*, which inhibit HIV-1 virus replication in the PM-1 cell line. A sulphated polysaccharide, naviculan, was isolated from deep sea diatom *Navicula directa* and shown to inhibit HSV-1 and HSV-2 (Lee *et al.*, 2006).

16.5.3 Microbial origin

The field of microbiology has become particularly important in recent years, with microorganisms being used in almost all arenas. The use of microbial-derived therapeutic agents began in the 20th century (Monaghan and Tkacz, 1990) and these bioactive compounds have become the foundation of modern pharmaceuticals (Capon, 2001). Major classes of microbes such as bacteria, actinobacteria and fungi are used in biomedical products. Antagonism is nature's method of survival and existence. Microbes produce some secondary metabolites as a form of defence against other micro-organisms, and these secondary metabolites serve as a source of bioactive compounds for use in human therapies. The marine environment also represents a largely unexplored source for isolation of new microbes (bacteria, fungi, actinomycetes, micro-algae – cyanobacteria and diatoms) that are potent producers of bioactive secondary metabolites (Bhatnagar and Kim, 2010).

Bacteria

Bacteria produce a wide variety of antagonistic factors that include primary and secondary metabolites (Piard and Desmazeaud, 1992). Primary metabolites are produced during the growth phase of an organism and play a vital role in active growth. Secondary metabolites can be described as compounds that are produced after active growth has taken place and which perform no vital function for the producing organism (Kleinkauf *et al.*, 1986). Inhibition may be due to the production of many metabolites such as organic acids, viz. lactic acid, acetic acid, hydrogen peroxide, diacetyl, lysozymes, proteases, siderophores and bacteriocins (Ennahar *et al.*, 2001). Marine bacteria are a rich source of potentially useful antimicrobial molecules; since 1990, the number of bioactive metabolites derived from marine bacteria has exponentially increased (Faulkner, 2000). These bacteria, such as *Altermonas, Pseudoalteromonas, Bacillus, Vibrio, Pseudomonas* and *Cytophaga*, were isolated from seawater, sediments, algae and marine invertebrates and were able to produce quinones, polyenes, macrolides, alkaloids, peptides and to a lesser extent terpenoids (Laatsch *et al.*, 1995). Bioactive substances derived from marine *Pseudomonas*, such as pyrroles, pseudopeptide pyrrolidinedione, phloroglucinol, phenazine, benzaldehyde, quinoline, quinolone, phenanthren, phthalate, andrimid, moiramides, zafrin and bushrin act as antimicrobial agents particularly against major aquatic pathogens (Romanenko *et al.*, 2008). A marine *Pseudomonas* strain I-2 produced inhibitory compounds against shrimp pathogenic vibrios including *V. harveyi, V. fluvialis, V. parahaemolyticus, V. damsela* and *V. vulnificus* (Chythanya *et al.*, 2002).

Fourteen Pseudomonas strains isolated from marine environments exhibited antagonistic action against a wide range of bacteria including *Vibrio* spp. (Than *et al.*, 2004) and extra cellular anti-*Vibrio* substances characterized from *Pseudoalteromonas* sp. A1-J11 (Castillo *et al.*, 2008). Gram *et al.* (1998) investigated Siderophore-producing *P. fluorescens* and found that it is able to suppress the mortality of rainbow trout infected with *V. anguillarum*. *Bacillus* is an interesting genus for the production of a diverse array of antimicrobial peptides with several different basic chemical structures (Bizani and Brandelli, 2002). Species of the genus act as a beneficial bacteria when used against bacterial or viral disease in shrimp aquaculture, and release antibacterial substances (Balcazar and Luna-Rojas, 2007). *Bacillus subtilis* helps to improve the immune system and bacterial clearance in shrimps against *V. alginolyticus* infection (Tseng *et al.*, 2009). A novel cyclic decapeptide antibiotic, Loloatin B, is derived from *Bacillus* sp. and inhibits the growth of methicillin-resistant *Staph. aureus* and vancomycin-resistant Enterococcus (Gerard *et al.*, 1999).

Recently, extracellular products (ECPs) of *Bacillus cereus* TC-1 and TC-2 isolated from coconut retting and solar salt pans respectively were shown to inhibit the pathogenic *V. harveyi* and *V. parahaemolyticus* isolated from infected *P. monodon* (Donio, 2009; Nair *et al.*, 2010). Phenyllactic acid (PLA), lactic acid, and acetic acid isolated from *Lactobacillus plantarum* act as antifungal agents (Prema *et al.*, 2010). A marine *Vibrio* (strain C33) that has an inhibitory effect on the growth of *V. anguillarum* was isolated from the north Chilean scallop *Argopecten purpuratus*, *V. parahaemolyticus* and *V. splendidus* (Jorquera *et al.*, 1999). EPS from marine bacteria such as *Alteromonas macleodii, Alteromonas fijiensis, Vibrio diabolicus* and

Alteromonas infernus act as sources for antiviral compounds. Sulphated EPS are known to interfere with the adsorption and penetration of viruses into host cells, as well as inhibiting various retroviral reverse transcriptases. Two new exopolysaccharides, EPS-1 and EPS-2, isolated from *Bacillus licheniformis* and *Geobacillus thermodenitrificans*, are very effective in the control of HSV (Arena *et al.*, 2009). Macrolactin A is another antiviral compound derived from marine bacteria that inhibited HIV replication (Gustafson *et al.*, 1989).

Actinobacteria

Actinobacteria are the most economically and biotechnologically valuable prokaryotes and are responsible for the production of bioactive secondary metabolites, notably antibiotics, antitumour agents, immunosuppressive agents and enzymes. Actinomycetes yields 45% of the bioactive secondary metabolites (Berdy, 2005). The marine environment represents a large untapped source for the isolation of new Gram-positive actinobacteria, which are known to produce chemically diverse compounds with a wide range of biological activities (Bredholt *et al.*, 2008). Several actinobacteria belonging to the family Micromonosporaceae were isolated from subtidal marine sediments collected from the Bismarck Sea and the Solomon Sea off the coast of Papua New Guinea, and demonstrated inhibitory activity against multidrug-resistant Gram-positive pathogens and vaccinia virus (Magarvey *et al.*, 2004). The akaliphilic actinobacteria *Streptomyces sannanensis*, strain RJT-1, isolated from alkaline soil in India, showed inhibitory activity against *Staph. aureus, B. cereus, B. megaterium, B. subtilis* and Gram-negative organisms such as *E. coli, Proteus vulgaris, Shigella dysentry, P. aeruginosa* and *Salmonella typhosa para* B (Vasavada *et al.*, 2006). Antagonistic *Nocadiopsis* sp., strain JAJ16, showed good antibacterial activity against bacteria such as *Staph. aureus, B. subtilis, S. typhi*, methicillin-resistant *Staph. aureus, K. pneumoniae, Enterobacter* sp. and *P. aeruginosa*. Good antifungal activity was also observed against fungi such as *C. albicans, Aspergillus flavus* and *Fusarium oxysporum* (Aruljose *et al.*, 2010). Atta *et al.* (2009) screened the inhibitory activity of five actinobacteria belonging to the genus *Streptomyces*, isolated from the river Nile, against *Staph. aureus, M. luteus, Bacillus pumilus, B. subtilis, E. coli, K. pneumonia, P. aeruginosa* and the fungi *C. albicans, A. niger, A. fumigatus, F. oxysporum* and *F. moniliform*.

Current attention should focus on the search for novel secondary metabolites and novel enzymes from Halophilic/alkali-tolerant and thermophilic actinobacteria from extreme environments. Novel alkaliphilic actinobacteria *Nocardiopsis alkaliphila, Nocardiopsis metallicus* and *Bogoriella caseilytica* are a valuable source of novel products of industrial interest, including enzymes and antimicrobial agents (Mitsuiki *et al.*, 2002; Tsujibo *et al.*, 2003). Pyrocoll, an antibiotic compound, was recently detected in a novel alkaliphilic *Streptomyces* strain (Dietera *et al.*, 2003).

Fungi

It is traditionally estimated that there are around 1.5 million species of fungi worldwide (Hawksworth and Rossman, 1987). Marine-derived fungi have been widely studied for their bioactive metabolites and have proven to be a rich and promising source of novel anticancer, antibacterial, antiplasmodial, anti-inflammatory and antiviral agents (Bhadury *et al.*, 2006; Newman and Hill, 2006). Although antagonistic fungal species are isolated from marine, terrestrial and soil sources, the majority show good antagonistic activity against phytopathogens. Baker *et al.* (2009) demonstrated the antagonistic activity of fungi such as *Agaricomycotina, Mucoromycotina, Saccharomycotina* and *Pezizomycotina* against *E. coli, Bacillus* sp., *Staph. aureus* and *Candida glabrata.* Byun *et al.* (2003) isolated a novel compound, diketopiperazine, from the fungal strain M-3 associated with the red alga *Porphyra yezoensis* which inhibited the fungus *Pyricularia oryzae.* Phomadecalins A, B, C and D, and Phomapentenone A, from cultures of *Phoma* sp. (NRRL 25697), a mitosporic fungal colonist isolated from the stromata of *Hypoxylon* sp. were found to be active against the Gram-positive bacteria *B. subtilis* (ATCC 6051) and *Staph. aureus* (ATCC 29213) (Che *et al.*, 2002). *Trichoderma* sp., soil isolated fungi, have been widely used as antagonistic fungal agents against several fungal pathogens such as *F. moniliforme, A. flavus, Pythium ultimum* and *Rhizoctonia solani.* They are also used for bioremediation purposes (Schirmböck *et al.*, 1994). The important metabolites are linear, amphipathic polypeptides, namely, peptaibols and peptaibiotics (Szekeres *et al.*, 2005).

Chaetomium cupreum, Chaetomium globosum, Trichoderma harzianum, Trichoderma hamatum and Penicillium chrysogenum extracts inhibited the growth of the grape pathogen *Colletotrichum gloeosporioides* (Soytong *et al.*, 2005). Fang and Tsao (1995) reported that *Penicillium funiculosum* could inhibit the growth of *Phytophthora cinnamoni, Phytophthora parasitica* and *Phytophthora citrophthora* (root rot of *Azalea* and orange). Metabolites such as clavacin and fumigacin from *Aspergillus clavatus* and *A. fumigatus* respectively inhibited the bacterial pathogens *Salmonella schottmuelleri, S. choleraesuis, Bacillus megatherium, B. cereus, B. subtilis* and *Staph. aureus* (Waksman *et al.*, 1942). *Muscodor albus* is a recently described endophytic fungus obtained from small limbs of *Cinnamomum zeylanicum*, which can kill a broad range of plant- and human-pathogenic fungi and bacteria (Strobel *et al.*, 2001). Gai *et al.* (2007) isolated Fusarielin E from *Fusarium* sp. (strain 05JANF165) which was bioactive against several pathogens. The marine-derived fungus *Ampelomyces* sp. has also been proven to have antimicrobial activity (Kwong *et al.*, 2006).

16.6   Plant antimicrobials

The use of herbal antimicrobials has a long history. Archaeologists discovered flower fragments from several different medicinal plants in

Neanderthal tombs in Iraq dating back some 60000 years, while various herbal remedies have been used in China for more than 8000 years (Shanidar, 1971; Emboden, 1979). Scientists have speculated that the birch fungus found in the stomach of a 5300-year-old mummified human discovered in 1991 was being used as a drug, possibly as a treatment for intestinal parasites (Capasso, 1998). Herbal medicine use reached a peak in United States at the end of the 19th and start of the 20th century, until the Food and Drugs Act was passed in 1906 (Tyler and Lydia, 1995). Herbal drug and phytoconstituents provide both safety and efficiency, with minimal or no side effects when compared to synthetic drugs. In the modern era many people believe that plants and phytoconstituents are preferable to allopathic drugs for the treatment of diseases (Sen *et al.*, 2009). Medicinal plants have rich, potent and powerful sources of antimicrobial compounds such as phenolics, alkaloids, flavanoids, terpenes, essential oils and lectins from various parts including the root, stem, bark, leafs, flower, fruit, seeds, gum and twigs. A wide range of medicinal plants are currently used to treat various bacterial, viral and fungal diseases in humans, animals and aquatic fish and shellfish species. Medicinal plants can thus be described as 'nature's gift' thanks to their versatile characteristics such as easy availability, low cost and biodegradability, and the absence of side effects.

16.6.1   Medicinal plants as antibiotics for the future

Medicinal plants synthesize antimicrobial compounds as part of their defence against invasion by microbial pathogens. It is estimated that almost 50% of synthetic medicines are derived from or patterned after phytochemicals (Canadian Pharmaceutical Association, 1988). In the medicinal plant family secondary metabolites such as alkaloids, phenolics and other compounds have contributed the largest number of antimicrobial drugs in the pharmacological industry. The safer, biodegradable plant-derived compounds offer a promising solution to the problem of resistant microbes (Citarasu, 2010).

16.6.2   Plant active compounds involved in microbial control

Plants produce primary and secondary metabolites with various functions. The primary metabolites – amino acids, simple sugars, nucleic acids and lipids – are compounds that are essential for their cellular processes. Secondary metabolites are compounds such as phenols, alkaloids, terpenoids and flavonoids, which are produced in response to stress.

Phenolics and phenolic acids
Phenolics make up a class of chemical compounds consisting of a hydroxyl group (–OH) attached to an aromatic hydrocarbon group; they serve as a defence against attack by microorganisms. Phenolic acids are widely distributed in plants and make up a diverse group that includes the widely

distributed hydroxybenzoic and hydroxycinnamic acids. Polyphenols, which can form heavy soluble complexes with proteins, may bind to bacterial adhesions thereby disturbing the availability of receptor on the cell surface (Perumalsamy and Gopalakrishnakone, 2008). Phenolic compounds are essential for the growth and reproduction of plants, and are produced as a response in order to defend injured plants against pathogens. Some of the simplest bioactive phytochemicals consist of a single substituted phenolic ring which is effective against viruses (Wild, 1994), bacteria (Brantner *et al.*, 1996) and fungi. Phenolic acids are very good antioxidants and reduce the formation of cancer-promoting nitrosamines from dietary nitrates and nitrites (Duke, 1985).

Quinones
Quinones are coloured aromatic rings with two ketone substitutions, having the molecular formula of $C_6H_4O_2$. They are ubiquitous in nature and are characteristically highly reactive. Pharmacologically, quinine is toxic to many bacteria, yeast, fungi and plasmodia. It also has antipyretic (fever-reducing), analgesic (pain-relieving) and local anaesthetic properties. Quinine concentrates in the red blood cells, and is thought to interfere with the protein and glucose synthesis of the malaria parasite (Woodward and Doering, 1944). The mode of antimicrobial action for quinones may be related to their ability to inactivate microbial adhesions, enzymes, cell envelope transport proteins and so on (Ya *et al.*, 1988).

Alkaloids
Alkaloids are a diverse group of heterocyclic nitrogen compounds with a bitter taste. The name derives from the word alkaline; originally, the term was used to describe any nitrogen-containing base. They are found in high concentrations in plants, and in fungi and marine organisms, and are pharmacologically important in humans and other animals. They have been found to have good antimicrobial effects against bacterial pathogens and protozoan parasites (Ghoshal *et al.*, 1996). Berberine is a particularly important member of the alkaloid group in this respect. Solamargine, a glycoalkaloid from the berries of *Solanum khasianum*, and other alkaloids, may be useful against HIV infection (McMahon *et al.*, 1995) as well as intestinal infections associated with AIDS. The mechanism of action of highly aromatic planar quaternary alkaloids such as berberine and harmane (Hopp *et al.*, 1976) is attributed to their ability to intercalate with DNA (Phillipson and O'Neill, 1987).

Tannins
Tannins are a group of polymeric phenolic substances with a molecular weight of between 500 and 3000 (Haslam, 1996). They are found in almost every plant part such as bark, wood, leaves, fruits and roots (Scalbert, 1991). They may be formed by condensations of flavan derivatives which have

been transported to the woody tissues of plants or by polymerization of quinone units (Geissman, 1963). Macrocyclic structures of bioactive ellagiannins (gluconic acid core) and oligomeric ellagitannins have been found in species of Myrtaceae and Elaeagnaceae; these also possess antibacterial activity against *Helicobacter pylori* (Yoshida *et al.*, 2000). Their antimicrobial activity is the result of the inactivation of microbial adhesins, enzymes such as protease activity, and cell envelope transport proteins. They are also toxic to filamentous fungi, yeasts and bacteria (Scalbert, 1991) and inhibit viral reverse transcriptase. They also stimulate phagocytic cells, host-mediated tumor activity, and a wide range of anti-infective actions in humans (Haslam, 1996).

Flavones, flavonoids and flavonols
Flavones are phenolic structures containing one carbonyl group; the addition of a 3-hydroxyl group yields a flavonol (Fessenden and Fessenden, 1982). Flavonoids are also hydroxylated phenolic substances but occur as a C_6–C_3 unit linked to an aromatic ring. Plant-derived flavones, flavonol and flavonoids have been found to be effective antimicrobial substances against a number of microorganisms. Their activity is probably due to their ability to complex with extracellular and soluble proteins and with bacterial cell walls (Tsuchiya *et al.*, 1996). Catechin, a flavonoid compound, inhibited *in vitro V. cholerae* O1 (Borris, 1996), *Streptococcus mutans* (Batista *et al.*, 1994), *Shigella* (Vijaya *et al.*, 1995), and other bacteria and microorganisms. Flavone and flavonoid derivatives such as swertifrancheside, glycyrrhizin, chrysin, quercetin, naringin and hesperetin exhibited inhibitory activity against various human viruses. A further important point is that flavonoids with no hydroxyl groups on their b-rings can be more active against microorganisms than those with the 2OH groups (Chabot *et al.*, 1992).

Terpenoids and essential oils
Terpenoids, sometimes referred to as isoprenoids, are a large and diverse class of naturally occurring organic chemicals similar to terpenes, derived from five-carbon isoprene units assembled and modified in thousands of ways. Most are multicyclic structures which differ from one another not only in functional groups, but also in their basic carbon skeletons (Yermakov *et al.*, 2010). They can be classified as monoterpenoids, sesquiterpenoids, diterpenoids, desterterpenoids, triterpenoids, tetraterpenoids and polyterpenoids. They are available in many herbal plants and are used in a number of pharmaceutical applications including as antimicrobial agents. In many plant species diterpenes and sesquiterpenes act as phytoalexins and are involved in the defence mechanism against fungal and bacterial pathogens. They are active against bacteria (Ahmed *et al.*, 1993), fungi, viruses and protozoa (Vishwakarma, 1990). The mechanism of action of terpenes is not fully understood but is speculated to involve membrane disruption by the lipophilic compounds (Mendoza *et al.*, 1997).

Lectins and polypeptides

Plant lectins have been found in many botanical groups including mono- and dicotyledons, moulds and lichens, but have most frequently been detected in Leguminoseae and Euphorbiaceae. They are glycoproteins with a molecular weight of 60 kDa–100 kDa, and are able to agglutinate erythrocytes *in vitro*. There are over 400 000 estimated binding sites for kidney bean agglutinin on the surface of each erythrocyte. Lectins are found in most types of beans, including soybeans. They play an important role in the defence mechanisms of plants against the attack of microorganisms, pests and insects; they have a specific interaction with certain carbohydrates. Lectins may bind with free sugar or with sugar residues of polysaccharides, glycoproteins or glycolipids which can be free or bound (Barondes, 1981), leading to antimicrobial action.

Polypeptides are positively charged and contain disulphide bonds (Zhang and Lewis, 1997) with antimicrobial activity (Balls *et al.*, 1942). The natural peptides that occur most widely are dipetideanhydrides, cyclic oligo- and polypeptides, depsipeptides, large peptides, large modified peptides and glycopeptides. They are able to control various Gram-positive and Gram-negative bacteria and viruses (Sewald & Jakubke, 2002). Their mechanism of action may be the formation of ion channels in the microbial membrane (Zhang and Lewis, 1997) or competitive inhibition of adhesion of microbial proteins to the polysaccharide receptors in the host (Sharon and Ofek, 1986).

16.7   Possible mode of action of herbal antimicrobials

Plant antimicrobial compounds generally inhibit nucleic acid synthesis leading to protein synthesis blocking, damage DNA and the cell membrane, arrest cell wall synthesis, and inactivate microbial adhesions. Antibacterial herbal compounds such as phenolics, alkaloids, tannins and flavanoids may lyse the cell wall, block protein synthesis and DNA synthesis, inhibit enzyme secretions and interfere with the signalling mechanism of the quorum sensing pathway of selected bacteria such as *Vibrio* sp. Antiviral herbal compounds may block the transcription of the virus to reduce replication in the host cells and enhance non-specific immunity. They also act as immunostimulants to the host immune system. In fungal species, the activity of the antimicrobial herbal compound involves cell wall lysis, altering the permeability, affecting the metabolism and RNA and protein synthesis.

16.7.1   Antimicrobial herbal compounds as immunostimulants

Antibacterial and antiviral herbal compounds are able to not only control microbial pathogens, but also boost the immune system in fish and shrimps against pathogenic infections. An immunostimulant is a chemical, drug, stressor or action that enhances the defence mechanisms or immune response (Anderson, 1992), thus rendering the animal more resistant to diseases. In cases where disease outbreaks are cyclical and can be predicted, immunostimulants may be used in anticipation of events to elevate the nonspecific defence mechanism, and thus prevent losses from diseases. It has been proven that immunostimulant herbal extracts such as *Cynodon dactylon, Phyllanthus niruri, Tridax procumbens, Zingiber officinalis, Ocimum sanctum, W. somnifera* and *Myristica fragrans* have improved the immune system in grouper *E. tauvina*, while also controlling *V. harveyi* (Sivaram *et al.*, 2004; Punitha *et al.*, 2008). Similarly, *C. dactylon, A. marmelos, T. cordifolia, Picrorhiza kurooa* and *Eclipta alba* improved the immune system in the shrimp *P. monodon* and reduced the WSSV load in the haemolymph as well as body tissue (Citarasu *et al.*, 2006).

16.7.2   Synergistic effect of antimicrobials

Use of a combination of antimicrobial herbal extracts provides enhanced effects against microbial pathogens when compared to the delivery of single herbal extracts. Treatment with poly-herbal formulations may confer synergistic, potentiative and agonistic/antagonistic pharmacological activity (Ebong *et al.*, 2008). Synergism is generally dependent on the number of extracts, the dose and the various active compound combinations such as phenolics, flavonoids, alkaloids, terpenoids and tannins. The correct doses and formulas ensure the safety and efficacy of the antimicrobial effect and in multi-herb formulas the ratio of each herb must be clearly specified. Most of our studies (Citarasu *et al.*, 2002, 2003, 2006; Punitha *et al.*, 2008) have shown that the application of a mixture of antimicrobial herbal extracts against pathogenic bacteria and WSSV is extremely beneficial, and that 800 mg/kg diets are the optimal dose. The addition of a mixture of Chinese herbs (*Rheum officinale, Andrographis paniculata, Isatis indigotica, Lonicera japonica*) to the feed of crucian carp resulted in increased phagocytosis of the white blood cells (Chen *et al.*, 2003). Ponpornpisit *et al.* (2001) fed a Chinese herbal mix, known as C-UPIII, to guppies (*Lebistes reticulata*) and observed an improved survival rate in fish infected with *Tetrahymena pyriformis*. Yuan *et al.* (2007) fed carp diets containing a mixture of *A. membranaceus, Polygonum multiflorum, Isatis tinctoria* and *Glycyrrhiza glabra* and observed significantly increased phagocytosis, respiratory burst activity and levels of total protein in the serum.

16.8   Routes of administration of antimicrobial herbal extracts to fish and shrimps

The easiest methods by which to administer antimicrobial herbal extracts are:

- Orally. The extracts can be added to feeds for fish and shrimp at the late larval, juvenile and adult stages.
- Bioenrichement or bioencapsulation through *Artemia nauplii* and rotifer.
- Direct injection of fish.

16.8.1   Administration through artificial feed

Delivery through artificial diet is one of the more conventional and easiest methods for successful treatment of bacterial infection. Herbal extracts at different percentages ranging from 100 to 1000 mg/g are incorporated into the artificial diets. The extracts are added along with fish oil such as cod liver oil after the feed ingredients have been cooked and allowed to cool. The mixture will be cold extruded, cut into pellets, air dried and stored in airtight bag or at 4 °C. In our experience 800 mg herbal extracts was the optimum level for best results (Citarasu *et al.*, 2002, 2006). The fish/shrimp will be fed three times a day with feed containing these antimicrobial extracts; virulent bacterial/viral pathogen will then be introduced at different intervals such as 20, 30, 40 and 60 days. The main advantage of this method is that it does not stress the fish.

16.8.2   Administration through bioenrichment

In this method, the antimicrobial extracts are incorporated into live feed organisms such as *Artemia* or rotifers either directly or indirectly to ensure the complete acceptability of the incorporated compounds. This is the simplest and most effective method of delivering antimicrobial extracts to the early fish and shrimp post-larvae of PL 1–20. The herbal extracts are emulsified by mixing herbal extracts, egg yolk and cod liver oil in a ratio of 1:1:1. The enrichment schedule is 200 mg of emulsified diet mixed with 100 ml water and the nauplii density is 100/ml. After enrichment at several intervals they will be ready to use for feeding.

16.8.3   Direct injection

Sometimes, the herbal extracts will be injected to the adult fish/shrimps by intramuscular injection. This is a crude method, mainly used for the purposes of research involving bacterial and viral infection, and a lower concentration, such as 100 μg, is normally used. The extracts are mixed with saline water, especially phosphate buffered saline (PBS) and injected into the fish/shrimps. The disadvantage of this method is that it creates stress.

Tables 16.1–16.3 summarise the impact of herbal antimicrobial active principles on pathogens of aquatic animals.

Table 16.1 List of highly influenced herbal natural antibacterial extracts on aquatic bacterial pathogens

Botanical name	Family	Useful parts	Major compounds	Reference
Andrographis paniculata	Acanthaceae	Whole plant	Andrographolid	Citarasu et al. (2003)
Solanum trilobatum	Solanaceae	Fruits and root	Solanine	
Psoralea corylifolia	Papilionaceae	Seeds	Psoralen	
Aegle marmelos	Rutaceae	Whole plant	Aegelin, aegelinine	Citarasu et al. (2003)
Solanum surattense	Solanaceae	Fruits and root	Solanine	
Terminalia bellirica	Combretaceae	Fruits	α-Glucosidase	Sivaram et al. (2004)
Myristica fragrans	Myristicaceae	Dried seeds	Beta-pinene borneol	
Leucus aspera	Labiatae	Whole plant	Saponin	Immanuel et al. (2004)
Cynodon dactycon	Gramineae	Leaf and root stalk	Cyanodin	Immanuel et al. (2009)
Terminalia chebula	Combretaceae	Fruits	α-Glucosidase	Rani (1999)
Tephrosia purpurea	Papilionaceae	Leaves and root	Isoflavone	Rani (1999)
Acalypha indica	Euphorbiaceae	Whole plant	Acalyphine	Citarasu et al. (1999)
Tridax procumbens	Asteraceae	Whole plant	β-Sitosterol	Punitha et al. (2008)
Piper longum	Piperaceae	Dried seeds	Piperine	Punitha et al. (2008)
Adathoda vasica	Acanthaceae	Whole plant	Vasicine and vasicinol	
Murraya koeniji	Rutaceae	Leaves	Beta-Ocimene 2,4-heptadienal	Velmurugan et al. (2010)
Ocimum basilicum	Labiatae	Whole plant	Cedrenol, β-ocimene	
Quercus infectoria	Cupuliferae	Galls and bark	Beta-D-Glucogallin gallic-acid Quercetin etc.	
Stellaria aquatica	Caryophyllaceae	Whole plant	Phenolics	
Oenothera biennis	Onagraceae	Whole plant	Quercetin; lignin	
Lonicera japonica	Caprifoliaceae	Whole plant	Flavanoids	
Impatiens biflora	Balsaminaceae	Whole plant	Dinaphthofuran, α-parinaric acid	Shangliang, et al. (1990)

Table 16.2 List of highly influenced herbal natural antiviral extracts on important aquatic viruses

Botanical name	Family	Useful parts	Major compounds	Reference
Cyanodon dactylon	Gramineae	Whole plant	Cyanodin	Citarasu *et al.* (2006)
Aegle marmelos	Rutaceae	Whole plant	Aegelin, Aegelinine	
Tinospora cordifolia	Menispermaceae	Whole plant	Berberine	
Picrorhiza kurroa	Scrophulariaceae	Whole plant	Piccrorhizin	
Eclipta alba	Asteraceae	Whole plant	Ecliptine	
Clinacanthus nutans	Acanthaceae	Whole plant	Phaeophytins	Direkbusarakom *et al.* (1998)
Phyllanthus sp.	Euphorbiaceae	Whole plant	Phyllanthin	Direkbusarakom *et al.* (1995)
Tinospora crispa	Menispermaceae	Whole plant	Tinocrisposid	
Momordica charantia	Cucurbitaceae	Whole plant	Cucurbitins, cucurbitacins,	
Psidium guajava	Myrtaceae	Bark, root	3β-*p-E*-Coumaroyloxy-2α-methoxyurs-12-en-28-oic acid	Direkbusarakom (2004)
Azadirachta indica	Meliaceae	Whole plant	Azadirachtinin	
Cassia fistula	Fabaceae	Seeds, fruits & bark	Phenolics	
Catharanthus roseus	Apocynaceae	Whole plant	Phenolics, alkaloids	
Curcuma longa	Zingiberaceae	Root	Curcumin, Curcumenol	
Melia azedarach	Meliaceae	Whole plant	Meliacine	Balasubramanian *et al.* (2007)
Ocimum americanum	Lamiaceae	Whole plant	Betulinic acid, Terpenes	
Solanum nigrum	Solanaceae	Fruits	β-Solamargine and solasonine	
Phyllanthus emblica	Phyllanthaceae	Fruits, seeds	Phyllanthin	
Tylophora indica	Asclepiadaceae	Leaf & flowers	Tylophorine	

Table 16.3 List of highly influenced herbal natural antifungal extracts on important aquatic fungi

Botanical name	Family	Useful parts	Major compounds	Reference
Azadirachta siamensis	Meliaceae	Whole plant	Azadirachtinin	Campbell *et al.* (2001)
Datura metel	Solanaceae	Leaf, Flower	Hyoscyamine, hyoscine and meteloidine	Dabur (2004)
Melaleuca alternifolia	Myrtaceae	Whole plant	oils	Campbell *et al.* (2001)
Terminalia catappa	Combretaceae	Bark, leaves and seed	β-Sitosterol	Chitmanat *et al.* (2005)
Tamarix dioica	Tamaricaceae	Leaf	Flavones – tamaridone	
Rhazya stricta	Apocynaceae	Root, stems, leaves and flowers	Alkaloids	Khan *et al.* (2004)

16.9   Conclusions

Natural alternative antimicrobials such as medicinal plants, marine invertebrates, seaweeds, micro-algae, antagonistic bacteria, actinobacteria and
fungi are highly useful in the control of pathogenic bacteria, viruses and
fungi involved in pathogenesis in aquaculture industry without any side
effects. They offer a suitable natural alternative to synthetic drugs and
antibiotics thanks to the antimicrobial activity of alkaloids, flavanoids, pigments, phenolics, terpenoids, starch, steroids and essential oils (medicinal
plants); lectin, antimicrobial peptides and other compounds (marine fish
and invertebrates); sterols, dibutanoids, peptides, sulphated polysaccharides,
terpenoids, phlorotannins, acrylic acid, phenolic compounds, halogenated
ketones, polysulphides and fatty acids (seaweeds); polyketides, amides, alkaloids and peptides (micro-algae); and various secondary metabolites, exopolysaccharides, antimicrobial peptides (microbes). The major advantages
of these natural alternative antimicrobials in the aquaculture industry are
(a) the ability to prevent emergence of resistant strains of bacteria, viruses
and fungi and to reduce the risks of non-target effects; (b) minimal or no
bio-magnification effect in the cultured shrimp/fish species, and reduction
in the expenses normally incurred by using commercial antimicrobial
compounds; (c) avoidance of any impact on human health caused by consumption of products from antibiotic-treated animals; (d) prevention of

consignment rejection due to the incorporation of antibiotics; (e) improvement in the survival and production rate and in the immunological response of the host system and (f) potential for reuse of shrimp/fish farms abandoned due to bacterial/viral diseases, improving the economic status of the shrimp farmers who incur substantial losses through these diseases. This practice also reduces the side effects observed with synthetic compounds. Hence, natural alternative antimicrobials prove to be very effective in aquaculture operations.

16.10   References

ADIGÜZEL A, MEDINE G, MERYEM B, HATICE U T C, FIKRETTIN A and ÜSA K (2005), 'Antimicrobial effects of *Ocimum basilicum* (Labiatae) extract', *Turk J Biol* **29**, 155–160.

AGUIRRE-GUZMÁN G, MEIJA R H and ASCENCIO F (2004), 'A review of extracellular virulence product of *Vibrio* species important in disease of cultivated shrimp', *Aquaculture Res* **35**, 1395–1404.

AHMED A A, MAHMOUD A A, WILLIAMS H J, SCOTT A I, REIBENSPIES J H and MABRY T J (1993), 'New sesquiterpene a-methylene lactones from the Egyptian plant *Jasonia candicans*', *J Nat Prod*, **56**, 1276–1280.

ANDERSON D P (1992), 'Immunostimulants, adjuvants, and vaccine carriers in fish: applications to aquaculture', *Annu Rev Fish Dis*, **2**, 281–307.

ARENA A, MAUGERI T L, PAVONE B, IANNELLO D, GUGLIANDOLO C and BISIGNANO G (2006), 'Antiviral and immunomodulatory effect of a novel exopolysaccharide from a marine thermotolerant *Bacillus licheniformis*', *Int Immunopharmacol*, **6**, 8–13.

ARENA A, GUGLIANDOLO C, STASSI G, PAVONE B, IANNELLO D, BISIGNANO G and MAUGERI T L (2009), 'An exopolysaccharide produced by *Geobacillus thermodenitrificans* strain B3-72: antiviral activity on immunocompetent cells', *Immunol Lett*, **123**, 132–137.

ARULJOSE P, SATHEEJASANTHI V and SOLOMON J R D (2010), '*In vitro* antimicrobial potential and growth characteristics of *Nocardiopsis* Sp.JAJ16 Isolated from crystallizer pond', *Int J Current Res*, **3**, 024–026.

ASTHANA R, DEEPALI K, TRIPATHI M K, SRIVASTAVA A, SINGH A P, SINGH S P, NATH G, SRIVASTAVA R and SRIVASTAVA B S (2009), 'Isolation and identification of a new antibacterial entity from the Antarctic cyanobacterium Nostoc CCC 537', *J Appl Phycol*, **21**, 81–88.

ATTA H M, DABOUR S M and DESOUKEY S G (2009), 'Sparsomycin antibiotic production by *Streptomyces* sp. AZ-NIOFD1: taxonomy, fermentation, purification and biological activities', *American-Eurasian J Agric Environ Sci*, **5**(3), 368–377.

AUSTIN B and AUSTIN D A (2007), *Bacterial Fish Pathogens – Diseases of Farmed and Wild Fish*, Chichester, Springer, in association with Praxis Publishing, 4–14.

AUSTIN B, BAUDET E and STOBIE M (1992), 'Inhibition of bacterial fish pathogens by *Tetraselmis suecica*', *J Fish Dis*, **15**, 55–61.

BAKER P W, KENNEDY J, DOBSON A D W and MARCHESI J R (2009), 'Phylogenetic diversity and antimicrobial activities of fungi associated with *Haliclona simulans* isolated from Irish coastal waters', *Mar Biotechnol*, **11**, 540–547.

BALASUBRAMANIAN G, SARATHI M, RAJESH KUMAR S and SAHUL HAMEED A S (2007), 'Screening the antiviral activity of Indian medicinal plants against white spot syndrome virus in shrimp', *Aquaculture*, **263**, 15–19.

BALCAZAR J L and LUNA-ROJAS T (2007), 'Inhibitory activity of probiotic *Bacillus subtilis* UTM 126 against *Vibrio* species confers protection against vibriosis in juvenile shrimp (*Litopenaeus vanamei*)', *Current Microbiol*, **55**, 409–412.

BALLS A K, HALE W S and HARRIS T H (1942), 'A crystalline protein obtained from a lipoprotein of wheat flour', *Cereal Chem*, **19**, 279–288.

BARONDES S H (1981), 'Lectins: their multiple endogenous cellular functions', *Annu Rev Biochem*, **50**, 207–231.

BATISTA O, DUARTE A, NASCIMENTO J and SIMONES M F (1994), 'Structure and antimicrobial activity of diterpenes from the roots of *Plectranthus hereroensis*', *J Nat Prod*, **57**, 858–861.

BERDY J (2005), 'Bioactive microbial metabolites. A personal view', *J Antibiot (Tokyo)*, **58**, 1–26.

BÉRESS A, WASSERMANN O, BRUHN T, BÉRESS L, KRAISELBURD E N, GONZALEZ L V, DE MOTTA G E and CHAVEZ P I (1993), 'A new procedure for the isolation of anti-HIV compounds (polysaccharides and polyphenols) from the marine alga *Fucus vesiculosus*', *J Nat Pros*, **56**, 478–488.

BERNAN V S, GREENSTEIN M and MAIESE W M (1997), 'Marine microorganisms as a source of new natural products', *Adv Appl Microbiol*, **43**, 57–90.

BEWLEY C A, GUSTAFSON K R, BOYD M R, COVELL D G, BAX A, CLORE G M and GRONENBORN A M (1998), 'Solution structure of cyanovirin-N, a potent HIV-inactivating protein', *Nat Struct Biol*, **5**, 571–578.

BHADURY P, MOHAMMAD B T and WRIGHT P C (2006), 'The current status of natural products from marine fungi and their potential as anti-infective agents', *J Ind Microbiol Biotechnol*, **33**, 325–337.

BHATNAGAR I and KIM S K (2010), 'Immense essence of excellence: marine microbial bioactive compounds', *Mar Drugs*, **8**, 2673–2701.

BIZANI D and BRANDELLI A (2002), 'Characterization of a bacteriocin produced by a newly isolated *Bacillus* sp. Strain 8A', *J Appl Microbiol*, **93**, 512–519.

BONAMI J R and LIGHTNER D V (1991), Unclassified viruses of crustacea, p. 597–622. In J.R. Adams and J.R. Bonami (eds.) *Atlas of Invertebrate Viruses*. CRC Press, Boca Raton.

BORRIS R P (1996), 'Natural products research: perspectives from a major pharmaceutical company', *J Ethnopharmacol*, **51**, 29–38.

BOYD M R, GUSTAFSON K, MCMAHON J and SHOEMAKER R (1996), 'Discovery of cyanovirin-N, a novel HIV-inactivating protein from *Nostoc ellipsosporum* that targets viral gp120', *Int Conf, AIDS*, **11**, 71.

BRANTNER A, MALES Z, PEPELJNJAK S and ANTOLIC A (1996), 'Antimicrobial activity of *Paliurus spina-christi* mill', *J Ethnopharmacol*, **52**, 119–122.

BREDHOLT H, FJOVIK E, JOHNSEN G and ZOTCHEV S B (2008), 'Actinomycetes from sediments in the Trondheim Fjord, Norway: diversity and biological activity', *Mar Drugs*, **6**(1), 12–24.

BROWN J (1989), 'Antibiotics: their use and abuse in aquaculture, *World Aquac*, **20**(2), 34–43.

BYUN H G, ZHANG H, MOCHIZUKI M, ADACHI K, SHIZURI Y, LEE W J and KIM S K (2003), 'Novel antifungal diketopiperazine from marine fungus', *J Antibiot (Tokyo)*, **56**, 102–106.

CAMPBELL R E, LILLEY J H, PANYAWACHIRA V and KANCHANAKHAN S (2001), '*In vitro* screening of novel treatments for *Aphanomyces invadans*', *Aquac Res*, **32**(3), 223–233.

CANADIAN PHARMACEUTICAL ASSOCIATION (CPA) (1988), *Self medication*, Ottawa, Canada.

CAPASSO L (1998), '5300 years ago the Ice Man used natural laxatives and antibiotics', *Lancet*, **352**, 1864.

CAPON R J (2001), 'Marine bioprospecting – Trawling for treasure and pleasure', *European J Org Chem*, 633–645.

CASTILLO C S D, WAHID M I, TAKESHI Y and TAIZO S (2008), 'Isolation and inhibitory effect of anti-*Vibrio* substances from *Pseudoalteromonas* sp. A1-J11 isolated from the coastal sea water of Kagoshima Bay', *Fisheries Sci*, **74**, 1, 174–179.

CHABOT S, BEL-RHLID R, CHENEVERT R and PICHE Y (1992), 'Hyphal growth promotion *in vitro* of the VA mycorrhizal fungus, *Gigaspora margarita* Becker and Hall, by the activity of structurally specific flavonoid compounds under CO_2-enriched conditions', *New Phytol*, **122**, 461–467.

CHATTOPADHYAY T and CHATTERJEE B P (1993), 'A low molecular weight lectin from the edible crab *Scylla serrata* hemolymph: purification and partial characterization', *Biochem Arch*, **9**, 65–72.

CHE Y, GLOER J B and WICKLOW D T (2002), 'Phomadecalins A-D and phomapentenone A: new bioactive metabolites from *Phoma* sp. NRRL 25697, a fungal colonist of *Hypoxylon stromata*', *J Nat Prod*, **65**, 399–402.

CHEN X, WU Z, YIN J and LI L (2003), 'Effects of four species of herbs on immune function of *Carassius auratus* gibelio', *J Fish Sci China*, **10**, 36–40.

CHITMANAT C, TONGDONMUAN K, KHANOM P, PACHONTIS P and NUNSONG W (2005), 'Antiparasitic, antibacterial, and antifungal activities derived from a terminalia catappa solution against some tilapia (*Oreochromis niloticus*) pathogens', *Acta Hortic*, **678**, 179–182.

CHOUDHURY S, SREE A, MUKHERJEE S C, PATTNAIK P and BAPUJI M (2005), '*In Vitro* antibacterial activity of extracts of selected marine algae and mangroves against fish pathogens', *Asian Fisheries Sci*, **18**, 285–294.

CHYTHANYA R, KARUNASAGAR I and KARUNASAGAR I (2002), 'Inhibition of shrimp pathogenic vibrios by a marine *Pseudomonas* I-2 strain', *Aquaculture*, **208**, 1–2, 1–10.

CITARASU T (2010), 'Herbal biomedicines: a new opportunity for aquaculture industry', *Aquacult Int*, **18**, 403–414.

CITARASU T, JAYARANI T V, BABU M M and MARIAN M P (1999), 'Use of herbal biomedicinal products in aquaculture of shrimp', *Aqua-Terr Annual Symposium*, School of Biological Sciences, M. K. University, Madurai.

CITARASU T, SEKAR R R, BABU M M and MARIAN M P (2002), 'Developing *Artemia* enriched herbal diet for producing quality larvae in *Penaeus monodon*', *Asian Fish Sci*, **15**, 21–32.

CITARASU T, RAJAJEYASEKAR R, VENKETRAMALINGAM K, DHANDAPANI P S and MARIAN M P (2003), 'Effect of wood apple *Aegle marmelos*, Correa (Dicotyledons, Sapindales, Rutaceae) extract as an antibacterial agent on pathogens infecting prawn (*Penaeus indicus*) lariviculture', *Indian J Marine Sci*, **32**(2), 156–161.

CITARASU T, SIVARAM V, IMMANUEL G, ROUT N and MURUGAN V (2006), 'Influence of selected Indian immunostimulant herbs against white spot syndrome virus (WSSV) infection in black tiger shrimp, *Penaeus monodon* with reference to haematological, biochemical and immunological changes', *Fish Shellfish Immunol*, **21**, 372–384.

COMMITTEE ON DRUG USE IN FOOD ANIMALS (1999), *The Use of Drugs in Food Animals: Benefits and Risks. Based on reports commissioned by the Panel on Animal Health, Food Safety, and Public Health (a joint activity of the [USA] National Research Council and the [USA] Institute of Medicine.* CABI Publishing, Wallingford, UK.

DABUR R (2004), 'A novel antifungal pyrrole derivative from Datura metel leaves', *Pharmazie*, **59**, 568–570.

DAS B K, PRADHAN J, PATTNAIK P, SAMANTARAY B R and SAMAL S K (2005), 'Production of antibacterials from the freshwater alga *Euglena viridis* (Ehren)', *World J Microbiol Biotechnol*, **21**, 45–50.

DECANO M M S, DEMULE M C Z, DE CAIRE C Z and DE HALPERIN D R (1990), 'Inhibition of *Candida albicans* and *Staphylococcus aureus* by phenolic compounds from the terrestrial cyanobacterium *Nostoc muscorum*', *J Appl Phycol*, **2**, 79–82.

DEFER D, BOURGOUGNON N and FLEURY Y (2009), 'Screening for antibacterial and antiviral activities in three bivalve and two gastropod marine mollusks', *Aquaculture*, **293**, 1–7.

DESBOIS A P, MEARNS-SPRAGG A and SMITH V J A (2009), 'Fatty acid from the diatom *Phaeodactylum tricornutum* is antibacterial against diverse bacteria including multiresistant *Staphylococcus aureus* (MRSA)', *Mar Biotechnol*, **11**, 45–52.

DIETERA A, HAMM A, FIEDLER H P, GOODFELLOW M, MULLER W E, BRUN R and BRINGMANN G (2003), 'Pyrocoll, an antibiotic, antiparasitic and antitumor compound produced by a novel alkaliphilic *Streptomyces* strain', *J Antibiot*, **56**, 639–646.

DIREKBUSARAKOM S (2004), 'Application of medicinal herbs to aquaculture in Asia', *Walailak J Sci Tech*, **1**(1), 7–14.

DIREKBUSARAKOM S, HERUNSALEE A, BOONYARATPALIN S, DANAYADOL Y and AEKPANITH-ANPONG U (1995), Effect of *Phyllanthus* spp. against yellow-head baculovirus infection in black tiger shrimp, *Penaeus monodon*. In: Shariff M, Arthur JR, Sub-asinghe RP (eds.) *Diseases in Asian Aquaculture* II. Fish Health Section, Asian Fisheries Society, Manila, pp. 85–92.

DIREKBUSARAKOM S, HERUNSALEE A, YOSHIMIZU M and EZURA Y (1996), 'Antiviral activity of several Thai traditional herb extracts against fish pathogenic viruses', *Fish Pathol*, **31**(4), 209–213.

DIREKBUSARAKOM S, RUANGPAN L, EZURA Y and YOSHIMIZU M (1998), 'Protective efficacy of *Clinacanthus nutans* on yellow-head disease in black tiger shrimp (*Penaeus monodon*)', *Fish Pathol*, **33**(4), 410–404.

DONIO M B S (2009), *Development and characterization of water probiotics from solar salt works, coconut retting and tannery wastes against pathogenic vibrios in shrimp aquaculture industry*, M. Phil dissertation, Manonmaniam Sundaranar University, India.

DUBBER D and HARDER T (2008), 'Extracts of *Ceramium rubrum*, *Mastocarpus stellatus* and *Laminaria digitata* inhibit growth of marine and fish pathogenic bacteria at ecologically realistic concentrations', *Aquaculture*, **274**, 196–200.

DUFF D C B, BRUCE D L and ANITA N J (1966), 'The antibacterial activity of marine planktonic algae', *Canadian J Microbiol*, **12**, 877–884.

DUFOSSE L, GALAUP P, YARON A, ARAD S M, BLANC P, MURTHY N C and RAVISHANKAR G A (2005), 'Microorganisms and microalgae as sources of pigments for food use: a scientific oddity or an industrial reality?', *Trends Food Sci Technol*, **16**, 389–406.

DUKE J A (1985), *Handbook of Medicinal Herbs*. CRC Press, Inc., Boca Raton, FL.

EBONG P E, ATANGWHO I J, EYONG E U and EGBUNG G E (2008), 'The antidiabetic efficacy of combined extracts from two continental plants: *Azadirachta indica* (A. Juss) (Neem) and *Vernonia amygdalina* (Del.) (African Bitter Leaf)', *American J Biochem Biotechol*, **4**(3), 239–244.

EL-SHEEKH M M, DAWAH A M, AZZA M, EL-RAHMAN A, EL-ADEL H M and EL-HAY R A A (2008), 'Antimicrobial activity of the cyanobacteria *Anabaena wisconsinense* and *Oscillatoria curviceps* against pathogens of fish in aquaculture', *Ann Microbiol*, **58**(3), 527–534.

EMBODEN W (1979), *Narcotic Plants*. Macmillan Publishing Co., New York.

ENNAHAR S, ASOU Y, ZENDO T, SONOMOTTOU K and ISHIZAKI A (2001), 'Biochemical and genetic evidence for production of enterocins A and B by *Enterococcus faecium* WHM 81', *Int J Food Microbiol*, **70**, 291–301.

FANG J G and TSAO P H (1995), 'Efficacy of *Penicillium funiculosum* as a biological control agent against *Phytophthora* root rot of Azalea and Citrus', *Phytopathology*, **85**, 871–878.

FAO/OIE/WHO (2006), *Antimicrobial Use in Aquaculture and Antimicrobial Resistance*: *Expert Consultation on Antimicrobial Use in Aquaculture and Antimicrobial Resistance*. Seoul, Republic of Korea, WHO Document Production Services, Geneva, Switzerland.

FARNSWORTH N R, AKERELE O, BINGEL A S, SOEJARTO D D and GUO Z (1985), 'Medicinal plants in therapy', *Bull WHO*, **63**(6), 965–981.

FAULKNER D J (2000), 'Marine pharmacology', *Antonie van Leeuwenhoek*, **77**, 135–145.

FESSENDEN R J and FESSENDEN J S (1982), *Organic Chemistry*, 2nd Edition, Willard Grant Press, Boston.

FISH S A and CODD G A (1994), 'Analysis of culture conditions controlling the yield of bioactive material produced by the thermotolerent cyanobacterium (blue green alga) *Phormidium*', *Eur J Phycol*, **29**, 261–266.

GAI Y, ZHAO L L, HU C Q, ZHANG H P and FUSARIELIN E (2007), 'A new antifungal antibiotic from *Fusarium* sp', *Chin Chem Lett*, **18**, 954–956.

GEISSMAN T A (1963), Flavonoid tannins, lignins and related compounds, In: M. Florkin and E. H. Stotz (ed.), *Pyrrole pigments, isoprenoid compounds and phenolic plant*, Elsevier, New York.

GERARD J, HADEN M T, KELLY M T and ANDERSON R J (1999), 'Loloatins – cyclic decapeptide antibiotics produced in culture by a tropical marine bacterium', *J Nat Prod*, **62**, 80–85.

GERBER P, DUTCHER J D, ADAM E V and SHERMAN J H (1958), 'Protective effect of seaweed extracts for chicken embryos infected with influenza B ormumps', *Proc Soc Exp Biol Med*, **99**, 590–593.

GHASEMI Y, YAZDI M T, SHAFIEE A, AMINI M, SHOKRAVI S and ZARRINI G (2004), 'Parsiguine, a novel antimicrobial substance from *Fischerella ambigua*', *Pharm Biol*, **42**, 318–322.

GHOSHAL S, KRISHNA PRASAD B N and LAKSHMI V (1996), 'Antiamoebic activity of *Piper longum* fruits against *Entamoeba histolytica in vitro* and *in vivo*', *J Ethnopharmacol*, **50**, 167–170.

GLOMBITZA K W (1979), Antibiotics from algae. In *Marine Algae in Pharmaceutical Science*. Waiter de Gruyter, Berlin.

GOWDA N M, GOSWAMI U and KHAN M I (2008), 'T-antigen binding lectin with antibacterial activity from marine invertebrate, sea cucumber (*Holothuria scabra*): possible involvement in differential recognition of bacteria', *J Invertebr Pathol*, **99**, 141–145.

GRAM L, MELCHIORSEN J, SPANGGAARD B, HUBER I and NIELSEN T F (1998), 'Inhibition of *Vibrio anguillarum* by *Pseudomonas fluorescens* AH2, a possible probiotic treatment of fish', *Appl Envi Microbiol*, **65**, 3, 969–973.

GRIGORAKIS K (2010), 'Ethical issues in aquaculture production', *J Agric Environ Ethics*, **23**, 345–370.

GUSTAFSON K R, ROMAN M and FENICAL W (1989), 'The macrolactins, a novel class of antiviral and cytotoxic macrolides from a deep-sea marine bacterium', *J Am Chem Soc*, **111**, 7519–7524.

GUSTAFSON K R, SOWDER R C, HENDERSON L E, CARDELLINA J H, MCMAHON J B, RAJAMANI U, PANNELL L K and BOYD M R (1997), 'Isolation, primary sequence determination, and disulfide bond structure of cyanovirin-N, an anti-HIV (human immunodeficiency virus) protein from the cyanobacterium *Nostoc ellipsosporum*', *Biochem Biophys Res Commun*, **238**(1), 223–228.

HASLAM E (1996), 'Natural polyphenols (vegetable tannins) as drugs: possible modes of action', *J Nat Prod*, **59**, 205–215.

HASUI M, MATSUDA M, OKUTANI K and SHIGETA S (1995), '*In vitro* antiviral activities of sulfated polysaccharides from a marine microalga (*Cochlodinium olykrikoides*) against human immunodeficiency virus and other enveloped viruses', *Int J Biol Macromol*, **17**, 293–297.

HAWKSWORTH D C and ROSSMAN A Y (1987), 'Where are the undescribed fungi?', *Phytopathology*, **87**, 888–891.

HAUG T, KJUUL A K, STENSVÅG K, SANDSDALEN E, and STYRVOLD O B (2002), Antibacterial activity in four marine crustacean decapods, *Fish and Shellfish Immunology*, **12**, 371–385.

HAYASHI T, HAYASHI K, MAEDA M and KOJIMA I (1996), 'Calcium spirulan, an inhibitor of enveloped virus replication, from a blue–green alga *Spirulina platensis*', *J Nat Prod*, **59**, 83–87.

HOPP K H, CUNNINGHAM L V, BROMEL M C, SCHERMEISTER L J and WAHBA KHALIL S K (1976), '*In vitro* anti trypanosomal activity of certain alkaloids against *Trypanosoma lewisi*', *Lloydia*, **39**, 375–377.

IMMANUEL G, VINCY BAI V C, PALAVESAM A and PETER MARIAN M (2004), 'Effect of butanolic extracts from terrestrial herbs and seaweeds on the survival, growth and pathogen (*Vibrio parahaemolyticus*) load on shrimp *Penaeus indicus* juveniles', *Aquaculture*, **236**, 53–65.

IMMANUEL G, UMA R P, IYAPPARAJ P, CITARASU T, PUNITHA S M J, BABU M M B and PALAVESAM A (2009), 'Effect of medicinal plant extracts on the growth, immune activity and survival of tilapia (*Oreochromis mossambicus*)', *J Fish Biol*, **74**, 1462–1475.

INGLIS V (1996), Antibacterial chemotherapy in aquaculture: review of practice, associated risks and need for action. In: *Proceedings of the Meeting on the Use of Chemicals In Aquaculture in Asia*, Philippines.

ISHIDA K, MATSUDA H, MURAKAMI M and YAMAGUCHI K (1997), 'Kawaguchipeptin B, an antibacterial cyclic undecapeptide from the cyanobacterium *Microcystis aeruginosa*', *J Nat Prod*, **60**, 724–726.

JAYAPRAKAS V and SAMBHU C (1996), 'Growth response of white prawn, *Penaeus indicus* to dietary L-carnitine', *Asian Fish Sci*, **9**, 209–219.

JORQUERA M A, RIQUELME C E, LOYOLA L A and MUÑOZ L F (1999), 'Production of bactericidal substances by a marine *Vibrio* isolated from cultures of the scallop *Argopecten purpuratus*', *Aquaculture International*, **7**, 433–448.

KAMEI Y and AOKI M (2007), 'A chlorophyll c2 analogue from the marine brown alga *Eisenia bicyclis* inactivates the infectious hematopoietic necrosis virus, a fish rhabdovirus', *Arch Virol*, **152**, 861–869.

KARUNASAGAR I, KARUNASAGAR I and UMESHA R K (2004), Microbial diseases in shrimp aquaculture. In: N. Ramiah (ed.) *Marine Microbiology: Facts and Opportunities*, National Institute of Oceanography, Goa, India, pp. 165–186.

KATIRCIOGLU H, BEYATLI Y, ASLIM B, YÜKSEKDAG Z and ATICI T (2006), 'Screening for antimicrobial agent production of some microalgae in freshwater', *Internet J Microbiol*, **2**, 2.

KELLAM S J and WALKER J M (1989), 'Antibacterial activity from marine microalgae in laboratory culture', *Br Phycol J*, **24**, 191–194.

KHAN S, KHAN G S, MEHSUD S, RAHMAN A and KHAN F (2004), 'Antifungal activity of *Tamarix dioica* – an *in vitro* study', *Gomal J Med Sci*, **2**, 2.

KLEINKAUF H, VON DHREN H, DORNAUER H and NESEMANN G (1986), *Regulation of Secondary Metabolite Formation*, Weinheim: VCH.

KOLANJINATHAN K, GANESH P and GOVINDARAJAN M (2009), 'Antibacterial activity of ethanol extracts of seaweeds against fish bacterial pathogens', *European Rev Medi Pharm Sci*, **13**, 173–177.

KWONG T F, MIAO L, LI X and QIAN P Y (2006), 'Novel antifouling and antimicrobial compound from a marine-derived fungus *Ampelomyces* sp', *Mar Biotechnol*, **8**, 634–640.

LAATSCH H, RENNERBERG B, HANEFELD U, KELLNER M, PUDLEINE H, HAMPRECHT G, KRAEMER H and ANKE H (1995), 'Structure–activity relationship of phenyl- and benzoylpyrroles', *Chem Pharm Bull*, **43**(4), 537–546.

LEE J B, HAYASHI K, HIRATA M, KURODA E, SUZUKI E, KUBO Y and HAYASHI T (2006), 'Antiviral sulfated polysaccharide from *Navicula directa*, a diatom collected from deep-sea water in Toyama Bay', *Biol Pharm Bull*, **29**, 2135–2139.

LEE S, NAJIAH M, WENDY W and NADIRAH M (2009), 'Chemical composition and antimicrobial activity of the essential oil of *Syzygium aromaticum* flower bud (Clove) against fish systemic bacteria isolated from aquaculture sites', *Front Agric China*, **3**(3), 332–336.

LI C, HAUG T, MOE M K, STYRVOLD O B and STENSVÅG K (2010), 'Centrocins: Isolation and characterization of novel dimeric antimicrobial peptides from the green sea urchin, *Strongylocentrotus droebachiensis*', *Dev Comp Immunol*, **34**, 9, 959–968.

LI S S and TSAI H J (2009), 'Transgenic microalgae as a non-antibiotic bactericide producer to defend against bacterial pathogen infection in the fish digestive tract', *Fish & Shellfish Immunol*, **26**, 316–325.

MA D, HU Y, WANG J, YE S and LI A (2006), 'Effects of antibacterials use in aquaculture on biogeochemical processes in marine sediment', *Sci Total Envi*, **367**, 273–277.

MAEDA M and LIAO C (1992), 'Effect of bacterial population on the growth of a prawn larvae, *Penaeus monodon*', *Bull Natl Res Inst Aquaculture*, **21**, 25–29.

MAGARVEY N A, KELLER J M, BERNAN V, DWORKIN M and SHERMAN D H (2004), 'Isolation and characterization of novel marine-derived actinomycete taxa rich in bioactive metabolites', *Appl Envi Microbiol*, **70**, 12, 7520–7529.

MANIVASAHAM S, SELVARAJ R, PURUSOTHAMAN A and SUBRAMANIAN A (1989), 'Antibacterial activity of *Nitzschia obtusata*', *Curr Sci*, **58**, 83.

MAYER A M S and HAMANN M T (2000), 'Marine pharmacology in 2000: marine compounds with antibacterial, anticoagulant, antifungal, anti-inflammatory, antimalarial, antiplatelet, antituberculosis, and antiviral activities; affecting the cardiovascular, immune, and nervous systems and other miscellaneous mechanisms of action', *Mar Biotechnol*, **6**, 37–52.

MCCUTCHEON A R, ROBERTS J E, ELLIS S M, BABIUK L A, HANCOCK R E W and TOWERS G H N (1995), 'Antiviral screening of British Columbian medicinal plants', *J Ethnopharm*, **49**, 101–110.

MCMAHON J B, CURRENS M J, GULAKOWSKI R J, BUCKHEIT R W J, LACKMAN-SMITH C, HALLOCK Y F and BOYD M R (1995), 'Michellamine B, a novel plant alkaloid, inhibits human immunodeficiency virus-induced cell killing by at least two distinct mechanisms', *Antimicrob Agents Chemother*, **39**, 484–488.

MENDOZA L, WILKENS M and URZUA A (1997), 'Antimicrobial study of the resinous exudates and of diterpenoids and flavonoids isolated from some Chilean *Pseudognaphalium* (Asteraceae)', *J Ethnopharmacol*, **58**, 85–88.

MEYER F P (1991), 'Aquaculture disease and health management', *J Anim Sci*, **69**, 4201–4208.

MITSUIKI S, SAKAI M, MORIYAMA Y, GOTO M and FURUKAWA K (2002), 'Purification and some properties of keratinolytic enzyme from an alkaliphilic *Nocardiopsis* sp. TOA-1', *Biosci Biotechnol Biochem*, **66**, 164–167.

MONAGHAN R L and TKACZ J S (1990), 'Bioactive microbial products: focus upon mechanism of action', *Annu Rev Microbiol*, **44**, 271–301.

MORITA T, OHTSUBO S, NAKAMURA T, TANAKA S, IWANAGA S, OHASHI K and NIWA M (1985), 'Isolation and biological activities of *Limulus* anticoagulant (anti-LPS factor) which interacts with lipopolysaccharide (LPS)', *J Biochem*, **97**, 1611–1620.

MORLEY P S, APLEY M D, BESSER T E, BURNEY D P, FEDORKA-CRAY P J, PAPICH M G, TRAUB-DARGATZ J L and WEESE J S (2005), 'ACVIM Consensus Statement', *J Vet Intern Med*, **19**, 617–629.

MTOLERA M S P and SEMESI A K (1996), Antimicrobial activities of extracts from six green algae from Tanzania. In M, Björk, A, Semesi, M, Pedersén, & B, Bergman, (eds). *Current Trends in Marine Botanical Research in the East African Region*, Gotab AB, Uppsala, Sweden, pp. 211–217.

MUNRO M H G, BLUNT J W, DUMDEI E J, HICKFORD S J H, LILL R E, LI S X, BATTERSHILL C N and DUCKWORTH A R (1999), 'The discovery and development of marine compounds with pharmaceutical potential', *J Biotechnol*, **70**, 15–25.

NAIR A G H, DONIO M B S, THANGAVIJI V, MICHAELBABU M and CITARASU T (2010), 'Isolation from coconut retting effluent of *Bacillus cereus* TC-2 antagonistic to pathogenic Vibrios', *Ann Microbiol*, **61**(3), 631–637.

NEWMAN D J and HILL R T (2006), 'New drugs from marine microbes: the tide is turning', *J Ind Microbiol Biotechnol*, **33**, 539–544.

NEWMAN D J, CRAGG G M and SNADER K M (2003), 'Natural products as source of new drugs over the period 1981–2002', *J Nat Prod*, **66**, 1022–1037.

NICOLAS J L, GATESOUPE F J, FROUEL S, BACHERE E and GUEGUEN Y (2007), 'What alternatives to antibiotics are conceivable for aquaculture?', *Health Aquaculture Special issue*, **20**, 3.

NOGA E J (1993), Fungal diseases of marine and estuarine fishes, In: *Pathophysiology of Marine and Estuarine organisms*, CRC Press, Boca Raton, FL, 85–110.

ÖZDEMIR G, CONK DOLAY M, KÜÇÜKAKYÜZ K, PAZARBANI B and YILMAZ M (2001), 'Determining the antimicrobial activity capacity of various extracts of *Spirulina platensis* produced in Turkey's conditions', *J Fisheries and Aquatic Sciences 1st. Algal Technology Symposium*, **18**(1), 161–166.

PALAVESAM A, SHEEJA L and IMMANUEL G (2006), 'Antimicrobial properties of medicinal herbal extracts against pathogenic bacteria isolated from the infected grouper *Epinephelus tauvina*', *J Biol Res*, **6**, 167–176.

PEKKARINEN M and LOTMAN K (2003), 'Occurrence and life cycles of *Dermocystidium* species (mesomycetozoa) in the perch (*Perca fluviatilis*) and ruff (*Gymnocephalus cernuus*) (Pisces: Perciformes) in Finland and Estonia', *J Nat Hist*, **37**, 1155–1172.

PEREIRA H S, LEAO-FERREIRA L R, MOUSSATCHE N, TEIXEIRA V L, CAVALCANTI D N, COSTA L J, DIAZ R and FRUGULHETTI I C (2004), 'Antiviral activity of diterpenes isolated from the Brazilian marine alga *Dictyota menstrualis* against human immunodeficiency virus type 1 (HIV-1)', *Antiviral Res*, **64**, 69–76.

PERUMALSAMY R and GOPALAKRISHNAKONE P (2008), 'Therapeutic potential of plants as anti-microbials for drug discovery', *eCAM*, **7**(3), 283–294.

PHILLIPSON J D and O'NEILL M J (1987), 'New leads to the treatment of protozoal infections based on natural product molecules', *Acta Pharm Nord*, **1**, 131–144.

PIARD J C and DESMAZEAUD M (1992), 'Inhibiting factors produced by lactic acid bacteria. Part 2. Bacteriocins and other antibacterial substances', *Lait*, **72**, 113–142.

PONPORNPISIT A, ENDO M and MURATA H (2001), 'Prophylactic effects of chemicals and immunostimulants in experimental *Tetrahymena* infection of guppy', *Fish Pathol*, **36**, 1–6.

PRASAD S and VARIYUR PADHYOY K B (1993), 'Chemical investigation of some commonly used spices', *Aryavaidyan*, **6**(4), 262–267.

PRATT R, MAUTNER R H, GARDNER G M, SHA Y and DUFRENOY F (1951), 'Report on the antibiotic activity of seaweed extracts', *J Amer Pharm Assoc Sci Edn*, **40**, 575–579.

PREMA P, SMILA D, PALAVESAM A and IMMANUEL G (2010), 'Production and Characterization of an antifungal compound (3-phenyllactic acid) produced by *Lactobacillus plantarum* strain', *Food Bioprocess Technol*, **3**, 379–386.

PROKSCH P, EBEL R, EDRADA R A, WRAY V and STEUBE K (2003), Bioactive natural products from marine invertebrates and associated fungi. In: W. E. G. Müller (ed.), *Progress in Molecular and Subcellular Biology, Sponges*, Springer-Verlag, Berlin, 117–142.

PUNITHA S M J, BABU M M, SIVARAM V, SHANKAR V S, DHAS S A, MAHESH T C, IMMANUEL G and CITARASU T (2008), 'Immunostimulating influence of herbal biomedicines on nonspecific immunity in grouper *Epinephelus tauvina* juvenile against *Vibrio harveyi* infection', *Aqua Int*, **16**, 511–523.

RAJASULOCHANA P, DHAMOTHARAN R, KRISHNAMOORTHY P and MURUGESAN S (2009), 'Antibacterial activity of the extracts of marine red and brown algae', *J Amer Sci*, **5**(3), 20–25.

RAMASAMY P, RAJAN P R, JAYAKUMAR R, RANI S and BRENNER G P (1996), '*Lagenidium callinectes* (Couch, 1942) infection and its control in cultured larval Indian tiger prawn, *Penaeus monodon* Fabricius', *J Fish Dis*, **19**, 75–82.

RANI T V J (1999), Fourth year annual report (CSIR Research Associateship), submitted to Council of Scientific and Industrial Research, New Delhi.

RAVEH A and CARMELI S (2007), 'Antimicrobial ambiguines from the cyanobacterium *Fischerella* sp. collected in Israel', *J Nat Prod*, **70**(2), 196–201.

REHULKA J (1991), Prevention and therapy of fish diseases: fungal diseases. In: J. Tesařeik and Z. Svobodová (ed.) *Diagnostics, Prevention and Therapy of Fish Diseases and Intoxications*, FAO, Rome, Italy, 270–307.

RESHEF V, MIZRACHI E, MARETZKI T, SILBERSTEIN C, LOYA S, HIZI A and CARMELI S (1997). 'New acylated sulfoglycolipids and digalactolipids and related known glycolipids from cyanobacteria with a potential to inhibit the reverse transcriptase of HIV-1', *J Nat Prod*, **60**, 1251–1260.

RIGOS G and TROISI G (2005), 'Antibacterial agents in Mediterranean finfish farming: a synopsis of drug pharmacokinetics in important euryhaline fish species and possible environmental implications', *Rev Fish Biol Fisheries*, **15**, 53–73.

ROBERTS R J (2003), *Fish Pathology*, W.B. Saunders, Edinburgh, 169–253.

ROMANENKO L A, UCHINO M, KALINOVSKAYA N I and MIKHAILOV V V (2008), 'Isolation, phylogenetic analysis and screening of marine mollusc-associated bacteria for antimicrobial, hemolytic and surface activities', *Microbiol Res*, **163**, 633–644.

ROSENFELD W D and ZOBELL C E (1947), 'Antibiotic production by marine microorganisms', *J Bacteriol*, **154**, 393–398.

SAITO T, KAWABATA S I, HIRATA M and IWANAGA S (1995), 'A novel type of *Limulus* lectin-L6 – Purification, primary structure, and antibacterial activity', *J Biol Chem*, **270**, 14493–14499.

SAMBHU C (1996), Effect of hormones and growth promoters on growth and body composition of pearlsport, *Etroplus suratensis* and white prawn *Penaeus indicus*. Ph.D. Thesis. University of Kerala, India.

SANANDAKUMAR S (2002), MPEDA asks aquafarms not to use banned antibiotics, *Times News Network*, 9 April.

SAPKOTA A, SAPKOTA A R, KUCHARSKI M, BURKE J, MCKENZIE S and WALKER P (2008), 'Aquaculture practices and potential human health risks: current knowledge and future priorities', *Env Int*, **34**, 1215–1226.

SCALBERT A (1991), 'Antimicrobial properties of tannins', *Phytochemistry*, **30**, 3875–3883.

SCHIRMBÖCK M, LORITO M, WANG Y L, HAYES C K, ARISAN-ATAC I, SCALA F, HARMAN G E and KUBICEK C P (1994), 'Parallel formation and synergism of hydrolytic enzymes and peptaibol antibiotics, molecular mechanisms involved in the antagonistic action of *Trichoderma harzianum* against phytopathogenic fungi', *Appl Environ Microbiol*, **60**(12), 4364–4370.

SCHRÖDER H C, USHIJIMA H, KRASKO A, GAMULIN V, THAKUR N L, DIEHL-SEIFERT B, MÜLLER I M and MÜLLER W E G (2003), 'Emergence and disappearance of an immune molecule, an antimicrobial lectin, in basal metazoa', *J Biol Chem*, **278**, 32810–32817.

SEN S, CHAKRABORTY R, DE B and MAZUMDER J (2009), 'Plants and phytochemicals for peptic ulcer: an overview', *Pharmacognosy Rev*, **3**(6), 270–279.

SERRANO P H (2005), *Responsible Use of Antibiotics in Aquaculture*, FAO Fisheries technical paper 469, Food and Agriculture Organization of the United Nations, Rome.

SEWALD N and JAKUBKE H (2002), *Peptides: Chemistry and Biology*, Wiley-VCH, Weinheim.

SHANGLIANG T, HETRICK F M, ROBERSON B S and BAYA A (1990), 'The antibacterial and antiviral activity of herbal extracts for fish pathogens', *J Ocean University of Qingdao*, **20**, 53–60.

SHANIDAR S M (1971), *The First Flower People*, Alfred Knopf, New York, 245–250.

SHARON N and OFEK I (1986), Mannose specific bacterial surface lectins. In: *Microbial Lectins and Agglutinins*, New York, John Wiley & Sons, Inc, 55–82.

SINDERMANN C J and LIGHTNER D V (1988), *Disease Diagnosis and Control in North American Marine Aquaculture*. New York, Elsevier.

SIPKEMA D, FRANSSEN M C, OSINGA R, TRAMPER J and WIJFFELS R H (2005), 'Marine sponges as pharmacy', *Mar Biotechnol*, **7**, 142–162.

SIVARAM V, BABU M M, CITARASU T, IMMANUEL G, MURUGADASS S and MARIAN M P (2004), 'Growth and immune response of juvenile greasy groupers (*Epinephelus tauvina*) fed with herbal antibacterial active principle supplemented diets against *Vibrio harveyi* infections', *Aquaculture*, **237**, 9–20.

SMITH P, HINEY M P and SAMUELSEN O B (1994), 'Bacterial resistance to antimicrobial agent used in fish farming: a critical evaluation of method and meaning', *Ann Rev Fish Dis*, **4**, 273–313.

SMITH V J A, DESBOIS P and DYRYNDA E A (2010), 'Conventional and unconventional antimicrobials from fish, marine invertebrates and micro-algae', *Mar Drugs*, **8**, 1213–1262.

SOYTONG K, SRINON W, RATTANACHERDCHAI K, KANOKMEDHAKUL S and KANOKMEDHAKUL K (2005), 'Application of antagonistic fungi to control anthracnose disease of grape', *J Agric Biotechnol*, **1**, 33–41.

STENSVÅG K, HAUG T, SPERSTAD S V, REKDAL Ø, INDREVOLL B and STYRVOLD O B (2008), 'Arasin 1, a proline–arginine-rich antimicrobial peptide isolated from the spider crab, *Hyas araneus*', *Dev Comp Immunol*, **32**, 275–285.

STROBEL G A, DIRKSE E, SEARS J and MARKWORTH C (2001), 'Volatile antimicrobials from *Muscodor albus*, a novel endophytic fungus', *Microbiology*, **147**, 2943–2950.

SZEKERES A, LEITGEB B, KREDICS L, ZSUZSANNA A, HATVANI L, MANCZINGER L and VAGVOLGYI C (2005), 'Peptaibols and related peptaibiotics of *Trichoderma*', *Acta Microbiol Immunol Hung*, **52**, 137–168.

THAN P P, CASTILLO C S D, YOSHIKAWA T and SAKATA T (2004), 'Extracellular protease production of bacteriolytic bacteria isolated from marine environments', *Fisheries Sci*, **70**, 4, 659–666.

TSAO R and DENG Z (2004), 'Separation procedures for naturally occurring antioxidant phytochemicals', *J Chromatogr*, **812**, 85–99.

TSENG D Y, HUANG P L, CHENG S Y, SHIU S C and LIU C H (2009), 'Enhancement of immunity and disease resistance in the white shrimp, *Litopenaeus vannmei*, by the probiotic *Bacillus subtilis* E 20', *Fish Shellfish Immunol*, **26**, 339–334.

TSUCHIYA H, SATO M, MIYAZAKI T, FUJIWARA S, TANIGAKI S, OHYAMA M, TANAKA T and IINUMA M (1996), 'Comparative study on the antibacterial activity of phytochemical flavanones against methicillin-resistant *Staphylococcus aureus*', *J Ethnopharmacol*, **50**, 27–34.

TSUJIBO H, KUBOTA T, YAMAMOTO M, MIYAMOTO K and INAMORI Y (2003), 'Characteristics of chitinase genes from an alkaliphilic actinomycete, *Nocardiopsis prasina* OPC-131', *Appl Environ Microbiol*, **69**, 894–900.

TYLER V E and LYDIA W E (1995), 'Pinkham's vegetable compound: an effective remedy?', *Pharm Hist*, **37**, 24–28.

TZIVELEKA L A, VAGIAS C and ROUSSIS V (2003), 'Natural products with anti-HIV activity from marine organisms', *Curr Top Med Chem*, **3**, 1512–1535.

VASAVADA S H J, THUMAR T and SINGH S P (2006), 'Secretion of a potent antibiotic by salt-tolerant and alkaliphilic actinomycete *Streptomyces sannanensis* strain RJT-1', *Curr Sci*, **91**, 10, 25.

VELMURUGAN S, PUNITHA S M J, BABU M M, SELVARAJ T and CITARASU T (2010), 'Select of Indian antibacterial medicine characteristics herbal to replace antibiotics for shrimp *Penaeus monodon* post larvae', *J Appl Aquaculture*, **22**, 230–239.

VIJAYA K, ANANTHAN S and NALINI R (1995), 'Antibacterial effect of theaflavin, poly-phenon 60 (*Camellia sinensis*) and *Euphorbia hirta* on *Shigella* spp. – a cell culture study', *J Ethnopharmacol*, **49**, 115–118.

VILLARREAL-GÓMEZ L J, SORIA-MERCADO I E, GUERRA-RIVAS G and AYALA-SÁNCHEZ N E (2010), 'Antibacterial and anticancer activity of seaweeds and bacteria associated with their surface', *Revista de Biología Marina y Oceanografía*, **45**, 2, 267–275.

VISHWAKARMA R A (1990), 'Stereoselective synthesis of a-arteether from artemisinin', *J Nat Prod*, **53**, 216–217.

WAKSMAN S A, HORNING E S and SPENCER E L (1942), *Two Antagonistic Fungi,* Aspergillus fumigatus *and* Aspergillus clavatus *and their Antibiotic Substances*, New Brunswick, Journal Series paper of the New Jersey Agricultural Experiment Station, Rutgers University.

WILD R (1994), *The Complete Book of Natural and Medicinal Cures*. Rodale Press, Inc., Emmaus, PA.

WITVROUW M, ESTE J A, MATEU M Q, REYMEN D, ANDREI G, SNOECK R, IKEDA S, PAUWELS R, BIANCHINI N V, DESMYTER J and DE CLERCQ E (1994), 'Activity of a sulfated poly-saccharide extracted from the red seaweed *Aghardhiella tenera* against human immunodeficiency virus and other enveloped viruses'. *Antiviral Chem Chemother*, **5**, 297–303.

WIUFF C, MALCOLM W, WILSON J, CROMWELL T, BENNIE M and EASTAWAY A (2010), *The Annual Surveillance of Healthcare Associated Infection Report January – December 2009*, Health Protection Scotland, Glasgow.

WOODWARD R and DOERING W (1944), 'The total synthesis of quinine', *J Am Chem Soc*, **66**(849).

WHO (1999), *Joint FAO/NACA/WHO Study Group on Food Safety Issues Associated with Products from Aquaculture*, WHO Technical Report Series, 883.

YA C, GAFFNEY S H, LILLEY T H and HASLAM E (1988), Carbohydratepolyphenol com-plexation, In: *Chemistry and Significance of Condensed Tannins*, New York, Plenum Press.

YERMAKOV A I, KHLAIFAT A L, QUTOB H, ABRAMOVICH R A and KHOMYAKOV Y Y (2010), 'Characteristics of the GC-MS mass spectra of terpenoids (C10H16)', *Chemi Sci J*, 7.

YOSHIDA T, HATANO T and ITO H (2000), 'Chemistry and function of vegetable poly-phenols with high molecular weights', *Biofactors*, **13**, 121–125.

YUAN C, LI D, CHEN W, SUN F, WU G, GONG Y, TANG J, SHEN M and HAN X (2007), 'Admin-istration of a herbal immunoregulation mixture enhances some immune param-eters in carp (*Cyprinus carpuio*)', *Fish Physiol Biochem*, **33**, 93–101.

ZHANG Y and LEWIS K (1997), 'Fabatins: new antimicrobial plant peptides', *FEMS Microbiol Lett*, **149**, 59–64.

17

The potential for antimicrobial peptides to improve fish health in aquaculture

A. Falco, A. Martinez-Lopez, Miguel Hernández University, Spain,
J. P. Coll, INIA-SIGT Biotechnology, Spain and A. Estepa,
Miguel Hernández University, Spain

Abstract: Infectious diseases cause severe problems in the aquaculture industry, with viral diseases being responsible for the greatest losses in production. Disease prevention strategies still have some limitations in terms of safety and efficacy. Antimicrobial peptides (AMPs) are molecules of the innate immune system, one of the first lines of defense against pathogens. Usually they not only exhibit antimicrobial activity, but also modulate the immune response. This review focuses on fish AMPs and their antiviral and immunoregulatory activities in order to assess their potential relevance to aquaculture. Since fish depend on their innate immune defenses more than mammals, they could be an alternative source of novel antiviral compounds.

Key words: antimicrobial peptides, AMPs, antimicrobial activity, antiviral, aquaculture, fish, disease, immune system, rhabdovirus.

17.1 Introduction

The OIE (Office International des Epizooties, now known as the World Organization for Animal Health) (http://www.oie.int) listed nine notifiable fish diseases in the 2009 Aquatic Animal Health Code. Among them, seven are viral in origin: *Koi herpesvirus disease* (KHVD) caused by herpesvirus, *epizootic haematopoietic necrosis* (EHN) and *red sea bream iridoviral disease* (RSIVD) caused by iridovirus, *infectious salmon anaemia* (ISA) caused by orthomyxovirus, and three diseases caused by rhabdovirus, namely *viral haemorrhagic septicaemia* (VHS), *infectious hematopoietic necrosis* (IHN) and *spring viremia of carp* (SVC). In general, viral diseases are responsible for the greatest losses in aquaculture production, since they affect fish at the early stages of development and produce an elevated percentage of mortality in the more economically valuable adult fish. In these terms, rhabdoviral diseases, caused by *viral haemorrhagic septicaemia virus* (VHSV), *infectious hematopoietic necrosis virus* (IHNV) (both belonging to the *Novirhabdovirus* genus) and *spring viremia of carp virus* (SVCV) (*Vesiculovirus-like* genus), represent the highest risk to worldwide aquaculture and they are commonly associated with frequent epizooties.[1–4] Also, within the last few years, severe mortality rates have been observed worldwide in farmed and wild carp as well as koi populations, with the deaths caused by *koi herpesvirus* (KHV), also known as *cyprinid herpesvirus-3* (CyHV-3). This is emerging as a serious problem for carp culture.[5–8]

There are two more diseases listed in the 2009 Aquatic Animal Health Code apart from those already mentioned: *epizootic ulcerative syndrome* (EUS) and *gyrodactylosis*. They are caused by the oomycetes *Aphanomyces invadans* or *Aphanomyces piscicida*, and the flatworm (Platyhelminthes) *Gyrodactylus salaris*, respectively. Dead fish from many different species presenting EUS signs (ulcerative lesions in the dermis caused by aggressive mycotic granulomas that can even penetrate into the skeletal muscle) have been found across a vast region that includes America, Asia and Australia.[9] Unlike EUS, *gyrodactylosis* is less widespread, limited so far to the Baltic countries, where it produces devastating epidemics, especially within Atlantic salmon (*Salmo salar*) populations.

Although they are not listed by the OIE in the recent Aquatic Animal Health Code, there are other fish pathogens that are potential risks for the aquaculture industry. The culture of eels, *Anguilla anguilla*, has recently increased in importance due to the decreasing wild populations and, hence, its rising market value. However, the eel aquaculture industry is suffering severe losses due to several viral infections, such as *eel virus B12* (EEV-B12), *eel virus C26* (EEV-C26) and *eel virus American* (EVA),[10] all of which are rhabdoviruses. Within the bacterial field we find *Ichthyophthirius multifiliis* (order Hymenostomatida, family Ichthyophthiriidae) and *Cryptocaryon irritans* (order Prorodontida, family Cryptocaryonidae), both ciliates that parasitize fresh water and marine fish. They are widely distributed and responsible for white spot disease.[11,12] *Aeromonas salmonicida* is the causative agent of furunculosis, which causes severe problems in marine and freshwater fish, especially salmonids, and is also related to carp *erythrodermatitis*.[13] Besides the genus *Aeromonas*, *Aeromonas hydrophila* is the dominant infectious agent of *bacterial haemorrhagic septicaemia*, a lethal disease, with significant outbreaks in China.[14] *Vibriosis* is one of the most prevalent fish diseases. It is caused by *Vibrio anguillarum*, a Gram-negative bacteria that is particularly devastating in the marine culture of gadoids, although it also seriously affects other fish species such as salmonids.[15]

17.2   Strategies for preventing disease in fish

The prevention of disease in farmed fish has been mainly focused on treatment with antibiotics/chemicals and/or vaccination. However, treatment with antibiotics can lead to antibiotic resistance in the pathogens, meaning that their safety needs to be continuously assessed (the therapeutic dosage and host tolerance threshold are often dangerously close) and vaccination may not be effective. Viral diseases are especially problematic since the only prevention method remains vaccination, but that is currently not always possible. In addition, this kind of treatment is ineffective against outbreaks of viral diseases, which have emerged as a serious problem because they not only cause severe losses, but also produce an enormous ecological impact.[16]

The prevention of viral diseases by means such as vaccination or immunostimulation is an attractive approach.[17,18] However, the development of cheap, effective and safe vaccines for the prevention of viral diseases in fish has proven to be a difficult task, and only a few commercial viral vaccines are available, despite extensive research over the years.[19] Immunostimulation seems to be an attractive approach, because the immunostimulants used are commonly of natural origin (bacterial or fungal products, plant extracts, nutritional factors and hormones, or even humoral components from the immune system itself, such as interferon, which may have an alternative use as adjuvants in vaccination) and do not produce resistance or contamination. On the other hand, the molecular basis underlying the activity of these compounds still remains unknown, as too are optimum treatment protocols and potential side effects. In an attempt to find antibiotic alternatives and to improve current vaccination and immunostimulation strategies, focus has shifted to increasing our understanding both of the molecules involved in triggering the host immune responses, and the pathogen-induced immune host molecules with antimicrobial activity. In this context, the development of techniques to enhance the innate immune system, particularly by using peptides produced by fish called antimicrobial peptides (AMPs), which have antimicrobial and also possibly immunostimulant properties, is an attractive challenge.

17.2.1   The innate immune system of fish

Fish represent the earliest class of vertebrates in which both innate and acquired, or adaptive, immune mechanisms are present. The innate immune system appears to play a central role in the response to infections in fish, whereas in mammals the adaptive immune system is more significant.[20] The intrinsic inefficiency of the adaptive immune response in fish is due to its evolutionary status – it only possesses IgM-like responses – and, moreover, due to environmental constraints such as temperature, because of the poikilothermic nature of fish.[21] These factors result in a limited antibody repertoire, poor affinity maturation and memory, slow lymphocyte proliferation, and a short-lived secondary response.[22]

The innate immune response is the first line of defense against infections, triggered immediately after the first encounter between host and pathogen. The innate defenses in fish comprise a wide repertoire of biological actions in which both cellular and humoral components are implicated. These components include macrophages, cytotoxic cells, complement, interferon (IFN) and AMPs. These actions are mainly initiated and driven by pattern-recognition receptors capable of recognizing, for instance, pathogen-associated molecular patterns (PAMPs) such as nucleic acids or surface glycoproteins, which thus give some specificity to a system previously referred to as non-specific.

It should be taken into account that microorganisms, including those that are pathogenic to fish, are widely distributed in aquatic environments; therefore, fish live in close contact with high concentrations of bacteria, viruses and parasites. Regarding viruses, it has been estimated that about 10^{10} virus particles per liter exist in aquatic habitats.[23] In addition, although fish seem to be highly vulnerable to pathogens, since they have large areas of delicate epithelium, such as gills, and allied to the fact that the epidermal surface of their skin is mostly composed of living cells, with low levels of keratin, under normal conditions, fish maintain their health through a complex network of defense mechanisms. Those include a mucosal barrier, involving AMPs as key components. AMPs offer strong immunological activity against a broad range of microbes.[24] In summary, it is believed that fish may possess a broader range of AMPs and other antimicrobial molecules in their mucosal sufaces than 'higher' vertebrates.[25,26]

17.3   Antimicrobial peptides (AMPs)

AMPs are present in almost all life forms, including unicellular organisms. Therefore, evolutionarily, AMPs may be among the earliest-developed molecular effectors of the innate immune system. Many different families of these host gene-coded defense molecules have been described. AMPs share several common properties, including: (i) having a cationic charge at physiological pH,[24,27–30] (ii) generally containing fewer than approximately 60 amino acid residues and (iii) showing broad-spectrum activity against bacteria, fungi and/or enveloped and non-enveloped viruses.[31–34] Many AMP families are expressed in more than one species and in more than one cell type, but AMPs are typically present in leukocytes or in epithelial surfaces.[27,35] Their expression can be constitutive and/or inducible.[20] They have been classified into at least three structural groups: linear/α-helical, disulfide stabilized/β-sheet, and extended structures rich in a single amino acid, such as tryptophan, proline, or histidine.[36]

17.3.1   Functions of AMPs: antimicrobial activity and immunomodulatory properties

In most cases, the cationic and amphipathic AMPs interact with the negatively-charged lipid cell membranes of pathogens, including those of enveloped viruses.[37–42] The mechanisms proposed to explain their mode of action include different pore-formation models.[43–50] With this general, but not exclusive mode of action, AMPs are frequently able to rapidly kill large numbers of pathogens by destabilization/permeabilisation of their membranes, and hence they do not easily select for resistant mutants.[20,51,52] This mechanism could explain why AMPs are more efficient against enveloped viruses than non-enveloped viruses. Thus, the activity of AMPs against enveloped viruses works by two mechanisms: direct inactivation of the viral particles via damage to their lipid membranes,[53] and/or inhibition of the membrane fusion step during the viral replication cycle, both of which will block viral spread.[54–56] AMPs show an extraordinary diversity in sequence and structure. Further to the pore-forming mechanisms mentioned above, there is evidence to suggest that AMPs might diffuse towards intracellular targets and inhibit the synthesis of pathogen cell walls, nucleic acids, or proteins, and that they may even reduce pathogen-induced enzymatic activity.[43,57–61]

In addition to their direct antimicrobial activity, AMPs also have immunomodulatory properties. The immunomodulatory ability of many AMPs may be even more important than their antimicrobial activity. AMPs' immunomodulation is produced by inducing cytokines and chemokines, changing the gene expression profile of host cells, inhibiting the host cells' proinflammatory response to pathogen components and/or stimulating monocyte chemotaxis.[33,62–64] In this regard, the functions of AMPs and chemokines are similar. It has even been proposed that some AMPs evolved from chemokines, since some chemokines exhibit antimicrobial activity,[65] and many antimicrobial peptides exhibit chemoattractant capacity.[66] This idea will remain controversial unless more evidence emerges.[67]

17.3.2 Antimicrobial peptides in fish

Research on fish AMPs began relatively recently, but the number of AMPs isolated from fish grows continuously.[68] Thus, large numbers of different AMPs have been identified within a broad spectrum of different fish species and tissues[26,29,69–71] including fish immune cells.[70,72–74] Taken together, the above-mentioned findings suggest the high importance of these molecules in the fish innate immune system. Fish AMPs are extremely diverse. They belong not only to families of AMPs present in other groups of vertebrates, and invertebrates, such as histone H2A-derived peptides, defensins, hepcidins and cathelicidins, but also to families of AMPs only present in fish, for instance, piscidins, pleurocidins and chrysophsins. A brief description of the best-characterized fish AMP families follows, in which they are classified as either linear (i) or disulfide-stabilized (ii) AMPs.

Linear fish AMPs

The main linear fish AMPs are piscidins, pleurocidins, cathelicidins and pardaxins (Table 17.1) and their classification in the above mentioned groups is based on sequence homology.

Table 17.1 Main characteristics of linear fish AMPs

Name	Mature peptide size (aas)	Genomic organization	Organisms	References
Piscidins	18–44	4 exons 3 introns	Acanthopterygii and Paracanthopterygii superorder	[70, 75–77, 79]
Pleurocidins	18–26	4 exons 3 introns	Pleuronectiformes order	[26, 29, 82, 83]
Cathelicidins	35–66	4 exons 3 introns	Salmonids, Atlantic cod and Atlantic hagfish	[87, 89, 91]
Pardaxins	33	?	Red Sea Moses sole and peacock sole of the western Pacific	[92–94]
Epinecidin	21	?	Grouper	[160]
Misgurin	21	?	Loach	[97]
Chrysophsins	20–25	?	Red sea bream	[98]
Hipposin	51	?	Atlantic halibut	[99, 100]
Parasin	19	?	Catfish	[101]

? = unknown.

Piscidins constitute a relatively new family of linear AMPs found in teleosts.[70] Nevertheless, they are the best-characterized fish AMPs thus far. In general, they are 18–26 residues long, although recently a novel type of piscidin (piscidin 4) of twice the length (44 aa) was isolated from hybrid striped bass (*Morone saxatilis* × *Morone chrysops*) gill.[75] All of them share a high proportion of basic amino acids. Piscidins were initially identified in the mast cells of the hybrid striped bass[70] and it is now known that they are widespread amongst Perciformes (Acanthopterygii superorder).[72,76–78] They have also been found in more basal teleosts of the Paracanthopterygii superorder, for example Atlantic cod (*Gadus morhua*).[79] Piscidins are also known as *moronecidins*, because piscidin-1 and piscidin-2 were discovered separately in two different laboratories, but at approximately the same time, in the skin and gills of hybrid striped bass.[80] *Dicentracins*, which were identified in European seabass (*Dicentrarchus labrax*) and assigned to the moronecidin group,[74] belong to the piscidin AMP family as well. Piscidin genes contain three introns and four exons that code for a putative precursor containing a signal peptide, the mature peptide and a C-terminal prodomain.[74,77]

Pleurocidins were first identified in Pleuronectiformes (Acanthopterygii superorder).[81] Several pleurocidins have been reported from other fish species such as winter flounder (*Pseudopleuronectes americanus*),[29,71,82]

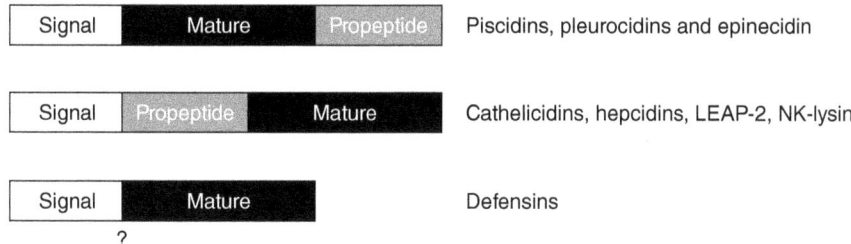

Fig. 17.1   Molecular organization of the AMP initial precursors.

Atlantic halibut (*Hippoglossus hippoglossus*), witch flounder (*Glyptoceph-alus cynoglossus*), yellowtail flounder (*Limanda ferruginea*), American plaice (*Hippoglossoïdes platessoïdes*)[81] and mud dab (*Limanda limanda*).[83] Pleurocidin expression has been described in epidermal mucus,[29] skin, gut[82] and eosinophils of the gills.[73] Developmental expression in winter flounder indicated that the pleurocidin gene is first expressed early in the larvae stage.[82] The gene structure of pleurocidin genes consists of four exons and three introns[26,29] coding for an initial precursor (signal peptide + mature peptide + propeptide) (see Fig. 17.1 for different initial precursor organiza-tions).[81,82] The structure is similar to those found in piscidins,[74,77] which after processing results in a bioactive mature peptide of 25 amino acids in length.[82] Significant homology in both peptidic and genomic organization has been found between pleurocidins and piscidins, which suggests an evolutionary relationship between the two AMP families.[77,80,82] However, the homology at the sequence level of mature peptide is weaker. Thus, pleurocidins do not express the highly conserved N-terminus of mature piscidins and, furthermore, unlike piscidins, no XQQ motif is present in pleurocidin prodomains.[77,80,82]

Members of the family *cathelicidins* are synthesized as precursors, in a similar manner to piscidins and pleurocidins, but their precursors are: signal peptide + propeptide + mature peptide. Cathelicidins are characterized by the presence of a highly conserved propeptide containing the cathelin-like domain (cathelin is an acronym for cathepsin-L-inhibitor protein), which has two disulfide bridges.[84,85] However, after endoproteolytic cleavage by elastase for propeptide removal, it is released as a mature peptide with antimicrobial activity, which is very heterogeneous in size and sequence, and may not contain cysteines.[34,85,86] As in mammals, fish cathelicidin genes contain four exons and three introns. The first three exons encode the signal peptide and the conserved propeptide, while the fourth exon encodes the cleavage site and the hypervariable mature peptide.[87] This last exon under-went a rapid diversification, based on the variation of the number of tandem repeats detectable in the sequences, and on insertion mutations in the coding sequence.[88] Cathelicidins have been identified in salmonid species such as rainbow trout (*Oncorhynchus mykiss*), Atlantic salmon (*S. salar*),[69,87] brown trout (*Salmo trutta fario*), brook trout (*Salvelinus fontinalis*), grayling (*Thymallus thymallus*),[88] Chinook salmon (*Oncorhynchus tshawytscha*) embryo cells (CHSE) and Arctic charr (*Salvelinus alpinus*),[89] but it has also been found in some non-salmonid species such as Atlantic cod[90] and Atlantic hagfish (*Myxine glutinosa*).[91] The expression of each cathelicidin gene has been shown to be constitutive and/or inducible either bacteria *A. salmonicida*, lipopolysaccharide (LPS), bacterial DNA or polyinosinic:polycytidylic acid (poly I:C), depending on the cathelicidin gene. Cathelicidins are present not only in mucosal tissues (gill, head and trunk kidney, gut, stomach, skin and spleen).[69,87–89]

Pardaxins have been isolated from the toxic shark-repelling secretions of Red Sea Moses sole (*Pardachirus marmoratus*) and the peacock sole of the western Pacific (*Pardachirus pavoninus*). They are 33 amino acids long and show a helix-(proline) hinge-helix structure, similar to the AMP melittin, a component of bee venom.[92–94] The 21-residue peptide *epinecidin*(-1), the gene for which was identified in grouper (*Epinephelus coioides*),[95] codes for an initial precursor similar to those found in pleurocidins and piscidins.[74,77,81,82] It exhibits broad tissue distribution (gill, head kidney, gut and skin) and can be upregulated by LPS and poly I:C.[96]

Other linear fish AMPs include the 21-residue peptide *misgurin*, isolated from homogenized loach (*Misgurnus anguillicaudatus*),[97] *chrysophsins* (-1, -2 and -3), which are 20–25-residue petides isolated from the gills of red sea bream (*Chrysophrys major*) and detected in some epithelial cells and eosinophilic-like granules from cells localated in the epithelium,[98] and, finally, the H2A amino terminus derived-peptides *hipposin*, a 51 residue AMP isolated from the skin mucus of Atlantic halibut,[99,100] and *parasin* (I), a 19-residue AMP isolated from the skin mucus of catfish (*Parasilurus asotus*),[101] which is produced by cathepsin D action.[102]

Several studies, mainly using circular dichroism and solid-state nuclear magnetic resonance methods, have tried to elucidate the molecular structure of linear peptides. The overall structure of these AMPs is markedly amphipathic, with well-defined hydrophobic and hydrophilic sectors, which are easily identified in a Shiffer–Edmundson helical wheel diagram (Fig. 17.2). Many of these peptides adopt a random coil structure in water but become α-helical upon interaction with dodecylphosphocholine (DPC) micelles (enhanced when anionic lipids are included in the membranes), which suggests that this last structure is the active conformation that can permeabilize the membrane. This has been shown to be the case for pleurocidin,[103] piscidin[104] and chrysophsin,[105] which were found to adopt an α-helical structure, lying parallel to the membrane surface (carpet-like model),[106] in agreement with models where permeabilization is a consequence of transient membrane disruption,[93,105,107–109] rather than these peptides forming ion channels via a barrel-stave mechanism.[50,110,111] Although some AMPs, such as pardaxin, do form ion channels via a barrel-stave

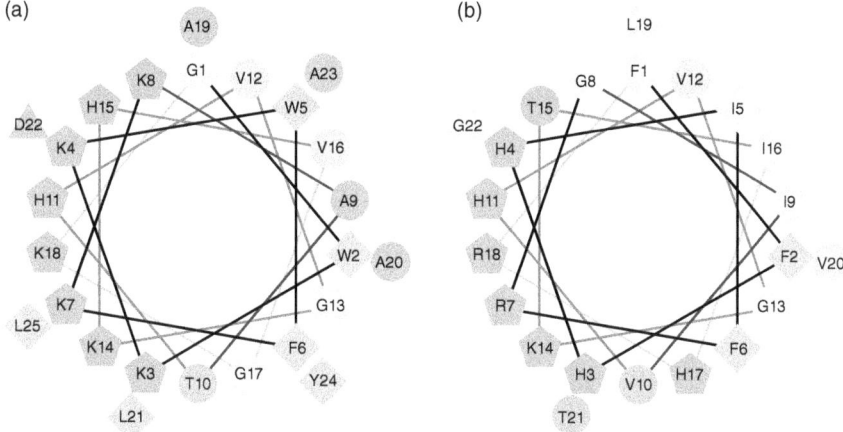

Fig. 17.2   Schiffer-Edmundson helical wheel representation of linear AMP mud dab pleurocidin (a) and hybrid striped bass piscidin-1 (b). Predicted α-helical structures adopt an amphipathic conformation for both AMPs. Hydrophilic charged residues are represented by pentagons and hydrophobic residues by diamonds. Other neutral or polar amino acids are pointed out by circles. Amino acid residues are designated using the standard single-letter abbreviated terminology and numbered starting from the amino terminus of the mature peptides. Mud dab pleurocidin sequence is GWKKWFKKATHVGKHVGKAALDAYL. Hybrid striped bass piscidin-1 sequence is FFHHIFRGIVHVGKTIHRLVTG. Software used from http://rzlab.ucr.edu/scripts/wheel/wheel.cgi.

mechanism, this may not be the mode of action that kills pathogens, and furthermore, the adoption of these conformations may be concentration dependent.[93]

Disulfide-stabilized fish AMPs
The main disulfide-stabilized fish AMPs are defensins, hepcidins and NK-lysins (Table 17.2). *Defensins* are probably the best studied family of AMPs since they are widely distributed within fungi, plants and animals (both vertebrates and invertebrates).[32] Vertebrate defensins are classified into three subgroups, α-defensins, β-defensins and θ-defensins, based on the differences in the location and distribution of the disulfide bonds between their six conserved cysteine residues; thus, in α-defensins the disulfide bond pattern is C1–C6, C2–C4 and C3–C5, in β-defensins is C1–C5, C2–C4 and C3–C6 and θ-defensins are cyclic peptides, also with three disulfide bonds, possibly derived from the splicing of two α-defensins.[112,113] To date, α- and θ-defensins are only known in mammals,[32,34,58] and θ-defensins are limited to primates.[113] Because the β-defensin subfamily is spread across all vertebrates, the other defensins might evolve from ancestral β-defensins.[114] In fish, β-defensin-like peptides have been identified in zebrafish (*Danio rerio*),

Table 17.2 Main characteristics of non-linear fish AMPs

Name	Mature peptide size (aas)	Number of Cys	Genomic organization	Organisms	References
β-defensins	38–45	6	3 exons 2 introns	Zebrafish, pufferfish, grouper, rainbow trout, medaka and olive flounder	[115–118, 161]
Hepcidins or LEAP-1	19–31	6–8	3 exons 2 introns	Hybrid striped bass, winter and Japanese flounder, Atlantic salmon, zebrafish, medaka, rainbow trout, long-jawed mudsucker, Japanese sea bass, tilapia and black porgy	[126, 135, 162, 163]
LEAP-2	35–66	4	3 exons 2 introns	Rainbow trout, channel catfish, blue catfish and grass carp	[138–140]
NK-lysins	About 100	6	5 exons 4 introns	Channel catfish, pufferfish, zebrafish and Japanese flounder	[141, 142]

pufferfish (*Tetraodon nigroviridis*), orange spotted grouper (*E. coioides*),[115] rainbow trout (*O. mykiss*),[116,117] medaka (*Oryzias latipes*)[118] and olive flounder (*Paralichthys olivaceus*).[119] Their genomic analysis revealed that they had the same gene organization, with three exons divided by two introns. The first exon encodes the signal peptide, while the other two exons encode the mature peptide, which ranges in size from 38 to 45 amino acid residues. Unlike upper vertebrate β-defensins, in which the precursor contains a negatively charged propeptide between the signal and mature peptide,[120] the presence of such a region has not yet been demonstrated in fish.

The other main family of disulfide-stabilized AMPs consists of the *hepcidins*, also known as liver-expressed AMPs (LEAP-1) due to the fact that they were originally identified in the liver by two different research groups.[121,122] Interestingly, hepcidin is not only an AMP, but also regulates intestinal iron absorption and releases iron from macrophages and blood cells[121,123,124] to reduce iron availability for invading bacteria and tumor cells.[125] From higher to lower vertebrates, hepcidin sequences have six to eight highly conserved cysteine residues.[126,127] Hepcidin AMPs have been identified in hybrid striped bass,[128] winter flounder,[129] Atlantic salmon,[126] zebrafish,[130] medaka, rainbow trout,[131] long-jawed mudsucker (*Gillichthys mirabilis*),[132] Japanese sea bass (*Lateolabrax japonicus*),[133] tilapia (*Oreochromis mossambicus*),[134] Japanese flounder (*P. olivaceus*)[135] and black porgy (*Acanthopagrus schlegelii*).[136] The genomic organization of fish hepcidins comprises three exons divided by two introns. The first exon encodes the signal peptide and the beginning of a highly negatively charged propeptide, the second exon encodes the rest of the propeptide and the third exon comprises the end of the propiece along with the entire mature peptide, which is over 20 amino acid residues in size.[128,130,133,135] As in upper vertebrates, fish hepcidin genes are mainly expressed in liver, and they can be highly up-regulated by some inducers, for example, LPS or iron over-load (iron-dextran). However, they can also be expressed in some other organs such as gill, kidney, heart, peripheral blood lymphocytes (PBLs), spleen and stomach at lower levels.[135]

LEAP-2 was the second liver-expressed AMP discovered in vertebrates.[137] LEAP-2 molecules are a distinct family of AMPs, since they show no similarity with hepcidins (or LEAP-1) or defensins. For instance, LEAP-2 has only four cysteines which form two disulfide bonds. To date, fish LEAP-2 AMPs have only been reported in rainbow trout,[138] channel catfish (*Ictalurus punctatus*), blue catfish (*Ictalurus furcatus*)[139] and grass carp (*Ctenopharyngodon idella*).[140]

NK-lysins are also a family of AMPs with a structure which is stabilized by three disulfide bonds. They have only been identified in channel catfish (*I. punctatus*), pufferfish, zebrafish[141] and Japanese flounder.[142] The genomic structure of NK-lysin genes comprises five exons and four introns, and they have been shown to be highly expressed in a wide variety of fish tissues.[141,142]

17.4   The potential role of antimicrobial peptides in preventing and treating fish diseases

17.4.1   Antimicrobial activity

Owing to increasing understanding of the importance of AMPs in fish immune systems and the increasing need to find safe, effective natural antimicrobial agents to combat serious disease outbreaks in fish the race to determine and characterize new AMPs has been accelerated and their antimicrobial effects on fish pathogens are being analyzed.

Several fish AMPs possess *in vitro* and *in vivo* antimicrobial activity against fish bacterial pathogens. For instance, piscidins and pleurocidins demonstrate potent, broad-spectrum, antibacterial activity against several fish and human pathogens, Gram-negative as well as Gram-positive, and including multi-drug resistant bacteria.[26,29,75,77,80,143] Synthetic epinecidin-1 derived-peptides inhibit *Propionibacterium acnes* growth *in vitro*.[144] Hepcidins also inhibit a broad array of Gram-negative and Gram-positive bacteria.[134,135,145] Moreover, a possible synergic effect between AMPs and other antimicrobial agents was shown for winter flounder pleurocidin and coho salmon histone H1 derived-peptide or lysozyme[146] and for hepcidin with piscidin from hybrid striped bass.[145] All the data cited above support the results obtained from *in vivo* experiments. For example, pleurocidin protected coho salmon from *in vivo V. anguillarum* infections,[31] transgenic zebrafish (*D. rerio*) and convict cichlid (*Archocentrus nigrofasciatus*) expressing tilapia hepcidin TH2-3[147] as well as intramuscularly injected or transgenic zebrafish expressing grouper epinecidin-1[148,149] showed significant clearance of the bacteria when they were challenged by *Vibrio vulnificus*. Other *in vivo* experiments with AMPs in fish have been carried out using non-fish AMPs. For instance, medaka expressing the *silk moth* (*Hyalophora cecropia*) and porcine (*Sus scrofa*) cecropin transgene exhibited higher resistance to *Pseudomonas fluorescens* and *V. anguillarum*[150] and so did transgenic channel catfish when challenged with *Edwardsiella ictalurii* and *Flavobacter columnare* using the same AMP.[151]

High *in vitro* fungicidal effects have been found for hybrid striped bass piscidin-1, -2 and -3 on several human pathogenic fungal strains. When piscidins were tested against *Candida albicans*, *Malassezia furfur* and *Trichosporon beigelii*, they were shown to be as potent inhibitors as melittin. The antifungal effect of piscidins was due to the damage to the membrane of the fungal cells they caused.[152] Piscidin-2 has·been also tested *in vitro* for antiprotozoan activity against fish protistan ectoparasites. Thus, the ciliates *Cryptocaryon irritans* and Trichodina sp., and the dinoflagellate *Amyloodinium ocellatum* (all of them affecting marine fish), and *Ichthyophthirius multifiliis* (another ciliate ectoparasite of freshwater fish), were susceptible to piscidin-2.[153] Grouper epinecidin-1 possessed high antifungal and antiprotozoan activity *in vitro* when it was tested against the human opportunistic pathogens *C. albicans* and *Trichomonas vaginalis*.[144] Activity against fish viral infections has been reported for some AMPs of fish origin such as epinecidin-1, hepcidins, piscidins and defensins.[76,116,154,155] Also, non-fish AMPs activity against fish viruses has been demonstrated for cecropins and defensins.[55,156]

17.4.2 Immunomodulatory properties

Several studies, some of them mentioned above, have suggested that some of the antiviral mechanisms of AMPs are due to their action on host cells, rather than in parallel with their direct action to inactivate viral particles. For example, they may make fish cells more resistant to and/or protected from viral infections. Cecropins are a family of linear AMPs initially discovered in giant silk moth (*H. cecropia*),[157,158] with no members found in fish. However, the antiviral effect of insect cecropin B (CecB) and a synthetic analogue (CF17) was reported *in vitro* against IHNV. Cecropin acted by triggering antiviral mechanisms in salmonid embryo-derived CHSE cell lines.[156] Also, the expression profiles of some representative immune genes (interleukin 1β, *il1β*, and cyclooxygenase, *cox2*) were modulated in a rainbow trout spleen (RTS) derived-macrophage RTS cell line in response to these AMPs and pleurocidin.[25] It has been also reported *in vivo* that mud dab pleurocidin significantly up-regulates the expression of some proinflammatory cytokines in rainbow trout.[159]

Amongst defensins, human α defensin 1 (HNP1) reduced VHSV infectivity to about 75% when it was added to epithelioma papulosum cyprini (EPC) and rainbow trout gonad (RTG) cells[55] and it was demonstrated that an IFN-related mechanism was activated as shown by the up-regulation of *mx3*, a marker of IFN-induction. Trout head kidney leucocytes were incubated with HNP1 as well and four IFN-related genes (*mx1*, *mx2*, *mx3* and *vig1*), *il1β* and inducible nitric oxide synthase (*inos*) genes were significantly modulated, thus confirming that HNP1 might trigger an antiviral response dependent on IFN induction. Moreover, HNP1 was able to modulate *in vivo* the expression profiles of some genes related to the innate immune response in rainbow trout.[159] The expression level of the proinflammatory cytokines *il1β*, tumour necrosis factor α1 (*tnfα1*) and interleukin 8 (*il8*) in muscle tissue, and also in blood in the case of *il1β*, and in head kidney in the case of *il8*, greatly increased in fish injected with this peptide. Moreover, it was found that HNP1 considerably attracted trout blood leukocytes showing an important role in chemotaxis.[159]

Preliminary results obtained with defensins from fish origin suggest they could also have immunomodulatory properties. It has been proven that EPC cells transfected with a DNA plasmid containing the sequence of a β-defensin-like peptide from rainbow trout (omBD1) are protected against VHSV infection and showed the type I IFN-mediated antiviral response is activated. These cells showed an up-regulation of type I IFN-mediated antiviral response by inducing the carp *mx1* gene. Moreover, the culture medium from omBD1-transfected EPC cells demonstrated acid and heat-stable antiviral activity.[116]

17.5   Future trends and conclusions

The teleosts are the most ancient and abundant group of vertebrates. They are found from the depths of the ocean to freshwater rivers and lakes, inhabiting many diverse niches. However, the pathogenic microorganisms that affect them have evolved simultaneously, and fish disease is the major risk factor in commercial aquaculture nowadays. The practice of aquaculture started a long time ago in territories known now as China; such simple practices have developed quite recently into a large industry with a world-wide presence. This has had a large impact on ecosystems, because intensive farming not only amplifies bacteria, virus and parasite diseases, but also spreads them to wild type populations.[16]

As mentioned in this chapter, the presence and widespread distribution of an abundant repertoire of AMPs in fish (which is continuously growing owing to new discoveries in the area), with a great variety of modes of action, not only provide fish with powerful immunological tools against microbial infections, but also give us the opportunity to design novel strategies to enhance the health status of aquacultured fish. The direct antimicrobial effects and/or immunomodulatory properties exerted by some of these AMPs against pathogens make them promising candidates as therapeutants. They can be administrated in different ways, for example as encapsulated food supplements or via injection. There is also the possibility of injecting a DNA plasmid containing their sequences. In this respect and in terms of toxicity, we know that only pardaxin and the three chrysophsins peptides have significant hemolytic activity, which might limit their applications as antimicrobial agents.[93,98]

AMP expression levels can be up-regulated by exposure to PAMPs (e.g. lipopolysaccharides). This is something to take into account because improving resistance during the periods in which the incidence of pathogens is known to be particularly high or when stressful events such as handling are going to be carried out, could help us to prevent potential outbreaks of infectious disease. Moreover, selective breeding of highly valuable species could lead to the development of fish populations that are genetically resistant to pathogens as they over-express AMPs.

Since AMP expression levels also vary depending on the health status of fish they also can be potentially used as accurate bio-markers of health status. They could be detected using methods such as enzyme-linked immunosorbent assay (ELISA). Certain AMPs which act as immunostimulants, owing to their ability to modulate the immune system and enhance the expression of important cytokines, also have the potential to be used as adjuvants to improve the vaccines already in existence.

17.6   Acknowledgements

This work was supported by the projects Consolider ingenio 2010 CSD2007–00002 and AGL2008–03519-C04 both from MICNN (Spain).

17.7   References

1   FU Z, HOFFMANN B, BEER M, SCHÜTZE H, METTENLEITER T. Fish rhabdoviruses: molecular epidemiology and evolution. *The World of Rhabdoviruses*: Springer, Berlin, Heidelberg, 2005: 81–117.

2   SKALL HF, OLESEN NJ, MELLERGAARD S. Prevalence of viral haemorrhagic septicaemia virus in Danish marine fishes and its occurrence in new host species. *Dis Aquat Organ* 2005 Sep 5;**66**(2):145–51.

3   ENZMANN PJ, KURATH G, FICHTNER D, BERGMANN SM. Infectious hematopoietic necrosis virus: monophyletic origin of European isolates from North American genogroup M. *Dis Aquat Organ* 2005 Sep 23;**66**(3):187–95.

4   AHNE W, BJORKLUND HV, ESSBAUER S, FIJAN N, KURATH G, WINTON JR. Spring viremia of carp (SVC). *Dis Aquat Organ* 2002 Dec 10;**52**(3):261–72.

5   AOKI T, HIRONO I, KUROKAWA K, FUKUDA H, NAHARY R, ELDAR A, *et al.* Genome sequences of three koi herpesvirus isolates representing the expanding distribution of an emerging disease threatening koi and common carp worldwide. *J Virol* 2007 May;**81**(10):5058–65.

6   GARVER KA, AL-HUSSINEE L, HAWLEY LM, SCHROEDER T, EDES S, LEPAGE V, *et al.* Mass mortality associated with koi herpesvirus in wild common carp in Canada. *J Wildl Dis* 2010 Oct;**46**(4):1242–51.

7   WALTZEK TB, KELLEY GO, STONE DM, WAY K, HANSON L, FUKUDA H, *et al.* Koi herpesvirus represents a third cyprinid herpesvirus (CyHV-3) in the family Herpesviridae. *J Gen Virol* 2005 Jun;**86**(Pt 6):1659–67.

8   MCGEOCH DJ, RIXON FJ, DAVISON AJ. Topics in herpesvirus genomics and evolution. *Virus Res* 2006 Apr;**117**(1):90–104.

9   SOSA ER, LANDSBERG JH, STEPHENSON CM, FORSTCHEN AB, VANDERSEA MW, LITAKER RW. *Aphanomyces invadans* and ulcerative mycosis in estuarine and freshwater fish in Florida. *J Aquat Anim Health* 2007 Mar;**19**(1):14–26.

10   ESSBAUER S, AHNE W. Viruses of lower vertebrates. *J Vet Med B Infect Dis Vet Public Health* 2001 Aug;**48**(6):403–75.

11   MATTHEWS RA. *Ichthyophthirius multifiliis* Fouquet and ichthyophthiriosis in freshwater teleosts. *Adv Parasitol* 2005;**59**:159–241.

12   YANG TB, CHEN AP, CHEN W, LI AX, YAN YY. Parasitic diseases of cultured marine finfishes and their surveillance in China. *Parassitologia* 2007 Sep;**49**(3):193–9.

13   WIKLUND T, DALSGAARD I. Occurrence and significance of atypical Aeromonas salmonicida in non-salmonid and salmonid fish species: a review. *Dis Aquat Organ* 1998 Feb 26;**32**(1):49–69.

14   NIELSEN ME, HOI L, SCHMIDT AS, QIAN D, SHIMADA T, SHEN JY, *et al.* Is *Aeromonas hydrophila* the dominant motile *Aeromonas* species that causes disease outbreaks in aquaculture production in the Zhejiang Province of China? *Dis Aquat Organ* 2001 Aug 22;**46**(1):23–9.

15   SAMUELSON OB, NERLAND AH, JØRGENSEN T, SCHRØDER MB, SVÅSAND T, BERGH O. Viral and bacterial diseases of Atlantic cod *Gadus morhua*, their prophylaxis and treatment: a review. *Dis of Aquat Organ* 2006 August 30;**71**(3):239–54.

16   SKALL HF, OLESEN NJ, MELLERGAARD S. Viral haemorrhagic septicaemia virus in marine fish and its implications for fish farming – a review. *J Fish Dis* 2005 Sep;**28**(9):509–29.

17   BIERING E, VILLOING S, SOMMERSET I, CHRISTIE KE. Update on viral vaccines for fish. *Dev Biol (Basel)* 2005;**121**:97–113.

18   SOMMERSET I, LORENZEN E, LORENZEN N, BLEIE H, NERLAND AH. A DNA vaccine directed against a rainbow trout rhabdovirus induces early protection against a nodavirus challenge in trout. *Vaccine* 2003;**21**:4661–7.

19   SALONIUS K, SIMARD N, HARLAND R, ULMER JB. The road to licensure of a DNA vaccine. *Curr Opinion Invest Drugs* 2007;**8**(8):635–41.

20   HANCOCK RE, DIAMOND G. The role of cationic antimicrobial peptides in innate host defences. *Trends Microbiol* 2000 Sep;**8**(9):402–10.

21   MAGNADOTTIR B. Innate immunity of fish (overview). *Fish Shellfish Immunol* 2006 Feb;**20**(2):137–51.

22   DU PASQUIER L. The immune system of invertebrates and vertebrates. *Comp Biochem Physiol B Biochem Mol Biol* 2001 May;**129**(1):1–15.

23   TORT L, BALASCH JC, MACKENZIE S. Fish health challenge after stress. Indicators of immunocompetence. *Contrib Sci*, 2004;**2**(4):443–54.

24   ZASLOFF M. Antimicrobial peptides of multicellular organims. *Nature* 2002;**415**:389–95.

25   BLY JE, CLEM LW. Temperature-mediated processes in teleost immunity: *in vitro* immunosuppression induced by *in vivo* low temperature in channel catfish. *Vet Immunol Immunopathol* 1991 Jul;**28**(3–4):365–77.

26   DOUGLAS SE, PATRZYKAT A, PYTYCK J, GALLANT JW. Identification, structure and differential expression of novel pleurocidins clustered on the genome of the winter flounder, *Pseudopleuronectes americanus* (Walbaum). *Eur J Biochem* 2003 September 15;**270**(18):3720–30.

27   PATRZYKAT A, DOUGLAS SE. Antimicrobial peptides: cooperative approaches to protection. *Protein Pept Lett* 2005 Jan;**12**(1):19–25.

28   BOMAN HG. Peptide antibiotics and their role in innate immunity. *Annu Rev Immunol* 1995;**13**:61–92.

29   COLE AM, WEIS P, DIAMOND G. Isolation and characterization of pleurocidin, an antimicrobial peptide in the skin secretions of winter flounder. *J Biol Chem* 1997 May 2;**272**(18):12008–13.

30   HANCOCK RE, LEHRER R. Cationic peptides: a new source of antibiotics. *Trend Biotechnol* 1998;**16**:82–8.

31   JIA X, PATRZYKAT A, DEVLIN RH, ACKERMAN PA, IWAMA GK, HANCOCK RE. Antimicrobial peptides protect coho salmon from *Vibrio anguillarum* infections. *Appl Environ Microbiol* 2000 May;**66**(5):1928–32.

32   GANZ T. Defensins: antimicrobial peptides of innate immunity. *Nat Rev Immunol* 2003 Sep;**3**(9):710–20.

33   MOOKHERJEE N, HANCOCK RE. Cationic host defence peptides: innate immune regulatory peptides as a novel approach for treating infections. *Cell Mol Life Sci* 2007 Apr;**64**(7–8):922–33.

34   YANG D, BIRAGYN A, HOOVER DM, LUBKOWSKI J, OPPENHEIM JJ. Multiple roles of antimicrobial defensins, cathelicidins, and eosinophil-derived neurotoxin in host defense. *Annu Rev Immunol* 2004;**22**:181–215.

35   ZASLOFF M. Antimicrobial peptides of multicellular organisms. *Nature* 2002 Jan 24;**415**(6870):389–95.

36   ZASLOFF M. Antimicrobial peptides in health and disease. *N Engl J Med* 2002 Oct 10;**347**(15):1199–200.

37   PARK Y, HAHM KS. Antimicrobial peptides (AMPs): peptide structure and mode of action. *J Biochem Mol Biol* 2005 Sep 30;**38**(5):507–16.

38   SHAI Y. Mode of action of membrane active antimicrobial peptides. *Biopolymers* 2002;**66**(4):236–48.

39   EPAND RM, EPAND RF. Modulation of membrane curvature by peptides. *Biopolymers* 2000;**55**(5):358–63.

40 JELINEK R, KOLUSHEVA S. Membrane interactions of host-defense peptides studied in model systems. *Curr Protein Pept Sci* 2005 Feb;**6**(1):103–14.

41 LOHNER K, BLONDELLE SE. Molecular mechanisms of membrane perturbation by antimicrobial peptides and the use of biophysical studies in the design of novel peptide antibiotics. *Comb Chem High Throughput Screen* 2005 May;**8**(3):241–56.

42 SALDITT T, LI C, SPAAR A. Structure of antimicrobial peptides and lipid membranes probed by interface-sensitive X-ray scattering. *Biochim Biophys Acta* 2006 Sep;**1758**(9):1483–98.

43 JENSSEN H, HAMILL P, HANCOCK RE. Peptide antimicrobial agents. *Clin Microbiol Rev* 2006 Jul;**19**(3):491–511.

44 BESSIN Y, SAINT N, MARRI L, MARCHINI D, MOLLE G. Antibacterial activity and pore-forming properties of ceratotoxins: a mechanism of action based on the barrel stave model. *Biochim Biophys Acta* 2004 Dec 15;**1667**(2): 148–56.

45 BOHEIM G. Statistical analysis of alamethicin channels in black lipid membranes. *J Membr Biol* 1974;**19**(3):277–303.

46 LEUSCHNER C, HANSEL W. Membrane disrupting lytic peptides for cancer treatments. *Curr Pharm Des* 2004;**10**(19):2299–310.

47 RAIMONDO D, ANDREOTTI G, SAINT N, AMODEO P, RENZONE G, SANSEVERINO M, *et al.* A folding-dependent mechanism of antimicrobial peptide resistance to degradation unveiled by solution structure of distinctin. *Proc Natl Acad Sci USA* 2005 May 3;**102**(18):6309–14.

48 HUANG HW. Action of antimicrobial peptides: two-state model. *Biochemistry* 2000 Jul 25;**39**(29):8347–52.

49 LUDTKE SJ, HE K, HELLER WT, HARROUN TA, YANG L, HUANG HW. Membrane pores induced by magainin. *Biochemistry* 1996 Oct 29;**35**(43):13723–8.

50 SHAI Y. Mechanism of the binding, insertion and destabilization of phospholipid bilayer membranes by alpha-helical antimicrobial and cell non-selective membrane-lytic peptides. *Biochim Biophys Acta* 1999 Dec 15;**1462**(1–2): 55–70.

51 HANCOCK REW, LEHRER R. Cationic peptides: a new source of antibiotics. *Trends Biotechnol* 1998;**16**(2):82–8.

52 SIMMACO M, MIGNONGNA G, BARRA D. Antimicrobial peptides from amphibian skin: what do they tell us? *Biopolymers* 1998;**47**(6):435–50.

53 DAHER KA, SELSTED ME, LEHRER RI. Direct inactivation of viruses by human granulocyte defensins. *J Virol* 1986 Dec;**60**(3):1068–74.

54 BAGHIAN A, JAYNES J, ENRIGHT F, KOUSOULAS KG. An amphipathic alpha-helical synthetic peptide analogue of melittin inhibits herpes simplex virus-1 (HSV-1)-induced cell fusion and virus spread. *Peptides* 1997;**18**(2):177–83.

55 FALCO A, MAS V, TAFALLA C, PEREZ L, COLL JM, ESTEPA A. Dual antiviral activity of human alpha-defensin-1 against viral haemorrhagic septicaemia rhabdovirus (VHSV): inactivation of virus particles and induction of a type I interferon-related response. *Antiviral Res* 2007 Nov;**76**(2):111–23.

56 OWENS RJ, TANNER CC, MULLIGAN MJ, SRINIVAS RV, COMPAS RW. Oligopeptide inhibitors of HIV-induced syncytium formation. *AIDS Research Human Retroviruses* 1990;**6**:1289–96.

57 HANCOCK RE, ROZEK A. Role of membranes in the activities of antimicrobial cationic peptides. *FEMS Microbiol Lett* 2002 Jan 10;**206**(2):143–9.

58 KLOTMAN ME, CHANG TL. Defensins in innate antiviral immunity. *Nat Rev Immunol* 2006 Jun;**6**(6):447–56.

59 LEIKINA E, DELANOE-AYARI H, MELIKOV K, CHO MS, CHEN A, WARING AJ, *et al.* Carbohydrate-binding molecules inhibit viral fusion and entry by crosslinking membrane glycoproteins. *Nat Immunol* 2005 Oct;**6**(10):995–1001.

60 GALLO RL, HUTTNER KM. Antimicrobial peptides: an emerging concept in cutaneous biology. *J Invest Dermatol* 1998;**111**:739–43.

61 YOUNT NY, BAYER AS, XIONG YQ, YEAMAN MR. Advances in antimicrobial peptide immunobiology. *Biopolymers* 2006;**84**(5):435–58.

62 BOWDISH DM, DAVIDSON DJ, HANCOCK RE. Immunomodulatory properties of defensins and cathelicidins. *Curr Top Microbiol Immunol* 2006;**306**:27–66.

63 TERRITO MC, GANZ T, SELSTED ME, LEHRER R. Monocyte-chemotactic activity of defensins from human neutrophils. *J Clin Invest* 1989 Dec;**84**(6):2017–20.

64 HOLZL MA, HOFER J, STEINBERGER P, PFISTERSHAMMER K, ZLABINGER GJ. Host antimicrobial proteins as endogenous immunomodulators. *Immunol Lett* 2008 Aug 15;**119**(1–2):4–11.

65 YANG D, CHEN Q, HOOVER DM, STALEY P, TUCKER KD, LUBKOWSKI J, *et al.* Many chemokines including CCL20/MIP-3alpha display antimicrobial activity. *J Leukoc Biol* 2003 Sep;**74**(3):448–55.

66 CHERTOV O, MICHIEL DF, XU L, WANG JM, TANI K, MURPHY WJ, *et al.* Identification of defensin-1, defensin-2, and CAP37/azurocidin as T-cell chemoattractant proteins released from interleukin-8-stimulated neutrophils. *J Biol Chem* 1996 Feb 9;**271**(6):2935–40.

67 DURR M, PESCHEL A. Chemokines meet defensins: the merging concepts of chemoattractants and antimicrobial peptides in host defense. *Infect Immun* 2002 Dec;**70**(12):6515–7.

68 NOGA EJ, SILPHADUANG U. Piscidins: a novel family of peptide antibiotics from fish. *Drug News Perspect* 2003 Mar;**16**(2):87–92.

69 CHANG TL, VARGAS J, JR., DELPORTILLO A, KLOTMAN ME. Dual role of alpha-defensin-1 in anti-HIV-1 innate immunity. *J Clin Invest* 2005 Mar;**115**(3): 765–73.

70 SILPHADUANG U, NOGA EJ. Peptide antibiotics in mast cells of fish. *Nature* 2001 Nov 15;**414**(6861):268–9.

71 COLE AM, GANZ T. Human antimicrobial peptides: analysis and application. *Biotechniques* 2000 Oct;**29**(4):822–31.

72 SILPHADUANG U, COLORNI A, NOGA EJ. Evidence for widespread distribution of piscidin antimicrobial peptides in teleost fish. *Dis Aquat Organ* 2006 Oct 27;**72**(3):241–52.

73 MURRAY HM, GALLANT JW, DOUGLAS SE. Cellular localization of pleurocidin gene expression and synthesis in winter flounder gill using immunohistochemistry and *in situ* hybridization. *Cell Tissue Res* 2003 May;**312**(2):197–202.

74 SALERNO G, PARRINELLO N, ROCH P, CAMMARATA M. cDNA sequence and tissue expression of an antimicrobial peptide, dicentracin; a new component of the moronecidin family isolated from head kidney leukocytes of sea bass, *Dicentrarchus labrax. Comp Biochem Physiol B Biochem Mol Biol* 2007 Apr;**146**(4): 521–9.

75 NOGA EJ, SILPHADUANG U, PARK NG, SEO JK, STEPHENSON J, KOZLOWICZ S. Piscidin 4, a novel member of the piscidin family of antimicrobial peptides. *Comp Biochem Physiol B Biochem Mol Biol* 2009 Apr;**152**(4):299–305.

76 CHINCHAR VG, BRYAN L, SILPHADUANG U, NOGA E, WADE D, ROLLINS-SMITH L. Inactivation of viruses infecting ectothermic animals by amphibian and piscine antimicrobial peptides. *Virology* 2004 Jun 1;**323**(2):268–75.

77 SUN BJ, XIE HX, SONG Y, NIE P. Gene structure of an antimicrobial peptide from mandarin fish, *Siniperca chuatsi* (Basilewsky), suggests that moronecidins and pleurocidins belong in one family: the piscidins. *J Fish Dis* 2007 Jun;**30**(6): 335–43.

78 CORRALES J, MULERO I, MULERO V, NOGA EJ. Detection of antimicrobial peptides related to piscidin 4 in important aquacultured fish. *Dev Comp Immunol* 2010 Mar;**34**(3):331–43.

79 FERNANDES JM, RUANGSRI J, KIRON V. Atlantic cod piscidin and its diversification through positive selection. *PLoS One* 2010;**5**(3):e9501.

80 LAUTH X, SHIKE H, BURNS JC, WESTERMAN ME, OSTLAND VE, CARLBERG JM, *et al.* Discovery and characterization of two isoforms of moronecidin, a novel antimicrobial peptide from hybrid striped bass. *J Biol Chem* 2002 Feb 15;**277**(7):5030–9.

81 PATRZYKAT A, GALLANT JW, SEO J-K, PYTYCK J, DOUGLAS SE. Novel antimicrobial peptides derived from flatfish genes. *Antimicrob Agents Chemother* 2003 August 1;**47**(8):2464–70.

82 DOUGLAS SE, GALLANT JW, GONG Z, HEW C. Cloning and developmental expression of a family of pleurocidin-like antimicrobial peptides from winter flounder, *Pleuronectes americanus* (Walbaum). *Dev & Comp Immunol* 2001;**25**(2): 137–47.

83 BROCAL I, FALCO A, MAS V, ROCHA A, PEREZ L, COLL JM, *et al.* Stable expression of bioactive recombinant pleurocidin in a fish cell line. *Appl Microbiol Biotechnol* 2006 Oct;**72**(6):1217–28.

84 RITONJA A, KOPITAR M, JERALA R, TURK V. Primary structure of a new cysteine proteinase inhibitor from pig leucocytes. *FEBS Lett* 1989 Sep 25;**255**(2): 211–14.

85 ZANETTI M, GENNARO R, ROMEO D. Cathelicidins: a novel protein family with a common proregion and a variable C-terminal antimicrobial domain. *FEBS Lett* 1995 Oct 23;**374**(1):1–5.

86 TOMASINSIG L, ZANETTI M. The cathelicidins – structure, function and evolution. *Curr Protein Pept Sci* 2005 Feb;**6**(1):23–34.

87 CHANG CI, ZHANG YA, ZOU J, NIE P, SECOMBES CJ. Two cathelicidin genes are present in both rainbow trout (*Oncorhynchus mykiss*) and atlantic salmon (*Salmo salar*). *Antimicrob Agents Chemother* 2006 Jan;**50**(1):185–95.

88 SCOCCHI M, PALLAVICINI A, SALGARO R, BOCIEK K, GENNARO R. The salmonid cathelicidins: a gene family with highly varied C-terminal antimicrobial domains. *Comp Biochem Physiol B Biochem Mol Biol* 2009 Apr;**152**(4):376–81.

89 MAIER VH, DORN KV, GUDMUNDSDOTTIR BK, GUDMUNDSSON GH. Characterisation of cathelicidin gene family members in divergent fish species. *Mol Immunol* 2008 Aug;**45**(14):3723–30.

90 MAIER VH, SCHMITT CN, GUDMUNDSDOTTIR S, GUDMUNDSSON GH. Bacterial DNA indicated as an important inducer of fish cathelicidins. *Mol Immunol* 2008 Apr;**45**(8):2352–8.

91 UZZELL T, STOLZENBERG ED, SHINNAR AE, ZASLOFF M. Hagfish intestinal antimicrobial peptides are ancient cathelicidins. *Peptides* 2003 Nov;**24**(11): 1655–67.

92 LAZAROVICI P, PRIMOR N, LOEW LM. Purification and pore-forming activity of two hydrophobic polypeptides from the secretion of the Red Sea Moses sole (*Pardachirus marmoratus*). *J Biol Chem* 1986 Dec 15;**261**(35):16704–13.

93 OREN Z, SHAI Y. A class of highly potent antibacterial peptides derived from pardaxin, a pore-forming peptide isolated from Moses sole fish *Pardachirus marmoratus. Eur J Biochem* 1996 Apr 1;**237**(1):303–10.

94 THOMPSON SA, TACHIBANA K, NAKANISHI K, KUBOTA I. Melittin-like peptides from the shark-repelling defense secretion of the sole *Pardachirus pavoninus. Science* 1986 Jul 18;**233**(4761):341–3.

95 PAN CY, CHEN JY, CHENG YS, CHEN CY, NI IH, SHEEN JF, *et al.* Gene expression and localization of the epinecidin-1 antimicrobial peptide in the grouper (*Epinephelus coioides*), and its role in protecting fish against pathogenic infection. *DNA Cell Biol* 2007 Jun;**26**(6):403–13.

96 PAN C-Y, CHEN J-Y, CHENG Y-SE, CHEN C-Y, NI IH, SHEEN J-F, *et al.* Gene expression and localization of the epinecidin-1 antimicrobial peptide in the grouper

(*Epinephelus coioides*), and its role in protecting fish against pathogenic infection. *DNA Cell Biol* 2007;**26**(6):403–13.

97   PARK CB, LEE JH, PARK IY, KIM MS, KIM SC. A novel antimicrobial peptide from the loach, *Misgurnus anguillicaudatus*. *FEBS Lett* 1997 Jul 14;**411**(2–3): 173–8.

98   IIJIMA N, TANIMOTO N, EMOTO Y, MORITA Y, UEMATSU K, MURAKAMI T, *et al.* Purification and characterization of three isoforms of chrysophsin, a novel antimicrobial peptide in the gills of the red sea bream, *Chrysophrys major*. *Eur J Biochem* 2003 Feb;**270**(4):675–86.

99   BIRKEMO GA, LUDERS T, ANDERSEN O, NES IF, NISSEN-MEYER J. Hipposin, a histone-derived antimicrobial peptide in Atlantic halibut (*Hippoglossus hippoglossus* L.). *Biochim Biophys Acta* 2003 Mar 21;**1646**(1–2):207–15.

100 BIRKEMO GA, MANTZILAS D, LUDERS T, NES IF, NISSEN-MEYER J. Identification and structural analysis of the antimicrobial domain in hipposin, a 51-mer antimicrobial peptide isolated from Atlantic halibut. *Biochim Biophys Acta* 2004 Jun 1;**1699**(1–2):221–7.

101 PARK IY, PARK CB, KIM MS, KIM SC. Parasin I, an antimicrobial peptide derived from histone H2A in the catfish, *Parasilurus asotus*. *FEBS Lett* 1998 Oct 23;**437**(3):258–62.

102 CHO JH, PARK IY, KIM HS, LEE WT, KIM MS, KIM SC. Cathepsin D produces antimicrobial peptide parasin I from histone H2A in the skin mucosa of fish. *Faseb J* 2002 Mar;**16**(3):429–31.

103 SYVITSKI RT, BURTON I, MATTATALL NR, DOUGLAS SE, JAKEMAN DL. Structural characterization of the antimicrobial peptide pleurocidin from winter flounder. *Biochemistry* 2005 May 17;**44**(19):7282–93.

104 CAMPAGNA S, SAINT N, MOLLE G, AUMELAS A. Structure and mechanism of action of the antimicrobial peptide piscidin. *Biochemistry* 2007 Feb 20;**46**(7):1771–8.

105 MASON AJ, CHOTIMAH IN, BERTANI P, BECHINGER B. A spectroscopic study of the membrane interaction of the antimicrobial peptide pleurocidin. *Mol Membr Biol* 2006 Mar-Apr;**23**(2):185–94.

106 CHEKMENEV EY, VOLLMAR BS, FORSETH KT, MANION MN, JONES SM, WAGNER TJ, *et al.* Investigating molecular recognition and biological function at interfaces using piscidins, antimicrobial peptides from fish. *Biochim Biophys Acta* 2006 Sep;**1758**(9):1359–72.

107 BECHINGER B, ZASLOFF M, OPELLA SJ. Structure and orientation of the antibiotic peptide magainin in membranes by solid-state nuclear magnetic resonance spectroscopy. *Protein Sci* 1993 Dec;**2**(12):2077–84.

108 POUNY Y, RAPAPORT D, MOR A, NICOLAS P, SHAI Y. Interaction of antimicrobial dermaseptin and its fluorescently labeled analogues with phospholipid membranes. *Biochemistry* 1992 Dec 15;**31**(49):12416–23.

109 SHAI Y. Molecular recognition between membrane-spanning polypeptides. *Trends Biochem Sci* 1995 Nov;**20**(11):460–4.

110 OJCIUS DM, YOUNG JD. Cytolytic pore-forming proteins and peptides: is there a common structural motif? *Trends Biochem Sci* 1991 Jun;**16**(6):225–9.

111 LAZAROVICI P, LELKES PI. Pardaxin induces exocytosis in bovine adrenal medullary chromaffin cells independent of calcium. *J Pharmacol Exp Ther* 1992 Dec;**263**(3):1317–26.

112 TRAN D, TRAN PA, TANG YQ, YUAN J, COLE T, SELSTED ME. Homodimeric theta-defensins from rhesus macaque leukocytes: isolation, synthesis, antimicrobial activities, and bacterial binding properties of the cyclic peptides. *J Biol Chem* 2002 Feb 1;**277**(5):3079–84.

113 TANG YQ, YUAN J, OSAPAY G, OSAPAY K, TRAN D, MILLER CJ, *et al.* A cyclic antimicrobial peptide produced in primate leukocytes by the ligation of two truncated alpha-defensins. *Science* 1999 Oct 15;**286**(5439):498–502.

114 SEMPLE CA, ROLFE M, DORIN JR. Duplication and selection in the evolution of primate beta-defensin genes. *Genome Biol* 2003;**4**(5):R31.

115 ZOU J, MERCIER C, KOUSSOUNADIS A, SECOMBES C. Discovery of multiple beta-defensin like homologues in teleost fish. *Mol Immunol* 2007 Jan;**44**(4): 638–47.

116 FALCO A, CHICO V, MARROQUI L, PEREZ L, COLL JM, ESTEPA A. Expression and anti-viral activity of a beta-defensin-like peptide identified in the rainbow trout (*Oncorhynchus mykiss*) EST sequences. *Mol Immunol* 2008 Feb;**45**(3): 757–65.

117 CASADEI E, WANG T, ZOU J, GONZALEZ VECINO JL, WADSWORTH S, SECOMBES CJ. Characterization of three novel beta-defensin antimicrobial peptides in rainbow trout (*Oncorhynchus mykiss*). *Mol Immunol* 2009 Oct;**46**(16): 3358–66.

118 ZHAO JG, ZHOU L, JIN JY, ZHAO Z, LAN J, ZHANG YB, et al. Antimicrobial activity-specific to Gram-negative bacteria and immune modulation-mediated NF-kappaB and Sp1 of a medaka beta-defensin. *Dev Comp Immunol* 2009 Apr;**33**(4):624–37.

119 NAM BH, YAMAMOTO E, HIRONO I, AOKI T. A survey of expressed genes in the leukocytes of Japanese flounder, *Paralichthys olivaceus*, infected with Hirame rhabdovirus. *Dev Comp Immunol* 2000;**24**:13–24.

120 VALORE EV, MARTIN E, HARWIG SS, GANZ T. Intramolecular inhibition of human defensin HNP-1 by its propiece. *J Clin Invest* 1996 Apr 1;**97**(7):1624–9.

121 KRAUSE A, NEITZ S, MAGERT HJ, SCHULZ A, FORSSMANN WG, SCHULZ-KNAPPE P, et al. LEAP-1, a novel highly disulfide-bonded human peptide, exhibits anti-microbial activity. *FEBS Lett* 2000 Sep 1;**480**(2–3):147–50.

122 PARK CH, VALORE EV, WARING AJ, GANZ T. Hepcidin, a urinary antimicrobial peptide synthesized in the liver. *J Biol Chem* 2001 Mar 16;**276**(11):7806–10.

123 WEINSTEIN DA, ROY CN, FLEMING MD, LODA MF, WOLFSDORF JI, ANDREWS NC. Inappropriate expression of hepcidin is associated with iron refractory anemia: implications for the anemia of chronic disease. *Blood* 2002 Nov 15;**100**(10): 3776–81.

124 VERGA FALZACAPPA MV, MUCKENTHALER MU. Hepcidin: iron-hormone and anti-microbial peptide. *Gene* 2005 Dec 30;**364**:37–44.

125 VYORAL D, PETRAK J. Hepcidin: a direct link between iron metabolism and immunity. *Int J Biochem Cell Biol* 2005 Sep;**37**(9):1768–73.

126 DOUGLAS SE, GALLANT JW, LIEBSCHER RS, DACANAY A, TSOI SC. Identification and expression analysis of hepcidin-like antimicrobial peptides in bony fish. *Dev Comp Immunol* 2003 Jun–Jul;**27**(6–7):589–601.

127 RODRIGUES PN, VAZQUEZ-DORADO S, NEVES JV, WILSON JM. Dual function of fish hepcidin: response to experimental iron overload and bacterial infection in sea bass (*Dicentrarchus labrax*). *Dev Comp Immunol* 2006;**30**(12):1156–67.

128 SHIKE H, LAUTH X, WESTERMAN ME, OSTLAND VE, CARLBERG JM, VAN OLST JC, et al. Bass hepcidin is a novel antimicrobial peptide induced by bacterial challenge. *Eur J Biochem* 2002 Apr;**269**(8):2232–7.

129 DOUGLAS SE, GALLANT JW. Isolation of cDNAs for trypsinogen from the winter flounder, *Pleuronectes americanus*. *J Mar Biotechnol* 1998 Dec;**6**(4): 214–19.

130 SHIKE H, SHIMIZU C, LAUTH X, BURNS JC. Organization and expression analysis of the zebrafish hepcidin gene, an antimicrobial peptide gene conserved among vertebrates. *Dev Comp Immunol* 2004 Jun;**28**(7–8):747–54.

131 BAYNE CJ, GERWICK L, FUJIKI K, NAKAO M, YANO T. Immune-relevant (including acute phase) genes identified in the livers of rainbow trout, *Oncorhynchus mykiss*, by means of suppression subtractive hybridization. *Dev Comp Immunol* 2001 Apr;**25**(3):205–17.

132 GRACEY AY, TROLL JV, SOMERO GN. Hypoxia-induced gene expression profiling in the euryoxic fish *Gillichthys mirabilis*. *Proc Natl Acad Sci USA* 2001 Feb 13;**98**(4):1993–8.

133 REN HL, WANG KJ, ZHOU HL, YANG M. Cloning and organisation analysis of a hepcidin-like gene and cDNA from Japan sea bass, *Lateolabrax japonicus*. *Fish Shellfish Immunol* 2006 Sep;**21**(3):221–7.

134 HUANG P, CHEN JY, KUO CM. Three different hepcidins from tilapia, *Oreochromis mossambicus*: analysis of their expressions and biological functions. *Mol Immunol* 2007;**44**:1932–44.

135 HIRONO I, HWANG JY, ONO Y, KUROBE T, OHIRA T, NOZAKI R, *et al.* Two different types of hepcidins from the Japanese flounder *Paralichthys olivaceus*. *FEBS J* 2005 Oct;**272**(20):5257–64.

136 YANG M, WANG KJ, CHEN JH, QU HD, LI SJ. Genomic organization and tissue-specific expression analysis of hepcidin-like genes from black porgy (*Acanthopagrus schlegelii* B). *Fish Shellfish Immunol* 2007 Nov;**23**(5):1060–71.

137 KRAUSE A, SILLARD R, KLEEMEIER B, KLUVER E, MARONDE E, CONEJO-GARCIA JR, *et al.* Isolation and biochemical characterization of LEAP-2, a novel blood peptide expressed in the liver. *Protein Sci* 2003 Jan;**12**(1):143–52.

138 ZHANG YA, ZOU J, CHANG CI, SECOMBES CJ. Discovery and characterization of two types of liver-expressed antimicrobial peptide 2 (LEAP-2) genes in rainbow trout. *Vet Immunol Immunopathol* 2004 Oct;**101**(3–4):259–69.

139 BAO B, PEATMAN E, XU P, LI P, ZENG H, HE C, *et al.* The catfish liver-expressed antimicrobial peptide 2 (LEAP-2) gene is expressed in a wide range of tissues and developmentally regulated. *Mol Immunol* 2006 Feb;**43**(4):367–77.

140 LIU F, LI JL, YUE GH, FU JJ, ZHOU ZF. Molecular cloning and expression analysis of the liver-expressed antimicrobial peptide 2 (LEAP-2) gene in grass carp. *Vet Immunol Immunopathol* 2010 Feb 15;**133**(2–4):133–43.

141 WANG Q, WANG Y, XU P, LIU Z. NK-lysin of channel catfish: gene triplication, sequence variation, and expression analysis. *Mol Immunol* 2006 Apr;**43**(10): 1676–86.

142 HIRONO I, KONDO H, KOYAMA T, ARMA NR, HWANG JY, NOZAKI R, *et al.* Characterization of Japanese flounder (*Paralichthys olivaceus*) NK-lysin, an antimicrobial peptide. *Fish Shellfish Immunol* 2007 May;**22**(5):567–75.

143 PATRZYKAT A, GALLANT JW, SEO JK, PYTYCK J, DOUGLAS SE. Novel antimicrobial peptides derived from flatfish genes. *Antimicrob Agents Chemother* 2003 Aug;**47**(8):2464–70.

144 PAN CY, CHEN JY, LIN TL, LIN CH. *In vitro* activities of three synthetic peptides derived from epinecidin-1 and an anti-lipopolysaccharide factor against *Propionibacterium acnes*, *Candida albicans*, and *Trichomonas vaginalis*. *Peptides* 2009 Jun;**30**(6):1058–68.

145 LAUTH X, BABON JJ, STANNARD JA, SINGH S, NIZET V, CARLBERG JM, *et al.* Bass hepcidin synthesis, solution structure, antimicrobial activities and synergism, and *in vivo* hepatic response to bacterial infections. *J Biol Chem* 2005 Mar 11;**280**(10):9272–82.

146 PATRZYKAT A, ZHANG L, MENDOZA V, IWAMA GK, HANCOCK REW. Synergy of histone-derived peptides of coho salmon with lysozyme and flounder pleurocidin. *Antimicrob Agents Chemother* 2001 May 1;**45**(5):1337–42.

147 HSIEH JC, PAN CY, CHEN JY. Tilapia hepcidin (TH)2–3 as a transgene in transgenic fish enhances resistance to *Vibrio vulnificus* infection and causes variations in immune-related genes after infection by different bacterial species. *Fish Shellfish Immunol* 2010 Sep;**29**(3):430–9.

148 LIN WJ, CHIEN YL, PAN CY, LIN TL, CHEN JY, CHIU SJ, *et al.* Epinecidin-1, an antimicrobial peptide from fish (*Epinephelus coioides*) which has an antitumor

effect like lytic peptides in human fibrosarcoma cells. *Peptides* 2009 Feb; **30**(2):283–90.

149 PENG KC, PAN CY, CHOU HN, CHEN JY. Using an improved Tol2 transposon system to produce transgenic zebrafish with epinecidin-1 which enhanced resistance to bacterial infection. *Fish Shellfish Immunol* 2010 May–Jun;**28**(5–6):905–17.

150 SARMASIK A, WARR G, CHEN TT. Production of transgenic medaka with increased resistance to bacterial pathogens. *Mar Biotechnol (NY)* 2002 Jun;**4**(3): 310–22.

151 DUNHAM RA, WARR GW, NICHOLS A, DUNCAN PL, ARGUE B, MIDDLETON D, *et al.* Enhanced bacterial disease resistance of transgenic channel catfish *Ictalurus punctatus* possessing cecropin genes. *Mar Biotechnol (NY)* 2002 Jun;**4**(3): 338–44.

152 SUNG WS, LEE J, LEE DG. Fungicidal effect and the mode of action of piscidin 2 derived from hybrid striped bass. *Biochem Biophys Res Commun* 2008 Jul 4;**371**(3):551–5.

153 COLORNI A, ULLAL A, HEINISCH G, NOGA EJ. Activity of the antimicrobial polypeptide piscidin 2 against fish ectoparasites. *J Fish Dis* 2008 Jun;**31**(6): 423–32.

154 WANG YD, KUNG CW, CHEN JY. Antiviral activity by fish antimicrobial peptides of epinecidin-1 and hepcidin 1–5 against nervous necrosis virus in medaka. *Peptides* 2010 Jun;**31**(6):1026–33.

155 WANG YD, KUNG CW, CHI SC, CHEN JY. Inactivation of nervous necrosis virus infecting grouper (*Epinephelus coioides*) by epinecidin-1 and hepcidin 1–5 antimicrobial peptides, and downregulation of Mx2 and Mx3 gene expressions. *Fish Shellfish Immunol* 2010 Jan;**28**(1):113–20.

156 CHIOU PP, LIN CM, PEREZ L, CHEN TT. Effect of cecropin B and a synthetic analogue on propagation of fish viruses *in vitro. Mar Biotechnol (NY)* 2002 Jun;**4**(3):294–302.

157 BOMAN HG, HULTMARK D. Cell-free immunity in insects. *Annu Rev Microbiol* 1987;**41**:103–26.

158 SALLUM UW, CHEN TT. Inducible resistance of fish bacterial pathogens to the antimicrobial peptide cecropin B. *Antimicrob Agents Chemother* 2008 September 1;**52**(9):3006–12.

159 FALCO A, BROCAL I, PEREZ L, COLL JM, ESTEPA A, TAFALLA C. *In vivo* modulation of the rainbow trout (*Oncorhynchus mykiss*) immune response by the human alpha defensin 1, HNP1. *Fish Shellfish Immunol* 2008 Jan;**24**(1):102–12.

160 PAN YL, CHENG JT, HALE J, PAN J, HANCOCK RE, STRAUS SK. Characterization of the structure and membrane interaction of the antimicrobial peptides aurein 2.2 and 2.3 from Australian southern bell frogs. *Biophys J* 2007 Apr 15;**92**(8):2854–64.

161 NAM BH, MOON JY, KIM YO, KONG HJ, KIM WJ, LEE SJ, *et al.* Multiple beta-defensin isoforms identified in early developmental stages of the teleost *Paralichthys olivaceus. Fish Shellfish Immunol* Feb;**28**(2):267–74.

162 DOUGLAS SE, GALLANT JW, BULLERWELL CE, WOLFF C, MUNHOLLAND J, REITH ME. Winter flounder expressed sequence tags: establishment of an EST database and identification of novel fish genes. *Mar Biotechnol (NY)* 1999 Sep;**1**(5):458–64.

163 HUANG PH, CHEN JY, KUO CM. Three different hepcidins from tilapia, *Oreochromis mossambicus*: analysis of their expressions and biological functions. *Mol Immunol* 2007 Mar;**44**(8):1922–34.

18

Advances in non-chemical methods for parasite prevention and control in fish

C. Sommerville, University of Stirling, UK

Abstract: The chapter sets out to introduce methods for the management of parasitic infection without the use of chemicals. It covers suggested preventative measures, management of disease in culture situations and non-chemotherapeutic interventions; measures to be used where available and measures which have potential for the future. Importantly, the methods outlined are intended to be used in concert as together they constitute ideas for an integrated pest management (IPM) strategy for parasite infection in aquaculture production where chemotherapy is the last resort. Though many methods require further development, the use of such a strategy is a sustainable way to manage disease so as to avoid or delay the use of chemicals and has the benefit of delaying parasite resistance to the chemicals.

Key words: integrated pest management (IPM) strategy, disease management, preventative measures, removal of infectious agents, biocontrol.

18.1 Introduction

In recent decades, the application of chemical treatments to control parasitic infection in fish culture has tended to become the dominant strategy, particularly in developed countries. This has undoubtedly been due to the increased value of cultured fish in the market economy. The increased production volume of cultured fish, particularly the salmonids, has attracted major pharmaceutical companies to invest in the development of selected effective products for use in fish parasite control and to obtain licences for their use. Despite the enormous costs of research and development for licensing new treatments, it has been cost-effective, particularly in the area of drug or chemotherapeutants against sea lice on salmon. There are many reasons, however, for choosing non-chemical methods for controlling a disease problem in aquaculture. In many cases there are no suitable or effective chemical methods available for controlling a specific disease, or it may be that conditions are such that a chemical method cannot be used in a particular situation. However, there are many arguments for trying to control disease without the use of chemicals or drugs which are related to sustainability and also, importantly, to cost.

All manipulations of the natural environment create imbalance in the ecosystem and the introduction of a chemical tends to cause a more severe and acute impact so that the return to equilibrium takes longer. Clearly, some chemicals are more damaging than others and a major effort is made by the producers of the chemicals and drugs, and by the regulators, to ensure minimal adverse impact. Precise details of administration and strict instructions on careful handling are presented with the chemical/drug by the producer and, provided these are followed, there should be no danger to the fish being treated or to the farm staff administering the treatment. There may be side effects of treatments on the fish, many of which are unknown, and a certain level of mortality may occur, particularly where the treatment dose has a low therapeutic index. This is because it is especially difficult to administer the same dose to all fish in a large rearing unit such as a sea cage containing tens of thousands of fish. For example, during in-feed treatments of a hierarchical population of fish such as salmon, dominant feeding fish may ingest more medicated food and thus there is a risk of overdose and an accompanying potential for the development of resistance in pathogens of under-dosed fish low in the hierarchy. There is also a risk with topical treatments. Obtaining an homogeneous distribution of the chemical during a bath treatment is difficult and may result in hotspots and consequent overdosing of some fish. As for the safety of the consumer, there are strict controls over the residue levels in fish and information regarding withdrawal times between treatment and harvest. The recovery of the environment is sometimes a slow process and residues inevitably find their way into the food chain whereby they become concentrated in organisms at the higher trophic levels. This is taken into account by the licensing authorities who require detailed information on the effect on non-target organisms in the environment. The dispersion of the drug or chemical is subjected to models and analysed prior to licensing, taking into account that elements of the product or its breakdown become bound in sediments at the fish farm site, e.g. Telfer *et al.* (2006).

All of these precautions apply but are monitored only in a limited number of countries where there is a constituted regulatory authority with well-defined, well-presented and well-policed regulations. However, aquaculture is practised worldwide (FAO records statistics from 186 countries) and even where regulations exist, safety depends on effective monitoring and surveillance of the regulations (FAO 2002). For example, Faruk *et al.* (2008) evaluated the current use of chemicals and antibiotics used in freshwater aquaculture in Bangladesh. They list a large number of these and reported that farmers in Bangladesh are using chemicals without knowing their necessity and effectiveness; they highlighted the lack of information accompanying chemical products on dose rate and methods of application for fish, some of which are products not developed for use in fish. Although Faruk *et al.* focused on Bangladesh, personal observations confirm that this is the case in a number of countries developing aquaculture, where there is a lack of expertise, information or regulation. Much of the production in the developing world is exported and, in terms of consumer safety, it is left to importing countries to test imports on a regular basis, although tests are often carried out for only a restricted number of known treatment chemicals, and sample sizes may be small (Johnston and Santillo 2002). The majority of aquaculture production is in countries which do not have elaborate drug approval processes, if any at all. Even those with drug approval have questionable surveillance and reporting.

There is a wide variety of ways in which disease can be prevented, have a minimised impact or be controlled without the use of drugs or chemicals, and advances in these will be explored in this chapter. Many of them represent cheaper alternatives to chemotherapy and it is unfortunate, and to some extent inexplicable, why the initial action of fish farmers appears to be to reach for a chemical solution before exploring the non-chemical alternatives. It is a common misconception that it is possible to eliminate a pathogen completely and that the instrument for this is chemical, the proverbial 'silver bullet'. This chapter does not promote the rejection of chemotherapeutic methods but the judicious use of them and only where other methods have failed. This will have the benefits of prolonging the useful life of a chemical by minimising the chances of development of resistance by the pathogen, maximising the chances of environmental recovery and avoiding consumer and public outcries, which have too often followed the development of new drug treatments against sea lice (*Lepeophtheirus salmonis*).

18.2   Principles of disease management without chemicals

Aquaculturalists need yet to fully embrace the usefulness of non-chemical control methods and to see them as essential components of an integrated pest management (IPM) system. Fortunately, there is now an increasing recognition that therapeutants need to be used more judiciously in order to prevent or delay the development of resistance to the licensed chemotherapies available. This realisation has come about the hard way and is particularly acute for the infection of salmon by sea lice (*L. salmonis* and *Caligus elongatus*) where epizootics occur over which there may be little control. Mortalities, together with costs of treatments, have resulted in losses reported to be on average 6% of the cost of production (Costello 2009). After more than 25 years and 7 or more licensed products, there is still no universally satisfactory chemotherapy available due to the development of resistance in many populations of lice, which has arisen often as a result of over-use, failure to rotate, or misuse. Yet the emphasis is still on the sole use of chemotherapies without full exploration and use of the non-chemical alternatives.

A good IPM strategy draws on a wide range of resources to combat disease and utilises non-chemical as well as chemical methods for control (Sommerville 2009). IPM strategies embrace the contribution of other control methods including biological, cultural/husbandry, regulatory and social resources, all used in concert. The 'integrated' element of pest management is often the one that is overlooked. Chemotherapy is only one component and should be used strategically in order to reduce the development of resistance in the pathogen and the health and safety of fish, consumers and environment. Although possibly more challenging, it is essential that the whole range of non-chemical methods available should be drawn on and incorporated into a comprehensive IPM, attacking the problem from many and varied angles to achieve best results (Fig. 18.1). There are

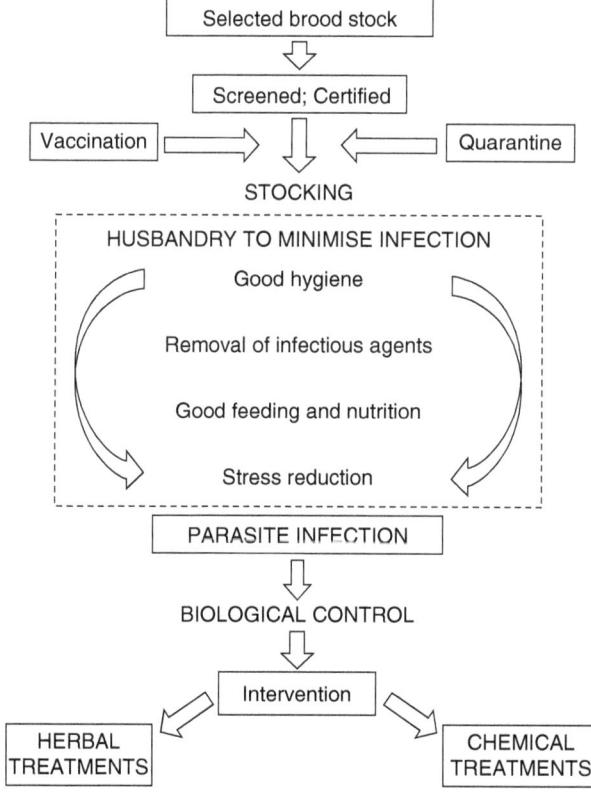

Fig. 18.1 Integrated pest management strategy for parasite infections in aquaculture: how the components described in this chapter can be integrated into a coherent strategy which will minimise the use of chemotherapeutants.

about 500 or more resistant insect species worldwide and the aquaculture industry could learn from successful IPM practices developed by horticulturalists which were brought about by the increase in pest resistance causing crises in crop production (Thomas 1999). A change of attitude is required where the aquaculturalist learns to live with parasites and considers parasite pathogen management and control rather than elimination. The alternative is to engage in a race with evolution; one which will always be lost by the fish farmer.

The following text attempts to draw together the non-chemical components of an IPM strategy in the expectation that all approaches are used simultaneously and chemical intervention is a last resort. Many of the components are elaborated fully elsewhere in this book but are brought together here as components of an IPM strategy as outlined in Fig. 18.1.

18.3   Preventative measures

Prevention is in the first line of disease management, and the development and implementation of biosecurity strategies for fish diseases is exercised at national level and is embodied in the codes of practice of several international organisations such as the EU 2006 Directive and the OIE code of practice. Oidtmann *et al.* (2011) provide an overview of international and national biosecurity strategies in aquatic animal health. Preventative measures on the whole require more risk analysis and closer adherence than they currently enjoy so that they may become a higher priority. Unfortunately, the preventative measures most closely adhered to tend to be largely those most closely regulated. Some of these preventative measures as outlined below are subjected to intensive research and are dealt with elsewhere in this book.

18.3.1   Selective breeding
Stocking a pathogen-resistant host population is the most desirable requisite for disease control, since clinical disease is prevented by the natural immune process of the fish. Resistant host populations can be achieved in a number of ways and research is advanced in some more than others.

The wide range of variation in disease susceptibility in the numerous offspring of a breeding pair and selectively breeding from the healthiest has been used traditionally for optimising performance under intensive rearing conditions. Oleson *et al.* (2003) discussed breeding goals and cautioned that increased susceptibility to disease may become an unwanted side effect of selection for other traits, for example, fish with a desirable growth trait might be more susceptible to certain virus diseases. Nevertheless, selection of brood stock for a variety of characters over many generations during the last four decades has produced some robust, semi-domesticated cultured fish species, for example Atlantic salmon, common carp and tilapia. Classical breeding programmes, that is, selective breeding, crossbreeding, and hybridisation, are the mainstream of finfish genetic improvement and these have largely been applied to salmonids, especially rainbow trout and salmon but also to Nile tilapia.

Genetic improvement using modern genetic technologies was recently reviewed by McAndrew and Napier (2011). Though the commercial use of these technologies is not as yet very widespread, it is on the increase, especially in developed countries. Most programmes still seek to improve growth rate and food conversion efficiency but as fish welfare is becoming more significant (Ashley 2007) there is now selection for animal welfare-related traits, and disease resistance has become a key driver. Pottinger and Carrick (1999) were able to show experimentally that the stress response could be manipulated by selection of rainbow trout. Several studies have been published showing the existence of genetically determined variation in resistance to infectious diseases within and between fish stocks (Chevassus and Dobson 1990; Gjedrem et al. 1991; Fjalestad et al. 1993; Midtlyng et al. 2002). These record specific traits for resistance to a variety of viral diseases (infectious pancreatic necrosis (IPN), infectious salmon anaemia (ISA), viral haemorrhagic septicaemia (VHS), infectious hematopoietic necrosis (IHN)) and bacterial diseases such as furunculosis, enteric redmouth (ERM) and rainbow trout fry syndrome (RTFS), indicating that resistance could be improved by selective breeding. Henryon (2005) showed that additive variation was also possible and selected for heritable traits conferring multiple resistances to ERM, RTFS and VHS. Some progress has been made in producing virus resistant stocks. It has been shown that major histocompatibility complex (MHC) variants may be associated with susceptibility or resistance to bacteria and viruses (Grimholt et al. 2003).

In general, however, there has been very little consideration of resistance to parasitic diseases with the exception of some myxosporidians (Dionne et al. 2009) and sea lice (Glover et al. 2005, 2007). Genomic analysis of salmon families of variable susceptibilities suggest that major genes may exist for sea lice resistance that could be used to select strains of salmon for breeding more resistant populations; this process would also enable the elimination of susceptible families in breeding programmes. Kolstad et al. (2005) demonstrated significant variation in heritability of resistance to sea lice following trials using both natural infection and experimental challenge infection. Commercial breeding programmes are still few but have resulted in the production of eggs of Atlantic salmon claimed to be resistant to sea lice now becoming available. There is as yet limited availability and only time will tell the extent to which they contribute towards sea lice control in the field.

The most notable application so far resulting from the use of molecular techniques is the breeding of lymphocystis disease-resistant Japanese flounder (*Paralichthys olivaceus*) (Fuji et al. 2006, 2007). This was not a selective breeding programme *per se* but researchers identified a major locus that is mapped to LG 15 of the *P. olivaceus* linkage map which is highly associated with resistance to lymphocystis disease. The lymphocystis resistance and the marker are inherited in Mendelian fashion with lymphocystis resistance being the dominant trait.

Fish have a high reproductive rate and, therefore, although generation time is relatively short, selecting fish with desirable traits is time-consuming and very labour-intensive. Selective breeding programmes have been greatly accelerated by molecular techniques (reviewed by Lui, 2009) in recent years which have overcome some of the time constraints, especially the technique of marker assisted selection (MAS). Hulata (2009) has recently reviewed the status of genetic improvement in finfish. Quantitative trait loci (QTLs) have been detected in various aquacultural species and amongst these traits are resistance to stress, pathogens and disease for salmon, char, tilapias and common carp. When evaluating traits it is necessary to consider them in the context of environmental variation so as to study the extent of genotype × environmental interaction to assess the value of the trait under the different rearing environments of varied geographical areas.

Genetic manipulation
There have been indications that triploidy may be linked to possible disease resistance in fish in some cases, though evidence is thin (Maxime 2008). Triploidy hybridisation of certain crosses produced disease resistance according to Parsons *et al.* (1986), suggesting that careful selection of resistant parents for the production of triploids is worthy of investigation. However, other studies have indicated that triploidy results in increased susceptibility and therefore should be avoided for stocking purposes in endemic areas. Tildesley (2008), for example, provided experimental evidence that triploid rainbow trout were significantly more susceptible to infection by the gill parasitic crustacean *Ergasilus sieboldi* than diploid or brown trout when stocked in a lake where this parasite was prevalent, and recommended stocking diploid rainbow or brown trout.

The use of transgenics for introducing genes for resistance to specific pathogens or genes which enhance innate immune mechanisms would appear to be some way in the future and more investment in research is required. In the meantime, innate susceptibility to specific disease agents should be taken into account when using recombinant technologies for other traits. There can however, be serendipitous incidental benefit when producing transgenic fish for other character traits as found by Ling *et al.* (2009). They found that the F4 generation of 'all fish' growth hormone transgenic carp (*Cyprinus carpio*) also resulted in resistance to the ciliate pathogen *Ichthyophthirius multifiliis*. One-year-old transgenic fish were significantly more resistant to the parasite infection as measured by number of trophonts per fish and the mortality associated with the disease. Jiang (1993) considered it likely that, in the long term, increasing the disease resistance in fish will be the greatest contribution gene manipulation will make in aquaculture.

18.3.2   Screening and certification

When salmonid aquaculture markets expanded in response to demand, this led to constraints on production due to seasonal factors and the length of the growth cycle. As a result, eggs are now transported across continents and equatorial divides with the consequent risk of disease spread. Mass screening of brood stock for specific exotic pathogens was established and many countries now operate surveillance programmes supported by legislation. The World Organisation for Animal Health sets out a code of practice for aquatic animals (OIE 2010) and incorporates tests for the most serious pathogens; the 'gold standard' methods for testing for these diseases are set out in the OIE Aquatic Manual of Diagnostic Tests for Aquatic Animals (OIE 2011). The most serious pathogens are listed in relevant country legislation and also, for example, the European Community Council Directive 2006/88 EC (EU 2006), Challenges to these biosecurity strategies are discussed by Oidtman *et al.* (2011).

Notwithstanding the legislation, the desirability of prevention of such dangerous pathogens into farm stocks by testing and certification is unquestioned, despite the high cost of testing in specialised laboratories. The lists consist largely of viral and bacterial diseases and only rarely do parasite pathogens appear. The myxosporean parasite pathogen *Myxosoma cerebralis = Myxobolus cerebralis*, the cause of whirling disease in salmonids, was placed on this list following serious outbreaks of disease in Europe and America in the 1960s. However, it was eventually withdrawn in the UK after major control efforts failed and the parasite was declared to be endemic in native trout populations. The methodology for testing was by a rather prolonged process which involved the digestion of the bony and cartilaginous tissues containing the myxospores, re-suspending them and then centrifuging to release spores into known volumes of aqueous medium, usually water. Despite there being no cure, the testing of brood stock seriously restricted its distribution until its life cycle was elucidated in 1983 and control became possible by the elimination of the alternative host, the oligochaete *Tubifex tubifex*. The monogenean *Gyrodactylus salaris* was added to the list in Europe after it was reported to seriously damage wild salmon parr in rivers in Norway. This parasite remains currently on the list and effective diagnostic methods have been the subject of much research as very few expert diagnosticians are available for the legislation to be effective (Shinn *et al.* 2010). There can be no doubt that extensive screening and certification have been on the whole effective in limiting the spread of highly pathogenic agents in Europe where there is high compliance with testing and certification. Though the compulsory testing is restricted to a limited number of highly pathogenic organisms, the testing and certification of brood stock for an even greater variety of pathogens would be highly beneficial for control, regardless of the legislation.

In addition to brood stock certification, fish farms are encouraged to routinely screen fish stocks from fry to harvest ideally on a biweekly basis as results can signal a problem before an epizootic arises and resultant crisis management ensues, usually with chemical intervention. Such ectoparasites as *Ichthyobodo necator* (agent of costiasis), trichodinids, *Chilodonella* spp., *I. multifiliis*, gyrodactylids, dactylogyrids and lice, as well as the intestinal flagellate *Spironucleus* sp., can cause large-scale mortalities, especially in fry and fingerlings in a short space of time. The methods for testing bacteria and viruses are well established, although not without problems; however, parasite pathogens in fish culture arise from a number of different phyla with a wide range of sizes and morphologies. They are often highly site-specific on the fish and skill, together with a microscope, is required to detect them. Methodologies are therefore difficult to standardise and a certain degree of site-staff training is required, which includes the use of light microscopy. The monogenean *G. salaris* is a case in point as it belongs to a genus which has over 400 described species, all of which are superficially very similar. The gyrodactylids are notoriously difficult to identify owing to the heavy dependence on hook morphology, mainly morphometrics, but made more difficult owing to a shape component (Harris *et al.* 2008). The *Gyrodactylus* species on salmonids are peculiarly difficult to distinguish from each other. Both automated classification systems based on morphometric data and molecular methods have been developed (Cunningham *et al.* 1995; Shinn *et al.* 2000).

The routine screening of fish for parasite pathogens is clearly an important parasite management tool regardless of legislation and is highly recommended. Unfortunately, routine examination is an activity sometimes lost or neglected when under pressure from other husbandry activities and put into operation only as a crisis measure after a parasite pathogen has been introduced to a site in an effort to prevent further introductions.

The potential for molecular methods for routine parasite diagnosis is still much underexplored but there is a great potential here for multiplex polymerase chain reaction (PCR) systems to operate which would have great benefits in preventing transfer of pathogens across natural barriers, countries, watersheds etc. It would be best targeted initially to detect microscopic parasites, especially the protozoa (amoebae, ciliates and flagellates), microsporidians, myxosporidians, intestinal pathogens such as helminths and coccidians. In some cases they are useful for a specific identification, e.g. for skin monogenea, it is useful to be able to distinguish *G. salaris* from the more common gyrodactylids which occur on salmonids such as *Gyrodactylus derjavini* and *Gyrodactylus truttae*. Gene sequence data are already available for many of these common parasites but to date are only routinely used for *Paramoeba* and *G. salaris* and this most often for research purposes. A variety of molecular methods are considered by Cunningham

(2002). Routine screening is only successful with a well-drawn up risk analysis and implementation of a clearly defined control strategy, which includes threshold levels appropriate to size, age and species of fish.

Many parasite pathogens are transferred to fish stocks via indigenous fish, ubiquitous in the local natural aquatic environment, and thus farm site selection should take this into account in disease prevention. Knowledge of pathogens in the ecosystem of the culture site and their seasonal dynamics would be a valuable weapon in the armoury against epizootics of cultured fish, yet this is rarely included in the preliminary surveys. Lakes, rivers, inshore and offshore waters have characteristic faunas which, known and understood, enable the analysis of risk of transfer of parasites into farm fish and for effective methods to be put into place to minimise or prevent introductions, e.g. fallowing, filters etc. to minimise the threat of epizootics.

It may not be possible to prevent the trans-boundary spread of pathogens entirely because, in addition to aquacultural practices, many transfers take place through ornamental and sport fish trading, which are less well regulated. The incentives for traders to exercise control measures for parasites and other pathogens is less because fish are rapidly moved on, together with the costs of morbidity and mortality, to importers, retailers and angling clubs. Even where legislation exists, it is not always effective (Yeomans *et al.* 1997). A greater utilisation of a wide variety of measures to prevent introduction of pathogens would pay high dividends. Oidtmann *et al.* (2011) present an overview of international and national biosecurity strategies in aquatic animal health.

18.3.3   Quarantine
The devastating effects of the introduction of exotic pathogens to naive indigenous stocks of fish, wild or cultured, has been witnessed many times over and is particularly evident in island habitats. Although it can occur accidentally through natural wild fish migrations, many unfortunate introductions causing major damage to native stocks have been anthropogenic and, therefore, avoidable. A cheap and effective quarantine system has been considered as an ideal by many countries but has not yet been fully realised. Legislation usually seeks to prevent introductions of serious pathogens by giving them notifiable or category status but controls are not always truly effective. For example in the UK, the legislation is largely focused on salmonids whereas a number of parasites were introduced with carps for pet or sport fish purposes. Good examples of these are the tapeworms *Bothriocephalus acheilognathi* (Andrews *et al.* 1981), *Khawia sinensis* (Yeomans *et al.* 1997) and the digenean *Sanguinicola inermis* (Iqbal and Sommerville 1986) in the 1980s. These were detected in consignments quarantined prior to stocking but were not rejected and stocking was permitted. The spread of *K. sinensis* in the British Isles since 1986 was documented by Yeomans *et al.* (1997). More recently there has been a recognition of the loopholes

and attempts to control the further spread of these and other parasites throughout the EU. The removal of trading barriers by expansion of the European Community has opened up more pathways by increasing and encouraging exchange and trade of fish between members. The extensive research and monitoring activity of UK Government to study the ability of the UK to reduce the risk of entry of *G. salaris* into the UK is an acknowledgement of the poor control barriers (Peeler and Thrush 2004).

Parts of Asia have struggled to develop quarantine systems for many years with the able assistance of the FAO. Further, and equally important, it has broadcast knowledge and awareness of the risks of importing exotic disease. A well-constructed quarantine policy, however, is of little use if it is not enforced and backed up at government level, as reported for Thailand by Tonguthai (1997). Whittington and Chong (2007) reported an appraisal of the effectiveness of risk analysis and quarantine controls in Australia as they are applied to the 'sanitary and phytosanitory (SPS) agreement' with the OIE. They found that ornamental fish represent a loophole in legislative controls in some countries. They went on to review the importation quarantine policies of a range of countries and classified them as stringent or non-stringent, based on levels of pre-border and border controls. Australia has pre-border and border controls with a quarantine period of 1 to 3 weeks and is thus classified as stringent. Nevertheless, they showed evidence of the establishment of viral, bacterial, fungal, protozoan and metazoan pathogens of ornamental fish in farmed native fish as well as free-living introduced specimens. They concluded that the agreement had not provided an acceptable level of protection and that the risk analysis process described by the OIE under the SPS agreement is not appropriate to the ornamental trade and recommended the OIE guidelines be reviewed and imports dramatically reduced.

At the local level, farms can make strategic use of sites in different geographic locations for quarantine purposes; they can request certification for specific pathogens and impose their own quarantine, isolating fish for a predetermined period of time based on their own local risk analysis which also involves water treatments for effluent. However, this is set against the extra cost and there is a reluctance to pursue it without legislative motivation. It is certainly feasible for ornamental and sport fish but, even when attempts to quarantine are made, they are often too short and are rarely accompanied by testing or screening for disease in a fully comprehensive manner.

18.3.4 Vaccination

Vaccination is a major tool in disease prevention and is dealt with elsewhere in this book. Multi-vaccination, together with genotype selection constitutes the major direction for the future management and control of disease for bacterial and viral pathogens. However, the future for parasite vaccines is less clear. The most advanced research is for a vaccine against *Cryptobia salmositica* (Tan *et al*. 2008). However, this parasite has a distribution restricted to the USA as a salmon pathogen. Also advanced is research for a vaccine for the ciliate protozoan *I. multifiliis*, but despite many years of research, a commercial vaccine is still obstinately 'round the corner'. Whereas the approach to the *I. multifiliis* vaccine has tended to pursue the ciliary i-antigen (Dickerson 2006) to the exclusion of others, the search for a recombinant vaccine against sea lice, particularly the salmon louse *L. salmonis* has seen several different approaches. Research into a sea lice vaccine is very active and now advanced to a patent which can be viewed online (Ross *et al*. 2008). Nevertheless, it still seems to be a number of years out of reach.

18.3.5 Immunostimulants

The ability to boost the natural defence mechanisms of cultured fish has major benefits and is the subject of a relatively new and highly active current research area, dealt with elsewhere in this volume. The main search has been for substances which can be incorporated in feed and delivered orally to fish but others may be injected along with vaccines. Many of the early reports of commercial benefits were not supported by investigations of the mechanism of action and evidence for involvement of the immune system could not be confirmed. More recent studies are now accompanied by data on the effects of treatment on a number of immune bioassays and, though the mode of action is unknown, there appears to be some form of immunomodulation. Whatever their action, immunostimulants directly or indirectly enhance the specific or non-specific defence mechanisms, or both. Virus and bacterial infections have been the major focus of the studies, some of which have involved experimental challenges, for example Ai *et al*. (2011) and, although a number of parasitic diseases have been reported to be improved following administration of immunostimulants, there are few detailed reports of successful treatments. Good results were achieved by intraperitoneal injection of beta glucans into rainbow trout exposed to infection by the microsporidian *Loma salmonae* (Guselle *et al*. 2006) but the timing of administration was considered to be important to achieve a reduction in spore xenomas in the gills (Guselle *et al*. 2007). However, administration by injection is not very practicable in the farm situation and ultimately an oral treatment was achieved (Guselle *et al*. 2010). Laurisden and Buchmann (2010) reduced levels of *I. multifiliis* in rainbow trout after feeding beta glucan in an experimental challenge and also recorded a slight but significant increase in lysozyme levels.

Hopes for the control of sea lice using immunostimulants have not yet been fully realised. Burrells *et al*. (2001) supplemented Atlantic salmon diet with dietary nucleotides for three weeks and achieved a 37% reduction in the mean number of attached lice per fish following an experimental challenge. Not all dietary supplements tried are benign and Refstie *et al.* (2010) found that health and FER (feed efficiency ratio) were affected by some formulations and, although they resulted in poor growth performance, the reduction in lice prevalence was 27%. Adding beta glucans in an attempt to ameliorate the negative growth effects reduced the lice prevalence by a further 28%.

Tests on cod infected with two species of renal myxosporeans showed no reduction in spore production after 6 weeks treatment with a beta glucans based immunostimulant (Gorgoglione 2009, personal communication). However, Harikrishnan *et al.* (2011b) significantly reduced the percentage mortality caused by *Uronema marinum* in the olive flounder, *P. olivaceus*, following 30 days of 50 and 100 mg per kilogram bodyweight of the traditional Korean medicine (TKM). They showed that the Korean herbs *Punica granatum*, *Chrysanthemum cinerariaefolium* and *Zanthoxylum schinifolium* had an effect on a range of immune parameters.

There may well be a contribution to parasitic disease control in dietary supplementation with immunostimulants, given the relative importance of the innate immune response in fish defences against parasites, but research is necessary to formulate and standardise the diets before they will be commercially useful and to establish that they are wholly beneficial without adverse side effects. As with other measures in this chapter, they will make a contribution to control when integrated with other measures and further investigations into the usefulness of immunostimulants in the control of parasitic diseases would be very fruitful.

18.4   Disease management

18.4.1   Husbandry activities

Many disease problems are basic husbandry problems and there are many, sometimes simple, easy and inexpensive techniques which contribute to the welfare of fish. On the whole, the factors which contribute to the objective of producing a strong, robust grower will also make a major contribution to the health of the fish, provided they do not cause physiological stress. Stocking density management is a primary point for disease control for all infectious diseases. A high stocking density encourages parasite transfer between hosts (resulting in a measure of high mean parasite intensity) and more frequent encounters of hosts by the parasite (resulting in a measure of high parasite prevalence). Lowering stocking densities is a very useful first step measure when ectoparasite infections break out, along with increasing water flow, to achieve a greater flushing effect on the parasites. For an interesting account of stocking density and fish welfare, see Turnbull *et al.* (2008).

Fallowing is an essential control measure which interrupts the parasite life cycle and, for parasite prevention, is based on the short-lived nature of the transmission stages of many parasites. The best example is for management of sea lice infections in marine cages of Atlantic salmon where it is now standard practice in sea sites and is very effective in the control of sea lice infection (Bron *et al.* 1993). In the case of sea lice, the fallow period is calculated on the basis of the longevity of the adult female which may be as long as 4–5 weeks plus the time to hatch and develop to the copepodid stage, which is the transmission stage of the progeny. Although fallowing is extremely useful it is, unfortunately, often relinquished in some culture systems owing to production pressure. It is particularly useful for ponds and can be accompanied by draining and drying. This removes leeches as well as parasite hosts, e.g. molluscs, crustaceans and oligochaetes. Fallowing should be accompanied by single year class stocking. Single year class stocking prevents hyper-infection of the more vulnerable, smaller fish which can occur as a result of exposure to the more tolerant and more heavily infected growers. This is now a universal practice in mariculture of salmon where the newly stocked smolts would otherwise be more vulnerable to sea lice originating from the second year growers (Bron *et al.* 1993). Area agreements which allow for coordination of husbandry activities can also have benefits (Rae 2002).

A key component of good husbandry is the monitoring of fish welfare, i.e. health, growth and the culture environment, and reacting to changes, e.g. in morbidity and mortality rates, parasite burden, and physicochemical properties of the environment.

18.4.2 Hygiene

The regular cleaning of ponds, tanks and nets is now a well-accepted method of disease control in aquaculture, even if it is not always practised, and is incorporated in the code of practice set out by OIE (2010). The cleaning, draining and drying of ponds done properly will remove any disease agents as well as invertebrate hosts of parasites. For example molluscs are host to digenean parasites and removal of the molluscs is the only effective method for the control of cataracts due to eye fluke infection caused by strigeid metacercariae, notably *Diplostomum* and *Tylodelphus* genera, and must be carried out annually to be effective. The blood fluke *Sanguinicola inermis* and a number of human zoonotic digeneans such as *Heterophyes* spp., *Haplorchis* spp., *Chlonorchis* spp. etc., all include molluscs in their life cycle. Oligochaete worms such as *Tubifex* host many species of myxosporeans, and plankton transmit cestode and nematode worms to fish. Thorough cleaning of ponds will kill or reduce numbers of the resistant stages of parasites, for example, spores of microsporeans, ciliates and flagellates, oocysts of coccidians, and tomonts which are the resistant stages of *I. multifiliis*, as well as of some parasite vectors such as weed fish. If drying is not possible, the addition of lime ($CaCO_3$) or disinfectants to the mud of the earth pond benthos can be effective. Tanks are more readily cleansed and disinfected prior to stocking and net pens must be freed of fouling organisms, some of which act as hosts for parasites and support facultative pathogens. If benthos is allowed to build up in tanks from waste food, fish excrement etc., then this is rapidly colonised by benthic organisms such as amoebae, scuticociliates, mobile and sessile peritrichous ciliates such as trichodinids and epistylids, and holotrichous ciliates such as *Tetrahymena* spp., all of which feed on the bacteria, accompanied by myxobacteria and fungi such as *Saprolegnia* and *Fusarium*. These can colonise gills and skin and the effects are readily visible as excess mucus production, skin and fin erosion and sloughing; a common feature of poorly managed fish tanks and ponds. Sometimes following chemotherapeutic treatments, an imbalance is set up in the benthic ecology and a single group of organisms, e.g. amoebae or yeasts, can dominate, becoming very difficult to control; such conditions are more common in flatfish tanks, e.g. turbot and olive flounder, but have been known also to occur in large inshore Atlantic salmon tanks.

The establishment of a high-quality environment with clean, well-aerated water is important in producing healthy fish and critical to those species native to oligotrophic waters such as the salmonids, and improvement is dealt with elsewhere in this book. In poor water environments, it is generally the nitrogen and phosphate levels which contribute to conditions in which parasites can abound. High organic content will produce a bacterial and fungal-rich environment where ectoparasitic and facultative protozoan pathogens will thrive, resulting in skin damage. Mixed flora and fauna infections with, for example, *Chilodonella* spp., trichodinids, dactylogyrids, myxobacteria and fungi, such as *Saprolegnia* sp., are common in these poor water conditions. In some cases, infections also include *Cryptobia* sp. and *I. multifiliis* and these mixed infections can signal a degraded, nitrogen-rich environment, which may be exacerbated by high temperatures and low oxygen levels. Such infections should be seen as an indicator of poor water quality and acted upon urgently.

Increased water flow, reduction in stocking density and aeration or oxygenation can provide immediate benefits, whereas chemical treatments may cause further deterioration of water quality. Formalin, for example, is a common treatment for ectoparasitic protozoa but it is a reducing agent and will cause further mortality as damaged gills are unable to support the oxygen requirements of the fish. Using chemical interventions without correction of the environmental conditions which engendered the poor water quality is wasted effort whilst those conditions prevail.

18.4.3   Removal of infectious agents

The use of disinfectant baths at the entrance to culture sites, for example of iodophors, has been used to good effect to prevent transfer of bacteria and viruses between farms but has little use for most parasite prevention. However, disinfection of nets and the use of individual nets for each aquaculture unit, i.e. tank, pond etc., will readily prevent transfer of many parasite infectious stages and ectoparasites from tank to tank and is especially useful for the control of the spread of parasitic protozoans.

Care when placing the mouth of the inlet pipe in rivers and lakes and inshore waters would involve sampling the water for parasite transmission stages before establishing its location, e.g. avoiding mollusc beds which may transmit glochidia or cercariae of, for example, eye fluke, or *Sanguinicola*. The removal of parasite agents from incoming water can be achieved by the use of particulate filters such as sand or a synthetic medium, e.g. cercariae of digenean parasites such as eye fluke and plankton vectors, which can carry intermediate cestode stages. Fine filters down to 10 µm may also be effective in removing infectious actinospores of myxosporeans; Heinecke and Buchmann (2009) recommend water filtration to remove all tomonts of *I. multifiliis*. It may be that filters, which incur costs, need not be in constant operation but activated during peak periods of parasite transmission, for example, cercariae of the eye fluke, *Diplostomum* spp., emerge from the mollusc (*Lymnaea* spp.) in a relatively restricted period, depending on the latitude. In Scotland, for example, this usually occurs in early June, with a further, smaller emission in September. A simple study of the local conditions would pay dividends where these parasites are problematic.

Removal by passing through ultraviolet sterilisers (Summerfelt 2003) is commonly recommended but can deal with only small amounts of water and is thus reserved for small tank use. It is also debatable how effective this is against parasites but evidence suggests it can be useful for small protozoa provided the distance is within a few microns. Subasinghe (1980, personal communication) showed that theronts of *I. multifiliis* could be destroyed in this way but would be suitable for use only on a small scale such as for small ornamental fish tank use. A number of methods should be combined to improve the efficiency of removal of potential pathogens.

Mechanical removal of *I. multifiliis* was achieved by Shinn *et al.* (2009) who developed a novel mechanical system for commercial trout raceways which, used daily, removed the settled reproductive cysts of *I. multifiliis*. The device involves suction and operates in the manner of a vacuum cleaner in association with a low adhesion polymer raceway lining. Daily cleaning not only reduced the number of parasites per fish but demonstrated a greater fish survival in the raceways where the device was in operation. McRobbie and Shinn (2011) devised a mechanical rotary device for cleaning commercial-scale circular tanks which they believe would perform a similar function of removal of encysted parasites. Mechanical removal of parasite agents of the freshwater louse *Argulus* have been known for some time to be useful in freshwater fisheries. This simply consists of boards suspended in the water column attached to buoys on which the *Argulus* lay their eggs. These are periodically removed and scraped clean and found to be very effective (Gault *et al.* 2002). A modification of this using invertible, vertical tubes was designed by Shinn (personal communication) which had the advantage of being self-cleaning as the tubes, when inverted, exposed the eggs to the drying and sterilising effect of air and sunlight and, at the same time, minimising interference with angling activity.

Culling of emaciated or chronically sick fish is very important, not only from a fish welfare point of view but to remove potentially high numbers of infectious agents. Examination of these will provide a quick identification of any causative agents present in the population as such fish will act as reservoirs of the disease agents present. Equally important for control, but of less value for diagnostic purposes, is the speedy removal of dead fish. This is important, of course, for any disease but is especially useful for controlling microsporidian and myxosporean parasites, the spores of which are produced in vast numbers in internal organs. Where they are in tissues with no exit, the death of the fish host is their only point of exit into the environment and the potential for exposure to the next host in the life cycle; they can only be released into the environment following decomposition of the dead fish.

Culling may be carried out in response to legislation in some countries as, at the government level, it is possible to see that eradication of the pathogen is still a goal. Eradication policies have been applied to virus diseases in several countries, for example to control ISA (Vagsholm *et al.* 1994), and Norway has also carried out a robust eradication policy to remove the parasite *G. salaris* from 46 Norwegian rivers using rotenone and aluminium sulphate (Mo 2007). Eradication policies are often controversial and evidence for their usefulness is not always available (Murray 2006).

18.4.4   Feeding and nutrition

Incorrect feeding is also a husbandry activity which can significantly contribute to disease. It is now recognised that there is a clear relationship between the nutritional status of the fish and the quality of their disease response. Whereas natural feeding may be considered to be nutritionally balanced, if it is allowed, there is a risk of ingestion of parasites which utilise prey items as intermediate or final hosts. For example, many helminths utilise invertebrates as first intermediate hosts, in which to grow to the next or final stage. Many of these invertebrates are important food items of prey for fish at some stage in their life, e.g., crustacean zooplankton, such as copepods and diaptomids which harbour procercoids of cestodes and larval nematodes. Benthic oligochaetes are an important source of food for some fish species but they also harbour cestodes which continue development following ingestion by the fish.

As invertebrate populations have a seasonal or diurnal dynamic, they may be ingested in large quantities by fish during a bloom. This is a common occurrence where fish feed on zooplankton populations which include copepod and diaptomid crustaceans, first intermediate hosts of some fish pathogenic cestodes, and fish can become heavily infected with these helminths. For example, Atlantic salmon held in net pens in large freshwater bodies can harbour large numbers of *Diphyllobothrium dendriticum* and *Diphyllobothrium ditremum* plerocercoids from high consumption of *Cyclops* and *Diaptomus* species. During plankton blooms in both fresh and seawater, salmonids held in net pens can become heavily infected with *Eubothrium crassum*. Young fish will take the plankton in preference to pelleted feeds in many cases; however, one might expect this to be mini-mised if the pelleted feed is an appropriate size and composition and fed to satiation. Stocking schedules, especially of small fish for which zooplank-ton is the natural diet, should avoid seasonal planktonic blooms where possible and site cages in locations away from where plankton accumulation occurs as a result of prevailing winds. Awareness of the seasonal abundance of benthic invertebrate prey is also necessary as they can act as intermediate hosts of parasites of fish, e.g. oligochaetes act as intermediate hosts of some seriously pathogenic cestodes such as *Khawia* spp. as well as actinospores of pathogenic myxosporeans. Such local knowledge assists in planning stocking strategies and general parasite management.

In some cases, where the cultured fish is a top predator, the intermediate hosts may be smaller fish which harbour intermediate stages of parasites which mature in the cultured fish. This is the main reason why by-catch or trash fish cannot be used unprocessed as fish feed. Anisakid nematodes are very widespread in many marine vertebrates and invertebrates, and readily re-establish in the predator's tissues. Deep freezing or silaging the fish is usually adequate to kill off these transmission stages; however, the inter-mediate stages of parasites are often in a dormant state and difficult to destroy so that many can survive chilling and refrigeration for very long periods and, as they may also be zoonotic, such as the anisakids and the heterophyid digenea, vigilance is necessary in freezing to −18 to 20 °C.

Poor feeding lowers resistance to infection and invariably the 'poor doers' become infected first, often leading to mistaken cause-and-effect conclusions, i.e. the assumption that the wasting or emaciated condition of the fish is due to the parasite infection rather than the reverse. This error may result in further chemical intervention when correction of diet or feeding regime is the root cause. A particularly critical stage for the fish is at weaning which, if not properly achieved, results in a widespread debilitat-ing disease condition of the early fry, which can succumb to many bacterial and viral infections as well as to ectoparasitic flagellate protozoa, e.g. *Ich-thyobodo* (costiasis). The intestinal flagellate *Spironucleus* (previously known as *Hexamita* and *Octomitis*) is commonly found in fry and fingerlings where nutrition is inadequate and results in a pinhead condition and what is sometimes called 'catarrhal enteritis'. As fry and fingerlings are growing rapidly, particular attention is necessary to the feed composition and pellet size. Better knowledge of diets means they can be boosted to optimise defence against disease, especially by correct dietary fatty acid balance and the inclusion of sufficient amounts of antioxidant compounds such as vita-mins C and E and minerals.

Gastrointestinal microbiota in the health of fish is an important aspect of nutrition, dealt with in other chapters. Efforts to investigate how microbiota can be altered to improve immunity and disease resistance have been carried out in terrestrial livestock and are now being applied to fish. Prebiotics used as dietary supplements benefit the terrestrial animal by stimulating growth and/or activity of a limited number of health-promoting bacteria e.g. *Lactobacillus* and *Bifidobacter* species in the intestine, at the same time possibly limiting pathogenic bacteria such as *Salmonella*, *Listeria* and *Escherichia coli*. There is, however, limited information on their activity and benefits in aquatic organisms.

Research with fish has focused more on the activity of probiotics and this is discussed extensively in other chapters. The primary application of microbial manipulation in aquaculture has been to alter the composition of the aquatic medium, and the microbiota in the fish gastrointestinal tract has not yet been fully characterised, especially the anaerobic microbiota.

18.4.5   Stress reduction

Fish respond to environmental challenges in a manner which is commonly called 'stress' but which constitutes a series of adaptive neuroendocrine processes that induce reversible changes both metabolic and behavioural. Most of the fish species cultured globally, with the possible exception of *C. carpio*, are still essentially wild animals and very prone to stress reactions to husbandry activity. An acute stress response is usually advantageous to the fish (Demers and Bayne 1997) and does not cause any long-term problems; however, prolonged activation of the stress response results in a chronic condition which is damaging. The artificial environment of the fish farm is at odds with the natural environment of most, if not all, fish species cultured and they are, therefore, under constant stress. Portz *et al.* (2006) provide a wide-ranging review of the stress-associated impacts of short-term holding of fish. Provided the stress is not harmful, the fish can adapt but many environmental stressors reach the limit of adaptability for fish. In addition to the culture conditions, culture activities such as handling, netting, grading and transporting are acute stressors.

It has been well reported that chronic stress leads to immunosuppression, reduced growth and reproductive dysfunction. The stress response in relation to fish welfare has been reviewed by Huntingford *et al.* (2006) and is such an important element in fish disease conditions that it is given a separate chapter in this book.

It has long been established that stress affects growth and lowers resistance to disease via the hypothalamic-pituitary-interenal (HPI) axis (Pickering 1993). Numerous studies have linked specific stressors to changes in immune parameters both innate and adaptive. An acute stress as simple as repeated netting can increase levels of cortisol, adrenaline and lysozyme in trout plasma (Demers and Bayne 1997). Such apparently minor stressors can affect parasite levels directly, for example an increase in a population of the gill monogenean, *Dactylogyrus* spp. was recorded in tank cultured carp by a single daily netting (Sommerville, unpublished data). Sunyer *et al.* (1995) showed the effect of stress on general immune competence. Natural cycles of activity of components of the immune system occur during seasonal and circadian rhythms mediated via the pineal gland and melatonin (Esteban *et al.* 2006; Cuesta *et al.* 2008) and these may readily be upset by changed photoperiods commonly employed in enclosed tank systems or cage systems using artificial lights to improve feeding and growth.

All aspects of aquaculture activity are potentially stressful for fish and prolonged stress will eventually impact on aspects of growth performance as well as disease. Controlling stress, therefore, is a major contribution to minimisation of the risk of disease outbreaks. Stress reduction has been given a lower priority than it deserves in the past but it is increasingly realised by fish farmers that the stress effect on many aspects of fish growth and welfare has significant economic impact. Identifying genes associated with the stress response in fish would inform breeding programmes and enable the production of populations of fish which are less inclined to stress and are therefore more domesticated. Minimising stress is a key component of any programme for disease reduction and thus a key platform of IPM.

18.5   Interventions for parasite prevention and control in fish

18.5.1   Herbal medicines and ethnopharmacology

The role of herbal substances, sometimes called botanicals, in delivering a boost to the immune system was noted above. Herbal medicines may have immunostimulant effects on a broad spectrum of diseases including parasites, or they may act as natural chemotherapeutants and, as such, may be used under the heading 'ethnopharmacology'. The research activity into the use of herbal products for disease prevention or treatment in intensive culture is as yet at a very early stage, although it has enjoyed a boost in activity in recent years.

Herbal medicines have been traditionally used in extensive systems although traditional Korean medicine (TKM) is used currently in intensive *P. olivaceus* culture as a methanol, ethanol or aqueous extraction against the ectoparasitic protozoan *U. marinum* (Harikrishnan *et al.* 2011b). Herbal treatments today tend to be considered as organics owing to their natural origins. However, they may well be delivering potent molecules such as insecticides, for example neem. Neem is a natural plant material from the neem tree, *Azadirachta indica*, originally from India but now distributed across the globe, which contains insecticidal properties. In traditional treatments, the leaves were used in carp ponds infected with *Argulus*. Research to date has shown that it has 140 active components, occurring in different parts of the tree, the most important of which are the tetratriterpenoids. An active ingredient, azadarachtin, has been identified which has potential as a powerful insecticide and thus as a treatment chemical for crustacean parasites of fish.

Although many licensed products originate from plant material, for the purpose of this account herbal treatments will be regarded as natural plant products with a largely unknown mode of action and their use commonly arising from traditional methods. The modes of action have been variously described as antistress, growth promotion, appetite stimulation, tonic, immunostimulation, or as having aphrodisiac or microbial properties in finfish or larviculture. Active principles such as alkaloids, flavonoids, pigments, phenolics, terpenoids, steroids and essential oils have been reported (Citarasu 2010) and herbal treatments have been better received in Asian and Oriental fish culture than in the intensive rearing systems practised in the West. A number of recent reviews (Harikrishnan *et al.* 2011a; Jeney *et al.* 2009), provide details of many studies, some of which have been used for ornamental fish rather than food fish. Direkbusakarom (2004) summarises herbs used for treatment of fish disease in Thailand, Vietnam and China and Citarasu (2010) elucidates some of the better researched herbals and their proposed activity. The vast majority of studies describe the treatment of viral, bacterial and, to a lesser extent, fungal diseases and only a few mention herbal treatments of parasitic diseases.

Trials of herbals against parasite infection have not been well documented, though reference has been made to treatments for myxoboliasis, trichodinids, gyrodactylosis, argulosis and scuticociliates. Direkbusakarom (2004) has identified a number of herbal remedies for the control of *Lernaea* sp., *Argulus* sp. and helminth infections which include *Bothriocephalus gowcongensis* [sic] (now *acheilognathii*). Other herbal extracts were reported by Auro de Ocampo and Jiminez (1993) to be active against the nematodes *Capillaria* and *Pseudocamallanus* (Zhou *et al.* 2003). Sánchez *et al.* (2000) reports the use of *Buddleja cordata* extract to control costiasis and Ekanem *et al.* (2004) used crude herbal extracts to reduce parasite-induced mortality in *Carassius auratus* infected with *I. multifiliis*. A preliminary study by Steverding *et al.* (2005) indicated that Australian tea tree, *Melaleucaalternifolia*, oil (TTO) was active against *Gyrodactylus* spp. Crude extracts of *Alium sativum* (garlic) and *Artemesia vulgaris* were claimed by Noor El Deen and Mohamed (2009) and Aboud (2010) to control both *Trichodina* epizootics and *Aeromonas* infection on *Oreochromis niloticus* and Madsen *et al.* (2000) found that squeezed garlic at a dose of 200 ppm effectively controlled *Trichodina jadranica* in eels. In their study they tried an arrangement of different treatments as potential alternatives to formalin but did not develop the garlic treatment further, favouring standardised chemical treatments instead. Thus, there would appear to be further potential in investigating the use of herbals as parasiticides.

The mode of action of some herbals is clear such as that of the tree *Azadirachta indica* and leguminous climber *Derris elliptica*, which have had their active ingredients determined and licensed as chemical insecticides, but the action of many herbals may well be in growth promotion or immunostimulation. Further research into the mode of activity of those traditional herbs which are active against parasites would be productive and those that support and strengthen the natural defence mechanisms may constitute a cost-effective, eco-friendly alternative to chemotherapy and vaccines. Research is hampered by a lack of standardisation and this must be solved before any meaningful data can be generated.

18.5.2   Biocontrol

Predator–prey–pathogen systems

The term biological control is very loosely used and may be utilised in different ways by different interests. The narrow definition is 'the practice by which an undesirable organism is controlled by means of another'. There are several approaches to biocontrol but the most common usually refer to the introduction of predators or pathogens of the 'pest' requiring control, ideally strategically in a controlled environment. There are many examples of success, mainly in horticulture where micropathogens, i.e. bacteria, fungi, viruses, are introduced to the pest population. It is likely that the definition will expand as more creative exploitations of biological interactions are discovered.

The most developed biological control method for fish parasites to date is the use of wrasse to control sea lice in sea cages of salmon (Sayer *et al.* 1996). It was first used in Norway and reported by Bjordal (1988) and has since been widely employed in Scotland, Shetland and Ireland as well as Norway. The method utilises the cohabiting of wrasse with salmon at variable ratios, and depends on the natural feeding ability of wrasse to take epibionts from surfaces such as rocks and shells in the natural environment. Whilst wrasse are not natural predators of sea lice, when hungry they will remove lice from the fish skin surface; thus, the system does not readily compare to a typical biocontrol method.

It is difficult to assess the efficacy of this method as the mix of species of wrasse, mainly goldsinny (*Ctenolabrus rupestris*), rockcook (*Centrolabrus exoletus*) and corkwing (*Crenilabrus melops*), size of wrasse and salmon and ratio of stocking varies in most cases. There have been many anecdotal reports and results of some experimental and commercial trials have been published. One of the more comprehensive studies was that of Deady *et al.* (1995) who described the results of a trial testing the use of wrasse to control lice in a commercial farm situation off the west coast of Ireland. Selected trial cages were followed through 1991 and 1992 after being stocked with wrasse in May through to November. Corkwing and goldsinny wrasse successfully controlled sea lice, keeping lice levels below five mobiles (pre-adults and adults) per fish, starting with an initial ratio of one wrasse to 250 salmon. It was calculated that individual wrasse consumed more than 50 lice each on average, though feeding declined in October possibly due to lowered temperature and shortened day length changes. Some 70% loss of wrasse from the cages was assumed to be due to escape or predation. Wrasse were found to be more effective in controlling lice on stressed, diseased fish than were oral chemotherapeutants where fish were unable to take the appropriate oral dose owing to their inappetence as a result of the disease. Tully *et al.* in 1996 used experimental and commercial-scale sea cages to evaluate the ability of wrasse to control *Caligus elongatus* in Ireland and showed that it was effectively removed by goldsinny and rockcook wrasse. Despite the inadequacies of many of the trials which lack appropriate controls and standardisation, the expansion of the use of wrasse and the anecdotal approval of the industry does suggest their usefulness, especially with newly stocked fish.

Unfortunately, the wrasse are all wild caught, although a breeding programme for wrasse is, belatedly, underway in Scotland and Norway. There are many problems associated with the approach relating to shortage of wrasse, their inability to overwinter in cages, their feeding where there are no or too few lice etc. The shortage of available wild wrasse has restricted their use to the newly stocked salmon cages where the small fish are most vulnerable and when only a few adult female lice can kill its host. Also, when stocked with the larger salmon the wrasse risk being predated by the salmon. The cost is also high as it becomes a capture fishery which is confined to coastal waters during certain seasons. There is a risk of introducing disease with the wild wrasse and they may also act as reservoirs of disease. They have been shown to be susceptible to IPN in salmon cages (Gibson *et al.* 1998), although studies are still few, and ultimately they will be subjected to the same disease legislation as their salmonid cohabitants. The best prospect is to make them commercially available through breeding programmes in pathogen-free hatcheries. However, there still remains an ethical issue relating to the death of the fish during the winter and the danger of treating them as a disposable commodity.

Wrasse have also been tried as biological control agents in attempts to remove large capsalid monogeneans from fish. Cowell *et al.* (1993) attempted to control *Neobenedenia meleni* from tilapia, whilst the bluestreak cleaner wrasse (*Labroides didimiatus*) was used by Grutter *et al.* (2002) to clean up the black eye thicklip *Hemigymnus melapterus* of *Benedenia lolo*. Caligid crustaceans such as sea lice and the large capsalid monogeneans are readily visible to the naked eye and attempts at smaller parasite control in this way would seem to be less likely. However, Picon-Comacho *et al.* (Shinn, personal communication) have investigated the efficacy of the biofilm grazer *Glyptoperichthys gibbiceps* (commonly known as the leopard pleco) to remove the pro-tomont and encysted stage of *I. multifiliis* in small-scale tank experiments with some success.

Marin *et al.* (2002) made a thorough study of the potential for *Udonella caligorum* as a biocontrol for *Caligus rogercresseyi* on farmed fish in southern Chile. *Udonella* spp. are hyperparasitic monogeneans widespread on caligid crustacea. Unfortunately they had to conclude that it had no significant effect on either the fecundity or survival of free-living stages of the lice and therefore could not be used as a biological control of sea lice.

Biopesticides
Biopesticides are largely microbial pathogens of the pest in need of control which can be processed and marketed in the same way as chemical pesticides and some have enjoyed considerable success in crop protection systems. Very little research has been done in this area on the pathogens or competitors of fish pathogenic organisms. There has been some preliminary research on the micropathogens of sea lice by Freeman (2002) who studied the pathogens of *L. salmonis* and assessed their potential as biocontrol agents. The epibionts and ectoparasites were considered to be nonpathogenic symbionts. However a hyperparasitic microsporean caused major pathology of internal organs and reduced fecundity. This parasite, described by Freeman and Sommerville (2009) as *Desmozoon lepeophtheirii*, has since been found to be widespread, in Norway (Nylund *et al.* 2010) as well as Scotland. The potential for this as a biocontrol agent was curtailed when it was also found within the host salmon tissue (Freeman *et al.* 2003; Nylund *et al.* 2010). Some progress has also been made in the search for viral and bacterial pathogens of *L. salmonis* (Sommerville and Harper, unpublished); however this necessarily is a long-term solution requiring major financial investment. Nevertheless, the search for biopesticides will produce beneficial solutions for the future as, although these have been slow to develop in horticulture, with the loss of so many chemical pesticides due to the development of parasite resistance, there is now a rush to find suitable biopesticides to fit in with integrated pest management programmes.

Parasite behaviour modification
The modification of the behaviour of the parasite pathogen has been a favoured approach in many host–parasite–pathogen systems of medical and veterinary importance. The use of naturally occurring, non-toxic substances that are able to influence the pest's behaviour by interfering with natural responses suggests an eco-friendly and therefore attractive solution to pest control. For biological control, the pheromones which orchestrate normal behaviour are exploited, the most potent being the sex pheromones, and mechanisms which may involve the use of lures and traps are designed to interfere with reproduction.

Many studies of the biology of parasites have investigated their chemosensory abilities and semiochemicals have been identified or proposed in a range of behaviours but mainly mate and or host location. There are fewer studies of fish as hosts than there are of homeotherms, and few of these studies have been taken to the point of developing a potential parasite control system. The principal use of semiochemicals for parasite control is in a stimulo–deterrent diversionary strategy (SD DS) or push–pull strategy. Push–pull strategies involve the behavioural manipulation of the pest in question by introduction of chemosensory cues into the culture environment which act to make the host unattractive or unsuitable (Push) while luring them towards an attractive source (Pull) from where they are subsequently removed. The push and pull components may consist of a number of stimuli having an additive or even synergistic effect and may be integrated to maximise efficacy (Cook *et al.* 2007). There are some successful methods for terrestrial pests composed of synthetic pheromones which are commercially available but research for aquatic pests has lagged behind and thus this approach is very hypothetical at the present time. There have been useful studies on the behaviour of some serious parasite pathogens of fish and the chemicals involved characterised but the identification and characterisation of the stimulatory molecules is relatively simple compared with the development of a suitable technology which will deliver a biocontrol system, thus it may be the lack of design technologies generally which have held back the use of this knowledge for practical purposes.

Buchmann and Nielsen (1999) studied the chemoattraction of *I. multifiliis* theronts to host molecules. Using *in vitro* experiments and a bioassay they found a high, unidirectional chemo-attractive effect using serum from a variety of potential hosts both freshwater and marine, reflecting the low host specificity exhibited by *I. multifiliis*. They determined through mucus analysis that theronts were specifically attracted to high molecular weight molecules in the mucus, one of them curiously reactively similar to host immunoglobulin.

In a study of the behaviour of cercariae (the transmission stage) of *Diplostomum spathaceum*, the eye fluke, Haas *et al.* (2002) investigated the specific stimuli used by the cercaria to identify the fish host, which he called the 'enduring' stimulus and which differed from the 'penetration' stimulus. The cercariae of *D. spathaceum* responded to a unique profile of cues which differed from other cercarial species and he suggested that cercariae may be interrupted by blocking, saturating or overstimulating the parasite's chemoreceptor. Haas (2003) went on to review the strategies of parasitic worms for host finding, recognition and invasion. It is well recognised that the miracidial stage of digenea locate snails in response to snail-emitted chemical compounds, and much work has been done to characterise these chemicals in snail tissues, generating the idea of miracidial attractants.

Semiochemicals associated with host and mate interaction have also been studied in the salmon louse *L. salmonis* and a host location molecule, isophorone, which activated adult male *L. salmonis* has been characterised (Ingvarsdottir *et al.* 2002). As a result, the authors proposed a slow release system for field trapping of lice. Moredue and Birkett (2009) reviewed the host finding behaviour of *L. salmonis* and proposed the potential for the use of kairomones in odour traps to reduce populations of sea lice on Atlantic salmon in cages. Finding the specific chemical attractants and repellents for a parasite is, unfortunately, a long way from finding a solution in the field, which may be the open ocean.

Taxes to physical stimuli may also be exploited in push–pull strategies. Many aquatic parasite transmission stages are phototactic, and sometimes strongly so, such that it would seem to be a straightforward method of trapping those that show positive phototaxis. However, attempts to design light lures in the past have failed (Gravil, 1996). This early prototype failed largely because the expense of each individual lure precluded properly illuminating sufficient area to make any difference to the sea lice population. Pahl *et al.* (1999a) developed a light trap for use in determining the distribution and abundance of marine larvae. This portable trap showed an efficiency of 46% in a laboratory static system and they suggested that it might be a viable option for removal of mobile sea lice stages. This device differed from the previous light lure which depended solely on phototaxis as it had an airlift system for filtering water. Pahl *et al.* (1999b) subsequently tested the photo mechanical sampling device which used halogen light and caught approximately 70% of the larval stages in a tank and approximately 24% of the adults it tested. It was field-tested in the ocean and the authors concluded it was an effective non-invasive and environmentally friendly method to monitor sea lice. Flamarique *et al.* (2009) have published the device on Free Patents Online as 'monitoring and potential control of sea lice using LED-based light trap'. Being a small device, it is less expensive than that previously designed and is intended that five units per net pen would be cost-effective. Although it is primarily a monitoring system, the authors have recognised its potential in sea lice control. Wahli *et al.* (1991) showed that theronts of *I. multifiliis* also exhibited positive phototaxis but no host finding behaviour. It does not appear that they have tried to develop this further as a control system.

Possibly one of the main reasons why lures of one kind or another have not been as successful as hoped is that an organism's response to stimuli is very complex and little understood. Aquatic organisms, as with terrestrial organisms, respond to a hierarchy of host cues (Lewis *et al.* 1995). Chemical cues can also act as modulatory stimuli affecting the magnitude of the response to other stimuli (Hölldobler 1999). Further, parasites with complex life cycles with a number of stages (*L. salmonis* has ten) exhibit a range of different behaviours and responses depending on the life stage. Mikheev *et al.* (1998, 2000) described the complexity of host finding cues involving olfaction and mechanoreception of *Argulus coregoni* which depended on whether the parasite was a juvenile or at the adult stage. It is clear that a greater understanding of the interactions between host/parasite and parasite/parasite, and their chemical ecology generally is essential if semio-chemicals are to be used for parasite population control in fish culture systems. Notwithstanding the difficulty in characterising, synthetically manufacturing and delivering such molecules in suitable traps, the signal strength of tens of thousands of fish hosts in a cage is always likely to outperform any possible, even very strong, signals from a manufactured odour or light trap.

It is unlikely that any single method of biological control can bring about control of a parasite epizootic or even sustain parasite populations at a manageable level. Nevertheless, their place is as components of an integrated pest management strategy with the great benefit of avoiding or delaying resistance development (Pickett *et al.* 1997).

18.6   Conclusions

The elements of non-chemical control outlined above are intended for use 'in concert'. The main objective is to bring about a change in the perspective of parasite pathogen control in aquaculture systems. There is a need to change from the single technology, pesticide-dominated approach towards a more sustainable strategy which is fully integrated and which will relieve the cycle of 'new product–resistance development–crisis' as illustrated with sea lice chemotherapy. Many of the technologies outlined will show major advances in the next decade and, together with the advances in diagnostic and vaccine technology, will bring about a more manageable disease situation; advances which are also ecologically more acceptable.

18.7   References

ABOUD O A E (2010), Application of some Egyptian medicinal plants to eliminate *Trichodina* sp. and *Aeromonas hydrophila* in tilapia (*Oreochromis niloticus*), *Researcher*, **2**, 12–16.

AI Q, XU H, MAI K, XU W, WANG J and ZHANG W (2011), Effects of dietary supplementation of *Bacillus subtilis* and fructooligosaccharide on growth performance, survival, non-specific immune response and disease resistance of juvenile large yellow croaker, *Larimichthys crocea*, *Aquaculture*, **317**, 155–161.

ANDREWS C, CHUBB J C, COLES T and DEARSLEY A (1981), The occurrence of *Bothriocephalus acheilognathi* Yamaguti, 1934 (*B. gowkongensis*) (Cestoda: Pseudophyllidea) in the British Isles, *J Fish Dis*, **4**, 89–93.

ASHLEY P J (2007), Fish welfare current issues in aquaculture, *Appl Anim Behav Sci*, **104**, 199–235.

AURO DE OCAMPO A and JIMENEZ E M (1993), Herbal medicines in the treatment of fish diseases in Mexico, *Veterinaria (Mex)*, **24**, 291–295.

BJORDAL A (1988), Cleaning symbiosis between wrasse (Labridae) and lice infested salmon (*Salmo salar*) in mariculture, *Int Coun Explor Sea, CM*, F17.

BRON J E, SOMMERVILLE C, WOOTTEN R and RAE G H (1993), Fallowing of marine Atlantic salmon, *Salmo salar* L., farms as a method for the control of sea lice, *Lepeophtheirus salmonis* (Krøyer, 1837), *J Fish Dis*, **16**, 487–493.

BUCHMANN K and NIELSEN M E (1999), Chemoattraction of *Ichthyophthirius multifiliis* (Ciliophora) theronts to host molecules, *Int J Parasitol*, **29**, 1415–1423.

BURRELLS C, WILLIAMS P D and FORNO P F (2001), Dietary nucleotides: a novel supplement in fish feeds: 1. Effects on resistance to disease in salmonids, *Aquaculture*, **199**, 159–169.

CHEVASSUS B and DOBSON M (1990), Genetics of resistance to disease in fishes, *Aquaculture*, **85**, 83–107.

CITARASU T (2010), Herbal biomedicines: a new opportunity for aquaculture industry, *Aquacult Int*, **18**, 403–414.

COOK S M, KHAN Z R and PICKETT J A (2007), The use of push–pull strategies in integrated pest management, *Annu Rev Entomol*, **52**, 375–400.

COSTELLO M J (2009), The global economic cost of sea lice to the salmonid farming industry, *J Fish Dis*, **32**, 115–118.

COWELL L E, WATANABE W O, HEAD W D, GROVER J J and SHENKER J M (1993), Use of tropical cleaner fish to control the ectoparasite *Neobenedenia melleni* (Monogenea:Capsalidae) on seawater-cultured Florida red tilapia, *Aquaculture*, **113**, 189–200.

CUESTA A, CEREZUELA R, ESTEBAN M Á and MESEGUER J (2008), *In vivo* actions of melatonin on the innate immune parameters in the teleost fish gilthead seabream, *J Pineal Res*, **45**, 70–78.

CUNNINGHAM C O (2002), Molecular diagnosis of fish and shellfish diseases present status and potential use in disease control, *Aquaculture*, **206**, 19–55.

CUNNINGHAM C O, MCGILLIVRAY D M, MACKENZIE K and MELVIN W T (1995), Identification of *Gyrodactylus* (Monogenea) species parasitising salmonid fish using DNA probes, *J Fish Dis*, **18**, 539–544.

DEADY S, VARIAN S J A and FIVES J M (1995), The use of cleaner-fish to control sea lice on two Irish salmon (*Salmo salar*) farms with particular reference to wrasse behaviour in salmon cages, *Aquaculture*, **131**, 73–90.

DEMERS N E and BAYNE C J (1997), The immediate effects of stress on hormones and plasma lysozyme in rainbow trout, *Dev Comp Immunol*, **21**, 363–373.

DICKERSON H W (2006), *Ichthyophthirius multifiliis* and *Cryptocaryon irritans* (Phylum Ciliophora) in Woo P T K, *Fish Disease Diseases and Disorders* Volume 1: *Protozoan and Metazoan Infections*, second edition. CAB International, 116–154.

DIONNE M, MILLER K M, DODSON J J and BERNATCHEZ L (2009) MHC standing genetic variation and pathogen resistance in wild Atlantic salmon. *Phil Trans R Soc B*, **364**, 1555–1565.

DIREKBUSARAKOM S (2004), Application of medicinal herbs to aquaculture in Asia, *Walailak J Sci & Tech*, **1**, 7–14.

EKANEM A P, OBIEKEZIE A, KLOAS W and KNOPF K (2004), Effects of crude extracts of *Mucuna pruriens* (Fabaceae) and *Carica papaya* (Caricaceae) against the protozoan fish parasite *Ichthyophthirius multifiliis*, *Parasitol Res*, **92**, 361–366.

ESTEBAN M Á, CUESTA A, RODRÍGUEZ A and MESEGUER J (2006), Effect of photoperiod on the fish innate immune system: a link between fish pineal gland and the immune system, *J Pineal Res*, **41**, 261–266.

EU (2006), European Community Council Directive 2006/88 EC http://eurlex.europa.eu/LexUriServ/LexUriServ.do?uri=OJ:L:2006:328:0014:0056:en:PDF

FAO (2002), Antibiotic Report: Basic overview of the regulatory procedures for authorisation of veterinary medicines with emphasis on residues in food animal species, Fisheries and Aquaculture Department. http://www.fao.org/DOCREP/004/AC343E/AC343E00.HTM. Last accessed 16 October 2011.

FARUK M A R, ALI M M and PATWARY Z P (2008), Evaluation of the status of use of chemicals and antibiotics in freshwater aquaculture activities with special emphasis to fish health management, *J Bangladesh Agril Uni*, **6**, 381–390.

FJALESTAD K T, GJEDREM T and GJERDE B (1993), Genetic improvement of disease resistance in fish: an overview, *Aquaculture*, **111**, 65–74.

FLAMARIQUE I N, GULBRANSEN C, GALBRAITH M and STUCCHI D (August 2009), Free patents online http://www.freepatentsonline.com/article/Canadian-Journal-Fisheries-Aquatic-Sciences/207283747.html. Last accessed 16 October 2011.

FREEMAN M A (2002), Potential biological control agents for the salmon louse *Lepeophtheirus salmonis* (Krøyer, 1837), PhD Thesis, University of Stirling.

FREEMAN M A and SOMMERVILLE C (2009), *Desmozoon lepeophtherii* n. gen., n. sp., (Microsporidia:Enterocytozoonidae) infecting the salmon louse *Lepeophtheirus salmonis* (Copepoda: Caligidae), *Parasites & Vectors*, **2**, 58.

FREEMAN M A, BELL A S and SOMMERVILLE C (2003), A hyperparasitic microsporidian infecting the salmon louse *Lepeophtheirus salmonis*: an rDNA-based molecular phylogenetic study, *J Fish Dis*, **26**, 667–676.

FUJI K, KOBAYASHI K, HASEGAWA O and COIMBRA M R M (2006), Identification of a single major genetic locus controlling the resistance to lymphocystis disease in Japanese flounder (*Paralichthys olivaceus*), *Aquaculture*, **254**, 203–210.

FUJI K, HASEGAWA O, HONDA K, KUMASAKA K, SAKAMOTO T and OKAMOTO N (2007), Marker-assisted breeding of a lymphocystis disease-resistant Japanese flounder (*Paralichthys olivaceus*), *Aquaculture*, **272**, 291–295.

GAULT N F S, KILLPATRICK D J and STEWART M T (2002), Biological control of the fish louse in a rainbow trout fishery, *J Fish Biol*, **60**, 226–237.

GIBSON D R, SMAIL D A and SOMMERVILLE C (1998), Infectious pancreatic necrosis virus: experimental infection of goldsinny wrasse, *Ctenolabrus rupestris* L. (Labridae), *J Fish Dis*, **21**, 399–406.

GJEDREM T, SALTE R and GJSEN H M (1991), Genetic variation in susceptibility of Atlantic salmon to furunculosis, *Aquaculture*, **97**, 1–6.

GLOVER K A, AASMUNDSTAD T, NILSEN F, STORSET A and SKAALA Ø (2005), Variation of Atlantic salmon families (*Salmo salar* L.) in susceptibility to the sea lice *Lepeophtheirus salmonis* and *Caligus elongatus*, *Aquaculture*, **245**, 19–30.

GLOVER K A, GRIMHOLT U, BAKKE H G, NILSEN F, STORSET A and SKAALA Ø (2007), Major histocompatibility complex (MHC) variation and susceptibility to the sea louse *Lepeophtheirus salmonis* in Atlantic salmon *Salmo salar*, *Dis Aquat Org*, **76**, 57–65.

GRAVIL H R (1996), Studies on the biology and ecology of the free swimming larval stages of *Lepeophtheirus salmonis* (Kroyer, 1838) and *Caligus elongatus* Nordmann, 1832 (Copepoda: Caligidae). PhD Thesis, University of Stirling.

GRIMHOLT U, LARSEN S, NORDMO R, MIDTLYNG P, KJOEGLUM S, STORSET A, SAEBØ S and STET R J M (2003), MHC polymorphism and disease resistance in Atlantic salmon (*Salmo salar*); facing pathogens with single expressed major histocompatibility class I and class II loci, *Immunogenetics*, **55**, 210–219.

GRUTTER A S, DEVENEY M R, WHITTINGTON I D and LESTER R J G (2002), The effect of the cleaner fish *Labroides dimidiatus* on the capsalid monogenean *Benedenia lolo* parasite of the labrid fish *Hemigymnus melapterus*, *J Fish Biol*, **61**, 1098–1108.

GUSELLE N J, MARKHAM R J F and SPEARE D J (2006), Intraperitoneal administration of β-1,3/1,6-glucan to rainbow trout, *Oncorhynchus mykiss* (Walbaum), protects against *Loma salmonae*, *J Fish Dis*, **29**, 375–381.

GUSELLE N J, MARKHAM R J F and SPEARE D J (2007), Timing of intraperitoneal administration of β-1,3/1,6 glucan to rainbow trout, *Oncorhynchus mykiss* (Walbaum), affects protection against the microsporidian *Loma salmonae*, *J Fish Dis*, **30**, 111–116.

GUSELLE N J, SPEARE D J, MARKHAM R J F and PATELAKIS S (2010), Efficacy of intraperitoneally and orally administered ProVale, a yeast β-(1,3)/(1,6)-D-glucan product, in inhibiting xenoma formation by the microsporidian *Loma salmonae* on rainbow trout gills, *North Amer J Aquacult*, **72**, 65–72.

HAAS W (2003), Parasitic worms: strategies of host finding, recognition and invasion, *Zoology*, **106**, 349–364.

HAAS W, STIEGELER P, KEATING A, KULLMANN B, RABENAU H, SCHÖNAMSGRUBER E and HABER L B (2002), *Diplostomum spathaceum* cercariae respond to a unique profile of cues during recognition of their fish host, *Int J Parasitol*, **32**, 1145–1154.

HARIKRISHNAN R, BALASUNDARAM C and HEO M-S (2011a), Impact of plant products on innate and adaptive immune system of cultured finfish and shellfish, *Aquaculture*, **317**, 1–15.

HARIKRISHNAN R, KIM J-S, KIM M-C, BALASUNDARAM C and HEO M-S (2011b), *Prunella vulgaris* enhances the non-specific immune response and disease resistance of *Paralichthys olivaceus* against *Uronema marinum*, *Aquaculture*, **317**, 1–15.

HARRIS P D, SHINN A P, CABLE J, BAKKE T A and BRON J (2008), GyroDb: gyrodactylid monogeneans on the web, *Trends Parasitol*, **24**, 109–111.

HEINECKE R D and BUCHMANN K (2009), Control of *Ichthyophthirius multifiliis* using a combination of water filtration and sodium percarbonate: dose–response studies, *Aquaculture*, **288**, 32–35.

HENRYON M, BERG P, OLESEN N J, KJAER T E, SLIERENDRECHT W J, JOKUMSEN A and LUNDE I (2005), Selective breeding provides an approach to increase resistance of rainbow trout (*Oncorhynchus mykiss*) to the diseases, enteric redmouth disease, rainbow trout fry syndrome, and viral haemorrhagic septicaemia, *Aquaculture*, **250**, 621–636.

HÖLLDOBLER B (1999), Multimodal signals in ant communication, *J Comp Physiol A*, **184**, 129–141.

HULATA G (2009), Genetic improvement of finfish, in Burnell G and Allan G, *New Technologies in Aquaculture, Improving Production, Efficiency, Quality and Environmental Management*, Woodhead, Cambridge, 55–72.

HUNTINGFORD F A, ADAMS C, BRAITHWAITE V A, KADRI S, POTTINGER T G, SANDØE P and TURNBULL J F (2006), Current issues in fish welfare, *J Fish Biol*, **68**, 332–372.

INGVARSDÓTTIR A, BIRKETT M A, DUCE I, MORDUE W, PICKETT J A, WADHAMS L J and MORDUE A J (2002), Role of semiochemicals in mate location by parasitic sea louse, *Lepeophtheirus salmonis*, *J Chem Ecol*, **28**, 10, 2107–2117.

IQBAL N A M and SOMMERVILLE C (1986), Effects of *Sanguinicola inermis* Plehn, 1905 (Digenea: Sanguinicolidae) infection on growth performance and mortality in carp, *Cyprinus carpio* L., *Aquacult Res*, **17**, 117–122.

JENEY G, YIN G, ARDÓ L and JENEY Z (2009), The use of immunostimulating herbs in fish. An overview of research, *Fish Physiol Biochem*, **35**, 669–676.

JIANG Y (1993), Transgenic fish – gene transfer to increase disease and cold resistance, *Aquaculture*, **111**, 31–40.

JOHNSTON P and SANTILLO D (2002), Chemical Usage in Aquaculture: Implications for Residues in Market Products, Technical Note 06/2002l Greenpeace Research Laboratories, http://www.greenpeace.to/publications/technical_Note_06_02.pdf. Last accessed 5 October 2011.

KOLSTAD K, HEUCH P A, GJERDE B, GJEDREM T and SALTE R (2005), Genetic variation in resistance of Atlantic salmon (*Salmo salar*) to the salmon louse *Lepeophtheirus salmonis*, *Aquaculture*, **247**, 145–151.

LAURIDSEN J H and BUCHMANN K (2010), Effects of short- and long-term glucan feeding of rainbow trout (Salmonidae) on the susceptibility to *Ichthyophthirius multifiliis* infections, *Acta Ichtyolet Pisc*, **40**, 61–66.

LEWIS E E, GREWAL P S and GAUGLER R (1995), Hierarchical order of host cues in parasite foraging strategies, *Parasitology*, **110**, 207–213.

LING F, LUO Q, WANG J-G, WANG Y-P, WANG W-B and GONG X-N (2009), Effects of the 'all-fish' GH (growth hormone) transgene expression on resistance to *Ichthyophthirius multifiliis* infections in common carp, *Cyprinus carpio* L., *Aquaculture*, **292**, 1–5.

LUI Z (2009), Genome-based technologies useful for aquaculture research and genetic improvement of aquaculture species in developments, in Burnell G and

Allan G, *New Technologies in Aquaculture, Improving Production, Efficiency, Quality and Environmental Management*, Woodhead, Cambridge, 3–41.

MADSEN H C K, BUCHMANN K and MELLERGAARD S (2000), Treatment of trichodiniasis in eel (*Anguilla anguilla*) reared in recirculation systems in Denmark: alternatives to formaldehyde. *Aquaculture*, **186**, 221–231.

MARIN S L, SEPÚLVEDA F, CARVAJAL J and NASCIMENTO M G (2002), The feasibility of using *Udonella* sp. (Platyhelminthes: Udonellidae) as a biological control for the sea louse *Caligus rogercresseyi*, Boxshall and Bravo 2000, (Copepoda: Caligidae) in Southern Chile, *Aquaculture*, **208**, 11–21.

MAXIME V (2008), The physiology of triploid fish: current knowledge and comparisons with diploid fish, *Fish and Fisheries*, **9**, 67–78.

MCANDREW B and NAPIER J (2011), Application of genetics and genomics to aquaculture development: current and future directions, *J Agric Sci*, **149**, 143–151.

MCROBBIE A S and SHINN A P (2011), A modular, mechanical rotary device for the cleaning of commercial-scale, circular tanks used in aquaculture, *Aquaculture*, **317**, 16–19.

MIDTLYNG P J, STORSET A, MICHEL C, SLIERENDRECHT W J and OKAMOTO N (2002), Breeding for disease resistance in fish, *Bull Eur Ass Fish Pathol*, **22**, 166.

MIKHEEV V N, VALTONEN E T and RINTAMAKI-KINNUNEN P (1998), Host searching in *Argulus foliaceus* L. (Crustacea: Branchiura): the role of vision and selectivity, *Parasitology*, **116**, 425–430.

MIKHEEV V N, MIKHEEV A V, PASTERNAK A F and VALTONEN E T (2000), Light-mediated host searching strategies in a fish ectoparasite, *Argulus foliaceus* L. (Crustacea: Branchiura), *Parasitology*, **120**, 409–416.

MO T A (2007), Surveillance and control programmes for terrestrial and aquatic animals in Norway, in Brun E, Jordsmyr H M, Hellberg H, Mørk T, *Surveillance and Control Programmes for Terrestrial and Aquatic Animals in Norway*. Annual Report 2007. Oslo: National Veterinary Institute, 143–149.

MORDUE A J and BIRKETT M A (2009), A review of host finding behaviour in the parasitic sea louse, *Lepeophtheirus salmonis* (Caligidae: Copepoda), *J Fish Dis*, **32**, 3–13.

MURRAY A G (2006), A model of the emergence of infectious pancreatic necrosis virus in Scottish salmon farms 1996–2003, *Ecol Model*, **199**, 64–67.

NOOR EL DEEN A I E and MOHAMED R A (2009), Application of some medicinal plants to eliminate *Trichodina* sp. in tilapia (*Oreochromis niloticus*), Report and Opinion 2009 Available at: http://www.sciencepub.net/report/report0106/01_1181_plants_report0106.pdf Last accessed 12 October 2011.

NYLUND S, NYLUND A, WATANABE K, ARNESEN C E and KARLSBAKK E (2010), *Paranucleospora theridion* n. gen., n. sp. (Microsporidia, Enterocytozoonidae) with a life cycle in the salmon louse (*Lepeophtheirus salmonis*, Copepoda) and Atlantic salmon (*Salmo salar*), *J Euk Microbiol*, **57**, 95–114.

OIDTMANN B C, THRUSH M A, DENHAM K L and PEELER E J (2011), International and national biosecurity strategies in aquatic animal health, *Aquaculture*, **320**, 22–33.

OIE (2010), World Organisation for Animal Health; Aquatic Animal Health Code 2010. Available at OIE, Paris. http://www.oie.int/en/international-standard setting/aquaticcode/access-online/. Last accessed 16 October 2011.

OIE (2011), World Organisation for Animal Health Manual of Diagnostic Tests for Aquatic Animals. Available at OIE, Paris. http://www.oie.int/eng/normes/fmanual/A_summry.htm. Last accessed 16 October 2011.

OLESEN I, GJEDREM T, BENTSEN H B, GJERDE B and RYE M (2003), Breeding programs for sustainable aquaculture, *J Appl Aquacult*, **13**, 179–204.

PAHL B C, COLE D G and BAYER R C (1999a), Sea lice control I, *J Appl Aquacult*, **9**, 85–96.

PAHL B C, COLE D G and BAYER R C (1999b), Sea lice control II, *J Appl Aquacult*, **9**, 75–88.

PARSONS J E, BUSCH R A, THORGAARD G H and SCHEERER P D (1986), Increased resistance of triploid rainbow trout × coho salmon hybrids to infectious hematopoietic necrosis virus, *Aquaculture*, **15**, 337–343.

PEELER E J and THRUSH M A (2004), Qualitative analysis of the risk of introducing *Gyrodactylus salaris* into the United Kingdom, *Dis Aquat Org*, **62**, 103–113.

PICKERING A D (1993), Growth and stress in fish production, *Aquaculture*, **111**, 51–63.

PICKETT J A, WADHAMS L J and WOODCOCK C M (1997), Developing sustainable pest control from chemical ecology, *Agric Ecosyst Env*, **64**, 149–156.

PORTZ D E, WOODLEY C M and CECH JR J J (2006), Stress-associated impacts of short-term holding on fishes, *Rev Fish Biol Fish*, **16**, 125–170.

POTTINGER T G and CARRICK T R (1999), Modification of the plasma cortisol response to stress in rainbow trout by selective breeding, *Gen Comp Endocrinol*, **116**, 122–132.

RAE G H (2002), Sea louse control in Scotland, past and present, *Pest Manage Sci*, **58**, 515–520.

REFSTIE S, BAEVERFJORD G, SEIM R R and ELVEBØ O (2010), Effects of dietary yeast cell wall β-glucans and MOS on performance, gut health, and salmon lice resistance in Atlantic salmon (*Salmo salar*) fed sunflower and soybean meal, *Aquaculture*, **305**, 109–111.

ROSS N W, JOHNSON S C, FAST M D and EWART K V (2008), Recombinant Vaccines Against Caligid Copepods (Sea Lice) and Antigen Sequences Thereof, United States Patent Application 20080003233. http://www.freepatentsonline.com/y2008/0003233.html. Last accessed 16 October 2011.

SÁNCHEZ B R D, JIMÉNEZ-ESTRADA M and AURO DE OCAMPO A (2000), Evaluación del efectoparasiticida de los extractosacuoso y metanólico de *Buddleja cordata* HBK (Tepozán) sobre *Costia necatrix* en tilapia (*Oreochromis* sp), *Veterinaria México*, **31**, 3, 1–9.

SAYER M D J, TREASURER J W and COSTELLO M J (1996), *Wrasse: Biology and use in Aquaculture*. Fishing News Books, Oxford.

SHINN A P, KAY J W and SOMMERVILLE C (2000), The use of statistical classifiers for the discrimination of species of the genus *Gyrodactylus* (Monogenea) parasitizing salmonids, *Parasitology*, **120**, 261–269.

SHINN A P, PICON-COMACHO S M, RAWDEN R and TAYLOR N G T (2009), Mechanical control of *Ichthyophthirius multifiliis* Fouquet, 1876 (Ciliophora) in a rainbow trout hatchery, *Aquacult Eng*, **41**, 152–157.

SHINN A P, COLLINS C, GARCÍA-VÁSQUEZ A, SNOW M, MATEJUSOVÁ I, PALADINI G, LONGSHAW M, LINDENSTRØM T, STONE D M, TURNBULL J F, PICON-CAMACHO S M, RIVERA C V, DUGUID R A, MO T A, HANSEN H, OLSTAD K, CABLE J, HARRIS P D, KERR R, GRAHAM D, MONAGHAN S J, YOON G H, BUCHMANN K, TAYLOR N G, BAKKE T A, RAYNARD R, IRVING S and BRON J E (2010), Multi-centre testing and validation of current protocols for the identification of *Gyrodactylus salaris* (Monogenea), *Int J Parasitol*, **40**, 1455–1467.

SOMMERVILLE C (2009), Controlling parasitic diseases in aquaculture: new developments, in Burnell G and Allan G, *New Technologies in Aquaculture, Improving Production, Efficiency, Quality and Environmental Management*, Woodhead, Cambridge, 215–237.

STEVERDING D, MORGAN E, TKACZYNSKI P, WALDER F and TINSLEY R (2005), Effect of Australian tea tree oil on *Gyrodactylus* spp. infection of the three-spined stickleback *Gasterosteus aculeatus*, *Dis Aquat Org*, **66**, 29–32.

SUMMERFELT S T (2003), Ozonation and UV irradiation – an introduction and examples of current applications, *Aquacult Eng*, **28**, 21–36.

SUNYER J O, GÓMEZ E, NAVARRO V, QUESADA J and TORT L (1995), Physiological responses and depression of humoral components of the immune system in gilthead seabream (*Sparus aurata*) following daily acute stress, *Can J Fish Aquat Sci*, **52**, 2339–2346.

TAN C W, JESUDHASAN R R R and WOO P T K (2008), Towards a metalloprotease-DNA vaccine against piscine cryptobiosis caused by *Cryptobia salmositica*, *Parasitol Res*, **102**, 265–275.

TELFER T C, BAIRD D J, MCHENERY J G, STONE J, SUTHERLAND I and WISLOCKI P (2006), Environmental effects of the anti-sea lice (Copepoda: Caligidae) therapeutant emamectin benzoate under commercial use conditions in the marine environment, *Aquaculture*, **260**, 163–180.

THOMAS M B (1999), Ecological approaches and the development of 'truly integrated' pest management, *Proc Natl Acad Sci USA*, **96**, 5944–5951.

TILDESLEY A S (2008), Investigations into *Ergasilus sieboldi* (Nordmann 1832) (Copepoda: Poecilostomatoida), in a large reservoir rainbow trout fishery in the UK. PhD Thesis, University of Stirling.

TONGUTHAI K (1997), Control of freshwater fish parasites: a Southeast Asian perspective, *Int J Parasitol*, **21**, 1185–1191.

TULLY O, DALY P, LYSAGHT S, DEADY S and VARIAN S J A (1996), Use of cleaner-wrasse (*Centrolabrus exoletus* (L.) and *Ctenolabrus rupestris* (L.)) to control infestations of *Caligus elongatus* Nordmann on farmed Atlantic salmon, *Aquaculture*, **142**, 111–124.

TURNBULL J F, NORTH B P, ELLIS T, ADAMS C E, BRON J, MACINTYRE C M and HUNTINGFORD F A (2008), Stocking density and the welfare of farmed salmonids, in Branson E J, *Fish Welfare*, Blackwell Publishing Ltd, Oxford, 111–120.

VAGSHOLM I, DJUPVIK H O, WILLUMSEN F V, TVEIT A M and TANGEN K (1994), Infectious salmon anaemia (ISA) epidemiology in Norway, *Prev Vet Med*, **19**, 277–290.

WAHLI T, MEIER W and SCHMITT M (1991), Affinity of *Ichthyophthirius multifiliis* theronts to light and/or fish, *Appl Ichthyol*, **7**, 244–248.

WHITTINGTON R J and CHONG R (2007), Global trade in ornamental fish from an Australian perspective: the case for revised import risk analysis and management strategies, *Prev Vet Med*, **81**, 92–116.

YEOMANS W E, CHUBB J C and SWEETING R A (1997), *Khawia sinensis* (Cestoda:Caryophyllidea) – indicator of legislative failure to protect freshwater habitats in the British Isles, *J Fish Biol*, **51**, 880–885.

ZHOU J, HUANG J and SONG X L (2003), Applications of immunostimulants in aquaculture, *Mar Fish Res*, **24**, 70–79.

Index

infectious haematopoietic necrosis
        virus (IHNV), 137, 261
infectious hypodermal and
        haematopoietic necrosis virus
        (IHHNV), 278, 280–1, 282,
        283–4, 296–7
infectious salmon anaemia, 10, 336
innate immune response, 459–60
innate immune system, 3–48, 459–60
    fish innate immune response, 14–32
        antimicrobial peptides, 21–4
        cellular components, 28–32
        fish complement system, 24–8
        innate immune cells, 31–2
        monocytes/macrophages, 28–30
        non-specific cell-mediated
            cytotoxicity cells, 30–1
        pattern recognition receptors,
            14–21
    immune cells and organs in fish,
        4–14
        gills, 10–11
        gut, 11–13
        head kidney, 7–8
        liver and integumentary surface,
            13–14
        spleen, 9–10
        teleost fish immune tissues, 6
        thymus, 5–7
innate immunity, 245
integrated pest management (IPM),
        190, 482–4
integumentary surface, 13–14
inter-laboratory variation, 174
interferon regulatory factors, 259
International Organisation for
        Standardisation (ISO), 148,
        157–8
interrenal cell, 8
intestinal bulb, 11
intestinal flora, 424
intra-laboratory variation, 183
Intralytix, 383
intubation, 220
invertebrates, 429–30

Japanese eel *see A. japonica*
Japanese flounder *see Paralichthys
        olivaceus*
Japanese oysters *see Crassostrea gigas*

Khawia sinensis, 489
Koch's postulates, 111–12
Kupffer cells, 29

lab seed, 281
Labeo rohita, 260
Lactobacillus, 403, 498
Lactococcus garvieae, 22, 252
Lactococcus garvieae infection, 371–2
Lateolabrax japonicus, 230
lateral-flow immunoassay, 135
'leaderless' secretory pathway, 16
LEAP-1, 21–2, 467
LEAP-2, 467
LEAPs, 21–2
lectin pathway, 27
lectins, 92–3, 430, 440
Lepeophtheirus salmonis, 197, 482
Lepeophtheirus spp., 202
leucocytes, 4
Limulus sp., 82
linear fish AMPs, 461–5
    main characteristics, 462
    molecular organisation of the AMP
        initial precursors, 463
    Schiffer-Edmundson helical wheel
        representation, 465
lipopeptides, 264–5
lipopolysaccharide, 77
Litopenaeus stylirostris, 81, 279, 281,
        282
    super-intensive culture systems, 284
    Taura syndrome virus (TSV) and
        IHHNV resistant, 283–4
Litopenaeus vannamei, 82, 278, 279, 281
    dominant species development in
        America, 281–5
    harvest-size pacific white and blue
        shrimp, 285
    pond studies at Texas A&M
        University, 282–3
    Ralston Purina, 281–2
    selection development in America,
        284–5
    super-intensive culture systems
        with *L. stylirostris*, 284
    Super Shrimp and SPR-43, 283–4
    Ralston Purina, 281–2
live attenuated vaccines, 219
live recombinant vaccines, 227–30
    Aeromonas hydrophila, 227
    Aeromonas salmonicida, 227
    Edwardsiella ictaluri, 227
    Edwardsiella tarda, 227–8
    Flavobacterium columnare, 228
    Flavobacterium psychrophilum, 228
    Francisella asiatica, 229
    Mycobacterium marinum, 229

Lightning Source UK Ltd.
Milton Keynes UK
UKHW022141251120
373986UK00012B/370